THE OXFORD HANDI

ENVIRONMENTAL HISTORY

MW00653265

THE OXFORD HANDBOOK OF

ENVIRONMENTAL

HISTORY

Edited by

ANDREW C. ISENBERG

OXFORD

UNIVERSITY PRESS

OXFORD
UNIVERSITY PRESS

Oxford University Press is a department of the University of Oxford.
It furthers the University's objective of excellence in research, scholarship,
and education by publishing worldwide.

Oxford   New York
Auckland   Cape Town   Dar es Salaam   Hong Kong   Karachi
Kuala Lumpur   Madrid   Melbourne   Mexico City   Nairobi
New Delhi   Shanghai   Taipei   Toronto

With offices in
Argentina   Austria   Brazil   Chile   Czech Republic   France   Greece
Guatemala   Hungary   Italy   Japan   Poland   Portugal   Singapore
South Korea   Switzerland   Thailand   Turkey   Ukraine   Vietnam

Oxford is a registered trade mark of Oxford University Press
in the UK and certain other countries.

Published in the United States of America by
Oxford University Press
198 Madison Avenue, New York, NY 10016

© Oxford University Press 2014

First issued as an Oxford University Press paperback, 2017

All rights reserved. No part of this publication may be reproduced, stored in a
retrieval system, or transmitted, in any form or by any means, without the prior
permission in writing of Oxford University Press, or as expressly permitted by law,
by license, or under terms agreed with the appropriate reproduction rights organization.
Inquiries concerning reproduction outside the scope of the above should be sent to the Rights
Department, Oxford University Press, at the address above.

You must not circulate this work in any other form
and you must impose this same condition on any acquirer.

Library of Congress Cataloging-in-Publication Data
The Oxford handbook of environmental history / edited by Andrew C. Isenberg.
pages cm. — (Oxford handbooks)
Includes index.
ISBN 978–0–19–532490–7 (hardcover : alk. paper); 978–0–19–067348–2 (hardcover : alk. paper)
1. Human ecology—History. 2. Environmental sciences—History. 3. Nature—Effect
of human beings on—History. I. Isenberg, Andrew C. (Andrew Christian),
editor of compilation. II. Title: Handbook of environmental history.
GF13.O84 2014
304.209—dc23
2014002933

CONTENTS

PART IV ENTANGLING ALLIANCES

ACKNOWLEDGMENTS

My foremost thanks are to the contributors to this book. I am grateful for how much I have learned from their essays, and for their patience and diligence in preparing this volume. I join the contributors in thanking the anonymous reviewer who provided detailed comments to all of the essayists.

Many of the participants in this volume sketched out some of the ideas for their essays in a workshop at Temple University in September 2007. In addition to the contributors who participated in the workshop, I would like to thank the scholars who attended the conference and offered their insights on the working papers, including Beth Bailey, David Biggs, Jeff Featherstone, Petra Goedde, Angela Gugliotta, Rob Mason, Mark Stoll, and Ellen Stroud.

The guiding force behind this volume has been Susan Ferber of Oxford University Press. She convinced me to edit this volume and has been unfailingly supportive throughout the process.

LIST OF CONTRIBUTORS

Thomas G. Andrews is Associate Professor of History at the University of Colorado at Boulder. He is the author of *Killing for Coal: America's Deadliest Labor War* (2008), winner of the Bancroft Prize and other honors.

Emily K. Brock is a historian of American science, global industry, and the environment. She is the author of the forthcoming *Money Trees: The Douglas Fir and American Forestry, 1900–1949* and is an assistant professor of history at the University of South Carolina.

Kathleen A. Brosnan is the Travis Chair of History at the University of Oklahoma. She authored *Uniting Mountain and Plain: Cities, Law and Environmental Change along the Front Range* (2002); edited the *Encyclopedia of American Environmental History* (2010); and co-edited *City Dreams, Country Schemes: Community and Identity in the American West* (2011) and *Energy Capitals: Local Impact, Global Influence* (2014).

Mark Carey is Associate Professor of History in the Clark Honors College at the University of Oregon. His book, *In the Shadow of Melting Glaciers: Climate Change and Andean Society* (2010), won the Elinor Melville Prize for the best book in Latin American environmental history.

Connie Y. Chiang is Associate Professor of History and Environmental Studies at Bowdoin College. She is the author of *Shaping the Shoreline: Fisheries and Tourism on the Monterey Coast* (2008).

Lawrence Culver is Associate Professor of History at Utah State University. He is the author of *The Frontier of Leisure: Southern California and the Shaping of Modern America* (2010).

Diana K. Davis, a geographer and veterinarian, is Associate Professor of History at the University of California, Davis. Her first book, *Resurrecting the Granary of Rome: Environmental History and French Colonial Expansion in North Africa* (2007), won the George Perkins Marsh Prize from the American Society for Environmental History. She is the recipient of Guggenheim and ACLS Ryskamp fellowships.

William Deverell is Director of the Huntington-USC Institute on California and the West and Professor of History at the University of Southern California. He is the author of *Whitewashed Adobe: The Rise of Los Angeles and the Remaking of Its Mexican Past*

(2005) and co-editor of the *Blackwell-Wiley Companion to California* (2008) and the *Blackwell-Wiley Companion to Los Angeles* (2010).

Kurk Dorsey is Associate Professor of History at the University of New Hampshire. He is the author of *The Dawn of Conservation Diplomacy: U.S.-Canadian Wildlife Protection Treaties in the Progressive Era* (1998) and *Whales and Nations: Environmental Diplomacy on the High Seas* (2013).

Andrew R. Graybill is Professor of History, department chair and Co-director of the William P. Clements Center for Southwest Studies at Southern Methodist University. He is the author or editor of three books: *Policing the Great Plains: Rangers, Mounties, and the North American Frontier, 1875–1910* (2007); *Bridging National Borders in North America: Transnational and Comparative Histories* (2010; ed., with Benjamin H. Johnson); and *The Red and the White: A Family Saga of the American West* (2013).

Marcus Hall is *Privat Dozent* (Senior Lecturer) of Environmental History at the University of Zurich. He is the author of the award-winning *Earth Repair: A Transatlantic History of Environmental Restoration* (2005), editor of *Restoration and History: The Search for a Usable Environmental Past* (2010), and co-editor (with Marco Armiero) of *Nature and History in Modern Italy* (2010).

Andrew C. Isenberg is Professor of History at Temple University. He is the author of *The Destruction of the Bison: An Environmental History, 1750–1920* (2000); *Mining California: An Ecological History* (2005); *Wyatt Earp: A Vigilante Life* (2013); and the editor of *The Nature of Cities: Culture, Landscape, and Urban Space* (2006).

Matthew Klingle is Associate Professor of History and Environmental Studies at Bowdoin College. He is the author of the award-winning *Emerald City: An Environmental History of Seattle* (2007).

Nancy Langston is Professor of Environmental History in the Great Lakes Research Center and the Department of Social Sciences at Michigan Technological University. She is author of several books, including *Toxic Bodies: Hormone Disruptors and the Legacy of DES* (2010).

Thomas Lekan is Associate Professor of History and a faculty associate in the Environment and Sustainability program at the University of South Carolina in Columbia. He is the author of *Imagining the Nation in Nature: Landscape Preservation and German Identity* (2004) and co-editor, with Thomas Zeller, of *Germany's Nature: Cultural Landscapes and Environmental History* (2005).

Michael Lewis is Professor of History and Chair of the Department of Environmental Studies at Salisbury University. He is the author of *Inventing Global Ecology: Tracking the Biodiversity Ideal in India, 1945–1997* (2004) and the editor of *American Wilderness: A New History* (2007).

Linda Nash is Associate Professor of History at the University of Washington. She is the author of *Inescapable Ecologies: A History of Environment, Disease, and Knowledge* (2006), which won the American Historical Association's John H. Dunning Prize, the American Historical Association-Pacific Coast Branch Book Prize, and the Western Association of Women Historian's Keller-Sierra Prize.

Sara B. Pritchard is Associate Professor in the Department of Science and Technology Studies at Cornell University. She is the author of *Confluence: The Nature of Technology and the Remaking of the Rhône* (2011) and co-editor of *New Natures: Joining Environmental History with Science and Technology Studies* (2013).

Steven Stoll is Associate Professor of History at Fordham University. He is the author of *The Fruits of Natural Advantage: Making the Industrial Countryside in California* (1998); *Larding the Lean Earth: Soil and Society in Nineteenth-Century America* (2002); *The Great Delusion: A Mad Inventor, Death in the Tropics, and the Utopian Origins of Economic Growth* (2008); and the editor of *U.S. Environmentalism Since 1945, A Brief History with Documents* (2006).

Paul S. Sutter is an Associate Professor of History at the University of Colorado at Boulder. He is the author of *Driven Wild: How the Fight against Automobiles Launched the Modern Wilderness Movement* (2002).

James Morton Turner is Associate Professor of Environmental Studies at Wellesley College. He is the author of *The Promise of Wilderness: American Environmental Politics since 1964* (2012).

Nancy C. Unger is Professor of History at Santa Clara University. She is the author of *Beyond Nature's Housekeepers: American Women in Environmental History* (2012) and *Fighting Bob La Follette: The Righteous Reformer* (2000).

Brett L. Walker is Professor of History at Montana State University. He is the author of *The Conquest of Ainu Lands: Ecology and Culture in Japanese Expansion, 1590–1800* (2001); *The Lost Wolves of Japan* (2005); *and Toxic Archipelago: A History of Industrial Disease in Japan* (2010), which won the George Perkins Marsh Prize from the American Society for Environmental History.

Louis Warren is Professor of History at the University of California, Davis. He is the author of *The Hunter's Game: Poachers and Conservationists in Twentieth-Century America* (1997); and *Buffalo Bill's America: William Cody and the Wild West Show* (2005), which won the Albert J. Beveridge Award of the American Historical Association, the Caughey Western History Association Prize, and the Great Plains Distinguished Book Prize.

Frank Zelko is Associate Professor of History and Environmental Studies at the University of Vermont. He is the author of *Make it a Green Peace! The Rise of Countercultural Environmentalism* (2013).

Thomas Zeller is Associate Professor of History at the University of Maryland, College Park. He wrote *Driving Germany: The Landscape of the German Autobahn, 1930–1970* (2007) and has co-edited the volumes *How Green Were the Nazis? Nature, Environment, and Nation in the Third Reich* (2005); *Germany's Nature: Cultural Landscapes and Environmental History* (2005); *Rivers in History: Perspectives on Waterways in Europe and North America* (2008); and *The World beyond the Windshield: Roads and Landscapes in the United States and Europe* (2008).

INTRODUCTION

A New Environmental History

ANDREW C. ISENBERG

THE EMERGENCE OF A NEW FIELD

WHEN the historian William H. McNeill's *Rise of the West* was published in 1963, it received widespread praise for its striking argument for the importance of cultural diffusion in world history. As McNeill himself recalled in 1991, the book challenged inward-looking historical orthodoxy by positing that "the principal factor promoting historically significant social change is contact with strangers possessing new and unfamiliar skills." *Rise of the West* won the National Book Award and firmly established McNeill's scholarly reputation. Yet McNeill, who thirteen years after writing *Rise of the West* wrote one of the founding texts of environmental history, *Plagues and Peoples*, later rued that in the writing of *Rise of the West* a number of subjects had "largely escaped my attention." By 1991, he had come to believe that in addition to inter-cultural encounters, historians also needed to take account of "our encounters and collisions with all the other organisms that make up the Earth's ecosystem." He pointed to the devasting plague of the fourteenth century (commonly known as the Black Death) and the transmission of diseases from Europe to the Americas after 1492 as important examples of such "collisions."[1]

When *Rise of the West* was first published, few historians would have seen the omission of "the Earth's ecosystem" in a historical study as worthy of regret. As recently as thirty years ago, textbooks in American, European, and world history contained little of what we would now call environmental history. Textbook authors, such as the historian R. R. Palmer, whose *History of the Modern World* went through ten editions between 1950 and 2006, sometimes included a few words at the outset that established a geographical setting for the narrative to follow. In the foreword to the 1983 edition of *History of the Modern World*, Palmer and his coauthor Joel Colton hinted that the environment could be understood historically, writing that "climate itself can change. The

Roman ruins in the interior of Morocco and Tunisia remind us that the climate there was once more favorable."[2] But the narrative following that short geographical and climatological introduction paid little attention to the environmental context of human history, and still less to the environment as an agent in history.

Indeed, few historians of McNeill's generation came to embrace environmental history as fully as he did. Nonetheless, textbook authors writing in the decades after McNeill's original study increasingly integrated the perspectives of environmental history into their narratives. In 1981, a leading textbook, *Western Civilization: A Concise History*, dealt with the famine that attended fourteenth-century Europe in a way typical of textbooks at the time. The authors attributed the famine to a number of factors, foremost among which was ignorance, including "limited knowledge of fertilizer." Among the other factors, none of which were deemed particularly important, was bad weather, particularly "intermittent bouts of prolonged heavy rains and frosts."[3] But twenty years later, climate had become an important explanatory factor. In 2001, Ronnie Hsia wrote in *The Making of the West* that "a cooling of the European climate also contributed to the crisis in the food supply. Modern studies of tree rings indicate that fourteenth-century Europe entered a colder period, with a succession of severe winters beginning in 1315. The extreme cold upset an ecological system already overtaxed by human civilization."[4]

Likewise, the authors of American history textbooks have done much to incorporate environmental history into their synthetic narratives. In the first edition of *America: A Narrative History*, published in 1984, the historian George Tindall wrote dismissively of Native Americans: "When the whites came, even the most developed Indian societies were ill-equipped to resist these dynamic European cultures. There were large and fatal gaps in their knowledge and technology."[5] By the eighth edition, published in 2010, Tindall and coauthor David Shi had revised their view to reflect scholarship in environmental history on the complexities of native land-use practices. "Over the centuries the Native Americans of North America had adapted to changing climates and changing environments. Their resilience was remarkable."[6]

The work of the environmental historian Alfred Crosby has been particularly influential. In 1984, Tindall mentioned disease as a contributing but relatively minor factor in the colonists' displacement of natives; he put far more emphasis on the colonists' violent conquest.[7] By 2010, however, Tindall and Shi attached more importance and causal weight to disease. "By far the most significant aspect of the biological exchange," they wrote, "was the transmission of infectious diseases from Europe and Africa to the Americas. European colonists and enslaved Africans brought with them deadly pathogens that Native Americans had never experienced...The results were catastrophic. Far more Indians—tens of millions—died from contagions than from combat. Major diseases such as typhus and smallpox produced pandemics in the New World on a scale never witnessed in history."[8]

Tindall devoted but a few sentences to the "dust bowl" of the North American Great Plains in his 1984 edition—and then only to note, in a waggish aside, that by destroying wheat, drought helped the goal of the Agricultural Adjustment Act, which was to stem the overproduction of staple crops. "In 1933, widespread drought in the wheat belt reduced production and removed any need to plow up growing wheat."[9] In 2010, by contrast, Tindall and Shi wrote that "a decade-long drought during the 1930s spawned an environmental and human catastrophe known as the dust bowl."[10]

The authors of *A People and a Nation: A History of the United States* had little to say about the environmental movement in their 1986 edition. In a section devoted primarily to a discussion of the social welfare legislation passed during the administration of Richard Nixon, the text's discussion of the environmental movement and the legislation it helped to create amounted to one sentence noting three bills that became law in 1970: "Congress responded to the growing environmental movement by passing the Clean Air Act [*sic*; in 1970 Congress passed an extension of the Act], the Water Quality Improvement Act, and the Resource Recovery Act."[11] The 2012 edition of the textbook, with a different group of authors responsible for writing the twentieth-century section, discussed the environment, pollution, and the environmental movement extensively. The authors noted not only Rachel Carson's 1962 book *Silent Spring* and the first Earth Day in 1970, but also the work of the environmentalist Barry Commoner, two infamous environmental disasters in 1969 (an oil spill in Santa Barbara, California, and the burning of the polluted Cuyahoga River in Cleveland Ohio), the nuclear accident at Three Mile Island in Pennsylvania in 1979, and the revelation of toxic pollution at Love Canal in New York in 1980. The revised text noted that Nixon's executive order creating the Environmental Protection Agency in 1970 was followed by the passage of eighteen major environmental laws during the 1970s.[12]

In short, since the early 1980s, the recognition of environmental history within the mainstream of academic history has been widespread. The work of what one might call the founding generation of environmental historians—William Cronon, Alfred Crosby, Thomas Dunlap, Samuel Hays, J. Donald Hughes, Carolyn Merchant, Martin Melosi, Arthur McEvoy, William McNeill, Roderick Nash, John Opie, Stephen Pyne, Hal Rothman, Susan Schrepfer, Joel Tarr, Richard White, and Donald Worster among them—had a profound impact on the practice of academic history.[13] By the beginning of the twenty-first century, most writers of broadly synthetic texts shared McNeill's view that neglecting the environment meant missing something important.

WHAT IS ENVIRONMENTAL HISTORY?

Mainstream historians, rushing to incorporate the insights of environmental history into their synthetic work, tended to think of the field as a novelty. "When early works in this new field first began to appear in the 1970s and 1980s," Cronon wrote in 2003, "the initial reaction from scholarly and lay readers alike was that these books represented something entirely new, a bold departure from traditional history."[14] Because of the sudden visibility of environmental history, many historians, including some environmental historians, emphasized both the field's novelty and its connection to the environmental movement. As Cronon argued in 1993, "like the several other 'new' histories born or reenergized in the wake of the 1960s—women's history, African-American history, Chicano history, gay and lesbian history, and the new social history generally—environmental history has always had an undeniable relation to the political movement that helped spawn it."[15]

Yet environmental history was neither entirely new nor entirely drawn from political environmentalism. While the environmental movement lent visibility and a sense of purpose to the field, the intellectual roots of environmental history can be traced well back into the nineteenth century. Those roots can be found in the work of writers such as George Perkins Marsh, the polymath who argued in *Man and Nature* (1864) that nature "avenges herself" on societies that degrade the environment.[16] The American historian Frederick Jackson Turner argued in 1893, in a widely influential essay, that "American development" could be explained by the progressive transformation of the "wilderness" to "civilization."[17] While Marsh and Turner had focused on people's transformations of the environment, the historian Walter Prescott Webb emphasized the limitations that nature placed on human endeavors. In his 1931 study, *The Great Plains*, Webb declared aridity to be the determining characteristic of the North American grasslands.[18] In France, Webb's contemporaries Marc Bloch and Lucien Febvre—the founders, in 1929, of the journal *Annales d'histoire économique et sociale*—rejected event-oriented history in favor of what Febvre's student Fernand Braudel called *la longue durée*: "man in his relationship to the environment, a history in which all change is slow, a history of constant repetition, ever-recurring cycles."[19] All of these scholars articulated, in one way or another, histories of human societies' interactions with nature.

Though different in many ways, Marsh, Turner, Webb, and the *Annalistes* all shared important similarities which they bequeathed to environmental history. All were expansively interdisciplinary in their analyses: Turner was one of the first historians to urge his fellow historians to adopt the methodologies and perspectives of emerging social sciences such as economics and sociology; the *Annalistes*, meanwhile, were deeply influenced by geography. Long before the emergence of social history that looked at the past from the bottom up, they all regarded history that merely recited the deeds of great men with skepticism. Their work was carried on in the middle of the twentieth century by scholars such as Clarence Glacken, James Malin, Karl Wittfogel, and Carl Sauer, all of whom contributed to the massive 1956 tome, *Man's Role in Changing the Face of the Earth*.[20] In short, environmental history was not simply an outgrowth of the environmental movement of the late twentieth century. Nor did it emerge in the 1970s among the founding generation of environmental historians *sui generis*. Rather, it had deep intellectual roots.[21]

In the 1980s, the founding generation's definition of environmental history reflected the field's interdisciplinary heritage. In 1988, in the preface to *The Ends of the Earth: Perspectives in Modern Environmental History*, Donald Worster summed up his scholarly generation's definition of environmental history as "the interactions people have had with nature in past times."[22] Reflecting environmental history's interdisciplinary origins, the definition made no real disciplinary distinctions: cultural geographers, human ecologists, and historians were invited to operate under its catholic definition. Indeed, contributors to *The Ends of the Earth* included an economist, an ecologist, and a geographer. In various permutations, Worster's pairing of the interactions of "people" with "nature" was the operative one for most environmental historians in the 1980s. In 1983, William Cronon defined the relationship between human society and the environment as "dialectical."[23] Richard White, writing in 1985, preferred the term "reciprocal."[24] At a time when environmental historians found their closest intellectual allies not in history departments but in departments of anthropology, geography, or biology, this simple binary between people and nature was a serviceable, umbrella-like definition for a field that integrated the material insights of the environmental sciences with an analysis of human societies and cultures.

Yet since 1988 there has been a burgeoning of subfields of environmental studies, each with its own methodologies and disciplinary perspectives. What once recommended Worster's definition—its intentional lack of disciplinary precision—now makes it seem vague. Absent the prepositional phrase "in past times," the definition does nothing to distinguish environmental history from fields, many of which did not exist in 1988, such as environmental philosophy, environmental sociology, environmental anthropology, environmental economics, geography, human ecology, political ecology, or ecocriticism. Because the past is not the exclusive province of historians— everything that anthropologists, sociologists, geographers, ecocritics and other scholars study happened in the past, after all—even the modifier "in past times" does little to define what disciplinary perspective or methodology historians bring to the study of human interactions with the environment.

What disciplinary perspectives do historians bring to the environmental humanities? Primarily, they offer a close attention to change, causation, contingency, and context.[25] Historians, including environmental historians, are drawn to explanations of change: explaining how, for instance, Old World fauna such as sheep and rabbits established themselves in Australia; or how New England colonists of the seventeenth century transformed the forested environment in which they settled into a landscape of farms and villages.[26] Beyond explanations of the causes of change, however, the perspective that perhaps most distinguishes history from other disciplines is the notion of historical context. Context presumes that every age has its own idioms, worldviews, meanings (and for environmental historians, climates, disease pools, and views of nature) that are specific to that time. Contexts are neither static nor uncontested; rather, they are, like everything else, subject to change. Change, in turn, is contingent upon context, but not bound by it. Thus, while historians believe that context shapes people, events, and interpretations, at the same time context does not determine history; individuals can transcend their contexts to exercise a decisive agency and thus influence historical change. As Karl Marx wrote in 1852, "Men make their own history, but they do not make it as they please; they do not make it under self-selected circumstances, but under circumstances existing already, given and transmitted from the past."[27] Since Marx's time, the complex relationship between context and agency has been an ongoing subject of inquiry and debate in historical methodology.[28]

Because a deep understanding of historical context in all its complexity, taking into account its influences and its limitations, is fundamental to understanding the meaning of the past, an understanding of context is both a tool of historical analysis and one of its goals. Historians' regard for the importance of context distinguishes them from, for instance, poststructural ecocritics who focus on the aesthetics of a text itself apart from its historical context, or political ecologists who make comparisons between societies that belong to different historical contexts. By contrast, the cultural historian Robert Darnton has pointed to the importance of historical context as well as the continued importance of the historian's task of analyzing evidence, in describing his method as "working back and forth between texts and contexts."[29]

When Darnton wrote those words, professional historians had already begun to expand their evidentiary base from traditional primary sources—wills, journals, letters, sermons—to include oral tradition, popular culture, and material culture (one of Darnton's own notable contributions was an analysis of early modern folklore). Environmental historians have pushed that inclusiveness in another direction, drawing non-human nature into historical analysis as both text and context. As text, historians "read" the environment for evidence about the past: an exotic plant or animal species found far from its place of origin, a stand of second-growth trees, or a toxic waste dump are each, in their own way, a text that provides insight into the past. At the same time, environmental historians believe that non-human nature is part of the context of historical change: climate, disease, and natural disasters influence human history. To put it succinctly, a definition of environmental history now, when the environmental humanities are so crowded, might read as follows: environmental history understands the environment in a historical context, while at the same time understanding human history in an environmental context.

The environmental historian Arthur McEvoy outlined a model for such reciprocally constituted contextualization in 1988, in the same volume in which Worster defined environmental history as "the interactions people have had with nature in past times." McEvoy argued that ecology, economy, and culture were equally important as interacting agents of change in environmental history. "Any explanation of environmental change should account for the inter-embeddedness and reciprocal constitution of ecology, production, and cognition," McEvoy wrote. "All three elements, ecology, production, and cognition, evolve in tandem, each partly according to its own particular logic and partly in response to changes in the other two. To externalize any of the three elements . . . is to miss the crucial fact that human life and thought are embedded in each other and together in the nonhuman world."[30] McEvoy's formulation anticipated in some ways the "actor-network theory" of the historian of science Bruno Latour, who has argued for the agency of nonhumans within a network in which human beings are bound to their natural and technological environments.[31] No consideration of historical context, McEvoy and Latour argued, could neglect the natural environment.

Numerous historians have adopted that insight. Just as environmental history has been integrated into widely read textbooks, historians specializing in other fields have used the methodologies of environmental history to inform their own work. Historians of medieval Europe such as William Jordan have integrated environmental history into an analysis of the subsistence crisis of the fourteenth century.[32] Scholars of early American history such as James Merrell have placed epidemic disease within a broader understanding of the encounter between colonists and natives in North America.[33] David Blackbourn, a historian of modern Germany, integrated a history of water engineering over two centuries with the history of the formation of the German state.[34] As a result of works such as these, which integrate the environment with the concerns of social, cultural, and political history, environmental history has come to look less like the sort of work that ecological anthropologists, human ecologists, and geographers do, and more like what other historians do. Yet as the field has become less novel and increasingly mainstream, it has struggled to define itself in a way that both reflects its acceptance within academic history yet maintains its old revisionist brashness.

The struggle over how to define a field once seen as revisionist and now suddenly mainstream came to the fore in 1990. That year, some of the field's founders—Worster, Crosby, Cronon, Merchant, White, and Pyne—engaged in a roundtable discussion in the pages of the *Journal of American History* in an attempt to chart a direction for the field. The discussion, however animated and replete with useful insights, produced more disagreement than consensus. Indeed, the exchange left the field deeply divided between materialist and idealist approaches. Worster, initiating the roundtable, offered a straightforwardly materialist agenda for environmental historians. He suggested that environmental history works on three levels: first, "the discovery of the structure and distribution of natural environments in the past"; second, a focus on "productive technology as it interacts with the environment"; and third, a "purely mental type of encounter in which perceptions, ideologies, ethics, laws, and myths have become part of an individual's or a group's dialogue with nature." Superficially, Worster's attention to ecology, economy, and culture seemed to echo McEvoy's model of ecology, production, and cognition as deeply intertwined. Yet, while McEvoy saw each of these three factors as equally important, Worster's view was hierarchical, with the environment as the base below a cultural superstructure. With this three-tiered approach in mind, he called for environmental historians to focus on "agroecological" history, that is, the shift from subsistence farming to "the capitalistic agroecosystem." The capitalist mode of production, Worster argued, not only organized human labor but transformed the environment. "The capitalist era in production introduced a new, distinctive relation of people to the natural world," Worster argued. "The *reorganization of nature*, not merely of society, is what we must uncover."[35]

Two of the respondents—White and Cronon—were particularly opposed to Worster's call for the field to devote itself to the study of agroecology. Cronon objected to Worster's "potentially excessive materialism" as an approach that was "needlessly closing doors to approaches different from his own." He critiqued Worster's argument as one that labeled "as capitalist or modern all forces for ecosystemic change, and as traditional or natural all forces for stability. The tautology of such an approach is too self-evident."[36] Like Cronon, White recoiled from the materialism that underscored Worster's perspective. He argued that Worster's emphasis on modes of production and the transition to capitalism, as well as his hierarchical "blueprint" grounded in an ecological-economic base, relegated culture to the superstructural periphery "in the old vulgar Marxist sense." Worster's model, White argued, failed "to recognize the role of value judgments and beliefs."[37]

The disputants had dug in their heels—and in doing so, not only neglected the work of their contemporaries such as McEvoy and Carolyn Merchant, who had suggested ways of integrating materialist and idealist approaches, but indeed ignored some of the insights of their own work, which had combined both perspectives. Worster, in his response to the critiques by Cronon and White, staked out a definition of environmental history that was—if possible—more materialist than his original call for a focus on agroecology. Cronon, he wrote, "would redefine environment as cultural landscape," a shift that would leave environmental history "on the same downward spiral that social history has taken toward fragmentation and a paralyzing fear of all generalization." As if in confirmation of the first part of this charge, over the next several years, Cronon issued a series of essays in which he indeed defined environment primarily as cultural landscape: of particular note, a 1996 essay argued that wilderness, "far from being the one place on earth that stands apart from humanity, is a profoundly human creation." In 2005, he lauded an essay entitled "Ideas of Nature" by the literary critic Raymond Williams as "the densest, richest, most suggestive 19 pages I know."[38] Worster was yet more strident in his rebuttal to White, whose refusal to "permit anything smacking of cultural materialism," he wrote, left him "in a confused, relativistic morass" that permits no "clear demonstration of causality."[39] In a 2004 essay, White dismissed such critiques of cultural history as "hysterical" and urged environmental historians to take the "cultural turn," or in other words, to follow the lead of cultural theorists who argued that in a postmodern world, all history is cultural history.[40]

And there the debate stagnated. The field, which had burst onto the scene of academic history with such promise in the 1980s, was by the end of that decade divided between materialist and idealist approaches. On one side was Worster, advocating a kind of base-superstructure Marxist-historical analysis that had gone out of fashion among most academic historians by the mid-1960s, when E. P. Thompson argued, in *The Making of the English Working Class*, that working-class consciousness was not simply a superstructural consequence of a capitalist mode of production, but a cultural construction created by workers themselves.[41] On the other side were Cronon and White, advocating a kind of cultural history of the environment that was in its own way just as dated as Worster's materialism; it had more than a passing resemblance to the work of 1950s and 1960s American Studies scholars such as Henry Nash Smith and Leo Marx, which pondered ideas of nature contained in the writings of literary luminaries such as Nathaniel Hawthorne, Ralph Waldo Emerson, and Henry David Thoreau.[42] What had made environmental history innovative in the 1980s was its integration of materialist and idealist approaches. Indeed, the discussants in the 1990s roundtable had, in their books, integrated both approaches, however much they emphasized one approach in their historiographical essays. Pulling the field apart into separate approaches in the

1990 roundtable merely emphasized that each perspective, on its own, had relatively little to offer that was new.

The materialist-idealist debate took on other dimensions as well: it became one between Worster's idea of environmental history operating on a grand scale, as a form of the *Annales* school's *histoire totale*, and Cronon and White's idea that works in the field should not only be more focused, but that the field should be a hyphenated history, in which environmental historians would also be, variously, cultural, social, labor, or gender historians.[43] It contrasted Worster—strongly supported by his colleague Alfred Crosby—who believed that in the final analysis the environment is in control and that human dominance of nature is illusory, with White and Cronon, who insisted that people are not simply buffeted by large forces beyond their control but rather exercise agency in historical change (here the ongoing problem about the boundary between context and agency, with which all historians must wrestle, came to the fore).[44] Most importantly, Worster wanted environmental history to retain its lively revisionism, to remain outside of mainstream history. Cronon and White, by contrast, did not want environmental history to be left behind as professional historians took the cultural turn. (Alas, however eager they were for environmental historians to embrace the cultural turn, cultural historians have not been, as of yet, nearly as enthusiastic about environmental history. A recent volume on the cultural turn in United States history, for instance, has almost nothing to say on the subject of environmental history.[45]) For environmental historians, there were virtues to each side in these disputes; most environmental historians found themselves agreeing, at least in part, with both sides.

Some scholars in the field were untroubled by the divergent understandings of the discipline. All historical subfields—active ones, at least—change as new scholarship redefines boundaries and categories of analysis. It is rare for a healthy field to arrive at a universally agreed-upon definition of itself—and the creative disagreement that results from divergence can be constructive. Others were encouraged enough by the remarkable growth of environmental history since the late 1970s to shrug off the dispute within the discipline; so long as the field seems to be thriving, why worry about it? Divergent understandings of the field did not dissuade several environmental historians from writing historiographical essays attempting to define and thus reconcile the field. The authors of these essays have variously emphasized culture, power, a transnational perspective, and evolutionary science, among other things.[46] Insights into the field abound in these essays—but the field remains too large, and is moving in too many directions simultaneously, to be bound easily by a single definition.

TOWARD A NEW ENVIRONMENTAL HISTORY

The profusion of essays over the last decade and a half attempting to define environmental history have indicated that environmental historians now face an entirely different challenge than the one that faced Cronon, Crosby, Merchant, White, Worster, and others at the time of the roundtable in 1990. The founding generation had to break into the textbooks—and they did so, brilliantly. The generation that has succeeded them—those engaged in a new environmental history—have had to integrate the insights of environmental history into a host of other subfields that are equally as complex as environmental history, including class, race, ethnicity, gender, consumption, borderlands, labor, law, and the history of science.[47] For the current generation of environmental historians, these other subfields—each with its own literature, questions, and debates about historical context—are as important as the environment in understanding environmental history. These subfields are some of the very subjects that are addressed in the chapters in this volume. Indeed, this volume is organized with the challenge of integrating environmental history with other subfields, methodologies, and historical contexts foremost in mind. In integrating environmental history with other, established subfields of history, the essayists in this volume have had to do justice to the historiographies of consumerism, or borderlands, or gender, while at the same time remaining faithful to the methodology of environmental history.

The essays in Part I, "Dynamic Environments and Cultures," though they branch out in many directions, share a root concern for the problem of environmental context. They thus address one of the biggest hurdles environmental historians face in integrating their methods with other historical fields: the problem and perception of environmental determinism. The perception that environmental history emphasizes determinism can be traced in large part to the influential 1986 book, *Ecological Imperialism*, in which Alfred Crosby recast the story of European imperialism as a tale of ecological accidents in which European colonists benefited from the microbes and weeds they inadvertently transported overseas.[48] Crosby argued that the Europeans who colonized the Americas, Australia, and New Zealand (places he blithely termed "Neo-Europes"), "were seldom masters of the biological changes they triggered.... They benefited from the great majority of these changes, but benefit or not, their role was less often a matter of judgment and choice than of being downstream of a bursting dam."[49] A minimizing of "judgment and choice" likewise characterized the work of the physiologist-turned-historian Jared Diamond, who argued in 1997 in his Pulitzer Prize-winning world history, *Guns, Germs, and Steel*, that the ecological advantages enjoyed by Europeans (such as easily cultivable plants including wheat and easily domesticable animals such as sheep and cattle) "were determined by geography, which sets ground rules for the biology of all plant and animal species, including our own. In the long run, and on a broad scale, where we lives makes us who we are."[50]

Rather than seeing environmental forces as wholly overpowering, most environmental historians now take an approach similar to that of J. R. McNeill, who emphasized in *Mosquito Empires*, his 2010 study of mosquito-borne illnesses in the colonial Caribbean, the adaptations of people to disease environments.[51] The essays in Part I of this book, which address the topics of climate, disease, fauna, and regional environments, make clear that few environmental historians nowadays subscribe to the environmental determinism of Crosby or Diamond. Certainly, ecological forces can have a profound influence on human societies. At moments of plague, volcanic eruption, earthquake, or tsunami, such forces can be, for a time, determining. Yet, as Mark Carey (in his essay on climate) and Linda Nash (in her essay on disease) show, human perception is as important a determining factor as ecological change in the response to climate or disease. Likewise, as Diana Davis and Paul Sutter argue, deserts or tropics can seem like determinative environments—until one considers the cultural contexts of such human perceptions. Moreover, environments and human societies are not only entangled in complex ways, but also both adaptable. The essays by Brett Walker (on animals), Andrew Isenberg (on grasslands), and Emily Brock (on forests) emphasize human and ecological adaptive processes. Together, these essays argue that adaptation provides a more satisfying narrative of historical change than ecological determination or despoliation.

The essays in Part II, "Knowing Nature," confront a problem with which current environmental historians must wrestle in their efforts to integrate their field with other approaches: namely, a changing understanding of scientific knowledge. Environmental historians must now address a more complex, dynamic idea of ecology than was accepted when the founding generation of environmental historians entered the scene. In his 1980 book *Land Use, Environment, and Social Change: The Shaping of Island County, Washington,* Richard White employed the concept of ecological "succession" leading toward a stable "climax" community to describe the environment of the Puget Sound region. In the years since that book appeared, environmental science has come to question the climax model, preferring to conceive of environments as prone to instability and change. When his book was reissued in 1992, White regretted that "I used ecological concepts like community, succession, climax, and ecosystem unproblematically even though within the profession those ideas had already come under attack."[52]

Among ecological scientists, a new science of nonlinear dynamics has complicated the older concepts of succession and climax. Environments, according to nonlinear dynamics, do not transcend historical change; rather, they change unpredictably, much as societies change over the course of time.[53] Most environmental historians have taken to the concept of dynamic, changeable environments eagerly, in part because it posits a historical understanding of environments. Indeed, a number of the essays in this volume—on climate, deserts, forests, and grasslands—treat these subjects in light of recent environmental science that emphasizes the dynamism in the non-human natural world.

Yet the science of nonlinear dynamics, like older models of the natural world, reflects its cultural context. While most environmental historians accept nonlinear dynamics in ecology, Worster has dismissed the concept as the scientific profession's reflection of the "chaos, uncertainty, and disintegration we find in our institutions and communities."[54] In at least one respect, Worster is correct: environmental science is merely one among many cultural representations of the natural world. As the essays in Part II by Michael Lewis, Sara Pritchard, and Nancy Langston demonstrate, environmental historians have adopted the methodology and perspective of historians of science, who place scientific inquiry in its historical context. Likewise, as the essays by Jay Turner,

Marcus Hall, and Thomas Lekan and Thomas Zeller show, when wilderness advocates, conservation biologists, nature restorationists, and landscape designers create parks or wilderness areas or attempt to restore past environments, they are deeply and unavoidably implicated in contemporary culture and politics.

The essays in Part III, "Working and Owning," address the intersection of environmental history and economic change—one of the central concerns of environmental historians. Perhaps the most well-known study of this type is Cronon's *Nature's Metropolis*, his 1991 Bancroft Prize-winning book about the ways in which nineteenth-century Chicago drew in natural resources—lumber, grain, and cattle—from its hinterlands and transformed those resources into marketable commodities. *Nature's Metropolis* is a brilliant exposition of the commodification of nature, commodity flows, and the often-obscure relations between city and countryside. Critics, however, noted that the book had relatively little to say about power relations in Chicago. The geography journal *Antipode* devoted part of an issue in 1994 to critiques of *Nature's Metropolis*, and many of the critics focused on Cronon's disregard for ethnicity, class, labor, and other common concerns of urban studies scholars. Cronon's response to these criticisms was, in part, that he deemphasized these concepts out of the fear that they would overwhelm the environment as a category of analysis. He wrote that class, labor, and ethnicity were well-established, and "so dominant that they trump all other analytical categories."[55] Faced with the problem of integrating the methodology of environmental history with longer-established fields of ethnicity, labor, and class, Cronon chose to minimize those other, established fields.[56]

In 1991, when environmental history was not nearly so firmly established as it is now, and when some of its practitioners feared that it might go the way of other once-popular methodologies born in the 1970s such as psychohistory or cliometrics, Cronon's choice was understandable. While Cronon's decision made sense for a member of the founding generation of environmental historians writing two decades ago, it is impossible to imagine a member of the new generation of environmental historians choosing to disregard important subfields in the way that Cronon did. Indeed, only four years after the publication of *Nature's Metropolis*, Andrew Hurley's exemplary study of industrial pollution in Gary, Indiana, *Environmental Inequalities*, made class and race central—yet no one could suggest that they trumped the environment as a category of analysis.[57]

The essayists in Part III follow Hurley's example as, from different perspectives, they integrate the environment with the study of labor, class, consumerism, and production. The material world matters, as the essays by Steven Stoll (on capitalism) and Thomas Andrews (on labor) strenuously remind us; the gritty details of material history—notably, the ways in which resource capitalism has inscribed changes on the land and on human bodies—transform our understanding of the rise of the state and wage labor. Environmental history further complicates our understanding of the history of capitalism by revealing that the path to resource capitalism was hardly linear. During the transition to capitalism, the privatization of natural resources, as Louis Warren argues, emerged alongside various forms of common property, including national parks. Similarly, the law, as Kathleen Brosnan argues, facilitated both the exploitation and protection of the environment. The essays by Matthew Klingle and Lawrence Culver carefully merge cultural and material history. Cities, as Culver points out, were not merely collection points where natural resources could be transformed into commodities; in places such as Los Angeles, cultural desires for leisure dictated the urban form. Likewise, consumers, as Klingle argues, were forces not only for wastefulness but for environmental politics.

Power is the central concern of the essays in Part IV, "Entangling Alliances." Several of the essayists in this final part of the volume tackle the intersection of environmental history with the history of the nation-state, one of the most powerful and well-established historical fields. The concept of the nation has long influenced historical practice. The historical profession emerged in the nineteenth century along with the rise nation-states; for most of its existence, academic history has taken the nation as its proper category of analysis.[58] Environmental history poses a profound challenge to national histories, both because environments transcend borders—environmental history lends itself in a unique way to transnational history—and because environmental historians, in their attention to non-human nature, force us to reconsider national identities and policies. As William Deverell demonstrates, American nationalists placed heavy demands on the environment; it was both a place where, through the exploitation of resources, the nation might become modern, and the place where the nation might seek anti-modern refuge, heal the divisions of the Civil War, and restore its national identity. Andrew Graybill finds, in the histories of borderland regions, national competition for resources, and yet in some cases the emergence of transnational ecological spaces that belong wholly to no single nation. Kurk Dorsey finds, likewise, that several decades of environmental diplomacy has left a variegated history: namely, examples of international cooperation amid a longer history of states pursuing their own competitive interests. It has often been left, as Frank Zelko shows, to non-governmental organizations to press for environmental regulations on the international stage. Connie Chiang and Nancy Unger analyze how the environment and environmental inequalities have manifested themselves in the history of race, ethnicity, and gender. All of these essays show how the history of the environment is deeply implicated in relations of power.

Altogether, the integration of environmental history with established fields distinguishes the work of the current generation of environmental historians represented in this volume. In the last two decades, environmental historians have increasingly incorporated the environment into the concerns of mainstream history. The essays in this volume, taken together, suggest that the environment is inextricably entangled in the human experience, just as our societies, cultures, and economies are embedded in the environment. Our laws, politics, markets, and property are bound to the environment; so, too, are our science, technology, and our conception of nations. The new environmental history must integrate both materialist and idealist perspectives with the ongoing work of the historical profession at large. The field has a long-standing intellectual vision, grounded in its interdisciplinary roots, that posits an interactive relationship between complex, changing human societies and an equally complex and dynamic environment. The founding generation of environmental historians demonstrated to their colleagues not only that such an interactive relationship existed, but that it mattered. The current generation represented in this volume, in continuing the intellectual project of environmental history, must (on one side of the old binomial definition of environmental history) integrate the environment into a host of complex subfields such as gender, labor, and borderlands. At the same time (on the other side of the old binary), they must understand the environment as neither a deterministic force nor as (absent human influence) inherently stable.

At the heart of environmental history's critical apparatus is a vision of the environment and human societies as inter-connected—a vision that McEvoy eloquently articulated in the late 1980s. That vision is inclusive—neither simply idealist nor only materialist, but always necessarily both. As the field moves increasingly into the mainstream of professional history, it is important to emphasize that environmental historians do not simply tack nature onto studies of gender, cities, or colonialism (among other subjects). Rather, they ask how the environment is implicated in these and other subjects as cause and context. The study of the environment in history is not an accessory to more familiar subjects of historical inquiry. Nor does the agency of the environment transcend social, cultural, or economic agency. Rather, it is bound to them, just as they are implicated in environmental change.

NOTES

1. William H. McNeill, *The Rise of the West: A History of the Human Community* (Chicago: University of Chicago Press, 1991 [1963]), xvi, xxix. McNeill, *Plagues and Peoples* (Garden City, NY: Doubleday, 1976).
2. R. R. Palmer and Joel Colton, *A History of the Modern World*, 6th ed. (New York: Alfred A. Knopf, 1983), 8.
3. Marvin Perry et al., *Western Civilization: A Concise History* (Boston: Houghton Mifflin, 1981), 247.
4. Lynn Hunt et al., *The Making of the West: Peoples and Cultures* (Boston: Bedford/ St. Martin's Press, 2001), 468.
5. George Brown Tindall, *America: A Narrative History*, 1st ed. (New York: W. W. Norton, 1984), 10.
6. Tindall and David Emory Shi, *America: A Narrative History*, 8th ed. (New York: W. W. Norton, 2010), 14.
7. Tindall, *America*, 1st ed., 21. Tindall drew on Alfred W. Crosby, Jr., *The Columbian Exchange: Biological and Cultural Consequences of 1492* (Westport, CT: Greenwood Press, 1972).
8. Tindall and Shi, *America*, 8th ed., 23–24. By this edition, Tindall and Shi were likely familiar with Crosby, *Ecological Imperialism: The Biological Expansion of Europe, 900–1900* (New York: Cambridge University Press, 1986).
9. Tindall, *America*, 1st ed., 1072.
10. Tindall and Shi, *America*, 8th ed., 1096. The revised accounting of the event seems to have been influenced by Donald Worster, *Dust Bowl: The Southern Plains in the 1930s* (New York: Oxford University Press, 1979).

11. Mary Beth Norton et al., *A People and a Nation: A History of the United States*, 2d ed. (Boston: Houghton Mifflin, 1986), 965.

12. Norton et al., *A People and a Nation: A History of the United States*, 9th ed. (Boston: Houghton Mifflin, 2012), 819–820, 878–879.

13. The work of the first generation of environmental historians includes William Cronon, *Changes in the Land: Indians, Colonists, and the Ecology of New England* (New York: Hill and Wang, 1983); Thomas Dunlap, *DDT: Scientists, Citizens, and Public Policy* (Princeton, NJ: Princeton University Press, 1981); Dunlap, *Saving America's Wildlife* (Princeton, NJ: Princeton University Press, 1988); Samuel Hays, *Conservation and the Gospel of Efficiency: The Progressive Conservation Movement, 1890–1920* (Cambridge, MA: Harvard University Press, 1959); Hays, *Beauty, Health, and Permanence: Environmental Politics in the United States, 1955–1985* (New York: Cambridge University Press, 1989); Arthur F. McEvoy, *The Fisherman's Problem: Ecology and Law in the California Fisheries, 1850–1980* (New York: Cambridge University Press, 1986); Carolyn Merchant, *Ecological Revolutions: Nature, Gender, and Science in New England* (Chapel Hill: University of North Carolina Press, 1989); Roderick Nash, *Wilderness and the American Mind* (New Haven, CT: Yale University Press, 1967); John Opie, *The Law of the Land: Two Hundred Years of American Farmland Policy* (Lincoln: University of Nebraska Press, 1987); Susan Schrepfer, *The Fight to Save the Redwoods: A History of Environmental Reform, 1917–1978* (Madison: University of Wisconsin Press, 1983); Richard White, *Land Use, Environment, and Social Change: The Shaping of Island County, Washington* (Seattle: University of Washington Press, 1980); White, *The Roots of Dependency: Subsistence, Environment, and Social Change among the Choctaws, Pawnees, and Navajos* (Lincoln: University of Nebraska Press, 1983); Worster, *Rivers of Empire: Water, Aridity, and the Growth of the American West* (New York: Pantheon, 1985); Worster, ed., *The Ends of the Earth: Perspectives in Modern Environmental History* (New York: Cambridge University Press, 1988).

14. Cronon, "Afterword," in *Changes in the Land* (2003), 172–173.

15. Cronon, "The Uses of Environmental History," *Environmental History Review* 17 (Fall 1993): 2.

16. George Perkins Marsh, *Man and Nature; or, Physical Geography as Modified by Human Action*, ed. David Lowenthal (Cambridge, MA: Belknap Press, 1965), 42–43.

17. Frederick Jackson Turner, "The Significance of the Frontier in American History," *American Historical Association Annual Report* (1893): 199–227.

18. Walter Prescott Webb, *The Great Plains* (Boston: Ginn, 1931). In 1957, Webb extended upon that insight, arguing that "the heart of the West is a desert, unqualified and absolute." Aridity determined that the American West would be an "oasis society" of defeated expectations. Webb, "The American West: Perpetual Mirage," *Harper's Magazine* 214 (1957): 25–31.

19. Fernand Braudel, *The Mediterranean and the Mediterranean World in the Age of Philip II*, vol. 1 (New York: Harper and Row, 1972 [1966]), 20. For the *Annalistes*, see Lucien Febvre, *A Geographical Introduction to History* (New York: Alfred A. Knopf, 1925 [1922]), 46; Marc Bloch, *French Rural History: An Essay on its Basic Characteristics* (Berkeley: University of California Press, 1966 [1931]).

20. See David Lowenthal, *George Perkins Marsh: Versatile Vermonter* (New York: Columbia University Press, 1958)Cronon, "Revisiting the Vanishing Frontier: The Legacy of Frederick Jackson Turner," *Western Historical Quarterly* 18 (1987): 157–176; Worster, "Doing Environmental History," in *The Ends of the Earth: Perspectives on Modern Environmental*

History, ed. Worster (New York: Cambridge University Press, 1988), 289; William L. Thomas, Jr., ed., *Man's Role in Changing the Face of the Earth* (Chicago: University of Chicago Press, 1956).

21. Andrew C. Isenberg, "Historicizing Natural Environments: The Deep Roots of Environmental History," in *Companion to Western Historical Thought*, ed. Lloyd Kramer and Sara Maza (Malden, MA: Blackwell, 2002), 372–389.

22. Worster, ed., *The Ends of the Earth: Perspectives in Modern Environmental History* (New York: Cambridge University Press, 1988), vii.

23. Cronon, *Changes in the Land*, 13.

24. White, "American Environmental History: The Development of a New Historical Field," *Pacific Historical Review* 54 (August 1985): 297–335. White's definition of the field, which he articulated on the first page of his essay, was much like Worster's: "the historical relationship between society and the natural environment."

25. See Thomas Andrews and Flannery Burke, "What Does It Mean to Think Historically?" *Perspectives on History* (January 2007). http://www.historians.org/publications-and-directories/perspectives-on-history/january-2007/what-does-it-mean-to-think-historically.

26. See Crosby, *Ecological Imperialism*; Cronon, *Changes in the Land*.

27. Karl Marx, *The Eighteenth Brumaire of Louis Bonaparte* (1852), quoted in Matt Perry, *Marxism and History* (New York: Palgrave Macmillan, 2002), 10.

28. Walter Johnson, "On Agency," *Journal of Social History* 37 (Fall 2003): 113–124; David Arnold, "Gramsci and Peasant Subalternity in India," and Rajnaranyan Chandavarkar, "'The Making of the Working Class': E. P. Thompson and Indian History," in *Mapping Subaltern Studies and the Postcolonial*, ed. Vinayak Chaturvedi (New York: Verson, 2000), 24–71; Gayatri Chakravorty Spivak, "Can the Subaltern Speak?" in *Marxism and the Interpretation of Culture*, ed. Cary Nelson and Larry Grossberg (Urbana: University of Illinois Press, 1988), 271–313.

29. Robert Darnton, *The Great Cat Massacre and Other Episodes in French Cultural History* (New York: Basic Books, 1984), 262. Professional historians' definition of their methodology has evolved, from a search for objective truths found in an analysis of primary documents, to an understanding that the past contains many meanings. Carl Becker, "Everyman His Own Historian," *American Historical Review* 37 (January 1932): 221–236. See also E. H. Carr, *What Is History?* (New York: Vintage, 1961); Peter Novick, *That Noble Dream: The "Objectivity Question" and the American Historical Profession* (Chicago: University of Chicago Press, 1988).

30. Arthur F. McEvoy, "Toward an Interactive Theory of Nature and Culture: Ecology, Production, and Cognition in the California Fishing Industry," *Environmental Review* 11 (Winter 1987): 289–305. McEvoy's colleague Carolyn Merchant added a fourth dimension: reproduction, an engine of change in all three areas. Moreover, as women have historically had different roles than men in economic production and cultural reproduction, for Merchant, environmental historians should be at all times gender historians as well. See Carolyn Merchant, "The Theoretical Structure of Ecological Revolutions," *Environmental Review* 11 (Winter 1987): 265–274; Merchant, *Ecological Revolutions*; Merchant, "Gender and Environmental History," *Journal of American History* 76 (March 1990): 1117–1121. See also Barbara Leibhardt, "Interpretation and Causal Analysis: Theories in Environmental History," *Environmental Review* 12 (Spring 1988): 23–36.

31. Bruno Latour, *Reassembling the Social: An Introduction to Actor-Network Theory* (Oxford: Oxford University Press, 2005).

32. William Chester Jordan, *The Great Famine: Northern Europe in the Early Fourteenth Century* (Princeton, NJ: Princeton University Press, 1997).

33. See James Merrell, *The Indians' New World: Catawbas and their Neighbors from European Contact though the Era of Removal* (New York: W. W. Norton, 1991).

34. David Blackbourn, *The Conquest of Nature: Water, Landscape, and the Making of Modern Germany* (New York: W. W. Norton, 2006).

35. Worster, "Transformations of the Earth: Toward an Agroecological Perspective in History," *Journal of American History* 76 (March 1990): 1087–1106. Worster had offered up a very similar definition of the field two years earlier in an appendix to *Ends of the Earth* entitled "Doing Environmental History," 293.

36. Cronon, "Placing Nature in History," *Journal of American History* 76 (March 1990): 1122–1131.

37. White, "Environmental History, Ecology, and Meaning," *Journal of American History* 76 (March 1990): 1111–1116. White was careful to note that Worster's approach was more "Braudelian" than Marxist. Worster does not judge himself a Marxist historian, either. He wrote in 2004 about his 1979 book, *Dust Bowl* (which begins on its first page with a quotation from Marx), that "I never intended…to offer a 'Marxist' interpretation of Great Plains history, for after all Marx missed quite a few things and turned out to be a bad prophet." Worster, "Afterword," in *Dust Bowl: The Southern Plains in the 1930s* (1979; New York: Oxford University Press, 2004), 246.

38. Cronon, "The Trouble with Wilderness; Or, Getting Back to the Wrong Nature," *Environmental History* 1 (January 1996): 7–28; Cronon, "The Densest, Richest, Most Suggestive 19 Pages I Know," *Environmental History* 10 (October 2005): 679–681. See also Cronon, "A Place for Stories: Nature, History, and Narrative," *Journal of American History* 78 (March 1992): 1347–1376.

39. Worster, "Seeing Beyond Culture," *Journal of American History* 76 (March 1990): 1142–1147.

40. White, "From Wilderness to Hybrid Landscapes: The Cultural Turn in Environmental History," *The Historian* 66 (September 2004): 557–564; David C. Chaney, *The Cultural Turn: Scene-Setting Essays on Contemporary Cultural History* (New York: Routledge, 1994), 1.

41. E. P. Thompson, *The Making of the English Working Class* (New York: Vintage, 1966). For a brief and excellent analysis of the significance of Thompson to Marxist historical scholarship, see Perry, *Marxism and History*.

42. See Henry Nash Smith, *Virgin Land: The American West as Symbol and Myth* (Cambridge, MA: Harvard University Press, 1950); Leo Marx, *The Machine in the Garden: Technology and the Pastoral Ideal in America* (New York: Oxford University Press, 1964).

43. See Ursula Lehmkuhl, ed., *Umweltgeschichte: Histoire Totale oder Bindestrich-Geschichte?* Erfurter Beiträger zur Nordamerikanischen Geschichte, 4 (Erfurt: University of Erfurt, 2002).

44. For Crosby's support of Worster, see Crosby, "An Enthusiastic Second," *Journal of American History* 76 (March 1990): 1107–1110.

45. James W. Cook, Lawrence B. Glickman, and Michael O'Malley, eds., *The Cultural Turn in U.S. History: Past, Present and Future* (Chicago: University of Chicago Press, 2009).

46. Alfred Crosby, "The Past and Present of Environmental History," *American Historical Review* 100 (October 1995): 1177–1189; Theodore Steinberg, "Down to Earth: Nature,

Agency, and Power in History," *American Historical Review* 107 (June 2002): 797–820; Edmund Russell, "Evolutionary History: Prospectus for a New Field," *Environmental History* 8 (April 2003): 204–228; J. R. McNeill, "Observations on the Nature and Culture of Environmental History," *History and Theory* 42 (December 2003): 5–43; Kristin Asdal, "The Problematic Nature of Nature: The Post-Constructivist Challenge to Environmental History," *History and Theory* 42 (December 2003): 60–74; Douglas R. Weiner, "A Death-Defying Attempt to Articulate a Coherent Definition of Environmental History," *Environmental History* 10 (July 2005): 404–419; Sverker Sörlin and Paul Warde, "The Problem of the Problem of Environmental History: A Re-reading of the Field," *Environmental History* 12 (January 2007): 107–130; Joseph E. Taylor III, "Boundary Terminology," *Environmental History* 13 (July 2008): 454–481.

47. These works include Kathleen Brosnan, *Uniting Mountain and Plain: Cities, Law, and Environmental Change along the Front Range* (Albuquerque: University of New Mexico Press, 2002); Mark Carey, *In the Shadow of Melting Glaciers: Climate Change and Andean Society* (New York: Oxford University Press, 2010); Connie Chiang, *Shaping the Shoreline: Fisheries and Tourism on the Monterey Coast* (Seattle: University of Washington Press, 2008); Mark Cioc, *The Rhine: An Eco-Biography, 1815–2000* (Seattle: University of Washington Press, 2006); Lawrence Culver, *The Frontier of Leisure: Southern California and the Shaping of Modern America* (New York: Oxford University Press, 2010); Diana K. Davis, *Resurrecting the Granary of Rome: Environmental History and French Colonial Expansion in North Africa* (Athens: Ohio University Press, 2007); Kurk Dorsey, *The Dawn of Conservation Diplomacy: U.S.-Canadian Wildlife Protection Treaties in the Progressive Era* (Seattle: University of Washington Press, 1998); Marcus Hall, *Earth Repair: A Translatlantic History of Environmental Restoration* (Charlottesville: University of Virginia Press, 2005); Andrew Hurley, *Environmental Inequalities: Class, Race, and Industrial Pollution in Gary, Indiana, 1945-1980* (Chapel Hill: University of North Carolina Press, 1995); David Igler, *Industrial Cowboys: Miller & Lux and the Transformation of the Far West, 1850–1920* (Berkeley: University of California Press, 2005); Isenberg, *The Destruction of the Bison: An Environmental History, 1750–1920* (New York: Cambridge University Press, 2000); Isenberg, *Mining California: An Ecological History* (New York: Hill and Wang, 2005); Ari Kelman, *A River and Its City: The Nature of Landscape in New Orleans* (Berkeley: University of California Press, 2003); Matthew Klingle, *Emerald City: An Environmental History of Seattle* (New Haven, CT: Yale University Press, 2007); Nancy Langston, *Forest Dreams, Forest Nightmares: The Paradox of Old Growth in the Inland West* (Seattle: University of Washington Press, 1995); Timothy J. LeCain, *Mass Destruction: The Men and Giant Mines that Wired America and Scarred the Planet* (Newark, NJ: Rutgers University Press, 2009); Thomas Lekan, *Imagining the Nation in Nature: Landscape Preservation and German Identity, 1885–1945* (Cambridge, MA: Harvard University Press, 2005); Michael Lewis, *Inventing Global Ecology: Tracking the U.S. Biodiversity Deal in India, 1947–1997* (Athens: Ohio University Press, 2004); J. R. McNeill, *Mosquito Empires: Ecology and War in the Great Caribbean, 1620–1914* (New York: Cambridge University Press, 2010); Linda Nash, *Inescapable Ecologies: A History of Environment, Disease, and Knowledge* (Berkeley: University of California Press, 2007); Jared Orsi, *Hazardous Metropolis: Flooding and Urban Ecology in Los Angeles* (Berkeley: University of California Press, 2004); Jennifer Price, *Flight Maps: Adventures with Nature in Modern America* (New York: Basic Books, 1999); Sara Pritchard, *Confluence: The Nature of*

Technology and the Remaking of the Rhône (Cambridge, MA: Harvard University Press, 2011); Adam Rome, *The Bulldozer in the Countryside: Suburban Sprawl and the Rise of American Environmentalism* (New York: Cambridge University Press, 2001); Edmund Russell, *War and Nature: Fighting Humans and Insects with Chemicals from World War I to Silent Spring* (New York: Cambridge University Press, 2001); Russell, *Evolutionary History: Uniting History and Biology to Understand Life on Earth* (New York: Cambridge University Press, 2011); Douglas Sackman, *Orange Empire: California and the Fruits of Eden* (Berkeley: University of California Press, 2007); Theodore Steinberg, *Nature Incorporated: Industrialization and the Waters of New England* (New York: Cambridge University Press, 1991); Mart A. Stewart, *"What Nature Suffers to Groe": Life, Labor, and Landscape on the Georgia Coast, 1680–1920* (Athens: University of Georgia Press, 1996); Steven Stoll, *The Fruits of Natural Advantage: Making the Industrial Countryside in California* (Berkeley: University of California Press, 1998); Stoll, *Larding the Lean Earth: Soil and Society in Nineteenth-Century America* (New York: Hill and Wang, 2002); Paul Sutter, *Driven Wild: How the Fight Against Automobiles Launched the Modern Wilderness Movement* (Seattle: University of Washington Press, 2004); Joseph E. Taylor III, *Making Salmon: An Environmental History of the Northwest Fisheries Crisis* (Seattle: University of Washington Press, 1999); Peter Thorsheim, *Inventing Pollution: Coal, Smoke, and Culture in Britain since 1800* (Athens: Ohio University Press, 2006); Conevery Bolton Valencius, *The Health of the Country: How American Settlers Understood Themselves and Their Land* (New York: Basic Books, 2002); Brett Walker, *The Conquest of Ainu Lands: Ecology and Culture in Japanese Expansion, 1590–1800* (Berkeley: University of California Press, 2001); Walker, *The Lost Wolves of Japan* (Seattle: University of Washington Press, 2005); Walker, *Toxic Archipelago: A History of Industrial Disease in Japan* (Seattle: University of Washington Press, 2010); Louis S. Warren, *The Hunter's Game: Poachers and Conservationists in Twentieth-Century America* (New Haven, CT: Yale University Press, 1997); and Thomas Zeller, *Driving Germany: The Landscape of the German Autobahn, 1930–1970* (Oxford: Berghahn Books, 2010).

48. Crosby, *Ecological Imperialism*. The notion of ecology as a dynamic agent of change was one that emerged in the middle of the twentieth century among biologists such as Charles Elton, who studied population dynamics and ecological invasion. Charles Elton, *Animal Ecology* (London: Sidgwick and Jackson, 1927); Elton, *The Ecology of Invasions by Animals and Plants* (London: Chapman and Hall, 1958), 15–32. See also Andrew Clark, *The Invasion of New Zealand by People, Plants, and Animals* (New Brunswick, NJ: Rutgers University Press, 1949).

49. Crosby, *Ecological Imperialism*, 192.

50. Jared M. Diamond, *Guns, Germs, and Steel: The Fate of Human Societies* (New York: W. W. Norton, 1997).

51. J. R. McNeill, *Mosquito Empires: Ecology and War in the Greater Caribbean, 1620–1914* (New York: Cambridge University Press, 2010). For adaptations, see also Isenberg, *Destruction of the Bison*, 7–8, 31–122; Isenberg, "Between Mexico and the United States: From Indios to Vaqueros in the Pastoral Borderlands," in *Mexico and Mexicans in the Making of the United States*, ed. John Tutino (Austin: University of Texas Press, 2012), 85–109.

52. White, *Land Use, Environment, and Social Change: The Shaping of Island County, Washington* (Seattle: University of Washington Press, 1992 [1980]), xviii, 8–11.

53. One of the first to embrace the idea was Cronon in his 1983 work, *Changes in the Land*, in which he understood that the ecological ideas of "climax" had given way to a more "historical" approach, "for which change was less the result of 'disturbance' than of the ordinary processes whereby communities maintained and transformed themselves." Cronon, *Changes in the Land*, 11. In this regard, Cronon's book, although published only four years after *Dust Bowl* and only three years after White's *Land Use, Environment, and Social Change*, feels as if it belongs to a different era.

54. Worster, *The Wealth of Nature: Environmental History and the Ecological Imagination* (New York: Oxford University Press, 1994), 177. For a recent volume in which historians of science adopted some of the methodology and perspectives of environmental history, see Jeremy Vetter, ed., *Knowing Global Environments: New Historical Perspectives on the Field Sciences* (New Brunswick, NJ: Rutgers University Press, 2010).

55. See Mary Beth Pudup et al., "William Cronon's *Nature's Metropolis*: A Symposium," *Antipode* 26 (April 1994): 113–176.

56. For an analysis of this aspect of *Nature's Metropolis* within the context of urban environmental history, see Isenberg, "Introduction: New Directions in Urban Environmental History," in *The Nature of Cities: Culture, Landscape, and Urban Space*, ed. Isenberg (Rochester, NY: University of Rochester Press, 2006), xi–xix.

57. Hurley, *Environmental Inequalities*. See also Klingle, *Emerald City*.

58. See Novick, *That Noble Dream*.

PART I

DYNAMIC ENVIRONMENTS AND CULTURES

CHAPTER 1

....

BEYOND WEATHER

The Culture and Politics of Climate History

....

MARK CAREY

In 1915, the Yale University geographer Ellsworth Huntington argued that climate was "the most fundamental" determinant of culture, and thus a powerful influence on human history. Climate, he went on to say, affected people's food, their natural resources, parasites and diseases, human occupations, livelihoods, and habits. It was also "the strongest [factor] in causing migration, racial mixture, and natural selection."[1] For Huntington, climate shaped world history and determined a society's level of civilization. While few scholars today would agree with his conclusions, Huntington nonetheless left a lasting legacy on the study of climate history.[2] Unfortunately, that legacy cast a dark, even reprehensible, shadow over climate research for several decades, as Huntington's (and others') climatic determinism helped justify colonial expansion and racist divisions in the world.

More recently, though, climate history scholarship has shed such racist undertones. Climatic determinism has not disappeared completely, as Mike Hulme suggests, for the predictive environmental sciences and climate modeling that try to predetermine future societal trajectories.[3] Nonetheless, research in climate history and historical climatology is rapidly evolving and has become quite sophisticated, with dramatic growth in the scholarship since I originally conceived and wrote this essay in 2007. Environmental historians have shown, for most topics, that environmental forces and ecological constraints can affect history without predetermining its path. Research on perceptions of nature and interpretations of environmental change within specific societal contexts have allowed scholars to blur supposed nature-culture dichotomies and conceptualize hybrid landscapes, thereby avoiding environmental determinism.[4] Studies on climate history conducted during the last several decades have followed this shift away from deterministic interpretations, revealing a complex dialectic between climate and society. Studies show how human factors worked alongside climate in shaping the past, thereby recognizing—as do most contemporary environmental historians—that nature and culture cannot be separated neatly into divided camps. Additionally, researchers are uncovering diverse sources and unique data to help reconstruct past climates. The history of climate science has also become more varied and goes back deeper in time, while increasingly scrutinizing climate science as a discursive construction by particular societies at specific points in history. Scholars have also examined the direct and indirect impacts of climate on societies. Yet climate affects different groups differently—and those holding power or those atop social hierarchies can often withstand climatic changes and weather meteorological disasters more easily than marginalized populations. In short, scholars of climate history now study climate not simply in its physical, measurable forms, such as temperature or precipitation, but also as a cultural construct. New research analyzes how societies have perceived, understood, and responded to climate based on their own worldviews. Increasingly, then, research is going beyond meteorological history and climate reconstructions to examine the cultural and political dimensions of climatic change in the past and present.[5]

This essay argues that climate history scholarship has become particularly strong in four areas: (1) the evaluation of diverse sources; (2) the analysis of historical knowledge systems; (3) the examination of the agency of non-human nature; and (4) the depiction of cultural values and perceptions of nature.[6] But achievements in these four areas remain incomplete, and researchers have historically faced—and still face—significant obstacles that they must overcome. Moreover, as this essay's concluding section points out, there are other areas and approaches that have yet to find their way into climate history scholarship. Examining these four areas of the scholarship fosters critical analysis of existing climate history research and thus highlights particular accomplishments and deficiencies. Moreover, the historiographical analysis of these four aspects of climate history scholarship—sources, knowledge, agency, and culture—parallel developments in environmental history scholarship more broadly. And running through all four of these themes are persistent problems with scale, both spatial and temporal, that continually confound researchers of climate history, as well as environmental historians in general. This study of climate historiography thus helps illuminate the successes and shortcomings of environmental history as a discipline.

THE PROBLEM OF CLIMATE IN ENVIRONMENTAL HISTORY

Climate is "average weather" over a period of time, generally at least a few decades. Or, as Mark Twain supposedly said more poetically, "climate is what we expect; weather is what we get." Although recent global warming discussions have focused primarily on temperature, climate also takes into account precipitation, wind, air pressure, humidity, sunshine, and other aspects of weather. Meteorology, on the other hand, is the study of weather, with a particular interest in prediction; it is thus more short-term

oriented than climate is. Climate history refers to the history of climate-society interactions rather than historical climatology, which is the compilation of past climate
data by climatologists and other scientists.[7] Scholars sometimes classify their work as
climate history, but focus on weather events, such as hurricanes, droughts, or El Niño
events. Consequently, this essay focuses on climate history but sometimes includes
investigations that might be more accurately classified as weather history. This incorporation of weather history and meteorological studies alongside climate histories
is useful, illuminating, and insightful for both climate and environmental history
scholarship more broadly. But, meteorological history is a major field too large to analyze here.[8]

Historians have long lamented that insufficient data exists for sophisticated analyses
of past climate-society relations. Yet, with increasingly detailed climatic data compiled
from diverse scientific and historical sources, our understanding of past climate has
grown markedly during the last few decades. Two major innovations helped refine and
expand knowledge about past climatic conditions: first, the availability of new or at
least increasingly detailed proxy data drawn from ice cores, tree rings, peat bogs, pollen, and ocean and lake sediment cores; second, the use of computers to organize and
analyze data and to run sophisticated climate models. Using these new data and analytical tools, scientists have reconstructed climate going back at least 540 million years,
through the Phanerozoic Eon.[9]

The climatic shifts that have proved most influential for human history occurred at
the end of the last Ice Age and during the Holocene (the last 11,500 years). Although
past data supported the idea of a relatively stable Holocene climate, more recent high
resolution data and computer models indicate multiple climate anomalies during
this period. Some of the most significant and well-documented of these include the
Younger Dryas (12,900 to 11,500 years ago), the onset of the Holocene (11,500 years
ago), a rapid-onset cool period from 8,200 to 8,000 years ago, the Medieval Warm
Period (from a few centuries after the birth of Christ to around 1350), the Little Ice
Age (roughly 1450 to 1850), and Global Warming (since the end of the Little Ice Age,
but accelerating after about 1980).[10] Many people today also identify the so called
Anthropocene, the current geological phase that started approximately two hundred years ago and is characterized by pronounced human impacts on Earth's ecosystems, climate, and geology.[11] These periods of dramatic or rapid climate change
affected societies in many ways, such as by facilitating (or impeding) innovations in
agriculture, by triggering major social changes and even societal collapse, or by making human migrations possible, such as the peopling of the Americas or the settlement of Greenland.[12] Although scientists cannot possibly provide climatic data for
every place at every time in human history, the ever-expanding and more accurate
data sets do allow for these broad generalizations about past climate and impacts on
societies. In some cases, however, climatic records offer extraordinarily specific data
for precise regions. And this information can illuminate both social and environmental history—thereby making possible some of the most innovative, interdisciplinary
climate history.[13] Although these broad periods offer some insight into large climatic

trajectories, they do not apply everywhere and they were driven by extremely complex forces. This essay is not an explanation of how the impossibly complex global, or even local, climate works, but rather an historiographical analysis of climate history research. Explaining the global climate in an essay like this would be akin to writing the history of the United States in three paragraphs.

Interestingly, environmental historians have not been the principal researchers of climate history. While many historians over several generations have grappled with climate-society interactions, climate history has not evolved into a subfield of environmental history in the same way that other areas have, such as conservation history or forest history, for example.[14] In fact, climate history never turned into what Robert Claxton predicted in the early 1980s: "a new focus within environmental history."[15] This was less the case in Europe, where the influence of historical climatologists such as Christian Pfister has continued from the 1970s to the present.[16] In the thirty years after the first issue of the journal *Environmental Review* in 1976, the principal environmental history journals—*Environmental Review, Environmental History Review, Environmental History*, and, more recently, *Environment and History*—published fewer than a dozen articles *total* on climate. The *Journal of Historical Geography* also has surprisingly few climate articles, though its 2009 special issue on historical narratives of climate change makes key contributions to the evolving field.[17] Of course, exceptions exist. Classic early environmental histories such as Emmanuel Le Roy Ladurie's *Times of Feast, Times of Famine* and Donald Worster's *Dust Bowl* underscored the importance of climate in history. Yet neither of these historians analyzed *interactions* between climate and human society: Le Roy Ladurie chronicled the Little Ice Age but did not emphasize climate's influence on human society; and Worster attributed the Dust Bowl primarily to human action rather than to drought.[18] Now, as global warming has triggered an avalanche of academic and popular studies on climate history, environmental historians could engage the topic more rigorously, especially to contextualize ongoing climate change or provide illuminating historical examples to guide future policies. In fact, environmental historians are uniquely positioned to add to the research by scientists, geographers, anthropologists, archaeologists, and, more recently, journalists on historical climate-society interactions.

There are, of course, many challenges environmental historians face when doing climate history. For one, climate research has a nefarious legacy dating to the nineteenth and early twentieth century. Huntington's environmental determinism, his racist classifications of civilization, and his overemphasis on climate as "the fundamental" historical agent shaping societies generated an enduring stigma against climate studies. And Huntington was not the first, nor the last, to explain history with climatic determinism. Others in Europe and North America used climatological justifications for colonialism and even slavery.[19] In some ways, the stigma against those types of conclusions about climate-society relations persisted for a long time. As J. Donald Hughes recently observed, "a charge often made against environmental historians is that of environmental determinism, which is the theory that history is inevitably guided by forces that are not of human origin or subject to human choice. Studies that emphasize the roles of climate and epidemic disease have been subjected in particular to this criticism."[20]

Other obstacles beyond environmental determinism can challenge environmental historians studying climate, and identifying them helps reveal some of the contributions and challenges to doing environmental history scholarship more generally. First, and as Hughes notes, the reliable scientific data needed to reconstruct precise climatic data simply does not exist. Despite tremendous accomplishments in climatology, most climatic data lacks information about the local and regional scales that most interest historians. But avoiding climate history because of a supposed lack of data points to a larger problem in environmental history: the overreliance on Western scientific data to provide a single environmental reality, rather than one among many representations of nature. As more environmental historians analyze science as discourse, as a cultural construct, they might free themselves from the supposed need for climatic data and thus produce more climate histories.[21]

Second, climate history has not made a good declensionist tale. Despite pleas to avoid the declensionist tendency that long characterized environmental history, stories about how people have destroyed their environments still pervade environmental historiography. Given the difficulty in pinpointing human impacts on climate change, however, climate history has not offered a clear declensionist history of human folly and arrogance in devastating the earth. [22] Third, and related, is the difficult task of identifying the historical agency of climate. Historians still need to theorize and navigate a balance between the environmental determinism of Huntington's era and the postmodern conceptualization of nature as a purely cultural construct.[23] The tendency, as with natural disaster scholarship, is to deemphasize the agency of climate and focus instead almost solely on human agency. But this approach runs the risk of abandoning the very foundations on which environmental history emerged, which traces the agency of physical environments and processes, even if nature and culture cannot be separated.

Fourth, most environmental histories have focused on ecology and the terrestrial world. Oceans, the cryosphere, the atmosphere, and other environments have received much less attention than forests, rivers, wildlife, and other land-based systems.[24] Fifth, climate history often forces researchers to think beyond the geographical restraints that the discipline of history imposes on its practitioners. To understand the relationship between poor harvests and weather in early seventeenth-century England, for example, a researcher might need to study a volcanic eruption in Peru and use climatic evidence gathered in Greenland, Antarctica, the Himalayas, and the Alps. For most historians, reading beyond one's geographical specialty is challenging, even impractical. Despite Richard White's plea for more transnational history, then, few scholars have taken up the challenge.[25] For climate history, a transnational approach is often a necessity, not a luxury.

SOURCES

Before the availability of detailed climatic data from ice cores, tree rings, and other proxy records, historians played a vital role in reconstructing past climates using sources untapped by most scientists. In fact, climate historians believed for several decades that their most important contributions could come from the unearthing of previously unknown climate records.[26] Documents containing weather details allowed historians to reconstruct past climates in particular regions before precise meteorological instruments were available.

Among the most famous works of climate reconstruction using innovative sources is Emmanuel Le Roy Ladurie's *Times of Feast, Times of Famine*. Le Roy Ladurie believed that historians who "burrow among archives" could produce climate data that was inaccessible to scientists but, at the same time, was both quantifiable and rigorously scientific. He sought to go beyond "the handwritten comments on climate from some parish register, or the worm-eaten and illegible records of some lawyer, [which] were too accidental and irregular to provide material for really organized knowledge." He also wanted to tease apart the human story from climatic details. His goal was to reconstruct climate rather than understand its impacts on people, though that of course mattered to him, too. Le Roy Ladurie thus compiled data on harvest dates, which depended on the year's weather, rather than on crop volumes or famines, which hinged on people's actions. Instead of summarizing effects of glacier advances on alpine pasture lands, he pinpointed the specific locations of glacier tongues through time. In short, Le Roy Ladurie used innovative sources and examined them in unique ways to provide the climatic details—the fluctuations and variability, the periods, and the origins and termination—of the Little Ice Age. In reconstructing Europe's climatic shifts over time, Le Roy Ladurie always remained cautious about attributing too much historical agency to climate. "I don't trust climatic explanations," he quipped.[27] His book largely avoids the human consequences of climate change, in part because he believed geographical variation and people's own behavior outweighed the force of climate in history. Though Le Roy Ladurie probably could have pushed his arguments further, as others have since, his work was sound, meticulous, and comprehensive, even path-breaking for environmental history.[28] After all, he analyzed grapes, glaciers, and flowers as historical texts.

Some historians continue to delve into archives, to scrutinize diaries and tithe records, to analyze paintings, and to examine almanacs and government files to accumulate information about past climatic conditions. A December 2006 environmental history podcast identified current climate history research that analyzes ship logs from 1750 to 1850. These logs recorded weather around the globe on every single day of the year.[29] Recent work from Peru and Jamaica to Western Europe, North America, and China has illuminated past climatic conditions by delving into unpublished personal, governmental, commercial, and church records.[30] Concerns about global warming perpetuate this interest in past climates, and continually drive researchers to find and analyze new sources in new ways.

KNOWLEDGE

There has been much fruitful intellectual exchange between environmental historians and historians of science—and more would certainly be useful in environmental history, as a means of analyzing environmental epistemologies, scrutinizing science (that is, not accepting it uncritically), and examining environmental sciences as discourses. The historical study of climate knowledge and knowledge systems, including the power embedded in certain scientific knowledge and the socio-historical contexts for climate knowledge, offer excellent insights into climate-society interactions and the possible ways of approaching environmental history research. There is also increasing research, though still mostly marginal to environmental history, on the role of Traditional Ecological Knowledge (TEK) or Indigenous Knowledge (IK) of climate change, especially in the Arctic region.[31] This type of climate history, probing the intersection of context, culture, race, ethnicity, and power, could significantly bolster current research in both environmental history and climate history.

The history of climatology has evolved and shifted in recent years as a result of both interest in global warming and trends in the history of science and technology.[32] Histories of climate science now emphasize two major areas: (1) the history of the scientific discipline itself, and (2) the applications of that knowledge to public policy, environmental control, and social relations. For several decades, history of science research on climate focused primarily on the progression of scientific breakthroughs, innovations in climatic reconstructions, explanations for climate change, technologies used for climatology, and historical meteorology.[33] More recently, the growing interest in global warming has inspired a host of studies on what Spencer Weart calls the "discovery of global warming."[34] These studies tend to implicitly or explicitly start with the recent debate about whether humans are causing global warming. The research then traces the historical trajectory of climate science since the nineteenth century, examining in particular the connections scientists came to see between atmospheric levels of carbon dioxide and global temperature. Some researchers have also illuminated the significant contributions of particular scientists, such as Guy Stewart Callendar, who understood the connection between carbon dioxide emissions and global climate, or Lonnie Thompson, who has demonstrated innovative methods of acquiring climatic data through the collection of dozens of ice cores from mountain glaciers in South America, Africa, Asia, and elsewhere.[35]

In addition to illuminating histories of scientific advances in climatology, research on climatology and meteorology has also exposed past uses and applications of climate science. One of the most direct applications of climate science has been on weather control. Historians like James Fleming offer key historical insights into the folly of climate control and its potential to produce dangerous long-term outcomes.[36] Kristine Harper calls some of these weather modification ideas "the lunatic fringe of scientific discussion." An American scientist in the late 1940s, for example, proposed using silver iodide smoke to produce more precipitation for the arid American West. To make rain, the scientist argued, a technician would just need to push a button when a suitable cloud passed nearby. By the early 1960s, Congress was providing millions of dollars to additional weather modification programs, such as the Office of Atmospheric Water Resources and Project Skywater, which were focused on creating rain, or Project Stormfury, which tried to use cloud-seeding techniques to dissipate or steer hurricanes.[37] President Lyndon Johnson even used weather modification science and technology to alleviate drought-induced famine in India. His administration saw climate control as a tool of the state—as a way to build international relations, to participate in a foreign government's policies, and to resolve environmental problems.[38]

Engineers in the Soviet Union also concocted plans to alter weather and stimulate human settlement in northern regions. They considered coating Arctic areas with soot dropped from planes to blacken the ice surface and hasten its melting, thereby changing both the landscape and the local weather. Soviet engineers also proposed damming the Bering Strait; they argued that they could then pump an astounding five hundred cubic kilometers of water per day from the Arctic Ocean to the Pacific, thereby warming the Arctic and melting its ice.[39] Not all weather control plans were so crazy, though. During World War II, British scientists devised a way to burn off clouds and fog so that warplanes could land on runways after returning from bombing missions in Germany.[40] While some of these weather modification programs might exemplify the post-World War II era of large-scale techno-scientific projects, they have not ended. Many scientists, engineers, and government officials still consider geoengineering, carbon sequestration, and other technological interventions as valid options to fight future global warming. Lessons from environmental history, whether from Fleming's work on weather modification or my own work on climate change adaptation technologies, suggest that even well-intentioned techno-scientific agendas (like climate geoengineering) can backfire, triggering far-reaching unintended consequences.[41]

Given the diverse uses of climate science and meteorology, researchers have increasingly examined the historical contexts and motivations for the production of particular scientific knowledge about weather and climate.[42] Building on advances in the history of science and technology, this research, as Fleming explains, generally seeks to "illustrate the cultural situatedness and historical contingency of our climate knowledge." Specifically, he demonstrates how the development of national systems of data collection and storage—and international cooperation to share that information—facilitated advanced storm warnings to aid citizens in Europe and the United States.[43] But as Fleming and others have shown, the concentration of that information in bureaucracies and among climate scientists has disempowered some people. The development of late nineteenth-century weather maps, for example, created a division between meteorology for the public—weather maps accessible to common citizens—and weather data produced for professional scientists, represented in complex ways that laypeople could not readily understand. As Katharine Anderson summarizes, weather maps "shaped the relationship between elite and popular forms of weather knowledge."[44] In another late nineteenth-century case, Erik Larson shows how power struggles ultimately killed thousands during the 1900 Galveston hurricane. Infighting within the US Weather

Bureau—combined with hubris and a dismissive attitude toward international relations with Cuba—caused meteorologists to overlook, even ignore, clear signs that a devastating storm was headed toward Texas.[45]

These power dimensions of climate science also exist within the present-day contexts of global warming. Clark Miller and Paul Edwards contend that climate change has created competitions about how to represent nature, and these power struggles influence governmentality, the distribution of power, access to natural resources, and the control of space. Climate science, they assert, "appears less an independent input to global governance than an integral part of it: a human institution deeply engaged in the practice of ordering social and political worlds." Consequently, computer modelers and laboratory or field scientists are deeply involved in global discussions about what serves as legitimate knowledge, who can speak for nature, and how science should inform public policy, laws, and international treaties.[46] Science and models, then, are more than just data.[47] Climate science is also about power—the power to define climatic discourse, to represent nature in certain ways, to possess (and use) data, and to project a certain view of the world. As Richard White has explained for similar arguments in environmental history, "who gets to define nature is an issue of power" with significant consequences for inhabitants of the represented regions.[48]

Beyond these recent power struggles and debates, climate knowledge has also played a role in more nefarious historical processes, such as imperialism, nation building, racism, and even slavery. On a practical level, climatic data has helped empires govern their colonial territories and subjects. Weather data gathered in nineteenth-century India, for example, advanced not only British knowledge of its colonial terrain, but also British science.[49] But views of climate—and especially the perceived relationships among climate, geography, race, health, and human character—also justified imperial expansion. Scientific discourse became a powerful tool for Europeans and North Americans in the eighteenth, nineteenth, and early twentieth centuries, allowing them to justify their subjugation of people worldwide, especially in the tropics. Science buttressed racist arguments because of its supposed objectivity and its statistical qualities. As David Livingstone puts it, "the language of science became the moralizing vehicle for conveying to the public the racial judgments of an intellectual elite."[50] In short, elite Europeans and North Americans believed that people from warm climates, which were associated with geographical places rather than climatic conditions, were inferior, lazy, and mentally unfit compared to the more vigorous and intelligent inhabitants of cold climates. Montesquieu outlined many of these theories in the eighteenth century, but they persisted into the twentieth—and were then reinforced by the eugenics movement.[51]

More disturbingly, ethnoclimatological justifications for European colonization of the tropics also helped plantation owners justify slavery. In the nineteenth-century American South, for example, slave owners maintained that the tropical origins of Africans had acclimated them to warm weather. In fact, plantation owners argued that only Africans or African Americans could labor in such a hot climate. Writing in 1858, Samuel Cartwright argued that "Negroes glory in a close, hot atmosphere. . . . This ethnical peculiarity is in harmony with their efficiency as labourers in hot, damp, close, suffocating atmosphere—where instead of suffering and dying, as the white man would, they are healthier, happier, and more prolific than in their native Africa—producing, under the white man's will, a great variety of agricultural products."[52] Clearly, ethnoclimatology, climatic determinism, and climate science have at times been associated with terrible trends, in which climate discourse fostered narratives that promoted and justified imperialism, racism, and slavery. In the twentieth century, climate science continued to serve governments, though generally through neoimperial and nation-building agendas, rather than the colonial empires of the nineteenth century.[53]

These historical studies that scrutinize climate knowledge—both to expose underlying social contexts and to identify the power dimensions embedded in the production and application of climatology and meteorology—represent a growing body of literature. Generally, historians of science and geographers dominate this analysis of scientific discourse. Yet environmental historians could also add their expertise by examining climate knowledge in the context of climate as a dynamic historical agent, rather than as a passive or static stage on which people act or as a consistent system scientists try to discern. Analyzing the relationship among environmental change, environmental knowledge, and power/politics/policies is, after all, part of the bedrock of environmental history as a field.

AGENCY

Environmental historians have long grappled with agency to discern the degree to which non-human nature shapes societies and the ways people affect species, landscapes, and the environment more broadly. Considering non-human nature as an agent of historical change does not, however, imply that there is any degree of environmental determinism, at least not in the sense of the scholarship a century ago when Huntington was writing. Today it is more common to conceptualize hybrid landscapes, social-ecological systems, and coupled natural-human systems that effectively blur the boundaries between nature and culture—and which thus recognize the combined agency of humans and non-human forces, such as in Actor Network Theory (ANT).[54] This literature in climate history increasingly recognizes the agency of climate but contextualizes it within socio-historical contexts, thereby avoiding climatic determinism while clearly illustrating the power of climate and weather to shape the past.

When Le Roy Ladurie concluded in *Times of Feast, Times of Famine* that "in the long term the human consequences of climate seem to be slight, perhaps negligible, and certainly difficult to detect," he was no doubt responding to previous critiques about both the pitfalls of climatic determinism and the overemphasis on climate as a mono-causal force in history.[55] While caution may have been necessary in 1970 to reverse the historiographical pendulum, by the late 1970s scholars began to challenge Le Roy Ladurie. Christian Pfister, for instance, sought to go beyond the physical dimensions of climate and delve into the effects of climate on crops and mortality. He thus brought climate history into agrarian and population history—and has ever since analyzed the effects of climate on European societies, especially during the Little Ice Age.[56] Hubert Lamb went further, contending that climate produces "shocks" on human societies.[57] Others, too, especially with the rising interest in global warming since the 1980s, have insisted that, "climate change and its human consequences stands on its own merits among the great dramas of human history."[58]

The difference between, on the one hand, recent scholarship that recognizes the historical agency of climate and, on the other, the climatic determinism of the nineteenth and early twentieth centuries, is the level of complexity evident in the illustration of how climate interacts with other historical forces. As David Hackett Fischer explained as early as 1980, "In all of those instances, the axial period and the dark ages, the era of oceanic expansion and the age of revolutions, important linkages may appear in the relationship between climate and culture—not in the form of a mindless, monistic determinism, but rather in the form of an intricate interaction of challenge and choice."[59] Finding the balance between climate as an historical agent and climatic determinism remains a challenge still today, as James McCann has pointed out for African historiography.[60] Nevertheless, successful research on the historical impacts of climate change on people—and the agency of climate—has emerged during the last three decades. It tends to concentrate in three general areas: (1) the collapse of civilizations; (2) social unrest; and (3) climate disasters.

Research on societal collapse has gained increasing popularity in recent decades, especially as researchers try to draw out "lessons from the past" to inform current policy on global warming.[61] Fortunately, most of these studies have increasingly rejected simplistic interpretations and have avoided climatic determinism. They accomplish this primarily by identifying climate as one among many historical forces that caused societies to collapse.[62] Research on the fate of the Classic Maya in the ninth century is perhaps the most well-known scholarship on societal collapse.[63] Over the last twenty years, explanations for the Maya collapse have increasingly focused on climate. Some of these scholars have argued for a predominant emphasis—perhaps even too much emphasis—on climate (drought in particular) as the most significant factor leading to Maya population decline and abandonment of advanced urban communities.[64] Others have identified a host of explanations, climate among them, for the Maya collapse.[65] Justine Shaw argues that localized climate evidence for lake cores reveals a mosaic of climatic conditions, rather than a uniform drought. Consequently, the Maya collapse resulted from the combined effects of Maya land use practices, local climate dynamics stemming from deforestation, and global climate change between 700 and 900 A.D.[66]

Jared Diamond's popular research on societal collapse—with case studies of climatic impacts on the Maya, Anasazi, Greenland Norse, and other groups—fits with historiographical tendencies to see climate change as a major influence on human history, but one that is always acting alongside other historical forces. While his book *Collapse* may be problematic due to its shaky attempt to draw lessons for today, Diamond nonetheless avoids some of the geographical determinism present in other works on climate history and in his own *Guns, Germs, and Steel*.[67] He argues in *Collapse* that civilizations fell not because of a single cause, but rather because of five interacting historical forces: hostile neighbors; isolation from trade; natural resource depletion; cultural inflexibility; and climate change. Drought for the Anasazi and Maya, then, may have pushed them over the edge and led to their downfall. But if drought had not accompanied and exacerbated warfare, social unrest, and the overuse of nearby resources, then drought alone likely would not have triggered societal collapse.

Social unrest, rebellion, and political instability can also stem from climate change—and research in this area forms a second major subfield in the study of climate impacts on societies. Richard Grove, for example, examines the global effects of the 1788–1795 El Niño event (which contributed to drought and famine in Africa and India), Native American warfare on the northern plains of the US, and revolution in France.[68] Others have shown how weather-induced crop failures, rising food prices, hunger and disease, and the formation of social movements led to rebellions against—or at least discontent with—governments.[69] Sam White has shown how climate during the Little Ice Age influenced the fate of the Ottoman Empire.[70] Climatic explanations for the birth of independence movements in Mexico also fall into this category of social unrest. Susan Swan, for example, attributes climatic variability in the late Little Ice Age to the massive uprisings that initiated Mexico's independence struggle from Spain. Perhaps exaggerating the singular role of climate, she suggests that eighty thousand people rebelled in 1810 because bad weather (a series of droughts) had destroyed their crops and raised prices. "In an agricultural economy," Swan maintains, "perhaps no other environmental factor is of such concern as bad weather."[71] Other scholars have expanded upon Swan's work to show both longer term climatic effects on the population of New Spain (colonial Mexico) and more precise connections between climate, agriculture, and social rebellion.[72] Georgina Endfield's research—in contrast to Swan's more climatically deterministic assertions—reveals how numerous other social and political forces were the root causes of discontent. Climate change exacerbated those preexisting conditions and worked alongside numerous societal forces to generate social change over time.

Beyond the collapse of civilizations and climate-induced social unrest, research on weather-related disasters, especially hurricanes, drought, and El Niño, accounts for a third principal area of this scholarship on societal impacts of climate. Whether studying the impacts of mega El Niños on Peru's Moche people or drought on the Maya or farmers in the American southern plains, the study of climatic catastrophes has become a rich subfield of climate history.[73] Though more often weather-related than climatic events, these disasters are part of climate history because they occur repeatedly through time or can change in response to climatic variability. They also effectively grapple with issues of agency and the capacity for weather and climate to influence historical trajectories. The circum-Caribbean basin, for example, has experienced periodic hurricanes for centuries, and these hurricane patterns may even be linked to El Niño events and other broader climatic conditions.[74] Residents of northeastern Brazil continue to deal with severe drought, as do people in China, northern

Africa, Australia, and many other world regions. El Niño has also attracted widespread and far-reaching attention from scholars, with a recent effort to understand El Niño as a global disaster.[75] Some scholars have linked climate not just to disasters in human history, but also to animal losses, such as during the Pleistocene Extinctions after the end of the last ice age.[76]

Scholarship on climatic catastrophes is vast and has, in general, remained more tied to natural disaster scholarship than to climate history.[77] Disaster scholarship tends to examine issues of power and social relations more than climate history does. As Nicholas Gabriel Arons explains in his analysis of repeated drought in northeast Brazil, "it is not droughts alone, but droughts within the context of social and political realities" that make them so deadly for some but not others, and thus so disproportionately experienced.[78] This conclusion parallels arguments among disaster scholars more broadly, who show that marginalized populations are generally the most vulnerable.[79] Others have drawn similar conclusions about drought in Brazil and elsewhere. One of the region's most deadly droughts (1877–1879), caused between 200,000 and 500,000 deaths. During this Great Drought, government leaders and other elites monopolized the discourse about the drought—about the climatic events, the state's relief efforts, and about the drought's victims, their homelands, and their behavior. Regional and national leaders actually provided little disaster relief, leaving many residents without political recourse who subsequently died by the thousands. But elites manipulated and controlled the discourse to convince other Brazilians, especially those far from the drought zone, that these leaders had in fact responded with compassionate charity. In reality, though, elite views of lower-class drought victims animated government response to the Great Drought. As Greenfield explains, the drought victims were powerless: "Their needs and best interests, then, would be determined by those at the top." Worse, elite rhetoric portrayed disaster victims through the same historical lens that had characterized lower classes as lazy and backward—a narrative that reinforced elite social standing in Brazil. The "discursive reality" of the Great Drought, as Greenfield summarizes, was not only death and displacement, but also the process by which elite discourse "transformed the *sertanejos* from innocent victims of a natural disaster into shiftless architects of their own misfortune."[80] In short, the power to define and control drought discourse— the narrative—shaped the societal impacts of climate on Brazilians. Timothy Finan's conclusions about the politics of drought in northeast Brazil go even further. "History," he explains, "shows that drought-related policy is not always motivated by humanitarian concerns or by constituency demand, but rather becomes both a discourse and a stratagem for consolidating political power and, in many cases, accumulating personal wealth."[81]

By examining the social history underlying climatic events, scholars can eschew environmental determinism and more effectively understand historical agency. They can also achieve what many historians have increasingly been urging since the late 1990s: a merger of social history with environmental history.[82] Eric Klinenberg, for example, explores the underlying social forces that produced a 1995 disaster in Chicago.

Hundreds of people died in that heat wave, he asserts, not because of the weather, but because of "social factors," namely the making of social outcasts. Elderly, poor, and other isolated residents became vulnerable to the weather through a host of economic, political, social, cultural, and environmental processes. Without social networks and money, and within a society that put political aspirations and economic gains ahead of people's well-being, the isolated residents could not withstand the heat.[83] The agency and causal factors underlying the calamity in Chicago were thus, for Klinenberg, centered on society.

In his analysis of El Niño events in the late nineteenth century, Mike Davis puts El Niño into the context of global politics and the international creation of weather disasters, and he thus offers a climate history on a radically different scale than Klinenberg's local analysis of Chicago.[84] El Niño events in the late nineteenth century killed 30 million people in Brazil, China, and India. To be sure, the physical aspects of droughts, epidemics, and crop failures caused by El Niño events generated tragic outcomes. But Davis argues that famines are political events more than climatic irregularities.[85] Millions died because of the "fatal meshing of extreme events between the world climate system and the late Victorian world economy."[86] In the late nineteenth century, global capitalism and imperialism had altered labor patterns, changed household structures, shifted connections between local and regional production systems, allowed global grain markets to influence local food prices, and made people dependent on the colonial state. El Niño events triggered regional famines because western European imperial powers had taken control of local people's land and labor, and thus their ability to grow and procure food. Understanding these interconnections on a global scale, and across multiple countries and regions, required Davis to ignore typical geographical boundaries in historical research and instead conduct an international, interdisciplinary analysis. This tackling of a new scale that is so challenging in historical research subsequently allowed him to draw connections and understand historical processes and contingencies unachievable without the climatic lens that forced him to reject the usual scales (primarily the nation state) that shape environmental history research.

What also makes Davis's book—and some other recent scholarship on societal impacts of climate—important for climate research is that it demonstrates how climate is not something "out there" that acts upon a society. Rather, as environmental historians have been trying to show for decades, nature and culture blur together. Climate is no exception: its impact on society cannot be divorced from the power structures and social divisions that exist on the ground. Much of the new research on societal impacts of climate—whether causing societal collapse, social unrest, or disasters—shows how social relations, political economy, and the social construction of science interacted with the agency of climate.[87] And some historical studies show agency in the other direction, plotting how people have affected the climate, such as William Ruddiman, who shows how humans have influenced the global climate for thousands of years.[88] This type of scholarship is obviously at the center of the more recent trend to consider the Anthropocene as the most recent geological epoch.

CULTURE

As climatological data becomes more precise and locally specific, and as studies of the social impacts of climate have become more varied and numerous, scholars increasingly recognize that distinct regions and people experience climate differently. Consequently, a growing number of studies examine ways in which culture shapes human-climate interactions both in the past and present. Aqqaluk Lynge, the president of the Inuit Circumpolar Council in Greenland, recently summed up the discrepancies and the subsequent need for more cultural analyses of climate. "We are all in this together," Lynge explains about global warming. "Yet our perceptions are different. This is the challenge."[89] More than a challenge for policymakers, cultural variation through time and space has also led scholars to focus considerably more attention on both perceptions of and responses to climate change. After all, different societies in distinct locations may perceive the same weather quite differently. Moreover, many researchers, especially historians of science and anthropologists, increasingly treat Western climate science not strictly as the objective facts on which to base arguments and draw conclusions, but rather as a discourse in and of itself—a cultural, social, and political construction worthy of scrutiny, not passive acceptance. Cultural approaches to climate history—what some scholars call "cultural climatology"—have thus become common.[90] As the anthropologists Sarah Strauss and Ben Orlove explain in the introduction to their provocative volume on culture-climate links, people's everyday conversations, their proverbs, oral histories, scientific discourse, and cultural constructions of time, all shape our understandings of climate.[91] Culture, then, acting in dialogue with the physical aspects of weather, shapes how people experience and respond to climatic change—and these understandings and responses vary not only throughout history, but also in diverse places around the world.

The cultural emphasis in climate history has emerged in recent decades for several reasons. First, climate data have become much more precise, complete, and far-reaching during the last several decades. Historians are thus borrowing climatic records from scientists and freeing themselves from their decades-long dedication to reconstructing past climates. Second, research now demonstrates clearly that different regions experience major climatic shifts differently. The Little Ice Age, for example, had different characteristics, as well as distinct timelines, in Western Europe, North America, and South America.[92] These geographical and chronological disparities produced unique climatic experiences for different societies. Third, as present-day climate change increasingly affects diverse populations such as the Inuit, these indigenous people are publicizing their experiences with and perceptions of global warming. These new voices have influenced the media and policymakers, as well as scholars.[93] Fourth, during the past few decades, postmodernism, poststructuralism, and the Cultural Turn have inspired researchers to delve more profoundly into issues such as culture, discourse, traditional ecological knowledge, and folk or indigenous science.[94] Fifth, research in science and technology studies increasingly scrutinizes science as a social construction, emerging historically from cultural, political, economic, and social conditions that are place-dependent rather than universal truths.[95] This scholarship thus analyzes climate science not only as one among many cultural discourses, but also as a historical narrative with the capacity to empower (or disempower) certain social groups. As a result of these historiographical, social, and scientific trends, climate history scholarship has increasingly turned toward cultural analyses that focus on three areas: (1) perceptions and understandings; (2) expressions of climate; and (3) responses to climate change.

Perceptions of climate vary through time and space.[96] Fortunately, most social scientists now recognize that indigenous or folk knowledge should not be thrown aside or neglected for its supposed lack of scientific rigor. Stories about the Inuit spirit master Narssuk, a "giant infant who was unpredictable in behavior and temperament, just like the Arctic weather," or about an animal-like glacier in northwestern Canada, barreling down on Tlingit people because they spoke too loudly or cooked with grease, represent valid views of weather and the consequences of climate change.[97] Moreover, as Julie Cruikshank contends, if researchers carefully analyze indigenous oral history, they may realize that Western thinkers and indigenous storytellers often say the same things, even if they express them in radically different ways. In her own studies of Tlingit and Athapaskan people's relationship with glaciers and climate change in northwestern North America, Cruikshank suggests that "Romantic poets, Tlingit elders, and prestigious geophysicists…bring different approaches to understanding climate change, but they no longer sound as incompatible as they once did."[98] All of these groups may have thought glaciers surged and retreated for different reasons. Yet, when compared, they all recognized the historical interplay between climate change and glacier behavior—and all three groups charted those advances and retreats historically, based on the climatic variability at the end of the Little Ice Age.

Scholars have also recognized that multiple and diverse understandings of climate exist in different world regions and throughout history. In his work on China since the seventh millennium B.C., for example, Cho-yun Hsu argues that, "social memory concerning climate is embedded in both high-level cosmology and local lore."[99] More than simply cultural constructions of climate, however, perceptions changed according to China's regional variations in geography and corresponding climate. While people in the north feared drought, for example, those in the south often prayed for relief from floods. Culture, climate, and space thus intersected to form people's attitudes about weather. These examples also show the importance of scale—both spatial and temporal—in historical studies of climate. Other forces shaped Chinese beliefs, too, such as the yin-yang dualism that influenced perceptions of seasons and seasonal change. Climate change, Hsu explains, was also deeply connected to human behavior because the equilibrium of the cosmic order could be upset by human actions, especially those of the state. Within this cultural framework, witchcraft and impeachment of rulers served as possible means of regulating the climate and restoring the cosmic equilibrium. In other contexts, perceptions of climate could also unite people as well. Jan Golinksi has suggested that shared climate—and, more importantly, shared understandings of climate—helped forge a British identity during the Enlightenment.[100] In short, physical climate may be grounded in specific locations and possess observable characteristics. But people's understanding of those physical conditions depend on their memory, their lore, their spiritual beliefs, their political structures, and their methods of observing and recording climatic data. Essentially, climate is as much culture as weather.

Culture also influences how people express their understandings of climate, and art offers one important way to present those perceptions. Recognizing the art-climate connection, Hans Neuberger analyzed more than 12,000 paintings, located in nine countries and dating back to 1400. He shows how artists represented climatic conditions during the Little Ice Age by examining cloud cover, visibility, and the amount of clothing people wore in each painting (to gauge temperature shifts).[101] While Neuberger's work fits most accurately into the climate sources category of the historiography, his meticulous and expansive research also indicates how paintings and art can reveal cultural perceptions and expressions of climate. Artists, after all, make decisions about what they paint based on artistic trends and their cultural understandings of the world around them. In another case, John Thornes and Gemma Metherell studied Monet's paintings of London in the late nineteenth century.[102] Combining an analysis of climate and art, as well as contemporary residents' fears of smog and epidemics, Thornes and Metherell place their study within Britain's unique historical context: the Victorian era, when London was among both the most advanced and powerful cities in the world, not to mention the most polluted. Perceptions of climate, then, evolved alongside political economy and scientific discourse—and Monet recorded this history in the one hundred paintings of the "London Series." In all of these cases, artists used their own judgments about how to depict nature. Their paintings thus not only represented climatic change over time—Neuberger found, for instance, that paintings chronicled Little Ice Age climate change quite accurately—but they also revealed human perceptions of and anxieties about those changes. While these studies of climate perceptions suggest innovative developments in the historiography, a great deal more research on climate expressions and cultural perceptions remains to be done, especially outside of Western Europe.

Scholars are increasingly examining how culture affects not only understandings and expressions of climate, but also societal responses to climate change. For environmental historians, Donald Worster's classic *Dust Bowl* may be one of the most well-known explanations for climate disaster. And Worster claims that cultural factors—the relentless clinging to capitalist agriculture and the belief in techno-scientific tools to overcome environmental obstacles—led southern plains farmers to over-exploit land, soil, and water. These farmers' belief that "rain follows the plow" also shows how folk knowledge should not just be attributed to or considered for indigenous societies, because even these Euro-American settlers held views that do not correspond with so-called Western science. When the Dust Bowl drought came in the early 1930s, it generated a catastrophe—a disaster that stemmed from customs, economy, beliefs, and an inability to adapt. In short, Worster argued that culture, not climate, caused the early Dust Bowl.[103]

Wolfgang Behringer provides one of the most fascinating historical analyses of human responses to climate change in his study of European witchcraft trials at the onset of the Little Ice Age. He argues that the erratic, cold, and stormy weather associated with the most extreme periods of the Little Ice Age were the same periods in which witch-hunts occurred. During the 1430s, 1450s, and 1480s–1490s, when hailstorms and other inclement weather were particularly destructive of crops, ecclesiastic leaders transformed weather-making into a crime. Most witches were charged with weather-related offenses, such as storms or crop failures. Moreover, the periods of the most aggressive witch hunts (the 1560s, 1580s, 1620s, and 1680s–1690s) also corresponded to some of the most devastating weather of the Little Ice Age. Behringer thus sees an "interdependence of meteorological disaster, crop failure and a popular demand for witch-hunts." Not only do his conclusions contribute to cultural studies of climate history, but, methodologically, Behringer challenges scholars' overreliance on Western scientific data to draw conclusions about past climate-society interactions. As he asserts, "Inasmuch as scientists and scholars have previously based their periodizations upon indicators drawn from the physical environment (dendrochronology, glaciology, etc.), this essay proposes another approach. I suggest taking into account the subjective factor and to consider human reactions to climatic changes as an important indicator for an assessment of the beginning, the duration and the end of the Little Ice Age."[104] As a reminder of how scholarship has changed over time, Le Roy Ladurie argued just the opposite in *Times of Feast, Times of Famine*. Christian Pfister has also shown that the exceptionally cold periods of the Little Ice Age, especially during the late sixteenth and early seventeenth-centuries, led to both food shortages and increased witchcraft trials in Europe. In fact, Pfister uses the example of this connection between witchcraft and climate change to make "a plea for bridging the gap separating studies of climate from those of culture."[105]

Cultural understandings of climate—and its relationship to geography and people—also affected European colonialism, especially in tropical regions. As Richard Grove summarizes, "Ideas about environmental influence, and about the influence of climate on culture in particular, increased in effectiveness as European expansion proceeded in all parts of the world."[106] Beliefs about colonists' ability to alter landscapes, change the weather, and improve their health in what they saw as hostile climates also shaped European views of the colonial process. In North America and the Caribbean, for example, English colonists worried about how the hot climate would affect their moral character and their health. Disgusted by the oppressive, deleterious heat of Barbados, they feared long-term settlement in tropical climates would erode their morality, not to mention their health. They surmised that living in hot or tropical climates for more than a few years would result in their adopting the negative characteristics of the native inhabitants of tropical America and Africa, people they saw as slothful. More practically, European colonists studied local and regional climates to learn which crops they could grow and export.[107] Karen Kupperman and Richard Grove, among others, offer excellent studies to show not only what Europeans thought about climate, but also their responses during the seventeenth and eighteenth centuries, such as the ways in which cultural perceptions shaped their actions, the manner of colonization, the location of colonies, the crops planted, the goods produced, and colonists' ability to adapt to new climates. Another recent study demonstrates how the culture of scientific discourse and colonial perceptions later interacted with trends in travel and tourism to transform the Caribbean climate from what British colonists construed as a deadly climate in the seventeenth century to the ideal tourist climate for sun, sand, and sea by the twentieth century. After all, the beaches of Barbados—the place European colonists previously feared—have for the last century ranked as one of the world's top tourist destinations, in part because of their enticing climate. The tropical weather itself changed little over those three hundred years from 1600 to 1900, but the science and discourse reversed course entirely—turning the climate from deadly to healthy, from dangerous to rejuvenating.[108] Clearly, cultural analyses of climate can illuminate as much about broader historical processes as they can reveal about climate and weather.

CONCLUSIONS

Climate history research has clearly moved beyond Ellsworth Huntington's era, when environmental determinism, simplistic mono-causal explanations, and racist interpretations of climate-geography-society interactions dominated social scientific analyses of climate. Significant contributions to climate history are now visible in four areas. First, historians have helped reconstruct past climates by augmenting scientific data with historical information from diaries, church records, government documents, and other sources normally overlooked by climatologists. Second, scholars have examined the history of climatology and meteorology to understand a great diversity of environmental knowledge. Much of this research historicizes the discovery of global warming or explains the evolution of knowledge about climate-emissions relationships. Additionally, many historians of science are now analyzing science as discourse, as a particular way of understanding climate. Scholars thus study climate scientists as historical actors whose scientific knowledge has not only illuminated atmospheric processes, but also has served political interests, affected social relations, and perpetuated or created power imbalances among social groups, classes, and nations. Third, a steadily growing body of literature analyzes agency to discover how climate affects societies. This research has become more nuanced and sophisticated since the days of climatic determinism. It now considers climate alongside many other forces in causing major societal transformations, social unrest, and climatic disasters. Fourth, climate researchers are increasingly studying culture to understand how perceptions of and responses to climate vary in space and time.

Existing climate historiography reveals key achievements and limitations that should inspire environmental historians not only to do more climate research but also to refine and broaden their approaches to environmental history in general. In other words, cross-pollination between environmental history and climate history could enrich all research fields that study climate. Four areas for innovation stand out: scale, boundaries, historical agency, and interdisciplinary research.

Climate research forces scholars to grapple with different scales because climate varies in both time and space.[109] To put global warming in context, for example, researchers extend their studies back hundreds of thousands, or even millions, of years. On another scale, historians studying weather and climate often find diaries that mention "the worst storm in my life" or other such comments that impose a generational timeframe. Weather also changes over the course of a year or a day, and devastating meteorological disasters such as a hurricane can last just hours. While time scales obviously vary from minutes to millions of years, spatial scales can also vary from a valley to a nation to the earth system. Smaller spatial scales exist, too, such as a climate-controlled office building. Scholars thus study a range of scales, from Brian Fagan's study across nearly 20,000 years to Mike Davis's global analysis of El Niño in the late 1800s to what Vladimir Jankovic calls the "intimate climates" of indoor air.[110] Researchers need both dexterity and flexibility in handling these often overlapping scales, which require historians to step outside a number of academic disciplinary boundaries that usually restrict them to narrow time periods in specific geographical regions.

The problem of scale also raises the issue of boundaries. When does weather turn into climate? Where does the atmosphere end, especially considering that we take air inside our bodies with every breath? Does the existence of a "heat island effect" mean that we should distinguish urban from rural climates? Is there a distinction between "natural" and human-caused climate change? Who has the power to answer these questions or to use the answers to make policies? Most of these questions, which remain insufficiently examined, point to a persistent reluctance among Westerners to blur nature-culture divisions. But breaking down these boundaries—or showing how and why past societies have imposed them—will illuminate important histories of science, culture, social relations, power dynamics, and environmental processes.

Unresolved questions about historical agency also emerge in the climate historiography. Environmental determinism no longer dictates our understanding of climate-society interactions. But, in some ways, fear of overemphasizing the role of climate in history has caused scholars to focus too exclusively on human agency, ignoring altogether the effects of climate on society, except in cases of climatic disasters. Although historians of science have recently produced some of the most innovative, compelling climate research, they very rarely treat climate or weather as an active historical actor. Environmental historians are uniquely positioned to study climate as a dynamic historical force without venturing into environmental determinism.[111]

The need for interdisciplinary research presents additional obstacles and opportunities for historians of climate. Climate histories have always stood out for their interdisciplinary breadth, especially the cross-disciplinary alliances bridging the social and environmental sciences.[112] Yet additional interdisciplinary links could better connect the social sciences and humanities. Environmental historians should thus read not only history and climatology, but also the history of science, political ecology, cultural ecology, and archaeology. They could also collaborate and coauthor research with scholars from this range of social and natural sciences to enhance dialogue, enrich conclusions, and reach broader audiences.

Issues related to scale, boundaries, agency, and interdisciplinarity point to achievements in the existing scholarship and areas for new innovation. Yet trends emerging in the climate historiography already reveal important transformations. The past dedication to reconstructing climate data and tracing the history of meteorology has given way to social and cultural analyses of climate. Historians of science have conducted some of the most innovative cultural critiques of climate, while a range of researchers—from geographers and anthropologists to journalists, sociologists, and environmental historians—have examined the social dimensions of historical climate change. Scholars now recognize that societies' experiences with climate have as much to do with physical weather over time as with power relations, social divisions, scientific discourse, and the cultural construction of climate knowledge. In other words, climate histories have gone beyond weather to focus on culture and power.

NOTES

1. Ellsworth Huntington, *Civilization and Climate*, 3d. ed. (New Haven, CT: Yale University Press, 1924 [1915]), 3.

2. For a good analysis of Huntington's climate theories, see James Rodger Fleming, *Historical Perspectives on Climate Change* (New York: Oxford University Press, 1998), chap. 8.

3. Mike Hulme, "Reducing the Future to Climate: A Story of Climate Determinism and Reductionism," *Osiris* 26 (2011): 245–266.

4. William Cronon, ed., *Uncommon Ground: Rethinking the Human Place in Nature* (New York: W.W. Norton, 1996); Richard White, "From Wilderness to Hybrid Landscapes: The Cultural Turn in Environmental History," *The Historian* 66, no. 3 (2004): 557–564; Mark Carey, "Latin American Environmental History: Current Trends, Interdisciplinary Insights, and Future Directions," *Environmental History* 14, no. 2 (2009): 221–252.

5. These cultural analyses of climate history often fulfill some of the objectives William Cronon outlined: namely, writing histories of past environmental stories and narratives. See William Cronon, "A Place for Stories: Nature, History, and Narrative," *Journal of American History* 78, no. 4 (1992): 1347–1376.

6. I have identified and discussed these categories elsewhere, though with a different analysis and body of literature. See Mark Carey, "Climate and History: A Critical Review of Historical Climatology and Climate Change Historiography," *Wiley Interdisciplinary Reviews: Climate Change* 3, no. 3 (2012): 233–249.

7. Confusingly, Europeans tend to use "historical climatology" to refer to what those in the US call "climate history." For this discussion and a more elaborate definition of this field that combines climatology with history, see Rudolf Brázdil et al., "Historical Climatology in Europe—The State of the Art," *Climatic Change* 70, no. 3 (2005): 363–430.

8. On the history of weather scholarship, see Brant Vogel, "Bibliography of Recent Literature in the History of Meteorology: Twenty Six Years, 1983–2008," *History of Meteorology* 5 (2009): 23–125.

9. For an overview of diverse data and methods used to reconstruct past climates, see Vogel, "Bibliography of Recent Literature in the History of Meteorology." For an overview of long-term climate change, see N. M. Chumakov, "Trends in Global Climate Changes Inferred from Geologic Data," *Stratigraphy and Geological Correlation* 12, no. 2 (2004): 117–138.

10. For overviews of major shifts in the history of global climate, see Eric J. Steig, "Mid-Holocene Climate Change," *Science* 286, no. 5444 (1999): 1485–1487; Richard Alley, *The Two Mile Time Machine: Ice Cores, Abrupt Climate Change, and Our Future* (Princeton, NJ: Princeton University Press, 2000); Paul Andrew Mayewski and Frank White, *The Ice Chronicles: The Quest to Understand Global Climate Change* (Hanover: University of New Hampshire/University Press of New England, 2002).

11. Paul J. Crutzen, "The 'Anthropocene,'" in *Earth System Science in the Anthropocene*, ed. E. Ehlers and T. Krafft (Berlin: Springer, 2006), 13–18.

12. For long-term analyses of historical climate-society interactions, see Wolfgang Behringer, *A Cultural History of Climate* (Malden, MA: Polity Press, 2010); Brian Fagan, *The Great Warming: Climate Change and the Rise and Fall of Civilizations* (New York: Bloomsbury Press, 2008); Brian Fagan, *The Long Summer: How Climate Changed Civilization* (New York: Basic Books, 2004); H. H. Lamb, *Climate, History and the Modern World*, 2d ed. (New York: Routledge, 1995 [1982]); H. H. Lamb, *Climate: Present, Past and Future, Vol. 2: Climatic History and the Future* (New York: Barnes and Noble Books, 1977); Arlene Miller Rosen, *Civilizing Climate: Social Responses to Climate Change in the Ancient Near East* (Lanham, MD: AltaMira Press, 2007).

13. J. A. Dearing, "Climate-Human-Environment Interactions: Resolving Our Past," *Climate of the Past* 2, no. 2 (2006): 187–203. For two excellent interdisciplinary social scientific studies that utilize climatological evidence, see Izumi Shimada et al., "Cultural Impacts of Severe Droughts in the Prehistoric Andes: Application of a 1,500-Year Ice Core Precipitation Record," *World Archaeology* 22, no. 3 (1991): 247–270; Justine M. Shaw, "Climate Change and Deforestation: Implications for the Maya Collapse," *Ancient Mesoamerica* 14 (2003): 157–167.

14. For other studies dealing with climate but not as the principal focus, see for example Walter Prescott Webb, *The Great Plains* (Lincoln: University of Nebraska Press, 1959); James Claude Malin, *The Grassland of North America, Prolegomena to Its History* (Gloucester, MA: P. Smith, 1967).

15. Robert H. Claxton, "Climate and History: From Speculation to Systematic Study," *The Historian* 45, no. 2 (1983): 220.

16. See Rudolf Brázdil et al., "Historical Climatology in Europe—The State of the Art"; Behringer, *A Cultural History of Climate*.

17. Stephen Daniels and Georgina H. Endfield, "Narratives of Climate Change: Introduction," *Journal of Historical Geography* 35, no. 2 (2009): 215–222.

18. Emmanuel Le Roy Ladurie, *Times of Feast, Times of Famine: A History of Climate Since the Year 1000* (Garden City, NY: Doubleday, 1971); Donald Worster, *Dust Bowl: The Southern Plains in the 1930s* (New York: Oxford University Press, 1979). Also see Richard Grove, *Ecology, Climate and Empire: Colonialism and Global Environmental History, 1400–1940* (Cambridge, UK: White Horse Press, 1997); Richard Grove, *Green Imperialism: Colonial Expansion, Tropical Island Edens and the Origins of Environmentalism, 1600–1860* (New York: Cambridge University Press, 1995).

19. See, for example, Mart A. Stewart, "'Let Us Begin with the Weather?': Climate, Race, and Cultural Distinctiveness in the American South," in *Nature and Society in Historical Context*, ed. Mikulás Teich, Roy Porter, and Bo Gustafsson (New York: Cambridge University Press, 1997); David N. Livingstone, "The Moral Discourse of Climate: Historical Considerations on Race, Place and Virtue," *Journal of Historical Geography* 17, no. 4 (1991): 413–434.

20. J. Donald Hughes, *What is Environmental History?* (Malden, MA: Polity Press, 2006), 97.

21. For provocative studies that treat environmental science as discourse, see, for example, Linda Nash, "The Changing Experience of Nature: Historical Encounters with a Northwest River," *Journal of American History* 86, no. 4 (2000): 1600–1629; Bruce Braun, "Producing Vertical Territory: Geology and Governmentality in Late Victorian Canada," *Ecumene* 7, no. 1 (2000): 7–46.

22. William Ruddiman does contend that humans have influenced climate for at least the last eight thousand years. See William Ruddiman, *Plows, Plagues, and Petroleum: How Humans Took Control of Climate* (Princeton, NJ: Princeton University Press, 2005).

23. For an excellent effort at this theorization, see Paul S. Sutter, "Nature's Agents or Agents of Empire? Entomological Workers and Environmental Change during the Construction of the Panama Canal," *Isis* 98 (2007): 724–754.

24. For discussions of these other realms of environmental history, see, for example, Mark Carey, "The History of Ice: How Glaciers Became an Endangered Species," *Environmental History* 12, no. 3 (2007): 497–527; W. Jeffrey Bolster, "Opportunities in Marine Environmental History," *Environmental History* 11, no. 3 (2006): 567–597.

25. Richard White, "The Nationalization of Nature," *Journal of American History* 86, no. 3 (1999): 976–986.

26. For a good representation of this historiographical thrust at its heyday, see the *Journal of Interdisciplinary History*, Special Issue on "History and Climate: Interdisciplinary Explorations," 10, no. 4 (1980).

27. Le Roy Ladurie, *Times of Feast, Times of Famine*, 2, 11.

28. Some today still share Ladurie's skepticism about the role of climate. See the discussion in, for example, Simon G. Haberle and Alex Chepstow Lusty, "Can Climate Influence Cultural Development? A View through Time," *Environment and History* 6 (2000): 349–369.

29. "Podcast 7: Climate History and a Forest Journey," 9 Dec. 2006. http://www.eh-resources.org/podcast/podcast2006.html.

30. Lizardo Seiner Lizárraga, *Estudios de historia medioambiental, Perú, siglos XVI–XX* (Lima, Peru: Universidad de Lima, 2002); Michael Chenoweth, *The 18th Century Climate of Jamaica Derived from the Journals of Thomas Thistlewood, 1750–1786* (Philadelphia: Transactions of the American Philosophical Society, 2003); Ioannis G. Telelis, "The Climate of Tübingen A.D. 1596–1605, on the Basis of Martin Crusius' *Diarium*," *Environment and History* 4, no. 1 (Feb. 1998): 53–74; Wang Shao-wu and Zhao Zong-ci, "Droughts and Floods in China, 1470–1979," in *Climate and History: Studies in Past Climates and Their Impact on Man*, ed. T. M. L. Wigley, M. J. Ingram, and G. Farmer (New York: Cambridge University Press, 1981), 271–288.

31. See, for example, Arun Agrawal, "Dismantling the Divide between Indigenous and Scientific Knowledge," *Development and Change* 26 (1995): 413–439; D. Green and G. Raygorodetsky, "Indigenous Knowledge of a Changing Climate," *Climatic Change* 100 (2010): 239–242.

32. For a compilation of the most recent history of science research on climate, see the special edition of *Osiris* 26 (Sept. 2011), ed. James Rodger Fleming and Vladimir Jankovic.
33. See, for example, Robert I. Rotberg and Theodore K. Rabb, eds., *Climate and History: Studies in Interdisciplinary History* (Princeton, NJ: Princeton University Press, 1981); Wigley, Ingram, and Farmer, eds., *Climate and History*.
34. Spencer R. Weart, *The Discovery of Global Warming* (Cambridge, MA: Harvard University Press, 2003).
35. James Rodger Fleming, *The Callendar Effect: The Life and Work of Guy Stewart Callendar (1898–1964), the Scientist Who Established the Carbon Dioxide Theory of Climate Change* (Boston: American Meteorological Society, 2007); Mark Bowen, *Thin Ice: Unlocking the Secrets of Climate Change in the World's Highest Mountains* (New York: Henry Holt, 2005).
36. James Rodger Fleming, *Fixing the Sky: The Checkered History of Weather and Climate Control* (New York: Columbia University Press, 2010).
37. Kristine C. Harper, "Climate Control: United States Weather Modification in the Cold War and Beyond," *Endeavour* 32, no. 1 (2008): 20–26, quote from p. 20.
38. Ronald E. Doel and Kristine C. Harper, "Prometheus Unleashed: Science as a Diplomatic Weapon in the Lyndon B. Johnson Administration," *Osiris* 21 (2006): 66–85.
39. Lamb, *Climate: Present, Past and Future*, 660–662.
40. Fleming, *The Callendar Effect*, chap. 4.
41. Fleming, *Fixing the Sky*; Mark Carey, Adam French, and Elliott O'Brien, "Unintended Effects of Technology on Climate Change Adaptation: An Historical Analysis of Water Conflicts Below Andean Glaciers," *Journal of Historical Geography* 38, no. 2 (2012): 181–91.
42. For an excellent example, see Paul N. Edwards, *A Vast Machine: Computer Models, Climate Data, and the Politics of Global Warming* (Cambridge, MA: MIT Press, 2010). Bruno Latour has stimulated some of these broader trends in this history of science and technology; see Bruno Latour, *The Pasteurization of France* (Cambridge, MA: Harvard University Press, 1988); Latour, *Science in Action: How to Follow Scientists and Engineers Through Society* (Cambridge, MA: Harvard University Press, 1987). Also see Wiebe E. Bijker, *Of Bicycles, Bakelites, and Bulbs: Toward a Theory of Sociotechnical Change* (Cambridge, MA: MIT Press, 1995).
43. Fleming, *Historical Perspectives on Climate Change*, chap. 3, quote from p. 9. For another good example of this trend in climate history, see Vladimir Jankovic, *Reading the Skies: A Cultural History of English Weather, 1650–1820* (Chicago: University of Chicago Press, 2001).
44. Katharine Anderson, "Mapping Meteorology," in *Intimate Universality: Local and Global Themes in the History of Weather and Climate*, ed. James Rodger Fleming, Vladimir Jankovic, and Deborah R. Coen (Sagamore Beach, MA: Science History Publications, 2006), 86.
45. Erik Larson, *Isaac's Storm: A Man, A Time, and the Deadliest Hurricane in History* (New York: Vintage Books, 2000).
46. Clark A. Miller and Paul N. Edwards, "Introduction: The Globalization of Climate Science and Climate Politics," in *Changing the Atmosphere: Expert Knowledge and Environmental Governance*, ed. Clark A. Miller and Paul N. Edwards (Cambridge, MA: MIT Press, 2001), 5.
47. Paul N. Edwards, *A Vast Machine: Computer Models, Climate Data, and the Politics of Global Warming* (Cambridge, MA: MIT Press, 2010).

48. Richard White, "From Wilderness to Hybrid Landscapes: The Cultural Turn in Environmental History," *The Historian* 66, no. 3 (2004): 560–561.

49. Katharine Anderson, *Predicting the Weather: Victorians and the Science of Meteorology* (Chicago: University of Chicago Press, 2005), 237.

50. Livingstone, "The Moral Discourse of Climate," 414. Also see Theodore M. Porter, *Trust in Numbers: The Pursuit of Objectivity in Science and Public Life* (Princeton, NJ: Princeton University Press, 1995).

51. Clarence J. Glacken, *Traces on the Rhodian Shore: Nature and Culture in Western Thought from Ancient Times to the End of the Eighteenth Century* (Berkeley: University of California Press, 1967), chap. 12; Philip D. Curtin, *Death by Migration: Europe's Encounter with the Tropical World in the Nineteenth Century* (New York: Cambridge University Press, 1989); Nancy Leys Stepan, *The Hour of Eugenics: Race, Gender, and Nation in Latin America* (Ithaca, NY: Cornell University Press, 1991); Stepan, *Picturing Tropical Nature* (Ithaca, NY: Cornell University Press, 2001); William G. Palmer, "Environment in Utopia: History, Climate and Time in Renaissance Environmental Thought," *Environmental Review* 8, no. 2 (Summer 1984): 162–178.

52. Quoted in Stewart, "'Let Us Begin with the Weather?'" 249.

53. Deborah R. Coen, "Scaling Down: The 'Austrian' Climate between Empire and Republic," in *Intimate Universality*, 115–140; Gregory T. Cushman, "The Struggle over Airways in the Americas, 1919–1945: Atmospheric Science, Aviation Technology, and Neocolonialism," in *Intimate Universality*, 175–222. For links between science and state-building more broadly, see James Scott, *Seeing Like a State: How Certain Schemes to Improve the Human Condition Have Failed* (New Haven, CT: Yale University Press, 1998); Raymond B. Craib, *Cartographic Mexico: A History of State Fixations and Fugitive Landscapes* (Durham, NC: Duke University Press, 2004); Stuart McCook, *States of Nature: Science, Agriculture, and Environment in the Spanish Caribbean, 1760–1940* (Austin: University of Texas Press, 2002); Gyan Prakash, *Another Reason: Science and the Imagination of Modern India* (Princeton, NJ: Princeton University Press, 1999); Braun, "Producing Vertical Territory."

54. C. Folke, "Resilience: The Emergence of a Perspective for Social-Ecological Systems Analyses," *Global Environmental Change* 16 (2006); B. L. Turner et al., "Illustrating the Coupled Human–Environment System for Vulnerability Analysis: Three Case Studies," *Proceedings of the National Academy of Sciences* 100, no. 14 (2003); White, "From Wilderness to Hybrid Landscapes"; Oran R. Young et al., "The Globalization of Socio-Ecological Systems: An Agenda for Scientific Research," *Global Environmental Change* 16 (2006). This methodological approach—the linking of climate change with other socio-historical forces—corresponds with environmental historians' efforts to blur nature-culture boundaries and historians' of science Actor Network Theory. See Kristin Asdal, "The Problematic Nature of Nature: The Post-Constructivist Challenge to Environmental History," *History and Theory* 42 (2003): 60–74; William Cronon, ed., *Uncommon Ground: Rethinking the Human Place in Nature* (New York: W.W. Norton, 1996); Timothy Mitchell, *Rule of Experts: Egypt, Techno-Politics, and Modernity* (Berkeley: University of California Press, 2002).

55. Le Roy Ladurie, *Times of Feast, Times of Famine*, 119.

56. See, for example, Christian Pfister, "Climate and Economy in Eighteenth-Century Switzerland," *Journal of Interdisciplinary History* 9, no. 2 (1978): 223–243; Christian Pfister, "Climatic Extremes, Recurrent Crises and Witch Hunts: Strategies of European Societies in Coping with Exogenous Shocks in the Late Sixteenth and Early Seventeenth

Centuries," *The Medieval History Journal* 10, no. 1–2 (2007): 33–73; Christian Pfister, "Little Ice Age-Type Impacts and the Mitigation of Social Vulnerability to Climate in the Swiss Canton of Bern Prior to 1800," in *Sustainability or Collapse? An Integrated History and Future of People on Earth*, ed. Robert Costanza, Lisa J. Graumlich, and Will Steffen (Cambridge, MA: MIT Press, 2007), 197–212; Christian Pfister and Rudolf Brázdil, "Social Vulnerability to Climate in the 'Little Ice Age': An Example from Central Europe in the Early 1770s," *Climate of the Past* 2, no. 2 (2006): 115–129.

57. Lamb, *Climate, History and the Modern World*.

58. Roderick J. McIntosh, Joseph A. Tainter, and Susan Keech McIntosh, "Climate, History, and Human Action," in *The Way the Wind Blows: Climate, History, and Human Action*, ed. Roderick J. McIntosh, Joseph A. Tainter, and Susan Keech McIntosh (New York: Columbia University Press, 2000), 1. Also see Neville Brown, *History and Climate Change: A Eurocentric Perspective* (New York: Routledge, 2001).

59. David Hackett Fischer, "Climate and History: Priorities for Research," *Journal of Interdisciplinary History*, Special Issue on "History and Climate: Interdisciplinary Explorations," 10, no. 4 (1980): 828.

60. James C. McCann, "Climate and Causation in African History," *The International Journal of African Historical Studies* 32, no. 2–3 (1999): 261–279. For a study on African history that achieves a good balance between ecology/climate and human agency, see James L. A. Webb, *Desert Frontier: Ecological and Economic Change Along the Western Sahel, 1600–1850* (Madison: University of Wisconsin Press, 1995).

61. For example, Jared Diamond, *Collapse: How Societies Choose to Fail or Succeed* (New York: Viking, 2004); Eugene Linden, *The Winds of Change: Climate, Weather, and the Destruction of Civilizations* (New York: Simon & Schuster, 2006).

62. Peter B. deMenocal, "Cultural Responses to Climate Change During the Late Holocene," *Science* 292 (2001): 667. Other discussions of climate-induced societal collapse include, for example, Thomas H. McGovern, "Management for Extinction in Norse Greenland," in *Historical Ecology: Cultural Knowledge and Changing Landscapes*, ed. Carole L. Crumley (Santa Fe, NM: School of American Research Press, 1994), 127–154; Carole L. Crumley, "The Ecology of Conquest: Contrasting Agropastoral and Agricultural Societies' Adaptation to Climatic Change," in *Historical Ecology*, 183–201.

63. For a good overview of the traditional explanations that did not (yet) embrace climate as a cause for collapse, see T. Patrick Culbert, ed., *The Classic Maya Collapse* (Albuquerque: University of New Mexico, 1973).

64. Richardson Benedict Gill, *The Great Maya Drought: Water, Life, and Death* (Albuquerque: University of New Mexico Press, 2000); Gerald H. Haug et. al., "Climate and the Collapse of Maya Civilization," *Science* 299, no. 5613 (2003): 1731–1735.

65. D. A. Hoddell, J. H. Curtis, and M. Brenner, "Possible Role of Climate in the Collapse of Classic Maya Civilization," *Nature* 375 (1995): 391–394.

66. Shaw, "Climate Change and Deforestation."

67. For a critical review of *Collapse*, see J. R. McNeill, "Diamond in the Rough: Is There a Genuine Environmental Threat to Security?," *International Security* 30, no. 1 (2005): 178–195.

68. Richard Grove, "Revolutionary Weather: The Climatic and Economic Crisis of 1788–1795 and the Discovery of El Niño," in *Sustainability or Collapse? An Integrated History and Future of People on Earth*, ed. Robert Costanza, Lisa J. Graumlich, and Will Steffen (Cambridge, MA: MIT Press, 2007), 151–168.

69. For example, J. Neumann, "Great Historical Events that Were Significantly Affected by the Weather: 2. The Year Leading to the Revolution of 1789 in France," *Bulletin of the American Meteorological Society* 58, no. 2 (1977): 163–168; Alan Taylor, "'The Hungry Year': 1789 on the Northern Border of Revolutionary America," in *Dreadful Visitations: Confronting Natural Catastrophe in the Age of the Enlightenment*, ed. Alessa Johns (New York: Routledge, 1999), 39–69.

70. Sam White, *The Climate of Rebellion in the Early Modern Ottoman Empire* (New York: Cambridge University Press, 2011).

71. Susan L. Swan, "Drought and Mexico's Struggle for Independence," *Environmental Review* 6, no. 1 (Spring 1982): 54; Susan L. Swan, "Mexico in the Little Ice Age," *Journal of Interdisciplinary History* 11, no. 4 (1981): 633–648.

72. For example, Georgina H. Endfield, *Climate and Society in Colonial Mexico: A Study in Vulnerability* (Oxford: Blackwell Publishing, 2008); Endfield, "Archival Explorations of Climate Variability and Social Vulnerability in Colonial Mexico," *Climatic Change* 83 (2007): 9–38; Endfield, "Climate and Crisis in Eighteenth Century Mexico," *The Medieval History Journal* 10, no. 1–2 (2007): 99–125.

73. Shaw, "Climate Change and Deforestation"; Shimada et al., "Cultural Impacts of Severe Droughts in the Prehistoric Andes"; Worster, *Dust Bowl*.

74. Louis A. Pérez Jr., *Winds of Change: Hurricanes and the Transformation of Nineteenth-Century Cuba* (Chapel Hill: University of North Carolina Press, 2001); Matthew Mulcahy, *Hurricanes and Society in the British Greater Caribbean, 1624–1783* (Baltimore: Johns Hopkins University Press, 2006); Walter J. Fraser Jr., *Lowcountry Hurricanes: Three Centuries of Storms at Sea and Ashore* (Athens: University of Georgia Press, 2006); Sherry Johnson, *Climate and Catastrophe in Cuba and the Atlantic World in the Age of Revolution* (Chapel Hill: University of North Carolina Press, 2011).

75. For studies on El Niño, see Cesar N. Caviedes, *El Niño in History: Storming Through the Ages* (Gainesville: University Press of Florida, 2001); Hallie Eakin, *Weathering Risk in Rural Mexico: Climatic, Institutional, and Economic Change* (Tucson: University of Arizona Press, 2006); Michael H. Glantz, *Currents of Change: Impacts of El Niño and La Niña on Climate and Society*, 2d ed. (New York: Cambridge, 2001). For recent studies demonstrating the global effects and cultural construction of El Niño disasters, see S. George Philander, *Our Affair with El Niño: How We Transformed an Enchanting Peruvian Current into a Global Climate Hazard* (Princeton, NJ: Princeton University Press, 2004); Gregory T. Cushman, "Enclave Vision: Foreign Networks in Peru and the Internationalization of El Niño Research During the 1920s," *Proceedings of the International Commission on History of Meteorology* 1, no. 1 (2004): 65–74; Mike Davis, *Late Victorian Holocausts: El Niño Famines and the Making of the Third World* (New York: Verso, 2001); Richard Grove, "The East India Company, the Raj and the El Niño: The Critical Role Played by Colonial Scientists in Establishing the Mechanisms of Global Climate Teleconnections 1770–1930," in *Nature and the Orient: The Environmental History of South and Southeast Asia*, ed. Richard Grove, Vinita Damodaran, and Satpal Sangwan (New York: Oxford University Press, 1998), 301–323.

76. For a good synthesis of this discussion about whether climate caused the Pleistocene Extinctions, see Shepard Krech III, *The Ecological Indian: Myth and History* (New York: W.W. Norton, 1999), chap. 1.

77. For a study linking disaster and climate historiographies, see Mark Carey, *In the Shadow of Melting Glaciers: Climate Change and Andean Society* (New York: Oxford University Press, 2010).

78. Nicholas Gabriel Arons, *Waiting for Rain: The Politics and Poetry of Drought in Northeast Brazil* (Tucson: University of Arizona Press, 2004), 5.

79. See, for example, Ted Steinberg, *Acts of God: The Unnatural History of Natural Disaster in America* (New York: Oxford University Press, 2000).

80. Gerald Michael Greenfield, *The Realities of Images: Imperial Brazil and the Great Drought* (Philadelphia: American Philosophical Society, 2001), 102, 106.

81. Timothy J. Finan, "Climate Science and the Policy of Drought Mitigation in Ceará, Northeast Brazil," in *Weather, Climate, Culture*, ed. Sarah Strauss and Ben Orlove (New York: Berg, 2003), 203.

82. Stephen Mosley, "Common Ground: Integrating Social and Environmental History," *Journal of Social History* 39, no. 3 (2006): 915–933; Alan Taylor, "Unnatural Inequalities: Social and Environmental Histories," *Environmental History* 1 (1996): 6–19.

83. Eric Klinenberg, *Heat Wave: A Social Autopsy of Disaster in Chicago* (Chicago: University of Chicago Press, 2003), 17.

84. "El Niño," of course, is my abbreviated reference to "El Niño-Southern Oscillation" (ENSO). Davis, *Late Victorian Holocausts.*

85. For another analysis of the political dimensions of famine, see Taylor, "'The Hungry Year': 1789 on the Northern Border of Revolutionary America."

86. Davis, *Late Victorian Holocausts*, 12.

87. For example, Carey, *In the Shadow of Melting Glaciers.*

88. Ruddiman, *Plows, Plagues, and Petroleum.*

89. Aqqaluk Lynge, "Foreword: Whose Climate Is Changing?" in *Thin Ice: Inuit Traditions Within a Changing Environment*, Nicole Stuckenberger (Hanover, NH: University Press of New England, 2007), 10.

90. J. E. Thornes and G. R. McGregor, "Cultural Climatology," in *Contemporary Meanings in Physical Geography*, ed. Stephen Trudgill and Andre Roy (London: Arnold, 2003), 173–197.

91. Sarah Strauss and Ben Orlove, "Up in the Air: The Anthropology of Weather and Climate," in *Weather, Climate, Culture*, ed. Sarah Strauss and Ben Orlove (New York: Berg, 2003), 3–14.

92. Julie Cruikshank, *Do Glaciers Listen?: Local Knowledge, Colonial Encounters, and Social Imagination* (Vancouver: University of British Columbia Press, 2005); Brian Fagan, *The Little Ice Age: How Climate Made History, 1300–1850* (New York: Basic Books, 2000); L. G. Thompson et al., "The Little Ice Age as Recorded in the Stratigraphy of the Tropical Quelccaya Ice Cap," *Science* 234, no. 4774 (1986): 361–364.

93. For example, Susan A. Crate and Mark Nuttall, eds., *Anthropology and Climate Change: From Encounters to Actions* (Walnut Creek, CA: Left Coast Press, 2009); Timothy B. Leduc, *Climate, Culture, Change: Inuit and Western Dialogues with a Warming North* (Ottawa: University of Ottawa Press, 2010).

94. For an overview related to environmental history, see White, "From Wilderness to Hybrid Landscapes."

95. For example, David N. Livingstone, *Putting Science in Its Place: Geographies of Scientific Knowledge* (Chicago: University of Chicago Press, 2003).

96. For overviews of changing Western understandings of climate over time, see Fleming, *Historical Perspectives on Climate Change*; Glacken, *Traces on the Rhodian Shore*, 551–622.

97. Cruikshank, *Do Glaciers Listen?*; Julie Cruikshank, "Glaciers and Climate Change: Perspectives from Oral Tradition," *Arctic* 54, no. 4 (2001): 377–393.

98. Cruikshank, "Glaciers and Climate Change," 378. Also see Anne Henshaw, "Climate and Culture in the North: The Interface of Archaeology, Paleoenvironmental Science, and Oral History," in *Weather, Climate, Culture*, 217–231.

99. Cho-yun Hsu, "Chinese Attitudes Toward Climate," in *The Way the Wind Blows: Climate, History, and Human Action*, ed. Roderick J. McIntosh, Joseph A. Tainter, and Susan Keech McIntosh (New York: Columbia University Press, 2000), 209.

100. Jan Golinski, *British Weather and the Climate of Enlightenment* (Chicago: University of Chicago Press, 2007).

101. Hans Neuberger, "Climate in Art," *Weather* 25 (1970): 46–56.

102. John E. Thornes and Gemma GMetherell, "Monet's 'London Series' and the Cultural Climate of London at the Turn of the Twentieth Century," in *Weather, Climate, Culture*, 141–160.

103. Donald Worster, "Climate and History: Lessons from the Great Plains," in *Earth, Air, Fire, Water: Humanistic Studies of the Environment*, ed. Jill Ker Conway, Kenneth Keniston, and Leo Marx (Amherst: University of Massachusetts Press, 1999), 51–77; Worster, *Dust Bowl*.

104. Wolfgang Behringer, "Climatic Change and Witch-Hunting: The Impact of the Little Ice Age on Mentalities," *Climatic Change* 43 (1999): 340, 336.

105. Christian Pfister, "Climatic Extremes, Recurrent Crises and Witch Hunts," 34.

106. Grove, *Green Imperialism*, 154.

107. Karen Ordahl Kupperman, "The Puzzle of the American Climate in the Early Colonial Period," *American Historical Review* 87, no. 5 (1982): 1262–1289; Kupperman, "Fear of Hot Climates in the Anglo-American Colonial Experience," *William and Mary Quarterly*, 3rd series, 41, no. 2 (1984): 213–240.

108. Mark Carey, "Inventing Caribbean Climates: How Science, Medicine, and Tourism Changed Tropical Weather from Deadly to Healthy," *Osiris* 26 (2011): 129–141.

109. For an excellent treatment of scale in weather and climate history, see Fleming, Jankovic, and Coen, eds., *Intimate Universality*.

110. Davis, *Late Victorian Holocausts*; Fagan, *The Long Summer*; Vladimir Jankovic, "Intimate Climates, from Skins to Streets, Soirées to Societies," in *Intimate Universality*, 1–34.

111. For a much deeper analysis and case study on this point, see, Carey, *In the Shadow of Melting Glaciers*.

112. For an excellent overview of these interdisciplinary achievements, see Brázdil et al., "Historical Climatology in Europe."

CHAPTER 2

··

ANIMALS AND THE INTIMACY
OF HISTORY

··

BRETT L. WALKER

THE tasteful austerity of Kenton Joel Carnegie's online memorial belied the gruesome complexities of his bloody death. The memorial contained traces of Carnegie's
twenty-two-year life: it provided links to his artwork, several family pictures, a discussion board, and a donation site at the University of Waterloo, where he was a third-year
geological engineering student. Donations supported the newly established Kenton
Carnegie Memorial Fund. Judging from the written content of the online memorial,
Carnegie was a person of "profound integrity" who possessed "an incredible understanding of the land." But the online memorial is silent on one matter: the cause of
Carnegie's death. All the memorial divulges is that he died "suddenly on Tuesday
November 8, 2005 as a result of a tragic incident in Points North, Saskatchewan."[1] News
reports proved more forthcoming with the grisly details: four wolves had killed and
eaten him on a trail near a uranium mine in Saskatchewan.

Wolf-conservation circles labeled the incident the "first documented case of healthy
wolves killing a human in North America."[2] Needless to say, the key word in this poorly
documented assertion is "documented," because it is hard to imagine that, given
wolves' opportunistic natures, unreported killings have not taken place. Barry Lopez
writes that both wolves and humans are social hunters, often seeking the same prey in
the same general locations. In such an environment, he concludes, confrontation was
probably inevitable.[3] If published accounts of Carnegie's death are reliable, however,
speculation regarding ancient hominid–lupine interaction is unnecessary. It appears
that, after a brief chase, the wolves dragged down and ate Carnegie near the shores of
Wollaston Lake.

Given that Carnegie was a geological engineering student, it is not surprising that
he was in the Points North Landing area. The *Duluth News Tribune* reported that the
"former wilderness area is a hotbed for uranium mining, as well as gold and diamond
exploration."[4] Carnegie was engaged in aerial surveys for an Ottawa company, Sander
Geophysics Ltd. The high number of miners, engineers, and support workers in the area

meant that some wolves, including the four that killed Carnegie, had seized the opportunity and started loitering around mining camps, eating garbage and food scraps. They had become habituated to people; they lived at the border of the uranium camp, or the industrial "ecotone" with its "edge effect," where ecological diversity occurs as a result of interplay between the wild and the domesticated.[5] Two wolves had been in the area for weeks, including just prior to the incident. More disturbing were indications that Carnegie and others had been "interacting with the wolves at close range, possibly feeding the animals."[6] Indeed, two days before his death, Carnegie, after showing photographs of wolves at a cafeteria, was warned against taking such photographs by trucker Bill Topping, who hauled supplies in the region. He relayed the story of a dog that had been "shredded" by wolves in the Paull River Wilderness Camp, south of Points North Landing. Topping cautioned Carnegie and another geology student that, "Wolves are the smartest creatures in the bush."[7] When Carnegie failed to return from a walk in the late afternoon on November 8, searchers discovered his body and chased off the four nearby wolves.[8] Andrew McKean, of *Field & Stream*, recreated the moment for his rifle-toting readers: "The footprints indicated that four wolves had shadowed Carnegie, who stopped, turned around and then tried to elude the animals breaking into a terrified sprint for safety. The tracks suggest that the man was knocked to the ground at least twice but struggled to his feet before he was taken down a final time."[9]

Topping, the trucker who had earlier warned Carnegie, remembered that the site where Carnegie's body was found "wasn't pretty." He recalled, "It was just as though those wolves had taken down a moose or a caribou." In *Field & Stream*, however, McKean presented a different interpretation and, by separating humans from other animals, forcefully reined in his readers from a dangerous philosophical abyss. He barked: "Only it wasn't an animal. The wolves' victim was a human."[10] On November 10, Saskatchewan conservation officers shot two of the suspected wolves at the dump. When necropsies were performed at the Prairie Diagnostic Services laboratories at the University of Saskatchewan, veterinarians discovered "hair and flesh in the large intestines that resembled human remains."[11] Paul Paquet, an ecologist from the University of Calgary, concluded his investigation in this manner: "I suspect that ultimately we will find that these are garbage-habituated wolves that are either being inadvertently fed or intentionally fed in the area.... That is the common thread to most wolf attacks that I've investigated."[12] Tim Trottier, a wildlife biologist for Saskatchewan Environment and Resource Management, explained that, "These wolves lived in a very unnatural state, so it's not that surprising that they might behave unnaturally."[13] Intimacy with humans is always unnatural and always dangerous. But it is also at this juncture, at the deadly intersection of the natural and unnatural, that these wolves entered history. In essence, by entering the dump, they became culturally visible.

Carnegie's kill site is not "pretty" for historians, either. It evokes the many tricky theoretical issues that we face when writing about nonhuman animals. Carnegie's mangled body outraged McKean, the *Field & Stream* journalist, precisely because the young man was "human," not an "animal" such as a moose or caribou. His anxieties exposed the carefully policed divide between humans, who have long fancied themselves as outside nature, and other animals.[14] Trottier, the Saskatchewan environmental official, likened the garbage dump to an "unnatural state" and characterized wolves that killed people as behaving "unnaturally." When humans manipulate and defile the sublime cathedral of wilderness—mining for uranium, surveying for diamonds, laying oil pipes, and logging forests—that pristine place falls from its natural grace, and, we must surmise, so do the other animals that live there.[15] In the case of wolves, they become, in the words of McKean, "junkyard dogs," living on the edge between civilization and wilderness. Real wolves do not kill and eat people, but junkyard dogs certainly do.

Carnegie's kill site was bloodied not only by the young man's torn body, but by the cruel reminder it provided: humans are indeed animals, sometimes even a meaty prey species, and as such, they are not external to nature or fundamentally different from other animals. Just as Carnegie did, they can have violently intimate relationships with other creatures. To be eaten by another animal is to become energy for that animal. It is to be forcefully pulled back into the metabolism of the natural realm, ripped from the safe confines of cultural dominion. One may contemplate the advanced technologies of the space shuttle, the lofty notes of Johann Sebastian Bach's *St. John Passion*, or the coarse brushstrokes of Vincent van Gogh's paintings; however, the debate over whether humans are anomalous, outside nature, exceptional, and separate from other animals abruptly ends when faced with the reality of a wolf's stomach acids dissolving the flesh of a young mining engineer.

Significantly, the cultural-constructionist arguments regarding "the animal"— namely, that our understanding of nonhuman animals is entirely culturally generated—border on the intellectually pedantic. Scholars claim that Michel Foucault, for example, the high-water mark of the "linguistic turn," remained "suspicious of claims to universal truths," and that for the French thinker there was "no external position of certainty, no universal understanding that is beyond history and society." For Foucault, all knowledge has a genealogy and nothing is "external" to culture and the history that generates it.[16] In many respects, Foucault was correct in thinking that truth-claims are historically created and function as technologies of social discipline. However, as we pause for a moment to contemplate Carnegie's ground flesh moving through the digestive tract of these edge-dwelling wolves, such claims to culture's dominion over all historical experiences and happenings succumb to the certainty of nature's hungry metabolism. The author Erica Fudge has insisted that, "The history of many of the ways in which we currently live with animals offers some sobering reminders that those ways have a source," a historical and cultural source, and that "they are not natural."[17] However, our reluctance to join our animal cousins on their terms, that is, on more natural terms, exposes a lingering devotion to human exceptionalism, one that is inherent in the humanities and social sciences and historically rooted in Cartesian philosophy. There is another "source" for the nature of our coexistence with animals operating in history, and it exists external to culture. If we must adhere to the analytical binaries of "culture" and "nature," rather than subscribe to what historian Paul Sutter cautions is the morally ambiguous worlds of "hybridity" and "second nature," then nature, as that which is not culture, also drives our historical interactions with other animals.[18]

David Quammen, in *Monster of God*, has written that "For as long as *Homo sapiens* has been sapient—for much longer if you count the evolutionary wisdom stored in our genes—alpha predators have kept us acutely aware of our membership within the natural world. They've done it by reminding us that to them we're just another flavor of meat."[19] This chapter investigates how some historians and other scholars have navigated the complex theoretical terrain of writing about nonhuman animals, including animals that kill and eat people. Over the past decade, writing on nonhuman animals has developed into a sizable literature. This article focuses on two broad themes, which, for our purposes, serve as subsections and represent some of the major areas of concentration in this subfield: "The Intimacy of Violence" and "The Intimacy of Transcendence." My use of the word "intimacy" will strike some as inappropriate, but I am interested in the closeness, familiarity, and private nature that the term evokes. Humans will always find familiarity as well as difference in animals, and we do so, I supposed, because we are animals.

The thread that holds this chapter together, with all its disparate references to important books and articles, is our shared intimacy with animals. They permeate our history and we theirs: tug at the threads and our stories, woven as they are into the same tightly knit tapestry, will not untangle.

THE INTIMACY OF VIOLENCE

Just like Carnegie, Val Plumwood was in the wrong place at the wrong time. In Kakadu National Park in Australia, the water lilies float on thick, slow-moving water; but the fiberglass canoe that Plumwood paddled through these lazy wetlands, meandering as they do through countless shallow channels with steep, muddy banks, seems in retrospect like a flimsy craft, given that, after decades of protection, the park belonged to hundreds of crocodiles. Plumwood had come to Kakadu because she wanted to view ancient Aboriginal rock art, and she ventured deep into crocodile country, despite cold temperatures, driving rain, and a stern warning from a park ranger. Finally, she arrived at the croc-infested main channel, but she grew nervous and decided to paddle back to the boat launch. It was then that a crocodile pursued her. She tried to paddle the canoe to avoid the half-submerged creature, but it effortlessly adjusted its course to intercept her. "For the first time," she recalled, "it came to me fully that I was prey."[20]

Plumwood was no longer the sole agent in this pending violent encounter; the crocodile had become an agent as well. For traditionalists such as R. G. Collingwood, historians look through, not at, certain happenings, and therefore episodes such as crocodile attacks are "mere events, not the acts of agents."[21] Such happenings just do not have the cognitive trappings to be considered historical. But when historians expand definitions of "agency" to include what Donald Worster described as those "independent energies that do not derive from the drives and intentions of any culture," nonhuman animals, from mega-fauna to microbes, become key agents in the unfolding of history around the world.[22]

Meanwhile, Plumwood, who would have been surprised to learn that her reptilian pursuer was not an agent of some sort, jumped from the craft and grabbed a low-hanging tree limb, but the crocodile launched from the water with a tremendous splash and grabbed her between the legs and pulled her into the water. Immediately, Plumwood found herself in the crocodile's "death roll," whirling violently in the frothing, bloody water. The creature was drowning her. Then, suddenly, the crocodile released her and, with what traces of power she could muster, she once more tried to climb the tree's draping limb. But just as before, the crocodile launched from the muddy depths of the channel and grabbed her by the thigh, dragging her into the water and then releasing her. Finally, she was able to scale the muddy bank of the channel by using her thumbs as dull, fleshy pitons. She tried to make it back to the boat launch, but she had lost a significant amount of blood and buckled under the excruciating pain and exhaustion. "I struggled on," she remembered, "through driving rain, shouting for mercy from the sky, apologizing to the angry crocodile, calling out my repentance to this place for the fault of my intrusion."[23] Ultimately, Plumwood was rescued from her harrowing experience. Later, she philosophized about the entire event. She pondered what she called the "hyperseparated" boundaries between the "sacred–human" and "profane-natural," boundaries that dissolve before one's eyes when one is trapped in the horrifically powerful jaws of a crocodile. Death at the jaws of one of these giant reptiles "multiplies these forbidden boundary breakdowns, combining decomposition of the victim's body with the overturning of the victory over nature and materiality that Christian death represents." She continued: "Crocodile predation on humans threatens the dualistic vision of human mastery of the planet in which we are predators but can never ourselves be prey. We may daily consume other animals in their billions, but we ourselves cannot be food for worms and certainly not meat for crocodiles." Indeed, the crocodile's "death roll" proved the inadequacies of culturally driven explanations for our encounters with other animals on Earth: "We live by illusion if we believe we can shape our lives, or those of the other beings with whom we share the ecosystem, in the terms of the ethical and cultural sphere alone."[24] Humans inhabit a vast living biosphere, teeming with creatures driven by biological needs and hungers, making us far from the sole agents of Earth's destinies. Given that, with the advent of climate change, Earth's destinies at a geologic planetary scale have become historically generated, animals not only create history, but they experience it as well, as all species do on our slowly warming planet.[25]

In many respects, the role of nature's agency in driving history has preoccupied environmental history from the field's inception. If history was once driven by "great men," but then broadened, at the behest of social history, to include more human actors, then one of environmental history's contributions has been to broaden historical agency to include nature. Linda Nash has argued that agency in history is generally defined as an "ability to convert ideas into purposeful actions," something that human architects do but bees, though builders of elaborate structures, do not. Of course, the distinction between bees and architects is pithy, but it is overly simplistic, because historians need to acknowledge different kinds of animals, and that some, such as Carnegie's wolves, daily "convert ideas into purposeful actions." To deny that they do is to ignore the science that demonstrates as much.[26] Rather, Nash argues that we need to recast our definitions of agency to be less about the "self-contained individual confronting the external world" and more about social milieus, such as Bruno Latour's actor-network theory or Tim Ingold's "organism-in-its-environment" perspective.[27]

However, after Lynn Margulis's theories of evolutionary symbiosis, and recent microbiological studies that demonstrate that the "individual" human body is better understood as a complex "microbiome," where some 100 trillion, or about two pounds, of animal microbes coexist everyday inside us, any discussion of the body as "self-contained" appears hopelessly outdated.[28] We share intimate relationships with all of these microscopic animals and they, as part of us, share in our historical agency, even when we "convert ideas into purposeful actions."

Like Carnegie's death, Plumwood's "death roll" with a crocodile served as a reminder that animals possess real agency in our world. Animals are not our technologies, though the case for viewing them in this manner, as works such as *Industrializing Organisms* propose, has proved compelling.[29] Industrializing organisms do serve as another example of our shared intimacy with animals, however: industrial culture has recrafted the bodies of animals, through what Edmund Russell has called "historical evolution," to serve our modern needs.[30] But every time the brainy hominid manipulates nature—birthing a "hemophiliac beagle," a large-breasted "chicken of tomorrow," or the "worker" hogs of industrial pharmaceutical settings—somewhere, hidden in some muddy channel, another animal is stealthily adjusting its course, always half submerged from our horizon of vision, glassy eyes probing, preparing a toothy interception.[31]

One compelling example of this interception is contagious diseases. In his classic *Plagues and Peoples*, William McNeill aptly characterized disease transfer as a kind of predation on humans by microparasites, microscopic meat-eaters that stalk the human herd. McNeill writes that, "one can properly think of most human lives as caught in a precarious equilibrium between the microparasitism of disease organisms and the macroparasitism of large-bodied predators."[32] Simply put, certain diseases are actually animals that exist at a microscopic level, or "tiny organisms—viruses, bacteria, or multi-celled creatures as the case may be—that find a source of food in human tissue suitable for sustaining their own vital processes."[33] It is a matter of scale; they are not all that different from wolves, crocodiles, and lions, just much smaller. Moreover, most killer pathogens, from smallpox to influenza, are the result of the domestication of livestock and, hence, transference between species. This is the price humanity has paid for the domestication and, later, industrialization of animals and animal breeding. The price for humanity's intimacy with livestock is the micro-predation of the human herd. McNeill explains that twenty-six known diseases have transferred from poultry to humans, thirty-two from rats, thirty-five from horses, forty-two from pigs, forty-six from sheep and goats, fifty from cattle, and sixty-five from dogs. Obviously, he writes, the "sharing of infection increases with the degree of intimacy that prevails between man and beast."[34] Man's best friend, the dog, has bestowed on the human species more infectious microparasites than any other domesticated creature. Like Plumwood's crocodile, this is nature adjusting course and intercepting. And, as Quammen has recently reminded us, nature is continuing to do so, thus shaping history in important ways.[35] In the case of dogs, however, it was probably a price worth paying for both species.

An Animal in the Bedroom

Dogs have deftly adjusted their evolutionary course throughout human history, serving as humanity's most intimate partners. When the first dogs departed the wolf tribe, they hitched their evolutionary wagon to the brainy hominid, which cleverly assured their survival; however, in doing so they surrendered themselves to a species that has inscribed its cultural and political desires, sometimes quite cruelly, on their bodies and behaviors. Archaeological sites reveal humans buried alongside wolves and dogs. Sites such as the Zhoukoudian in North China (300,000 B.C.), Lazeret in the south of France (150,000 B.C.), and Boxgrove near Kent, England (400,000 B.C.) all yielded wolf bones in close association with hominid bones. As mentioned earlier, these early sites of intimate association should not surprise us. As Juliet Clutton-Brock has written, "the sites of occupation and hunting activities of humans and wolves must often have overlapped." Archaeologists unearthed a dog mandible from a late Paleolithic gravesite at Oberkassel in Germany (14,000 B.C.). It is likely that dogs emerged as humanity's subsistence partner, when hunting techniques shifted from direct impact with stones and axes to arrows tipped with microliths in the Epipaleolithic or Natufian Age. Dogs could help track down and dispatch wounded game.[36] In these hunting fields, our intimate relationship with dogs began. The wolf tribe, by contrast, has been forced to struggle on its own.

Splitting with wolves and joining humans was evolutionarily wise for dogs: many dogs share beds with people in opulent homes; while wolves are chased down or shot from low-flying aircraft. Humans have recrafted the bodies of dogs to advertise class difference and display other social signals, sculpting them, through selective breeding, to play into human social needs as they shift over historical time. In *The Beast in the Boudoir*, Kathleen Kete observes, "When bourgeois people spoke of their pets, as they loquaciously did, they pointedly spoke also of their times, and above all else of themselves." The "bourgeois dog" was the carefully sculpted product of a Parisian fantasy, whereas "working-class" dogs and "Oriental" dogs "led unstructured, more natural, less cultured lives." Class anxieties paralleled the pet-keeping fantasy; consistent with the theme of intimacy in this chapter, the "bourgeois dog" was protected and invited into the bedroom, the "working-class" and "Oriental" dog ostracized, chased down, and killed, because it was closer to wolves.[37]

This was certainly true in Victorian England. In *The Animal Estate*, Harriet Ritvo writes that similar to the "bourgeois dog" of Paris, "good animals" never challenged human superiority. She continues, "The best animals were those that displayed the qualities of an industrious, docile, and willing human servant; the worst not only declined to serve, but dared to challenge human supremacy."[38] In this hierarchy (one challenged in this chapter by wolves, crocodiles, and microparasites), "Eating human flesh symbolized the ultimate rebellion, the radical reversal of roles between master and servant."[39] While dog fanciers systematized pedigrees, participated in dog shows, and fussed over portraits with their prized pets, the unlicensed mongrels of the working class in London were hunted down and clubbed to death because of anxieties over rabies. But the real anxieties were over the working class itself. Power over nature (and people) was at the heart of Victorian England's "cult of the pet." As Ritvo explained, the goal of pet fanciers "was to celebrate their desire and ability to manipulate, rather than to produce animals that could be measured by such extrinsic standards as utility, beauty, or vigor."[40] Careful manipulation safely transforms animals from the "profane-natural" to the "sacred-human."

These attitudes were quickly projected into Victorian Britain's imperial designs as well. In Japan, for example, though not part of the empire, indigenous breeds were labeled "pariah dogs" or "kaffir," because of their unstructured, wolfish behavior, whereas English dogs became symbols of "civilization." Immediately after the Meiji Restoration of 1868, indigenous breeds, because of rabies and the threat to livestock, were rounded up and shot; foreign breeds were often protected as emblems of Japan's desired Western-style modernity. Starting in 1877, on the northern island of Hokkaido, in such cities as Sapporo, Hakodate, and Nemuro, policed tracked and clubbed to death all unlicensed dogs because they reportedly harassed livestock and allegedly carried disease. Police dispatched hundreds of what were labeled "wild dogs," "bad dogs," and "mad dogs" in this brutal manner.[41] Only with the rise of Japanese ethnic nationalism in the early twentieth century were indigenous Japanese breeds rescued from the brink of extinction. Within the context of Japanese ethnic nationalism and other distinctly Japanese cultural values, the "Japanese dog" was born and celebrated, resculpted to fit the fascist political climate of its day.[42]

Holding Humanity Down

In all these histories, the human need to conquer nature through subduing and resculpting animals glares like translucent crocodile eyes in a dark Australian wetland. In *Eyelids of Morning*, Alistair Graham wrote of his and Peter Beard's crocodile research on Lake Rudolf in Kenya in the mid-1960s. Graham, too, fixated on the theme of human dominance over this particularly toothy representation of nature. He wrote, "In the face of man's inexorable expansion, Lake Rudolf will one day fall and its dragons be subdued, for civilized man will not tolerate wild beasts that eat his children, his cattle, or even the fish he deems to be his. That would be regression into barbarism."[43] Being eaten by a crocodile, or even a wolf for that matter (as Carnegie's tragedy instructs), is to confront our shared animal nature with other organisms, to surrender being human, a crafter of culture and artifice, and to regress into animal "barbarism."

This is the "hyperseparation" that Plumwood identified in the aftermath of her encounter with a crocodile's "death roll." Graham told many grisly stories of crocodile attacks, but the most gripping was the death of William Olsen, a Peace Corps volunteer in Ethiopia, in 1966. He, too, was in the wrong place at the wrong time, and a crocodile killed and ate him for his error. Later, police shot the reptile and a field necropsy produced Olsen's torn body. Karl Luthy, a safari-outfitted big-game hunter, witnessed the slicing open of the croc's belly. "We found his legs," Luthy recalled, "intact from the knees down, still joined together at the pelvis. We found his head, crushed into small chunks, a barely recognizable mass of hair and flesh; and we found other chunks of unidentifiable tissue."[44] Graham wrote, "So long as one is constantly threatened by savage brutes one is to some extent bound in barbarism; they hold you down. For this reason there is in man a cultural instinct to separate himself from and destroy wild beasts such as crocodiles."[45] The hallmark of the human relationship with other animals is our need to separate ourselves from them, lest they "hold us down." Wild, toothy animals deny us our right to heavenly immortality by holding us down; they deny us our divine origin myths by holding us down. Crocodiles and other predators force us to confront our shared fleshy nature with other organisms on Earth, which flies in the face of our deepest cultural myths of monotheistic transcendence and sacred difference. Look around carefully; look at your fingernails, hair, and incisors. We were not built in the likeness of gods, but in the likeness of the other organisms with whom we share Earth.

Predators such as wolves and crocodiles do not eat people in chance encounters, either. The fate of becoming prey, or participating in the crunchiest of animal intimacies, is motivated by biological necessity, but often made possible by humanity's social drivers. Quammen labeled the social and cultural forces that render people prey as "the muskrat conundrum."[46] In brief, the "muskrat conundrum" refers to the marginalization of the socially vulnerable in a community. His principal example is the Maldharis of India, whose livestock and selves often serve as meat for the Asiatic lion of the Gir Wildlife Sanctuary and National Park. For nearly a century and a half, the Maldharis (a composite of several older pastoral communities) were nomadic, traveling the Kathiawar with their cattle, gaining access to pasturelands for the manure that their cattle would inevitably deposit there. As the Kathiawar landscape became more restricted due to private ownership, they established makeshift camps in the forests around Gir. In 1972, however, all Maldhari families living within the Gir forests—some 845 of them—were to be forcibly resettled to make way for what the Gujarat government called the Gir Lion Sanctuary Project, a project undertaken at the behest of such international conservation groups as the World Wildlife Fund. The Maldharis were slated to become farmers. Nonetheless, by the 1980s, only about six hundred families had left the Gir for surrounding farmlands, while some three hundred Maldharis families remained, scattered throughout the Gir forests in makeshift camps. These Maldharis and their livestock now share the edges of the Gir with lions and leopards. Living outside mainstream Indian society (a phenomenon shaped by India's rigid caste system), the Maldharis and their livestock fend for themselves against lions and leopards. They live in close proximity to large predators, and their poverty and pastoral lifestyle render them vulnerable.[47]

Quammen likened this situation to the "muskrat conundrum" because forcing the Maldharis to the edges of Indian society, where big cats lurk in dark forested places, is similar to how muskrats treat their furry outcasts. Muskrat populations are notoriously density dependent, and only a certain number of individuals can find safe, suitable den sites and adequate food within a given territory. The elderly, outcast, or those weakened from disease are forced from the safe den sites and become what scientists call the "wasted parts" of the population. Predators such as minks lurk on the edges, preying on muskrat "wasted parts," or those forced to the dangerous edges of muskrat territories. Quammen likened the Maldharis to the "wasted parts" of Indian society, bound to the edges of the Gir forest by pastoral tradition, botched government relocation programs, poverty, and the desire for charismatic lions and leopards by the wealthy in industrialized nations.[48] So that "haves" in the safe center can enjoy the spiritual and aesthetic pleasure of the Gir Sanctuary's majestic lions, "have-nots" such as the Maldharis wander with their cattle through lion and leopard-infested forests, losing livestock and members of their community in the process. Those closest to the Gir lions, the Maldharis, understandably resent the large cats. Quammen wrote, "No one wants to be among the 'wasted parts' of a population." He asked, "Is it inevitable that the costs exacted by alpha predators be borne disproportionately by poor people...while the spiritual and aesthetic benefits of those magnificent beasts are enjoyed from afar? He concluded that, "it's a matter that we cozier muskrats need to address."[49]

However, "cozier muskrats" in the US, as portrayed by Mike Davis in *Ecology of Fear*, have their own edgy, predator-infested problems to deal with.[50] If Indian society has pushed pastoral peoples such as the Maldharis to the edges of the Gir, where lions and leopards wait for them and their livestock, violent and impoverished inner cities in the US have pushed the "haves" to the suburbs. In California and across the US, as suburban neighborhoods encroach on once wilder lands, crossover ecologies have emerged, places where wildlife, such as the four wolves that killed and ate Carnegie, becomes more habituated to people. That is, whether in India or California, nonhuman animals, such as Asiatic lions or mountain lions, rarely if ever hunt people in the heart of major cities or metropolitan areas—this is where the "have-nots" hunt members of their own species. Mostly, these animals hunt people on the edges, where the green Kentucky bluegrass yields to arid sage lands. Predation occurs at these edges, just as it does along the Gir forest's edges. Only it is not herdsmen who suffer, but rather the mobile, well-heeled members of American society, people who yearn to retrieve a small, carefully manicured piece of subdued nature. But here again, animals just beyond our horizon, just like Plumwood's overly curious crocodile, have adjusted their course. In an edge ecosystem, Davis recounts how a mountain biker was torn from his bike in the San Gabriel Mountains, and how a coyote ripped the head from a young girl in a Glendale, California, suburb. Most predators kill people along the edges of suburban areas, mines, wildlife sanctuaries, and logging camps.[51] These suburbanites live in a new kind of intimacy with wildlife, creatures that have come to appreciate the benefits of living near more vulnerable people.

Intimate Affinities

However, sharing cultural and biological affinities and intimacies with the brainy hominid can be risky business. For these creatures, humans reserve a level of violence they normally focus on members of their own species. The more similar to the human species other animals are—whether socially, physiologically, or behaviorally—the more menacingly they challenge the human/animal distinction which humans so carefully police. It is hard to separate (let alone "hyperseparate"), when everywhere similarities, not differences, abound between humanity and other animals. Chickens, for example, are not like people: they lay eggs, cluck, have feathers, roost, and scratch and forage for grasshoppers and grains. But monkeys are quite like us, so much so that Japanese hunters, when asked by angry persimmon fruit growers to cull them, refuse to do so. Japanese hunters recount how if you point a rifle at them, they gesture and beg for their lives; if you do shoot them, they grab their wound in pain, like a scene in a spaghetti western.[52]

Some primates are so closely related to us that they serve as a kind of "natural other" to humans: the study of certain species, such as chimpanzees, can be cast in a manner that underscores and legitimizes patriarchy and capitalism in certain societies. In *Simians, Cyborgs, and Women*, Donna Haraway argues that natural knowledge regarding animals has often been used to buttress social domination over women and reinforce forms of political-economic control. "Women know very well that knowledge from the natural sciences has been used in the interest of our domination and not our liberation," writes Haraway. She continues: "natural knowledge is reincorporated covertly into techniques of social control instead of being transformed into sciences of liberation." Indeed, the study of animal groups, observes Haraway, has proved "unusually important in the construction of oppressive theories of the body political."[53] Monkeys, too, "hold us down," because primatology can serve the purposes of the architects of our patriarchal, capitalistic, scientific realities. Clearly, primatology and other animal sciences are the subject of the history of science more than environmental history, but it is through the lens of such science that environmental historians often see the animals they write about, so studies of the social and cultural construction of science are critical aspects of the environmental historian's toolkit. We increasingly see ourselves within the context of this history of science. With "deep history," historians have started pushing further back in time to uncover Paleolithic precedents for our historical-behavioral convergences. But when historians do so, they eventually stumble on our ape ancestors. As Daniel Lord Smail acknowledges, "A deep history demands that we acknowledge a genetic and behavioral legacy from the past." But Smail also concedes that many historians assume that, "history begins when humans ceased being animals and became people," and, therefore, there is a reluctance to acknowledge that the neurology that we share with other animals, such as psychotropic mechanisms, created the intricacies of human civilization.[54]

In this regard, animals serve as symbolic markers between historical humans and ahistorical apes. Indeed, they can signify many different scientific, political, social, and cultural forces in human societies around the globe.[55] In pre-revolutionary France, cats signified bourgeois luxury, witchcraft, and vaginas. In Japan's folk beliefs, raccoon-dogs signified neglected debts and large testes, whereas deer signified gratitude, the continuum of life, and the transmigration of the soul in Buddhist theologies.[56]

Wolves do not look like us, but they live in complex societies, nurture and educate their young for years, communicate through complex vocalizations and body languages, grieve and sacrifice themselves, violently defend their territories, eat meat, cooperate, and live within elaborate social hierarchies, which they occasionally challenge. They really do compete with humans on an ecological and economic level. Wolves view the livestock that humans defend for their livelihood as part of their own livelihood, too.[57]

The manner in which nature's economic competition between wolves and humans spirals out into the cultural realm of myth-making and ritualized violence is the theme that Jon Coleman explores in *Vicious*.[58] Principally, Coleman seeks to explain the violence inflicted on wolves by humans, violence that travels beyond simple competition for calories. Even if the two species did compete over territory, livestock, and other sources of meat, that competition still fails to explain the meticulous cruelty that humans have unleashed on the wolf tribe. In the 1660s, for example, along the Maine coast, John Josselyn and his hunting partners twice captured live wolves and tortured them for sheer enjoyment. Once, Josselyn's mastiff pinned a wolf by its throat in a low tide, so that the hunters could bind the wolf and carry it back home, "like a Calf upon a staff between two men." Later that night, they let the wolf loose in the living room. "The beast sank to the floor," narrated Coleman. "No biting, no snarling, he just slouched there, staring at the door." Even the mastiffs proved uninterested in riling up the wolf. Coleman explains, "Their evening's entertainment ruined, the hunters took the wolf outside and crushed his skull with a log."[59] As Coleman documents, the violence inflicted on wolves by Americans was ghoulish. The title *Vicious*, it turns out, refers to the hominid species, not the lupine one. Coleman's prose on this topic is as eloquent as his examples are grisly:

> Euro-Americans fractured wolf skulls and shot-gunned wolf puppies. They set the animals on fire and dragged them to pieces behind horses. They destroyed wolves for a host of pragmatic reasons: to safeguard livestock, to knit local ecosystems into global capitalist markets, to collect state-sponsored bounties, and to rid the world of beasts they considered evil, wild, corrupt, and duplicitous. Their motives appear as blunt as a gunshot to the head, but wolves' deaths were neither that quick nor that straightforward. They died with fractured spines and severed hamstrings, gifts from a predator dissatisfied with mere annihilation. The brutality of wolf killing transformed bloody-but-understandable acts of agricultural pacification into deeds as inexplicable as they were horrendous.[60]

Coleman argues that wolves and humans fought for one common goal: "transcendence." He writes, "Both struggled to pass down genetic, cultural, and material legacies to their offspring, and this conquest of time, played out over history, culture, and biology, explained the longevity and intensity of the species' conflict."[61] Coleman observes that all creatures seek transcendence at some level, but humans also seek to pass along their "possessions and ideas." He continues, "Humanity's quest to reproduce ideas and possessions clashed with wolves' mission to survive as a species through sexual reproduction. Wolves were formidable biological competitors."[62] Coleman integrates biological analysis into his portrait of the viciousness marking the relationship between wolves and humans, but he carefully makes one distinction between humans and other nature. "Humans set conditions on life while biology sets none," he explains.[63]

Perhaps the most persuasive example of humans setting "conditions on life" through biological and cultural transcendence are colonialism and empire. Coleman observes, "One of the ways Europeans coped with their reproductive bonanza was to pack up their progeny and leave. Population fueled European colonization, and colonization rearranged biological communities throughout the world."[64] Importantly, if colonialism delivered biological and cultural transcendence for some members of humanity and their allied organisms, it set the stage for mayhem and genocide for others. Though some humans vigorously competed with wolves for transcendence, they enlisted allies from throughout the natural world to assist with ensuring the survival of their offspring, possessions, and ideas. Humanity needed imperial partners to craft empires, and they received substantial help from nonhuman animals. This is the intimacy of transcendence.

THE INTIMACY OF TRANSCENDENCE

Human expansion created opportunities for biological transcendence. Many animals, even the smallest creatures with whom we share the planet, have nurtured biological advantages through allying themselves with humankind. In *The Fire Ant Wars*, for example, Joshua Blu Buhs argues that fire ants exploited what historians call the "bulldozer revolution" in the American South, following humans as they disturbed the landscape.[65] The ant, because it evolved in the floodplains of South America, thrived in areas of upheaval. Buhs submits that fire ants "exploited this revolution to spread across the region. Thus it was a combination of the ant's natural history and human action that caused the insect's irruption." Buhs carefully pointed out that the "ant is nature independent of humans."[66] That is, the ant acts outside the realm of our minds. But, like most nonhuman animals explored in this chapter, ants established an ecological intimacy with humans, particularly exploiting the human penchant to expand and disrupt distant landscapes. Most invasive insect species hitched their evolutionary wagons to humans: the Japanese beetle, for example, stowed away in the root bundles of a batch of azaleas shipped to Riverton, New Jersey in 1916. It proved an advantageous decision for the hungry chafers, as they managed to escape several species of predatory flies and wasps, not to mention several deadly diseases, which had kept their numbers down in

Japan.[67] Similarly, in the mid-nineteenth century, Etienne Leopold Trouvelot imported European gypsy moths to the Boston area and, after they mysteriously escaped, they became an agricultural scourge and led to the "gypsy moth wars."[68] In every pesky case, though, whether fire ants, Japanese beetles, or gypsy moths, it was the intimate ecological partnerships with humanity that ensured these insects' survival and their expansion to new corners of the globe. Indeed, humanity can be a good partner when it comes to expanding your species.

When Iberians set out for the New World in 1492—seeking, to evoke Coleman's words, biological and cultural transcendence—they brought their ships, navigational sciences, military hardware, and lots of critters. Wooden ships rocked and jolted in stormy seas; they creaked and groaned while their cargo holds echoed with the agonizing bellows of livestock. But livestock had to come: these allied organisms proved instrumental in the conquest of the Native American civilizations. Allied organisms such as pigs, sheep, cows, dogs, and other creatures devoured native agriculture, transformed landscapes, introduced noxious weeds, out-competed indigenous fauna, and provided other imperial services critical in the European conquest of the New World. While pigs rummaged through native gardens, posing the threat of starvation, two-hundred-pound English mastiffs chased Indians from newly established plantations. When Captain Martin Pring landed at Provincetown Harbor near Cape Cod in 1603, he was accompanied by Fool and Gallant, two large mastiffs. When Pring wanted to be rid of local Nauset traders, "wee would let loose the Mastives, and suddenly with out-cryes they would flee away."[69] Fool and Gallant, and a host of other creatures, were allies in the European colonial project.

The Intimacies of Empire

Though snarling mastiffs can prove helpful, Alfred Crosby argues that the real key to Europeans' success in the New World was their ability to "Europeanize" it, a process that was underway by 1500 and, in Crosby's estimation, irreversible by 1550. Within decades of 1492, disease had ravaged the Antillean peoples of the Caribbean and had made headway on the American mainland. "As the number of humans plummeted," writes Crosby, "the population of imported domesticated animals shot upwards." The land was emptied of Indians, and European livestock replaced them. The first wave of allied organisms arrived in 1493, and populations of these horses, sheep, dogs, pigs, cattle, chickens, and goats exploded given the lack of natural predators and diseases, and an abundance of grasses, roots, wild fruits, and carefully tended Indian gardens. Crosby explains: "Their numbers burgeoned so rapidly, in fact, that doubtlessly they had much to do with the extinction of certain plants, animals, and even the Indians themselves, whose gardens they encroached upon." Virtually everywhere Iberian conquerors went, they deposited pigs and cattle from their ships, and the animals' numbers multiplied and displaced or destroyed local fauna and human populations.[70]

As pigs busily displaced peoples, Europeans saddled and rode horses, herding cattle whose meat could be consumed to provide colonial calories. In the New World, the Iberian rancher became iconic: Spaniards, astride their equine allies, drove their herds onto the plains, the llanos, and the pampas. Crosby estimates that by the seventeenth century there were more cattle in the New World than any other type of "vertebrate immigrant."[71] Sheep, too, accompanied Iberian conquerors, though they acclimated more slowly to the New World. Sheep not only brought wool to the New World, but microparasites, which, once transmitted to llama and alpaca populations, swiftly devastated those species. Elinor Melville, in *A Plague of Sheep*, likens the increase in sheep populations in the New World to "virgin soil" irruptions of diseases such as smallpox. She calls the increase in sheep populations an "ungulate irruption," whereby sheep contributed to the creation of a "conquest landscape" in Mexico. Sheep did not simply replace men," writes Melville. Rather, "they displaced them—ate them, as the saying goes." She succinctly defines the conquest landscape as "The process by which sheep grazing displaced agriculture, and sheep displaced humans, resulted in the formation of a new and far less hospitable landscape within which the indigenous populations were marginalized and alienated, their traditional resources degraded or lost, and their access to the means of production restricted."[72]

The conquest landscape was not unique to the Iberian experience in the New World, as animals often competed with humans for calories in certain disrupted environments. In northeastern Japan, for example, similar environmental processes occurred: after the brutal unification wars in the sixteenth century that ended centuries of civil war (a period known as the Era of the Warring States), a domestic conquest landscape emerged when that region was shoehorned into the new political order emerging in such metropolitan areas as Kyoto and, later, Edo (present-day Tokyo). When the warlords militarily subdued essentially autonomous northeastern domains, the region was, over the centuries, transformed from a landscape characterized by horse pastures (which supplied some of Japan's most famous samurai mounts), to a landscape dominated by soybean monoculture. In the eighteenth century, with neo-Boreal (Little Ice Age) events and increased demand for soybeans in Edo, colder and windier weather ruined crops, forcing farmers into the mountains to search for yams, arrowroot, and other wild vegetables. However, the slash-and-burn agricultural techniques that had made large-scale soybean cultivation possible in the mountains also created large swaths of habitat for wild boars, whose populations grew rapidly. When farmers felled forests to make soybean fields, they inadvertently created ideal habitat for calorie competitors. With the threat of famine in 1749, farmers retreated to the mountains to harvest naturally occurring vegetables, but they found none; meanwhile, marauding boars devoured what little remained of their harvests. Thus, Japan's sixteenth-century unification wars—conflicts that paved the way for the establishment of the modern state—colonized once-autonomous domains, restructured biological communities, changed landscapes, and sparked an "ungulate irruption" that contributed to the death of thousands of people in 1749 in areas such as Hachinohe. Several brutal famines struck eighteenth-century Japan, but chroniclers called this one the "wild boar famine" of 1749, because of the competition between hominids and ungulates.[73]

A Stockyard on the Hill

Europeans' allied organisms crafted the conquest landscape in the Americas. In the Caribbean and Mexico, there were few predators and plenty of food, and livestock marauded through Indian gardens or turned feral and competed with indigenous fauna. Ultimately, these allied animals proved intimate partners in the conquest of South America. They partnered with hominids in the "Europeanization" of the New World, but they also had their own agendas to attend to, such as reproducing, eating, bedding down, rummaging, marauding, and otherwise imprinting their rather large hoof-prints on the savage history of the conquest of the New World. As Virginia DeJohn Anderson argues in *Creatures of Empire*, European livestock, once in seventeenth-century New England, shouldered much of the burden of creating and expanding the English colonies.[74] These creatures not only grazed the grasslands of the conquest landscape, but harassed, challenged, and ultimately brutally displaced Indians from their homes as part of the slow, bloody march of white settlers across the North American continent. These beasts of colonial burden served in the vanguard.

If Europeans' allied organisms intruded into the "virgin soil" of the Caribbean and Mexico, essentially as invasive species transported in ships by an invasive civilization, livestock in New England worked in a closer empire-building partnership with European settlers. Anderson argues that the English colonists' livestock not only reconfigured New World environments, but changed the "hearts and minds and behavior of the peoples who dealt with them" in a cultural clash that left New England-area Indians in tatters.[75] Specifically, although Europeans categorized all nonhuman creatures as "beasts" or "animals," Indians viewed the natural world differently, ignoring such strict dichotomies. They neglected to police divisions between the "sacred-human" and "profane-natural." Famously, the Algonquian peoples believed that a spiritual power, known as *manitou*, manifested itself in human and nonhuman animals. This belief contrasted sharply with Christian conceptions of animals in the natural order of Creation. Anderson explains: "The assumption that animals could possess *manitou* literally empowered nonhuman creatures in a manner foreign to the colonists' Christian beliefs. Access to spiritual protection gave animals a special status in native societies. Indians certainly recognized differences between people and animals, but could not regard animals as lesser beings, defined from the moment of Creation as invariably subordinate to humans."[76] Therefore, the importation of European livestock to the New World sparked cultural shifts, not simply ecological ones. The husbandry practices of English settlers challenged the natural order of Indians. Indians adorned themselves with animal motifs, dressed in animal skins, refrained from killing certain animals for deep cultural reasons, and "negotiated" with animals as they hunted and killed them. Anderson writes that, "Native peoples conceived of their connections with animals

HANDBOOK OF ENVIRONMENTAL HISTORY

in terms of mutual support rather than human dominance and shaped their behavior accordingly. In this, as in so many other aspects of Indian culture, the principle of reciprocity structured both thought and conduct."[77]

Respect for animals and reciprocity toward their slaughter should not be interpreted as suggesting that Indians were closer to animals, however. Indians might have respected animals, but it was English colonists who actually lived with them: husbandry is an intimate relationship with organisms that would have proved unfamiliar to Native Americans. Domestication, observes Anderson, "is as much a relationship as a condition, signaled by the frequent association of people with certain animals."[78] She argues that the key to understanding the colonists' relationship with their allied organisms is that the animals were "property" (or part of human material dominion or ownership). If Indians made little distinction between wild animals and domesticated ones, Europeans did, precisely because domesticated animals were the property of colonists. Foreshadowing the role that domesticated animals played in the history of Early America, Anderson submits: "The Indians could not have known that a distinction that had little or no meaning in their culture would loom so large in the colonists' minds." Much as they did in the conquest landscape in the Caribbean and Mexico, cattle became key players in the empire-building project in New England. Cattle as agents in the New World operated in an ecological, legal, and symbolic manner. Simply, cattle's symbolic association with civility and their material association with property made the "land Christian as well as English." "As agents of empire," Anderson writes, "livestock occupied land in advance of English settlers, forcing native peoples who stood in their way either to fend the animals off as best they could or else move on." Even though Indians eventually started to incorporate livestock into their culture, their lands became pastures for these hoofed agents of empire. "Indians found room in their world for livestock," Anderson concludes, "but the colonists and their descendants could find no room in theirs for Indians."[79]

Nor could they find room for wild animals that threatened their livestock. Importantly, the relationship between English colonists and their livestock was a reciprocal one—a little like the reciprocity between Indians and their game, only more intimate. That is, livestock provided calories and served in the vanguard for white settlers' cruel march across the North American continent. White settlers reciprocated by protecting them. Later, in the eighteenth and nineteenth centuries, this meant eliminating Indians who raided cattle, killing ungulates such as bison that competed for grass with livestock, and exterminating predators such as wolves that tracked, killed, and ate them. Cattle replaced bison on the American frontier and, as Richard White explains, "Americans came to think that they were living in the 'Golden Age of American Beef.'"[80] Some western cities were little but giant slaughterhouses. After the 1860s, observes William Cronon, Chicago's South Side became famous for its unified stockyards and slaughterhouses, where the meat market was concentrated, railroads loaded and unloaded squealing animals, and gawking tourists waded through the thick, sticky blood from stuck pigs and cows bleeding to death from hooks thrust through their bodies. In this brutal, mechanical, "death factory," as Rudyard Kipling described it, livestock were sacrificed for their hungry hominid overseers. It was in the Chicago stockyards, Upton Sinclair writes, that one heard "the hog-squeal of the universe."[81] That squeal, as recent environmental history has shown, echoes throughout human history as well.

Sharing the Kills of Another

Empire-builders around the globe left rotting bodies in their wake, and animals were, once again, intimately involved. Violence was perpetrated to protect organisms such as livestock, but it was also perpetrated by certain animals against hominid invaders. In India, Africa, and parts of Asia, where nineteenth- and early twentieth-century El Niño droughts laid waste to entire communities and, as Mike Davis explained, provided a "green light for an imperialist landrush," dogs, jackals, wolves, and other animals hunted down weakened people. Animals shared the kills of others—in this case, the kills of Europeans—gorging themselves on the "wasted population" left as carrion on global killing fields of biological transcendence. In August 1877, for example, in India's Nellore and other districts in the Madras Deccan, pariah dogs feasted on the famine dead. "I came upon two dogs worrying over the body of a girl about eight years old," wrote one British official. "They had newly attacked it, and had only torn one of the legs a little, but the corpse was so enormously bloated that it was only from the total length of the figure one could tell it was a child's."[82] That same year, El Niño droughts ravaged parts of China's Qing Empire. In Shaanxi, for example, the Qing allowed two representatives of the British Inland Mission to tour the Wei River Valley, where they discovered to their horror hundreds of thousands of dead bodies. Those lucky enough to survive were weakened by malnutrition and, according to reports, were hunted down by wolves that prowled the outskirts of town, "gorged and stupid from the fullness of many ghastly meals."[83]

In some regions of China, locusts devoured what remained of crops, magpies and "pariah dogs" picked at bloated bodies, and people—the most un-animal of all the animals—reportedly sold and consumed their own children in acts of grave desperation, blurring the contrived line between human and nonhuman animals. Chinese officials reported in Shanxi that "children abandoned by their parents...were taken to secret locations, killed and consumed." One Welsh missionary described stories of "parents exchanging young children because they could not kill and eat their own."[84] The British Empire preyed on societies weakened from El Niño droughts, and different varieties of animals—from pariah dogs to wolves—preyed on the native bodies left in the wake of cruel British misrule.

Dead animals often provide markers for the triumph of expanding civilizations. Throughout history, as Chinese dynasties and their allied organisms expanded across North and Southeast Asia, dead elephants were left in the wake of the march of Confucian civilization. Indeed, Mark Elvin, in *The Retreat of the Elephants*, begins by explaining: "Four thousand years ago there were elephants in the area that was later to become Beijing (in the Northeast), and in most of the rest of what was later to be China. Today, the only wild elephants in the People's Republic are those in a few protected enclaves in the Southwest, up against the border with Burma."[85] Elvin's book is not about elephants, but the pachyderms provide a fascinating historical indicator for China's biological and cultural transcendence. Similar to Euro-American monotheistic beliefs, pacifying animals proved central to Confucian notions of civilization. Mencius, one of the great Confucian philosophers, boasted that the sage-ruler, the Duke of Zhou, rid the landscape of animals. "He drove the tigers, leopards, rhinoceroses, and elephants far away," wrote Mencius, "and the world was greatly delighted."[86] Ridding the landscape of elephants was cruel business, however. In 1547, villagers from Hepu County, in the southwest, killed a herd of marauding elephants that had trampled their gardens by surrounding them with barricades, trapping them, felling all the trees from within the barricades, and then roasting them to death in the scorching sunlight. A Ming Dynasty chronicler wrote, "People were also told to wait for a moment when they could cut down the trees that grew within the barricades, so the herd could be attacked by the heat of the midday sun. Elephants are afraid of heat; and in three or four days all of them were dead."[87] As the Chinese hunted elephants as pests and for their ivory tusks, and enlisted them as allied organisms in warfare, the creatures gradually disappeared. In time, the landscape was depopulated of elephants and repopulated by Chinese people: the destruction of elephant herds became an indicator of the advance of the human one.

The elephant's destruction by the Chinese reminds one of the ruthless annihilation of the bison in the US. If elephants were killed as agricultural pests, sources of ivory, and as evidence of Confucian pacification of the landscape, bison were principally killed for hides that provided leather belts for industrial machinery. In *The Destruction of the Bison*, Andrew Isenberg traces the multi-causal nature of the bison's near extinction: the changing nature of the Great Plains ecosystem, eastern markets and consumer demands for hides, the emergence of equestrian nomadism among certain Native American societies, wasteful white hunting practices, and the spread of disease all conspired to destroy the bison. The nuances of this complex debate aside (that is, was it overzealous native hunting after the enlistment of horses as allied organisms, or white hunters slaughtering for eastern markets that ultimately devastated herds from an estimated 30 million to about one thousand?), in the wake of the European invasion of the New World, the bison's near-extinction became a potent indicator of the biological mayhem that followed.[88]

In the Pacific Northwest, salmon became a compelling indicator of biological disruption on the Columbia River. Just as bison served as organisms that transferred the sun's energy from the grasses of the Great Plains to Native American and white settler bodies, salmon transported energy from the Pacific to the bodies of those humans and other organisms who relied on the Columbia. Richard White has written about the role of salmon, before their virtual decimation, in bringing energy upstream. "During their time at sea Columbia salmon harvest the far greater solar energy available in the Pacific's food chain and, on their return, make part of that energy available in the river," observes White. "By intercepting the salmon and eating them, other species, including humans, in effect capture solar energy from the ocean. Salmon thus are a virtually free gift to the energy ledger of the Columbia."[89] Energy from organisms such as bison, salmon, and later cattle fueled humanity's biological and cultural transcendence. To date, environmental historians have focused mostly on the relationship between human transcendence and terrestrial animals, but compelling research by Arthur McEvoy, David Helvarg, and Mark Kurlansky is redirecting attention to the hidden environmental histories of the ocean and the myriad animals that live in its depths.[90]

CONCLUSION

When Plumwood paddled through those eerie channels of the Kakadu wetlands, she confronted a crocodile swimming just below the muddy water's surface. She changed course, trying to get around the animal; but, with a flip of its tail, it adjusted its course, too, successfully intercepting her. This is what animals represent to environmental historians: mobile, thinking, feeling—cleverly adjusting course and intercepting us—nature with a profoundly important form of agency. We ignore this agency at our own peril, however, because of our shared intimacy. Animals permeate global histories, their crocodile eyes glaring back from within virtually every story we tell. They are not separate from humanity, but rather an intimate partner in humankind's biological and historical transcendence. This is the principal lesson of writings on animals in environmental history.

NOTES

A version of this chapter was previously published in *History and Theory*, Theme Issue (December 2013): 45–67.
1. For the Kenton Joel Carnegie Memorial website, see http://www.mtechservices.ca/Kenton/Kenton.html.
2. Sara Gilman, "First Fatal Wolf Attack Recorded in North America?" *High Country News* (February 6, 2006). http://www.hcn.org/issues/315/16084.
3. Barry Holstun Lopez, *Of Wolves and Men* (New York: Touchstone, 1978), 69–73.
4. John Myers, "Death Could Be Rare Wolf Attack," *Duluth News Tribune* (December 21, 2005). http://www.freerepublic.com/focus/f-news/1545040/posts.
5. Mike Davis, *Ecology of Fear: Los Angeles and the Imagination of Disaster* (New York: Metropolitan Books, 1998), 206–207.
6. Myers, "Death Could Be Rare Wolf Attack."
7. Jamie Kneen, "Saskatchewan Gov't Denies Wolf Attack," *MiningWatch Canada*. http://www.northernminer.com/issuesV2/VerifyLogin.aspx?&er=NA (accessed March 30, 2014).

8. Darren Bernhardt, "Wolves Undergo Tests," *Saskatchewan News Service; CanWest News Service* (November 16, 2005).

9. Andrew McKean, "Manhunters: A Killing in Canada Puts an End to the Myth That Wolves Won't Harm Humans," *Field & Stream* (April 2007). http://www.outdoorlife.com/articles/hunting/2007/09/manhunters.

10. Ibid.

11. John Myers, "Death Could Be Rare Wolf Attack."

12. "Wolves Killed in Wollaston Lake Area," Canadian Broadcasting *Corporation CBC.CA News* (November 16, 2005).

13. McKean, "Manhunters."

14. For more on the history of the human–animal divide, see Richard W. Bulliet, *Hunters, Herders, and Hamburgers: The Past and Future of Human–Animal Relationships* (New York: Columbia University Press, 2005), 47–70.

15. William Cronon, "The Trouble with Wilderness; or, Getting Back to the Wrong Nature," in *Uncommon Ground: Rethinking the Human Place in Nature*, ed. William Cronon (New York: W.W. Norton, 1996), 69–90.

16. Michel Foucault, *The Foucault Reader*, ed. Paul Rabinow (New York: Vintage Books, 2010), 4.

17. Erica Fudge, *Animal* (London: Reaktion Books, 2002), 10.

18. Paul S. Sutter, "The World with Us: The State of American Environmental History," *Journal of American History* 100, no. 1 (June 2013): 96.

19. David Quammen, *Monster of God: The Man-Eating Predator in the Jungles of History and the Mind* (New York: W. W. Norton, 2003), 13.

20. Val Plumwood, "Being Prey," *Terra Nova* 1, no. 3 (Summer 1996): 34.

21. R. G. Collingwood, *The Idea of History*, rev. ed., ed. Jan Van Der Dussen (Oxford: Oxford University Press, 1994), 214.

22. Donald Worster, "Transformations of the Earth: Toward an Agroecological Perspective in History," *Journal of American History* 76, no. 4 (March 1990): 1089–90.

23. Ibid., 37.

24. Ibid., 42.

25. Dipesh Chakrabarty, "The Climate of History: Four Theses," *Critical Inquiry* 35 (Winter 2009): 201.

26. For a summary of this perspective on wolves, see Brett L. Walker, *The Lost Wolves of Japan* (Seattle: University of Washington Press, 2005).

27. Linda Nash, "The Agency of Nature or the Nature of Agency," *Environmental History* 10, no. 1 (2005): 67–69.

28. Michael Pollan, "Some of My Best Friends Are Germs," *The New York Times Magazine* (May 15, 2013). See also Lynn Margulis, *Symbiotic Planet: A New Look at Evolution* (New York: Basic Books, 1998).

29. Susan R. Schrepfer and Philip Scranton, eds. *Industrializing Organisms: Introducing Evolutionary History*. Hagley Perspectives on Business and Culture, vol. 5 (New York: Routledge, 2004).

30. Edmund Russell, "Evolutionary History: Prospectus for a New Field," *Environmental History* 8 (April 2003): 205–206.

31. On the industrialization of chickens, see Steve Striffler, *Chicken: The Dangerous Transformation of America's Favorite Food* (New Haven, CT: Yale University Press, 2005).

32. William H. McNeill, *Plagues and Peoples* (Garden City, NY: Anchor Books, 1979), 24.

33. Ibid.

34. Ibid., 70.

35. David Quammen, *Spillover: Animal Infections and the Next Human Pandemic* (New York: W.W. Norton, 2012).

36. Juliet Clutton-Brock, "Origins of the Dog: Domestication and Early History," in *The Domestic Dog: Its Evolution, Behavior and Interactions with People*, ed. James Serpell (Cambridge: Cambridge University Press, 1995), 8–10. For more on the origins of dogs and their behavior, see Raymond Coppinger and Lorna Coppinger, *Dogs: A Startling New Understanding of Canine Origin, Behavior and Evolution* (New York: Scribner, 2001). See also James Serpell, *In the Company of Animals: A Study of Human–Animal Relationships* (Cambridge: Cambridge University Press, 1996).

37. Kathleen Kete, *The Beast in the Boudoir: Petkeeping in Nineteenth-Century Paris* (Berkeley: University of California Press, 1994), 2.

38. Harriet Ritvo, *The Animal Estate: The English and Other Creatures in the Victorian Age* (Cambridge, MA: Harvard University Press, 1987), 17.

39. Ibid., 29.

40. Ibid., 115.

41. Walker, *The Lost Wolves of Japan*, 118–119.

42. Aaron Skabelund, "Can the Subaltern Bark? Imperialism, Civilization, and Canine Cultures in Nineteenth-Century Japan," in *JAPANimals: History and Culture in Japan's Animal Life*, ed. Gregory M. Pflugfelder and Brett L. Walker (Ann Arbor: Center for Japanese Studies, University of Michigan, 2005), 195–243.

43. Alistair Graham and Peter Beard, *Eyelids of Morning: The Mingled Destinies of Crocodiles and Men* (San Francisco: Chronicle Books, 1990 [1973]), 11.

44. Ibid., 68–69.

45. Ibid., 200–201.

46. Quammen, *Monster of God*, 119–124.

47. Ibid., 79–118.

48. Ibid., 123–124.

49. Ibid., 102, 123–124.

50. Mike Davis, *Ecology of Fear: Los Angeles and the Imagination of Disaster* (New York: Metropolitan Books, 1998).

51. Ibid., 200–201, 206–207, 237–238.

52. John Knight, *Waiting for Wolves in Japan: An Anthropological Study of People–Wildlife Relations* (Oxford: Oxford University Press, 2003), 84–122. For more on Japanese macaques, see Pamela J. Asquith, "Primate Research Groups in Japan: Orientations and East-West Differences," in *The Monkeys of Arashiyama: Thirty-Five Years of Research in Japan and the West*, ed. Linda M. Fedigan and Pamela J. Asquith (Albany: State University of New York Press, 1991), 81–98.

53. Donna J. Haraway, *Simians, Cyborgs, and Women: The Reinvention of Nature* (New York: Routledge, 1991), 10–11. See also Shirley C. Strum and Linda Marie Fedigan, eds., *Primate Encounters: Models of Science, Gender, and Society* (Chicago: University of Chicago Press, 2000).

54. Daniel Lord Smail, *On Deep History and the Brain* (Berkeley: University of California Press, 2008), 84, 118, 188–89.

55. Roy Willis, ed., *Signifying Animals: Human Meaning in the Natural World* (London: Routledge, 1990); Aubrey Manning and James Serpell, eds., *Animals and*

Human Society: Changing Perspectives (London: Routledge, 1994); Jennifer Wolch and Jody Emel, eds., *Animal Geographies: Place, Politics, and Identity in the Nature–Culture Borderlands* (London: Verso, 1998).

56. Robert Darnton, "Workers Revolt: The Great Cat Massacre of the Rue Saint-Séverin," in *The Great Cat Massacre and Other Episodes in French Cultural History* (New York: Vintage Books, 1984); Karen A. Smyers, *The Fox and the Jewel: Shared and Private Meanings in Contemporary Japanese Inari Worship* (Honolulu: University of Hawai'i Press, 1999); Jane Marie Law, "Violence, Ritual Reenactment, and Ideology: The Hōjō-e (Rite for Release of Sentient Beings) of the Usa Hachiman Shrine in Japan," *History of Religions* 33, no. 4 (1994): 325–357; Duncan Ryuken Williams, "Animal Liberation, Death, and the State: Rites to Release Animals in Medieval Japan," in *Buddhism and Ecology: The Interconnection of Dharma and Deeds*, ed. Mary Evelyn Tucker and Duncan Ryuken Williams (Cambridge, MA: Harvard University Press, 1997), 149–157.

57. Walker, *The Lost Wolves of Japan*, 181.

58. Jon Coleman, *Vicious: Wolves and Men in America* (New Haven, CT: Yale University Press, 2004).

59. Ibid., 69. Obviously, Jon Coleman is not the first to document the pogrom (as Barry Lopez called it) perpetrated against wolves. See also Stanley Paul Young and Edward A. Goldman, *The Wolves of North America: Part I, Their History, Life Habits, Economic Status, and Control; Part II, Classification of Wolves* (Washington DC: The American Wildlife Institute, 1944); Stanley Paul Young, *The Wolf in North American History* (Caldwell, ID: The Caxton Printers, Ltd., 1946); Lopez, *Of Wolves and Men; War against the Wolf: America's Campaign to Exterminate the Wolf*, ed. Rick McIntyre (Stillwater, OK: Voyageur Press, Inc., 1995); Karen R. Jones, *Wolf Mountains: A History of Wolves along the Great Divide* (Calgary: University of Calgary Press, 2002); Michael J. Robinson, *Predatory Bureaucracy: The Extermination of Wolves and the Transformation of the West* (Boulder: University of Colorado Press, 2005).

60. Coleman, *Vicious*, 71–72.

61. Ibid., 5.

62. Ibid., 6.

63. Ibid., 9.

64. Ibid., 8.

65. Joshua Blu Buhs, *The Fire Ant Wars: Nature, Science, and Public Policy in Twentieth-Century America* (Chicago: Chicago University Press, 2005).

66. Ibid., 4.

67. Brett L. Walker, "Sanemori's Revenge: Insects, Eco-System Accidents, and Policy Decisions in Japan's Environmental History," *Journal of Policy History* 19, no. 1, Special Issue: New Perspectives on Public Health Policy, ed. James Mohr (2007): 130–131.

68. Robert J. Spear, *The Great Gypsy Moth War: A History of the First Campaign in Massachusetts to Eradicate the Gypsy Moth, 1890–1901* (Amherst: University of Massachusetts Press, 2005).

69. Coleman, *Vicious*, 33. For more on dogs in the early Americas, see Marion Schwartz, *A History of Dogs in the Early Americas* (New Haven, CT: Yale University Press, 1997).

70. Alfred W. Crosby, *The Columbian Exchange: Biological and Cultural Consequences of 1492* (Westport, CT: Greenwood Press, 1972), 75, 77.

71. Ibid., 85.

72. Elinor G. K. Melville, *A Plague of Sheep: Environmental Consequences of the Conquest of Mexico* (Cambridge: Cambridge University Press, 1994), 39–40.
73. Brett L. Walker, "Commercial Growth and Environmental Change in Early Modern Japan: Hachinohe's Wild Boar Famine of 1749," *Journal of Asian Studies* 60, no. 2 (May 2001): 329–351.
74. Virginia DeJohn Anderson, *Creatures of Empire: How Domestic Animals Transformed Early America* (Oxford: Oxford University Press, 2004).
75. Ibid., 5.
76. Ibid., 21.
77. Ibid., 31.
78. Ibid., 38.
79. Ibid., 246.
80. Richard White, "Animals and Enterprise," in *The Oxford History of the American West*, ed. Clyde A. Milner, Carol A. O'Connor, and Martha A. Sandweiss (New York: Oxford University Press, 1994), 256.
81. Quoted in William Cronon, *Nature's Metropolis: Chicago and the Great West* (New York: W. W. Norton, 1991), 208.
82. Mike Davis, *Late Victorian Holocausts: El Niño Famines and the Making of the Third World* (London: Verso Books, 2001), 34.
83. Ibid., 71.
84. Ibid., 76.
85. Mark Elvin, *The Retreat of the Elephants: An Environmental History of China* (New Haven, CT: Yale University Press, 2004), 9.
86. Ibid., 11.
87. Ibid., 13.
88. Andrew C. Isenberg, *The Destruction of the Bison: An Environmental History, 1750–1920* (Cambridge: Cambridge University Press, 2000).
89. Richard White, *The Organic Machine: The Remaking of the Columbia River* (New York: Hill and Wang, 1995), 15.
90. Arthur F. McEvoy, *The Fisherman's Problem: Ecology and Law in the California Fisheries, 1850–1980* (Cambridge: Cambridge University Press, 1986); Mark Kurlansky, *Cod: A Biography of the Fish that Changed the World* (New York: Penguin, 1997); and David Helvarg, *Blue Frontier: Dispatches from America's Ocean Wilderness* (San Francisco: Sierra Club Books, 2006).

CHAPTER 3

....

BEYOND VIRGIN SOILS

Disease as Environmental History

....

LINDA NASH

IN 1972, Alfred Crosby published *The Columbian Exchange*, in which he chronicled the devastating impact of introduced infectious disease on Native Americans in the wake of Spanish colonization.[1] Arguably this story has been the single most influential contribution of environmental history to the larger discipline. It is now a standard part of US history textbooks, and ranks as one of the seminal stories of American and world history.

But beyond the impact of introduced pathogens in colonial North America, environmental historians largely ignored the topic of disease for the next two decades, only turning to it in the 1990s as part of a broader interest in the body and its relationship to "nature." Yet the movement of microorganisms and chemicals through landscapes, animals, insects, and human bodies offers one of the densest and most obvious links between human beings and their environments. Why, then, was disease not more central to the field in its formative decades?

Perhaps to focus on disease is to court a kind of anthropocentrism that an earlier generation of environmental historians wanted to avoid.[2] But more mundane professional considerations are not irrelevant. The various subfields of history lay claim to certain kinds of material; historians define themselves, in part, by their archives. Environmental historians, by and large, have laid claim to the material generated by the environmental sciences—fields such as ecology, geology, and marine biology—and have, at the same time, left other sources—those generated by doctors, medical biologists, and even patients—to those who identify themselves as historians of medicine and public health. Moreover, environmental historians are more likely to interpret the effects of ecological change as broadly social rather than intimately biological. Following the path set by so much of twentieth-century biomedicine, historians have, with some notable exceptions, overlooked or downplayed the environmental dimensions of disease. That is, twentieth-century historians largely wrote the history of disease to accord with the laboratory model of twentieth-century biomedicine. And even as a few maverick historians of medicine pointed to the environmental determinants of illness, their histories for too long remained outside the sphere of most environmental historians.[3] Thus, much of the literature on environment and disease lies either outside the subfield of environmental history, or geographically outside of North America and Europe, where the divisions among these historical subfields are not so stark. This essay ranges across a disciplinarily diverse literature, gathering together work on disease and environment in order to survey its intellectual antecedents and political implications, and to consider some of its current directions. It looks at the nexus between these two topics within three broad historical themes: conquest and colonization, urbanization and industrialization, and the so-called "re-emergence" of infectious diseases in the late twentieth century.

DISEASE ECOLOGIES AND COLONIAL HISTORIES

In US history, discussions of disease and environment inevitably begin with the effects of introduced pathogens on Native Americans in the wake of Columbus. The "Columbian Exchange"—the transfer of plants, animals, and pathogens between the Old and New Worlds—was unquestionably a watershed in the history of the world, and the epidemics subsequently unleashed in the Americas marked a pivotal moment in the global history of disease. Native societies suddenly experienced multiple out-breaks of many diseases, most of which they had never before encountered—including smallpox, measles, malaria, scarlet fever, yellow fever, and whooping cough. In many regions, catastrophic population losses followed. Sickness, along with warfare and the ensuing social disorganization, played a crucial role in the European invasion of the Americas. Alfred Crosby's telling of these events, first in a 1967 article and later in his groundbreaking books, has made this story central to any contemporary understand-ing of North American history.[4]

The interest in disease history sparked by Crosby's work was magnified by the subsequent publication of an equally influential volume just four years later. In 1976, William McNeill's retelling of world history as epidemic history appeared as *Plagues and Peoples*. McNeill emphasized the overriding role of infectious disease in the course of human events, finding in disease history not only the reason for European success in the Americas, but also the primary explanation for the devolution of the Roman Empire, the timing of Chinese and European imperial expansion, and the rise of both Christianity and Buddhism.[5] Taken together, these two works forced histori-ans to reckon with pathogens as crucial factors in social and political history, and they spurred new interest in the biological interactions between human beings and their environments. But what about the origins of these works? Although environmental historians are quite familiar with these authors, the intellectual contexts that produced their work have been far less studied. How had it become possible in the late 1960s to cast pathogens as the major shapers of history? And what were the broader implications of these disease narratives?

Both Crosby and McNeill drew upon the field of disease ecology, a subfield of medical research that took shape during the first decades of the twentieth century. Whereas the discoveries of germ theory had encouraged bacteriologists to regard pathogens as the essential cause of infectious disease, some researchers saw the emphasis on pathogens as overly narrow. Instead they insisted that the insights of germ theory had to be situated within a more complex framework that encompassed environmental and genetic factors. This emerging field had strong links to tropical and veterinary medicine, specialties that had never abandoned an environmental outlook.

The inspiration to pursue a more ecological understanding of disease typically came from the periphery rather than the colonial center, perhaps because in those lands where Europeans were still relative newcomers the ecological impacts of colonization were both more recent and more obvious, and thus the relationship between disease organisms and environmental change easier to recognize. The interest in disease ecology was, in Warwick Anderson's phrase, a "legacy of colonial settler anxieties."[6] In the US, the crucial figure was Theobald Smith, whose sensitivity to environmental factors was nurtured by his study of animal diseases in the American South and West on behalf of the US Department of Agriculture in the late nineteenth century. Best known for his discovery of the first insect vector—the tick that transmits Texas cattle fever— Smith offered a radically new description of disease, casting it not as an invader of an otherwise healthy organism but merely as a disturbance of the equilibrium between host and parasite. In Africa, British colonial scientists working to control sleeping sickness (a disease of both cattle and humans carried by the tsetse fly) perforce adopted an ecological perspective on disease, linking fly populations to landscape ecology. Shortly after World War II, the Paris-trained surgeon Jacques May drew on his experience in Bangkok and French Indochina to argue that infectious disease had to be situated in relation to place; he would make his life's work the publication of a fourteen-volume opus on modern medical geography. And in Australia, Frank Macfarlane Burnet's investigation of psittacosis among parrots led him to the conclusion that unfavorable environmental conditions had activated an otherwise latent infection. With the publication of *Biological Aspects of Infectious Disease* (1940), Burnet would offer the most sophisticated articulation of the ecology of pathogens in the period before World War II.[7]

The first histories to mobilize this ecological perspective on disease were penned by scientists. In 1935, Hans Zinsser, a bacteriologist and colleague of Theobald Smith's at Harvard, published a popular history of typhus that drew explicitly on Smith's ideas about parasitism. A decade later, Percy Ashburn, an American physician who had worked in the colonial Philippines, published his popular account of the effects of introduced diseases on Native American populations in the sixteenth and seventeenth centuries. Rene Dubos, aided by his wife Jean, followed in 1952 with a pathbreaking history of tuberculosis, emphasizing how both biological and social factors influenced the waxing and waning of tuberculosis mortality in Europe and America.[8] Despite this spate of innovative work, however, ecological approaches to both disease and history remained somewhat marginal in the immediate postwar years, overshadowed by the discovery of antibiotics and the belief, shared by both doctors and lay people, that infectious disease would soon be conquered, at least in the industrialized world. Mid-century medical successes prompted triumphalist accounts of bacteriology and scientific medical progress while pushing more nuanced ecological accounts of both disease and disease history to the margins.

It was in the 1960s that environmental concerns combined with the postcolonial critique of Western "progress" to generate new interest in the ecology of disease. Rene Dubos helped popularize the field in two best-selling books, becoming something of a guru of the nascent environmental movement in the process.[9] Disease ecology also fit well with the revisionist trajectory of historical writing in that period. For historians of the late 1960s and 1970s, disease ecology offered another means of questioning western assumptions about technological superiority, while underwriting a historical narrative that sidelined powerful individuals. Like the broad social and economic histories pioneered by French historians associated with the Annales School, disease history downplayed the agency of traditional elites in favor of larger structural and ecological forces. Moreover, disease history brought some of the key intellectual positions of modern environmentalism into professional history writing. For Crosby, McNeill, and the generation of environmental historians who followed, the insights of disease ecology implied that human history had to be understood in biological as well as social terms. Pathogens, plants, and animals were as—and likely more—important than the decisions of political elites or the strategies of invading armies; human beings were part of a larger ecology that they could affect—often negatively—but could by no means control. History was thus best understood as a story of ecological change and adaptation, rather than as a tale of intellectual and technological progress.

The new disease histories drew not only on the science of disease ecology but also on the emerging discipline of historical demography, a set of methodologies typically associated with Annales historians where they represented one more facet of the quantitative and structural turn in postwar historical writing. However, historical demography also arose independently in the US among anthropologists and historians of Native America.[10] Crucial to Alfred Crosby's work were the population histories of central Mexico produced in the 1950s and early 1960s by Sherburne Cook and Woodrow Borah at the University of California at Berkeley. Cook, a physiologist whose appointment was in the university's medical school, brought his training in medicine and statistics to the study of California and Mexico during the Spanish colonial period. Cook had begun to study the history of the California Indians in the 1930s, employing the techniques of animal ecology to estimate past human populations. In the 1950s, he began a decades-long collaboration with Borah, an economic historian of Mexico. Together, they pioneered the use of new historical sources, such as tax rolls and mission registries, to estimate colonial and precontact populations. Their estimate of Mexico's precontact native population—25 million—was a stunning claim when it was published in 1963. Soon after, Borah would propose 100 million as a likely figure for the pre-Columbian population of the Americas, a number twelve times greater than Alfred Kroeber's then-authoritative figure of 8.4 million.[11] Cook and Borah's population estimates had many radical implications: foremost among them was the assertion that Native American mortality in the post-contact period had been far higher than acknowledged by previous scholars, potentially as high as 90 percent on the Mexican plateau. Although Cook and Borah pointed to the pivotal role played by disease, they offered a nuanced account of population decline that encompassed multiple factors; Cook, especially, emphasized the disruption of a broader ecological equilibrium rather than solely the differential effects of disease.[12]

It was other writers—first Ashburn, later Crosby and McNeill—who placed greater emphasis on differential disease immunities to explain the European conquest. These authors pointed to the fact that most Europeans who came to the New World were immune to smallpox, having contracted and survived the disease during childhood ("acquired immunity"); many had similarly acquired immunity to measles. They also suggested that Europeans likely had inherited immunities to other deadly diseases to which their ancestors had been heavily exposed—the result of natural selection over many generations. Both Crosby and McNeill drew on several hypotheses about disease evolution to ground their claims about differential immunities. The first was crowd theory, which asserted that many of the most deadly human diseases (e.g., smallpox, measles, mumps) required a relatively dense human population to sustain themselves, and therefore likely emerged in Asia and the Middle East, areas with long histories of settled agriculture and urbanization. From there, trade carried them to Europe and parts of Africa; the Americas, however, remained isolated and too sparsely populated to evolve similar diseases. Secondly, most human pathogens are assumed to have evolved from similar animal pathogens and thus domesticated animals, because of their close proximity to humans, are the likely source of most human diseases. Thus, scholars argued that the majority of human infectious diseases evolved in the dense, agricultural societies of Europe, Asia, and the Middle East.[13]

A third element crucial to these histories was the malaria hypothesis. In the late 1940s, the British geneticist J. S. Haldane speculated that the prevalence of thalassemia (a form of anemia) among southern European immigrants in the US might be the result of natural selection for resistance to malaria. However, the theory did not gain popular attention until nearly decade later when another researcher, A. C. Allison, used it to explain the evolution of the sickle-cell trait in African and African-derived populations. In other words, the presence of these hemoglobin anomalies—thalassemia and the sickle-cell trait—could be explained by the survival advantage that these conditions conferred to those whose ancestors had lived in malaria-endemic regions of southern Europe and western Africa. Despite the speculative nature of the evidence, the story of African "resistance" to malaria quickly gained widespread attention in the 1960s and proved irresistible to scholars across several fields.[14] Allison's work allowed a much older colonial discourse that associated race with disease susceptibility to be reframed in modern scientific terms: as the process of natural selection operating on human populations who lived in malaria-endemic West Africa.

The concept of differential immunity transformed studies of European colonialism beginning in the late 1960s. Because of their long history of agricultural settlement, old world populations were assumed to have a long experience with a multitude of pathogens. North American natives, on the other hand, were presumed to have had little or none. Massive Native American mortality due to smallpox could be explained by their lack of exposure to the disease in childhood (that is, their lack of acquired immunity), while their apparent susceptibility to tuberculosis could be ascribed to their isolation from the TB pathogen and consequent genetic makeup. To put it crudely (as several medical researchers and some historians did), the American Indian population had not endured the same "weeding out" of susceptible individuals and thus had not reaped the benefits of natural selection with respect to diseases like TB; since the host (Native Americans) and the parasite (*M. tuberculosis*) had not yet evolved an equilibrium relationship, the virus killed off Native Americans in dramatic numbers. It was Crosby who brought the potent if problematic term "virgin soil epidemic" into the historical lexicon, defining it as "those in which the populations at risk have had no previous contact with the diseases that strike them and are therefore immunologically almost defenseless."[15]

In a seminal article published in 1968, historian Philip Curtin drew on the concept of differential immunity to argue that some Africans held a slight survival advantage over Amerindians and Europeans in the disease environment of the Caribbean, and that this was one factor that encouraged the expansion of the Atlantic slave trade throughout the seventeenth and eighteenth centuries. Although Curtin was judicious in his use of the limited and inconclusive biological evidence, others were less so. For instance, in the 1980s, Kenneth Kiple claimed that genetics and evolution were *the* crucial factors shaping the Caribbean region's demographic history. Deeply wedded to the concept of "race" as a meaningful biological category and to the notion of "blacks' biological distinctiveness," Kiple claimed that Africans in the New World were a "biological elite," shaped by generations of deprivation in the harsh, disease-ridden environment of West Africa and the horrors of the middle passage. Vastly overstating the case for African "immunity" to malaria and insisting on an unproven inherited immunity to yellow fever, Kiple asserted that black slaves had a substantial survival advantage in the Caribbean, that they were in fact "inexorably selected" by their biology as the region's workforce. Kiple's work was notable for his efforts to bring human biology and disease into the study of African slavery, but in his case it came at the cost of ignoring, or at least severely downplaying, the social realities that crucially shaped slaves' biological existence. Perhaps most obviously, Kiple ignored the excessive morbidity and mortality experienced by slaves in the Caribbean (roughly one-third of slaves died within three years of arrival, with "fevers" among leading causes of death in the British Caribbean during the early nineteenth century), while ignoring mortality variations across the region (e.g., the startlingly different demographic patterns among slaves on sugar versus coffee plantations) and among different African ethnic groups—facts which undermined his insistence on the meaningfulness of "race," as well as his astounding and unsupportable conclusion that racial slavery emerged as a result of putative biological differences.[16]

Moreover, despite his interest in how environments shaped both pathogens and human biology, Kiple's narrow emphasis on the evolutionary history of humans led him away from any careful consideration of historical environments and environmental change. Treating West African environments as both spatially homogeneous and temporally static, he upstreamed postcolonial conditions and downplayed the effects of colonial rule and cash-crop agriculture in transforming the region's lifeways and disease environments, while similarly ignoring how human actions, especially the extension of sugar plantations, had transformed the Caribbean's disease ecology. Yet the seeming pathology of tropical environments and the rapid spread of disease were often direct outcomes of colonial activity—whether the introduction of clay-bottomed sugar pots which created an ideal habitat for *Aedes* mosquitoes (vectors of yellow fever), the clearing of forests for cash-crop agriculture, or the unparalleled regime of work and inadequate diets that characterized slave life on Caribbean sugar plantations. That is, Caribbean and African environments changed radically over time, and their propensity to foster certain diseases increased markedly in the face of European colonialism, if not before.

The story is similar in the Bengal region of India, where malaria and cholera emerged as epidemics in the nineteenth century, generating massive mortality among populations that were in no way isolated or bacteriologically inexperienced. In this case, the wide-ranging environmental changes instigated by British colonial authorities (namely, the construction of embankments and dams as part of a colonial transportation network) provided new habitat for anopheline mosquitoes, while creating conditions favorable to the transmission of highly pathogenic cholera. [17] That is, epidemic disease in colonial situations has frequently been the consequence of radical environmental change as much as, and likely far more than, any difference in biological immunity between European and native populations.

Nonetheless, understanding colonization as a biological and ecological process, rather than solely as an economic and political one, has yielded important insights. Nowhere has this been more true than in the history of Africa, where attention to disease has helped to enable not only an environmental but a postcolonial reading of that continent's history. Many historians have long taken the disease environment of Africa as the central issue shaping the continent's history: disease posed immense challenges to the establishment of civilizations, limited contacts with other regions, discouraged European colonization, and, as discussed above, encouraged the slave trade to the Caribbean. But such readings have tended to ignore the historicity of disease environments. It is now widely understood that first the slave trade and later European colonialism introduced several new diseases (e.g., smallpox, TB, cholera, bubonic plague) while disrupting ecological and social relationships throughout most of the continent.[18] These biological developments had far-reaching social effects.

What is perhaps the seminal study of disease and environment in Africa was written not by a historian, but by a natural scientist. An entomologist by training, John

Ford brought decades of field experience with the tsetse fly to his analysis of trypanosomiasis (sleeping sickness)—a disease indigenous to the continent and also one of the most devastating maladies during the colonial period. Although grounded in the science of disease ecology, *The Role of Trypanosomiasis in African History* (1971) is deeply historical in its insistence that disease cannot be understood apart from how societies have responded to it. Moreover, Ford directly challenged the prevailing notion that sleeping sickness had limited African development for centuries and that colonial eradication efforts aimed at the tsetse fly (the vector of trypanosomiasis) were unambiguously progressive. Instead, he argued that precolonial African communities had successfully held sleeping sickness in check by achieving an ecological balance with both parasite and fly through limited but persistent interaction. That balance was destroyed in the late 1880s, when rinderpest irrupted among Eritrean cattle—the result of the Italian military's importation of infected Indian cattle to supply their campaign against Somalia. The rinderpest outbreak in East Africa was quickly followed by smallpox, locust invasions, and jiggers. As rinderpest spread across the continent, it set off a catastrophic feedback loop in which cattle and wildlife die-offs were followed by high human mortality, which resulted in the resurgence of bush, wildlife, and tsetse flies. Skyrocketing fly populations brought epidemics of sleeping sickness; colonial wars furthered this cycle of sickness, devastation, and depopulation. When new cattle were imported, they quickly succumbed to a disease to which they were not adapted. Colonial control efforts focused on regrouping and relocating African people only exacerbated the situation. By Ford's reading, Africans in the 1930s faced a much more terrible disease environment than they had fifty years earlier. The ensuing emphasis on tsetse eradication was one outcome, and a very expensive one, of Europeans' misreading of the continent's disease history. Ford insisted that, contrary to popular belief, European science had not conquered African tropical disease; rather, colonial scientists had left African countries "a legacy of ideas that had little relevance to the biological processes with which they had unwittingly interfered." Subsequent historical studies of African sleeping sickness have elaborated Ford's basic insight that the ecological and social changes accompanying colonialism created the conditions for epidemic trypanosomiasis, although none have attended so closely as Ford to the historical ecology.[19]

Disease has also been prominent in studies of the US South, where the near-tropical environment was assumed to set the region apart from the rest of North America. Like the Caribbean, much of the colonial South was plagued by malaria and yellow fever, while the region's climate also fostered the hookworm parasite. The environmental dimensions of disease did not go unnoticed by contemporaries. High morbidity and mortality affected regional demography and migration patterns, with the most sickly areas, such as the South Carolina low country, registering substantial population declines among white settlers in the late sixteenth and early seventeenth centuries. Though the social effects of disease are harder to trace, scholars have suggested that disease increased economic polarization while fostering more closely-knit communities in some areas such as the colonial Chesapeake. Moreover, during the nineteenth

century, repeated epidemics of yellow fever became the impetus for urban environmental reform across the South.[20]

Disease also shaped the colonization of western North America, though the topic
has been noticeably less prominent in Western than in Southern US history. In a region
repeatedly extolled for its healthfulness, stories of disease have perhaps been harder
to see. However, in a study published in 1949, medical historian Erwin Ackerknecht
chronicled the effect of malaria on the history of the upper Mississippi Valley; his work
was pathbreaking in that he pointed specifically to the role of human-induced environmental change in the spread of that disease and its eventual decline. In the late 1960s
and early 1970s, the historical geographer Kenneth Thompson attended to the disease
environments of California—a region long assumed to be a paradise of health. Among
other things, Thompson described the prevalence and rapid spread of malaria during the chaotic years of the Gold Rush. More recently, Conevery Valencius and Linda
Nash have detailed how concerns over disease and health shaped settlement practices
in the Arkansas-Missouri region and California, respectively. And although told primarily as a story of medical discovery, Victoria Harden's history of Rocky Mountain
Spotted Fever points to the unexpected results of human intrusion into the habitat of a
micro-organism that otherwise inhabits the bodies of ticks and small mammals. All of
these works illustrate how the environmental changes consequent to Western colonization altered disease patterns and prevalence, typically for the worse.[21]

Recently, historians have focused more closely on the ideological dimensions of disease environments in different times and places. Historians of the Southern US have
long recognized the ways in which the experience of disease contributed to an ideology
of regional distinctiveness that would become a central component of the pro-slavery
argument. The fusion of race, geography, and medicine reached an apotheosis of sorts
in the antebellum South, but it was part of a much larger colonial discourse. Recovering
the field of imperial medical geography from the late eighteenth and nineteenth centuries, scholars have shown how then-scientific concepts of disease and environment
were tightly interwoven with European ideologies of race and the project of colonial
expansion. Western scientific medicine was part and parcel of the colonial project. And
yet, the unpredictable and uncontrollable bodily outcomes could challenge as well as
underwrite European colonialism. Whether in Southern Africa, India, the Philippines,
tropical Australia, New England, or California's Central Valley, the prevalence of disease fostered the persistent anxiety that white bodies in the colonies might be "out of
place," and underscored the sense that colonial settlement was a gamble in both economic and physical terms.[22] To contemporaries, it was not always clear that nature was
on the European side.

The politics of writing histories of colonialism based on disease ecology could cut
either way. On the one hand, disease history decentered individual human actors
and, much like the social history of the 1960s, undercut the political-intellectual history that had for so long dominated historical writing. Suddenly it was possible to cast
Variola major and *Aedes aegypti* as agents of historical change. To write, for example,
that the conquest of South and Central America was not the result of Spanish strategy

or superior technology, but rather the unwitting outcome of epidemics, was a kind of liberal revisionism. In this sense, disease ecology appealed to those historians intent on assailing traditional Eurocentric narratives of history.

And yet, as David Jones has argued, the concern with disease has often served to rewrite Euro-American elites' unending concern with race and racial history in ostensibly more—or at least newer—scientific terms. The basic insight that historians derived from disease ecology was a rather simplistic one concerning the differential immunities of racialized populations. When Percy Ashworth penned his popular account of Native American disease history in the 1940s, his conclusions were unabashedly racist. As he put it, "Perhaps the white race has been and is great because of the difficulties it has surmounted, the diseases and habits it has mastered or partly mastered in the past." Though later writers were more circumspect in their rhetoric, the sentiment of biological superiority, albeit accidental, persisted among many Anglo-American authors, where Europeans' racial superiority was replaced by a "biological weapon...implanted in the bloodstreams of civilized peoples" (McNeill) or "an immune system of such experience and sophistication" (Crosby). A similar if more extreme reliance on immunological differences underpins the popular arguments of Jared Diamond in his Pulitzer prize-winning *Guns, Germs, and Steel* (1997). Though the emphasis of each author is different (Crosby, in particular, is the most circumspect about the ultimate significance of immunological differences), in all of these stories the outcomes of colonial struggles were largely determined in the proto-historic past, and pathogens become the unwitting allies of Europeans. For Europeans and their descendants in North America, the story was one of evolutionary immunity from a host of infectious diseases that rationalized, even if it did not endorse, the colonial order. In the case of Africans, disease ecology offered at least a partial explanation for their enslavement, and for Native Americans it offered the crucial explanation for their defeat and political marginalization. Despite the novelty of putting disease at the center of history, many of the resulting stories have a familiar teleological ring.[23]

Part of the problem is one of emphasis. Without denying the very real impacts of disease and variable immunities, it is impossible to parse the biological from the social in these histories. Where historians like Crosby, McNeill, and Kiple stressed the effects of pathogens, others have given more attention to the social practices and cultural responses that accompanied colonial invasions and settlement. Paul Kelton, for instance, stresses the effects of the Native American slave trade in fostering the circumstances that set off deadly epidemics.[24] Yet another part of the problem lies in using science to write history without subjecting scientific work to the same historicizing process. Science itself is part of a larger social context, and thus always subject to revision. The questions that medical researchers ask, the hypotheses that gain attention and credibility, the research that is marginalized or not even conducted—these are undeniably social phenomena. It is worth recalling that imperial medical geography was a cutting-edge science in its day. Even modern medicine, as Warwick Anderson and Keith Wailoo have shown, has played no small part in constructing and sustaining race as a social category. Moreover, the existing scientific literature is more diverse and complex than historians have generally acknowledged. There is strong evidence to suggest, for instance, that tuberculosis and influenza, diseases that are presumed to have originated in the Old World, may have existed in pre-Columbian America; that is, the Americas and their native inhabitants may not have been as free from infectious disease as has typically been assumed. Moreover, recent genetic studies of *Shigella*, a dysentery-causing bacterium, date its evolution to no later than 50,000 years ago, long before the Neolithic revolution. Similarly, molecular genetic studies of smallpox, tuberculosis, and cholera have cast doubt on the assumption that domestic animals were the source of most of the important human diseases. These findings all suggest that historians need to at least reconsider the basic story of disease emergence on which they have so heavily relied. And although the classic model of disease ecology postulates that disease virulence necessarily decreases over time, as pathogens and hosts adapt to one another, this is not always the case; in some cases, natural selection may work to increase virulence. Medical understanding of immunity has also grown far more complex over the last three decades, rendering simplistic accounts of "acquired" and "inherited" immunity anachronistic. The immense advances in genetics have told us much about the evolutionary history of both malaria and human hemoglobin anomalies, but they have also revealed just how difficult it is to substantiate the influence of natural selection on human populations.[25] Yet *these* scientific developments have yet to influence historical writing. Is it perhaps because they might complicate a simple and cherished colonial narrative?

DISEASES OF URBANIZATION
AND INDUSTRIALIZATION

As historians have long recognized, disease is a crucial part of the story of urbanization and industrialization. Denser concentrations of humans inevitably create new ecological conditions, and both scientists and historians have chronicled the association between urban environments and infectious disease. No disease has received more attention than Europe's Black Death, which killed more than one-third of the population between the mid-fourteenth and mid-fifteenth centuries, prompting one historian to call it "Europe's greatest known ecological disaster." The basic story identifies the Black Death as an outbreak of the bacillus *Yersinia pestis* that had been transported to Europe from Asia, where it existed among wild rodents in non-epidemic form. Once present in a local rodent population, bubonic plague may spread to humans through the bite of infected fleas—and the dense populations of humans, fleas, and rats in medieval European cities made epidemic plague possible. Scholars have suggested that the rapid spread of disease and its exceedingly high mortality rate were affected by ecological factors—including higher-than-normal precipitation that triggered an explosion in rat (and flea) populations.[26]

But it was the nineteenth century, which witnessed the rise of industrial urban centers and far more intensive colonial connections, that saw the most dramatic global epidemics. Bubonic plague re-emerged in 1894 in Hong Kong, spreading to every continent, where it disproportionately affected the poor and marginalized. In telling these stories, historians have typically drawn less on disease ecology than on the traditions of social medicine and the literature of public health. And while environmental historians have contributed much to the story of late nineteenth-century urbanization, they have emphasized its effects on the non-human environment: smoke-laden air, polluted rivers, and the enormous quantities of resources necessary to sustain urban life.[27] Yet as public health historians have long known, the process of industrialization also brought tremendous costs in health and multiple epidemics of infectious disease. The mortality rates in urban centers rose throughout most of the nineteenth century before leveling off and then declining in that century's final decades. Many diseases raged through crowded cities in addition to bubonic plague: typhoid and other diarrheal diseases, tuberculosis, whooping cough, measles, diphtheria, cholera, and smallpox. There are many excellent histories of both disease and environment under regimes of industrialization, but the literatures have been, until quite recently, surprisingly separate.

Tuberculosis was the principal killer in industrializing cities. The story of TB in America and Europe was critically told by Rene and Jean Dubos in their pathbreaking study, *The White Plague*, in which they described TB as "the first penalty that capitalistic society had to pay for the ruthless exploitation of labor." The Duboses refuted arguments about racial susceptibility and insisted that TB was essentially a social disease—that is, a disease whose principal cause lay in human social organization; or, more specifically, a disease whose cause could be found in the conditions of urban-industrial poverty. In their case, the term "social" encompassed the environmental conditions that characterized urban slums: overcrowded, garbage-strewn areas with poorly ventilated buildings that constantly exposed individuals to infection and reinfection. At the same time, the harsh conditions and poor rewards of industrial labor left most workers malnourished, fatigued, and more susceptible to infection. In industrial cities, environmental and social conditions were so thoroughly intertwined that they cannot be untangled.[28]

The history of tuberculosis has, like capitalist development, been uneven. In the modern period, it first reached epidemic proportions in England at the end of the eighteenth century, and then engulfed the industrializing populations of Germany, France, and the United States. In rapidly growing industrial capitals like London and Paris, TB infection rates may have reached nearly 100 percent. The disease came later to some Native American societies—typically only after they were forced onto crowded reservations—and later still to Alaska, where it remains a significant problem. Urban TB rates in the US were high throughout the early twentieth century. However, they were anything but uniform: African Americans suffered from TB at rates seven times greater than their white counterparts. At the end of the nineteenth century, as European and American imperialism helped spread the industrial revolution outward, tuberculosis followed. In fact, TB remains the principal killer worldwide. Randall

Packard's brilliant study of tuberculosis in South Africa links the disease to the specific working and environmental conditions that inhered in large-scale mining operations. Inadequate ventilation, dust exposure, high heat and humidity, wide temperature fluctuations, long hours of continuous work, and the risk of hookworm infection all contributed to extraordinarily high TB rates among African miners. Moreover, high labor turnover combined with the policy of repatriating sick black workers to their rural homes ensured that TB would spread rapidly through the South African countryside. Although colonial officials interpreted the higher TB incidence and mortality among black (versus white) miners to their supposed "virgin soil" status, Packard argues that declining disease rates among white miners (and industrial workers in Europe and America for that matter) were less the result of genetic characteristics than of public health reforms that were implemented in a racially selective way. While white workers saw improvements in working and living conditions, blacks did not. (Nor did Native Americans in the US.) Similarly, although tuberculosis rates declined dramatically in the industrialized world throughout most of the twentieth century, developing countries experienced no such reprieve. Then, in the mid-1980s, TB experienced a resurgence in places such as the former Soviet Union and in America's inner cities—a response to the spread of HIV, the development of antibiotic resistant strains, and the decline of infrastructure and health services in poor, urban communities. Even today, despite TB's global reach, it retains a social and geographic specificity. It is found, for instance, among the rural poor of Haiti, the urban working-class of Lima, the prisons and homeless shelters of North America.[29]

Horrific pollution often exacerbated the prevalence of disease. The literature on urban pollution and health in nineteenth-century Europe and North America is vast, most of it penned by historians of public health. Industrialization in Europe and the US was driven by the consumption of coal, and over the course of the nineteenth century the urban atmosphere was increasingly characterized by a thick, acrid smog. Historical accounts of industrializing cities like Pittsburgh and Manchester remark on the fact that the skies were gloomily dark even at midday. Smoke-laden skies obviously increased respiratory disease; by the 1870s, bronchitis had become the most common cause of death in England's factory towns, killing between 50 and 70 thousand people each year. However, the effects of air pollution ramified in bodies in less obvious ways. The ever-thickening clouds of soot and smoke that poured out of factories and residential chimneys kept the sun from reaching residents' skin. Since sunlight is the principal source of vitamin D, one result was an epidemic of rickets—a childhood bone disease. A British study conducted in 1889 revealed the geographic distribution of rickets to be highest in industrial and mining centers, and within the London suburbs the incidence of rickets correlated inversely with distance from the city center. Those most affected were the working classes, who could not afford the dietary sources of vitamin D (dairy products) that might have compensated for the lack of sunlight. Not only did rickets affect children's skeletal development, however, it also increased their mortality from whooping cough and measles and decreased their resistance to TB and other respiratory diseases.[30]

Yet nineteenth-century environments were implicated not only in the causes of disease, but also in its cure. Before and even after the advent of germ theory, the medical recommendation for "phthisis" (a historical diagnosis that often encompassed the contemporary diagnoses of TB, bronchitis, and other respiratory problems) was to seek out a good and healthful environment, which meant a non-urban environment—whether in Davos, the mountains of Silesia, New York's Saranac Lake, or Southern California. If the urban slums were perceived as the source of disease, dry mountain climes offered the best hope for recovery in the era before antibiotics. Nineteenth-century medicine fostered an appreciation for the therapeutic qualities of particular landscapes and furthered the interest in "wilderness" and outdoor recreation as crucial elements of health. "Climatotherapy" was its own medical specialty, sanatoria and health resorts sprang up across much of Europe and the western US, and a wave of early twentieth-century "health seekers" helped populate locations like Denver, San Diego, and Phoenix. John Bauer's classic book on the health migration to California chronicles one piece of this migration, while Gregg Mitman's innovative history of asthma and allergy, *Breathing Space*, wonderfully combines social, medical, and environmental history to illustrate not only how ideas about disease shaped treatment and migration, but also how concerns over health shaped particular landscapes. In Mitman's telling, diseases, environments, and bodies evolve together, as do our understandings of all three.[31]

The study of sanitation and public health reform has often brought the histories of disease and environment together. Nineteenth-century medicine typically located the cause of epidemic disease in the broader environment—where miasma, noxious winds, poor soils, and unhealthy water all threatened to push a body out of balance and into illness. Edwin Chadwick's landmark report on London, *The Sanitary Condition of the Labouring Population* (1842), argued for the remediation of "noxious physical circumstances" in the interests of health and helped launch the British sanitary movement. Over the next several decades, the effort to control epidemic diseases such as typhoid and cholera would prompt the reshaping of city environments across the globe. The municipal collection of garbage, the protection and filtration of water supplies, the installation of sewerage services, the paving of streets and sidewalks, the construction of parks, the filling of wetlands, and the regulation of urban animals—along with the disciplining of citizens through instruction in "hygiene"—were efforts to construct an urban environment that was less prone to disease. Nearly three decades ago, Judith Walzer Leavitt demonstrated how easily the public health history of this period bled into urban environmental history in her pioneering study of Milwaukee, while Joel Tarr, a historian of technology, called attention to the urban infrastructure—including the changing approaches to garbage and water supply—and its implications for health. More recently, Martin Melosi's epic survey, *The Sanitary City*, recounts the evolution of urban environmental and infrastructure reforms in the US and their health effects from the colonial period through the twentieth century. In an incredibly detailed and ambitious study of urban pollution and reform in two key cities—Chicago and Manchester—Harold Platt offers a transatlantic comparison of the environmental and health costs of industrialization and efforts at amelioration, arguing that Progressive reformers achieved significant improvements in both environmental equality and health—though the results were mixed. Whereas they were largely successful in curbing waterborne disease through the adoption of new technologies, air pollution and respiratory disease proved less tractable, and reformers largely accepted the spatial segregation of the city, which ensured that the costs of pollution would be disproportionately borne by the poor and nonwhite. [32] Sanitary reform also crossed oceans in the form of colonial public health administrations, where the new health experts read the pathology of foreign environments as a legitimation of both existing racial hierarchies and the colonial control of land and bodies. Endemic disease—often the result of colonial disruption—became the justification for colonial control.[33]

Industrial conditions fostered new diseases as well as old, particularly among workers. Between 1870 and 1900, as technology and capital transformed American industry, the total number of manufacturing, mining, and quarry workers in the US quadrupled. Germany and Great Britain had already experienced a similar expansion. Accompanying this wave of industrial expansion and technological elaboration were a growing number of ailments associated with particular trades: "painter's colic" (lead poisoning), "miner's asthma," (pneumoconiosis), "phossy jaw" (phosphorous poisoning). Though many of the diseases associated with particular forms of work had been known for generations, if not centuries, these diseases now appeared much more frequently and affected many more people; moreover, bacteriological techniques increasingly allowed doctors to distinguish between infectious diseases like tuberculosis and those that had their source in the environment of the modern factory. The disturbing association between disease symptoms and certain forms of industrial work led in turn to the new profession of industrial hygiene.[34]

What allowed reformers both in Europe and the United States to diagnose occupationally caused disease was their clinical and environmental approach. The new industrial hygienists visited factories, talked to workers about their illnesses, tracked the occurrence of observable symptoms, and then linked what they knew about workers' diseases to conditions present in particular settings. British physician Thomas Oliver published his seminal volume on the topic, *The Dangerous Trades*, in 1902. In the US, the key figure was Alice Hamilton—daughter of an elite mid-Western family, a bacteriologist by training, and a member of Jane Addams' Hull House. Commissioned in 1910 by the state of Illinois to study disease among workers in lead-based industries, Hamilton exhaustively documented the massive scope of the problem. Her report put the problem of occupational disease squarely before both legislators and public health officials, and helped generate some of the first worker health and safety reforms in the US.[35]

For the most part, the history of occupational disease has been told from the perspectives of labor and social history, focusing on capitalist exploitation of workers' health and workers' struggles to secure better working conditions and compensation for their illnesses.[36] Far less attention has been paid to the specific material environments that gave rise to symptoms. Moreover, labor historians have typically treated occupational diseases as more or less transhistorical elements and have not questioned the authority or genesis of modern scientific accounts. Michelle Murphy, however, has traced the complex processes through which one disease—sick building syndrome—emerged at the end of the twentieth century. In Murphy's account, a new disease was made manifest not by a straightforward process of discovery but by a congeries of unlikely forces: feminist activism, toxicology, ventilation engineering, ecology, and new chemical monitoring technologies. Occupational disease, like disease more broadly, can be both made and unmade through a variety of scientific and popular practices.[37]

Christopher Sellers has contributed to this critical rethinking of disease and its parameters though his history of toxicology—the laboratory-based study of chemically induced disease—tracing how this discipline emerged in the 1920s in response to heightened concern among both physicians and employers over workplace-based illness. As Sellers explains, toxicology served the interests of new scientific professionals, by granting them the authority of the laboratory, and the interests of corporate managers, by helping to normalize chemical exposures in the workplace. Toxicology may be a science, but it was also a political-economic compromise born of the interwar period. Of equal importance is the link that Sellers traces between Progressive-era industrial hygiene and post-World War II environmental policy by showing how this new toxicological science moved out of the factory to become the basis for regulating chemicals in the broader environment. Christian Warren's study of lead poisoning illustrates how the health problems faced by early twentieth-century workers would be echoed in the bodies of consumers in the postwar decades. Both environmental and labor historians have begun to make important contributions to this emerging history of chemicals and their regulation (see Chapter 10). In particular, the literature has begun to take a much-needed transnational turn because, as recent scholarship has shown, international organizations—both governing organizations such as the World Health Organization and scientific and activist organizations such as the International Union Against Cancer—played an important role in establishing and extending regulatory frameworks in the post-World War II period. Moreover, regulations were often driven and shaped by concerns over international trade—a topic that deserves closer investigation.[38]

Our understanding of non-infectious disease and its environmental causes is bound up not only with the history of toxicology, but also with the history of postwar epidemiology. These are the two scientific disciplines upon which contemporary systems of chemical regulation are based. Epidemiologists study the occurrence of disease across groups rather than in individuals; by observing patterns of disease across a population, epidemiologists try to identify the variables most strongly associated with illness. The discipline's roots are typically dated to John Snow's famous investigation of a cholera outbreak in mid-nineteenth-century London, and epidemiology would underwrite much of the modern public health movement. However, by World War II the field had been pushed to the margins of biomedicine, eclipsed by bacteriology and the rise of antibiotics. Then, in the 1950s, the field attained new prominence when epidemiologic studies identified key risk factors for both lung cancer (smoking) and heart disease (serum cholesterol)—the most prevalent forms of chronic illness in the United States.[39]

It was concern over the health effects of air pollution that first led epidemiologists to apply their methods to broad-based environmental exposures in the postwar period, a story explored by epidemiologist Devra Davis in her popular history, *When Smoke Ran Like Water*. Many epidemiologists were guided by their belief, shared with disease ecologists, that the origins of disease were environmental in a broad sense—that is, external to the organism. In effect, epidemiology recovered some of the insights of an earlier medical geography, successfully linking respiratory disease to its social and environmental contexts. Some clinicians and medical researchers had also become convinced of the links between organic chemicals and rising rates of many cancers in the industrialized world, and they marshaled epidemiologic evidence along with toxicological studies to support these connections.[40]

Taken as a whole, however, epidemiological studies of cancer and other chronic illnesses in the postwar period have been much more likely to validate those risk factors that can be directly ascribed to individuals (e.g., smoking cigarettes), while environmental and social factors that might contribute to disease (e.g., polluted environments) have tended to fall out of the analyses. Although many people have taken this as evidence that individual risk factors (such as whether a person smokes or what she eats) are more "real" or significant in some sense, it is also true that given the methods and assumptions of modern epidemiology, environmental and social causes of disease are much less likely to be sustained from the outset—because they cannot be adequately quantified, because they are viewed as something less than truly causal, or because they are subordinated to other individual factors.[41] The resulting tendency to ascribe disease to individuals while calling into question the relevance of noxious environmental conditions has stoked vehement public controversies over the last several decades, and frequently has given rise to a kind of grassroots guerilla science. While studying the efforts of activists in Woburn, Massachusetts to link chemically contaminated groundwater to their illnesses, sociologist Phil Brown coined the term "popular epidemiology"—a reference to the techniques through which ordinary people and sympathetic scientists have reshaped research agendas and techniques in response to local people's felt connections between place and disease. And although historians have chronicled several controversies in which epidemiology played a crucial role, they have yet to examine critically the history and evolution of the discipline and its implications for environmental knowledge. To date most of the relevant histories and critiques of postwar epidemiology have been penned by practitioners.[42] The field still awaits its historians.

While we might, following Rene and Jean Dubos, label tuberculosis as the first "environmental" epidemic of industrial society, obesity is among the latest. To be clear, the definition of "obesity" and the existence of an "epidemic" are contemporary cultural questions rather than simple facts. But what is demonstrably true is that in late twentieth-century America something called an "obesity epidemic" had become an object of public health research and intervention. What has been documented is an increase in body size (as commonly measured by the "body mass index") among industrialized populations, with the United States leading the way. What such a shift means in terms of health and disease is far from clear. For both American and international public health officials, obesity has become a primary public health concern because of its statistical correlation with Type II diabetes, coronary heart disease, and even some forms of cancer. The World Health Organization (WHO) has declared "overweight" as one of the top ten health risk conditions in the world and one of the top five in developed nations. Worldwide more than 1 billion adults are overweight by WHO standards, while 300 million are obese. Once considered a problem only in high-income countries, overweight and obesity are now rising dramatically in low- and middle- income countries, especially in urban areas.[43]

Although there has been a great deal of research into the genetic factors that contribute to obesity, the rapid onset of the problem points, once again, to socioeconomic and environmental changes as the cause. When narrowly construed, overweight and obesity in an individual stem from an imbalance between energy intake and energy expenditure; in other words, eating too many calories and exercising too little. However, decisions about what foods to eat and how much to exercise are not simply individual decisions. As every historian knows, environmental and social factors shape individual actions, and the factors relevant to obesity operate at multiple scales: from the size of spoons to an industrialized food supply to the unequal distribution of food and resources across the globe. Researchers have identified a plethora of factors that affect food intake and energy expenditure including food packaging, the proliferation of inexpensive "energy-dense" foods (e.g., soda pop and Big Macs), the relative availability of low-quality versus high-quality foods in specific communities, the layout and safety of neighborhoods, the entry of more women into paid labor outside the home, longer working days and commute times, the increase in sedentary labor and leisure, cutbacks in mandatory physical education and disinvestment in public space, and agricultural subsidies that favor the production of massive amounts of corn over other vegetables and grains. There is a broad recognition among scholars that the obesity epidemic is quite literally embedded within the landscape of late capitalist industrialization, prompting some to speak of "obesogenic" environments.[44] More provocatively, recent research has suggested a link between environmental chemical exposures (including bisphenol-A and organocholorine compounds) and obesity. According to a 2012 review by the US National Toxicology Program, developmental exposures may act synergistically with high-fat, high-calorie diets in adulthood to promote obesity and diabetes.[45]

Interest in obesity has shifted the focus of some epidemiologists away from individual risk factors and toward the social and environmental conditions that determine health. And yet, contemporary epidemiologic methods do not handle social and structural variables all that well, nor is the discipline well-suited to spanning the relevant scales of analysis (home, neighborhood, community, nation, etc.). Despite their statistical sophistication, cross-sectional and longitudinal studies may be less informative than careful historical work in revealing the multiple and overlapping causes behind shifts in bodily size, shape, and disease. For instance, cross-sectional studies that examine whether the proximity of supermarkets affects the amount and quality of food consumed are confounded by the difficulty of incorporating into a statistical model all the relevant factors that affect people's daily shopping behavior. Though these studies succeed in foregrounding environmental and social factors, in the end they will likely tell us far less than we can learn from nuanced community histories linking changes in the urban-environmental landscape to local culture and bodily change, or from histories that illuminate the connections among public policy, the corporatization of agriculture, advertising, and the ubiquity of corn syrup in the modern American diet. Studies of obesity, like studies of trypanosomiasis, need to take into account not only material environmental changes but also the multiple ways in which humans respond to those changes both individually and collectively; this is the terrain of environmental history.

EMERGING INFECTIOUS DISEASES

The 1990s saw the rise of "emerging" diseases as a new public health issue. Driven by the sudden irruption of the global AIDS pandemic during the 1980s and the utter failure of the public health establishment to either predict its magnitude or stem its spread, many people began to rethink the teleological narrative of the "epidemiologic transition"—the assumption that technological and economic development inevitably produces a change in disease pattern: from one dominated by infectious to one dominated by chronic diseases.[46] It has now become commonplace to observe the persistence of infectious disease and to argue that emerging and re-emerging infectious diseases (EIDs) will constitute a major health concern throughout the twenty-first century.

Although AIDS was by far the most significant new disease in terms of the deaths and disability it inflicted, dozens of novel pathogens infected human populations during the twentieth century's final decades.[47] Among these were Ebola virus in Central Africa, several hantaviruses (Sin Nombre virus in the western US, Andes virus in South America), Nipah virus in Malaysia, Rift Valley Fever in Mauritania and Egypt, and Creutzfeldt-Jakob disease (known in its zoonotic form as bovine spongiform encephalopathy or mad cow disease) in Great Britain. At the same time, familiar diseases became much more prevalent (e.g., malaria) or developed new, drug-resistant strains (e.g., *M. tuberculosis, P. falciparum*), extremely virulent forms (*E. coli* 0157:H7), or expanded into new geographic areas (West Nile Virus in North and Central America, dengue fever in the southern US and South America, Rift Valley Fever in the Middle East).[48]

From the outset, many observers connected EIDs to environmental change. In his sensational account of a hemorrhagic fever outbreak among a group of lab monkeys in Virginia published in 1994 (*The Hot Zone*), journalist Richard Preston somewhat blithely attributed the problem of emerging diseases to the human trashing of the planet. Two decades later, David Quammen offered a similar thesis in his dramatic account (*Spillover*) of several recent epidemics, including AIDS, whose origins lay in animal infections. Such accounts suggest the popular appeal of a broadly environmentalist—if overly simplistic—narrative of disease emergence; both Preston's and Quammen's books were bestsellers.[49]

Yet the scientific research on EIDs takes a necessarily ecological approach. Roughly 60 percent of known human pathogens have an animal origin, including 75 percent of the identified EIDs. One distinguishing factor of most viral EID pathogens is that they have very high mutation rates, allowing them to evolve quickly to survive in new host species and to adapt to, and even exploit, rapid environmental changes. However, the genetic properties of these viruses do not explain their emergence as potent human diseases; even highly adaptable viruses typically remain stable in their given ecological niche. Rather, it is frequently environmental change—much of it anthropogenic—that triggers new epidemics. Climate—especially changes in temperature and rainfall—is among the most potent drivers of epidemic outbreaks, but many other environmental changes contribute to disease emergence. Deforestation has brought humans and domestic animals into contact with new reservoirs of disease, such as the fruit bats that carry Nipah virus and the sandflies that carry leishmaniasis. Conversely, in the northeastern US, reforestation associated with the suburban development of agricultural land has created edge environments that favor the ticks which carry Lyme disease, while also bringing humans into frequent contact with deer, the tick's preferred host. Dam and irrigation projects have fostered the explosive growth of snails that carry schistosomiasis, flies that carry onchocerciasis (river blindness), and mosquitoes that carry malaria, dengue, and Rift Valley Fever. Industrialized agriculture has created the conditions necessary for the production and spread of toxic *E. coli*, mutations of otherwise harmless bacteria that have infected meat and produce in Europe and North America; the most common pathogenic strain of *E. coli* (O157:H7) is formed primarily in the gut of grain-fed (as opposed to grass-fed) cattle, and its proliferation owes much to the shift away from small ranches and toward the industrial feedlot system. Across much of Asia the decline of natural wetlands for wild ducks and geese has brought them into closer contact with industrialized poultry farms, creating conditions that foster epidemic pathogenic influenza that is, in turn, spread through globalized trade networks. The bacterium *Legionella*, discovered in 1976 when a mysterious pneumonia-like ailment struck the American Legion convention in Philadelphia, was able to proliferate only because of the ecological conditions found in the plumbing and air-conditioning systems of large, modern buildings. And this is only a partial list. Environmentally dependent disease—which western scientists once confidently believed to be confined to tropical, undeveloped countries—is again a major concern in Europe and North America.[50]

Even cholera—a disease whose cause was once narrowly construed as a water-borne germ (*Vibrio cholerae*) that was carried solely by human hosts—has recently been re-narrativized along environmental lines. For roughly four decades, public health professionals were confident that they not only fully understood the cause of cholera but that improvements in urban water supplies had rendered global pandemics a thing of the past. Then, in 1961, the seventh cholera pandemic emerged on the island of Celebes in Indonesia, spreading throughout southern Asia and into Africa. To everyone's surprise, the disease erupted in South America in 1991, after more than a century of absence from that continent, and it has recurred there every year since.

Although public health officials initially assumed that cholera had been transmitted by humans (through the discharge of either ship sewage or contaminated ballast water), the near-simultaneous appearance of the disease in several South American coastal locations suggested the pathways of the disease were more complex. *Vibrio cholerae* is now known to be ubiquitous in many regions (including North America), residing for long periods of time in both temperate and tropical coastal waters in a dormant, non-virulent state. However, the bacterium can be aroused to virulence by warm temperatures and shock. Moreover, when ingested with certain kinds of seafood, the pathogen becomes more toxic. In other words, cholera is not caused merely by a "germ," nor is it transmitted only by humans; rather it emerges out of the interaction of certain vibrios with environmental conditions and cultural practices.[51]

These technical facts about *V. cholera* exist alongside other relevant social and environmental facts. From 1990 to 1995, Latin America experienced an unusually long El Niño–Southern Oscillation (ENSO) event that increased both seawater temperatures and flooding. At the same time, Peru and other South American countries also saw a decline in the quality and availability of water-supply and sewage services as Latin nations, under pressure from Western banks and international aid agencies, retreated from earlier commitments to preventative medicine and public health and adopted the policies of neoliberal "structural adjustment." Thus, many residents continued to rely upon untreated water supplies. At the same time, increasingly polluted rivers and streams fed into Pacific coastal areas, contributing to large-scale plankton blooms. Floods were more frequent and larger due to both climatic factors (ENSO) and anthropogenic changes in upstream land use. Moreover, in Peru, the locus of the South American epidemic, ceviche—a dish made from raw fish—was a popular part of the regional diet.[52]

The complex mix of social, biological, and environmental factors driving EIDs like cholera calls out for more historical and anthropological work. One of the best histories to emerge thus far is Randall Packard's short history of malaria, which insists on the environmental and social conditions that have fostered the disease's re-emergence. Nicholas King has offered a trenchant analysis of the policy discourse surrounding EIDs; King usefully counterposes the discourse of the threat, which is construed as broadly global and frequently dire, to the proffered solutions, which are quite local and less than radical: laboratory investigation and information management that will presumably take place in the advanced research facilities of North America and western Europe. As both Packard and King emphasize, the turn toward neoliberal economic policies at the end of the twentieth century has been accompanied by a retreat on the part of Western countries and aid organizations to providing preventative medical care and decreasing health disparities across the globe.[53] Instead the answer to epidemic disease is said to lie in more vaccines.

Moreover, even though the research consistently recognizes the relevance of the environment to disease emergence, the analysis has not been taken very far in the scientific literature. Environmental causes are typically treated in a very generic way. Oftentimes, the cause of environmental change—whether deforestation, dam construction, industrial agriculture, or the displacement of populations due to famine or war—is simply equated with "population growth," "migration," or "development" rather than with specific regional histories and national and international policies. The result is often a reprise of the all-too-familiar Malthusian narrative. What is lacking is close attention to the specific social, economic, and environmental contexts that have produced the conditions conducive to outbreaks of infectious disease in discrete locations; and here environmental history can make an important contribution. Also needed is closer attention to the emergence of emerging disease science—its intellectual antecedents in disease ecology, its social and political locations, its continuities and discontinuities with earlier conceptualizations of disease (racial and otherwise), its larger discursive alliances and effects. Emerging diseases are real enough, but they are also social constructions. How we conceptualize disease will determine the possibilities for responding to it (e.g., through environmental reform, new economic policies, and preventative health measures? Or through additional investments in bio-surveillance and pharmaceuticals?), and critical histories of our current biological situation are vitally important to the contemporary conversation.

Conclusion

As contemporary debates over emerging diseases make clear, there are compelling political reasons for telling the history of disease as an environmental and social story, rather than simply as a medical and personal one. In a society marked by enormous inequality and an ideology of obsessive individualism, it is imperative to uncover the structural and environmental causes of illness and to insist upon the collective responsibility for care and remedies. From this perspective, epidemic disease can in many cases best be understood as an issue of environmental justice. Thus, the question is not whether disease is a topic for environmental historians, but how to address it. How can historians write not only landscapes but animals, microbes, and chemicals into their stories in a critical and compelling way? How should scholars grapple with and use the insights of an increasingly complex biomedicine? How can historians do justice to the multiple perspectives on disease and its causes in a global world? How do the traditional archives limit the stories that can be told, and what new archives might be tapped?

These are open questions that await future authors. But what is clear is that the intervention should not simply take the form of selectively raiding the latest scientific findings to analyze the "role" of disease in history. Pathogens and diseases are not simply agents or actors whose effects historians can recount; the effects of disease do not exist apart from the way that humans respond to them. And the ways that people respond depends, in turn, on how diseases are understood. Diseases are at once material realities and social creations. They exist apart from human control, and yet—like the "nature" that environmental historians are so familiar with—disease is constructed from within the fields of culture. Even within the field of biomedicine, there are many avenues and practices for understanding a given set of agents and symptoms, each with its own effects.[54] Our definitions and delineations of disease are inherently unstable, however coherent they may appear. They exist within fields of knowledge that are highly contingent, partial, and always changing. Consequently, scholars need to attend to the social contexts and contingent histories that shape biomedicine, while resisting the temptation to merely write histories of science and scientists, or to grab the latest scientific finding and run with it. The best histories will bring together an understanding of the most recent science, knowledge of the crucial social and environmental changes that contribute to health and disease in any moment, and—perhaps most importantly—sensitivity to the cultural and political contexts that have produced both past and present understandings of disease. Thus while the subject of disease takes environmental historians into the field of medicine, it should also take them beyond. Although biomedicine has laid a privileged claim to knowledge about human bodies, that is a claim that historians need to contest. It has been not only doctors and patients who have encountered and constructed disease, but also ecologists, veterinarians, workers, engineers, entomologists, chemists, philosophers, activists, and policymakers. In any moment and place, differing constructions of bodies and disease have competed and coexisted with another. How and when certain shared bodily conditions can be connected to environmental conditions, and why they coalesce into an "epidemic" are always questions. There are many archives to consult, and many more histories of environment and disease that can and need to be written.

NOTES

1. Alfred W. Crosby, *The Columbian Exchange; Biological and Cultural Consequences of 1492* (Westport, CT: Greenwood, 1972).
2. Gregg Mitman, "In Search of Health: Landscape and Disease in American Environmental History," *Environmental History* 10, no. 2 (2005): 184–210.
3. Andrew Cunningham, "Transforming Plague: The Laboratory and the Identity of Infectious Disease," in *The Laboratory Revolution in Medicine*, ed. A. Cunningham and P. Williams (Cambridge: Cambridge University Press, 1992), 209. For exceptions to this trend, see Erwin H. Ackerknecht, "Malaria in the Upper Mississippi Valley, 1760–1900," *Supplements to the Bulletin of the History of Medicine No. 4* (1945); Mirko D. Grmek, "Geographie Medicale et Histoire des Civilisations," *Annales ESC* 18 (1963): 1071–1097. Also, Charles E. Rosenberg, "Erwin H. Ackerknecht, Social Medicine, and the History of Medicine," *Bulletin of the Hisory of Medicine* 81 (2007): 511–532.
4. Alfred W. Crosby, "Conquistador y Pestilencia: The First World Pandemic and the Fall of the Great Indian Empires," *Hispanic American Historical Review* 47 (1967); Crosby, *Columbian Exchange*; Crosby, *Ecological Imperialism: The Biological Expansion of Europe, 900–1900* (Cambridge: Cambridge University Press, 1986).
5. William H. McNeill, *Plagues and Peoples* (Garden City, NY: Anchor Press, 1976).
6. Warwick Anderson, "Natural Histories of Infectious Disease: Ecological Vision in Twentieth-Century Biomedical Science," *Osiris* 19 (2004): 59.

7. Anderson, "Natural Histories"; Claude E. Dolman, *Theobald Smith, Microbiologist: Suppressing the Diseases of Animals and Man* (Boston: Boston Medical Library, 2003); Theobald Smith, *Parasitism and Disease* (Princeton, NJ: Princeton University Press, 1934); Helen Tilley, "Ecologies of Complexity: Tropical Environments, African Trypanosomiasis, and the Science of Disease Control Strategies in British Colonial Africa, 1900–1940," *Osiris* 19 (2004): 21–38; Jacques M. May, "Medical Geography: Its Methods and Objectives," *Geographical Review* 40, no. 1 (1950): 9–41; Frank M. Burnet, *Biological Aspects of Infectious Disease* (Cambridge: The University Press, 1940); Hyung Wook Park, "Germs, Hosts, and the Origin of Frank Macfarlane Burnet's Concept of 'Self' and 'Tolerance,' 1936–1949," *Journal of the History of Medicine and the Allied Sciences* 61, no. 4 (2006): 492–534.

8. Anderson, "Natural Histories"; Hans Zinsser, *Rats, Lice and History* (Boston: Little, Brown, and Company, 1935); Percy M. Ashburn, *The Ranks of Death, a Medical History of the Conquest of America* (New York: Coward-McCann, 1947); René J. and Jean Dubos, *The White Plague; Tuberculosis, Man and Society*, 1st ed. (Boston: Little, Brown, 1952).

9. René J. Dubos, *Mirage of Health: Utopias, Progress, and Biological Change* (Garden City, NY: Anchor Books, 1959); René J. Dubos, *Man Adapting* (New Haven, CT: Yale University Press, 1965).

10. Annales historians had first emphasized the study of prices. In the 1950s, the demographer Louis Henry, a leading figure of the Annales School, turned from the study of contemporary population problems to historical demography; he would influence several historians of the subsequent generation. Peter Burke, *The French Historical Revolution: The Annales School, 1929–89* (Stanford, CA: Stanford University Press, 1991), 56–57.

11. Sherburne F. Cook, *Population Trends Among the California Mission Indians* (Berkeley: University of California Press, 1940), 1–2; Woodrow H. Borah and Sherburne F. Cook, *The Aboriginal Population of Central Mexico on the Eve of the Spanish Conquest*, Ibero-Americana Series No. 44 (Berkeley: University of California Press, 1963); Woodrow H. Borah, "America as Model: The Demographic Impact of European Expansion upon the Non-European World," *35th Congreso Internacional de Americanistas Actas y Memorias* 3 (1964): 379–387; Albert L. Hurtado, "California Indian Demography, Sherburne F. Cook, and the Revision of American Indian History," *Pacific Historical Review* 58, no. 3 (1989): 323–343.

12. Sherburne F. Cook, *The Extent and Significance of Disease Among the Indians of Baja California, 1697–1773* (Berkeley: University of California Press, 1937); Cook, *Soil Erosion and Population in Central Mexico* (Berkeley: University of California Press, 1949).

13. Burnet, *Biological Aspects of Infectious Disease*; Aidan Cockburn, *The Evolution and Eradication of Infectious Diseases* (Baltimore: Johns Hopkins Press, 1963), 68–102.

14. J. B. S. Haldane, "The Rate of Mutation of Human Genes," *Herditas* 35 (1949): 267–273; A. C. Allison, "Sickle Cell Anemia and Evolution," *Scientific American* 195 (August 1956): 87–88. The immediate fascination with this discovery, despite its speculative nature, was likely related to the fact that it, in Keith Wailoo's phrase, "endorsed and legitimated cherished racial categories." Wailoo, *Drawing Blood: Technology and Disease Identity in Twentieth-Century America* (Baltimore: Johns Hopkins University Press, 1997), 141–149. Although hemoglobinopathies such as the sickle-cell trait do provide a strong survival advantage vis-à-vis malaria, true malarial immunity does not exist. Rather, scientists use the term "premonition" to describe a state of relative immunity to severe infection that results from the continued presence of small numbers of parasites in the body.

Premonition is a transient condition and will abate if one leaves an endemic malarial region for a year or two. Moreover, premonition is specific to a particular malaria strain; it offers no protection against other species or even other strains within the same species. Thus even endemic malarial areas can suffer from malaria epidemics when a new strain is introduced. Lorena Madrigal, *Human Biology of Afro-Caribbean Populations* (New York: Cambridge University Press, 2006), 78–82; Denise L. Doolan, Carlota Dobano, and J. Kevin Baird, "Acquired Immunity to Malaria," *Clinical Microbiology Reviews* 22, no. 1 (2009): 13–36, doi:10.1128/CMR.00025-08.

15. Alfred W. Crosby, "Virgin Soil Epidemics as a Factor in the Aboriginal Depopulation in America," *William and Mary Quarterly*, 3rd ser., 33, no. 2 (1976): 289. Christian W. McMillen notes that the term itself had been in use among medical professionals since at least 1903. See McMillen, "'The Red Man and the White Plague': Rethinking Race, Tuberculosis, and American Indians, ca. 1890–1950," *Bulletin of the History of Medicine* 82, no. 3 (2008): 622. Michael Worboys and Randall Packard have traced the concern with so-called virgin soil populations to British colonial scientists and their concern with tuberculosis and its effect on "primitive" races. Michael Worboys, "Tuberculosis and Race in Britain and Its Empire, 1900–1950," in *Race, Science and Medicine, 1700–1960*, ed. Waltraud Ernst and Bernard Harris (London: Routledge, 1999), 144–166; Randall M. Packard, *White Plague, Black Labor: Tuberculosis and the Political Economy of Health and Disease in South Africa* (Berkeley: University of California Press, 1989), 22–32. McMillen, cited above, has traced a similar trajectory for US physicians examining TB among Native Americans. Rene and Jean Dubos forcefully refuted the differential immunity argument for TB in *The White Plague*.

16. Philip D. Curtin, "Epidemiology and the Slave Trade," *Political Science Quarterly* 83 (1968): 190–216; Kenneth F. Kiple, "Survey of Recent Literature," in *The African Exchange: Toward a Biological History of Black People* (Durham, NC: Duke University Press, 1987); Kenneth F. Kiple, *The Caribbean Slave: A Biological History* (Cambridge: Cambridge University Press, 1984); Kenneth F. Kiple, *Another Dimension to the Black Diaspora: Diet, Disease, and Racism* (Cambridge: Cambridge University Press, 1981). While no scientific work has yet supported inherited immunity for yellow fever, Francis Black's work on the isolated Yanomani tribe in South America has indicated that indigenous group mounted a typical immunological response to the disease. Francis Black, Gerald Schiffman, and Janardan P. Pandey, "HLA, Gm, and Km Polymorphisms and Immunity to Infectious Diseases in South Amerinds," *Experimental and Clinical Immunogenetics* 12 (1995): 214, cited in Jones, "Virgin Soils Revisted," 730 n87. For a different interpretation and impassioned refutation of Kiple, see Sheldon Watts, "Yellow Fever Immunities in West Africa and the Americas in the Age of Slavery and Beyond: A Reappraisal," *Journal of Social History* 34, no. 4 (2001): 955–967 and Sheldon Watts, *Epidemics and History: Disease, Power, and Imperialism* (New Haven, CT: Yale University Press, 1997). For more judicious historical analyses of disease and epidemiology among Caribbean slaves, see B. W. Higman, *Slave Population of the British Caribbean, 1807–1834* (Baltimore: Johns Hopkins University Press), 340–341 on causes of death; Richard B. Sheridan, *Doctors and Slaves: A Medical and Demographic History of Slavery in the British West Indies, 1680–1834* (Cambridge: Cambridge University Press, 1985); Madrigal, *Afro-Caribbean Populations*.

17. A. Ramenofsky, "Diseases of the Americas, 1492–1700," in *Cambridge World History of Human Disease*, ed. Kenneth F. Kiple (New York: Cambridge University Press, 1993), 325;

Harold F. Gray and Russell E. Fontaine, "A History of Malaria in California," *Proceedings of the California Mosquito Control Association* 25 (1957): 18–37; Paul S. Sutter, "Nature's Agents or Agents of Empire? Entomological Workers and Environmental Change during the Construction of the Panama Canal," *Isis* 98, no. 4 (2007): 724–754. John McNeill attends carefully to changes in the disease environment of the Caribbean, though he also draws large conclusions from fragmentary evidence and speculative arguments about differential immunity. McNeill, *Mosquito Empires: Ecology and War in the Greater Caribbean, 1620–1914* (New York: Cambridge University Press, 2010). On Bengal, see Ira Klein, "Development and Death: Reinterpreting Malaria, Economics and Ecology in British India," *Indian Economic & Social History Review* 38, no. 2 (2001): 147–179; Ira Klein, "Imperialism, Ecology and Disease: Cholera in India, 1850–1950," *Indian Economic & Social History Review* 31, no. 4 (1994): 491–518.

18. For authors that emphasize the role of disease in shaping the long history of Africa, see John Iliffe, *Africans: The History of a Continent* (Cambridge: Cambridge University Press, 2007); James L.A. Webb, *Desert Frontier: Ecological and Economic Change Along the Western Sahel, 1600–1850* (Madison: University of Wisconsin Press, 1995); James L.A. Webb, "Malaria and the Peopling of Early Tropical Africa," *Journal of World History* 16, no. 3 (2005): 270–291.

19. John Ford, *The Role of the Trypanosomiases in African Ecology; a Study of the Tsetse Fly Problem* (Oxford: Clarendon Press, 1971). For a summary of Ford's contribution and its subsequent neglect by other scholars, see James Giblin, "Trypanosomiasis Control in African History: An Evaded Issue?" *The Journal of African History* 31, no. 1 (1990): 59–80. Other important studies of African trypanosomiasis include Helge Kjekshus, *Ecology Control and Economic Development in East African History: The Case of Tanganyika 1850–1950* (London: Heinemann, 1977); James Leonard Giblin, *The Politics of Environmental Control in Northeastern Tanzania, 1840–1940* (Philadelphia: University of Pennsylvania Press, 1992); Maryinez Lyons, *The Colonial Disease: A Social History of Sleeping Sickness in Northern Zaire, 1900–1940* (Cambridge: Cambridge University Press, 1992).

20. Peter A. Coclanis, *The Shadow of a Dream: Economic Life and Death in the South Carolina Low Country, 1670–1920* (New York: Oxford University Press, 1989); H. Roy Merrens and George D. Terry, "Dying in Paradise," *The Journal of Southern History* 50, no. 4 (1984): 533–550; Darrett B. Rutman and Anita H. Rutman, "Of Agues and Fevers: Malaria in the Early Chesapeake," in *Biological Consequences of European Expansion, 1450–1800*, ed. Kenneth F. Kiple and Stephen V. Beck (Aldershot, UK: Ashgate, 1997); Margaret Humphreys, *Yellow Fever and the South* (New Brunswick, NJ: Rutgers University Press, 1992).

21. Ackerknecht, *Malaria in the Upper Mississippi Valley*; Kenneth Thompson, "Insalubrious California: Perception and Reality," *Annals of the Association of American Geographers* 59, no. 1 (1969): 50–64; Conevery Bolton Valencius, *The Health of the Country: How American Settlers Understood Themselves and Their Land* (New York: Basic Books, 2002); Linda Nash, *Inescapable Ecologies: A History of Environment, Disease, and Knowledge* (Berkeley: University of California Press, 2006); Victoria A. Harden, *Rocky Mountain Spotted Fever: History of a Twentieth-Century Disease* (Baltimore: Johns Hopkins University Press, 1990).

22. Mart A. Stewart, "'Let Us Begin with the Weather?': Climate, Race, and Cultural Distinctiveness in the American South," in *Nature and Society in Historical Context*, ed. Mikulas Teich, Roy Porter, and Bo Gustafsson (Cambridge: Cambridge University Press, 1997), 240–256; John H. Warner, "The Idea of Southern Medical Distinctiveness: Medical

Knowledge and Practice in the Old South," in *Science and Medicine in the Old South*, ed. Ronald L. Numbers and Todd Lee Savitt (Baton Rouge: Louisiana State University Press, 1989), 179–205; Nicolaas A. Rupke, ed., *Medical Geography in Historical Perspective* (London: Wellcome Trust Center for the History of Medicine, 2000); Dane Kennedy, "The Perils of the Midday Sun: Climatic Anxieties in the Colonial Tropics," in *Imperialism and the Natural World*, ed. John M MacKenzie (Manchester: University of Manchester, 1990), 118–140; Warwick P. Anderson, *The Cultivation of Whiteness: Science, Health and Racial Destiny in Australia* (New York: Basic Books, 2003); Mark Harrison, *Climates and Constitutions: Health, Race, Environment, and British Imperialism in India, 1600–1850* (New York: Oxford University Press, 1999); Karen Ordahl Kupperman, "Fear of Hot Climates in the Anglo-American Colonial Experience," *William and Mary Quarterly*, 3rd ser., 41, no. 2 (1984): 213–240; Nash, *Inescapable Ecologies*, 10–81.

23. David S. Jones "Virgin Soils Revisited," *William and Mary Quarterly*, 3rd ser., 60, no. 4 (2003): 703–742; Ashburn, *Ranks of Death*, 211; McNeill, *Plagues and Peoples*, 87; Crosby, *Ecological Imperialism*, 34; Jared M. Diamond, *Guns, Germs, and Steel: The Fates of Human Societies* (New York: W.W. Norton, 1997). For a similar interpretation of this literature, see Gregg Mitman, "In Search of Health: Landscape and Disease in American Environmental History," *Environmental History* 10, no. 2 (2005): 184–210.

24. Paul Kelton, *Epidemics and Enslavement: Biological Catastrophe in the Native Southeast, 1492–1715* (Lincoln: University of Nebraska Press, 2007); Suzanne Austin Alchon, *A Pest in the Land: New World Epidemics in a Global Perspective* (Albuquerque: University of New Mexico Press, 2003).

25. On race and medicine, see Anderson, *Cultivation of Whiteness*; Wailoo, *Drawing Blood*. On TB, influenza, and cholera as present in the Americas, see A. F. Ramenofsky, A. K. Wilbur, and A. C. Stone, "Native American Disease History: Past, Present, and Future Directions," *World Archaeology* 35, no. 2 (2003): 241–257; Robert Fortune, *"Must We All Die?": Alaska's Enduring Struggle with Tuberculosis* (Fairbanks: University of Alaska Press, 2005), 1–6.; Atiqul Islam et al., "Indigenous Vibrio Cholerae Strains from a Non-Endemic Region Are Pathogenic," *Open Biology* 3, no. 2 (2013), doi:10.1098/rsob.120181.; Jessica M. C. Pearce-Duvet, "The Origin of Human Pathogens: Evaluating the Role of Agriculture and Domestic Animals in the Evolution of Human Disease," *Biological Reviews* 81, no. 3 (2006): 369–382, doi:10.1017/S1464793106007020. On the complexity of natural selection and immunity, Jones, "Virgin Soils Revisited"; Jean Langhorne et al., "Immunity to Malaria: More Questions Than Answers," *Nature Immunology* 9, no. 7 (2008): 725–732, doi:10.1038/ni.f.205. On pathogen evolution, see Paul W. Ewald and George E. Burch, *Evolution of Infectious Disease* (Oxford: Oxford University Press, 1994). For a similar argument about historians' selective reading of the medical literature, see Bruce S. Fetter, "History and Health Science: Medical Advances Across the Disciplines," *Journal of Interdisciplinary History* 32, no. 3 (2002): 423–442.

26. Quote from David Herlihy, *The Black Death and the Transformation of the West*, ed. Samuel Kline Cohn (Cambridge, MA: Harvard University Press, 1997). The Black Death has spawned an immense literature, and much (arguably too much) effort has gone toward trying to definitively identify the illness as bubonic plague or something else (both anthrax and hemorrhagic fever have been proposed). Yet much of this medically focused literature has paid close attention to the environmental factors that may have allowed the epidemic to emerge. For a useful overview, see Ole Georg Mosang, "Climate,

Ecology and Plague: The Second and the Third Pandemic Reconsidered," in *Living with the Black Death: A Medieval Symposium* (Odense: University Press of Southern Denmark, 2004). Also, Samuel Kline Cohn, *The Black Death Transformed: Disease and Culture in Early Renaissance Europe* (New York: Oxford University Press, 2002); Michael McCormick, "Rats, Communications, and Plague: Toward an Ecological History," *Journal of Interdisciplinary History* 34, no. 1 (2003): 1–25, doi:10.1162/002219503322645439.

27. Martin V. Melosi, *Pollution and Reform in American Cities, 1870–1930* (Austin: University of Texas Press, 1980); Melosi, *Garbage in the Cities: Refuse, Reform, and the Environment: 1880–1980* (College Station: Texas A & M University Press, 1981); Joel A. Tarr, *The Search for the Ultimate Sink: Urban Pollution in Historical Perspective* (Akron, OH: University of Akron Press, 1996); Tarr, *Devastation and Renewal: An Environmental History of Pittsburgh and Its Region* (Pittsburgh, PA: University of Pittsburgh Press, 2003); David Stradling, *Smokestacks and Progressives: Environmentalists, Engineers, and Air Quality in America, 1881–1951* (Baltimore: Johns Hopkins University Press, 1999); William Cronon, *Nature's Metropolis: Chicago and the Great West* (New York: W.W. Norton, 1992); Matthew W. Klingle, *Emerald City: An Environmental History of Seattle* (New Haven, CT: Yale University Press, 2007).

28. Dubos and Dubos, *White Plague*, 207; Friedrich Engels, *The Condition of the Working Class in England* (New York: Macmillan, 1958), 111.

29. Dubos and Dubos, *White Plague*; Matthew Gandy, "Life Without Germs: Contested Episodes in the History of Tuberculosis," in *The Return of the White Plague: Global Poverty and the "New" Tuberculosis*, ed. Matthew Gandy and Alimuddin Zumla (New York: Verso, 2003), 15–38; F. B. Smith, *The Retreat of Tuberculosis, 1850–1950* (London: Croom Helm, 1988); Fortune, *"Must We All Die?"*; Paul Farmer, *Infections and Inequalities: The Modern Plagues* (Berkeley: University of California Press, 1999).

30. Stephen Mosley, *The Chimney of the World: A History of Smoke Pollution in Victorian and Edwardian Manchester* (Cambridge: White Horse Press, 2001), 60–65; Anne Hardy, *The Epidemic Streets: Infectious Disease and the Rise of Preventive Medicine, 1856–1900* (Oxford: Oxford University Press, 1993).

31. David L. Ellison, *Healing Tuberculosis in the Woods: Medicine and Science at the End of the Nineteenth Century* (Westport, CT: Greenwood Press, 1994); John E. Baur, *The Health Seekers of Southern California, 1870–1900*, Huntington Library Publications (San Marino, CA: Huntington Library, 1959); Billy Mac Jones, *Health-Seekers in the Southwest, 1817–1900*, 1st ed. (Norman: University of Oklahoma Press, 1967); Kenneth Thompson, "Climatotherapy in California," *California Historical Quarterly* 50 (1971): 111–130; Gregg Mitman, *Breathing Space: How Allergies Shape Our Lives and Landscapes* (New Haven, CT: Yale University Press, 2007).

32. Edwin Chadwick, *Report on the Sanitary Condition of the Labouring Population of Gt. Britain*, ed. Michael W. Flinn (Edinburgh: University Press, 1965); Joanna Dyl, "The War on Rats versus the Right to Keep Chickens: Plague and the Paving of San Francisco, 1907–1908," in *The Nature of Cities: Culture, Landscape and Urban Space*, ed. Andrew C. Isenberg (Rochester, NY: University of Rochester Press, 2006), 38–61; Judith Walzer Leavitt, *The Healthiest City: Milwaukee and the Politics of Health Reform* (Princeton, NJ: Princeton University Press, 1982); Joel A. Tarr, *The Search for the Ultimate Sink: Urban Pollution in Historical Perspective* (Akron, OH: University of Akron Press, 1996); Martin V. Melosi, *The Sanitary City: Urban Infrastructure in America from Colonial Times to the Present* (Baltimore: Johns Hopkins University Press, 2000); Harold L. Platt, *Shock*

Cities: The Environmental Transformation and Reform of Manchester and Chicago (Chicago: The University of Chicago Press, 2005).

33. John and Jean Comaroff, "Medicine, Colonialism, and the Black Body," in *Ethnography and the Historical Imagination* (Boulder, CO: Westview Press, 1993), 215–234; Myron J. Echenberg, *Black Death, White Medicine: Bubonic Plague and the Politics of Public Health in Colonial Senegal, 1914–1945* (Portsmouth, NH: Heinemann, 2002); Ken De Bevoise, *Agents of Apocalypse: Epidemic Disease in the Colonial Philippines* (Princeton, NJ: Princeton University Press, 1995); David Arnold, *Colonizing the Body: State Medicine and Epidemic Disease in Nineteenth-Century India* (Berkeley: University of California Press, 1993); David Arnold, ed., *Imperial Medicine and Indigenous Societies: Disease, Medicine, and Empire in the Nineteenth and Twentieth Centuries* (Manchester: Manchester University Press, 1988); Ruth Rogaski, *Hygienic Modernity: Meanings of Health and Disease in Treaty-Port China* (Berkeley: University of California Press, 2004).

34. Jacqueline K. Corn, *Response to Occupational Health Hazards: A Historical Perspective* (New York: Van Nostrand Reinhold, 1992); Christopher C. Sellers, *Hazards of the Job: From Industrial Disease to Environmental Health Science* (Chapel Hill: University of North Carolina Press, 1997); Gerald Rosner and David Markowitz, *Dying for Work: Workers' Safety and Health in Twentieth-Century America* (Bloomington: Indiana University Press, 1987).

35. Thomas Oliver, *Dangerous Trades; the Historical, Social, and Legal Aspects of Industrial Occupations as Affecting Health, by a Number of Experts* (London: J. Murray, 1902); Madeleine P. Grant, *Alice Hamilton; Pioneer Doctor in Industrial Medicine* (London: Abelard-Schuman, 1967); Alice Hamilton, *Exploring the Dangerous Trades; the Autobiography of Alice Hamilton, M.D* (Boston: Little, Brown and Company, 1943).

36. Daniel M. Berman, *Death on the Job: Occupational Health and Safety Struggles in the United States* (New York: Monthly Review Press, 1978); Alan Derickson, *Black Lung: Anatomy of a Public Health Disaster* (Ithaca, NY: Cornell University Press, 1998); David Rosner, *Deadly Dust: Silicosis and the Politics of Occupational Disease in Twentieth-Century America* (Princeton, NJ: Princeton University Press, 1991). For an account of workplace health and disease that takes gender into account, see Laurie Mercier, *Anaconda: Labor, Community, and Culture in Montana's Smelter City* (Urbana: University of Illinois Press, 2001).

37. Michelle Murphy, *Sick Building Syndrome and the Problem of Uncertainty: Environmental Politics, Technoscience, and Women Workers* (Durham, NC: Duke University Press, 2006).

38. Christopher C. Sellers and Joseph Melling, *Dangerous Trade: Histories of Industrial Hazard Across a Globalizing World* (Philadelphia: Temple University Press, 2011); Soraya Boudia and Nathalie Jas, eds., *Toxicants, Health and Regulation Since 1945* (London: Pickering & Chatto, 2013).

39. Neil Pearce, "Traditional Epidemiology, Modern Epidemiology, and Public Health," *American Journal of Public Health* 86 (1996): 678–683; Mervyn Susser, "Choosing a Future for Epidemiology: I-Eras and Paradigms," *AJPH* 86 (May 1996): 668–673; Nancy Krieger, "Epidemiology and the Web of Causation: Has Anyone Seen the Spider?," *Social Science and Medicine* 39 (1994): 887–903; William G. Rothstein, *Public Health and the Risk Factor: A History of an Uneven Medical Revolution* (Rochester, NY: University of Rochester Press, 2003); Walter W. Holland, Jørn Olsen, and Charles du V. Florey, eds., *The Development of Modern Epidemiology: Personal Reports from Those Who Were There* (Oxford: Oxford University Press, 2007).

40. Devra Lee Davis, *When Smoke Ran Like Water: Tales of Environmental Deception and the Battle Against Pollution* (New York: Basic Books, 2002); Lynne Page Snyder, "The Death-Dealing Smog over Donora, Pennsylvania: Industrial Air Pollution, Public Health Policy, and the Politics of Expertise, 1948–1949," *Environmental History Review* 18, no. 1 (1994): 117–139; Christopher C. Sellers, "Discovering Environmental Cancer: Wilhelm Hueper, Post-World War II Epidemiology, and the Vanishing Clinician's Eye," *American Journal of Public Health* 87 (1997): 1824–1835; Robert Proctor, *Cancer Wars: How Politics Shapes What We Know and Don't Know About Cancer* (New York: Basic Books, 1995); John Higginson and Calum S. Muir, "The Role of Epidemiology in Elucidating the Importance of Environmental Factors in Human Cancer," *Cancer Detection and Prevention* 1 (1976): 79–105.

41. Mervyn Susser, "Epidemiology Today: 'A Thought-Tormented World'," *International Journal of Epidemiology* 18 (1989): 481–488; Mervyn Susser, "Epidemiology in the United States After World War II: The Evolution of Technique," *Epidemiologic Reviews* 7 (1985): 147–77. On individualism in epidemiology and the loss of the "population perspective," see Sylvia Noble Tesh, *Hidden Arguments: Political Ideology and Disease Prevention Policy* (New Brunswick, NJ: Rutgers University Press, 1988); Steve Wing, "Limits of Epidemiology," *Medicine and Global Survival* 1 (1994): 74–86; Robert A. Aronowitz, *Making Sense of Illness: Science, Society, and Disease* (New York: Cambridge University Press, 1998).

42. Nash, *Inescapable Ecologies*, 192–202; Phil Brown, "Popular Epidemiology: Community Response to Toxic Waste-Induced Disease in Woburn, Massachusetts," *Science, Technology, and Human Values* 12 (1987): 78–85; Phil Brown and Edwin J. Mikkelsen, *No Safe Place: Toxic Waste, Leukemia, and Community Action* (Berkeley: University of California Press, 1990); Barbara L. Allen, *Uneasy Alchemy: Citizens and Experts in Louisiana's Chemical Corridor Disputes* (Cambridge, MA: MIT Press, 2003). The best review of the field by a practitioner is Nancy Krieger, *Epidemiology and the People's Health: Theory and Context* (New York: Oxford University Press, 2011). See also Mervyn Susser, *Eras in Epidemiology: The Evolution of Ideas* (New York: Oxford University Press, 2009).

43. For a critique of the discourse around obesity, see Julie Guthman, *Weighing In: Obesity, Food Justice, and the Limits of Capitalism* (Berkeley: University of California Press, 2011). Statistics from World Health Organization, "Obesity and Overweight," Fact Sheet No. 311 (September 2006). http://www.who.int/mediacentre/factsheets/fs311/en/.

44. James O. Hill and John C. Peters, "Environmental Contributions to the Obesity Epidemic," *Science* 280, no. 5368 (May 29, 1998): 1371–1374, doi:10.1126/science.280.5368.1371; World Health Organization, "Obesity and Overweight," Fact Sheet No. 311, (September 2006), http://www.who.int/mediacentre/factsheets/fs311/en/; James O. Hill et al., "Obesity and the Environment: Where Do We Go from Here?," *Science* 299, no. 5608 (February 7, 2003): 853–855, doi:10.1126/science.1079857; Steven Cummins, "Commentary: Investigating Neighbourhood Effects on Health—Avoiding the 'Local Trap'," *International Journal of Epidemiology* 36, no. 2 (2007): 355–357; Steven Cummins and Sally Macintyre, "Food Environments and Obesity—Neighbourhood or Nation?," *International Journal of Epidemiology* 35, no. 1 (2006): 100–104, doi:10.1093/ije/dyi276. Barry M. Popkin, Kiyah Duffey, and Penny Gordon-Larsen, "Environmental Influences on Food Choice, Physical Activity and Energy Balance," *Physiology & Behavior* 86, no. 5 (2005): 603–613, doi:S0031-9384(05)00388-4; Jeffery Sobal and Brian Wansink,

"Kitchenscapes, Tablescapes, Platescapes, and Foodscapes: Influences of Microscale Built Environments on Food Intake," *Environment and Behavior* 39, no. 1 (2007): 124–142, doi:10.1177/0013916506295574; Mia A. Papas et al., "The Built Environment and Obesity," *Epidemiologic Reviews* 29, no. 1 (2007): 129–143, doi:10.1093/epirev/mxm009; Julie Guthman and Melanie DuPuis, "Embodying Neoliberalism: Economy, Culture, and the Politics of Fat," *Environment and Planning D: Society and Space* 24 (2006): 427–448. There have also been discussions of the environmental-evolutionary basis of the epidemic, i.e., the hypothesis that humans are well adapted to survive in food-scarce environments but poorly adapted for the food-rich environments of the contemporary era. See, for instance, Peter Brown, "Culture and the Evolution of Obesity," *Human Nature* 2, no. 1 (1991): 31–57, doi:10.1007/BF02692180. Although the evolutionary arguments are highly speculative, they also point to the role of the contemporary environment in fostering the epidemic.

45. Kristina A. Thayer et al., "Role of Environmental Chemicals in Diabetes and Obesity: A National Toxicology Program Workshop Review," *Environmental Health Perspectives* 120, no. 6 (June 2012): 779–789, doi:10.1289/ehp.1104597; J. L. Tang-Péronard et al., "Endocrine-Disrupting Chemicals and Obesity Development in Humans: A Review," *Obesity Reviews* 12, no. 8 (2011): 622–636, doi:10.1111/j.1467-789X.2011.00871.x; Jerrold J. Heindel and Frederick S. vom Saal, "Role of Nutrition and Environmental Endocrine Disrupting Chemicals During the Perinatal Period on the Aetiology of Obesity," *Molecular and Cellular Endocrinology* 304, no. 1–2 (2009): 90–96, doi:10.1016/j.mce.2009.02.025.

46. Abdel R. Omran, "The Epidemiologic Transition: A Theory of the Epidemiology of Population Change," *Milbank Memorial Fund Quarterly* 49, no. 4 (1971): 509–538.

47. According to the UN AIDS organization, approximately 35 million people had died from AIDS and AIDS-related causes, while 34 million people were living with HIV worldwide at the end of 2011. UNAIDS, *Global Report: UNAIDS Report on the Global AIDS Epidemic 2012* (Geneva: Joint United Nations Program on HIV/AIDS (UNAIDS), 2012), http://www.unaids.org/en/resources/publications/2012/name,76121,en.asp.

48. Ronald Barrett et al., "Emerging and Re-Emerging Infectious Diseases: The Third Epidemiologic Transition," *Annual Review of Anthropology* 27 (1998): 247–271; David M. Morens, Gregory K. Folkers, and Anthony S. Fauci, "The Challenge of Emerging and Re-Emerging Infectious Diseases," *Nature* 430 (July 8, 2004); K. E. Jones et al., "Global Trends in Emerging Infectious Diseases," *Nature* 451 (February 21, 2008): 990–993.

49. Richard Preston, *The Hot Zone* (New York: Random House, 1994); David Quammen, *Spillover: Animal Infections and the Next Human Pandemic* (New York: W. W. Norton, 2012).

50. S. Cleaveland, D. T. Haydon, and L. Taylor, "Overviews of Pathogen Emergence: Which Pathogens Emerge, When and Why?" *Current Topics in Microbiology and Immunology* 315 (2007): 85–111, doi:10.1007/978-3-540-70962-6_5; Eric Ka-Wai Hui, "Reasons for the Increase in Emerging and Re-Emerging Viral Infectious Diseases," *Microbes and Infection* 8, no. 3 (2006): 905–916, doi:10.1016/j.micinf.2005.06.032; Jonathan D. Mayer, "Geography, Ecology and Emerging Infectious Diseases," *Social Science & Medicine* 50, nos. 7–8 (2000): 937–952, doi:10.1016/S0277-9536(99)00346-9; Jonathan A. Patz et al., "Unhealthy Landscapes: Policy Recommendations on Land Use Change and Infectious Disease Emergence," *Environmental Health Perspectives* 112, no. 10 (2004): 1092–1098; K. E. Jones et al., "Global Trends in Emerging Infectious Diseases,"

Nature 451 (February 21, 2008): 990–993; Robert G. Wallace, "Breeding Influenza: The Political Virology of Offshore Farming," *Antipode* 41, no. 5 (2009): 916–951, doi:10.1111/j.1467-8330.2009.00702.x.; Robert F. Brieman, "Impact of Technology on the Emergence of Infectious Disease," *Epidemiologic Reviews* 18, no. 1 (1996): 4–9. Although many of the emerging and re-emerging diseases are fostered by environmental changes, AIDS is something of an exception. Although HIV was also zoonotic in origin, the rise of the AIDS epidemic was likely much more strongly affected by rapid urbanization, changing social practices, and iatrogenic infection than by more narrowly "environmental" factors like changes in land use, climate, pollution, etc.; however, the co-variance of AIDS with malaria infection does suggest an indirect environmental driver. Jacques Pepin, *The Origins of AIDS* (Cambridge: Cambridge University Press, 2011); John Iliffe, *The African AIDS Epidemic: A History* (Athens: Ohio University Press, 2006).

51. Rita R. Colwell, "Global Climate and Infectious Disease: The Cholera Paradigm," *Science* 274 (December 20, 1996): 2025–2031; Christopher Hamlin, *Cholera: The Biography* (Oxford: Oxford University Press, 2009), 267–300; Andrew E. Collins, "Vulnerability to Coastal Cholera Ecology," *Social Science & Medicine* 57, no. 8 (2003): 1397–1407, doi:10.1016/S0277-9536(02)00519-1.

52. Collins, "Vulnerability to Coastal Cholera Ecology"; Charles L. Briggs, *Stories in the Time of Cholera: Racial Profiling During a Medical Nightmare* (Berkeley: University of California Press, 2003); Marcos Cueto, "Stigma and Blame during an Epidemic: Cholera in Peru, 1991," in *Disease in the History of Latin America: From Malaria to AIDS*, ed. Diego Armus (Durham, NC: Duke University Press, 2003), 268–289.

53. Randall M. Packard, *The Making of a Tropical Disease: A Short History of Malaria* (Baltimore: Johns Hopkins University Press, 2007); Nicholas B. King, "The Scale Politics of Emerging Diseases," *Osiris* 19 (2004): 62–78.

54. Annemarie Mol, *The Body Multiple: Ontology in Medical Practice* (Durham, NC: Duke University Press, 2002).

CHAPTER 4

••

DESERTS

••

DIANA K. DAVIS

BARREN, void, parched, a wasteland. In the popular imagination, deserts are where the trees are not—where nothing grows, where there is no shade, where everything bakes in the sun and withers (see Figure 4.1). For most people, especially those not born and raised in the desert, this is troublesome. For a small minority of people, the desert is sometimes conceived as a sublime space, a place of spirituality or a proving ground for self-worth. Most, however, see deserts as ruined, forbidding wastes very much in need of improvement to make them more productive for more people. This global view of deserts today, embodied in the UN Convention to Combat Desertification (UNCCD) and many national development agendas, is primarily an Anglo-European conception and has deep roots in the Western imaginary of deserts stretching back at least two thousand years.[1] This essay explores some of the origins and evolution of our thinking about deserts, how this thinking has influenced our actions, and why the problematic notion of spreading deserts continues to be so ubiquitous and tenacious, despite plentiful scientific evidence to the contrary.[2]

DESERTS, FEAR, AND POLICY
••

The creation in 1997 of the global mechanism, a subsidiary body of the 1994 United Nations Convention to Combat Desertification (UNCCD), represents in numerous ways the apogee of contemporary global concern with deserts and their spread, usually called "desertification." Premised on the claim that desertification is a significant and growing problem, in large part caused by human action, the global mechanism distributes UN funds to try to halt this menace. Yet, even as millions of dollars are spent every year to try to combat allegedly growing deserts, a debate continues over the very existence of desertification, and a growing body of scientific evidence undermines the common claims of its supposedly catastrophic spread. Indeed, within the scientific and academic community there is a broad consensus that the extent and severity of

FIGURE 4.1 Camel bones in Southern Morocco.

Photo by James E. Housefield.

desertification has been greatly exaggerated.[3] This is not to say that land degradation, especially from over-irrigation and ploughing marginal soils, has not taken place. Such degradation has occurred—but it is most often limited to discrete locations, usually in the semi-arid zones and not frequently in deserts proper. Many contend that this kind of land degradation in arid zones should not be called desertification. The common anxious assumption that the word's deserts are inexorably spreading, however, has been repeatedly demonstrated to be false, and a recent "greening" has in fact been detected in many desert regions over the last two decades.[4]

Most deserts, like the Sahara, are not continually expanding; rather, they appear to grow and shrink according to rainfall patterns more than any other single factor, and have done so over the last several millennia.[5] Deserts, according to most scientists, are the arid and hyper-arid regions covering about 20 percent of the globe that receive very little rainfall of a highly unpredictable nature.[6] Average annual rainfall is low and temperatures are usually high (except in the cold deserts), although no universally accepted definition of "desert" exists.[7] Moreover, annual and interannual rainfall in deserts is highly variable, leading to widespread non-equilibrium ecological conditions.[8] There is little debate that deserts are naturally occurring physical features. Another 20 to 30 percent of the earth's surface, though, is variously categorized as semi-arid, and together deserts and semi-arid areas are often called "drylands." Most of the debate about desertification concerns the semi-arid zones and fears that they will become arid deserts, especially along desert margins. Located primarily in the low humidity, high-pressure zones of 30 degrees north and south latitude, desert environments are highly variable, ranging from sand seas to stony pavements to arid mountains to desert crusts and varnishes.[9] The majority of arid regions where deserts are located contain vegetable and animal life highly adapted to heat, aridity, and drought.[10] In many deserts numerous plants, such as annual grasses and wildflowers, are only visible for short periods following rainfall. They spend the majority of their life cycles as below ground biomass in seed banks and in bulbs or other forms. Deserts are not dead, lifeless zones as so often assumed, but rather most are rich in biodiversity.[11] Moreover, many of the plants, including a majority in places such as the Mediterranean basin, are also well-adapted to fire and grazing.

Given the nature of deserts and their biophysical components, it is perhaps not surprising that well-documented cases of large-scale desertification are relatively few.[12] In the words of W. M. Adams, a geographer with many years of experience working in drylands, "perceptions of the severity of recent desertification are to a large extent self-confirming, but have a scanty scientific base …and the urgency of policy concern about 'desertification' is based on loosely conceived concepts that provide no clear and consistent theoretical basis" for future work.[13] A significant problem is that desertification has over one hundred definitions, and no agreement has been reached in the scientific or policy communities on how to measure and monitor desertification.[14] Arid lands scientists S. M. Herrmann and C. F. Hutchinson have insightfully noted recently that, "we have a non-equilibrium world that is saddled with an overriding equilibrium mindset and policies that reflect it."[15] They conclude that, "policies that affect people on the ground may be formulated largely independent of science that is current and thus may serve to degrade rather than enhance the lives of people most affected."[16] One of the biggest points of contention in the "desertification debates," then, appears to be between the scientific/academic community and the global policy community, especially at the levels of international agencies like the UN and many national governments.

Despite the absence of an agreed definition for, and the paucity of documentation of, desertification, the crisis narrative of desertification is consistently invoked for a variety of reasons.[17] In the words of geographer Andrew Warren, desertification has been "used as a deliberate strategy to attract attention or funding."[18] As other scholars have repeatedly pointed out, claiming an environmental crisis such as desertification is an efficient way for governments to raise international development funds, spur action, and to justify social policies that control "difficult" populations.[19] For the approximately 38 percent of the world's population who live in drylands (which, according to a recent estimate, currently cover an estimated 41 percent of the globe), the fight against desertification is often intensely personal.[20] All too many projects to halt desertification, while enriching bureaucrats, have disrupted livelihoods, especially for pastoralists who are often incorrectly blamed for overgrazing and desertification, and have driven already poor populations deeper into poverty.[21]

In part because the perceived crisis of desertification has been beneficial to so many, from agencies seeking international funds, to scientists seeking grant money, to governments seeking control over recalcitrant populations, the popular conception of the encroaching desert has been tenacious despite the data undermining it. This imaginary of the desert as an aberration, as innately threatening, is usually dated to the early twentieth century, to the writings of the French colonial scientists Louis Lavauden and André Aubréville, and to the British forester E. P. Stebbing.[22] It has, however, a much longer history in Western thinking and can be traced back several hundreds of years. An analysis of the deep history of our ideas about deserts reveals that the notion of deserts as worthless wastelands, as formerly forested or fertile regions that were ruined by wanton deforestation, burning, and overgrazing by "non-civilized" peoples, and that thus urgently need "restoration," is deeply entrenched and exerts a continuing detrimental effect on policy decisions in arid lands.

Deserts in the Ancient and Early Christian Worlds

Deserts as we know them today, in distribution and extent, were largely formed by the time Greek civilization began to produce writers of geography and history.[23] The word *desert* itself may be one of the oldest written words, probably originating in the Egyptian hieroglyph, which was pronounced "tésert."[24] *Tésert* was likely filtered through Latin with the word *deserere*, meaning to abandon or forsake, to obtain its contemporary meaning.[25] It is in the writings of the ancient Greeks and Romans that we find some of the earliest articulations of thinking about deserts—those around the Mediterranean and in Asia. Many of the desert regions around the Mediterranean were familiar to the ancient Greeks, especially Egypt and the North African littoral, as well as the semi-arid areas of southern Europe and the eastern Mediterranean. They also had some knowledge of the Arabian peninsula and the desert regions of south and central Asia.

Some of the earliest descriptions of deserts are to be found in the writings of the Greek Herodotus (484-c. 425 BCE). He combined information from his own travels to places such as Egypt with what he gleaned from earlier Greek writings, as well as from hearsay.[26] Deserts were well-recognized as being places with extreme heat, sand, aridity and a dearth of vegetation. Herodotus describes parts of Libya, Egypt, and Syria in this way, as well as parts of India and central Asia.[27] Such seemingly barren landscapes were often contrasted with the agricultural fertility and lushness of, for example, the Nile Valley. Importantly, for these early Greeks, deserts were portrayed as simple facts of life on earth, not as aberrations to be corrected. The significance of deserts for the Greeks appears to lie in the difficulties they presented to travel as well as in the cultural and physical traits they believed they caused in the peoples who inhabited them.[28] It was widely believed at this time that different climates (such as wet, dry, cold, or hot) caused differences in the various peoples of the world.[29] By the turn of the millennium, such thoughts had coalesced into the Greek conception of the five different climate zones of the known world, and their inhabitants. Some early articulations of these determinist ideas are provided in the works of Aristotle (384–322 BCE) and of Posidonius (135–50 BCE), and they were later taken up by many Greek and Roman writers, including Strabo (64 BCE–20 CE).

Of these five zones, the "best" was the zone in which Greece and its self-proclaimed advanced civilization was located, the middle of the middle zones. The two extreme zones—the frigid Arctic and the torrid equatorial zones—were considered uninhabitable.[30] The more northerly temperate zone was cold and produced people who "are spirited but deficient in skill and intelligence," while the warm southerly temperate zone produced people who "are intelligent and inventive but lack[ing] in spirit and are in subjection and servitude."[31] The Greeks, being in the middle of these extremes, benefited from the advantages of each: "they are high-spirited and intelligent."[32] The desert zones, being on the edge of habitability, produced, according to this worldview, people who were decidedly different from, and mostly inferior to, the Greeks. For example, Strabo and later the Roman Pliny (23–79 CE), along with several writers before them, both attribute Ethiopians' skin color and hair texture to their having been "scorched" by the sun in their hot desert location.[33] Many other examples of early environmental determinism may be found in such classical writings.

In many places in the Hebrew Bible (the Christian Old Testament), the desert—rendered in several different Hebrew terms—is mentioned simply as a natural part of the earth (much as it was in classical writings).[34] In other parts, though, such as the wanderings of the Israelites in the Sinai desert following their exodus from Egypt, or the sufferings meted out to Job, the desert is a place of torment. That idea appears as well in the Christian Gospels, which tell of Jesus being led by the devil to the desert to be tempted. Moreover, there is the strong suggestion that the desert is the result of environmental deterioration caused by the expulsion of Adam and Eve from the Garden of Eden.[35] After the expulsion, "the earth mourns and *withers*," "it lies polluted," "a curse devours the earth and its inhabitants suffer for their guilt."[36] Indeed, as geographer Jeanne Kay has noted, "throughout the Bible, when people disobey, God causes nature to deteriorate as a tool of punishment and sends severe drought, plagues, pests, warfare or diaspora."[37] This particular line of Christian thinking and its relation to deserts became especially relevant during the colonial period in Africa and elsewhere.

Perhaps more significant during the early Christian period was the idea that, as the creation of God, humans had dominion over nature and, for some religious thinkers, a responsibility to improve the environment, including deserts. For example, Tertullian (ca. 160–240 CE), a Christian Roman from Carthage, wrote that "most pleasant farms have obliterated all traces of what were once dreary and dangerous wastes; cultivated fields have subdued forests; flocks and herds have expelled wild beasts; sandy deserts are sown."[38] This idea was echoed a thousand years later by Albert the Great, a theologian and German writer, who wrote in the thirteenth century that nature could be improved and that "the plow and the hoe transform the desert wild to cultivated land."[39] Jeanne Kay summarizes this by explaining that "God himself seems to place enormous value on sustained and fertile landscapes" throughout the Bible.[40]

A third significant change in thinking about deserts during this period resulted from the rise of monasticism in the third and fourth centuries CE. Known as the "Desert Fathers" or the "Desert Saints," an assortment of religious men and some women withdrew to the Egyptian desert, away from the Nile River, in order to test or prove their religious piety. Modeled on the example of Jesus fasting in the desert, religious figures such as Saint Anthony retreated to the desert and lived an ascetic life which, they believed, would bring them closer to God.[41] This way of living became so popular among Christians that it spread rapidly and the desert became "the assumed location of ascetic perfection."[42] So influential was this "Christian myth of the desert" that it influenced theological writers to change their texts to make the environment described therein less "Eden like" and more desert like, even in forest settings.[43] The popularity of the Desert Fathers and their writings led to the widespread belief that the desert landscapes of Egypt and the Levant were the most perfect location for Christian withdrawal from the world. In the words of one scholar, "the desert becomes, in fact, a heaven on earth."[44] This interpretation has had a long-lasting influence on Western thinking about deserts.

New knowledge and new thinking about deserts during the European Middle Ages (roughly 500–1400 CE) was scant. Most of what was written related to studies and interpretations of the Bible and related holy writings. Some limited early travel writing resulted from the Christian Crusades to the Holy Land and from other adventurers such as Marco Polo. Outside of the Christian context, such descriptive travel writing primarily portrayed deserts as magnificent and exotic. Marco Polo's recounting of his adventures while traveling to and from China between 1260 and 1295, for example, contains several descriptions of large and wondrous deserts like the Taklimakan and other desert areas in Central Asia. His descriptions are strikingly similar to those of the ancient writers, noting the difficult physical environments and the strange peoples living therein. Near Pamir, Polo describes an area of forty days' travel, with many deserts containing neither vegetation nor human life.[45] At the edge of this region are found a mountain people who he describes as savage, violent, and cruel. Later during his journey, Polo encounters the Taklimakan Desert of western China. He marvels that this desert "is so extensive that if a man were to travel through its whole length, it would employ a year … It consists altogether of mountains and valleys of sand, and nothing is got to eat; but after traveling a day and night, you find sweet water sufficient for from fifty to a hundred men, with their animals."[46] Marco Polo's book narrating his travels was quite popular in certain circles in Europe after it was published. Christopher Columbus owned a copy in which he made notes.[47] Indeed, much of this Western thinking about deserts would be carried with Europeans as they began to travel farther from home in their explorations during the next several centuries.

EARLY EXPLORATIONS, DESERTS, AND DESICCATION

The "age of exploration" ushered in some different thinking about deserts as parts of the world new to Europeans were "discovered." This period in the history of the West coincided with several important intellectual developments that bear directly on changing Anglo-European views of deserts, their nature, and their evolution. It is during this period that thinking about deserts becomes so intricately related to ideas about forests and their destruction, and to notions of humans as agents of environmental transformation.

Early in the fifteenth century, Latin translations of Greek and Roman texts, long lost to European readers but carefully conserved in Arabic translation, became available to those interested and able to read. In addition to introducing European readers to ancient conceptions of deserts in North Africa and Asia, these translations introduced ideas linking deforestation to desiccation and to the general idea of human-generated environmental change. The Greek naturalist Theophrastus, for instance, had pondered the relationship he perceived in the draining of a large marsh in Thessaly and the subsequent increase in freezing temperatures, which he attributed to causing a decrease in grape vines and olive trees.[48] He likewise mused about how the clearing of some woodlands near Philippi seemed to make the climate warmer in that area.[49] Plato, similarly, was convinced that deforestation in Attica had caused significant erosion, and lamented the lost fertility of former times.[50]

It was Theophrastus, though, who may have been the earliest thinker to draw a connection between deforestation and decreases in rainfall.[51] This idea became, by the mid-eighteenth century, a fairly coherent body of thought commonly called "desiccation discourse." Elegantly detailed by Richard Grove in his numerous publications, desiccation theory included the idea that deforestation causes the climate to dry out and also the corollary, that reforestation restores lost rainfall.[52] From the sixteenth through the eighteenth centuries, colonial interactions with nature in parts of the world such as the Caribbean, the South Pacific, and the Indian Ocean, places far removed from previous European sites of experience, helped to form such conceptions. These experiences, mostly of colonial plantation agriculture in tropical island settings, revealed serious consequences in the wake of deforestation such as soil erosion.[53] As Grove has so persuasively argued, the image of the tropical island, with its lush vegetation and association with the Garden of Eden, came to represent an ideal landscape with very positive moral connotations during the eighteenth century.[54] The fear that deforestation was causing desiccation in such settings therefore took on an urgency it might not have otherwise. Those concerned with the effects of deforestation were worried not only about desiccation and erosion but also, and perhaps more significantly, about these Edenic islands becoming arid desert wastes. In the words of Joseph Banks, who visited some of these tropical islands in 1771, "the paradise" of St. Helena had been allowed, through deforestation and overgrazing, to "become a desert."[55] After the mid-eighteenth century, desiccation theory was so dominant in Europe, especially in Britain and France, that it formed the basis for a growing number of forest protection and reforestation laws and plans in colonial territories.

Related notions of environmental determinism, environmental degradation, and human culpability were developing in other realms of European thinking during this period. The age of exploration had opened European minds in several respects. One significant example is that European voyages to the remote parts of the world disproved the previously common belief, developed by the ancient Greeks, that humans could not live in the "torrid" zone of the tropics or the "frigid" zone of the polar regions. By the sixteenth century, it was becoming increasingly clear that humans could and did live in such environments. In fact, according to Clarence Glacken, "men in the 16th century were more impressed with the discovery that the equatorial regions were habitable than they were with the discovery of the new World."[56] Glacken was analyzing the works of Jean Bodin, whom he considered "the most important thinker of the Renaissance" on the relations between history and geography.[57] Bodin's writings also reveal, however, that ancient Greek notions of environmental determinism—that is, that people were formed and significantly influenced by their environment—were still widespread. He wrote, for example, of the "savagery" of the North Africans and explained that, "it happens that those who are in the furthest regions are more inclined to vices."[58]

Especially important for thinking about deserts was the growing sentiment that humans could cause environmental change on a large scale. We see this idea articulated by George Hakewill in 1627, when he wrote that, "God has not everywhere ordained either fruitfulness or barrenness....Present wastelands and sandy deserts were fertile lands in former ages [and] the condition might be reversed."[59] Although humans could cause degradation, we see here that some began to believe that humans could also improve the environment through their works. This line of thinking, when combined with the common acceptance of desiccation theory, led many to believe, as did the eighteenth century author Count Buffon, that in desert lands such as Arabia, planting "a forest in the midst of [a] scorching desert might bring rain, fertility, and a temperate climate."[60] A century later, in 1862, the French scientist Jules Clavé summarized the prevailing environmental thinking by writing that "the terrible droughts which desolate the Cape Verde islands must also be attributed to the destruction of the forests. In the island of St. Helena, where the wooded surface has been considerably extended within a few years, it has been observed that the rain has increased in the same proportion....In Egypt, recent plantations have caused rains, which hitherto were almost unknown."[61]

Although few Europeans during the period from the fifteenth through the eighteenth centuries had much actual experience in deserts, environmental ideas formed during these four centuries were increasingly applied to desert regions. Primarily theoretical until the nineteenth century, such thinking about deserts—namely, that they were aberrations, ruined and deforested wastelands—grew and became increasingly influential. In essence, a growing number of people began to perceive deserts as former forests that must be made forests again. There were some who paired these environmental conceptions with the predominantly Christian belief that only sinful people lived in desert wastelands. Related to early Christian notions of deserts as the result of original sin, and the expulsion from the Garden of Eden, such environmental imaginaries would have serious consequences during the heyday of colonialism in the drylands in the nineteenth and early twentieth centuries.

COLONIALISM AND THE DESERT BLAME GAME

By the nineteenth century, deserts were increasingly conceptualized by those in the West as spaces of fear, as ruined, deforested landscapes in urgent need of repair. Culpability for the apparent destruction began to be apportioned to local populations with disturbing regularity. Whereas earlier, in the seventeenth and eighteenth centuries, blame was placed on European colonial activities such as plantation agriculture and logging operations nearly as frequently as on "native" agricultural and grazing practices, in the nineteenth century, the balance shifted. George Perkins Marsh exemplified the thinking of most Anglo-Europeans of the mid-nineteenth century with respect to deserts and their origins. In his 1864 book, *Man and Nature*, drawing on a wide range of European and other sources, he wrote that, "I am convinced that forests would soon cover many parts of the Arabian and African deserts, if man and domestic animals, especially the goat and the camel, were banished from them."[62] Indigenous peoples and local practices were the biggest cause of deserts in this increasingly common view.

Early in the nineteenth century, deserts came to be more strongly associated with "evil" just as forests were vigorously associated with "good." As Grove has explained, there were even those who "conceived of drought as the wages of environmental sin or sins of moral disorder … [who saw] drought as a form of moral retribution" against the indigenous population.[63] Writers including François René Vicomte de Chateaubriand, believed that, "everywhere the trees have disappeared, man has been punished for his improvidence. I can tell you better than others, sirs, what is produced by the presence or absence of forests, since I have seen … the deserts of old Arabia where nature appears to have breathed its last breath."[64] Desert landscapes, in this view then, were signs of divine retribution against those destructive and immoral "natives" who had ruined the land.

One of the earliest examples of the results of this kind of thinking about deserts in a colonial setting comes from southern Africa. In 1820 a Scottish missionary, Robert Moffat, arrived in Cape Colony and proceeded to the northern area of Latakoo, just south of the Kalahari Desert.[65] The local people, the Tswana agro-pastoralists, were then trying to survive a severe drought that lasted from 1820 to 1823. Moffat believed that the region had previously been lush and verdant, but that deforestation and burning by the Tswana had degraded the land. Moreover, Moffat believed that the Tswanas' "sinfulness" in destroying vegetation and in their general way of life (notably, rejecting Christianity, theft, and witchcraft) had caused the drought. In Moffat's eyes, the Tswanas' transgressions, particularly veld-burning, had "brought about the drought and the arid landscape of divine retribution."[66]

A quarter century later, Moffat's writings and ideas were taken up by another Scottish missionary, who had a more significant impact on the region and the colony. John Croumbie Brown visited the Cape as a missionary from 1844 to 1848, during another period of drought in southern Africa. He returned in 1862 as the Colonial Botanist during yet another severe drought. He subscribed to Moffat's views and mirrored other European thinking about deserts of the mid-nineteenth century in much of his work. He wrote many reports in his official capacity that laid blame for increased heat and drought, and the creation of desert-like conditions, on the destruction of vegetation by those living in the region.[67] Brown concluded that it was urgently necessary to proceed with "conservation and [the] extension of forests as a means of counteracting the evil referred to [creating deserts]."[68] Brown's many reports and books provided long-lived justifications for the creation of official policies in southern Africa that controlled lands by criminalizing the activities of local Africans in the name of preventing the "desert's spread."

About the same time, at the other end of the African continent, the French were developing the concept of desertification during their occupation of North Africa. Armed with the belief that North Africa had been the fertile granary of Rome, they mistakenly thought that the arid and semi-arid landscapes they encountered in 1830s Algeria were severely deforested and desertified. They blamed the local North Africans, especially the nomadic pastoralists whom they assumed were the destructive descendants of the tribes of the eleventh-century "Hillalian invasions" for the assumed environmental ruin.[69] As early as 1834, nomads were being blamed for destroying the environment.[70] Such claims were used to justify confiscating land and sedentarizing nomads from this point onward during the colonial period. Importantly, beginning in the 1830s, this view of the desert as land ruined by the "natives" was written into sedentarization policies and into laws such as the colonial forest and land codes for the Maghreb and related policies governing environmental management, in Algeria (and later Tunisia and Morocco).[71] These laws and policies restricted and criminalized many local land uses appropriate for arid lands that had been used sustainably for centuries by the North Africans while they facilitated colonial activities and enriched European settlers.

Although the word "desertification" was not used until 1927, when the French scientist Louis Lavauden wrote that "desertification...is uniquely the act of humans.... [T]he nomad has created what we call the pseudo-desert zone,"[72] the concept of desertification was widespread in French thought much earlier. In 1906, for example, Augustin Bernard described what came to be called desertification when he wrote that due to burning and overgrazing by nomads, "the forest gives way to scrub, the scrub to herbaceous vegetation, the herbaceous vegetation to bare soil, which finished by being itself detached and which becomes the victim of the wind."[73] Half a century earlier, in 1865, a powerful French doctor and politician in Algeria, August Warnier, voiced the common opinion that due to the Arabs' environmental destruction, Algeria, "long ago a sort of terrestrial paradise covered in groves…is [now] a sterile desert, bare and without water, that all call the *land of thirst*."[74] The fear of desert spread and related levels of xenophobia reached a fever pitch in the last half of the nineteenth century in French Algeria.[75] Warnier, for example, wrote in his typical style that, "France itself, so prosperous, would soon become a desert if it were in the hands of the Arabs."[76] Warnier was responsible for writing and passing into law extensive legislation that destroyed indigenous communal land tenure and instituted private property rights (and a booming market in property). In French North Africa, the locals and the desert they had apparently created were the enemy against which a continual war was waged. The French were thus the first to use and define "desertification," and their colonial definition of desertification as deserts created by nomads, and other pastoralists, burning, deforesting, and overgrazing has been very long-lived and influential, especially at the global policy level.

In British India similar desiccationist arguments were used from the 1830s to institute increasingly draconian policies restricting indigenous land uses, especially in the area of forestry. Colonial surgeons employed in the East India Company were the primary advocates of desiccation theory in India; they lobbied long and hard for forest conservation and reforestation in order to prevent what they perceived as further spread of desert-like conditions.[77] Their efforts culminated in the passing of the "notoriously unjust" Forest Act of 1878, which followed, in many ways, ideas and policies that had been developed by the French. In fact, as Richard Grove has noted, "by the late 1880s the typologies of anti-desiccation forest policy in the French and British colonial states were so closely inter-related that they can be said to have constituted a single ruling philosophy rather than two separate traditions."[78]

What was new in British India was the extent to which water was manipulated to transform arid and semi-arid areas. The British idea that in India deserts were abnormal and in need of correction engendered extensive efforts to increase irrigation. Many arid "wastes" that were previously primarily the domain of pastoralists who subsisted on the natural pastures, particularly in the Punjab, were canalized and irrigated to produce rational, productive agricultural fields rife with malaria and suffering salinization.[79] This highlights the European notion that only agricultural land has value and that land not cultivated with crops, or being used for raising cattle in the European style of intensive production, is being wasted. In the words of Neeladri Bhattacharya, "the extension of cultivation was synonymous with progress, and the 'reclamation of waste land' was a civilizing project."[80] Along the way, as in French Algeria, desert denizens, primarily "primitive" nomadic pastoralists, were marginalized and sedentarized.[81] Arid wastelands, like forests, were declared the property of the state in British India, and land rights were redefined.[82] Grazing itself was officially declared "undesirable," and those who "wandered," or migrated, were criminalized with the 1871

Criminal Tribes Act, which required a permit just to leave one's village.[83] In British India also, then, deserts were defined as aberrations and the local inhabitants were blamed for their creation. Both deserts and their inhabitants thus were "rationalized and civilized" with the enforcement of sedentary agriculture.

However, a radically different view of the desert and its inhabitants, especially nomads, was held by a group of British subjects in the early twentieth century in the "holy land." As explained by historian Priya Satia, on the eve of World War I, a group of idealistic British government and military personnel raised on the tales of the *Arabian Nights* and visions from the Bible went to the deserts of Mesopotamia, known variously as "Arabia," "the holy land," and "the cradle of civilization." Inspired by common Victorian notions of "Arabia" as a mysterious desert idyll, a biblical land, a place of miraculous convictions and extremes, they saw the desert environment as absolutely foreign, dangerously monotonous, and ultimately unknowable.[84] At first, many of these people saw "Arabia" as a kind of extraterrestrial utopia and romanticized the local peoples, in particular the nomads. The desolation of the landscape was blamed on the oppressive Ottoman tyrants and previous barbaric conquerors. As fighting got underway, however, it did not take long for perceptions of the environment to become more negative.

The desert idyll transformed, in the eyes of most involved, into an "autarkic wasteland, a fallen Eden" in need of redemption by the British.[85] Thanks in large part to the influence of T. E. Lawrence, though, a shred of romantic feeling was retained for the nomads, especially by those in the Air Force. Lawrence, filled with ideas of the desert sublime and respect for much of the nomads' way of life, thought that the Air Force could conquer the desert, much like the Arab nomads had in the past. Deeply ingrained views shaped by environmental determinism, though, twisted this strain of respect for the "noble nomads" and portrayed them as tough inhabitants of a harsh environment that "could tolerate random acts of violence" in ways that others could not.[86] Combined with an "environmental imaginary of land so barren that bombardment could not possibly worsen it," this view of the desert and its people coalesced into seeing "Arabia" as a "state of exception" that allowed, and even facilitated, excessive brutality in the name of "developing" the region.[87] This particular view of the desert has had a long-lasting and disturbing impact on Western thinking about and action in the Middle East.[88]

THE DUST BOWL, THE SAHELIAN FAMINE, AND DESERT ANGST TODAY

Following World War I, during the interwar period, increasing concern about the spread of existing deserts became more global in scope and more widespread. William Mcdonald's alarmist *Conquest of the Desert*, on southern African deserts and drylands, was published in 1914, and was followed by E. H. Schwartz's 1923 book *The Kalahari or Thirstland Redemption*. Echoing the earlier sentiments of the likes of Moffat and Brown, the conclusions of these and similar books and articles were included in influential policies in South Africa.[89] In North Africa very similar fears were expressed about the spreading Sahara by many authors during the 1920s, including Louis Lavauden, Henri Hubert, and Edward Bovill.[90] The French, of course, had been warning about the spread of the Sahara in dire terms since the mid-nineteenth century, based largely on their experiences in Algeria and later in Tunisia and Morocco. These fears also found their way into many laws and policies, especially in the French territories ringing the Sahara.

Such worries about desertification were not limited to the African continent; they also included South, Central, and East Asia, as well as North America.[91] In the United States, thinking about deserts had reflected most of the negative stereotypes and fears ubiquitous in Europe since at least the early nineteenth century, when Zebulon Pike wrote that the land of the plains and into the inter-mountain West was "a sterile waste like the sandy deserts of Africa."[92] Early travelers, traders, and trappers had even described the Great Plains as a desert before many had yet experienced the true deserts of the Southwest and West.[93] Once travel became more common west of the Rocky Mountains, through the actual deserts, negative views of deserts very similar to those Europeans had been voicing about African and Asian deserts became widespread. Throughout the nineteenth century, American deserts were primarily categorized as barren, sterile, savage, and altogether unnatural—as places to be conquered.[94] As in India, irrigation was seen as a way to redeem the deserts and make them useful.[95] Irrigated crop agriculture was not the only way to try to create a "garden in the desert" of the US West, however. Ranching, that is, extensive livestock production, also boomed in the late nineteenth century and into the twentieth century in the arid and semi-arid regions. Using a hodgepodge of land laws that ill-fitted the land, climate, and ecology of the region, many different people and corporations grabbed Western land quickly in order to try make the region "productive" and thus turn a profit in one of several different ways.[96]

The boom in ranching, for a variety of reasons, appears to have led to some serious overgrazing in parts of the West. This was the reason, in part, for the creation of the 1934 Taylor Grazing Act, which regulated grazing. This Act also, however, allowed cattle ranchers to gain grazing land at the expense of others, especially sheep ranchers, just as the 1891 law allowing the creation of forest reserves had disadvantaged other marginalized groups, such as the small ranchers in northern New Mexico.[97] This is quite similar to what the European colonial powers had done in many colonial territories where traditional forest grazing and other forest activities were restricted and criminalized. Many have, in fact, argued that the US West has been treated as a colony of the more influential Eastern powerbrokers.[98] Thinking about, and policies in, American deserts were similar to those in Europe and its colonies in disturbing ways, especially with respect to the inhabitants of deserts. The treatment of Native Americans in the West during the nineteenth and early twentieth centuries mirrored surprisingly closely the racist and determinist treatment of local peoples by European imperial powers in the arid parts of Asia and Africa.[99] These local peoples with long histories of sustainably living in the arid lands of the American West were sedentarized and corralled in reservations, partly in the name of environmental protection.

The Dust Bowl of the 1930s brought a heightened level of worry to existing thinking about deserts. Whereas in the United States, the new conservationist ideology condemned modern agricultural techniques and a rapacious view of an "unending frontier" in American thought, the effect of Dust Bowl thinking in much of the rest of the world was different.[100] In places such as the drylands of Africa and Asia, American prescriptions for soil conservation were taken as gospel and applied to heterogeneous environments, where they frequently criminalized local land use practices while introducing ecologically inappropriate techniques that often caused more harm than good.[101] This process was aided and abetted by the widely influential writings of authors such as the British forester Edward Stebbing, who wrote a 1935 article entitled "The Encroaching Sahara" and many others similar in tone and content. Stebbing's writings about the Sahara were highly alarmist and predicted that West Africa was going to become like the Sahara if action was not taken.[102] The continuing influence of writers like these cannot be overstated, despite the fact that scientific research based on field studies as early as the mid-1930s was demonstrating that the Sahara was not encroaching in any significant way on surrounding territories.[103] A spate of popular publications in French and English, with titles like *The Rape of the Earth* and *Africa: The Land that Dies*, continued to inflame these fears into the 1940s and later.[104]

As such perceptions of menacing, spreading deserts took root in a much broader portion of the scientific, bureaucratic, and lay populations in the 1940s, 1950s, and 1960s, a majority of land management policies in colonial—and later independent—countries around the world came to be based on them. Such crisis narratives, which blamed local peoples for desertification, were immensely useful to colonial powers and newly independent rulers for justifying many political, social, and economic changes in the name of environmental protection. In Mandate Palestine, for instance, this Anglo-European imaginary of deserts and their spread led the British-run forest department to report, incorrectly, that half of the habitable area of Palestine had become an artificial desert due to overgrazing.[105] The presumed overgrazing in this case, as in so many places around the globe, was blamed on nomads whose actions and movements nearly all governments have wanted to control.[106] Reducing livestock numbers, sedentarizing nomads, fire suppression, and planting trees were seen as the answers and such approaches have become nearly universal policy prescriptions in such cases.

The drought and concomitant famine in the African Sahel in the early 1970s generated an unprecedented level of concern about desertification around the globe. Images of millions of livestock and hundreds of thousands of Africans dead and dying from starvation filled television screens and newspapers during the early 1970s. The drought and desertification, the spread of the Sahara, were identified as the primary culprits. The "natural disaster" of drought was believed to have pushed environments and burgeoning communities already weakened by decades, if not centuries, of desertification through improper land use, over the edge of survival. This particular drought was perhaps more dramatic because it followed a couple of decades of above normal rainfall in much of Africa.[107] The change in vegetation and decline in nutrition, which occurred quickly, was highly visible. This drought also followed several decades of colonial rule in most of the region, which had disrupted precolonial social and agricultural systems to a great degree. Although such precolonial systems were often far from equitable and sometimes based on slavery, they did nonetheless provide several mechanisms for mitigating drought so that it was unusual to have people die from drought-related famine in the large numbers that were seen in the early 1970s.[108]

The Sahelian drought captured the attention, and to a certain degree, the sympathy of the world. It also had a significant impact on our thinking about deserts. It was determined by a number of powerful global actors that something should be done to stop the menace of encroaching deserts, which allegedly had created such a disaster. The outcome of this concern was the 1977 United Nations Conference on Desertification (UNCOD), which generated a detailed plan of action to combat desertification.[109] Deserts came to be seen as problems that could be solved with money, technology, and social engineering. UNCOD also ushered in a new era of bureaucratic management that required more personnel, more monitoring, more money, and more action plans. The UN Environment Programme (UNEP) created a desertification branch, which established the publication of the *Desertification Control Bulletin* and the *World Atlas of Desertification*, now in its second edition. As Jeremy Swift and others have pointed out, deserts and desertification have become big money-makers.[110] UNCOD's perspectives and prescriptions were reaffirmed in the 1994 UN Convention to Combat Desertification (UNCCD) and further institutionalized with the creation of its subsidiary body, "the global mechanism," designed to disperse anti-desertification funds, in 1997.

The hegemonic perception of deserts and desertification embodied in UNCOD/UNCCD, though, was deeply flawed. As noted by David Thomas and Nick Middleton, "it seems to have triggered many problems, of understanding and of action, that manifested themselves in the ways in which desertification was conceptualized, represented and approached as an environmental, social and political issue."[111] According to these two geographers who specialize in arid lands, desertification itself is in large part a myth, because desertification "data are at best inaccurate and at worst centred on nothing better than guesswork" while dryland ecosystems "appear to be well-adapted to cope with and respond to disturbance, demonstrating good recovery characteristics."[112] In other words, most deserts around the world are resilient, not suffering significant degradation, and not expanding, and moreover, "desertification exists in the eye of the beholder and particularly the land manager."[113]

CONCLUSION

The thinking about deserts embodied today by UNCOD, UNCCD, and innumerable environment and development projects around the world is not new. As this essay has shown, it incorporates, to a significant degree, Anglo-European perceptions dating back many hundreds of years, which assume deserts to be deforested and overgrazed wastelands that hold promise for improvement and productivity. The contemporary policy prescriptions generated by this view of deserts and desertification—as aberrations in need of correction—are also disturbingly similar to those put in place during the colonial period, to the detriment of many local peoples and all too often to the detriment of the environment. Based as they are on the tenacious ideas of desertification as solely anthropogenic, and the "pseudo-desert zone" being the result of nomad abuse so well expressed by Lavauden nearly a century ago, these policies most usually consist of a combination of the sedentarization of mobile pastoralists, destocking, fire suppression, and frequently "reforestation," all of which have been shown repeatedly by contemporary ecological science to be environmentally harmful in many arid, non-equilibrium environments. Extensive pastoralism (moving livestock frequently over large expanses of land), is actually the best use of a majority of land in the drylands.[114]

Side by side with such policies, most governments in arid lands promote irrigation and the extension of dryland agriculture into marginal areas (areas with annual average precipitation below that which can reliably support most grains, usually defined as below the 400 mm. isohyet). This has resulted in growing problems with salinization of agricultural soils, declining aquifers, and the destruction of native vegetation, especially range vegetation, by ploughing, which further desiccates soils in arid regions. Many policymakers, however, appear to be blind to these important and widespread causes of land degradation in arid lands. Development in the drylands has been greatly confused by these very old notions of deserts and more contemporary ideas of desertification—and this has resulted in misguided policies and prevented understanding the actual causes (and potential remedies) for land degradation in arid regions. Millions of dollars are spent every year to "improve deserts" and "fight desertification," disrupting many lives, when much of the time such interventions are not ecologically warranted. Just as our perceptions of past damage to deserts are usually flawed, so too are our aspirations for the future. Why should the desert bloom? This is a question urgently worth asking—and answering with very careful social and natural history.

To some extent, the strong bias against deserts in Western cultures is due to the "forest-centric" mood that has gripped the Anglo-European environmental imaginary since at least the nineteenth century.[115] Our preoccupation with trees is hindering our understanding of other environments such as deserts and leading to policies that not only frequently harm marginalized peoples but also all too often fail ecologically. We need to step outside these constraining preconceptions and try to see deserts for what they are: unpredictable, arid, and yet incredibly rich, beautiful, and productive for those who can appreciate them for what they are. In the recent words of arid lands expert Farouk El-Baz, professor of remote sensing at Boston University, we need to "let deserts be."[116] The "problem" of deserts is found mostly in our imaginations.

NOTES

1. I use the term "Anglo-European" here to denote the broad array of Western countries with long-standing interests in deserts and whose scholars, artists, and travelers made significant contributions to its representation over time. These countries include, but are not limited to: France, Britain, the United States of America, Germany, and Italy. Because global development policy in arid lands has been dominated by Western governments and policies, this essay focuses on Western governments and their colonial experiences and does not explore the development of thinking about deserts in other countries or regions where deserts are found, such as China or Central Asia, to any great degree. The focus is also on "expert" rather than "indigenous/local knowledge" about deserts because this is what has driven international policy.

2. For more details and extended analysis, see my forthcoming book, *Wasteland: An Environmental History of Deserts from the Divine to Desertification*.

3. Sharon E. Nicholson, *Dryland Climatology* (Cambridge: Cambridge University Press, 2011); S. M. Herrmann and C.F. Hutchinson, "The Changing Context of the Desertification Debate," *Journal of Arid Environments* 63, no. 3 (2005): 538–555; William M. Adams, *Green Development: Environment and Sustainability in a Developing World*, 3d ed. (London: Routledge, 2009), chapter 8; Sharon E. Nicholson, "Desertification," in *Encyclopedia of World Environmental History*, ed. Shepard Krech, John R. McNeill, and Carolyn Merchant (New York: Routledge, 2003), 297–303; J. F. Reynolds and D. M. Stafford Smith, eds., *Global Desertification: Do Humans Cause Deserts?* (Berlin: Dahlem University Press, 2002); L. C. Stringer, "Reviewing the International Year of Deserts and Desertification 2006: What Contribution Towards Combating Global Desertification and Implementing the United Nations Convention to Combat Desertification?" *Journal of Arid Environments* 72, no. 11 (2008): 2065–2074; David S. G. Thomas, "Science and the Desertification Debate," *Journal of Arid Environments* 37, no. 4 (1997): 599–608; and David S. G. Thomas and Nicholas Middleton, *Desertification: Exploding the Myth* (West Sussex, UK: John Wiley & Sons, 1994).

4. Ulf Helldén and Christian Tottrup, "Regional Desertification: A Global Synthesis," *Global and Planetary Change* 64, no. 3 (2008): 169–176. These authors note, however, that the cause and nature of the greening observed is not yet well understood.

5. Sharon E. Nicholson, C. J. Tucker, and M. B. Ba, "Desertification, Drought, and Surface Vegetation: An Example from the West African Sahel," *Bulletin of the American Meteorological Society* 79, no. 5 (1998): 815–829; Nicholson, *Dryland Climatology*; and Julie Laity, *Deserts and Desert Environments* (Oxford: Wiley-Blackwell, 2008), 67–69.

6. Andrew Goudie and John Wilkinson, *The Warm Desert Environment* (Cambridge: Cambridge University Press, 1977) and Julie Laity, *Deserts and Desert Environments* (Oxford: Wiley-Blackwell, 2008).

7. Nick Middleton, *Deserts: A Very Short Introduction* (Oxford: Oxford University Press, 2009), 2.

8. The high annual and interannual variability of rainfall in the majority of deserts creates conditions of non-equilibrium that render many conventional development approaches, such as the fenced ranching model for livestock production, invalid. The indigenous systems of extensive pastoralism (moving livestock over large spaces), is frequently the best use of land and resources in deserts. See Adams, *Green Development*, 228–232; K. Booker, L. Huntsinger, J. W. Bartolome, N. Sayre, and W. Stewart, "What Can Ecological Science

Tell us About Opportunities for Carbon Sequestration on Arid Rangelands in the United States?" *Global Environmental Change* 23, no. 1 (2013): 240–251; and James F. Reynolds, D. Mark Stafford Smith, Eric F. Lambin, et al., "Global Desertification: Building a Science for Dryland Development," *Science* 316, no. 5826 (2007): 847–850.

9. See Goudie, *Warm Desert,* for an excellent and concise overview of biophysical environments of deserts.

10. See Yitzchak Gutterman, *Regeneration of Plants in Arid Ecosystems Resulting from Patch Disturbance* (Dordrecht, Netherlands: Kluwer Academic Publishers, 2001); and Kamal A. Batanouny, *Plants in the Deserts of the Middle East* (Berlin: Springer-Verlag, 2001).

11. Laity, *Deserts,* 237.

12. Nicholson, "Desertification," see especially her detailed discussion on pp. 299–300. According to this eminent meteorologist, "the extent of desertification has never been adequately assessed" and in many cases "it is virtually impossible to separate the impact of drought from that of desertification." She points out that "attempts to portray desertification on a global scale have been made but, like the UNEP map, all lack readily measured, objective indicators. ... [and] are based on expert opinion rather than hard data." Nicholson, *Dryland Climatology,* 441. She cautions that "no techniques exist that can actually measure desertification on a global scale and objectively" (Ibid.). Thomas adds that the 1977 UN map of desertification "is in fact simply a map of the extent of dryland environments." Thomas, "Science," 604. This map, however, has been widely used to generate alarmist statistics that indicated that approximately 23 percent of "land had been affected by desertification and that perhaps another 35% of the earth's surface was at risk of undergoing similar changes." Nicholson, *Dryland Climatology,* 432.

13. Adams, *Green Development,* 216. Chapter 8 of this book provides an excellent, timely, and detailed analysis of desertification.

14. Thomas, "Science," and Nicholson, *Dryland Climatology.*

15. Herrmann, "Changing Contexts," 38.

16. Ibid.

17. Jeremy Swift, "Desertification: Narratives, Winners & Losers," in *The Lie of the Land: Challenging Received Wisdom on the African Environment,* ed. Melissa Leach and Robin Mearns (London: The International African Institute, 1996), 73–90.

18. A. Warren and L. Olsson, "Desertification: Loss of Credibility Despite the Evidence," *Annals of the Arid Zone* 42 (2003): 271–287, p. 275 cited.

19. See for example Diana K. Davis, *Resurrecting the Granary of Rome: Environmental History and French Colonial Expansion in North Africa* (Athens: Ohio University Press, 2007) and Thomas and Middleton, *Desertification.*

20. Reynolds, "Global Desertification."

21. For examples, see Adams, *Green Development,* Chapter 8; Thomas and Middleton, *Desertification*; and Diana. K. Davis, "Indigenous Knowledge and the Desertification Debate: Problematising Expert Knowledge in North Africa," *Geoforum* 36, no. 4 (2005): 509–524. Anti-desertification projects frequently fail to achieve their objectives, and have been known to harm the environment. Native vegetation, for example, is destroyed when plantations of atriplex are created. When attempted on the wrong soils, as has happened in southern Morocco, the introduced atriplex does not grow and the disrupted native vegetation can be extremely slow to become re-established. See Ibid.

22. See Diana K. Davis, "Desert 'Wastes' of the Maghreb: Desertification Narratives in French Colonial Environmental History of North Africa," *Cultural Geographies* 11, no. 4 (2004): 359–387 and Adams, *Green Development*, Chapter 8.

23. After the end of the last ice age, in the early to mid Holocene, roughly 9,000—5,500 years before the present, many deserts such as the Sahara became much wetter and more vegetated due to changing climate conditions. Much of the contemporary Sahara and Sahel regions had large lakes, tropical mammals, and much more luxuriant vegetation. Between 6,500 and 4,500 years before the present, natural climatic desiccation occurred which created, by 2,500 BCE, many of the great deserts we know today, including the Sahara, the Arabian Desert, and the Thar Desert. See Neil Roberts, *The Holocene: An Environmental History* (Oxford: Blackwell Publishers, 1998), especially 115–117, 162–163.

24. Farouk El-Baz, "Origin and Evolution of the Desert," *Interdisciplinary Science Reviews* 13, no. 4 (1988): 331–347. I am grateful to Nick Middleton for helping me to track down this reference.

25. Middleton, *Deserts*, 2. In Greek, the word *eremos* was widely used for desert.

26. Preston E. James and Geoffrey J. Martin, *All Possible Worlds: A History of Geographical Ideas* (New York: John Wiley & Sons, 1981), 21–24.

27. See Herodotus, *The History*, trans. David Grene (Chicago: The University of Chicago Press, 1987).

28. For example, see Herodotus, *History*, 143–144, where he discusses the abnormally small black men of the interior deserts of Libya. See also Eireann Marshall, "Constructing the Self and the Other in Cyrenaica," in *Cultural Identity in the Roman Empire*, ed. Ray Laurence and Joanne Berry (London: Routledge, 1998), 49–63, for many examples of how the Greeks and Romans saw the inhabitants of Libyan deserts as strange, barbaric, and "other."

29. Clarence Glacken, *Traces on the Rhodian Shore: Nature and Culture in Western Thought from Ancient Times to the End of the Eighteenth Century* (Berkeley: University of California Press, 1967), 8.

30. See Glacken, *Traces on the Rhodian Shore*, 98–100 for a discussion of these zones.

31. Glacken, *Traces on the Rhodian Shore*, 93.

32. Ibid.

33. See Benjamin Isaac, *The Invention of Racism in Classical Antiquity* (Princeton, NJ: Princeton University Press, 2006), 80. Glacken points out, however, that Strabo actually had a more complex view of environmental determinism and was not a complete determinist. He credits Strabo with a more sophisticated view and lauds his seventeen-volume book on geography as "the high point of cultural-geographical theory" in Greek and Roman writing. Glacken, *Traces on the Rhodian Shore*, 105.

34. See, for example, a passage in Job, quoted in Glacken, *Traces on the Rhodian Shore*, 156.

35. Glacken, *Traces on the Rhodian Shore*, 162.

36. Isaiah 24:4–6, quoted in Glacken, *Traces on the Rhodian Shore*, 162, emphasis mine.

37. Jeanne Kay, "Human Dominion over Nature in the Hebrew Bible," *Annals of the Association of American Geographers* 79, no. 2 (1989): 214–232, p. 217 cited.

38. Tertullian quoted in Glacken, *Traces on the Rhodian Shore*, 296.

39. Glacken, *Traces on the Rhodian Shore*, 315. This quote reveals the thorny problem of translation and the multiple meanings of "desert" in many different languages. It is highly unlikely that Albert the Great was writing of sandy, arid deserts, as was Tertullian. From Hebrew to Latin, from French to English (and likely many other languages), desert can

mean deserted of people, a wilderness, or a hot, arid desert in the modern sense. It has parallels with "waste" in the sense that it can mean an uncultivated area, but a waste (or wilderness) can apply equally to a forest, for example. Thus, problems of etymology and translation make tracing the history of deserts in Western thought challenging.

40. Kay, "Human Dominion," 223.
41. See Peter Brown, *The World of Late Antiquity, AD 150–750* (New York: W. W. Norton, 1989), 96–102.
42. James E. Goehring, "The Dark Side of Landscape: Ideology and Power in the Christian Myth of the Desert," *Journal of Medieval and Early Modern Studies* 33, no. 3 (2003): 437–451, p. 445 cited.
43. See Ibid., 445–446.
44. Ibid., 447.
45. Hugh Murray, *The Travels of Marco Polo* (New York: Harper & Brothers, 1858), 210.
46. Ibid, 216.
47. James, *All Possible Worlds*, 46.
48. Glacken, *Traces on the Rhodian Shore*, 130. Large bodies of water are known to have a moderating effect on local temperatures.
49. Ibid. Microclimate changes such as those described by Theophrastus are not unusual.
50. See Glacken, *Traces on the Rhodian Shore*, 121. Recent research, however, has demonstrated that Plato was likely incorrect in his assumptions and that subsequent interpretations of his writing on this subject were exaggerated. See A. T. (Dick) Grove and Oliver Rackham, *The Nature of Mediterranean Europe: An Ecological History* (New Haven, CT: Yale University Press, 2003), 288.
51. For a detailed discussion, see Richard H. Grove, *Green Imperialism: Colonial Expansion, Tropical Island Edens and the Origins of Environmentalism, 1600–1860* (Cambridge: Cambridge University Press, 1995), 20–21.
52. See Richard H. Grove, *Ecology, Climate and Empire: Colonialism and Global Environmental History 1400–1940* (Cambridge: The White Horse Press, 1997), Chapter One, for a concise and detailed analysis of the development of desiccation theory. His *Green Imperialism* goes into much more detail. For an analysis of some of the errors of desiccation theory and why it does not apply in large parts of the world, see Davis, *Resurrecting the Granary of Rome*, 78.
53. See Grove, *Ecology, Climate and Empire*, and *Green Imperialism*. The French led the way in much of the early thinking and action on desiccation theory. The work of Pierre Poivre in Mauritius in the late eighteenth century, as Grove details, was especially formative for later French and British conservation thought and planning.
54. This image was fostered by the artistic works of J. J. Rousseau, J. H. Bernadin de St. Pierre, and D. Defoe. See Grove, *Green Imperialism*, 62–63.
55. Ibid., 52.
56. Glacken, *Traces on the Rhodian Shore*, 437. For an excellent discussion of "tropicality" and the development of Western thinking about tropical regions since the fifteenth century, see David Arnold, *The Problem of Nature: Environment, Culture and European Expansion* (Oxford: Blackwell, 1996).
57. Glacken, *Traces on the Rhodian Shore*, 434. Bodin was a French lawyer and economist whose many books were widely read in Europe.
58. Bodin quoted in Glacken, *Traces on the Rhodian Shore*, 438.

59. Hakewill paraphrased in Glacken, *Traces on the Rhodian Shore*, 388. Hakewill was an English pastor whose writings on senescence in nature garnered the attention of many in the seventeenth century.

60. Buffon paraphrased in Glacken, *Traces on the Rhodian Shore*, 669–670. Count Buffon was an eighteenth-century French naturalist who became the director of the Jardin du Roi and was elected a member of the French Academy of Sciences. His extensive writings, especially the thirty-five-volume *Natural History*, were widely influential in Europe and elsewhere.

61. Jules Clavé, *Études sur l'économie forestière* (Paris: Guillaumin, 1862), as quoted in George Perkins Marsh, *Man and Nature: Or, Physical Geography as Modified by Human Action*, ed. David Lowenthal (Cambridge, MA: The Belknap Press of the Harvard University Press, 1965), 160.

62. Marsh, *Man and Nature*, 117. Marsh, an American, lived for many years in Europe and read widely in many languages. His book is a very useful synthesis of Anglo-European thinking about the environment in the mid-nineteenth century and was widely influential overseas as well as in the US.

63. Richard Grove, "Scottish Missionaries, Evangelical Discourses and the Origins of Conservation Thinking in Southern Africa 1820–1900," *Journal of Southern African Studies* 15, no. 2 (1989): 163–187, 180–181 cited. See also Georgina H. Endfield and David J. Nash, "Drought, Desiccation and Discourse: Missionary Correspondence and Nineteenth-Century Climate Change in Central Southern Africa," *The Geographical Journal* 168, no. 1 (2002): 33–47.

64. René Chateaubriand, "Opinion sur le projet de loi relatif aux finances," in *Oeuvres complètes de Chateaubriand*, vol. 14 (Paris: Administration de Libraire, 1852), 71. This opinion paper was read in 1817. Later an advisor to government, Chateaubriand had traveled to Egypt and the Levant in 1806.

65. Grove, *Ecology, Climate and Empire*, Chapter 3.

66. Ibid., 97.

67. Ibid., 109–111. It should be noted that Brown tended to blame settler agriculture as well as indigenous land use practices for ruining the land and the climate, especially later in his career.

68. Brown, *Crown Forests of the Cape of Good Hope*, 1887, quoted in Grove, *Ecology, Climate and Empire*, 109.

69. For details, including why this environmental narrative is incorrect, see Davis, *Resurrecting the Granary of Rome*.

70. Ibid., p. 33.

71. For details on these laws and policies, including the forest and land laws, and related policies and their effects on the local populations, see Davis, *Resurrecting the Granary of Rome*.

72. Louis Lavauden, "Les Forêts du Sahara," *Revue des Eaux et Forêts* 65, no. 6 (1927): 265–277 and 329–341, p. 267 cited.

73. Augustin Bernard and Napoléon Lacroix, *L'Évolution du nomadisme en Algérie* (Alger: Adolphe Jourdan, 1906), 42–43.

74. Auguste Warnier, *L'Algérie devant l'empereur, pour faire suite à L'Algérie devant le sénat, et à L'Algérie devant l'opinion publique* (Paris: Challamel Ainé, 1865), 28, emphasis in original.

75. Such environmental imaginaries and xenophobia were related to generalized degeneration anxieties that connected deforestation and environmental degradation with

questions of climate change, race, hygiene, and the (im)possibilities of European colonization. Quite common throughout the colonial world, such ideas led many to believe that deforestation had led to an unhealthy desiccation and a torrid climate that threatened the survival of European civilization. See Derek Gregory, "(Post)Colonialism and the Production of nature," in *Social Nature: Theory, Practice, and Politics*, ed. Noel Castree and Bruce Braun (Malden, MA: Blackwell Publishers, 2001), 84–111; Grove, *Green Imperialism*; Davis, *Resurrecting the Granary of Rome*, 72, 102, 103–106, 109–122, 149–150; Arnold, *Problem of Nature*; Nancy Leys Stepan, *Picturing Tropical Nature* (Ithaca, NY: Cornell University Press, 2001); and Caroline Ford, "Reforestation, Landscape Conservation, and the Anxieties of Empire in French Colonial Algeria," *American Historical Review* 113, no. 2 (2008): 341–362.

76. Warnier, *L'Algérie*, 28.

77. For details, see Grove, *Ecology, Climate and Empire*, Chapter 2. The surgeon Hugh Cleghorn was especially influential in these efforts.

78. Ibid., 30.

79. For a good discussion of this process, see David Gilmartin, "Models of the Hydraulic Environment: Colonial Irrigation, State Power and Community in the Indus basin," in *Nature, Culture, Imperialism: Essays on the Environmental History of South Asia*, ed. David Arnold and Ramachandra Guha (Delhi: Oxford University Press, 1998), 210–236. For details on the disastrous consequences of colonial irrigation in India, see Elizabeth Whitcombe, "The Environmental Costs of Irrigation in British India: Waterlogging, Salinity, Malaria," in *Nature, Culture, Imperialism: Essays on the Environmental History of South Asia*, ed. David Arnold and Ramachandra Guha (Delhi: Oxford University Press, 1998), 237–259.

80. Neeladri Bhattacharya, "Pastoralists in a Colonial World," in *Nature, Culture, Imperialism: Essays on the Environmental History of South Asia*, ed. David Arnold and Ramachandra Guha (Delhi: Oxford University Press, 1998), 49–85, p. 72 cited. On British ideas of improvement, see also Richard Drayton, *Nature's Government: Science Imperial Britain, and the "Improvement" of the World* (New Haven, CT: Yale University Press, 2000).

81. Western thinking about deserts is tightly bound to related negative notions about pastoralism and the primitive level of nomads in the "stages of civilization" that have a deep history. For more details, see Davis, *Resurrecting the Granary of Rome*, 65–66, and John Noyes, "Nomadic Fantasies: Producing Landscapes of Mobility in German Southwest Africa," *Ecumene* 7, no. 1 (2000): 47–66.

82. Bhattacharya, "Pastoralists," 56.

83. Ibid, 69, 83. Not only nomadic pastoralists were criminalized with this act, but also merchants and others who needed to be mobile for their livelihoods.

84. For more details, see the excellent discussion in Priya Satia, "'A Rebellion of Technology': Development, Policing, and the British Arabian Imaginary," in *Environmental Imaginaries of the Middle East and North Africa: History, Policy, Power & Practice*, ed. Diana K. Davis and Edmund Burke (Athens: Ohio University Press, 2011).

85. Ibid.

86. Ibid.

87. Ibid.

88. Diana K. Davis, "Power, Knowledge and Environmental History in the Middle East," *International Journal of Middle East Studies* 42, no. 4 (2010): 657–659.

89. For more details see Grove, *Ecology, Climate and Empire*, 32–36.

90. Edward W. Bovill, "The Encroachment of the Sahara on the Sudan," *Journal of the Royal African Society* 20, no. 79 (1921): 174–185, 259–269; Henri Hubert, "Le désseche-ment progressive en Afrique Occidentale," *Bulletin de la Comité d'Études Historiques et Scientifiques de l'Afrique Occidentale Française 1920*, (1920): 401–437; and Lavauden, "Les Forêts."

91. For Asia, see C. Coching, "Climatic Pulsations During Historic Times in China," *Geographical Review* 16 (1926): 274–282; and Peter Kropotkin, "The Desiccation of Eur-Asia," *Geographical Journal* 23 (1904): 722–741.

92. Henry Nash Smith, *Virgin Land: The American West as Symbol and Myth* (New York: Vintage Books, 1950), 202.

93. See, for example, Alexis de Tocqueville, *Quinze jours dans le désert américain* (Paris: Mille et Une Nuits, 1998 [1860]).

94. See the interesting discussion in Patricia N. Limerick, *Desert Passages: Encounters with American Deserts* (Albuquerque: University of New Mexico Press, 1985), Chapters 1–4. See also the more detailed discussion of the evolution of American views of deserts and the West in general in Smith, *Virgin Land*. As Smith points out, not all views of the arid lands of the American West were negative, especially when the person had a vested interest in promoting settlement in the region, as was the case with William Gilpin. Smith, *Virgin Land*, 38–46. There were similar boosters who proclaimed even the desert regions of Algeria as fertile lands for settlement in the nineteenth century. Davis, *Resurrecting the Granary of Rome*.

95. For excellent analyses of irrigation in the American West, see Donald Worster, *Rivers of Empire: Water, Aridity and the Growth of the American West* (New York: Pantheon Books, 1985) and Tyrrell, Ian, *True Gardens of the Gods: Californian-Australian Environmental Reform, 1860–1930* (Berkeley: University of California Press, 1999). In the American West (and Australia), reforestation was often seen as a key part of "reclaiming desert wastes" along with irrigation, because desiccation theory was widely embraced. See Tyrrell, *True Gardens*, 107–109. There are fascinating parallels between European colonial forestry and American forestry, especially in the West. See Davis, *Resurrecting the Granary of Rome*; Ravi Rajan, *Modernizing Nature: Forestry and Imperial Eco-Development, 1800–1950* (Oxford: Clarendon Press, 2006); F. B. Hough and N. H. Egleston, *Report on Forestry* (Washington, DC: Government Printing Office, 1878); and Chapter 6 in this volume.

96. These laws included the 1862 Homestead Act, the 1873 Desert Lands Act, and the 1902 Newlands Reclamation Act, among others. See Paul F. Starrs, *Let the Cowboy Ride: Cattle Ranching in the American West* (Baltimore: The Johns Hopkins University Press), especially chapter 3, and also Worster, *Rivers of Empire*.

97. See Starrs, *Let the Cowboy Ride*, 58–59. See also Jake Kosek, *Understories: The Political Life of Forests in Northern New Mexico* (Durham, NC: Duke University Press, 2006).

98. Ibid., 36.

99. For example, compare Chapters 7 and 8 of Karl Jacoby, *Crimes Against Nature: Squatters, Poachers, Thieves, and the Hidden History of American Conservation* (Berkeley: University of California Press, 2001) with Davis, *Resurrecting the Granary of Rome*. For similar (colonial) treatments of indigenous peoples in the American West, see also Marsha Weisiger, *Dreaming of Sheep in Navajo Country* (Seattle: University of Washington Press, 2009) and Andrew Isenberg, *The Destruction of the Bison: An Environmental History, 1750–1920* (Cambridge: Cambridge University Press, 2000). Common policies included sedentarization of mobile peoples (e.g. nomads), state appropriation of forests, criminalization

of grazing or gathering in forests, outlawing the use of fire in forests or for agricultural purposes, and land privatization, among others. Many if not most of these actions were justified with arguments that the local people were ruining the environment in one way or another (overgrazing, deforestation, desertification, etc.).

100. For the American Dust Bowl, see Donald Worster, *Dust Bowl: The Southern Plains in the 1930s* (Oxford: Oxford University Press, 1979) and William Cronon, "A Place for Stories: Nature, History and Narrative," *The Journal of American History* 78, (1992): 1347–1376. There is an extensive literature on the Dust Bowl that is very interesting and worth exploring.

101. See, for instance, Kate B. Showers, *Imperial Gullies: Soil Erosion and Conservation in Lesotho* (Athens: Ohio University Press, 2005). For related discussions, see William Beinart, "Soil Erosion, Conservation and Ideas about Development: A Southern African Exploration," *Journal of Southern African Studies* 11 (1984): 52–84, and David M. Anderson, "Depression, Dust Bowl, Demography and Drought: The Colonial State and Soil Conservation in East Africa during the 1930s," *African Affairs* 83 (1984): 321–344.

102. See Edward P. Stebbing, "The Encroaching Sahara: The Threat to the West African Colonies," *Geographical Journal* 85, no. 6 (1935): 506–524. In the United States, Paul Sears, who wrote *Deserts on the March*, performed a similar function in raising the public's concern about desertification. Sears roundly condemned nearly every major culture group around the globe for environmental degradation and upsetting the "balance of nature." See Paul B. Sears, *Deserts on the March* (Norman: University of Oklahoma Press, 1935). This widely read book went through four editions, and the original 1935 edition was reprinted with a eulogistic preface in 1988.

103. See, for example, some of the results of the Anglo-French Forestry Commission in Brynmor Jones, "Desiccation and the West African Colonies," *Geographical Journal* 41, no. 5 (1938): 401–423. See also the excellent discussion in Adams, *Green Development*, Chapter 8.

104. See Graham V. Jacks and Robert O. Whyte, *The Rape of the Earth: A World Survey of Soil Erosion* (London: Faber and Faber, 1939), and Jean-Paul Harroy, *Afrique terre qui meurt: La dégradation des sols africains sous l'influence de la colonisation* (Bruxelles: Marcel Hayez, 1944).

105. See Diana K. Davis, "Scorched Earth: The Problematic Environmental History that Defines the Middle East," in *Is There a Middle East*, ed. Abbas Amanat, Michael Bonine, and Michael Gasper (Stanford, CA: Stanford University Press, 2011), 183.

106. James C. Scott, *Seeing like a State: How Certain Schemes to Improve the Human Condition have Failed* (New Haven, CT: Yale University Press, 1998). See also Weisiger, *Dreaming of Sheep*, for a detailed account of forced destocking and effective sedentarization in the American West.

107. Thomas and Middleton, *Desertification*, 23.

108. For an excellent explanation, see Michael J. Watts, *Silent Violence: Food, Famine and Peasantry in Northern Nigeria* (Berkeley: University of California Press, 1983).

109. For fascinating details, see Thomas and Middleton, *Desertification*, and Adams, *Green Development*, Chapter 8. Nicholson points out that climate was intentionally excluded from the UNCOD statements, and from most research on, and definitions of, desertification from the 1970s until 1994, when the UNCCD revised their definition of desertification. See Nicholson, *Dryland Climatology*, 434–435. This helps to explain the overwhelming focus of anti-desertification research on curtailing the "damage" of

traditional land management techniques despite the fact that, as has been pointed out by many researchers, "it is virtually impossible to separate the impact of drought from that of desertification." Ibid., 433.

110. Swift, "Desertification."

111. Thomas and Middleton, *Desertification*, 49.

112. Ibid., 160. Other scientists, including meteorologist Sharon Nicholson, have corroborated the significant lack of data with which to assess desertification. Nicholson notes that, "no techniques exist that can actually measure desertification on a global scale directly and objectively." Nicholson, *Dryland Climatology*, 441. See also S. R. Verón, J. M. Paruelo, and M. Oesterheld, "Assessing Desertification," *Journal of Arid Environments* 66, no. 4 (2006): 751–763.

113. R. H. Behnke, P. M. Doll, J. E. Ellis, et al. "Responding to Desertification at the National Scale," in *Global Desertification: Do Humans Cause Deserts?*, ed. J. F. Reynolds and D. M. Stafford Smith (Berlin: Dahlem University Press, 2002), 357–385, p. 360 cited.

114. Key to the success of extensive pastoralism, however, are social, political, and econo-moc mechanisms that support the decision-making of the herders, rather than policies that hinder their best use of arid rangelends. See Emery Roe, Lynn Huntsinger, and Keith Labnow, "High Reliability Pastoralism," *Journal of Arid Environments* 39, no. 1 (1998): 39–55.

115. See Davis, *Resurrecting the Granary of Rome*, 63–66.

116. Farouk El-Baz, "Let Deserts Be," *Nature* 456, no. 30 (2008), doi: 10.1038/twas08.30a.

CHAPTER 5

....

SEAS OF GRASS

Grasslands in World Environmental History

....

ANDREW C. ISENBERG

I.

....

THANKS to the work of William McNeill, Alfred Crosby, and other scholars, we have long understood that European navigators' trans-oceanic crossings, beginning in the late fifteenth century, marked a decisive moment in world environmental history. As these leading environmental historians have shown, the voyages of Christopher Columbus, John Cabot, Ferdinand Magellan, and others initiated exchanges of germs, plants, and animals that transformed formerly isolated environments. European colonists' introduction of smallpox, measles, and other deadly crowd diseases devastated the indigenous populations of the Americas, Australia, and New Zealand; the wheat, sheep, horses, and other domesticated flora and fauna the Europeans introduced transformed overseas environments. Explorers returned home with commodities including potatoes, beans, maize, cacao, and tobacco that changed Europeans' diets, health, and environments in significant ways.[1]

Yet Columbus, Cabot, and other late fifteenth-century sailors were not the first voyagers to transgress the environmental obstacles that had long separated human societies. Between the third and fourteenth centuries, waves of nomads on horseback sailed across the Old World on a sea of grass between China in the east and Hungary in the west—the nomads' mastery of the horse and of the semi-arid grassland environment presaged sailors' mastery of the winds and the ocean. Like Columbus and other seafarers, mounted nomads exerted a decisive influence in world history. Between late antiquity and the early modern age, they battered the empires of China, India, Persia, and Europe. The nomads' accidental ecological influences were arguably as significant as their conquests; just as a member of Hernán Cortés's retinue inadvertently introduced smallpox to Mexico in 1519, two centuries earlier, a caravan accidently transmitted bubonic plague from the Asian Steppes to Europe. The resulting Black Death killed perhaps one-third of Europe's population.[2] In light of these influences, some historians, especially those striving for a less Eurocentric vision of world history, mark the beginning of the modern age not with the trans-Atlantic crossings of the late fifteenth century, but with the trans-Steppes movements of the fourteenth.[3]

Yet the Steppes and other temperate grasslands have been more than just pathways between environments; they have been the scenes of critical—even iconic—episodes in environmental history, including, in North America, the near-extinction of the bison in the nineteenth century and the Dust Bowl of the 1930s. These North American episodes were by no means unique to the world's grasslands: while the bison nearly became extinct in the Great Plains, European colonists destroyed the quagga and blue buck in South Africa's grasslands; and the wheatlands of South Africa, the Russian Steppes, and the Argentine Pampas have suffered droughts and soil erosion much like in the North American Dust Bowl. As these cases exemplify, grasslands epitomize the complex interplay of dynamic environments and human societies—a critical, if not the central, problem in environmental history. When things have gone awry in that complex interaction, dust storms, erosion, and species extinction have resulted.

Much of what environmental historians have written on the subject of grasslands concerns those catastrophes. They have drawn different conclusions about them—differences that reflect conflicting cultural ideas of grasslands. Like the grasslands themselves, which can be verdant one year and dust-choked the next, conceptions of grasslands in Old and New World cultures have varied widely. Algerians divide the North African grassland lying between the desert and the Mediterranean Sea into *tell* (well-watered farmland) and *sahara* (dryland suitable only for pastoralism). Yet the division is based as much on perception and prevailing land use as on climate.[4] In the United States, the Great Plains have been understood variously as "America's Breadbasket," the "Great American Desert," and as the potential site of a "Buffalo Commons." These cultural constructions of nature shape perceptions of ecological change and support some types of resource use while contesting others; at they same time they are in part derived from a changing environment and shifting land use. The Dust Bowl is a case in point: prevailing cultural perceptions of the southern Great Plains as a wheatland merely waiting for farmers to unlock its potential encouraged the influx of farmers into the region in the early twentieth century; the dust storm of the 1930s encouraged the perception of the region as a desert wasteland.

The importance of grassland environments has drawn many environmental historians to the subject. Grasslands, particularly the North American Great Plains, have been the subjects of influential histories that have addressed the problem of understanding the impact of human habitation on a dynamic environment. One of the first historians to write about grasslands, Walter Prescott Webb, completed his study, *The Great Plains*, in 1931, on the eve of the dust storms in the southern Great Plains. For Webb, aridity was the region's determining characteristic. His work paralleled that of his contemporaries, the geographers Carl Ortwin Sauer and Lucien Febvre, who emphasized the inherent conditions of regional environments. In Webb's view, the semi-arid environment imposed harsh limits on human habitation of the grasslands. According to Webb, the Great Plains' suffocating aridity stalled the development of native societies and frustrated Spanish and Mexican attempts at settlement, leaving the region to the Anglo-Americans, whom he saw as more technologically sophisticated.[5]

Succeeding generations of historians have taken different lessons from Webb's *Great Plains*. The historian James Malin took up as his main subject the last part of Webb's narrative: the arrival of Anglo-Americans in the Great Plains. For Malin, the semi-arid environment did not impose limits to Anglo-American settlement; rather, it posed a challenge to settlers' ingenuity. It was not the land that made settlers in the grasslands prosperous, Malin argued, but human innovation; the semi-arid environment of the Great Plains simply called for greater innovation. When Red May wheat failed in the dry climate, for instance, settlers switched to Turkey wheat; when the use of moldboard plows led to wind erosion of the soil, farmers shifted to listers and created high ridges of soil crosswise to prevailing winds. In the 1930s, Malin was a strident critic of New Deal policies that sought to restrain farmers from cultivating marginal lands in the Great Plains—to him, this was a stifling of human innovation and technology. Later, during the Cold War, he extended his objection to other forms of state regulation, arguing that such limitations would lead inevitably to some form of totalitarianism. These views never found much purchase among environmental historians, but some, such as Matthew Paul Bonnifield in his 1979 study, *The Dust Bowl*, have endorsed his view that the semi-arid grassland environment is not a limit to settlement, but rather a challenge.[6]

The environmental historian Donald Worster embraced a different lesson from Webb's work: the semi-arid grassland environment imposed limits that people ignore at their peril. While Webb maintained that aridity frustrated all those who preceded Anglo-Americans, Worster argued that the Great Plains climate was too severe even for Anglo-American industry; indeed, in a rebuke to Malin's faith in technology, he maintained that tractors and other farm machinery only made the grassland more inhospitable. For Malin, mechanized farming was an innovation that was the source of human prosperity and freedom; for Worster, it was a capitalist culture's destructive, short-sighted way of maximizing the resources of the soil. The capitalist ethos, Worster argued, has no view of nature other than as a source of wealth, no restraint in exploiting that wealth, and no plan for what to do when that wealth was exhausted. This culture, in the drought-prone Great Plains, produced the Dust Bowl of the 1930s.[7]

Two decades ago, the environmental historian William Cronon cast the competing histories of Malin/Bonnifield and Worster, in light of then-current postmodern theory, as a choice between progressive and declensionist narratives.[8] Indeed, many historians of the Great Plains and other grasslands have constructed their narratives as stories of progress or decline. Yet these are not the only ways to conceive of changes in the grasslands. Looking not to narrative theory but to recent work in the ecology of grasslands, one finds not unidirectional trajectories of progress or decline, but dynamic environments that shifts unpredictably and without regard for human purpose.

A scientific understanding of grasslands as dynamic places was not always the case. Before ecologists came to think of the grasslands as changeable, they saw them as inherently stable. According to the "climax community" model that ecologists developed in the early twentieth century, every regional environment proceeds through a succession of stages toward an ideal steady state, in which its plant and animal constituencies are at equilibrium. The primary developer of the climax model, Frederic Clements of the University of Nebraska, based much of his work on his observations of the grasslands. He also drew on the German botanist Oscar Drude's 1896 study, *Deutschlands Pflanzengeographie*, which emphasized the suitability of particular plants to their environments. Clements's successor at the University of Nebraska, John Weaver, believed like Clements that in the absence of human interference, grasslands would remain almost entirely unchanged. The prairie, Weaver wrote, "is a slowly evolved, highly complex organic entity, centuries old. It approaches the eternal."[9]

Clements and Weaver were never completely unopposed; working at the same time as Clements, Henry Chandler Cowles posited an ever-changing grassland environment in which constituent parts struggled for dominance. By the later twentieth century, most ecologists had discarded Weaver's belief that regional environments strive toward stability and equilibrium and had adopted the idea that disequilibrium is inherent in nature. Theoretical ecologists have a term for this unpredictable change: *chaos*. Drawing attention to the complexity and interconnectedness of natural systems, chaos theory emphasizes the multiplicity of ever-changing factors that influence outcomes, producing systems that are unpredictably complex. The meteorologist Edward Lorenz, one of the founders of chaos theory, focused much of his work on the problem of weather forecasting, which is famously impossible to do accurately because weather systems are complex, interconnected, and sensitively dependent on small changes. Pleasant weather on a spring day in the Great Plains might be followed by a similarly pleasant day; or, owing to just a few differences of micro-degrees in temperature, barometric pressure, and wind speed, the next day might bring a tornado. At the same time, chaos theoreticians insist that natural systems, however complex and unpredictable, adhere to certain rules and order—chaos is not to be confused with randomness. Also called non-equilibrium dynamics, adherents of chaos theory are found not only in ecology but in mathematics, physics, economics, chemistry, and biology as well.[10]

Some of the formative work within theoretical ecology on chaos theory dealt with how climatic change and reproduction dynamics produce disequilibrium between grazers and grassland forage. Rather than proceeding through a series of stages toward equilibrium, ecologists who studied the "irruptive paradigm" in grassland-herbivore relations, for example, suggested that grasslands and the mammals that graze on them are inherently at disequilibrium. When cattle or sheep are introduced to a previously ungrazed pasture, for instance, their populations, if unmanaged, rise sharply in the resource-rich environment, only to crash when populations exceed the carrying capacity of the range. Even in the absence of such introductions, ungulate populations are always moving upward or downward—sometimes quite rapidly so—and never at equilibrium.[11]

Despite the emergence of chaos theory within academic ecology, with a few exceptions environmental historians have been slow to embrace chaos theory.[12] That reluctance is understandable; as William H. McNeill, one of the first environmental historians, wrote in 1976, "We all want human experience to make sense, and historians cater to this universal demand by emphasizing elements in the past that are calculable, definable, and, often, controllable as well."[13] Yet chaos does make sense for environmental historians, if they think of it as *contingency*, a principle of academic history that reminds us that many factors contribute to historical change, that context influences the direction of change, and above all that outcomes are unpredictable. For environmental historians, chaos theory steers us away from arbitrarily imposed, uni-directional trajectories or progress and decline and lets us take account of the contingencies of historical change. It offers something richer and more complex than the paradigms of progress or decline; it acknowledges the agency of the environment without veering into environmental determinism; it recognizes the adaptability of human societies while also recognizing that adaptation is neither limitless nor a guarantor of a happy outcome; it understands that however ill-advised some schemes have been (one thinks of wheat farming in the southern Great Plains), catastrophe is not inevitable.

II.

Grasslands are found the world over; together with deserts (biomes with which grasslands often overlap and that many geographers and climatologists categorize together) they are the world's most common clime. By one estimate, grasslands cover one-third of the Earth's landmasses.[14] The largest grasslands are found in Asia (from Ukraine in the west to Mongolia in the east) and North America (from Canada south along the eastern foothills of the Rocky Mountains to Mexico); other grassland expanses are found in Argentina, Australia, and Africa.[15]

The trophic dynamics of grasslands constitute what may be called a *solar economy*. Grasses absorb the energy of the sun and, through photosynthesis, transform part of that energy into carbohydrates. Indigenous grazing animals (bison in North America, guanaco in South America; wildebeests and zebras in Africa), transform some of those carbohydrates into protein. Predators (wolves in North America, lions and their relatives in Africa, human beings throughout) hunt grazing animals and consume their protein. The introduction of domesticated grasses such as wheat, rye, and alfalfa and grazing animals such as cattle have altered and domesticated the floral and faunal constituency of grasslands without changing their basic trophic structure.[16] Some energy from the sun is lost as it moves from one trophic level to another—from grasses to herbivores to carnivores—so at each trophic level above grasses there exists less chemical energy in the form of food. A grassland ecosystem is thus pyramidal in structure: extensive grasses support a sizeable but smaller biomass of grazers, who in turn support a smaller number of predators. Nonetheless, because the energy of the sun feeds the grasses at the base of the pyramid, theoretically the system is indefinitely renewable, so long as the population of grazing animals does not outstrip the supply of forage, and predators (human or otherwise) confine their harvest of grazing animals to a sustainable level. A domesticated version of a grassland's trophic system, with domesticated plants such as wheat and alfalfa substituted for native grasses, domesticated livestock in place of wildlife, and herders and farmers in place of hunters and gatherers, is, theoretically at least, equally sustainable.[17]

Yet grasslands worldwide—both those characterized by indigenous grasses and wildlife and those transformed by ranching and farming—have been characterized not by renewable stability, but by unpredictable change. In the North American Great Plains alone, the near-extinction of the bison in the early 1880s was followed in short order by the collapse of the free-range cattle industry later in the decade and the Dust Bowl of the 1930s. Likewise, in the Argentine Pampas, Asian Steppes, and in Africa and Australia, disappointment and defeated expectations have punctuated the history of the grasslands. Why has this been so?

Three factors stand out. First, although the great extent of some grasslands (notably the Eurasian Steppes and the North American Great Plains) and the millions of grazing animals that inhabit grasslands make them seem almost inexhaustibly productive, they are climatologically marginal places for both grazing animals and human beings looking to eke out a living from the land. As anyone who has gardened or tended a houseplant knows, plants require not only solar energy but water. Grasslands are, by definition, places that receive relatively little precipitation. The North American Great Plains, the Argentine Pampas, and large expanses of the Asian Steppes are located on the leeward side of mountain ranges; the high mountains cast so-called *rain shadows* that inhibit the flow of moisture-bearing air currents. In such regions, where average annual precipitation is 500 mm (20 inches) or less, grasses are the dominant vegetation because their above-ground structures are minimal (and thus require minimal moisture to maintain) while their dense roots close to the surface take advantage of available soil moisture. Trees are rare outside of river valleys. Grasslands, defined by their stunted vegetation, demand that expectations be scaled down.[18]

A second important reason why grasslands have so often failed to meet people's expectations is that, as ecologists who have embraced the lessons of chaos theory have shown, grassland environments are not as regular as the solar economy model implies. Precipitation in grasslands typically is not only scarce but unpredictable. In the Great Plains, a few rainy years in a row are as likely as not to be followed by a period of drought. Intervals between droughts can be anywhere from fifteen to thirty years; the droughts themselves can last anywhere from a year to ten years or longer. The infamous drought in North America in the 1930s that helped cause the Dust Bowl is perhaps the most well-known, but it was by no means an anomaly in the Great Plains. Dendrochronological studies dating to the mid-sixteenth century show that precipitation in the Great Plains has been extremely variable; the region has been characterized by many periods of low precipitation lasting ten years or longer.[19] "Herein lies the major problem of steppe regions," wrote the geographers Robert Gabler, Robert Sager, Sheila Brazer, and Daniel Wise. They "seem like better-watered deserts at one time and like slightly subhumid versions of their humid climate neighbors at another."[20]

Thirdly, grasslands have been places of defeated expectations because they are not isolated from other environments. Nomads and emigrants, from third-century Turkic raiders in the Eurasian Steppes to nineteenth-century Argonauts in the North American Great Plains, traversed these inland seas of grass. As different human populations passed through or invaded grasslands, they brought with them their own labor power, new technologies to exploit the resources of the grasslands, and exotic ecological elements, both floral and faunal, which often displaced indigenous plants and animals. These ecological and economic invasions have sometimes destabilized regions that are already, given their marginal and drought-prone character, inherently unstable.[21]

By and large, people who relocated to grasslands saw themselves not as invaders but as tamers. Beginning with the domestication of the horse in the Eurasian Steppes roughly six thousand years ago, human societies across the world embarked upon a long and still ongoing effort to domesticate grasslands, or in other words, to harness the trophic system of grasslands to the regular production of grain, meat, milk, hides, and other products. Domestication of the horse enabled pastoralists to bring other domesticated grazing animals—primarily cattle and sheep—deep into the Steppes. First in the Asian Steppes and later in other grasslands of the world, settlers decimated wildlife populations, replacing them with domesticated species. Eventually, those settlers replaced native grasses with domesticated plants—oftentimes domesticated grasses such as wheat and alfalfa designed to produce, respectively, grain for people and feed for livestock. Increasingly elaborate technologies aided the domestication effort: wheels, saddles, plows, railroads, central pivot irrigation, chemical fertilizers. Yet the domestication of inherently dynamic environments proved elusive. Technologies intended to regularize and pacify grasslands have sometimes only further destabilized already marginal biomes. In a large sense, the disappointments and defeated expectations that have often accompanied human habitation of grasslands have represented failures to domesticate inherently chaotic environments.

III.

Before the domestication of the horse, the big mid-continental grasslands—the Steppes, Plains, and Pampas, for instance—defied most human settlers. These grasslands were too unpredictable and moreover too vast for continuous human habitation. The grasslands abounded with mammals—bison and pronghorn antelope in the Great Plains, and Saiga antelope and Przewalski's gazelle in the Eurasian Steppe, for example—but pursuing such game on foot across extensive grasslands was an effective strategy for only the most skilled hunters. Rather, people rimmed the edges of grasslands, making seasonal forays into them to gather and hunt (usually when grazing mammals assembled for their mating season) but always returning home to the forests and farms on the margins of the grasslands.[22] They were not unlike maritime peoples who made regular expeditions from their coastal homes to fish.

The conquest of the world's grasslands began when human beings domesticated the horse somewhere in the Eurasian Steppes—perhaps in Ukraine, perhaps in Kazakhstan—between 4000 and 3500 BCE; domestication likely happened independently in several locales. Both the place(s) and the date(s) of the domestication of the horse made it distinctly unlike the domestication of sheep, goats, and cattle, all of which were tamed between 8000 and 6000 BCE in Southwest Asia.[23] Initially, horse husbandry was part of a mixed economy including farming; horses in such an economy were a source of transport but also of milk and meat. As mixed farmers, those who raised horses continued to inhabit the margins of the Steppes. They could not branch out into the interior grasslands until a technological innovation—the wheel—enabled them. The invention of wheeled transport followed shortly after the domestication of the horse, sometime after 3500 BCE. Wagons opened up the Steppes: Eurasian pastoralists who had grazed their herds of sheep and goats on the edges of the grassland could, with the aid of wagons, penetrate the Steppes and transport shelter and supplies to pastures deep in the interior; just as importantly, they could shift to new pastures when their herds exhausted the available forage. Access to the extensive pastures of the Steppes meant, in turn, larger herds.[24]

The domestication of the horse and the technology that opened the resources of the grasslands drew numerous people from the fringe of the Steppes—Chinese farmers and Siberian hunters, for instance—into the grasslands to become nomadic pastoralists. The Eurasian Steppes were, in many ways, an ethnic melting pot. The grasslands were also an incubator of technology: by 900 B.C., people of the Steppes had developed the bit and bridle to guide a horse, as well as the horned saddle and stirrups, and a short bow to use from horseback.[25] While these technologies helped make the grasses of the Steppes accessible to mounted herders, it did not make the grasslands a predictable or reliable resource. Superficially, at least, the Steppe grasses provided a seemingly inexhaustible supply of forage for what the Mongols called "the five animals": sheep, goats, cattle, horses, and camels. (Of these, the horse and sheep were most important: the sheep as a source of meat, milk, and wool; the horse for transportation.) Yet seasonal fluctuations in the grassland environment imposed limits on the number of grazers. As a Kazakh proverb puts it, "Sheep are fat in the summer, strong in the autumn, weak in the winter, and dead by spring"—reflecting how the thick Steppe grasses of the summer withered in the colder months. Taking account of these seasonal limits, nomadic groups dispersed in the winter and spring in the hope that by reducing the density of their herds, they would lessen the likelihood of overgrazing.[26]

The limits of the Steppe environment also compelled pastoral nomads to draw on resources outside the grasslands, either through commerce or conquest. Almost from the time they first emerged, pastoral nomads of the Steppes exchanged their products—meat, hides, and wool—for the products of agricultural communities on the periphery of the Steppes. By the first century BCE, Steppe traders had elaborated these trans-ecological exchanges into a commercial network that extended from Europe to China—the so-called "Silk Road." (The term is a misnomer: it was in fact many roads, and most traders operated caravans only along one stretch.) Trade caravans across the Steppes were conduits not only for commodities but for germs; trade routes across the central Asian grasslands were the primary pathways for the microbes that cause smallpox, measles, and bubonic plague to move from formerly isolated disease pools to virgin populations of new human hosts. Inhabitants of Europe and China, according to William McNeill, "were in an epidemiological position analogous to that of Amerindians in the later age: vulnerable to socially disruptive attacks by new infectious diseases." Epidemics—probably smallpox, perhaps measles—struck both China and the Roman Mediterranean in the second, third, and early fourth centuries CE.[27]

The ecological limits of the Steppes, which encouraged nomads to seek other resources, impelled them not only to trade but to conquest. The people of more densely populated regions in China, India, and Southwest Asia disparaged the Steppe pastoralists as barbarians, but as early as the first millennium BCE the nomads possessed distinct military advantages over their sedentary neighbors. On horseback, nomads were more mobile than any other military force at the time; they could outflank other armies, or charge through infantry. Their military advantage was nearly absolute so long as horses remained rare and horsemanship relatively inferior outside of the Steppes. Accordingly, the military prowess of the central Asian nomads was infamous: Attila, Genghis Khan, and Tamerlane assembled cavalries that overawed peripheral regions.[28]

There were two great waves of nomadic invasions of city-states on the periphery of the Steppes. The first was between 200 and 550 CE. In China, Turkic and Mongol nomads conquered the Han Empire in 222; the invaders ruled until 316, when another wave of nomads invaded. The Huns crossed the Volga in 372; under their leader Attila they pillaged Gaul, the Balkans, and the Italian Peninsula between 434 and 453. While the western Huns moved into Europe, the eastern Huns moved eastward and southward toward India and conquered the Gupta Empire between 500 and 550. In short, the nomads of the Steppes helped to finish off the Roman, Han, and Gupta empires.

In the second wave of conquest between 1000 and 1500 CE, Turkic and Mongol nomads came to rule the wealthy societies on their fringes. Those conquests emerged, in part, from the ecological limitations of the Steppes. In the late twelfth and early thirteenth centuries—at the beginning of a period many environmental historians call the "Little Ice Age"—a prolonged period of cooler temperatures stunted grasses and impelled Mongols under Genghis Khan to conquest.[29] Genghis Khan, with his Mongol army, overran China north of the Great Wall in 1211; in 1215 he captured Beijing; he conquered Korea in 1218; then his army moved westward and conquered northern India; between 1219 and 1223 he conquered the Khwarizm sultanate (modern Iran and Iraq); in 1223 the Mongols defeated a Russian army in the Ukraine. Genghis Khan died in 1227 but his successors continued the conquests, conquering Moscow in 1237, Kiev in 1238, and Zagreb, the modern capital of Croatia, in 1241. At the other end of the empire, Genghis Khan's grandson, Kublai Khan, consolidated the conquest of China and campaigned against Indochina, Burma, Java, and Japan. Descendants of Genghis Khan ruled China, Persia, and most of Russia until the late fourteenth century.[30]

As the nomads gradually moved from the grasslands into the settled periphery, Russian tsars began to reassert their control of the Steppes beginning in the mid-sixteenth century. By the beginning of the eighteenth century, thousands of Russians had settled in the Steppes, plowed up native grasses, and sowed wheat.[31] Russia's colonization of the Steppes was part of a global expansion of European colonists into the world's temperate grasslands beginning in the sixteenth and seventeenth centuries, as Europeans sought to control more resources to produce commodities to supply an expanding population. Since the onset of European colonization in the sixteenth century, both the Eurasian Steppes and New World grasslands have reeled from a series of invasions of flora and fauna: horses, cattle, sheep, and wheat among them.[32] The invasion of the grasslands was not merely ecological. Just as wheeled transport enabled access to the resources of the ancient Steppes, so too new technologies—plows, steam, hydraulic engineering—opened up the resources of grasslands in the early modern and modern age.

The Pampas of South America was the first of the overseas grasslands to feel the impact of European colonization. The Pampas (from a Quechua word meaning "plains") overspread much of Argentina from the Atlantic Ocean to the foothills of the Andes. To the east, around Buenos Aires, the region receives considerable rainfall; most of Argentina's population is concentrated in this humid flatland. Inland, nearer to the Andes, the Pampas become a semi-arid grassland that receives only about 250 mm (10 inches) of precipitation a year. While most Spanish settlers avoided the dry Pampas, Spanish livestock, which missionaries first introduced in 1580, thrived in the semi-arid grassland. By 1700, millions of wild or semi-wild cattle and horses grazed in the Argentine grasslands. Heavy trampling and grazing by these exotic fauna destroyed many native grasses, which were soon replaced by Old World imports better adapted to livestock; by the 1920s, only one-quarter of the Pampas grasses were native species. Such an explosion of exotic grasses in the wake of the disturbances caused by Old World livestock was a hallmark of European ecological imperialism in New World grasslands.[33]

This ecological conquest decimated native flora, but native societies proved more adaptable and resilient. Indeed, the ecological incursion of horses and cattle into the Pampas following European colonization created opportunities for native groups in Argentina such as the Tehuelches, Puelches, and Querandi analogous to the domestication of the horse in the Eurasian Steppes: the natives abandoned former land use strategies and became mounted nomads. By the early seventeenth century, native groups possessed more horses than the Spanish. They subsisted by hunting cattle which had gone feral; the guanaco, a relative of the Andean llama; and the rhea, a large, flightless bird. The natives adopted some Spanish equestrian tools: the lasso and the hocking blade to hamstring cattle. They also contributed their own technologies, notably the bola. Like the Mongols and the natives of the North American Great Plains, the adoption of the horse momentarily lifted the natives of the Pampas to regional dominance. Until the last quarter of the nineteenth century, native groups controlled the dry Pampas (sometimes known as the Araucanian region after its largest native inhabitants) and its horses and feral cattle.[34]

In the North American Great Plains, natives adapted to the introduction of the horse in ways strikingly similar both to the changes in Argentina between the late sixteenth and late nineteenth centuries, and to the emergence of nomadic societies in the Eurasian Steppes thousands of years earlier. When Europeans first came to North America, the Blackfeet, Cheyenne, Comanche, Crow, Sioux, and others subsisted on the fringes of the grasslands. While they traveled seasonally to the Great Plains to hunt bison from foot, they relied primarily on hunting and gathering (and in some cases, farming) in the regions outside the grasslands. As in the Argentine Pampas, it was Europeans' introduction of the horse—diffusing into the Great Plains from Mexico largely along trade routes between 1700 and 1750—that prompted the Sioux, Cheyennes, and others to reinvent themselves as equestrian nomads of the grasslands. Like the Steppe pastoralists, the nomads of the Great Plains augmented the products of the grasslands by raiding or trading with those outside the region. Like the Argentine nomads, the control of Old World livestock and New World pasturage raised the natives of the Great Plains to regional dominance for much of the nineteenth century. Like them, the nomads of the Great Plains resisted conquest longer and more success-fully than many of their neighbors.[35]

In the Steppes, the nomads were pastoralists who herded domesticated livestock; in the Pampas, the nomads combined herding with the hunting of feral cattle. The Great Plains were different: the horse regularized natives' access to and procurement of a native grazing animal, the bison. Millions of bison, the largest land mammal in North America, inhabited the Great Plains in the late eighteenth century. Some Euroamerican observers in the nineteenth century estimated that the total bison population in North American might be as high as 75 or 100 million. For most of the twentieth century, ecol-ogists and historians believed that the historic bison population probably numbered no less than 60 million.[36] In more recent years, however, more sober estimates of the historic bison population, based on realistic assessments of the carrying capacity of the grasslands, have led scholars to the conclusion that between 24 and 30 million bison grazed in the Great Plains in the late eighteenth century.[37]

Whatever the upward limit of the bison population might have been, in the dynamic, unpredictable Great Plains environment, the bison population was rarely stable. Bison, like other ungulates, are usually at disequilibrium with their forage: their populations are apt to irrupt, place too much pressure on the range, and then crash.[38] Moreover, unpredictable ecological factors imposed further variables on the bison's numbers. In some years, wolf predation, competition from other grazers (increasingly, the natives' own extensive herds of horses), and blizzards depressed the bison's natural increase. Above all, drought periodically caused significant declines in range carrying capac-ity and consequently in the bison population.[39] In the southern Great Plains, in places where prehistoric American hunters drove bison to their deaths, there were two long periods—from 6000 to 2500 BCE and from 500 to 1300 CE—when bison were absent from the kill site remains, indicating steep declines in the bison population.[40]

The near extinction of the bison in the last quarter of the nineteenth century resulted from a combination of these inherent ecological pressures on the bison popu-lation together with the new pressures, both economic and ecological, that followed in the wake of the horse: commercial hunting, exotic bovine disease, and habitat degradation.[41] In a larger sense, those factors converged in the *domestication* of the western plains environment, particularly the displacement of wildlife by domesticated animals—thus extending into the North American grasslands the process that had characterized the Eurasian Steppes in the fourth millennium BCE.

Domestication began with the emergence of nomadic societies who relied on a domesticated animal, the horse, to procure the bison. By the middle of the nineteenth century, the nomads not only hunted bison for their own subsistence and for intertribal exchange, but increasingly to supply Euroamerican traders with bison robes. Traders ascended the Missouri River by steamboat into the heart of the northern Great Plains beginning in 1831. By the end of the 1850s, native commercial bison hunters had largely driven the bison from the vicinity of the Missouri.[42]

The nomads' commercial pressure on the bison paled in comparison to that of Euroamerican hunters, thousands of whom arrived in the Great Plains to hunt bison in the 1870s. Technological innovations made the hunters' invasion of the Great Plains possible: accurate, powerful rifles developed during the American Civil War; the extension of railroads into the plains to ship bulky bison hides; and new chemical methods to transform spongy, pliable bison hides into leather. The hunters slaughtered over three million bison—most only for their hides—between 1872 and 1874, and perhaps as many as six million. By 1881, the bison population had been reduced to fewer than one thousand. The United States government considered legislation to preserve the bison, but decided against it: bison, key policymakers reasoned, were the wild resource of the savage Indians, and their destruction would open the way for domesticated livestock and Euroamerican settlement.[43]

Indeed, the encroachment of domesticated livestock into the Great Plains contributed significantly to the destruction of the bison. In the southern Great Plains, a bovine disease, Texas fever, may have contributed to the demise of the bison. The disease, which affects grazing animals in warm climates, is caused by parasites transmitted by certain ticks. Texas longhorns carried the tick but were immune to the disease. When cattle were driven to market, however, the ticks sometimes detached themselves from their old hosts and attached themselves to new ones, such as bison, who were not immune. In the central Great Plains, Euroamerican overland emigrants herded hundreds of thousands of cattle and sheep to the Pacific West, denuding the central Great Plains river valleys of forage. In the northern Great Plains between the early 1870s and the early 1880s, the number of cattle increased from roughly one hundred thousand to over one million. Domestic cattle, moving into the northern plains from the south and east, occupied the regions into which bison might have otherwise dispersed during the droughts that struck the northern plains in the 1870s and early 1880s.[44]

The fate of the bison was predictive of the fate of indigenous grazing animals in the grasslands that Europeans colonized. Throughout the temperate grasslands of the world, European colonists decimated indigenous grazing animals and introduced in their stead domesticated livestock from the Old World. In the southern African veld, Dutch and later British colonists nearly exterminated the blue buck and the quagga and significantly depleted the numbers of elands and springboks. In their place they introduced merino sheep: twelve million sheep grazed in the Cape colony in 1890; there were 24 million by 1930. Both cultivation and increased grazing pressure led to soil erosion in the veld by the end of the nineteenth century. Nowhere in European colonial grasslands were sheep more important than in Australia, where legislation in 1843 made sheep and wool legal security for loans. A few sheep introduced into New South Wales in 1792 had become nine million by 1845 and twelve million by 1854. Squatters—half of the Australian population in 1840—let their herds overgraze available pasturage and then moved on. By the end of the 1850s, as a result of the sheep population, the Australian grasslands were characterized by loss of vegetative cover, loss of plant and animal diversity, soil compaction, and soil erosion. Even as the most heavily pastured areas were being abandoned, the expansion of sheep-herding continued. By the end of the nineteenth century, tens of millions of sheep grazed in Australia. By the 1930s, there were 300 million sheep in the Southern Hemisphere—most of them in the grasslands of Australia, New Zealand, South Africa, and South America.[45]

In North America, not sheep but cattle rapidly overtook the bison's former range. On the eve of the American Civil War there were between three and four million cattle in west Texas—the descendants of cattle introduced by Franciscan missionaries and Mexican settlers. By the end of the war, the west Texas cattle population, left largely untended during the conflict, had ballooned to over five million. The end of hostilities coincided with the extension of railroads to the central Great Plains, and Texas ranchers soon began to drive their cattle north to Kansas, where the animals were shipped by rail to the stockyards of Chicago. Financing poured in, much of it from British investors; by the early 1880s between seven and eight million cattle grazed in the Great Plains.[46] Overstocking characterized large areas of the grasslands, which in turn led to a deterioration of the pasturage; in one range, five acres were sufficient for one head of cattle in 1870; by 1880, the range was so trampled and overgrazed that each head of cattle required fifty acres.[47]

While ranchers may have owned their homesteads or well-located springs, the federal government owned most of the rangeland grazed by cattle. To prevent farmers from settling on this public land—as they were entitled to do under the auspices of the Homestead Act of 1862—as well to prevent competing ranchers from introducing their own herds, large ranchers increasingly turned to enclosing large areas of the public domain as their private pasturage, particularly following the invention of barbed wire in 1874. The practice of enclosure was a longstanding one in industrializing societies; the enclosure of rural commons had been instrumental in early modern England to the creation of a market society. There, landed gentry forced peasants off of ancestral commons to make room for larger herds of sheep, and wool became a profitable commodity on the emerging textile market. Newly landless peasants became agricultural wage laborers or moved to cities.[48] When enclosure reached large continental grasslands of the New World, it meant a change in the way societies sought to raise domesticated livestock on an enormous scale. For centuries, pastoralists had grazed livestock collectively, though the animals were owned privately. Enclosure privatized grasses. Just as enclosure in early modern Europe helped to create a class of wage laborers, in the

North American Great Plains, large-scale ranches not only pushed out or prevented the establishment of homesteads within the cattle range, but were staffed by itinerant, wage-earning herdsmen. The region's dependence on transient labor only became more pronounced in the first years of the twentieth century, with the establishment of large-scale grain farms.[49]

The irruption of cattle in the newly enclosed Great Plains was followed by decline— the usual parabola when domesticated grazing animals are introduced to virgin pasturage.[50] Cattle are both far less efficient grazers than bison and less well-suited to the bitter, snowy winters in the northern Great Plains. By the late 1870s, many cattle were dying each winter as they searched for forage. In areas where enclosure was common, many cattle died along fences, unable to reach grass a few feet away. Devastating winters in the mid-1880s destroyed about 15 percent of the herds. In the aftermath of the blizzards, ranchers shifted away from allowing cattle to fend for themselves in the winter; instead, they planted hay to see their reduced herds through the winter. Acres of hay harvested in Montana and Wyoming rose from eighty thousand in 1880 to over one million in 1900.[51] Likewise, in Argentina, where cattle herds doubled to 8.5 million during the 1880s, hectares sown with alfalfa rose from just over seven hundred thousand in 1895 to over seven million in 1914.[52]

In the first decades of the twentieth century, in grasslands from the Americas to Eurasia to Australia, the cultivation of feed gave way to the cultivation of grain to supply ever-larger urban populations. In Argentina between 1890 and 1914, the cattle population declined from 8.5 to 7.5 million, while hectares sown with wheat rose from 1.2 million to 6.2 million. In the Russian Steppes, cultivated hectares rose fivefold to 34 million between the beginning of the eighteenth and the end of the nineteenth century, while hectares of pasturage fell by one-quarter.[53] In the North American Great Plains, the surge in farming was yet more abrupt. In twenty-seven counties in the southern Great Plains, the amount of land in farms rose from 10 million acres in 1910 to 17.6 million in 1920; the total land area in farms rose from 38.9 percent to 68.4 percent; acreage of harvested corn, oats, wheat, barley, rye and sorghum rose from just under one million acres in 1910 to 2.2 million in 1920. Altogether, Great Plains farmers broke 32 million acres of sod between 1909 and 1929. New technologies fueled the expansion of farming: in 1915, there were fewer than three hundred tractors in the southern Great Plains. By the early 1930s, there were ten thousand.[54]

The expansion of wheat cultivation to new lands helped to raise the production of wheat to new levels. In the first years of the twentieth century in Buenos Aires province, the average farmer harvested a whopping 1200 kilograms of wheat per hectare. In 1919, United States farmers harvested 952 million bushels of wheat, a 38 percent improvement over the period between 1909 and 1913. In both the Pampas and the Great Plains, of course, there were particular economic pressures encouraging higher production. In Argentina, many wheat farms were cultivated by short-term tenants who had no stake in maintaining the long-term viability of the soil and thus did not hesitate to plow up marginal lands and sow it with wheat. In the North American Great Plains, where tenant farming was also common, farmers were urged to plow up new lands to produce wheat for Americans and their allies in the First World War. Apart from these economic pressures, high yields in the first years after plowing represented what the Argentine historian Adrián Gustavo Zarrilla has called *ecosystem harvest*: in those initial years of high yields, farmers captured soil nutrients that had taken centuries to accumulate.[55] The burst of high yields in newly cultivated soils was akin to the irruption of domesticated livestock in previously ungrazed rangelands. Just as irruptions of livestock were followed by population crashed, ecosystem harvests were followed by erosion and the contraction of farming.

The ecological consequences of destroying native grasses in drought-prone regions and replacing them with domesticated crops arrived by the late nineteenth century in the Steppes and in the 1930s and 1940s in the Great Plains, the Pampas, and South Africa. In Russia, drought-induced crop failures in the late 1840s, 1855, 1873–1874, and 1885 only foreshadowed the "catastrophic drought, harvest failure, and famine" of 1891–1892, according to the historian David Moon.[56] In the Pampas, the removal of native grasses led to significant water erosion in Buenos Aires province: 18 million hectares suffered some runoff damage, while 3.6 million suffered "severe" or "critical" water erosion, according to a 1950s study.[57] In southern Africa, the same wheat-growing techniques used in the southern Great Plains led to significant wind erosion by the 1950s.[58] In North America, drought beginning in 1932 shriveled wheat and left soil exposed. Over the next few years, wind periodically kicked up the exposed soil in dust storms that rained Great Plains soil across the eastern half of the continent. A dust storm in November 1933 spread soil from the Great Lakes to the lower Mississippi River valley. In early May 1934 a dust storm rained soil over Chicago, New York, Boston, and Atlanta, before proceeding out to sea to blanket ships three hundred miles off the Atlantic coast with Great Plains soil.[59]

Rationalizing schemes of the mid-twentieth century left much of the world's grasslands looking like the southern Great Plains in the wake of the dust bowl. In the 1940s, South African wheat farmers persisted in applying the dry-farming techniques of the southern Great Plains despite the North American dust storms, so confident were they that the techniques were sound. Between 1954 and 1965, the Soviets plowed up some 40 million acres in the Steppes; ensuing drought created wind erosion in almost half of that area.[60] As the historian Kate Brown has noted, as a result places such as Montana and Kazakhstan, despite their different histories, had striking resemblances by the end of the twentieth century: eroded grassland environments laid out in rectangular grids.[61]

In the last century, catastrophe has beset the world's grasslands: the destruction of native wildlife, overgrazing by herds of domesticated animals, and soil erosion caused by the cultivation of subhumid lands. Yet the history of grasslands is not simply one of unremitting decline. Their prospect, like their ecology, is unpredictable. And their histories have been multivalent. Before the European ecological invasion of the Americas led to the destruction of native animals and the degradation of the soil, it enabled natives to establish regional dominance as pastoral nomads in both the Pampas and the Great Plains. In the United States, the number of bison reached its nadir at the end of the nineteenth century, but in recent decades the population of bison in the region has risen while the population of ranchers and farmers has fallen precipitously. The bison's nineteenth-century chapter was a story of decline—a decline emblematic of the world's grasslands during that era—but it was not the last chapter.

The problem environmental historians face in interpreting episodes such as the destruction of the bison or the Dust Bowl is that their legacies remain open-ended. Was the Dust Bowl, as Worster maintained, the result of a modern, mechanized economy's assault upon the environment? Or will it be a lesson in the process of becoming what Malin called "grass men"—people who have learned to live within the ecological confines of the grasslands? Becoming "grass men" was what the pastoralists of the Eurasian Steppes set out to do five or six thousand years ago; the horse nomads of the Great Plains embarked on a similar project two and a half centuries ago. Despite setbacks, people have long inhabited grasslands—they have, in fact, constructed grassland societies that sustained themselves for long periods—but they never fully domesticated the grasslands environment. In this sense, grasslands are indeed like the seas: even the most knowledgeable and experienced "grass man" may be defeated by the semi-arid environment, just as even an experienced sailor may be shipwrecked in a sudden storm. The seas of grass are, like the oceans, unpredictable places liable to tempests. An essential aspect of becoming "grass men," it would seem, is an understanding of that dynamism.

NOTES

1. William H. McNeill, *Plagues and Peoples* (New York: Anchor, 1976); Alfred W. Crosby, *The Columbian Exchange: Biological and Cultural Consequences of 1492*, 2d ed. (Westport, CT: Praeger, 2003); Russell Thornton, *American Indian Holocaust and Survival: A Population History since 1492* (Norman: University of Oklahoma Press, 1990); Elinor G.K. Melville, *A Plague of Sheep: Biological Consequences of the Conquest of Mexico* (New York: Cambridge University Press, 1997); Noble David Cook, *Born to Die: Disease and New World Conquest, 1492–1650* (New York: Cambridge University Press, 1998); Virginia DeJohn Anderson, *Creatures of Empire: How Domestic Animals Transformed Early America* (New York: Oxford University Press, 2004).

2. For Eurasian pastoralists, see Owen Lattimore, *Inner Asian Frontiers of China* (New York: American Geographical Society, 1940). A recent study on the importance of Steppe pastoralists is David W. Anthony, *The Horse, the Wheel, and Language: How Bronze-Age Riders from the Eurasian Steppes Shaped the Modern World* (Princeton, NJ: Princeton University Press, 2007). For epidemics in late antiquity and medieval Europe, see McNeill, *Plagues and Peoples*; see also David Herlihy, *The Black Death and the Transformation of the West* (Cambridge, MA: Harvard University Press, 1997). For the eventual European conquest of the Steppes, see McNeill, *Europe's Steppe Frontier, 1500–1800: A Study of the Eastward Movement in Europe* (Chicago: University of Chicago Press, 1964); Michael Khodarkovsky, *Russia's Steppe Frontier: The Making of a Colonial Empire, 1500–1800* (Bloomington: Indiana University Press, 2002).

3. See Jack Weatherford, *Genghis Khan and the Making of the Modern World* (New York: Three Rivers Press, 2004); Robert Tignor et al., *World Together, Worlds Apart: A History of the Modern World, 1300 to the Present* (New York: Norton, 2002).

4. R. I. Lawless, "The Concept of Tell and Sahara in the Maghreb: A Reappraisal," *Transactions of the Institute of British Geographers* 57 (November 1972): 125–137. For another example, see Dee Mack Williams, "The Desert Discourse of Modern China," *Modern China* 23 (July 1997): 328–355. See also Chapter 4 in this volume.

5. Walter Prescott Webb, *The Great Plains* (Boston: Ginn, 1931); Carl O. Sauer, "The Morphology of Landscape," *University of California Publications in Geography* 2 (1925): 19–53; Lucien Febvre, *A Geographical Introduction to History* (New York: Knopf, 1925 [1922]). For Webb's place in the evolution of environmental history, see Andrew C. Isenberg, "Historicizing Natural Environments: The Deep Roots of Environmental History," in *A Companion to Western Historical Thought*, ed. Lloyd Kramer and Sarah Maza (Malden, MA: Blackwell, 2002), 372–389.

6. James C. Malin, "The Adaptation of the Agricultural System to the Sub-Humid Environment: Illustrated by the Activities of the Wayne Township Farmers' Club of Edwards County, Kansas, 1886–1893," *Agricultural History* 10 (July 1936): 118–141; Malin, *The Grassland of North America: Prolegomena to its History* (Lawrence, KS: 1947). See also Allan C. Bogue, "The Heirs of James C. Malin: A Grassland Historiography," *Great Plains Quarterly* 1 (Spring 1981): 105–131; Bogue, "Tilling Agricultural History with Paul Wallace Gates and James C. Malin," *Agricultural History* 80 (Autumn 2006): 436–460. See also Matthew Paul Bonnifield, *The Dust Bowl: Men, Dirt, and Depression* (Albuquerque: University of New Mexico Press, 1979); R. Douglas Hurt, *The Dust Bowl: An Agricultural and Social History* (Chicago: Nelson-Hall, 1981); Pamela Riney-Kehrberg, *Rooted in Dust: Surviving Drought and Depression in Southwestern Kansas* (Lawrence: University Press of Kansas, 1994); Geoff Cunfer, *On the Great Plains: Agriculture and Environment* (Lubbock: Texas A&M University Press, 2005).

7. Donald Worster, *Dust Bowl: The Southern Plains in the 1930s* (New York: Oxford University Press, 1979). For Worster's place in the environmental history of semi-arid regions, see Isenberg, "Environment and the Nineteenth-Century West; or, Process Encounters Place," in *A Companion to the History of the American West*, ed. William Deverell (Malden, MA: Blackwell, 2004), 77–92.

8. William Cronon, "A Place for Stories: Nature, History, and Narrative," *Journal of American History* 78 (March 1992): 1347–1376.

9. John Weaver, quoted in Ronald Tobey, *Saving the Prairie: The Life Cycle of the Founding School of American Plant Ecology* (Berkeley: University of California Press, 1981), 3; see also Frederic Clements, "Nature and Structure of the Climax," *Journal of Ecology* 24 (February 1936): 252–284.

10. For chaos theory, see Edward Lorenz, *The Essence of Chaos* (Seattle: University of Washington Press, 1993); James Gleick, *Chaos: Making a New Science* (New York: Penguin, 1988); Daniel B. Botkin, *Discordant Harmonies: A New Ecology for the Twenty-first Century* (New York: Oxford University Press, 1990); Robert V. O'Neill, "Is It Time to Bury the Ecosystem Concept? (With Full Military Honors, of Course!)" *Ecology* 82 (December 2001): 3275–3284; Jianguo Wu and Orie L. Loucks, "From Balance of Nature to Hierarchical Patch Dynamics: A Paradigm Shift in Ecology," *Quarterly Review of Biology* 70 (December 1995): 439–466; John Kricher, "Nothing Endures but Change: Ecology's Newly Emerging Paradigm," *Northeastern Naturalist* 5 (1998): 165–174.

11. For chaos theory in the grasslands, see Graeme Caughley, "Plant-Herbivore Systems," in *Theoretical Ecology: Principles and Applications*, ed. Robert M. May (Philadelphia: W. H. Saunders, 1976); David M. Forsythe and Peter Caley, "Testing the

Irruptive Paradigm of Large-Herbivore Dynamics," *Ecology* 87 (Febriary 2006): 297–303; D. D. Briske, S. D. Fuhlendorf, and F. E. Smeins, "Vegetative Dynamics in Rangelands: A Critique of Current Paradigms," *Journal of Applied Ecology* 40 (August 2003): 601–614.

12. Elinor Melville adopted the irruptive paradigm in 1994 for *A Plague of Sheep*. For an integration of chaos theory into a grassland environmental history, see Andrew C. Isenberg, *The Destruction of the Bison: An Environmental History, 1750–1920* (New York: Cambridge University Press, 2000). See also a work by a political scientist, Andrew T. Price-Smith, *Contagion and Chaos: Disease, Ecology, and National Security in the Era of Globalization* (Cambridge, MA: MIT Press, 2009).

13. McNeill, *Plagues and Peoples*, 4.

14. H. L. Schantz, "The Place of Grasslands in the Earth's Cover of Vegetation," *Ecology* 34 (1954): 143–145.

15. More than a century ago, the Russian climatologist Wladimir Köppen divided the planet into five basic environments based on climate: tropical (type A), dry (B), temperate (C), continental (D), and polar (E). In more recent years geographers have tried to make Köppen's broad categories more sophisticated. A revised version of Köppen's scheme—the so-called Köppen-Geiger system—has retained Köppen's basic categorization while subdividing his climate types into thirty categories according to more nuanced standards. Type B climates are divided into deserts (W for German *Wüste* or desert) and steppes (S); steppes are further divided into those where the coldest month has an average temperature below freezing (k, such as the northern reaches of the Canadian Great Plains), and those that do not (h, such as the grasslands of New South Wales in Australia). The apparent sophistication of the scheme does not change the fact that categorizing regions is an inherently subjective exercise. See M. C. Peel, B. L. Finlayson, and T. A. McMahon, "Updated World Map of the Köppen-Geiger Climate Classification," *Hydrology and Earth System Sciences Discussions* 4 (2007): 439–473.

16. For grassland ecology, see Howard B. Sprague, ed., *Grasslands* (Washington, DC: American Association for the Advancement of Science, 1959); Paul Sears, *Lands Beyond the Forest* (Englewood Cliffs, NJ: Prentice-Hall, 1969); Robert T. Coupeland, ed., *Grassland Ecosystems of the World: Analysis of Grasslands and their Uses* (Cambridge: Cambridge University Press, 1979); Lauren Brown, *Grasslands* (New York: Knopf, 1985); Maria Shahgedanova, *The Physical Geography of Northern Eurasia* (Oxford: Oxford University Press, 2003); Andrew S. Goudie, *The Nature of the Environment*, 4th ed. (Oxford: Wiley-Blackwell, 2001).

17. For the concept of "solar economy," see Christian Pfister, "The Early Loss of Ecological Stability in an Agrarian System," in *The Silent Countdown: Essays in European Environmental History*, ed. Christian Pfister and Peter Brimblecomb (Berlin: Springer-Verlag, 1990), 37–55. The concept is treated extensively in Crosby, *Children of the Sun: A History of Humanity's Unappeasable Appetite for Energy* (New York: Norton, 2006), and Richard White, *The Organic Machine: The Remaking of the Columbia River* (New York: Hill and Wang, 1995).

18. See Isenberg, *Destruction of the Bison*, 17–19.

19. Dan Flores, "Bison Ecology and Bison Diplomacy: The Southern Plains from 1800 to 1850," *Journal of American History* 78 (September 1991): 465–85; Isenberg, *Destruction of the Bison*, 17–18. See also Isenberg, *Mining California: An Ecological History* (New York: Hill

and Wang, 2005), 103–130; Scott Stine, "Extreme and Persistent Drought in California and Patagonia during Mediaeval Time," *Nature* 369 (June 16, 1994): 546–549.

20. See Robert E. Gabler, Robert J. Sager, Sheila M. Brazer, and Daniel L. Wise, *Essentials of Physical Geography*, 3d ed. (Philadelphia: Saunders, 1987), 210.

21. For a good discussion of the impact of overland emigrants on the North American Great Plains, see Elliott West, *The Way to the West: Essays on the Central Plains* (Albuquerque: University of New Mexico Press, 1995).

22. For hunting in the North American Great Plains before the horse, see Isenberg, *Destruction of the Bison*, 31–44; for a Eurasian example see Tim Ingold, *Hunters, Pastoralists, and Ranchers: Reindeer Economies and their Transformations* (Cambridge: Cambridge University Press, 1980), 69–75.

23. For an overview of human domestication of animals, see Jared Diamond, *Guns, Germs, and Steel: The Fates of Human Societies* (New York: Norton, 1997), 157–175. A recent intervention into the debate over the site and date of horse domestication is Alan K. Outram et al., "The Earliest Horse Harnessing and Milking," *Science* 323 (March 6, 2009): 1332–1335. Domesticated goats diffused to North Africa by 7700 B.C. Both goats and cattle were in East Africa by 4000 B.C. and had diffused to southern Africa by 2000 B.C. See Andrew B. Smith, "Origins and Spread of Pastoralism in Africa," *Annual Review of Anthropology* 21 (1992): 125–141.

24. See Anthony, *The Horse, the Wheel, and Language*, 72–73; William H. McNeill, "The Eccentricity of Wheels, or Eurasian Transportation in Historical Perspective," *American Historical Review* 92 (December 1987): 1111–1127.

25. Thomas J. Barfield, *The Nomadic Alternative* (Englewood Cliffs, NJ: Prentice-Hall, 1993), 133–134.

26. Barfield, *Nomadic Alternative*, 137–142.

27. David Christian, "Silk Roads or Steppe Roads? The Silk Roads in World History," *Journal of World History* 11 (Spring 2000): 1–26; William McNeill, *Plagues and Peoples*, 69–175.

28. Luc Kwanten, *Imperial Nomads: A History of Central Asia, 500–1500* (Philadelphia: University of Pennsylvania Press, 1979); Christopher Kaplonski, "The Mongolian Impact on Eurasia: A Reassessment," in *The Role of Migration in the History of the Eurasian Steppe: Sedentary Civilization vs. "Barbarian" and Nomad*, ed. Andrew Bell-Fialkoff (New York: St. Martin's Press, 2000), 251–274.

29. See Gareth Jenkins, "A Note on Climate Cycles and the Rise of Chinggis Khan," *Central Asiatic Journal* 18 (1974): 217–226; Anatoly Khazanov, "Ecological Limitations of Nomadism in the Eurasian Steppes and their Social and Cultural Implications," *Asian and African Studies* 24 (1990): 1–15; Joseph Fletcher, "The Mongols: Ecological and Social Perspectives," *Harvard Journal of Asiatic Studies* 46 (June 1986): 11–50.

30. Weatherford, *Genghis Khan*, 50–125; Owen Lattimore, "The Geography of Chingis Khan," *Geographical Journal* 129 (March 1963): 1–7.

31. David Moon, "Agriculture and the Environment on the Steppes in the Nineteenth Century," in *Peopling the Russian Periphery: Borderland Colonization in Eurasian History*, ed. Nicholas Breyfogle, Abby Schroeder, and Willard Sunderland (London: Routledge, 2007), 83–84.

32. See Crosby, *Ecological Imperialism*.

33. Crosby, *Ecological Imperialism*, 161, 178.

34. Ronald E. Gregson, "The Importance of the Horse in Indian Cultures of Lowland South America," *Ethnohistory* 16 (Winter 1969): 33–50; Richard O. Perry, "Warfare on the Pampas in the 1870s," *Military Affairs* 36 (April 1972): 52–58.

35. See Isenberg, "Between Mexico and the United States: From Indios to Vaqueros in the Pastoral Boderlands," in *Mexico and Mexicans in the Making of the United States*, ed. John Tutino (Austin: University of Texas Press, 2012), 83–109; Isenberg, *Destruction of the Bison*, 31–62; Richard White, "The Winning of the West: The Expansion of the Western Sioux in the Eighteenth and Nineteenth Centuries," *Journal of American History* 65 (September 1978): 319–343; Pekka Hämäläinen, *The Comanche Empire* (New Haven, CT: Yale University Press, 2008). For the Navajo, see James F. Brooks, *Captives and Cousins: Slavery, Kinship, and Community in the Southwest Borderlands* (Chapel Hill: University of North Carolina Press, 2002). The Great Plains nomads might also be compared to the Zulu of southern Africa, who dominated their region well into the nineteenth century as well; their source of subsistence was cattle that had long preceded European colonization. See James O. Gump, *The Dust Rose Like Smoke: The Subjugation of the Zulu and the Sioux* (Lincoln: University of Nebraska Press, 1996).

36. For high estimates of the historic bison population, see Ernest Thompson Seton, *Lives of Game Animals*, vol. 3, pt. 2 (New York: Doubleday, 1929), 654–656. Seton estimated that there were 75 million bison in North America; Frank Gilbert Roe, *The North American Buffalo* (Toronto: University of Toronto Press, 1951), implied that Seton's estimate might be too low; Donald Worster endorsed these high estimates as recently as 1994 in *An Unsettled Country: Changing Landscapes of the American West* (Albuquerque: University of New Mexico Press, 1994).

37. For lower estimates, each employing a different methodology to arrive at their estimates, see Tom McHugh, *The Time of the Buffalo* (Lincoln: University of Nebraska Press, 1972); Dan Flores, ""Bison Ecology and Bison Diplomacy: The Southern Plains from 1800 to 1850," *Journal of American History* 78 (September 1991): 465–485; Isenberg, *Destruction of the Bison*, 28–30.

38. For the fluctuations of grazing populations, see Graeme Caughley, "Wildlife Management and the Dynamics of Ungulate Population," in *Applied Biology*, vol. 1, ed. T. H. Coaker (London: Academic Press, 1976), 180, 240. The environmental historian Elinor Melville applied the concept of ungulate irruption to her study of sheep in central Mexico and cattle in Australia. See Melville, *A Plague of Sheep*, 6–7.

39. Isenberg, *Destruction of the Bison*, 83–84; Flores, ""Bison Ecology and Bison Diplomacy," 465–485. For drought see Robert T. Coupeland, "The Effects of Fluctuations in Weather upon the Grasslands of the Great Plains," *Botanical Review* 24 (May 1958): 284–86; Carl Friedrich Kraenzel, *The Great Plains in Transition* (Norman: University of Oklahoma Press, 1955), 17–23. For the loss of carrying capacity during droughts, see Charles Rehr, "Buffalo Population and Other Deterministic Factors in a Model of Adaptive Processes on the Shortgrass Plains," *Plains Anthropologist* 23 (November 1978): 23–39.

40. Tom Dillehay, "Late Quaternary Bison Population Changes in the Southern Plains," *Plains Anthropologist* 10 (August 1974): 180–196.

41. Rudolph W. Koucky, "The Buffalo Disaster of 1882," *North Dakota History* 50 (Winter 1983): 23–30; Richard White, "Animals and Enterprise," in *The Oxford History of the American West*, ed. Clyde Milner et al. (New York: Oxford, 1994), 247–249; West, *Way to the West*, 72–83.

42. Isenberg, *Destruction of the Bison*, 93–122.

43. See Isenberg, *Destruction of the Bison*, 133–140; Richard I. Dodge, *The Plains of North America and Their Inhabitants* (Newark: University of Delaware Press, 1989), 155.

44. For the northern and southern plains, see Isenberg, *Destruction of the Bison*, 140–142. For the central plains, see West, *Way to the West.*

45. William Beinart and Lotte Hughes, *Environment and Empire* (Oxford: Oxford University Press, 2007), 71–72, 94–105; Melville, *A Plague of Sheep*, 60–77; Kate B. Showers, "Soil Erosion in the Kingdom of Lesotho: Origins and Colonial Response, 1830s–1950s," *Journal of Southern African Studies* 15 (January 1989): 263–286.

46. Terry Jordan, *North American Cattle-Ranching Frontiers: Origins, Diffusion, and Differentiation* (Albuquerque: University of New Mexico Press, 1993); Wiliam Cronon, *Nature's Metropolis: Chicago and the Great West* (New York: Norton, 1991).

47. Worster, *Dust Bowl*, 83.

48. For enclosure, see E. P. Thompson, *The Making of the English Working Class* (New York: Vintage, 1963); J. L. and B. Hammond, *The Village Labourer, 1760–1832: A Study in the Government of England Before the Reform Bill* (London: Longman's, 1912); J. M. Neeson, *Commoners: Common Right, Enclosure, and Social Change in England, 1700–1820* (Cambridge: Cambridge University Press, 1993); Robert C. Allen, *Enclosure and the Yeoman* (London: Clarendon, 1992). For the extension of enclosure to the ranchlands of the nineteenth-century United States West, see Isenberg, *Mining California*, 131–162.

49. David Lopez, "Cowboy Strikes and Unions," *Labor History* 18 (1977): 325–340. For transient labor in the North American West, see Mark Wyman, *Hoboes: Bindlestiffs, Fruit Tramps, and the Harvesting of the West* (New York: Hill and Wang, 2010).

50. See Melville, *Plague of Sheep*, 6–7.

51. Richard White, "Animals and Enterprise," in *The Oxford History of the American West*, ed. Clyde A. Milner II, Carol A. O'Connor, and Martha Sandweiss (New York: Oxford University Press, 1996).

52. Jeremy Adelman, *Frontier Development: Land, Labour, and Capital in the Wheatlands of Argentina and Canada, 1890–1914* (Oxford: Clarendon, 1994), 70–73.

53. Adelman, *Frontier Development*, 70–74; Moon, "Agriculture and the Environment on the Steppes," 85.

54. Hurt, *The Dust Bowl*, 23–24.

55. Adrián Gustavo Zarrilli, "Capitalism, Ecology, and Agrarian Expansion in the Pampaean Region, 1890–1950," *Environmental History* 6 (October 2001): 570; Worster, *Dust Bowl*, 89.

56. Moon, "Agriculture and the Environment on the Steppes," 85, 95.

57. Zarrilli, "Capitalism, Ecology, and Agrarian Expansion in the Pampaean Region," 561–583.

58. Sara T. Phillips, "Lessons from the Dust Bowl: Dryland Agriculture and Soil Erosion in the United States and South Africa, 1900–1950," *Environmental History* 4 (April 1999): 245–266.

59. Hurt, *Dust Bowl*, 34–35; Worster, *Dust Bowl*, 13–14.

60. Phillips, "Lessons from the Dust Bowl"; Worster, *Dust Bowl*, 4.

61. Kate Brown, "Gridded Lives: Why Kazakhstan and Montana Are Nearly the Same Place," *American Historical Review* 106 (February 2001): 17–48.

CHAPTER 6

....

NEW PATTERNS IN OLD PLACES

Forest History for the Global Present

....

EMILY K. BROCK

For centuries, the Mayan city of Palenque thrived, supported by a network of surrounding villages and a domesticated landscape of intensive agriculture. As the city was abandoned, forest grew over the urbanized areas, obliterating traces of human endeavor. When the Spanish arrived, the tropical forests at Palenque appeared the perfect image of pristine nature. The monumental temples and humble homes that had made up the city were hidden from view, and have only been uncovered with painstaking archaeological effort. The unbroken green of cedar and mahogany that had seemed to be a primordial forest was in fact no more than four centuries old. Similar cases of ruined cities found deep within vast tropical forests can be found from Angkor in Cambodia to Great Zimbabwe in Zimbabwe. These remnants pose what Emmanuel Kreike has termed the "Palenque Paradox."[1] If wild forests are full of the vestiges of human activity, then there is no clear line between culture and nature, and no clear direction of environmental change. The dominant historical models of environmental change have all implied a cumulative and irreversible gradient from nature to culture, with nature the victim of human agency. But the history of Palenque and its forest challenges historians' attempts to depict unilinear environmental change.

Environmental historians often adopt a narrative of decline and despoliation. Yet the history of forests is not just the history of deforestation. Forests are complex and dynamic systems, and have served as sustaining shelters for human life since our species first appeared among the trees. If history is the chronicle of change, then forest history should encompass all types of change in forests, in which both humans and nature are actors. People cut trees down, and people plant trees. Fire sweeps through forests, and forests regrow. Species die out, and species invade. The climate changes. When historians look at forests, we should consider the full tapestry of these interactions.

Histories of forests often follow a common template, chronicling ever-escalating deforestation in the face of the pressures of global demand for wood and agricultural land. This essay questions the universality, even the value, of that pattern. If the forest is merely the mute backdrop for human activity, the stories told are declensionist patterns of humans dismantling wild nature. If, however, the forest itself is an actor, a dynamic assemblage of growing things and changing forces, then its histories may become much more diverse and intriguing. Writing history means finding meaning in patterns of change over time. Making a decision to include non-human actors in the histories we write does not mean we lose the ability to ascribe meaning to the changes we detect. The history of forests should incorporate the ecological changes forests endure. It should broaden in scope to include changes occurring over many centuries, and those happening so subtly as to be overlooked. By decentering the human within the forest, forest history is reframed.[2]

Forest history should encompass the full scope of both ecological and anthropogenic change, not just logging and other intentional, economically-driven changes to forests. We see our own species' impact on forest ecosystems clearly in the often-abrupt changes brought about by industrial logging, and may naturally focus our attention on those changes. The most abrupt and recognizably anthropogenic environmental changes are more easily encompassed historically than are those subtle, long-term shifts which are also unfolding. Broadening the scope of forest history does not imply denial of the severity of deforestation. Rather, complexifying forest history places deforestation within its broader context. Disparate cultural, economic, and ecological forces come into contact in the world's forests. The forest has always been a source of important resources that have fueled economic development, but has also impacted many other facets of human life. As historians we should question whether forests always dwindle, whether human greed always results in ecological destruction, and whether the world's forests are doomed to extinction.

This essay is organized in a manner intended to undermine the declensionist mode of forest history. It is organized around different types of relationships between humans and forests, offering examples grouped to transect standard historical categories. By doing so, it moves our attention to the interface between people and forests rather than imposing a misleading chronological or geopolitical framework onto the world's forests. This essay thus invites the reader to foreground human interaction and non-unilinear environmental change rather than look for expected patterns. The first section covers changes to forests, ranging from ecologically driven forest changes to the interventions of economically motivated exploitation. The second section covers forests as both players in and pawns of societal goals through military action, colonial rule, state-sponsored conservation, and scientific study. The third section considers intentional interventions and reversions of forest loss through tree plantations and reforestation. My intention in this essay is not only to point out interesting patterns of historical inquiry, but also to suggest fruitful paths forward for further study of the forest and its role in human history.

ECOLOGICAL FORCES FOR FOREST CHANGE

A forest, like any ecosystem, is a dynamic entity, responding to widely varying conditions and events. Changes in climate, topography, or other abiotic factors may shift an area's species assemblage dramatically. Tracing a particular landscape through periods of forest cover, felling, agricultural use, and reversion to forests can reveal a narrative of fluctuations over centuries.[3] In a classic example, Richard White's study of two forested islands in Washington's Puget Sound showed cascading ecosystem responses to the social and cultural changes among the islands' inhabitants. Technological improvements in logging, farming, and irrigation allowed humans to alter the environments of the islands in ever more disruptive ways. Likewise, as technologies of transportation improved, an increasing diversity of exotic species made their way to the islands' forests in the company of humans. Both intentionally and unintentionally, these developments in human culture and society changed Puget Sound forests in innumerable ways. To oversimplify such complex changes into a mere case of decline in forest cover would have meant ignoring the full spectrum of changes in the land.[4] While much of this essay concerns instances where humans controlled forest change, not all important changes in forests are primarily the result of human influence. This section examines ecological changes in which human agency played little role, as with prehistoric climate shifts, and ecological changes in which human agency was coupled with strong non-human forces, as with forest fires, species invasions, and ecological reversion of previously cleared areas.

Just by existing, we change a place. Since the rise of modern human cultures in the wake of the last Ice Age, humans have manipulated ecosystems, both intentionally and unintentionally. While the human species cannot help but be a factor in any ecosystem in which it is present, not all large-scale changes are the intentional work of humans. A multitude of ecological changes—many forces all acting at once—occurs in concert with a cacophony of human pressures. The most dramatic changes in global forest cover occurred with minimal human intervention, following the last Ice Age. Ten millennia ago, as the climate warmed, the surface area of the globe that was covered by forests increased rapidly. Boreal forests developed in the Northern Hemisphere as the ice retreated. As the planet warmed, many ecosystems in more temperate zones changed radically. Grasslands became forest in West Africa and South America. Forest became grasslands in North America, and became desert in North Africa.[5] Examining this prehistory broadens the historian's view of forest change, challenging presuppositions of a pristine and unchanging primeval forest. Forests, like any ecosystem, are not and have never been static. The forests in existence during human history should be recognized as the most recent unfoldings of a much longer ebb and flow.

While long-term, slow-moving shifts in the global climate cause long-term, widespread changes in forests, sudden ecological pressures can also cause significant, even severe, localized changes to forests. Hurricanes, storms, floods, and other non-anthropogenic events are all parts of the endless cycles of change in forest land-scapes. Over millennia, these chronic and acute pressures help shape the life of the forest. Individual species within the forest ecosystem evolve, due both to these abi-otic pressures and to inter- and intraspecies competition. Some species may die off, while others may increase their ranges as conditions become amenable to their needs. Invading species often dramatically change the character of an ecosystem over a rela-tively short period of time. These effects are often most acutely felt in the low-diversity, fragile environments of small islands, where the arrival of a new terrestrial species may result in profound ecological change. Animal and plant species have colonized remote islands through non-anthropogenic vectors, causing change on an evolutionary times-cale.[6] Anthropogenic introductions of exotic tree species can establish significant for-est cover in previously treeless environments. One of the most closely observed cases of this came when the Red Quinine tree, introduced to the Galapagos for the purposes of agricultural cultivation, soon established itself on the naturally treeless highlands of Santa Cruz Island. In less than twenty years, this species had spread to create a new, non-native forest.[7]

Many forest types are ecologically dependent on periodic fire, but human disrup-tion of normal fire patterns can have dramatic effect. While lightning strikes create forest fires independent of any human need, fire can also be a form of technology, made and controlled by humans who wield it for their own aims. Periodic uninten-tional fire can cause dramatic ecological effects, as with the evolution of many grass-land and savanna ecosystems. Forest fire has been traditionally used for calculated objectives—natives of eastern North America, for instance, periodically burned forest underbrush to create open, park-like conditions suitable to hunting deer and other game. Light, seasonal fires could limit underbrush and encourage growth of edible green plants. When such forms of extensive use are in place, those in control often limit access and strictly control use. The subtle effects of periodic fire may be most apparent following periods of fire suppression. For example, following several disastrous Western forest fires early in the twentieth century, the US Forest Service adopted a three-part strategy: prevent ignition, modify fire environments so that fires would burn with less intensity, and suppress fires while they were still small. In 1935, after a series of drought years led to more forest fires, the Forest Service adopted its "every fire out by 10 a.m." policy: to prevent catastrophic fires, all fires, no matter how small or remote, were to be suppressed by ten in the morning. Decades of fire suppres-sion transformed many forest types which were ecologically dependent on periodic fires, and led to large, hot, and uncontrollable burns in later years. Examining forests through the lens of fire foregrounds the ecological dynamics of the biotic and abiotic factors in their changes over short- and long-time scales.[8]

Abandoned agricultural lands may revert to forest without human intervention, creating dramatic and unanticipated ecological shifts as disturbance-tolerant spe-cies overtake former farmland. Agriculture expands and contracts on the map, gov-erned by the forces of a society's culture and economy as much as by ecology. Soils are depleted, market forces shift, politics change. Farmers who once worked to keep encroaching forests at bay relent, and new trees creep back onto old fields. This reversion to forest is not a universal response to abandonment, and in some places originally covered in forest, soil exhaustion and erosion has led to post-agricultural grassland or desert. Further, the new forest may contain a much different species array than the old, as changed conditions create different ecological niches. Parts of the American South have also seen a dramatic decline in agricultural land, and a corresponding spike in forest cover. Following the abolition of slavery and the large-scale movement of rural African Americans to cities, agricultural labor shortages forced a transformation of the Southern economy. Significant amounts of unmanaged forest established itself on formerly agricultural acreage, much of which was then subsequently converted to highly managed industrial tree farms in the twentieth century. The "Cutover" areas of the Upper Midwest were abandoned by lumber companies following the collapse of the white pine lumber industry in the early twentieth century. While the area was initially thought to be permanently ruined, with time forest did regenerate on the Cutover, although with aspen and hardwoods dominant in place of white pine. While the ecological quality of post-abandonment forest may be marginal, such landscape shifts reflect changing economic and cultural structures of the regions in which they occur.[9]

CULTURAL FORCES FOR FOREST CHANGE

Humans have long looked to the forest to provide resources for their very survival: shelter, fuel, crops, game. Indeed, the forest is one of the earliest places where human culture used nature and created wealth. Demands on forest resources are apparent in the earliest records of human history. Historians have focused much energy on the history of commercial logging, while often neglecting the rest of the economic spectrum. Yet forests host a wide spectrum of economic activity, varying from place to place as well as evolving over time. The same forest can be changed in multiple ways by multiple pressures, either sequentially or simultaneously. Some forms of economic use of forests, such as logging, are *intensive*, causing concentrated shifts in the structure of the forest. Other forms of economic use, such as hunting or gathering wild foodstuffs, are *extensive*, causing less perceptible but more widespread human impact on the landscape. For example, William Cronon's seminal *Changes in the Land* examines multiple intensive and extensive economic uses of a forested landscape. Amerindians modified New England forests to create edge habitats amenable to their resource strategies. As Amerindian populations decreased, the forest recovered, and was then significantly altered by colonists, who created fields and harvested lumber for local use. Although export logging in the forests of New England had a significant effect, it was only one of many intensive and extensive pressures creating the forest's pattern of change over time.[10] Deforestation for lumber is simply one particularly intensive method by which the forest is altered for economic gain. In this section I examine different human uses of forests, both intensive and extensive. First, I consider the roles of fuelwood use, charcoal production, and forest products harvested with little or no tree felling, then move on to the development of logging. Within each sort of economic use, colonial power and private industry can amplify impact and transform forests.

Populations may use nearby forests in ways whereby exploitation's impact is subtle but extensive, and such seemingly low-impact use may accumulate, causing enduring change. The use of wood for heating purposes has often had this effect. Fuelwood gathering need not involve felling trees, but it cannot be done without affecting the ecosystem, sometimes significantly. The activity could contribute to severe deforestation, as localized dependence on the resource increases. This pattern has not been completely eliminated by the trend toward other energy sources in the twentieth century. Especially vivid examples of the link between fuelwood use and deforestation can be found in India, Brazil, China and elsewhere in recent decades. However, firewood use is not always a cause of degradation. The English practice of coppicing trees proved fairly stable, lasting from the Neolithic Era until the rise of the coal furnace. Coppicing essentially turned a forest into an orchard for fuelwood, by repeatedly and habitually pruning back the large branches of the tree to encourage small, fuelwood-sized new sprouts from the trunk. Coppicing allowed people to maximize the amount of fuelwood a particular tree could provide, while allowing the tree to live. Charcoal production has been another longstanding method of energy production and of great use for industries such as metalsmithing, glass blowing, and industrial food production. The process has been understood since the prehistoric era and has often been thought to have created a significant impact on forests. However, the effects of charcoal production appear to have been far less severe in sixteenth and seventeenth century Europe than many have thought. Many food production practices are based in and around forests. Some, such as maple syrup or pinyon nut collection, involve human control and modification of natural trees themselves to maximize their yield. Crops such as rattan and bamboo are often cultivated within natural forests. Livestock grazing in forested areas can transform landscapes, leading to significant deforestation in fragile forests.[11]

While extensive forest use practices are often small-scale, at times they can also become large-scale industrial enterprise. Beginning with the colonial naval stores industry, the forests of the coastal Carolinas and Georgia became a major site of turpentine production. The naval stores industry gleaned resin from pines without felling trees, then distilled it to create turpentine, rosin, and related substances.[12] With the widespread shift away from wooden ships in the nineteenth century, the southeastern production of naval stores declined dramatically. In the twentieth century, the extensive light use of the forests in the naval stores industry was replaced by intensive use for logging in some areas and complete abandonment in other areas.[13] A similar process transformed Brazil's forests for the sake of a different substance. Following that nation's 1822 independence from Portugal, the development of an industrial rubber industry in the Amazon basin caused a wave of change in the forests. The rubber tree (*Hevea brasiliensis*) creates a latex sap which dries to a unique flexible consistency. The sap is tapped from living trees, then purified and treated to create natural rubber. With the invention of the vulcanization process in 1839, more stable and durable latex rubber could be made. Demand for *Hevea* soon exploded, as applications for vulcanized rubber were found in innumerable consumer and industrial purposes. Latex production grew rapidly into a large-scale industrial forest industry. Rubber plantations in Brazil became the site of a struggle for political control and economic dominance among local and foreign entities. *Hevea* grew wild only in South America, but the burgeoning rubber industry moved away from wild-grown trees and outside the species' original ecological range. Industrialists responded to the struggle for control within Brazil by creating rubber plantations in India, Indonesia, Malaysia, the Belgian-controlled Congo Free State, and elsewhere, reproducing their production capabilities in settings more amenable to their industrial control.[14]

The most dramatic and rapid changes humans can make to forests are to clear them entirely. Clearing forests for agricultural use has been one of the most longstanding patterns of human modification of the landscape. The ancient Sanskrit text of the *Mahābhārata* describes the use of prescribed fire to manage the Indian forest, clearing it of wild animals and creating agricultural fields. This pattern has been repeated worldwide. For example, the forests of the Caribbean and South America were often significantly affected by the imposition of European colonial agriculture. Much of the region's forest was cleared to make way for agricultural lands for sugar, coffee, cacao, tobacco, and other export crops. Likewise, most of the Caribbean saw an almost complete decimation of its natural forests under colonial rule. As sugar plantations in the region became one of the most significant sources of wealth in the colonies, the land was almost completely converted to agricultural use. Indeed, wood was so scarce on the islands at the height of the colonial sugar economy that it was necessary to import wood from New England to fabricate barrels for molasses and rum. While traditional slash-and-burn rotations often had little long-term impact on the environment, the soil erosion from intensive agriculture on former tropical forest may cause long-term declines in productivity, precluding recovery. By the twentieth century, potential and actual impacts of such agricultural activity in many tropical forest regions became a concern for international environmentalism.[15]

The production of dyes from the heartwood of certain tree species was once an industry of major importance. Dye has long been extracted from barks and heartwoods to color wool, cotton, and other fibers. Pliny describes several dyes derived from oak heartwood, sumac, and other trees, yielding reddish-browns cheaper than the cherished, marine-sourced *Murex* purple. The dyestuff from the Insular Southeast Asian sapang tree (*Caesalpinia sappan*) became a lucrative trade commodity in the Pacific and a common coloring for Indian trade cloths. Brazilwood or pau Brasil (*Caesalpinia echinata*) is congeneric with sapang but grows in the mixed hardwood coastal Atlantic Forest of Brazil. Brazilwood became one of the earliest European exports from the New World, beginning in 1501, and its dye proved even more vivid and colorfast than sapang. The lucrative Portuguese-controlled brazilwood dye market grew quickly, as did a smaller market for its ornamental red wood. While still a luxury, the brazilwood dye was still more easily obtained by Europeans than most others of a similar shade. European demand for the red dye derived from brazilwood was immense, resulting in widespread selective logging of that species. The market for the dyewood tapered in the late nineteenth century with the invention of cheaper synthetic dyes. However, by this time brazilwood had already been logged extensively, and the greater Atlantic Forest ecosystem almost completely obliterated. This species, the national tree of Brazil, remains threatened, with little chance of natural recovery.[16]

The historical linkages between forests and paper production are complex, with paper manufacture developing its dependency on wood pulp only in the mid-nineteenth century. The first true papers, developed in China in the second century A.D., were made from recycled rags and the bast fiber of trees and plants. Each region of the country used those species accessible to them in their locality, mainly paper mulberry, hemp, ramie, and rattan. The papermaking process developed in China was efficient and inexpensive, and created a more useable writing material than anything that had come before it. Official demand increased during the Sung Dynasty, especially after the invention of printing. Manufacture became standardized as demand increased, with the easily harvested and found bamboo the most common source of paper fiber, often combined with paper mulberry. Papermaking technology came to Europe, but the lack of Asian species meant European paper would be made mainly out of expensive cotton rag. The use of wood pulp, rather than bast fiber or cotton rag, was not possible until the development of pulping and bleaching methods in the 1830s. This cheaper method of manufacture initially relied on the otherwise noncommercial willow and poplar, but softwoods yielded a better finished product. As the paper industry grew, the demand for ideal softwood pulp soon prompted establishment of dedicated paper pulp plantations.[17]

Timber was one of the first globalized industrial products, as European empires exploited their colonial holdings worldwide. Large-scale industrial logging has shifted in intensity and location as economy, technology, and demand have evolved. Colonial forests were often an important resource for maintaining both economic and political power. Indeed, the desire to create a reliable source for wood to export to England drove the early stages of that nation's acquisition of lands in northeastern North America. The value of North American pine forests as a source for masts, other naval timber, and non-timber forest products was widely acknowledged at the time of the colonies' founding. However, while the tactical importance of maintaining the supply of potential mast timber remained great, lumber exports from New England were also important for building and other purposes in England. Lumber also generated capital through export to timber-poor European nations. As they had in North America earlier, the British exploited their colonial forest resources in India. Beginning in the mid-1860s, the subcontinent's forest cover changed dramatically due to Indian logging for foreign export, to fuel a growing iron smelting industry, and for railroad ties to build the Indian Railways.[18]

As colonial empires waned, large European and American corporations increasingly dominated the world lumber industry. The Pacific coastal region of North America became the location of extensive logging by the early twentieth century, both for domestic use and for export to timber-poor countries around the world. The region's industry developed in conjunction with the adaptation of complex logging technology and centralized management.[19] The establishment of the first outposts of the lumber industry in the late eighteenth century developed into a major source of income and employment for the region. Lumber harvesting in the region developed as late nineteenth-century industrial capitalism was already in full bloom, creating a different pattern than in regions where it had begun earlier. Technologies to efficiently cut and process the large trees of the Pacific coast, such as the steam-powered "donkey engine" to haul logs, developed in the redwood forests of northern California in the late nineteenth century, and their use spread quickly. As the region's lumber industry grew, government attempted to control the forests through regulating industrial harvest and predicting future growth.[20]

THE INTELLECTUAL HISTORY OF FORESTS: CONSERVATION AND FORESTRY

Our interactions with forests comprise more than just resource extraction. Examining economic use of forests, the topic of the previous section, is one way to approach forest history, but just as important are explorations of how humans have defined forests intellectually, managed them, and understood them. Management of forests has evolved in conjunction with definitions and redefinitions of the forest and its value to a society. Scientific knowledge about forests has been leveraged to govern individual and collective decisions about use and preservation. In this section I first discuss control and management of forests in military, colonial, and political contexts, and the development of government forest conservation. Then I discuss the professionalization of forest management and science, and the rise of preservation movements.

Forests have long had military importance, not just through supplying ship timbers but in facilitating strategic goals during times of war and of occupation. The exertion of one nation's power over the ecosystems of another is a manifestation of control, and an essential and palpable method of declaring dominance by destroying economic livelihood and national pride. Much military strategy hinges on knowledge of landscapes, including forests, as seen from the Roman era to the protracted American efforts to disrupt the Ho Chi Minh Trail during the Vietnam War. The impenetrability of forests by overland armies was a considerable factor in battle plans during both World Wars. The use of wood for war materiel can be traced through millennia, beginning with Bronze Age use of fuelwood in forging metal weapons. Wood was the only way to meet transportation requirements for war vehicles and fortification from enemy attack in Mediterranean, Asian, and other cultures worldwide. As the reach of European and Asian trade, exploration, and warfare increased, nations built larger and more complex wooden ships which required not simply a high volume of wood, but very specific varieties, sizes, and shapes of lumber. Both naval power and merchant trade depended upon wooden-hulled ships which could withstand not just the rigors of open-ocean sailing, but also pitched battles at sea.[21]

Fear of timber shortage, of so-called "timber famines," was especially pronounced during times of naval conflict, as the loss of ships in battle underscored the importance of maintaining supplies of timber suitable for masts and other ship-framing timbers. The decline of Venice as a naval power has been attributed to that water-bound state's difficulty in accessing sufficient timber for maintaining a fleet. As early as 1350, Venetian laws were in place to conserve oak for the Arsenal's use. By the turn of the sixteenth century, nearly all oak had been depleted from Venetian-controlled land.[22] In the naval battle between the Spanish Armada and an English-Dutch alliance in 1588, Spain lost approximately fifty ships. Replacing the ships became a burden on the Spanish forest, as the availability of adequate trees dwindled. England searched for a reliable source of naval timber as domestic supplies of mast timber were quickly depleted by the growing navy. Timber trees for naval stores, especially the long, straight-trunk timbers needed for masts, were an important consideration in the establishment of many European colonies around the world.[23] In England, responses to fears of a timber famine developed over time, as the nation struggled to maintain its military and economic dominance on the world stage. First, trade increased between European naval powers and the timber-rich Baltic states for mast timbers and other naval stores, including pitch for waterproofing. By the late nineteenth century, wooden warships were no longer being built and the linkage between forest conservation and naval power dissolved.

National fears of timber shortages were articulated in military contexts, but such worries were not limited to wartime. As the use of any resource intensifies, those who depend on it anticipate future scarcity. Forecasts of timber shortages have been a periodic phenomenon across cultures. The fear of timber famine is, in one sense, simply a fear of disruption of supply of a needed resource, the worry that a population may outstrip its provisions. The long time to maturity of timber trees, however, meant that addressing any true timber shortage could take decades. Discourses on timber famine express recognition of both environmental degradation and the finite nature of natural resources. State interventions to prevent timber famine can be viewed as early manifestations of the modern conservationist impulse. Maintaining a constant supply of wood required not just a judicious hand in the present but prediction of future use and needs of the forest as well. Timber famine fears led to attempted curtailment or elimination of hunting, grazing, and foraging activities within many European forests. Historically, foragers and fuelwood gatherers have been especially prone to regulation, since they were often low in socioeconomic status and without recourse when access was curtailed. The forest has long been seen as a peripheral place where outcasts find refuge, and managing forests had the added effect of imposing order on places governments perceived as lawless. The creation of forest preserves and royal hunting grounds both reserved forests for elite leisure, and ensured emergency supplies of timber for the crown.[24]

The profession of forestry has deep roots in conservation, but has always drawn on economics and public policy as well as scientific understanding. Government-financed professional forest management constitutes not just the manipulation of forested environments, but also an assertion of state control over natural landscapes.[25] The gamekeepers and woodswards of English royal forests were tasked with ensuring a continuous supply of game animals for royal hunts, maintaining a healthy forest cover, and controlling access. The management of forests for optimum use grew to be a more specialized profession in Europe by the seventeenth century. Forestry first became a defined profession in the eighteenth century, as the Prussian government established schools of professional training for foresters. The development of professional standards for forestry followed a pattern of professionalization across many disciplines in eighteenth- and nineteenth-century Europe. Other European countries, most notably France, followed the German example by creating forest schools to train professionals to manage their forests. During the eighteenth and nineteenth centuries, German forest schools were the primary source of forest management ideas throughout Europe and America. The German forest schools developed a rigorous methodology of management intended to create maximum timber yield.[26] Other nations, following the European model, also linked professionalization of forestry training to intensification of government management of forests. The Biltmore Forest, on the North Carolina estate of George W. Vanderbilt, provided the United States with its first case of modern systematic forest management as well as the first professional training school for foresters. The Biltmore Forest School's role in American forestry education was short-lived, however, and degree-granting forest schools associated with Cornell and Yale were also founded in 1898 and 1900, respectively. [27] British management and harvest of Indian forests constituted a complete restructuring of management of Indian forests around a European scientific-theoretical basis. Foresters in the US, India, and elsewhere started by modeling their efforts on the European schools and used examples of European forests as objects of study. Within a few decades, however, these foresters acknowledged the insurmountable differences between European forests and their local conditions, and broke away to tailor their efforts to the local biology and economy. [28] Soviet forest conservation efforts, within much different political, economic, and professional contexts, demonstrate the extent to which government-organized conservation has always entailed the assertion of state power over land and resources.[29]

Forest science has not only been about discovering how best to manage resources for profit, but also includes forest ecology and natural history. Study of particular tree species, especially those with important cultural uses, has been a constant interest of natural history since antiquity. Pliny the Elder wrote at length about tree species and the attributes of forests.[30] Charles Darwin used forest trees to illustrate the dynamics of competition between individuals of a species for scarce resources.[31] In the twentieth century the scientific study of forests became increasingly allied with the emerging discipline of ecology even as the management, economics, and industrial use of forests remained the purview of professional forestry. The concept of a succession of plant types leading to a stable, indefinitely maintained climax forest was an early, important articulation of ecological relationships for the burgeoning field. Frederic Clements's classification of forest and plant communities and concept of the forest as a climax plant community in a stable equilibrium would heavily influence scientific research in forest ecology for many decades.[32] By the middle of the twentieth century, forest biology became both more scientifically rigorous and more specialized. Studies of the diversity of tree species and the ecology of forests were most often undertaken by scientists trained in plant biology rather than forestry.[33] Ecologists have turned to forests for important long-term ecological studies, including the groundbreaking long-term ecological research on the Hubbard Brook site in New Hampshire, La Selva Biological Station in Costa Rica, and the Wind River Experimental Forest in the Cascade Mountains of Oregon.[34]

Maintaining forests as preserved landscapes, as parks or wilderness, has proved to be a challenging proposition. The first wave of American preservation efforts focused primarily on sublime and unique land formations: rugged canyons and dramatic mountain ranges. However, the forests of some areas of the country were understood to have an intrinsic value and to be worthy of preserving from possible economic activity. The 1892 creation of the Adirondack State Park in New York was an early example of preserving an American forest for recreational, aesthetic, and watershed protection purposes. The Mariposa Big Tree Grove near Yosemite Valley, its immense ancient trees appealing to the preservationist impulse, was deemed so important that in 1864 it was preserved within the boundaries of the new park, despite not being contiguous with Yosemite Valley. A number of other parks were created in the twentieth century with the intention of preserving intact forests, including the Everglades, Congaree, Great Smoky Mountains, Sequoia-Kings Canyon, and Redwoods. As the acreage of intact ancient forest dwindled in the mid-twentieth century, forest itself has become of more interest to new generations of preservationists and wilderness advocates worldwide.[35] However, when their natural ecological dynamism is arrested, the results may be unforeseen by those responsible. Due in part to the region's settlement patterns and in part to its ecology and topography, much of the forest acreage of the American West is public land under state forest, national forest, or other designation. The sites of much of the industrial logging of the twentieth century, these areas became the focus of some of the most hard-fought battles to preserve their remaining untouched forests from commercial activity. Forest conservation, for the federal government, centered on reserving natural resources for the benefit of the economy; tourism and environmental health were secondary goals. The need to commodify Western forests for tourists led to debate on the building of roads into intact forests. Mismanagement in post-World War II federal forest policy resulted in the destruction of much of the American West's forests by under-regulated industrial logging. Relatively small Wilderness Areas were established within the region's otherwise industry-oriented National Forests—isolated preserves of natural forest within a patchwork of industrial development.[36]

ADDRESSING FOREST CHANGES: REPAIRING DAMAGE AND SUSTAINING INDUSTRY

As discussed earlier, some forest types may regrow simply through the neglect of the cleared fields that had replaced them. In other instances, however, a forest's ecological demands mean reestablishing satisfactory forest cover only comes with great cost and effort. Sometimes voluntarily, sometimes compelled by law, the lumber industry has established industrial tree farms on many post-logging landscapes. In 1864 George Perkins Marsh advocated transforming deeply degraded natural places into idealized, semi-natural landscapes modified and designed especially for optimum human use. In the twentieth century, Gifford Pinchot adapted and extended Marsh's conception of reparative land use to create a utilitarian protocol of forest management. As Chief Forester, Pinchot adhered to a utilitarian view of conserving the nation's forest resources, stating that "conservation is the foresighted utilization, preservation and/ or renewal of forests, waters, lands and minerals for the greatest good of the greatest number for the longest time."[37] The slow recovery of the Cutover region of the Upper Midwest, and the resultant decimation of local economies, showed that maintaining the lumber economy of the United States would be difficult without intervention. By the late 1930s, that Forest Service ethos was extended by the lumber industry to imagine that following logging, the denuded land itself would hold potential for further use as the site for a future crop.[38]

The multiple motivations behind large-scale reforestation are often apparent, and most industrial reforestation differs greatly from the ecology of a natural forest. The extensive redistribution of Australian tree species around the world for plantations demonstrates well the breadth and success of technoscientific forest reshaping for industrial goals.[39] By the mid-twentieth century, foresters had developed silvicultural regimes to grow trees quickly for industrial use worldwide. Younger trees and more marginal wood fibers can be used in paper and wood products due to increasingly sophisticated pulp treatment and technological innovations such as plywood and chipboard. In many places the native forest, once logged, has been replaced with faster-growing or more lucrative exotic commercial species. While the work of non-governmental organizations in restoring public lands is the focus of most studies of restoration, much restorative work is taking place on private lands at the instigation of large corporate entities. Industrial tree farms are inspired by economics, public relations, and politics, not environmentalism or aesthetics. As lumber companies involved in sawlog harvest in the southeastern United States shifted into pulpwood harvest, they created a region dominated by the paper industry. These highly concentrated sites of short-rotation plantation forestry mainly consist of young stands of fast-growing, non-native pines. The replanting of tree cover on clear-cut areas slowed soil erosion, protected streams from silt runoff, and provided continued structure for wildlife habitat. Industrial tree plantations for paper or rubber have become widespread around the globe. The lumber industry's foresters have developed methods of industrial reforestation not meant to restore a forest ecosystem, but rather to reconstruct a forest resource for future logging. In its rhetoric as well as its methods, tree farming was a marriage of the lumber industry's practical needs with the larger cultural impetus to curtail forest loss.[40]

Trees are also planted for motives that have little to do with economics, though such plantations rarely mark the significant acreage of industrial tree farms. Tree planting has often memorialized and celebrated human endeavor, or, as with Arbor Day, served civic and political motives. Tree plantings can carry political weight, as with the Liberty Trees planted in town squares as demonstrations of revolutionary sentiment in eighteenth-century Europe and America.[41] Larger-scale plantations may have political and aesthetic overtones as well. The New Deal era's Shelterbelts on the Great Plains were designed to reduce wind and conserve soils and waters in the agricultural areas. The massive reforestation projects undertaken by the Civilian Conservation Corps addressed the damage of the Midwest Cutover region while also providing work and training to the unemployed.[42] Large-scale afforestation has been a facet of the Israeli government's claim to its landscape, especially during the early years of settlement establishment. Along with memorial groves and windbreaks, Israeli tree plantations on previously Arab-owned lands were often intended to "hold the land" and prevent the reestablishment of Arab pasture or farmland on those sites. [43]

Industrial tree farms are undeniably significant, but tree planting may be motivated by reasons other than industrial production. The English undertook a widespread tree-planting project to restore the supply of wood on domestic land during the wooden-ship era. This interest in replanting English forests is expressed most famously in John Evelyn's *Sylva*. The widely read *Sylva*, financed by the Royal Society, served to call attention to England's perceived timber famine. Part arboreal appreciation, part handbook for planting, *Sylva* was widely read and even more widely acclaimed. Evelyn advocated planting forest reserves on private estates as well as maintaining existing wood supplies. Establishing forested acreage became a way for wealthy Englishmen to demonstrate their patriotism. The turn toward the "picturesque" style of European estate design was in part a manifestation of the elite's fears of timber famine. However successful the drive to plant new trees for future naval use may have been, the newly-planted timber did little to solve the immediate needs of the Royal Navy. By the eighteenth century, timber extraction from colonial forests increased to fill the needs of the navy, especially for the difficult-to-obtain mast timber, and the rage for tree-planting abated.[44]

Japan has a long history of persistent and intensive reforestation for political, aesthetic, and environmental goals. Japan experienced severe deforestation on a national scope earlier than most places in the world, and also moved toward government-coordinated reforestation earlier. The demands of timber-intensive building styles and an increasing population began to affect the forests surrounding Japanese cities as early as the seventh century. Throughout the Japanese medieval period, expanding agricultural lands pushed the edges of the forest farther back from human settlement, and increasing use of forests for timber caused large-scale deforestation in Japan by the late sixteenth century. Amid concerns about damage to watersheds, loss of aesthetically and historically important forests, and dwindling domestic timber supply, the Tokugawa-era government enforced both the protection of remaining intact forests and the reforestation of logged areas. By the eighteenth century, in an attempt both to remedy the loss of forests through logging and to ensure forest availability for future generations' consumption, forest managers developed a strict silvicultural regime. The government quickly established plantations througout the country, and began government-run programs of sustained-yield forestry.[45] Japan began to rely on imported timber from around the world to supply its inhabitants with wood. The nation still maintains significant domestic forest cover and Japanese forest ecosystems have largely recovered from their earlier deforestation. The ultimate success of Japan's reforestation efforts, though, must be attributed in large part to the shift toward reliance on imported wood when trade reopened during the post-Sakoku Imperial period of the late nineteenth century.[46]

Organized ecological restoration of forest landscapes recovering from human impact has proven highly complex. In many cases restoring a forest for non-economic reasons has meant not only replanting trees, but also painstakingly reintroducing ecosystem function. Ecologists' efforts to restore fire to forest ecosystems have proven both ecologically difficult and politically fraught.[47] Reintroduction of fire has often occurred in conjunction with increased ecological understanding of a forest type. The longleaf pine-grassland ecosystems of the Southeastern US, for example, evolved in response to periodic burning by humans. While paper plantations and fast-growing exotics now dominate the region's forest cover, areas of natural longleaf pine forest still remain. Their rarity has made them more celebrated, and ecologists have endeavored to restore them by reestablishing their ecological function and biodiversity. Recognition of the fire dependence of natural forests also led to new ecological management policies, including prescribed burns. This has facilitated the creation of effective management of naturally fire-dependent forest ecosystems in the Southeastern United States and elsewhere.[48]

CONCLUSION: GLOBAL FORESTS AND HUMAN IDEALS

Since the 1990s, under the flag of sustainability, internationally coordinated regeneration of tropical forests has often become a hybrid improvement plan, with stated goals of both environmental and social well-being. The term "sustainable forestry" first became prevalent following the 1992 Earth Summit in Rio de Janeiro. The Rio statements on forestry declared that "efforts to maintain and increase forest cover and forest productivity should be undertaken in ecologically, economically and socially sound ways through the rehabilitation, reforestation and re-establishment of trees and forests on unproductive, degraded and deforested lands, as well as through the management of existing forest resources."[49] From this beginning, sustainable forestry has been connected to the regeneration of forests and to large-scale tree planting. In 1994 a consortium of American forest industry representatives established the Sustainable Forestry Initiative, declaring it the industry's voluntary and direct response to the statements at Rio. The SFI provided an impetus and an administrative framework for its lumber industry partners to provide for the reforestation of vast areas of American landscape affected by logging while simultaneously bidding for public approval.[50] Carbon offset planting—the attempt to offset carbon emissions and combat global warming through large-scale tree planting— has brought new interest and scrutiny to large-scale reforestation. While carbon offset planting could potentially repair human-caused damage to the global climate as well as correct human-caused loss of global forest cover, in practice such schemes have often been undertaken without proper foundation in either climate science or forestry.[51] Progress in the field of conservation biology, as well as lessons from past mistakes, has led to new formulations of government policies to preserve species diversity in concert with ecosystem functionality. The United States federal management plans for the endangered Northern Spotted Owl (*Strix occidentalis caurina*) integrated ecological analysis to force a revision of industry-oriented regional forest use. Likewise, plans for continent-spanning wildlife corridors show the potential for North American forest conservation to move toward a scientific sophistication and ecosystem integration missing in earlier eras. Focusing less on pristine wilderness and more on ecosystem function and linkage, such plans steer current and future forest preservation toward ecologically pragmatic ends.[52]

The ruins of Palenque remind us that there is no straightforward linear progression from nature to culture, and that forest transformations are not necessarily historically indelible.[53] The forest is an important locus for the intersection of human impulses with ecological constraints. Political, cultural, economic, and intellectual facets of human behavior are all affected and changed by interaction with forests. Deforestation creates devastating change, but it should not be seen as the only story to tell about human interaction with the forest landscape. The environmental history of forests does not have to be told simply as declensionist parables about human greed and ignorance. While constructing more nuanced forest history may mean sacrificing some polemical value, forests have played far more complex roles in history than as mere backdrops or passive victims of human greed. Indeed, perhaps only through understanding the long history of the fundamentally interwoven forces of human culture and forest ecology can global deforestation really be slowed.

Notes

1. Emmanuel Kreike, "The Palenque Paradox: Beyond Nature-to-Culture," in *Deforestation and Reforestation in Namibia: The Global Consequences of Local Contradictions*

(Leiden: Brill, 2009); Emmanuel Kreike, "The Nature-Culture Trap: A Critique of Late 20th Century Global Paradigms of Environmental Change in Africa and Beyond," *Global Environment* 1 (2008): 114–145; Emmanuel Kreike, "The Palenque Paradox: Bush Cities, Bushmen, and the Bush," in *The Nature of Cities: Culture, Landscape, and Urban Space*, ed. Andrew C. Isenberg (Rochester, NY: University of Rochester Press, 2006), 159–174.

2. On this point, see William Cronon, "A Place for Stories: Nature, History, and Narrative" *Journal of American History* 78, no. 4 (March 1992): 1347–1376.

3. William Cronon, *Changes in the Land: Indians, Colonists, and the Ecology of New England*, 2d ed. (New York: Hill and Wang, 2003); Brian Donahue, *The Great Meadow: Farmers and the Land in Colonial Concord* (New Haven, CT: Yale University Press, 2007). See also Steven Stoll, *Larding the Lean Earth: Soil and Society in Nineteenth Century America* (New York: Hill and Wang, 2003); Ellen Stroud, *Nature Next Door: Cities and Trees in the American Northeast* (Seattle: University of Washington Press, 2012).

4. Richard White, *Land Use, Environment, and Social Change: The Shaping of Island County, Washington* (Seattle: University of Washington Press, 1992).

5. Michael Williams, *Deforesting the Earth: From Prehistory to Global Crisis* (Chicago: University of Chicago Press, 2003).

6. The primary non-anthropogenic vector is usually thought to be floating rafts of drift-wood during hurricanes. For the development of theories of island biogeography and non-anthropogenic ecological invasions see Alfred Russel Wallace, *The Geographical Distribution of Animals* (New York: Harper & Brothers, 1876); Charles S. Elton, *The Ecology of Invasions by Animals and Plants* (New York: Wiley, 1958); Robert H. MacArthur, *The Theory of Island Biogeography* (Princeton, NJ: Princeton University Press, 1967). See also E. Alison Kay, "Darwin's Biogeography and the Oceanic Islands of the Central Pacific, 1859–1909" in *Darwin's Laboratory: Evolutionary Theory and Natural History in the Pacific*, ed. Roy MacLeod and Philip F. Rehbock (Honolulu: University of Hawaii Press, 1994), 49–69.

7. I. A. W. MacDonald, L. Ortiz, J. E. Lawesson, and J. B. Nowak, "The Invasion of Highlands in Galapagos by the Red Quinine-Tree *Cinchona succirubra*," *Environmental Conservation* 15 (1988): 215–220; Heinke Jäger, Alan Tye, and Ingo Kowarik, "Tree Invasion in Naturally Treeless Environments: Impacts of Quinine (*Cinchona pubescens*) Trees on Native Vegetation in Galapagos," *Biological Conservation* 140 (2007): 297–307.

8. Stephen J. Pyne, *Fire in America: A Cultural History of Wildland and Rural Fire* (Seattle: University of Washington Press, 1982); Mark Hudson, *Fire Management in the American West: Forest Politics and the Rise of Megafires* (Boulder: University Press of Colorado, 2011); William M. Denevan, "The Pristine Myth: The Landscape of the Americas in 1492," *Annals of the Association of American Geographers* 82, no. 3 (1992): 369–385. See also Stephen J. Pyne, *Burning Bush: A Fire History of Australia* (Seattle: University of Washington Press, 1998); Johann Georg Goldammer, *Fire in Tropical Biota: Ecosystem Processes and Global Challenges* (Basel: Berkhauser Verlag, 1992). For an early articula-tion of this idea see Carl Sauer, "The Agency of Man on Earth," in *Man's Role in Changing the Face of the Earth*, ed. William L. Thomas (Chicago: University of Chicago Press, 1956), 49–69.

9. Stoll, *Larding the Lean Earth*; R. Harold Brown, *The Greening of Georgia: The Improvement of the Environment in the Twentieth Century* (Macon, GA: Mercer University Press, 2002); Michelle M. Steen-Adams, Nancy Langston, and David J. Mladenof, "White Pine in the Northern Forests: An Ecological and Management History of White Pine on the

Bad River Reservation of Wisconsin," *Environmental History* 12, no. 3 (July 2007): 614–648; James Kates, *Planning a Wilderness: Regenerating the Great Lakes Cutover Region* (Minneapolis: University of Minnesota Press, 2001).

10. Cronon, *Changes in the Land.*

11. Clarence Glacken, *Traces on the Rhodian Shore: Nature and Culture in Western Thought from Ancient Times to the End of the Eighteenth Century* (Berkeley: University of California Press, 1967); John F. Richards, *The Unending Frontier: An Environmental History of the Early Modern World* (Berkeley: University of California Press, 2006). For an analysis and overview of the impact of charcoal production, see Williams, *Deforesting the Earth.* For bamboo, see Mark Elvin, *The Retreat of the Elephants: An Environmental History of China* (New Haven, CT: Yale University Press, 2004). For maple syrup, see Stroud, *Nature Next Door.*

12. Robert B. Outland III, *Tapping the Pines: The Naval Stores Industry in the American South* (Baton Rouge: Louisiana State University Press, 2004). For the global naval stores industry see also Robert Greenhalgh Albion, *Forests and Sea Power: The Timber Problem of the Royal Navy, 1652–1862* (Cambridge, MA: Harvard University Press, 1926); Joseph J. Malone, *Pine Trees and Politics: The Naval Stores and Forest Policy in Colonial New England* (Seattle: University of Washington Press, 1964); Sven Erik Åström, *From Tar to Timber: Studies in Northeast European Forest Exploitation and Foreign Trade, 1660–1860* (Helsinki: Finnish Society of Sciences and Letters, 1988).

13. Williams, *Americans and their Forests*; Albert E. Cowdrey, *This Land, This South: An Environmental History* (Lexington: The University Press of Kentucky, 1983); Timothy Silver, *A New Face on the Countryside: Indians, Colonists, and Slaves in South Atlantic Forests, 1500–1800* (Cambridge: Cambridge University Press, 1990).

14. Colin Barlow, *The Natural Rubber Industry:Its Development, Technology, and Economy in Malaysia* (Kuala Lumpur: Oxford University Press, 1978); Warren Dean, *Brazil and The Struggle for Rubber: A Study in Environmental History.* (Cambridge: Cambridge University Press, 1987); Austin Coates, *The Commerce in Rubber: The First 250 Years* (Singapore: Oxford University Press, 1987); Colin Barlow, Sisira Jayasuriya, and C. Suan Tan, *The World Rubber Industry* (London: Routledge, 1994). See also Greg Grandin, *Fordlandia: The Rise and Fall of Henry Ford's Forgotten Jungle City* (New York: Picador, 2010); William Beinart and Lotte Hughes, "Rubber and the Environment in Malaysia," in *Environment and Empire* (New York: Oxford University Press, 2009), 233–250.

15. Richard Grove, *Green Imperialism: Colonial Expansion, Tropical Island Edens, and the Origins of Environmentalism, 1600–1860* (New York: Cambridge University Press, 1995); Dennis M. Roth, "Philippine Forests and Forestry: 1565–1920," in *Global Deforestation in the Nineteenth-Century World Economy*, ed. Richard P. Tucker and J. F. Richards (Durham, NC: Duke University Press, 1983); Williams, *Deforesting the Earth*; Warren Dean, *With Broadaxe and Firebrand: The Destruction of the Brazilian Atlantic Forest* (Berkeley: University of California Press, 1997); Richard Dunn, *Sugar and Slaves: The Rise of the Planter Class in the English West Indies 1624–1713* (Chapel Hill: University of North Carolina Press, 2000); Thomas D. Rogers, *The Deepest Wounds: A Labor and Environmental History of Sugar in Northeastern Brazil* (Chapel Hill: University of North Carolina Press, 2010). For a sweeping discussion of traditional slash-and-burn agriculture, see Joseph Earle Spencer, *Shifting Cultivation in Southeastern Asia* (Berkeley: University of California Press, 1966). For discussion of fire in the *Mahābhārata*, see Stephen J. Pyne, *World Fire: The Culture of Fire on Earth* (Seattle: University of Washington Press, 2010).

16. Pliny the Elder, *The Natural History*, especially 16.53, 22.3, and 30.18. For sappanwood see John Guy, *Woven Cargoes: Indian Textiles and the East* (New York: Thames and Hudson, 1998); note also that sappanwood is sometimes incorrectly termed brazilwood. For brazilwood see Dean, *Broadaxe and Firebrand*; Gustavo A. B. da Fonseca, "The Vanishing Brazilian Atlantic Forest," *Biological Conservation* 34 (1985):17–34. For the rise of synthetic dyes see Carsten Reinhardt and Anthony Travis, *Heinrich Caro and the Creation of Modern Chemical Industry* (New York: Springer, 2001); Anthony Travis, *The Rainbow Makers: The Origins of the Synthetic Dyestuffs Industry in Western Europe* (Bethlehem, PA: Lehigh University Press, 1993).

17. Joseph Needham with Tsien Tsuen-Hsuin, *Paper and Printing*, vol. 5, pt. 1 of *Science and Civilization in China* (Cambridge: Cambridge University Press, 1985); Ricardo Carrere and Larry Lohmann, *Pulping the South: Industrial Tree Plantations and the World Paper Economy* (London: Zed Books, 1996).

18. Ramachandra Guha and Madhav Gadgil described British colonialism in India as "an ecological watershed," and highlighted forest management as "possibly the most important aspect of the ecological encounter" between the two countries. Madhav Gadgil and Ramachandra Guha, *The Use and Abuse of Nature* [an omnibus edition incorporating *This Fissured Land: An Ecological History of India* (1992) and *Ecology and Equity* (1995)] (New Delhi: Oxford University Press, 2000), 116.

19. William G. Robbins, *Landscapes of Promise: The Oregon Story 1800–1940* (Seattle: University of Washington Press, 1997); Richard Rajala, *Clearcutting the Pacific Rainforest: Production, Science, and Regulation* (Vancouver: University of British Columbia Press, 1998); Thomas R. Cox, *The Lumberman's Frontier: Three Centuries of Land Use, Society, and Change in America's Forests* (Corvallis: Oregon State University Press, 2010).

20. Thomas R. Cox, *Mills and Markets: A History of the Pacific Coast Lumber Industry to 1900* (Seattle: University of Washington Press, 1975); Robert E. Ficken, *The Forested Land: A History of Lumbering in Western Washington* (Seattle: University of Washington Press, 1987); Robert Bunting, *The Pacific Raincoast: Environment and Culture in an American Eden, 1778–1900*; William G. Robbins, *Hard Times in Paradise: Coos Bay, Oregon, 1850–1896.* (Seattle: University of Washington Press, 1988); William G. Robbins, *Landscapes Of Conflict: The Oregon Story, 1940–2000* (Seattle: University of Washington Press, 2004); Andrew C. Isenberg, *Mining California: An Ecological History* (New York: Hill and Wang, 2005). W. Scott Prudham, *Knock on Wood: Nature as Commodity in Douglas-fir Country* (New York: Routledge, 2005). See also White, *Land Use, Environment, and Social Change*.

21. Richard Stevens, *The Trail: A History of the Ho Chi Minh Trail and the Role of Nature in the War in Vietnam* (New York: Garland Publishing, 1993); J. R. McNeill, "Woods and Warfare in World History," *Environmental History* 9, no. 3 (July 2004): 388–410; Steward Gordon, "War, the Military, and the Environment: Central India, 1560–1820" in *Natural Enemy, Natural Ally: Toward an Environmental History of War*, ed. Richard P. Tucker and Edmund Russell (Corvallis: Oregon State University Press, 2004); David Biggs, "Managing a Rebel Landscape: Conservation, Pioneers, and the Revolutionary Past in the U Minh Forest, Vietnam," *Environmental History* 10, no. 3 (July 2005): 448–476 ; David Zierler, *The Invention of Ecocide: Agent Orange, Vietnam, and the Scientists Who Changed the Way We Think About the Environment* (Athens: University of Georgia Press, 2011); Lisa Brady, *War upon the Land: Military Strategy and the Transformation*

of Southern Landscapes during the American Civil War (Athens: University of Georgia Press, 2012).

22. John Perlin, *A Forest Journey: The Role of Wood in the Development of Civilization* (New York: W. W. Norton, 1989); Richard P. Tucker, "The World Wars and the Globalization of Timber Cutting," in *Natural Enemy, Natural Ally: Toward an Environmental History of War*, ed. Richard P. Tucker and Edmund Russell (Corvallis: Oregon State University Press, 2004), 110–141; Karl Appuhn, *A Forest On the Sea: Environmental Expertise in Renaissance Venice* (Baltimore: Johns Hopkins University Press, 2009).

23. Albion, *Forests and Sea Power*; Malone, *Pine Trees and Politics*; Perlin, *Forest Journey*; Graeme Wynn, *Timber Colony: A Historical Geography of Nineteenth-Century New Brunswick* (Toronto: University of Toronto Press, 1981); Simon Schama, *Landscape and Memory* (New York: Alfred A. Knopf, 1995).

24. At one time the word "forest" had a more specific meaning related to extensive use. Strictly speaking, in medieval England the Royal Forest denoted a controlled-access deer park tended to maximize the population of game animals. The upper class could hunt at leisure and be assured of plentiful game, while lower-class access, both for game and for firewood, was restricted. Landscapes designated as forest could be virtually tree-less, with the primary landscape being heath, moor, or even fen, if they were managed with this protocol. Land-use policies and cultural practices developed to manage this sort of state-owned resource. The literature on the history of the English forest is extensive, see especially the classic texts, W. G. Hoskins, *The Making of the English Landscape* (London: Hodder and Stoughton, 1963); Oliver Rackham, *Trees and Woodland in the British Landscape* (London: J. M. Dent & Sons, 1976); Robert Pogue Harrison, *Forests: The Shadow of Civilization* (Chicago: University of Chicago Press, 1992); Oliver Rackham, *The History of the Countryside* (London: Weidenfeld & Nicolson, 1995).

25. A point made well in the "parable" of the state and scientific forestry which opens James C. Scott's influential *Seeing Like a State: How Certain Schemes to Improve the Human Condition Have Failed* (New Haven: Yale University Press, 1998). See also Gregory Barton, *Empire Forestry and the Origins of Environmentalism* (Cambridge: Cambridge University Press, 2002).

26. Henry Lowood, "The Calculating Forester: Quantification, Cameral Science, and the Emergence of Scientific Forest Management in Germany," in *The Quantifying Spirit in the Eighteenth Century*, ed. Tore Frangsmyr, J. L. Heilbron, and Robin E. Rider (Berkeley, University of California Press, 1990), 315–342.

27. While a number of universities and agricultural colleges offered coursework in forest topics in the mid- to late nineteenth century, professional degrees in forestry were not offered in the US until the turn of the century. On the history of American forestry see Henry Clepper, *Professional Forestry in the United States* (Baltimore: Johns Hopkins University Press, 1983); William G. Robbins, *American Forestry: A History of National, State, and Private Cooperation* (Lincoln: University of Nebraska Press, 1985); Char Miller, *Gifford Pinchot and the Making of Modern Environmentalism* (Washington: Island Press, 2001). For Biltmore, see Carl Schenck *Cradle of Forestry in America: The Biltmore Forest School, 1989–1913*, ed. Ovid Butler (Durham, NC: Forest History Society, 1998). See also Mark Kuhlberg, *One Hundred Years and Counting: Forestry Education and Forestry in Toronto and Canada, 1907–2007* (Toronto: University of Toronto Press, 2009).

28. On British forestry on the Indian subcontinent, see especially Ramachandra Guha, *The Unquiet Woods: Ecological Change and Peasant Resistance in the Himalaya*, expanded

ed. (Berkeley: University of California Press, 2000). See also Raymond L. Bryant, "Rationalizing Forest Use in British Burma 1856–1942" in *Global Deforestation in the Nineteenth-Century World Economy*, ed. Richard P. Tucker and J. F. Richards (Durham, NC: Duke University Press, 1983); S. S. Negi, *Sir Dietrich Brandis: Father of Tropical Forestry* (Dehra Dun, India: Bishen Singh Mahendra Pal Singh, 1991); Benjamin Weil, "Conservation, Exploitation, and Cultural Change in the Indian Forest Service, 1875–1927," *Environmental History* 11, no. 2 (2006): 319–343.

29. Douglas Weiner, *A Little Corner of Freedom, Russian Nature Protection from Stalin to Gorbachev* (Berkeley: University of California Press, 2002); Stephen Brain, *Song of the Forest: Russian Forestry and Stalinist Environmentalism, 1905–1953* (Pittsburgh: University of Pittsburgh Press, 2011).

30. The natural history of forests and forest species was a major subject of both Pliny the Elder and Theophrastus. Theophrastus, *De Historia Plantarum*, books 2–5; Pliny the Elder, *The Natural History*, Book 16, "The Natural History of the Forest Trees." See also Plato, *Critias*, 110–112.

31. Charles Darwin, *On the Origin of Species by Means of Natural Selection*, 1st ed. facsimile (Cambridge, MA: Harvard University Press, 2001). See especially 74–75.

32. Frederic E. Clements, "Nature and Structure of the Climax," *The Journal of Ecology*, 24, No. 1 (February 1936): 252–284; Frederic E. Clements, *Plant Succession: An Analysis of the Development of Vegetation* (Washington, DC: Carnegie Institution of Washington, 1916). In the 1890s Henry Chandler Cowles studied the plant communities of the Sand Dunes on the southern shores of Lake Michigan, describing successive replacements of arrays of plant species with woodier, more complex tree species. Cowles argued for a parallel between walking inland from loose sand into increasingly dense vegetation, and traveling forward in time through ecological generations. Henry Chandler Cowles, "The Ecological Relations of the Vegetation on the Sand Dunes of Lake Michigan," *Botanical Gazette* (1899): 27, 95–117, 167–202, 281–308, 361–388. Much has been written on the history of ecology; of some relevance to the current discussion are Ronald C. Tobey, *Saving the Prairies: The Life Cycle of the Founding School of American Plant Ecology, 1895–1955* (Berkeley: University of California Press, 1981); Eugene Cittadino, "Ecology and the Professionalization of Botany in America, 1890–1905," *Studies in the History of Biology* 4 (1980): 171–198; Frank Golley, *A History of the Ecosystem Concept in Ecology: More than the Sum of its Parts* (New Haven, CT: Yale University Press, 1993); Peder Anker, *Imperial Ecology: Environmental Order in the British Empire, 1895–1945* (Cambridge, MA: Harvard University Press, 2001); Sharon Kingsland, *The Evolution of American Ecology, 1890–2000* (Baltimore: Johns Hopkins University Press, 2005); Nancy Slack, *G. Evelyn Hutchinson and the Invention of Modern Ecology* (New Haven, CT: Yale University Press, 2011). See also Emily K. Brock, "The Challenge of Forest Restoration: Experiments in the Douglas Fir Forest, 1920–1940," *Environmental History* 9, no. 1 (January 2004): 57–79.

33. As just three quite different examples of important midcentury forest-based ecological papers, see J. Roger Bray and J. T. Curtis, "An Ordination of the Upland Forest Communities of Southern Wisconsin," *Ecological Monographs* 27 (1957): 325–349; Robert H. MacArthur, "Population Ecology of some Warblers of Northeastern Coniferous Forests," *Ecology* 39 (1958): 599–619; and Jerry S. Olson, "Energy Storage and the Balance of Producers and Decomposers in Ecological Systems," *Ecology* 44 (1963): 322–331.

34. Gene E. Likens and F. Herbert Bormann, *Pattern and Process in a Forested Ecosystem: Disturbance, Development, and the Steady State based on the Hubbard Brook*

Ecosystem Study (New York: Springer-Verlag, 1979); Margaret Herring and Sarah Greene, *Forest of Time: A Century of Science at Wind River Experimental Forest* (Corvallis: Oregon State University Press, 2007); Susan M. Pierce, "Environmental History of La Selva Biological Station," in *Changing Tropical Forests: Historical Perspectives on Today's Challenges in Central and South America*, ed. Harold K. Steen and Richard P. Tucker (Durham, NC: Forest History Society, 1992). The development of the Long Term Ecological Research program of the National Science Foundation has also nurtured a number of such forest studies. For an analysis of the limitations of forest ecology see Nelson G. Hairston, "Experiments in Forests," in *Ecological Experiments: Purpose, Design, and Execution* (Cambridge: Cambridge University Press, 1989), 67–127.

35. Alfred Runte, *Yosemite: The Embattled Wilderness* (Lincoln: University of Nebraska Press, 1990); Roderick Nash, *Wilderness and the American Mind*, 4th ed. (New Haven, CT: Yale University Press, 2001); Donald Worster, *A Passion for Nature: The Life of John Muir* (New York: Oxford University Press, 2008); James Morton Turner, *The Promise of Wilderness: American Environmental Politics since 1964* (Seattle: University of Washington Press, 2012). For another example of forests as subject of wilderness activism, see Robert Marshall, *The People's Forests* (H. Smith and R. Haas, 1933).

36. The history of American forest conservation and preservation in the nineteenth and twentieth centuries has been a subject of deep study by environmental historians, including Susan R. Schrepfer, *The Fight to Save the Redwoods: A History of Environmental Reform, 1917–1978* (Madison: University of Wisconsin Press, 1983); Thomas R. Cox, *The Park Builders: A History of State Parks in the Pacific Northwest* (Seattle: University of Washington Press, 1988); Paul Hirt, *A Conspiracy of Optimism: Management of the National Forests Since World War Two* (Lincoln: University of Nebraska Press, 1994); Nancy Langston, *Forest Dreams, Forest Nightmares: The Paradox of Old Growth in the Inland West* (Seattle: University of Washington Press, 1995); Samuel P. Hays, *Conservation And The Gospel Of Efficiency: The Progressive Conservation Movement, 1890–1920* (Pittsburgh: University of Pittsburgh Press, 1999); Paul Sutter, *Driven Wild: How the Fight Against Automobiles Launched the Modern Wilderness Movement* (Seattle: University of Washington Press, 2002); David Louter, *Windshield Wilderness: Cars, Roads, and Nature in Washington's National Parks* (Seattle: University of Washington Press, 2006); Kevin Marsh, *Drawing Lines in the Forest: Creating Wilderness Areas in the Pacific Northwest* (Seattle: University of Washington Press, 2007); and Samuel P. Hays, *Wars in the Woods: The Rise of Ecological Forestry in America* (Pittsburgh: University of Pittsburgh Press, 2007).

37. Gifford Pinchot, *Breaking New Ground* (New York: Harcourt Brace, 1947).

38. George Perkins Marsh, *Man and Nature, or Physical Geography as Modified by Human Action*, ed. David Lowenthal (Cambridge, MA: Harvard University Press, 1965); David Lowenthal, *George Perkins Marsh: Prophet of Conservation*, 2d ed. (Seattle: University of Washington Press, 2003). See also Henry Clepper, *Professional Forestry in the United States* (Baltimore: Johns Hopkins University Press, 1971); Char Miller, *Gifford Pinchot and the Making of Modern Environmentalism* (Washington: Island Press, 2001); Marcus Hall, *Earth Repair: A Transatlantic History of Environmental Restoration* (Charlottesville: University of Virginia Press, 2005).

39. Brett M. Bennett, "A Global History of Australian Trees," *Journal of the History of Biology* 44 (2011): 125–145. See also Thomas Dunlap, *Nature and the English Diaspora: Environmental History in the United States, Canada, Australia, and New Zealand* (Cambridge: Cambridge University Press, 1999).

40. Emily K. Brock, "Tree Farms on Display: Presenting Industrial Forests to the Public in the Pacific Northwest, 1941–1960" *Oregon Historical Quarterly* 113, no. 4 (Winter 2012): 526–559. For replacement of native species with exotics in industrial reforestation see Ricardo Carrere and Larry Lohmann, *Pulping the South: Industrial Tree Plantations and the World Paper Economy* (London: Zed Books, 1996). For a case study of large-scale cooperative reforestation see Gail Wells, *The Tillamook: A Created Forest Comes of Age* (Corvallis: Oregon State University Press, 1999).

41. Schama, *Landscape and Memory*; Gayle Samuels, *Enduring Roots: Encounters with Trees, History, and the American Landscape* (New Brunswick, NJ: Rutgers University Press, 1999); Shaul Cohen, *Planting Nature: Trees and the Manipulation of Environmental Stewardship in America* (Berkeley: University of California Press, 2004); Henry W. Lawrence, *City Trees: A Historical Geography from the Renaissance through the Nineteenth Century* (Charlottesville: University of Virginia Press, 2006). See also "Liberty Tree," a poem by Revolutionary War figure Thomas Paine, in which the tree, planted by the "Goddess of Liberty," is praised highly. It is worth noting that Thomas Jefferson's oft-quoted declaration that "the tree of liberty must be refreshed from time to time with the blood of patriots & tyrants" [in a letter from Thomas Jefferson to William Smith on Nov. 13, 1787] alluded not to those Liberty Trees of the Revolution years, but to the cynical resurrection of that symbol by the participants in Shay's Rebellion.

42. William G. Robbins, *American Forestry: A History of National, State, and Private Cooperation* (Lincoln: University of Nebraska Press, 1985); Neil M. Maher, *Nature's New Deal: The Civilian Conservation Corps and the Roots of the American Environmental Movement* (New York: Oxford University Press, 2009); Joel Orth, "Directing Nature's Creative Forces: Climate Change, Afforestation, and the Nebraska National Forest," *The Western Historical Quarterly* 42, no. 2 (2011): 197–217; see also Robert Gardner, "Constructing a Technological Forest: Nature, Culture, and Tree-Planting in the Nebraska Sand Hills," *Environmental History* 14 (April 2009): 275–297.

43. Shaul Cohen, *The Politics of Planting: Israeli-Palestinian Competition for Control of Land in the Jerusalem Periphery* (Chicago: University of Chicago Press, 1993).

44. John Evelyn, *Sylva: Or, a Discourse on Forest Trees, and the Propagation of Timber in His Majesty's Dominion* (York: A. Ward, 1786 [1666]); Blanche Henrey discusses Sylva and related works at length in *British Botanical and Horticultural Literature before 1800* (New York: Oxford, 1975). See Schama, *Landscape and Memory*, for discussion of the picturesque.

45. Masako M. Osako, "Forest Preservation in Tokugawa Japan," in *Global Deforestation in the Nineteenth-Century World Economy*, ed. Richard P. Tucker and J. F. Richards (Durham, NC: Duke University Press, 1983); Conrad Totman, *The Green Archipelago: Forestry in Preindustrial Japan* (Berkeley: University of California Press, 1989); Conrad Totman, *The Lumber Industry in Early Modern Japan* (Honolulu: University of Hawaii Press, 1995). See also John F. Richards, *The Unending Frontier: An Environmental History of the Early Modern World* (Berkeley: University of California Press, 2006).

46. Thomas R. Cox, "The North American-Japanese Timber Trade," in *World Deforestation in the Twentieth Century*, ed. Richard P. Tucker and John F. Richards (Durham, NC: Duke University Press, 1989); M. Patricia Marchak, *Logging the Globe* (Montreal: McGill-Queens University Press, 1995); Nigel Sizer, "Trade, Transnationals, and Tropical Deforestation," in *Footprints in the Jungle: Natural Resource Industries, Infrastructure, and Biodiversity Conservation*, ed. Ian A. Bowles and Glenn T. Prickett (Oxford: Oxford University Press, 2001), 115–133.

47. For a valuable discussion of the reintroduction of fire see Eric Higgs, *Nature by Design: People, Natural Process, and Ecological Restoration* (Cambridge, MA: MIT Press, 2003). See also Joanne Vining, Elizabeth Tyler, and Byoung-Suk Kweon, "Public Values, Opinions, and Emotions in Restoration Controversies," in *Restoring Nature: Perspectives from the Social Sciences and Humanities*, ed. Paul H. Gobster and R. Bruce Hull (Washington: Island Press, 2000), 143–162.

48. Albert G. Way, *Conserving Southern Longleaf: Herbert Stoddard and the Rise of Ecological Land Management* (Athens: University of Georgia Press, 2011).

49. "Non-Legally Binding Authoritative Statement of Principles for a Global Consensus on the Management, Conservation and Sustainable Development of all Types of Forests (Annex III)," principle 8(b), in United Nations Document A/CONF.151/26 (Vol. III), Report of the United Nations Conference on Environment and Development (Rio de Janeiro, June 3–14, 1992). See also Roger A. Sedjo, Alberto Goetzl, and Steverson O. Moffat, *Sustainability of Temperate Forests* (Washington DC: Resources for the Future, 1998); David Humphreys, *Logjam: Deforestation and the Crisis of Global Governance* (London: Earthscan, 2006).

50. Sustainable Forestry Initiative, "2005–2009 Standard." See also Jerry F. Franklin, "The Fundamentals of Ecosystem Management with Applications in the Pacific Northwest," in *Defining Sustainable Forestry*, ed. Gregory H. Aplet, Nels Johnson, Jeffrey T. Olson, and Al Sample (Washington DC: Island Press, 1993), 127–144; Benjamin Cashore, Ilan Vertinsky, and Rachana Raizada, "Firms' Responses to External Pressures for Sustainable Forest Management in British Columbia and the US Pacific Northwest," in *Sustaining the Forests of the Pacific Coast: Forging Truces in the War in the Woods*, ed. Debra J. Salazar and Donald K. Alper (Vancouver: University of British Columbia Press, 2000), 80–119; Cohen, *Planting Nature*.

51. "Rio Declaration on Environment and Development (Annex I)," principle 1, in United Nations Document A/CONF.151/26 (Vol. I), Report of the United Nations Conference on Environment and Development (Rio de Janeiro, June 3–14, 1992); "The Role of Land Carbon Sinks in Mitigating Global Climate Change," Royal Society Policy Document 10/01(July 2001); see also Stephen Bocking, *Nature's Experts: Science, Policy and the Environment* (New Brunswick, NJ: Rutgers University Press, 2004); Cohen, *Planting Nature*.

52. Russell Lande, "Demographic Models of the Northern Spotted Owl (*Strix occidentalis caurina*)," *Oecologia* 75 (1988): 601–607. For discussion of the spotted owl case see Barry R. Noon and Kevin S. McKelvey, "Management of the Spotted Owl: A Case History in Conservation Biology," *Annual Review of Ecology and Systematics* 27 (1996): 135–162; Susan Harrison, Andy Stahl, and Daniel Doak, "Spatial Models and Spotted Owls: Exploring Some Biological Issues Behind Recent Events," *Conservation Biology* 7, no. 4 (1993): 950–953. The literature on ecoforestry and sustainable forestry is vast, for some entry points see Stephanie Mansourian, Daniel Vallauri, and Nigel Dudley, eds., *Forest Restoration in Landscapes: Beyond Planting Trees* (New York: Springer, 2005); Alan Drengson and Victoria Stevens, "Ecologically Responsible Restoration and Ecoforestry," in *Ecoforestry: The Art and Science of Sustainable Forest Use*, ed. Duncan M. Taylor and Alan R. Drengson (Gabriola Island, BC: New Society Publishers, 1997), 68–74; Reed Noss, "Wilderness Recovery: Thinking Big in Restoration Ecology," in *Restoration Forestry: An International Guide to Sustainable Forestry Practices*, ed. Michael Pilarski (Durango, CO: Kivaki Press, 1994), 92–101; Dean Apostol and Marcia Sinclair, *Restoring the Pacific Northwest: The Art and Science of Ecological Restoration in Cascadia* (Washington: Island Press, 2006).

53. Kreike, "Palenque Paradox."

....

THE TROPICS

A Brief History of an Environmental Imaginary

....

PAUL S. SUTTER

IN 1898, the British sociologist Benjamin Kidd published a short book titled *The Control of the Tropics*. Kidd's treatise appeared not only at the height of the New Imperialism among European powers, but also as the United States was expanding its territorial holdings and commercial interests beyond continental bounds. How peoples from temperate regions would control the tropics in the twentieth century was Kidd's great concern. Kidd saw the tropics as "the region most richly endowed by nature on the face of the globe—a region possessing capacities of production probably beyond any that have been imagined." And yet he averred that the nonwhite natives of the tropics, the region's "*natural* inhabitants," were incapable of developing their resources or even of achieving meaningful self-government: "there never has been, and never will be, within any time with which we are practically concerned, such a thing as good government, in the European sense, of the tropics by the natives of those regions." Thus, white temperate peoples were best situated to bring tropical development to fruition. But there was a catch. According to Kidd, the tropics was a region of "the most unhealthy physical and moral conditions" for white peoples from temperate latitudes, and he insisted on "the innate unnaturalness of the whole idea of acclimatization in the tropics, and of every attempt arising out of it to reverse by any effort within human range the long, slow process of evolution which has produced such a profound dividing line between the inhabitants of the tropics and those of the temperate regions." In the tropics, Kidd evocatively concluded, "the white man lives and works only as a diver lives and works under water."[1]

Benjamin Kidd's *The Control of the Tropics* was a vivid example of a much larger tradition within Western thinking of seeing the tropics as a place distinct from the temperate (and polar) regions of the globe—a place that, most importantly, seemed to resist all efforts by temperate Europeans and North Americans at settlement, control, and development. Not everyone who wrote, spoke, or acted within this genre agreed with Kidd's characterization of the tropics, and, as we will see, the twentieth century saw an environmentalist reconceptualization of the tropics, one that began not long after Kidd's book appeared. But almost all have agreed that the tropics stood in sharp contrast to the temperate world. My major focus in this chapter is to trace this history of tropical thinking. I am interested in how theorists such as Kidd came to see the tropics as a discrete and sharply delineated geographical, ecological, medical, and racial space, and the instrumental purposes that such sharp delineations served. This, then, is a brief history of the tropics as a Western environmental imaginary. I do not deny that real material differences existed between the places here described as temperate and tropical; indeed, I will often point out those differences. But I am less interested in those material differences than I am in the intellectual history of the tropics as a category for defining and containing such difference. If this brief history of tropical thinking has an argument, it is that environmental historians need to engage in the same kind of critical reconsideration of "the tropics" as an environmental category as they have for wilderness.[2]

While there is a deep history to tropical thinking, discussed briefly below, this chapter focuses on the emergence of the tropics as a modern environmental category. As a number of scholars have noted, tropicality (a term that indicates a bundle of qualities commonly ascribed to the region) took on a distinctive cast in the process of European expansion, and gained greater scientific precision as an organizing category beginning in the late eighteenth century, when Europeans, North Americans, Australians, and other so-called temperate peoples attempted to explore, exploit, and settle tropical places.[3] Indeed, it was during this period that tropicality took shape as a modern discourse akin to the "Orientalist" discourse Edward Said described—a discourse aimed not merely at understanding, but also at producing, a region called the tropics, and building up bodies of expertise designed to study, discipline, and control it.[4] Seeing tropicality as a discourse does not imply that all differences between tropical and temperate places are socially constructed; rather, it insists that we see the artifice of, and the instrumental intention behind, starkly dividing these differences into the dual categories of tropical and temperate. Said's analysis, although too monolithic in its attention to the power of a projected discourse, has helped a number of scholars to understand the tropics as a category created by outsiders to the region with interests in it—a category that both organized disparate phenomena in ideological ways and constantly needed defense against instabilities and anomalies.[5]

TROPICAL PRE-HISTORY

What do we mean when we talk about the tropics? In strict geographical terms, the tropics is the region that lies between the Tropics of Cancer and Capricorn, which circumscribe the Earth at 23° 27' north and south latitudes. The equator marks the region's center. These tropic lines delineate the farthest points, north and south, at which the sun can be directly overhead during the earth's seasonal tilting on its axis. In this sense, the "tropics" is a term that links the terrestrial to the celestial. The lands within these tropic lines include: large portions of the Americas, extending from central Mexico and the Caribbean South to just inside the northern border of Argentina and excluding most of the Southern cone nations; the majority of Africa, excluding only Mediterranean North Africa and the continent's southern tip; portions of the Arabian Peninsula, including Yemen, most of Oman, and parts of Saudi Arabia and the United Arab Emirates; Southern India as well as Sri Lanka; most of Southeast Asia, including Indonesia and the Philippines; the northern third of Australia; most of Oceania or the South Pacific; and significant portions of the Atlantic, Pacific, and Indian Oceans. Of the populated continents, only Europe lacks tropical territory. But the qualities that came to define tropicality in the modern era were not always confined within the tropic lines, and large areas within those lines—particularly its desert and savanna regions—have evaded the tropical label. The archetypal tropics are the humid lowlands and palm-lined tropical islands.

One can trace the identification of discrete geographical zones of climate and ideas about their medical and constitutional effects back to the ancient world.[6] In his *Geography*, Strabo (64/63 BCE—ca. 24 CE) attributed the division of the earth into five zones (polar, temperate, torrid, temperate, polar) to Parmenides (fifth century BCE), and such geographical divisions were well established by Aristotle (384–322 BCE), who wrote of the Earth as consisting of a "Torrid Zone," confined within the tropic lines and flanked by "Temperate" and then "Frigid" Zones. Aristotle insisted that only temperate areas were habitable by humans; "the lands beyond the tropics are uninhabitable," he wrote.[7] While others of his and later eras disagreed with his thinking on the Torrid Zone, Aristotle's formulation was nonetheless a critical one—not only because his argument about tropical habitability would be at the heart of debates over European exploration and expansion, but also because he mixed his geographical categories with chauvinistic political and ethnographic observations that have had tremendous lasting power. In particular, he applied his theory of the golden mean—that virtue is found in a balance between opposite extremes—to geographical and medical thinking. For Aristotle, the temperate was not merely a habitable geographic zone; it was a philosophical ideal that stood in opposition to the frigid and the torrid.

Another body of ancient theorizing that influenced the modern notion of the tropics (and, for that matter, Aristotle) was the Hippocratic corpus, and particularly *Airs, Waters, Places*, a work associated with the physician Hippocrates of Cos (ca. 460– ca. 370 BCE) and his followers. The Hippocratic corpus not only introduced humoral medicine—the notion that health rests in the states and balances of the four bodily humors: blood, black bile, yellow bile, and phlegm—but it also pioneered medical geography, the study of the ways in which particular places, environments, and climates influenced bodily health. Importantly, *Airs, Waters, and Places* not only offered specific advice about the local disease-environment relationship, but it also reached for broader distinctions that we might today call anthropological—insisting, for instance, that Asians and Europeans differed in fundamental ways for reasons that were environmental. Here was an important—and, again, a surprisingly lasting—precedent for the co-mingling of the micro- and macro-territorial conclusions that would be a basis for modern tropical discourse. As the geographer Clarence Glacken aptly put it, *Airs, Waters, and Places* "is responsible for the fallacy that, if environmental influences on the physical and mental qualities of individuals can be shown, they can by extension be applied to whole peoples."[8] Galen later offered a synthesis of humoral medicine that lasted in various forms until the germ theory displaced it in the late nineteenth century, while the Roman natural historian Pliny made the dangerous connection between climates and racial formation by correlating skin color with latitude and, again, privileging the temperate. Well into the Middle Ages, theories of geographical influence remained derivative of classical thinking, and modern ideas of tropicality were influenced by the rediscovery of these deep historical sources.[9]

THE EMERGENCE OF THE TROPICS AS AN IMAGINATIVE REGION, 1400–1800

Despite these classical roots, a coherent discourse on the tropics emerged only during the early period of European exploration in the fifteenth century, and it assumed its modern form in the eighteenth and nineteenth centuries. In a remarkable reinterpretation of Columbus's voyages, Nicolás Wey Gómez has argued that the Genoese mariner was as intent on sailing south as he was east, and that his voyages into the torrid zone fundamentally challenged the geographic assumptions of the ancients by showing the equatorial latitudes not only to be habitable but also rich in resources and economic possibility. In the process, Columbus invented what Gómez calls "the tropics of empire."[10] But, as Nancy Stepan has argued, it still took several centuries for the emergence of modern notions of the tropics, which relied on a rigorous and geographically informed natural history, the development of modern human sciences such as anthropology, and a revival of medical geography. Together these disciplinary efforts made sense of the natural, human, and medical tropics, assembling them into a single geographical category at a moment when temperate peoples from various parts of the globe had turned their attention to these regions. Modern tropical thinking was the product of several coalescing efforts at scientific exploration and categorization in an expansionistic and imperial age.[11]

Early modern tropical thought, from the fifteenth through the early nineteenth centuries, was characterized by the anxieties and possibilities felt by Europeans moving out into an alien world.[12] One aspect of this new experience involved contending with the maritime tropics. Histories of tropicality generally confine their analyses to the terrestrial tropics, but for many, entering the tropics was a gateway experience that occurred with land nowhere in sight. Mariners used celestial calculations to mark the tropical transit with precision. Often the trade winds, which propelled sailing ships while tempering the rising temperatures, welcomed the maritime entry into the tropics. Many also experienced the physical hardships of a hot vertical sun and, in lower latitudes, the doldrums, which could strand boats and sailors in an oceanic tropical purgatory. To mark the tropical transit, mariners during this period commonly practiced "tropical baptisms," whereby experienced sailors dunked those entering the tropics for the first time into tubs of seawater, a ritual often accompanied by strong drink. A similar ritual sometimes accompanied the crossing of the equator, or "crossing the line," such as the one Charles Darwin described during his voyage on the *Beagle*. As such rituals make clear, experiencing the maritime tropics—the qualities of its waters, its weather, its place in relation to the sun and stars, even its sea life—was critical to broader perceptions of the region as discrete and exotic. It was on the high seas that expanding Europeans most precisely honored the tropics as a geographical abstraction.[13] On land the boundaries blurred, and the evidence of tropicality was more sensory.

European maritime explorers and traders often first encountered the terrestrial tropics, and constructed their initial images of the region, as they used islands for staging and reprovisioning, and from that set of early encounters sprang one of the most durable environmental idylls within the discourse: the tropical island Eden. From the Fortunate Isles (the Canaries, Azores, and Madeiras) and the Cape Verde Islands, which served as launching points for European expansion, to the West Indies, St. Helena in the South Atlantic, and Mauritius and other Indian Ocean islands, merchant seamen sought out tropical islands as supply depots for fresh water, food, and other supplies, avoiding the tropical mainland for fear of disease. Tropical plantation societies also got their starts on these islands. As Richard Grove notes, citing Shakespeare's *The Tempest* (ca. 1611) as but one example, "the tropical environment was increasingly utilised as the symbolic location for the idealised landscapes and aspirations of the western imagination." Indeed, these islands served as imaginative resources for those searching for Eden, and many sailors greeted them, after exhausting sea voyages, as terrestrial paradises filled with exotic flora and fauna. But the use of such islands also had dramatic material consequences, so much so that Grove makes the provocative argument that Western environmentalism originated among a small cadre of colonial officials—mostly physician-naturalists—who from their peripheral positions began to observe the steady decline of tropical island environments at the hands of the various agencies of European expansion. His is a salvo in a broader debate about the relationship between imperialism, science, and tropical conservation, and an important caution that, when writing the history of ideas such as the tropics, we need to take seriously not only the imaginative capacities of the metropole, but also the empiricism of the periphery. From the earliest European expansion through the publication of *Robinson Crusoe* (1719) and Cook's South Pacific voyages (1760s–1770s), the reef-protected and palm-lined island redolent with tropical fruit became a key symbol of beneficent tropicality—one that remains alive and well among twenty-first-century tourists.[14]

Scholars writing about early American exploration and colonization have also noted that Europeans encountered new lands with anxieties about how their bodies would respond to these places, particularly those with warm climates. Such concerns about "warm climates" presaged the modern tropical discourse, but they were not bound by the tropics and could be longitudinal and seasonal as well as latitudinal. Europeans arriving in the North American environments of the Chesapeake, the Lowcountry, and the West Indies became preoccupied with "seasoning," an expected period of physiological adjustment to the new environment that some would not survive. This was also true for British colonists arriving in Australia in the late eighteenth and early nineteenth centuries. In general, colonists and colonial officials mixed optimism about the generative capacities of warmer climes with fears of disease and degeneration. These hopes and fears were heightened within the geographic tropics, but they spilled beyond them and could even extend into ostensibly temperate areas—as with settler anxieties about the wide temperature swings in New England, or the hot winds of otherwise temperate Victoria and New South Wales. As these examples suggest, the loosely constructed paradigm of warm climates largely prevailed before 1800; the rigid geography of tropicality emerged as concerns about bodies out of place migrated toward the equator.[15]

This discourse on the constitutional effects of warm climates mixed with perceptions about tropical peoples to shape the development of race slavery in the Americas. Before the emergence of modern medicine in the late nineteenth century, Western medical tradition conceptualized most tropical diseases as miasmatic—as literally emanating from tropical environments. And because humoral medicine still dominated, Westerners understood disease not as the workings of a discrete biological agent but, rather, as a body out of balance with the climate and environment. Disease was a relation, not a thing.[16] The question of which bodies were best suited to particular climates was compelling as Europeans built an expansive plantation economy. Part of the rationale Europeans offered for employing African slaves on tropical and subtropical plantations was their alleged suitability to hot, humid, and diseased environments that killed or enervated temperate whites. The emergence of the New World plantation complex thus not only helped to solidify a tropical-temperate dichotomy, but it racialized the tropics as a place of nonwhite labor. This correlation between dark skin and suitability to living and laboring in tropical places, though a deep historical one, served the interests of New World planters in novel ways. Looking for a labor force to grow tropical crops such as sugar, planters turned to African slaves in part under the dubious assumption that their tropical Old World origins matched them well to harsh laboring conditions in the New World, from British North America to Portuguese Brazil. If nothing else, this environmentalist logic was a convenient justification for an institution that developed for a series of reasons well beyond the environmental. Similar tropical encounters between Europeans and tropical "natives" in other parts of the world produced similar conclusions that "tropical peoples," assumed to be nonwhite, were uniquely suited to labor under hot and humid conditions. Tropicality thus emerged in part to naturalize race slavery, and race slavery served to embody tropicality. Put another way, tropicality was an environmental discourse for defining an emergent ideal of temperate whiteness. Indeed, by the end of the eighteenth century, Johann Friedrich Blumenbach's formative anthropological treatise, *On the Natural Variety of Mankind*, had helped to establish a strong connection between geography, climate, and color or race.[17]

European impressions that Africans and other tropical peoples were better adapted to tropical conditions were not entirely the result of pernicious imaginings. In certain cases, peoples of African origins did have advantages in New World plantation disease environments over white temperate peoples, though for reasons that had to do with previous experience with similar disease environments and not with so-called racial physiology. West Africans who lived in regions where malaria was endemic tended to have high rates of blood polymorphisms, for instance, that lent them resistance to malaria, though often at a cost. The best-known example here is the sickle cell trait, which can cause debilitating anemia, but, in its recessive form, provides some resistance to malaria. But this was in no way a "racial" trait—it was a genetic marker of a long community history with malaria. Acquired immunities also could seem racial, even though they resulted from lived experience with diseases, such as yellow fever, associated with the tropics. But planters shoehorned these differential immunities into racial containers. As a result, race and the geographical tropics matured together as ideological categories.[18]

The disease environments on New World plantations were themselves often the products of tropical biological exchanges and the substantial ecological transformations involved in creating plantation landscapes. John McNeill has shown how the creation of Caribbean sugar plantations provided ideal conditions—what he calls a "creole ecology"—for the spread of yellow fever to and throughout the region. Both the yellow fever virus and its mosquito vector, the *Aedes aegypti*, were imports from Africa, and while the Aedes mosquito found near-perfect breeding conditions in landscapes transformed for sugar production, the increased commerce between islands and the introduction of large numbers of non-immunes (in this case, people who had not survived a mild inoculating childhood case) created ideal epidemiological conditions for frequent epidemics. Malaria, too, was an Old World import that thrived in—and well beyond—the disturbed neotropics. In other words, Old World colonization of the New World tropics, and of warm regions north and south of the tropic lines, often created the disease conditions that planters and other colonial observers came to naturalize as tropical, and that they then used to justify race slavery as a natural response to said environmental conditions.[19]

During European expansion, tropicality also invoked a set of agricultural, horticultural, and forest commodities that could not be grown or gathered in most temperate regions (except under the most controlled conditions). These commodities—coffee, tea, sugar, cacao, bananas, pineapples, mangoes, coconuts, tropical woods, rubber, all sorts of spices, and even animals such as parrots or monkeys—came to function as shorthand representations of the tropical, signifiers of a dense set of imagined tropical environments and societies crafted in the minds of temperate world consumers. This was particularly true as these commodities became widely available, and as more people had the capacity to travel to the tropical places from which they originated. People from outside the region constructed the tropics in countless acts of consumption, and those acts, as the anthropologist Sidney Mintz argued, structured relations of power across the imagined tropical-temperate divide.[20] Moreover, many of these commodities worked their own fetishistic magic, obscuring relations of production and leading to another common assumption offered by temperate travelers in the tropics: that tropical commodities were the products of a fecund nature and could be had with little or no labor.

As Europeans moved into these warmer climates and environments to experiment with growing crops or extracting natural resources, they also began moving plants and animals around the world, between tropical regions as well as from the tropical periphery to the temperate core. This massive and diffuse process of transplanting flora and fauna occurred both along and across latitudinal gradients, and it coincided with, and profoundly informed, the anxious process of colonists transplanting themselves in new soil. Out of these twin processes grew a discourse on acclimatization—the process of moving living things from, and adapting them to, new environments—that helped to define the tropical and the temperate as discrete categories dividing plants, biotic communities, and the conditions under which they survived. Europeans found that they thrived in what Alfred Crosby famously (if too facilely) called the "neo-Europes" of North America, temperate South America, Australia, New Zealand, and southern Africa. Where Europeans and their biota did less well, many soon concluded, was between the tropic lines. To adapt, Europeans moved plants and peoples around the world's equatorial belt, creating pan-tropical networks of circulation and exchange that fed the growth of plantation economies.[21]

Acclimatization was critical to the history of the tropics as an environmental category in several ways. First, in suggesting the possibility of adapting to new climates and environments, acclimatization theories often ran counter to rigid theories of geographic racial differentiation. Indeed, while some argued that temperate peoples could never adapt themselves to tropical conditions and that they would degenerate racially and sexually (tropicality was often a discourse about gender as well as race) if they tried to do so, others insisted on the possibility of acclimatization: that is, that white people, under certain circumstances and with proper care, could live and even thrive in equatorial regions. Debate over the possibility of tropical acclimatization was one of the most fractious parts of the discourse, and while it was rooted in this early period of European expansion, it persisted into the twentieth century. Second, acclimatization efforts led to the creation of a series of institutions, such as acclimatization societies and botanical gardens, which not only helped to direct scientific attention to the tropics, but also brought glimpses of the tropics back to the temperate world. Third, the dissemination of tropical crops across the global tropics, which would accelerate after 1800, led increasingly to a pan-tropical awareness of the region's suitability to the production of certain commodities that could not be produced in temperate regions, and of the region's hostility to temperate laborers.[22] Finally, debates about acclimatization led natural historians and biologists to ask a set of questions about the relations between living things and their physical environments, which led to a revolution in biological understanding, one to which travel and research in the tropics would be key. Together, these developments provide a good bridge to a discussion of the modern period in the history of tropicality.

CONSTRUCTING THE MODERN TROPICS

If the first phase of tropical exploration and colonization was mostly on islands and along coasts, the late eighteenth and nineteenth centuries saw more thorough exploration and, in some cases, settlement of inland tropical areas of Amazonia, West Africa, Central America, South and Southeast Asia, and northern Australia.[23] And as more Western peoples moved deeper into the global tropics and developed ambitious plans for the region, they also shifted their general impression of the tropics. While an Edenic view of the tropics had prevailed from Columbus on (despite substantial experiences and anxieties to the contrary), by the nineteenth century the tropics took on a more ominous cast, particularly as disease decimated outsiders on a larger scale. Well into the twentieth century, a negative tropicality dominated the discourse.[24]

Several developments in the late eighteenth and early nineteenth centuries provide rough markers of a critical transition to a brand of thinking that more rigidly separated the tropical from the temperate. The founding and evolution of London's Kew Gardens during this period was a watershed development, as Kew became a showplace of exotic flora, an instrument of imperial botanizing, and the centerpiece of a growing network of colonial botanical gardens that sought to find agriculturally useful crops and move them throughout Britain's tropical holdings. The French, Dutch, Germans, and other imperial powers followed suit, developing botanical gardens to collect and array the floral tropics, and to figure out how they might profit from moving biota from one portion of the tropics to another. Moreover, as these institutions opened themselves to public view in the nineteenth century, and as advances in the use of architectural steel and glass and steam heating allowed for controlled artificial environments that mimicked the climatic consistency of the equatorial regions, conservatories became widely influential microcosms of tropical nature. Kew Gardens' Palm House was the most famous, but other public and private conservatories, orangeries, and hothouses proliferated in Europe, while tropical plants made their way into temperate homes as ornaments. For many who first visited the lowland tropics in the nineteenth century, the region seemed like nothing more than a vast and chaotic hothouse.[25]

The turn of the nineteenth century was also a critical moment in the history of tropical travel, with an accelerating number of Europeans and other temperate peoples traveling to tropical areas and writing about their travels. Tropicality, then, grew up in a literate age of travel. No traveler was more influential in defining the modern tropics, and spurring scientific and exploratory interest in the region, than the Prussian scientist-explorer Alexander von Humboldt. Between 1799 and 1804, Humboldt and his traveling companion, the botanist Aimé Bonpland, traveled throughout tropical America, along the Orinoco and Amazon Rivers as well as in the Andean highlands. Humboldt's *Travels* and subsequent publications helped to define the modern tropics. First, Humboldt greeted the tropics with enthusiasm, celebrating the region's natural beauty and biological richness in innovative ways. His writings, which were tremendously influential, helped to paint imaginative pictures not only of the beauty and fecundity of tropical lowlands, but also of tropical highland scenes—and the region's volcanoes in particular. Bucking the trend of the era, his was, to borrow a phrase from David Arnold, an "affirmative tropicality."[26] Humboldt and Bonpland also assembled a massive collection of tropical plant specimens that formed the foundation for a revolution in biological knowledge of the tropics and a nascent appreciation of the region's biological diversity. Humboldt offered an isothermal theory of plant geography, in which he correlated and mapped the distribution of plants with climate and other environmental factors, and pointed out how latitude and altitude had similar effects on plant distribution. For Humboldt, then, the tropics was not merely one biotic zone among many; rather, it was the only geographical zone where the full order of nature was viewable in its entirety.[27] While some scholars have argued that Humboldtean natural history worked in imperial ways to privilege the natural over the social tropics, other have pointed out that Humboldt offered passionate critiques not only of plantation slavery but also of the cultural and environmental chauvinisms that were so dismissive of tropical peoples. It was Humboldt who made the tropics a truly scientific category and a region for heroic exploration, and his tropical travels motivated many an important figure to follow in his footsteps.[28]

Scientific naturalists of the nineteenth century kept alive an affirmative tropicality even as they pushed well beyond Humboldtian plant geography. The most famous naturalist celebrants of the biological tropics after Humboldt were Charles Darwin and Alfred Russell Wallace, who separately pieced together the theory of natural selection during their tropical journeys. Darwin carried with him on the *Beagle* a copy of Humboldt's *Travels*, and he went to the American tropics—newly opened to travelers with the decline of the Iberian empires—with a set of expectations honed by Humboldt and the tropical displays of British botanical gardens. Wallace also traveled extensively in the Brazilian Amazon (with the naturalist Henry Walter Bates, whose 1863 memoir *The Naturalist on the River Amazons* did much to conceptualize and publicize Amazonia as a tropical space) and then in Malaya, and while he was less impressed aesthetically with tropical nature than had been Darwin and Humboldt before him, Wallace did write extensively about the tropics, most notably in *Tropical Nature, and Other Essays* (1878). In particular, Wallace provided an influential explanation of tropical biological diversity as the product of the climatic and geological stablility of the tropics. Such stability, Wallace theorized, allowed for uninterrupted speciation and biological development, as compared to those regions where glaciations had repeatedly wiped the biological slate clean. While today there are numerous and sometimes competing theories about what has made tropical environments so biodiverse, Wallace's theory of climatic stability remains an important one. Moreover, his belief in the tropics as a space of deep-temporal constancy fed into twentieth-century notions of the pristine wilderness tropics.[29]

The late eighteenth and early nineteenth centuries were also critical in the construction of tropical Africa and South Asia. While early exploration and the development of the slave trade had contributed to an understanding about African environments and peoples in the first several centuries of European expansion, it was not until the slave trade, mostly a coastal phenomenon for Europeans, came under attack that Europeans began looking anew at Africa as a field of opportunity.[30] From the 1780s through the middle of the nineteenth century, the British struggled to make sense of West Africa and the influences of race and environment on its development. Two variants on prominent myths emerged out of this experience. First, there was the myth of the tropics as the "white man's grave," a region where European outsiders perished in startlingly large numbers. There was nothing new to notions that the tropics would be a challenge to temperate peoples and constitutions, but the British experience in West Africa during this period gave them grave empirical and statistical support. As British travelers, explorers, scientists, and soldiers moved inland, they died in large numbers, sparking a debate not only over how race and disease mortality correlated, but also over just what it was about the tropical environment that produced disease so prolifically. One product of this push into tropical West Africa was the development of a more empirical brand of tropical medicine, which, though still rooted in miasmatic and humoral theories, did pay specific attention to medical topography and resulted, by mid-century, in several important improvements in tropical public health—the widespread use of quinine to combat malaria, for instance—for white outsiders entering the tropics.[31]

The second myth about tropical West Africa that developed during this period, and that had wide currency in broader discussions of the tropics, was the myth of tropical exuberance: that the tropics encapsulated tremendous biological productivity which, when harnessed to the temperate work ethic, would yield unprecedented bounty. "Luxurious" was the de rigueur adjective. We now realize that this estimate of the agricultural productivity of tropical soils was overly sanguine, but such fantasies were a core part of tropical thinking well into the twentieth century. And they in turn raised a set of questions about tropical peoples that troubled temperate observers. Why, if the tropics were so bountiful, had tropical peoples done so little to develop the agricultural possibilities of the region? Was the climate too enervating? Or were the racial deficiencies of tropical peoples to blame? Many even saw the tropics as a challenge to temperate masculinity; tropical peoples, they believed, were particularly effeminate. But the myth of tropical exuberance offered another answer to the question of why tropical peoples had allegedly underachieved: that the tropical environment, in its very fecundity, so easily provided for the basic needs of tropical peoples that there was no need to work for a living. Why cultivate the earth when one could simply and easily pluck the spontaneous productivity of the tropics? By the early nineteenth century, then, the tropics as a laborless Eden had become a developmental problem, and the myth of tropical exuberance lent an environmental explanation to the archetype of the lazy native.[32]

Just as tropical environments were coming under the gaze of scientists and others traveling to West Africa, so race was becoming an increasingly scientific category. Indeed, many temperate peoples defined their white racial identities in contradistinction not only to blackness, but also to a geographical tropics that seemed hostile to their presence. But modern racial and tropical thinking could also work at cross-purposes. Where some saw the tropical environment as the shaping influence on African societies and cultures, for instance, another brand of racial theorizing argued that African underdevelopment was fundamentally rooted in racial physiology, not environment. In the early nineteenth century, this was the case among polygenists, who believed that the black and white races resulted from separate lineages. Monogenists, on the other hand, were more likely to use the shaping influence of tropical climates and environments to explain and account for racial and developmental differences. This debate played out with particular clarity in encounters with and plans for the West African tropics.[33]

South Asia provides a unique case study of how tropicality came to the fore in colonial discourse in the first half of the nineteenth century. As in Africa and the Americas, South Asia between 1800 and 1850 saw a dramatic increase in Western travel and travel writing, a growing interest in mapping and the natural sciences, more careful attention to medical geography and topography, an enthusiasm about the potential for tropical export agriculture, and a growing European imperial military presence. But India was (and is) a subcontinent with stunning topographical and environmental diversity, and a rich political and cultural history that had long been the domain of Orientalist scholars. India was thus a difficult place to tropicalize. Nonetheless, as the British exercised more control over India in the first half of the nineteenth century, tropicality came to dominate colonial discourse about India, largely displacing Orientalism by mid-century. This shift occurred despite the fact that, amid both a growing pan-tropical consciousness and certain assumptions about tropical places, India consistently disappointed visitors who had expected something more like the Amazon basin or equatorial West Africa—or, for that matter, Ceylon. To make India tropical took some work.[34]

As the British tropicalized India, medical theorizing was particularly important. While India did adhere to the growing nineteenth-century perception that tropical places were also deadly ones for white outsiders, it nevertheless offered anomalies. While malaria, often conceptualized as a "tropical fever," was a major cause of morbidity and mortality, other causes of disease and death seemed less tropical—cholera, for instance, and dysentery, which could just as easily be seen as moral or sanitary problems. As in other tropical areas subject to imperial advances, India saw in the early nineteenth century growing attention to medical geography and topography, a prototype of the tropical medicine that would flourish in the late nineteenth century. These inquiries were susceptible to tropical generalizing, but they also paid careful attention to local conditions that did not always map easily onto tropicality. Some medical theorists worked within the older, looser paradigm of "warm climates" that correlated disease with heat and humidity, but not the tropics per se. Nonetheless, the era also produced works such as James Johnson's *The Influence of Tropical Climates, more especially the Climate of India, on European Constitutions* (1813), which did truck in such generalization. "By the 1840s," David Arnold concluded, "the representation of India's diseases as 'tropical' or the products of a 'tropical climate' was so commonplace as to require little explanation." Such terms had hardly been used as late as 1800.[35]

Botanists also did a lot of the work of tropicalizing India in the early nineteenth century. Since Humboldt, botanical distinctiveness and diversity had marked the tropics as separate from temperate regions, giving it a distinctive "floristic identity," and the work of botanists in India furthered that separation, though again with some telling anomalies. Many botanists were deeply disappointed with large swaths of India's landscape, which seemed too arid, barren, or otherwise divergent to meet tropical expectations, while others—such as Joseph Hooker, one of the great British botantists of the nineteenth century and the long-time director of Kew Gardens—found some of the subcontinent's most "tropical" landscapes in the shadows of the Himalaya, well north of the Tropic of Cancer. The tropicality of India, then, was also often at odds with the geographical tropics.[36]

For the British in India in the early 1800s, tropicality came to naturalize a set of differences and developmentalist objectives. Where the work of Orientalist scholars had focused on the complex cultural, linguistic, religious, and historical aspects of Indian life, those who constructed a tropical India emphasized the dominance of nature on the subcontinent, a view that served the engineering mindset which came with modernization and tighter imperial administrative control. The British construction of India as tropical obscured the subcontinent's humanity, just as the wilderness ideal had in North America, and suggested that the work of the British in improving agriculture and modernizing the landscape involved the conquest of nature rather than the control of people. By the mid-nineteenth century, in India as elsewhere in the tropics, the power of tropicality to naturalize imperial actions—to make them appear heroic conquests of nature rather than exercises in social control—was growing.[37]

Europeans led the way in crafting this tropical discourse, but by the middle of the nineteenth century North Americans, too, showed a marked interest in the tropics as an environmental space. Nineteenth-century Americans understood the particular health risks of tropical commerce, travel, and military engagement—risks that had been inherent in Caribbean geopolitics for a couple of centuries.[38] Nonetheless, there was also a growing fascination with tropical nature and exploration. Travel writers such as John Lloyd Stephens, who "discovered" the lost cities of the Mayas, wrote extensively about their neotropical explorations.[39] The young John Muir, soon to become a canonical figure in the American wilderness tradition, dreamed of following in the footsteps of Humboldt and visiting the Brazilian tropics.[40] Indeed, the Amazon Basin was Muir's ultimate destination when he embarked on his thousand-mile walk to the Gulf of Mexico, and only a case of malaria contracted deep in the American South shunted him to the Sierras of California. The American South also provided an arena for fleeting tropical encounters, and there were growing concerns, beginning in the late 1800s, that Americans faced a "tropical" problem within their own borders.[41] A number of America's most famous landscape artists, including Frederic Church and Martin Johnson Heade, painted the American tropics to popular acclaim and created what art historian Katherine Manthorne has aptly referred to as a "unified pictorial consciousness" of the tropics.[42] Few moments were more important to North Americans' growing acquaintance with the tropics than the Gold Rush and the Isthmian transit period that followed. Between 1848 and 1869, half a million people crossed Central America on their way west, and another 300,000 took the same routes back east. This was a massive influx of human traffic through tropical America, and thousands of migrants produced letters, journals, or published accounts of their tropical transit that were widely read. Most hewed to a tropical logic and vocabulary well established by the most influential publications informing this wave of tropical travel, with Bayard Taylor's *Eldorado* (1850) being perhaps the single most cited source. The Isthmian transit period, then, was an intense moment of tropical learning for North Americans.[43] Moreover, both before and after the Civil War, the US government sponsored a series of scientific expeditions to Central America to scout out possible locations for a canal—at the very moment when better-known expeditions were charting the landscapes of the American West. Americans also watched intently as the French floundered in their efforts to build a sea-level canal in Panama in the 1880s, losing more than 20,000 workers to an environment that one prominent Frenchman described as "literally poisoned."[44] Finally, the late nineteenth century saw American capital moving into tropical regions and Americans gaining commercial access to tropical consumer commodities that were among the most potent markers of tropicality. Even before the United States became a self-conscious imperial power concerned about tropical environmental management, then, Americans knew the tropical.

The Tropics and the Twentieth Century

The power of the tropics as a conceptual space peaked during the late nineteenth and early twentieth centuries, when imperialist nations sought to control and develop tropical colonies, and contended with the difficulties they encountered in doing so. Out of this effort emerged several tropical scientific and social scientific subdisciplines that worked to describe the characteristics of the region and unlock it for development.[45] Those disciplinary efforts were essential to modern tropical discourse, but they also produced critical instabilities within it. As a result, the second half of the twentieth century saw a dramatic shift in the popular Western assessment of the tropics as an environmental space. While many continued to see tropical regions as backward, and some even engaged in new types of environmental determinism to explain that backwardness, the racial and civilizational chauvinisms that defined tropicality for most of its history faded, and in their place an affirmative environmentalist tropics arose. Indeed, as we will see, the scientific efforts to discipline tropicality in the first part of the twentieth century propelled the reassessment that marked the second part.

The most important scientific discipline in the transition to twentieth-century tropicality was tropical medicine. Tropical medicine came into being as an imperial discipline largely aimed at the problem of settling whites in tropical areas, but it was also a discipline shrouded in the heroic narrative of modern medicine. By most accounts, the "father" of tropical medicine was Patrick Manson, who, while working in Asia in the late 1870s, figured out that the Filarial worm that causes elephantiasis was spread between human beings by Culex mosquitoes. This discovery of an insect vector for a disease was a critical one in the larger realm of tropical medicine, and Manson would go on to found the London School of Hygiene and Tropical Medicine in 1899. Other vector discoveries were also critical to tropical medicine, most notably the discovery that *Anopheles* mosquitoes spread malaria, confirmed by Ronald Ross, who was working in British India, in 1897 (Giovanni Battista Grassi made a similar discovery in Italy and contested Ross's primacy), and the work of Walter Reed and the Reed Commission in 1900 in Havana, Cuba, who confirmed experimentally that the *Aedes aegypti* mosquito spread yellow fever (Carlos Finlay, a Cuban physician, had made the same suggestion almost two decades earlier).[46] These vector discoveries worked a number of transformations. First, they undermined older miasmatic theories of medicine, and thus notions that particular environments, such as the lowland tropics, emanated diseased air. In fact, in most areas of medicine, the discovery that discrete germs caused diseases shifted medical attention away from environmental explanations of disease and toward the body. But tropical medicine was an exception because so many of the diseases that fell under its rubric were vector diseases. Even with the decline of the miasmatic theory, the tropics thus remained a redoubt for environmental medicine, and the tropical environment retained its reputation for ill health. Tropical medicine's vector discoveries also led to effective—if often limited—techniques for tropical public-health interventions, and thus gave doctors, sanitarians, and other environmental experts a central role in imperial success. In many cases, such as the United States' efforts to rid the Panama Canal Zone of malaria and yellow fever, environmental management in the form of mosquito control worked effectively, though most of the effort was focused on protecting white outsiders. The British attempted similar approaches at controlling mosquito vectors in British India and Malaya. But in other settings tropical public-health efforts followed a more pernicious path. Under the miasmatic paradigm, the understanding was that tropical peoples were largely immune to tropical diseases but not themselves a public health threat (except, perhaps, with regard to basic sanitary practices). But under the vector theory, imperial powers and their medical experts increasingly saw tropical peoples as disease reservoirs who threatened white outsiders. As a result, in some settings—British West Africa, for instance—public health officials urged racial segregation as a tropical sanitary solution.[47] Modern medicine, then, could increase fears that tropical peoples spread disease.

However tropical medical officials approached the problem of sanitation and disease prevention, the efficacy of new sanitary approaches made tropical medicine a celebrated profession and led to tropical triumphalism. Medical officials claimed that the scientific and sanitary advances of tropical medicine had, in effect, solved the problem of the tropics as an environmental space inaccessible to temperate peoples. In the wake of these discoveries, medical and sanitary officials of all sorts—those in the employ of colonial and national governments; those working with universities and non-governmental organizations such as the Rockefeller Foundation; and those working for private corporations such as the United Fruit Company and the Firestone Rubber Company—fanned out across the global tropics to protect officials, settlers, global commerce, and workers from tropical diseases and the inefficiencies that came with them.[48] As the twentieth century proceeded, however, tropical med-icine would become less imperial and more capacious in its concerns. As empires declined and the Cold War restructured relations between the developed and developing worlds, tropical sanitation became an imperative of development and a domain of increasingly powerful non-governmental organizations. Indeed, improv-ing the health of tropical peoples has become one of the most important interna-tional mobilizations of recent decades, particularly with the resurgence of malaria and the emergence of new diseases.

Tropical medicine, however imperial its origins, was not only a discipline exercised by imperial and first world officials. Some in the tropical world used "tropicality" to claim their own medical expertise and to build important institutional strengths. This was particularly true in Brazil, where a group of doctors founded the Escola Tropicalista Bahiana in the 1860s and attempted not only to adapt European medicine to the public health problems of Brazil, but also to oppose tropical fatalism and to claim Brazilian expertise and primacy in the treatment of their own medical problems. Their approach challenged imperial tropical medicine with a nationalist variant.[49] Indeed, perhaps more than any other nation, Brazil historically has adopted and embraced the tropical mantle with pride. Particularly notable in this vein are Gilberto Freyre's cel-ebration of Brazil's multiracial Luso-tropicalism in his *New World in the Tropics* (1960), and Robert Burle Marx's tropical landscape designs.[50] As the case of Brazil suggests, tropicality could be contested from within.

When medical and other imperial officials engaged in tropical triumphalism dur-ing the early twentieth century, claiming that they had made the tropics safe for white settlement and enterprise, much of what they had in mind was a potential explosion of tropical agricultural production. Indeed, by 1900 it was a commonplace that, once rid of disease, the tropics would become a cornucopia of global food production. Benjamin Kidd made just such an assumption. Agriculture in the tropics did explode across the twentieth century, and tropical agriculture itself became a specialized subdiscipline. At one end of the spectrum, tropical plantation agriculture expanded rapidly to accom-pany expanding mass consumer demand for tropical commodities such as bananas, sugar, coffee, pineapples, rubber, and palm oil. In the process, these enterprises cleared vast expanses of tropical forest. Tropical forest products themselves became important as commodities, a development that, while not entirely new, accelerated in line with industrial timber cutting in other parts of the world. Finally, there was the growth of cattle ranches and cattle export from the tropics, resulting in the conversion of tropical forests to grasslands. These operations were large-scale, often financed by outside capi-tal, increasingly reliant upon scientific research and expertise, often labor intensive, and environmentally destructive. Moreover, many of these crop sectors were prone to boom-bust cycles as a result of crop pests and diseases, market vicissitudes, and even the discovery of synthetic alternatives. As scholars such as Richard Tucker and John Soluri have shown, the US was the preeminent consumer of tropical commodities—and thus the preeminent driver of tropical environmental change—across much of the twentieth century, although European and Asian consumption grew toward the century's end. During the twentieth century, tropical regions provided the consuming world with a distinctive suite of agricultural commodities.[51]

Even as commercial plantation agriculture spread across the tropics, agronomists closely studied—and usually fretted about—the techniques and conditions of the small-scale agricultural production that characterized much of the tropical world. Several "problems" with small-scale tropical agriculture dominated this literature. First, there was the paradox of tropical soils. While many early tropical observers had assumed that the "luxuriant" vegetation of the hot, humid tropics reflected the high productivity of tropical soils, twentieth-century soil scientists quickly concluded that tropical soils were poor as a rule. Indeed, this dramatic reversal of the assessment of tropical soils could become its own variant on tropical environmental determinism.[52] Rather than the fecundity of the tropics promoting sloth, poor soils retarded economic development. Second, there was the problem of livestock in the tropics. Most attempts to move temperate breeds and pasture crops to tropical regions ended in failure as a result of diseases, pests, and poor fodder. As one early twentieth-century expert put it: "As a rule, the animals imported from cold countries rapidly deteriorate."[53] Indeed, theories of white degeneration in the tropics likely had one source of inspiration in the efforts of temperate peoples to acclimatize their livestock to tropical places. Thus, the difficulty of practicing mixed husbandry or even intensive pastoralism became another tropical agricultural "problem" in the twentieth century. Perhaps most tellingly, tropical agronomists studied, and often lamented, the tendency of small-scale tropical producers to engage in shifting cultivation, whereby they would clear small pieces of land, often with the assistance of fire, and then farm them for a short period of time, until the cleared land lost fertility. Such systems had different names in different parts of the world—such as "milpa" in Latin America and "chena," in South Asia—but they struck many outside observers as not only uniform across the tropics but also uniformly inefficient and destructive. In his *Agriculture in the Tropics: An Elementary Treatise* (1909), J. C. Willis, a British agricultural expert who had mostly worked in Ceylon (Sri Lanka), not only considered tropical peoples to be "naturally indolent," but he also criticized chena production as "an exceedingly vicious mode of production, and wasteful and destructive beyond measure."[54] This was a common assessment at the beginning of the twentieth century, though as the century went on agricultural experts increasingly recognized the virtues of such production in a region with nutrient-poor soils and a rich population of potential crop pests.[55] Nonetheless, for many agricultural experts—and, in the second half of the twentieth century, for many environmentalists—shifting cultivation, in its apparent primitiveness and seeming destructiveness, was a tropical bane. After World War II, in a postcolonial world increasingly shaped by Cold War competition for allegiances, and in an environmentalist climate in which the apparent inefficiencies of subsistence tropical agriculture seemed to be colliding with a demographic explosion, the tropics became the central theater of the Green Revolution—the effort by First World agronomists to increase the efficiency of Third World agriculture through improved crop breeding and intensive inputs.[56] "Tropical agriculture" persisted through the twentieth century as a category for geographically containing and explaining a particular set of agricultural problems.[57]

Another disciplinary effort critical to the persistence of the tropics as a conceptual space was modern geography. To a degree, geographers extended the geopolitical theorizing of scholars such as Benjamin Kidd, who insisted that temperate peoples could never acclimatize to the tropics. American geographers such as Ellsworth Huntington often supported this line of reasoning, and they developed grand climatic theories to undergird what Huntington called "the new science of geography." For Huntington, climate largely (though not wholly) determined civilization. In his masterwork, *Civilization and Climate* (1915), he argued that temperate climates produced great mental activity and that tropical climates produced "tropical inertia," which he defined as a lack of willpower and self-control and a general "deteriorating force" on temperate whites. Huntington's description of the effects of tropical climate on whites resembled the increasingly popular medical diagnosis of "tropical neurasthenia," first postulated by Charles Woodruff in his 1905 treatise, *The Effects of Tropical Light on White Men*.[58] For Huntington, climate made the tropics an "unstimulating environment" hostile to white settlement. Indeed, Huntington redefined the tropics in isothermal rather than latitudinal terms, insisting that the region was climatically confined within the mean annual temperature isotherm of 70° F. Many other prominent geographers in the early twentieth century shared Huntington's dim view of the tropics as a space of civilization and potential white enterprise, including Ellen Churchill Semple, who, in *Influences of Geographic Environment* (1911) concluded: "the conquering white race of the Temperate Zone is to be excluded by the adverse climatic conditions from the productive but undeveloped Tropics, unless it consents to hybridization." Geographical thinkers such as Huntington and Semple were widely influential in attaching social-scientific grand theory to ideas that were, as the historian David Livingstone has suggested, deeply moralistic. But while their ideas might sound familiar, it is important to note that their notion of climate had changed. Whereas nineteenth-century medical thinkers had seen climate and disease as effectively the same thing, the claims of these climatic determinists were largely focused on the physiological effects of temperature, humidity, and the tropical sun. They thus stood in opposition to those medical officials who claimed that disease was the central obstacle to white colonization of the tropics, and that tropical sanitation had solved that problem.[59]

Other geographers were not as bullish on climatic determinism. A strong counter-narrative, which was decidedly more possibilist, strengthened during the middle of the twentieth century. One important practitioner of this school was the Australian geographer A. Grenfell Price, whose *White Settlers in the Tropics* (1939) rejected the climatic determinism of Huntington and others in favor of a more sanguine assessment of white tropical settlement possibilities—albeit one that fell in line with the racist tropical optimism of the "White Australia" policy (which saw Australia's tropical regions as not only amenable to, but also demanding of, white settlement). Nonetheless, Price still worried about the capacity of women and children to thrive in tropical environments, and he insisted that successful white tropical settlement would only come through modern scientific administration. Another important tropical theorist at mid-century was the French Indochina expert, Pierre Gourou, whose *Les Pays Tropicaux* (1947; translated into English by E. D. Laborde as *The Tropical World* in 1953) rejected the climatic determinism of Huntington et al., the triumphalism of tropical medicine, and even the racism of theorists such as Price. Gourou defined the tropics in environmental terms, as the globe's "hot, wet lands," which excluded much of the land base within the tropic lines and included some lands without them. He insisted that two main environmental forces shaped these "hot, wet lands": their poor soils, which limited agricultural productivity and thus development, and the persistent unhealthiness of the tropics. Climate remained a determinant for Gourou, but it was an indirect one—it made the tropics unhealthy by nurturing public health problems among tropical peoples, such as malaria and intestinal diseases, and it had a major effect on soil quality through the processes of rapid decomposition and weathering. Gourou could also be quite critical of European interventions in the tropics, which he saw as generally creating havoc. But while Gouroru's analysis represented a transition from the chauvinisms of the imperial age to the technocratic and humanitarian emphases prevalent since the end of World War II, he was still a tropical exceptionalist who insisted that the "hot, wet lands have their own special physical and human geography."[60]

By the second half of the twentieth century, as grand climatic theorizing and patronizing racial and civilizational generalizations lost their appeal, other descriptors such as "Third World," "Developing World," and "Global South" replaced the geographically and racially distinctive tropical world. Most of these new categories emphasized socioeconomic and other human explanations for tropical underdevelopment rather than climate, environment, or geography. And as they did, the environmental dimensions of the discourse migrated into an increasingly insular effort to define and protect tropical nature as a storehouse for biodiversity. Many of those efforts emerged out of research efforts in tropical botany, tropical agriculture, and even tropical medicine. The result would be another tropical scientific subdiscipline: tropical ecology.

Tropical ecologists were beneficiaries of empire. In the case of the US, one of the most important sites for tropical research developed in the Panama Canal Zone. In the early 1920s, a group of American biologists formed the Institute for Research in Tropical America, and they created a research station at Barro Colorado Island in the middle of Lake Gatun. Originally designed to be supported by a consortium of universities and museums in the United States, the station became the Smithsonian Tropical Research Institute (STRI) in 1946, and has since been one of the most important centers for tropical biological research in the world. Its founding administrator, James Zetek, was an entomologist who had participated in mosquito control efforts in the Canal Zone. During this same period, the US conducted research in tropical forestry in the Philippines and created the Caribbean National Forest (now known as El Yunque National Forest) in Puerto Rico, which has also been home to critical research in tropical forestry and ecology. Even the US Department of Agriculture was busy developing tropical agricultural expertise and sending its plant explorers all over the world during the first decades of the twentieth century.[61] Other imperial nations similarly used their imperial holdings to conduct tropical ecological research.

As modern scientists made tropical nature the subject of ecological study, there also emerged in early twentieth-century Western popular culture a fascination with the "jungle" as a tropical eco-type. From Rudyard Kipling's *Jungle Book*, Edgar Rice Burroughs's Tarzan books, and Theodore Roosevelt's adventures in Africa and the Amazon, to the Disney films and amusement-park rides of the immediate postwar years, tropical jungles came to represent what the historian Kelly Enright has called "maximum wilderness." Scientists working in the tropics helped to construct cultural representations of the jungle by publishing memoirs and producing popular films that bridged the divide between science and society. The jungle, as a product of popular culture, became a tropical extension of the American wilderness ideal, a place of great natural diversity as well as a testing ground for scientific explorers.[62]

In the second half of the twentieth century, research in tropical ecology merged with a growing global environmental consciousness to transform the jungle into the rainforest, and to laud the tropics as a storehouse of biodiversity. Although the idea of biodiversity had deep roots, use of the term exploded in the 1980s, when biodiversity emerged as a new conservation ideal. Through the work of scientists such as Norman Myers and Edward O. Wilson, biodiversity also was attached to tropical moist forests and their accelerating transformation during the 1970s and 1980s. Increasingly, conservation biologists made tropical environmental preservation a central goal. At the same time, the 1980s saw the growth of international environmental groups— including the Rainforest Alliance and the Rainforest Action Network—and international environmental governance, and tropical conservation was central to those efforts. But as the tropics became vital to international conservation, critical debates emerged over whether the rainforest ideal, which mixed older notions of pristine wilderness with more recent notions about biological diversity, privileged the protection of tropical nature to serve and soothe First World peoples at the expense of the people that lived in these tropical places. As the twentieth century ended, the relationship between species richness (the dominant, though not only, measure of biodiversity) and latitudinal gradients continued to be an area of extensive research and broad fascination for tropical ecologists, even though, as one practitioner put it, "the latitudinal gradient in species richness is a gross abstraction."[63] But tropical conservation efforts also have, in a number of ways, absorbed the critiques of those who charged that First World conservationists valued the nature of the tropics and its salvation over the people who lived there: by democratizing and decentralizing conservation and empowering local peoples, for instance, and by examining ways in which conservation and development might be compatible. As has happened in the temperate world with the critiques of wilderness, scholars have also discovered that tropical places many assumed to be pristine in fact have a long history of human land use and transformation—and that some of their diversity might even spring from the effects of human disturbance.[64]

Finally, in the last fifty years, as air travel has put distant tropical places within the orbits of First World travelers, the tropics has matured as a vacation zone, with many heading toward the equator in search of the tropical island idyll in modern form or opting for tropical ecotourism in new national parks and rainforest reserves. The tourist tropics, like the environmentalist tropics, is a powerful imaginary largely imposed on the region from the temperate world, and like the environmentalist tropics, it comes with its share of contradictions. It is a vision of the tropics as something to be consumed, one that retains a sense of the region as an Edenic place where nature gives and labor is unnecessary. Yet, as Catherine Cocks has argued, the reconceptualization of the tropics as a tourist paradise—an effort pioneered by tourism promoters in the early twentieth-century Americas—was also built upon a fundamental rethinking of tropical nature and its social and cultural power. For "tropical whites," temperate tourists in the tropics, tropical environments became less culturally limiting and racially degenerative and more therapeutic and culturally fluid as the twentieth century progressed.[65] Meanwhile, back in the temperate zones, contemporary consumers not only continue to make extensive use of tropical products in their daily lives, but even the modern meaning of the tropics is available for consumption—in, for instance, the many locations of the Rainforest Café, a restaurant that "re-creates a tropical rainforest, with waterfalls, lush vegetation, and indigenous creatures."[66] Similar representations exist in rainforest exhibits in zoos and aquariums, which distill the tropics for the visiting public. This ersatz tropics is but one piece of evidence to suggest that the discourse is alive and well in the twenty-first century.

Tropicality has been one of the most powerful environmental discourses of the modern era, and while how we think about the tropics has changed dramatically since Benjamin Kidd penned his lament about the impossibility of white temperate peoples ever making a permanent home between the tropic lines, persistent tropical generalizing still bears many of the vestigial marks of this earlier discourse. We thus need to be careful not to uncritically adopt a progressive environmentalist narrative of the tropics as an historical idea, one whose central lesson is that while we used to revile tropical nature as threatening and in need of human conquest, we now appreciate its manifest virtues. Rather, we ought to appreciate the several centuries of imperial history that have shaped the use of the category, and question the purposes that bifurcating the world at the tropic lines have served and continue to serve. The tropics, to borrow the language that William Cronon used two decades ago to urge a rethinking of wilderness, "is quite profoundly a human creation—indeed, the creation of very particular human cultures at very particular moments in human history."[67] The tropics will almost certainly remain a utilized category, but, as I have suggested throughout this chapter, that category ought not to be deployed uncritically as an objective descriptor of a singular geographical and environmental space. Rather, the tropics ought to be a term deployed with a sense of its significant historical and cultural content, and particularly with a sense of the ways in which the notion historically worked to organize, make sense of, and even naturalize human power relations.

Notes

1. Benjamin Kidd, *The Control of the Tropics* (London: The MacMillan Company, 1898): 39, 51, 30, 54.
2. The classic essay in the reconsideration of wilderness in a North American context is William Cronon's "The Trouble with Wilderness; or, Getting Back to the Wrong Nature," in *Uncommon Ground: Rethinking the Human Place in Nature*, ed. William Cronon (New York: W. W. Norton, 1996), 69–90. The North American environmental history literature influenced by Cronon is too lengthy to be cited here.
3. I am indebted to a number of scholars who have explored the history of tropicality. Among the most useful sources in constructing this essay have been: David Arnold, "Inventing Tropicality," in *The Problem of Nature: Environment, Culture, and European Expansion* (Cambridge, MA: Blackwell, 1996), 141–168; Arnold, "Illusory Riches: Representation of the Tropical World, 1840–1950," *Singapore Journal of Tropical Geography* 21, no. 1 (2000): 6–18; Arnold, *The Tropics and the Traveling Gaze: India, Landscape, and Science, 1800–1856* (Seattle: University of Washington Press, 2006); Nancy Leys Stepan, *Picturing Tropical Nature* (Ithaca, NY: Cornell University Press, 2001); Felix Driver and Luciana Martins, eds., *Tropical Visions in an Age of Empire* (Chicago: University of Chicago Press, 2005); Felix Driver, "Imagining the Tropics: Views and Visions of the Tropical World," *Singapore Journal of Tropical Geography* 25, no. 1 (2004): 1–17; and Warwick Anderson, "The Natures of Culture: Environment and Race in the Colonial Tropics," in *Nature in the Global South*, ed. Paul Greenough and Anna Lowenhaupt Tsing (Durham, NC: Duke University Press, 2003): 31–46. I have also come rather late to Nicolás Wey Gómez, *The Tropics of Empire: Why Columbus Sailed South to the Indies* (Cambridge, MA: MIT Press, 2008), which provides a wonderful bridge between the ancient and modern intellectual traditions.
4. Edward Said, *Orientalism* (New York: Vintage Books, 1978), 3.
5. For other scholars of tropicality who have invoked Said and Orientalism, see Arnold, *The Problem of Nature*, 141; Felix Driver and Luciana Martins, "Views and Visions of the Tropical World," in *Tropical Visions in an Age of Empire*, 4–5; Stephen Frenkel, "Cultural Imperialism and the Development of the Panama Canal Zone, 1912–1960," (PhD diss., Syracuse University, 1992).
6. My major source for this section has been Clarence Glacken, *Traces on the Rhodian Shore: Nature and Culture in Western Thought from Ancient Times to the End of the Eighteenth Century* (Berkeley: University of California Press, 1967). Another useful discussion of the Ancient origins of tropicality can be found in Gary Okihiro, *Pineapple Culture: A History of the Tropical and Temperate Zones* (Berkeley: University of California Press, 2009): 5–9.
7. Strabo, *Geography*, Book II, Chapter 2. http://penelope.uchicago.edu/Thayer/E/Roman/Texts/Strabo/2B*.html#ref96. Nicolás Wey Gómez suggests that modern scholars accept Parmenides as the originator of this schema. See Gómez, *The Tropics of Empire*, xv, 233–235. Aristotle, *Meteorologica*, translated by H. D. P. Lee (Cambridge, MA: Harvard University Press, 1952), Book II, Chapter V, 181.
8. On the Hippocratic corpus and its continuing influence, see Glacken, *Traces on the Rhodian Shore*, 80–115; Conevary Bolton Valenčius, *The Health of the Country: How American Settlers Understood Themselves and Their Land* (New York: Basic Book, 2002): 59, 182–83.

9. Glacken, *Traces on the Rhodian Shore,* 107–112, 256.

10. Wey Gómez, *Tropics of Empire.*

11. Stepan, *Picturing Tropical Nature,* 16–17.

12. Arnold, "Inventing Tropicality," 143.

13. On tropical baptisms, see Srinivas Aravamudan, *Tropicopolitans: Colonialism and Agency, 1688–1804* (Durham, NC: Duke University Press, 1999), 238–240; Robert Robinson, *The History of Baptism* (Boston: Lincoln and Edmands, 1817), 361–62. On Darwin's equatorial baptism, see Janet Browne, *Charles Darwin: A Biography—Voyaging* (Princeton, NJ: Princeton University Press, 1995), 192–193.

14. Richard Grove, *Green Imperialism: Colonial Expansion, Tropical Island Edens, and the Origins of Environmentalism, 1600–1860* (New York: Cambridge University Press, 1995), 3. See also Alfred Crosby, *Ecological Imperialism: The Biological Expansion of Europe, 900–1900* (New York: Cambridge University Press, 1986): 70–103.

15. Karen O. Kupperman, "The Fear of Hot Climates in the Anglo-American Colonial Experience," *William & Mary Quarterly* 41, no. 2 (April 1984): 213–240; Joyce Chaplin, *Subject Matter: Technology, the Body, and Science on the Anglo-American Frontier, 1500–1676* (Cambridge, MA: Harvard University Press, 2001); Warwick Anderson, *The Cultivation of Whiteness: Science, Health, and Racial Destiny in Australia* (New York: Basic Books, 2003).

16. I borrow this phrasing from Gregg Mitman, *Breathing Space: How Allergies Shape Our Lives and Landscapes* (New Haven: Yale University Press, 2007), 252.

17. On Blumenbach and his importance, see Okihiro, *Pineapple Culture,* 11–12.

18. Philip D. Curtin, "Epidemiology and the Slave Trade" *Political Science Quarterly* 83, no. 2 (June 1968): 190–216; Curtin, "Disease Exchange across the Tropical Atlantic," *History and Philosophy of the Life Sciences* 15 (1993): 329–356; John McNeill, *Mosquito Empires: Ecology and War in the Greater Caribbean, 1620–1914* (Cambridge, MA: Cambridge University Press, 2010); James L. A. Webb, Jr., *Humanity's Burden: A Global History of Malaria* (New York: Cambridge University Press, 2009).

19. McNeill, *Mosquito Empires.*

20. Sidney Mintz, *Sweetness and Power: The Place of Sugar in Modern History* (New York: Penguin, 1985).

21. On "Neo-Europes," see Crosby, *Ecological Imperialism,* On acclimatization, see Thomas Dunlap, *Nature and the English Diaspora: Environment and History in the United States, Canada, Australia, and New Zealand* (New York: Cambridge University Press, 1999); Michael Osborne, *Nature, the Exotic, and the Science of French Colonialism* (Bloomington: Indiana University Press, 1994); Harriet Ritvo, "Going Forth and Multiplying: Animal Acclimatization and Invasion," *Environmental History* 17, no. 2 (2012): 404–414.

22. Arnold, *The Tropics and the Traveling Gaze,* 143.

23. Arnold, *The Tropics and the Traveling Gaze,* 113.

24. See Philip Curtin, *The Image of Africa: British Ideas and Action, 1780–1850* (Madison: University of Wisconsin Press, 1964); Curtin, *Death by Migration: Europe's Encounter with the Tropical World in the Nineteenth Century* (New York: Cambridge University Press, 1999); Arnold, "Inventing Tropicality."

25. On Kew Gardens and the larger network of botanical gardens, see Grove, *Green Imperialism*; Lucille Brockaway, *Science and Colonial Expansion: The Role of the British Royal Botanic Gardens* (New Haven, CT: Yale University Press, 2002 [1979]); Richard

Drayton, *Nature's Government: Science, Imperial Britain, and the 'Improvement' of the World* (New Haven, CT: Yale University Press, 2000); Stuart McCook, *States of Nature: Science, Agriculture, and Environment in the Spanish Caribbean, 1760–1940* (Austin: University of Texas Press, 2002).

26. Arnold, "Inventing Tropicality," 146.

27. Michael Dettlebach, "Global Physics and Aesthetic Empire: Humboldt's Physical Portrait of the Tropics," in *Visions of Empire: Voyages, Botany, and Representations of Nature*, ed. David Philip Miller and Hans Peter Reill, (Cambridge: Cambridge University Press, 1996), 258–292; Driver and Martins, "Views and Visions of the Tropical World," 12.

28. On Humboldt's importance more generally, and debates about that influence, see Aaron Sachs, *The Humboldt Current: Nineteenth Century Exploration and the Roots of American Environmentalism* (New York: Viking, 2006); Mary Louise Pratt, *Imperial Eyes: Travel Writing and Transculturation* (New York: Routledge, 1992); Arnold, *The Tropics and the Traveling Gaze*, 113–114.

29. Arnold, "Inventing Tropicality," 148–149, 156; Luciana L. Martins, "A Naturalist's Vision of the Tropics: Charles Darwin and the Brazilian Landscape," *Singapore Journal of Tropical Geography* 21, no. 2 (2000): 19–33; David Taylor, "A Biogeographer's Construction of Tropical Lands: A.R. Wallace, Biogeographical Method, and the Malay Arhcipelago," *Singapore Journal of Tropical Geography* 21, no. 2 (2000): 63–75; Arnold, "'Illusory Riches.'" For an interesting reading of Bates, see Hugh Raffles, *In Amazonia: A Natural History* (Princeton, NJ: Princeton University Press, 2002), Chapter 5.

30. Curtin, *Image of Africa*, v-vi.

31. Curtin, *Death by Migration*; Curtin, *Image of Africa*; Daniel Headrick, *Tools of Empire: Technology and European Imperialism in the Nineteenth Century* (New York: Oxford University Press, 1981).

32. Curtin, *Image of Africa*. These perceptions and questions were rife in other parts of the tropical world as well.

33. Curtin, *Image of Africa*.

34. Arnold, *The Tropics and the Traveling Gaze*. On Ceylon (Sri Lanka), see James Webb, *Tropical Pioneers*.

35. Arnold, *The Tropics and the Traveling Gaze*, 142.

36. Arnold, *The Tropics and the Traveling Gaze*, 142–3.

37. Here again I am echoing the argument of Arnold's *The Tropics and the Traveling Gaze*.

38. McNeill, *Mosquito Empires;* Mariola Espinosa, *Epidemic Invasions: Yellow Fever and the Limits of Cuban Independence, 1878–1930* (Chicago: University of Chicago Press, 2009).

39. On Stephens, see Larzer Ziff, *Return Passages: Great American Travel Writing, 1780–1910* (New Haven, CT: Yale University Press, 2000), 58–117.

40. Sachs, *The Humboldt Current*, 27. See also Michael P. Branch, ed., *John Muir's Last Journey: South to the Amazon and East to Africa* (Washington, DC: Island Press, 2001).

41. On the tropical South, Natalie J. Ring, "Mapping Regional and Imperial Geographies: Tropical Disease in the U.S. South," in *Colonial Crucible: Empire in the Making of the Modern American State*, ed. Alfred W. McCoy and Francisco Scarano (Madison: University of Wisconsin Press, 2009), 297–308.

42. Katherine Manthorne, *Tropical Renaissance: North American Artists Exploring Latin America, 1839–1879* (Washington, DC: Smithsonian Institution Press, 1989), 60. See also David C. Miller, *Dark Eden: The Swamp in Nineteenth-Century American Culture* (New York: Cambridge University Press, 1989).

43. On the Panama Gold Rush transit, see Aims McGuiness, *Path of Empire: Panama and the California Gold Rush* (Ithaca, NY: Cornell University Press, 2008); Brian Roberts, *American Alchemy: The California Gold Rush and Middle-Class Culture* (Chapel Hill: University of North Carolina Press, 2000); Albert Hurtado, *Intimate Frontiers: Sex, Gender, and Culture in Old California* (Albuquerque: University of New Mexico Press, 1999). On the broader history of US perceptions of Latin American nature and society, see Frederick Pike, *United States and Latin America: Myths and Stereotypes of Civilization and Nature* (Austin: University of Texas Press, 1992).

44. David McCullough, *The Path between the Seas: The Creation of the Panama Canal, 1870–1914* (New York: Simon and Schuster, 1978), 80.

45. Driver and Martins, "Views and Visions of the Tropical World," 4.

46. On the history of tropical medicine, see Douglas Haynes, *Imperial Medicine: Patrick Manson and the Conquest of Tropical Disease* (Philadelphia: University of Pennsylvania Press, 2001); Roy Porter, "Tropical Medicine, World Diseases," in *The Greatest Benefit to Mankind: A Medical History of Humanity* (New York: W. W. Norton, 1997): 462–492; Michael Worboys, "Manson, Ross and Colonial Medical Policy: Tropical Medicine in London and Liverpool, 1899–1914," in *Disease, Medicine, and Empire: Perspectives on Western Medicine and the Experience of European Expansion*, ed. Roy MacLeod and Milton Lewis (London: Routledge, 1988), 21–37; Warwick Anderson, "Immunities of Empire: Race, Disease, and the New Tropical Medicine, 1900–1920," *Bulletin of the History of Medicine* 70, no. 1 (1996): 94–118; Espinosa, *Epidemic Invasions*; Randall M. Packard, *The Making of a Tropical Disease: A Short History of Malaria* (Baltimore: Johns Hopkins University Press, 2007; Warwick Anderson, *Colonial Pathologies: American Tropical Medicine, Race, and Hygiene in the Philippines* (Durham, NC: Duke University Press, 2006).

47. See Paul Sutter, "Nature's Agents or Agents of Empire? Entomological Workers and Environmental Change during the Construction of the Panama Canal," *Isis* 98 (December 2007): 724–754; Philip Curtin, "Medical Knowledge and Urban Planning in Tropical Africa," *American Historical Review* 90, no. 3 (June 1985): 594–613; John Cell, "Anglo-Indian Medical Theory and the Origins of Segregation in West Africa," *American Historical Review* 91, no. 2 (April 1986): 307–335.

48. On Rockefeller Foundation efforts, see Marcus Cueto, "Sanitation from Above: Yellow Fever and Foreign Intervention in Peru, 1919–1922," *Hispanic American Historical Review* 72, no. 1 (1992): 1–22; John Farley, *To Cast Out Disease: A History of the International Division of the Rockefeller Foundation, 1913–1951* (New York: Oxford University Press, 2003); Steven Palmer, *Launching Global Public Health: The Caribbean Odyssey of the Rockefeller Foundation* (Ann Arbor: University of Michigan Press, 2010). On United Fruit, see Aviva Chomsky, *West Indian Workers and the United Fruit Company in Costa Rica, 1870–1940* (Baton Rouge: Louisiana State University Press, 1996). On Firestone, see Gregg Mitman and Paul Erickson, "Latex and Blood: Science, Markets, and American Empire," *Radical History Review* 107 (Spring 2010): 45–73.

49. Julyan Peard, *Race, Place, and Medicine: The Idea of the Tropics in Nineteenth-Century Brazilian Medicine* (Durham, NC: Duke University Press, 1999).

50. Gilberto Freyre, *New World in the Tropics: The Culture of Modern Brazil* (New York: Alfred Knopf, 1960); see also Stanley Stein, "Freyre's Brazil Revisited: A Review of *New World in the Tropics: The Culture of Modern Brazil*," *Hispanic American Historical Review* 41, no.

1 (February 1961): 111–113; Driver and Martins, "Views and Visions," 4; Stepan, *Picturing Tropical Nature*.

51. This discussion summarizes Richard Tucker's *Insatiable Appetite: The United States and the Ecological Degradation of the Tropical World* (Berkeley: University of California Press, 2000). See also John Soluri, *Banana Cultures: Agriculture, Consumption, and Environmental Change in Honduras and the United States* (Austin: University of Texas Press, 2006); Daniel Headrick, "Botany, Chemistry, and Tropical Development," *Journal of World History* 7, no. 1 (1996): 1–20.

52. Raffles, *In Amazonia*, 7.

53. J. C. Willis, *Agriculture in the Tropics: An Elementary Treatise* (Cambridge: Cambridge University Press, 1909): 139.

54. Willis, *Agriculture in the Tropics*, 153–4. Willis's book provides support for many of the above statements as well.

55. For an example of a mid-twentieth-century reconsideration of shifting cultivation, see P. H. Nye and D. J. Greenland, *The Soil Under Shifting Cultivation* (Farnham Royal, UK: Commonwealth Agricultural Bureaux, 1960).

56. On the Green Revolution, see Nick Cullather, *The Hungry World: America's Cold War Battle against Poverty in Asia* (Cambridge, MA: Harvard University Press, 2010).

57. C. C. Webster and P. N. Wilson, *Agriculture in the Tropics*, 2d ed. (New York: Longman, 1980).

58. On Woodruff, "tropical neurasthenia," and modern notions of tropical degeneration, see Kennedy, "The Perils of the Midday Sun"; David Livingstone, *The Geographical Tradition: Episodes in the History of A Contested Enterpise* (Oxford: Blackwell, 1992), 236–237.

59. Ellsworth Huntington, *Civilization and Climate* (New Haven, CT: Yale University Press, 1915); Ellen Churchill Semple, *Influence of the Geographic Environment, on the Basis of Ratzel's System of Anthro-Geography* (New York: Henry Holt and Company, 1911), 628; Warwick Anderson, "Geography, Race, and Nation: Remapping 'Tropical' Australia, 1890–1930," *Historical Records of Australian Science* 11, no. 4 (December 1997): 457–468; Anderson, *The Cultivation of Whiteness*; Livingstone, *The Geographical Tradition*, 222–240.

60. Pierre Gourou, *The Tropical World: Its Social and Economic Conditions and Its Future Status*, trans. E. D. LaBorde (London: Longmans, Green, and Company, 1953). On Gourou's tropicality, see Arnold, "'Illusory Riches.'" On Gourou's work in Indochina, see David Biggs, *Quagmire: Nation-Building and Nature in the Mekong Delta* (Seattle: University of Washington Press, 2010), 91–126.

61. On the history of STRI, see Catherine Christen, "At Home in the Field: Smithsonian Tropical Science Field Stations in the U.S. Panama Canal Zone and the Republic of Panama," *The Americas* 58, no. 4 (April 2002): 537–575; Joel Hagen, Problems in the Institutionalization of Tropical Biology: The Case of the Barro Colorado Island Biological Laboratory," *History and Philosophy of the Life Sciences* 12 (1990): 225–247; Elizabeth Royte, *The Tapir's Morning Bath: Mysteries of the Tropical Rainforest and the Scientists Who Are Trying to Solve Them* (Boston: Houghton Mifflin, 2001). On other efforts in tropical ecology and environmental management, see McCook, *States of Nature*; Tucker, *Insatiable Appetite*, 345–416; Greg Bankoff, "Conservation and Colonialism: Gifford Pinchot and the Birth of Tropical Forestry in the Philippines," in *Colonial Crucible*, 479–488.

62. Kelly Enright, *The Maximum of Wilderness: The Jungle in the American Imagination* (Charlottesville: University of Virginia Press, 2012). See also Gregg Mitman, *Reel Nature: America's Romance with Wildlife on Film* (Cambridge, MA: Harvard University Press, 1999).

63. Kevin J. Gaston, "Global Patterns in Biodiversity," *Nature* 405 (May 11, 2000): 220–227.

64. Timothy Farnham, *Saving Nature's Legacy: Origins of the Idea of Biodiversity* (New Haven, CT: Yale, 2007); David Takacs, *The Idea of Biodiversity: Philosophies of Paradise* (Baltimore: Johns Hopkins University Press, 1996); Candace Slater, "Amazona as Edenic Narrative" in *Uncommon Ground: Rethinking the Human Place in Nature*, ed. William Cronon (New York: Norton, 1996), 114–131; Slater, *Entangled Edens: Visions of the Amazon* (Berkeley: University of California Press, 2002); Raffles, *In Amazonia*.

65. Catherine Cocks, *Tropical Whites: The Rise of the Tourist South in the Americas* (Philadelphia: University of Pennsylvania Press, 2013).

66. Rainforest Café website, http://www.rainforestcafe.com/. See also Jennifer Price, *Flight Maps: Adventures with Nature in Modern America* (New York: Basic Books, 1999), 167–206.

67. Cronon, "The Trouble with Wilderness," 69.

PART II

KNOWING NATURE

CHAPTER 8

....

AND ALL WAS LIGHT?—SCIENCE AND ENVIRONMENTAL HISTORY.

....

MICHAEL LEWIS

ENVIRONMENTAL historians, as much as any other group of scholars, find themselves straddling the famous if simplistic divide between the "two cultures" of the sciences and the humanities.[1] Arguably, it is this attempt to combine science and the humanities that provides environmental history with its energy and intellectual niche. Environmental history works against the compartmentalization of knowledge that so many disciplinary traditions rely upon; nowhere does it do this more directly than in its integration of the insights of the sciences into its scholarship.[2] This position can be uncomfortable, though, and not without intellectual costs. Unlike our fellow travelers in the humanities and social sciences (the historians and sociologists of science, for instance), environmental historians rely upon scientific information in order to make sense of our object of study: namely, the history of interactions between the human and non-human world. Scientific studies using methods such as dendochronology, pollen analysis, satellite imagery, GIS mapping, ice cores, or changes in mitochondrial DNA, among others, allow environmental historians to consider material factors key to the human past that otherwise leave no evidence in the documentary record. With few exceptions, environmental historians do not practice sciences such as historical ecology—the scientific reconstruction of past ecosystems—but instead cite scientists who do so (articles that are more properly based upon the methodologies of the natural sciences are sometimes published in *Environmental History*, but scientists write these articles).[3] For some of our colleagues in the humanities and social sciences, especially in science studies, environmental historians' practice of *using* science, as well as studying it, is disconcerting.

This is because, unlike our fellow travelers in the natural sciences (the ecologists, geologists, and climatologists, among others), environmental historians, along with colleagues in the history of science and science studies, study science as a culturally mediated system of knowledge.[4] Environmental historians are not just interested in the best current science and what it tells us about the material reality of the world; rather, we are also interested the history of the science, including missteps and now invalidated ideas, and their implications for environmental history. Environmental historians believe that science (whether accurate in its claims or not) tells us something useful about the human cultures that produced and believed it, and plays a key role in human impacts on the material world. Many scientists find this baffling, or even, when historians focus upon science as a series of culturally mediated practices, vaguely anti-scientific.

In other words, at different moments, sometimes in the same article or book, and at other times in different publications, environmental historians use science as a tool (a source of information about the natural world in the present and past), or as an object of study itself (an intellectual system for producing culturally situated knowledge). To many scholars outside of environmental history who focus on only one of these two scholarly approaches to science, such moves run the risk of seeming confused or intellectually inconsistent. But these two different ways of interacting with science (as a source of information, and as a cultural product) need not be incompatible. When environmental historians discuss "nature," they typically use science as a screen to make this discussion as materially accurate as possible. Environmental historians also analyze that screen with an eye to why and how it was produced, what it helps them to see, and what it obscures, ultimately historicizing not just the method, but also the evidence produced.

How a scholar approaches science can lead to completely different questions and research agendas—and these agendas might be at odds. Consider, for example, how environmental historians have approached the scientific studies of the Indian ecologist Madhav Gadgil, specifically his work on sacred groves (fragments of habitat that local people protect due to varied religious beliefs and practices). By choosing either to use the science uncritically, or to analyze the science as a culturally situated set of practices and agendas, an environmental historian would be led in different directions.

In 1972 Mahav Gadgil returned to India after receiving his PhD in biology from Harvard University. Although he trained as a theoretical ecologist, and used early computers to analyze the survival strategies of fish, Gadgil soon found himself conducting field studies on previously undescribed (by science) sacred groves in the hills of Maharashtra. As Gadgil and his colleague V. D. Vartak wrote in 1974 in the *Journal of the Bombay Natural History Society,* these sacred groves—ranging in size from a few trees to several acres—harbored important remnants of Indian biodiversity. Subsequent field studies in these and other Indian sacred groves revealed many new or rare species, particularly of plants and insects. In subsequent decades Gadgil's early studies provided the impetus for his formulation of village-level biodiversity activism, and led to his conviction that for conservation to succeed in India it must begin with poor villagers. This also would eventually lead Gadgil into a collaboration with the historian Ramachandra Guha; they cowrote a synthetic environmental history of India, *This Fissured Land: An Ecological History of India* (1992). More broadly, Gadgil's field studies (as well as those of other Indian ecologists) provided the basis for a number of other historical studies, most notably those of Donald Hughes and M. D. Subhash

Chandran, which discussed sacred groves in India (as well as elsewhere in the world) as a successful indigenous form of nature conservation.[5] Over the last decade, Indian sacred groves have also become central to environmental advocates' arguments about the stability and potential sustainability of non-Western forms of nature use, sometimes juxtaposed with Western models of top-down, national-park-based conservation. Without science cataloguing the long-standing and contemporary ecological health (at least in terms of biodiversity) of these remnants, the historical argument would be far less compelling.

Instead of, or in addition to, focusing upon what Gadgil's science illustrated about the environment, the past, and the success of indigenous forms of nature conservation, my own work on conservation in India considered the cultural production of Gadgil's science. My study considered Gadgil's personal history, his training, and the political context that made his science both possible and publishable: essentially, the knowledge regimes in which his sacred grove studies occurred. In this analysis, Gadgil's decision to return home rather than pursue a career as a US academic (he was offered positions at both Princeton University and Harvard University) was linked to his analysis of a specifically Indian form of conservation—a nature nationalism. In other writings, including *This Fissured Land*, Gadgil argued that the Indian caste system had evolved following a process much like that of speciation in animals: different groups of people had focused upon different niches and resources, creating an effective and environmentally sustainable division of resources. How might this sort of socio-biological argument (which in effect naturalized the caste system and implied that it had less to do with human power than with natural processes of differentiation) have influenced Gadgil's understanding of sacred groves? Why focus upon sacred groves, rather than non-religious sites where forests were still somewhat intact? And what made Indian sacred groves so appealing to some Westerners—did they conform to a gentle Orientalism mixed with a little bit of Judeo-Christian self-loathing? This type of environmental history looked at Gadgil's sacred groves studies—and the resulting popularity of Indian sacred groves in international environmentalist discourse—as reflective of a certain current in contemporary Indian, and global, environmentalism. In this study, the accuracy (or not) of his science mattered less than what his science revealed about historically and culturally situated ways of seeing and knowing nature. This form of environmental history sees science as another—albeit powerful—way of understanding nature, and as a product of historical (social) forces, not a neutral recorder of fact.[6]

Environmental historians such as Stephen Pyne, with his extensive work on fire history, have succeeded in integrating both of the above approaches: ably using the knowledge generated by science to enrich our understanding of the past, while also historicizing that information in a rich cultural and social context. Fire ecology is crucial to how historians understand the long history of many terrestrial ecosystems. The attempt to parse the history of fire-induced modifications to ecosystems, and the role of humans in setting or managing those fires, has yielded dozens of research projects in the sciences and in environmental history. But the science of fire ecology is strongly culturally mediated and how a society understands fire—and the related concepts of nature, risk, and settlement, among others—plays a role in how their fire science develops. Pyne's work on fire has gracefully combined the need to use science to reconstruct what happened in the past, and the need to analyze science as a specific and powerful form of human interactions with the natural world. Pyne's work on fire has spanned his career, and included several articles and books—it has taken him a certain amount of space and time to pull off these career-defining histories. And while his strategy is not unusual among environmental historians, at the very least, the skill needed to combine these two different ways environmental history uses science does not always come quickly or easily—or in a single essay.[7]

Any attempt at listing or describing the role of science in environmental history is bound to be partial, as science (as both tool and object of study) is so omnipresent in environmental history scholarship. Several other chapters in this volume deal specifically with subjects and themes that are linked to science: technology, restoration, chemicals, health, and climate, among others—these topics, as well as the many intriguing studies by scientists in historical ecology are all being set aside. Of the vast terrain that still remains, this chapter focuses upon only a few examples of the uses (and only rarely abuses) of science as a tool within environmental history. Then we turn to a distinct variety of environmental history that is closely linked to the history of science and cultural and intellectual history: the history of scientific ideas and how they have shaped and reflected human interactions with the non-human world. The chapter closes with a brief consideration of three different calls for "new directions" in environmental history and its relationship to science.

USING SCIENCE TO READ THE PAST

The exchange of species around the globe, from smallpox to pigs to dandelions, has been one of the earliest and most productive avenues of environmental history scholarship. Alfred Crosby's pathbreaking works on the Columbian Exchange, beginning in 1972, retold the story of European expansion into the "neo-Europes" by focusing on the role of flora, fauna, and microbes in aiding colonization. Scientists had been concerned about biological "invasions" for decades; Crosby added these ecological concepts and data to the documentary record of settlement and adaptation. Many other scholars have followed suit, tracing the spread of plants, animals, and microbes around the globe, the resulting ecological impacts, and the roles of humanity in facilitating and responding to this process. Subsequent scholarship also called into question the language used to describe these processes—"aliens," "natives," "invasions," "exotics"— and the social and cultural history of species exchange.[8] At times, environmental histories of biological exchange have leaned close to environmental determinism, but this

has been more typical of misunderstandings and popularized versions of this scholarship, rather than the academic work produced by environmental historians.[9]

A second crucial application of scientific information by environmental historians
has been the use of scientific data on the composition (fauna, flora, soils, hydrology,
and climate) of past landscapes. This has been particularly attractive as a way of charting the interplay of human decisions (recorded through more traditional documentary
history) and changing landscapes. Classic early works of environmental history such
as William Cronon's *Changes in the Land: Indians, Colonists, and the Ecology of New
England* (1983) and Richard White's *Land Use, Environment and Social Change: The
Shaping of Island County, Washington* (1980) relied upon the work of scientists (through
tools such as pollen analysis) to characterize historical landscapes.[10] As mentioned
earlier, with very rare exception, environmental historians were not conducting this
historical ecology (or climatology, or geography) themselves, but relying upon science
done by others. Many early environmental historians were self-taught in the sciences;
the official history of the American Society for Environmental History records that its
first members worked at "acquiring a dual literacy in environmental science and ecology as well as history."[11] One tangible effect of this was that, at times, environmental historians used scientific information and ideas that were no longer considered accurate
by practicing scientists. A lack of formal scientific training left environmental historians at greater risk of crediting bad science, using weak science, or of not using helpful
science of which they simply were not aware. Insofar as this was true, this scientific illit-
eracy—or sub-literacy—would also corrupt the peer-review process in environmental
history journals, as reviewers would be less likely to catch scientific errors or omissions.
This was especially true in the earliest years of environmental history. Both Cronon
and White later considered their earlier works, mentioned above, to be too heavily
influenced by what they came to recognize as outdated notions of climax communities,
popularly referred to as the idea of a "balance of nature." These ecological ideas, based
in part on the work of pioneering plant biologist Frederic Clements, were the ecological
state-of-the-art up until the 1940s, when ecologists began to move to an understanding of ecological communities that were in a constantly shifting flux: change without
climax. As White wrote, "In [*Land Use*] . . . I used ecological concepts like community,
succession, climax and ecosystem unproblematically . . . I did this even though within
the discipline of ecology these ideas had already come under attack. Looking back now,
I realize that this book and other historical studies were themselves undermining such
ecological concepts even as they relied upon them." White went on to call for environmental historians to "pay more attention to newer ecological constructions in which
stability plays little part."[12] Within popular environmentalism, discussions of the "balance of nature" and unchanging climax communities still are common, but not within
environmental history. Instead, nuanced work, such as the award-winning *The Great
Meadow* (2004) by Brian Donahue, is more typical, with carefully constructed maps
and characterizations of agricultural practices that maintained tenuous ecological stability only with careful effort.[13]

On some occasions, historians use the best science available to understand a past landscape or event, but that science might not be the last word on the subject. New scientific breakthroughs or models allow for—and on occasion necessitate—new interpretations of historical events. For example: environmental historians working on the history of the American Dust Bowl must attempt to determine the exact boundaries of unproductive land, of drought, and the ecological causes of the disaster. In a 2012 article in *Environmental History*, Kenneth Sylvester and Eric Rupley used new ArcGIS mapping technologies to create a spatial model of a portion of the Dust Bowl landscape, and then overlay onto that model improved soil classification and survey data, as well as airborne photography. This new scientific modeling technology did not completely invalidate earlier scholarship on the Dust Bowl, but it did make certain claims about the material reality of the event that were new, and that had to be incorporated into subsequent historical scholarship.[14] In other cases, new data or a new modeling technique allows a scholar to study topics that had, to that point, been beyond the reach of historians. One notable example of this was Nicholas Robins's 2011 use of air-pollution modeling (developed by the American Meteorological Society and the EPA) to estimate air quality and human health impacts of colonial silver mining in the Andes. As is almost always the case when historians collaborate with scientists in this fashion, the science operated as a "black box," with Robins using the model, but not participating in its construction. It yielded intriguing and historically useful results—but if scientists construct a newer and more accurate model, another historian will have to revisit Robins's conclusions.[15]

ENVIRONMENTAL HISTORIES
OF ECOLOGICAL IDEAS

These few examples have only suggested the many usages of science as a tool for revealing the material past; while the details of other studies vary, the basic premises are consistent. We now turn our attention to the scholarly tradition within environmental history of considering the history of scientific ideas and practices, and how they have shaped and reflected human interactions with the non-human world. At first glance, this might seem to be a subfield of the history of science. An ever-increasing number of environmental historians have been trained in that tradition, but there are noticeable differences in the practices of environmental historians and historians of science. This can be easily seen in comparing four different books focused upon the history of the ecological sciences: historian of science Sharon Kingsland's *Modeling Nature: Episodes in the History of Population Ecology* (1985), and the environmental histories of Donald Worster (*Nature's Economy: A History of Ecological Ideas* [1977]), Susan Flader (*Thinking Like a Mountain: Aldo Leopold and the Evolution of an Ecological Attitude towards Deer, Wolves, and Forests* [1974]), and

Carolyn Merchant (*Death of Nature: Women, Ecology, and the Scientific Revolution* [1980]).

Environmental historians hoping to learn or understand basic ecology could use Kingsland's book to do so; it is an accurate and technical summation of the development of key ecological theories throughout the twentieth century, with particular attention to population ecology. Kingsland's book would look familiar to an ecologist: she discusses the theories and scientists that ecologists recognize as central to their discipline. In contrast, Worster focuses upon the cultural context (specifically the emergence of environmentalism) for the development of ecological science more than upon the technical details. *Nature's Economy* includes key environmentalists (such as Thoreau and Muir) who were interested in ecology, but who would never be included in a more formal history of that science. Worster also focuses upon some figures, such as James Lovelock (with his Gaia hypothesis), who have received more attention outside of scientific circles than within them. The Gaia hypothesis (which proposes that the totality of the earth be studied as a self-regulating system) has considerable cultural importance, but most practicing ecologists dismiss it as either non-scientific, or too unwieldy to offer a coherent line of scientific investigation. In short, Worster's book is more useful as a history of the cultural climate in which some ecological ideas arose and were popularized, as opposed to a history of ecology.[16]

Occupying something of a middle ground between *Nature's Economy* and *Modeling Nature* is Susan Flader's *Thinking Like a Mountain*. Unlike Worster's book, Flader focuses more carefully upon the changing ecological theories and implications of one specific set of ecological ideas: predator-prey interactions and populations (a subject that Worster considers for thirty-three pages, but for Flader is central to her entire text). And unlike Kingsland, Flader ties her analysis to her central subject, Aldo Leopold, and his experiences in specific landscapes, where he observed specific environmental transformations. Flader, in other words, includes both the cultural and material contexts of ecological ideas, and offers a more careful discussion of the changes in the science itself. The end result is a book that does not look as tangential to mainstream history of science as Worster's, but which nevertheless still shares environmental history's materiality and focus on cultural impact.[17]

Unusual among environmental historians for its focus upon science prior to the eighteenth century, Carolyn Merchant's *Death of Nature* is one of the most widely recognized books about science published by an environmental historian, and it offers additional insights into the differences between the two subfields of history. Merchant's polarizing thesis was that the scientific revolution (linked with the capitalist revolution) succeeded in transforming conceptions of nature from an organic, female, spirit-filled entity into a machine open to investigation and domination. Merchant then linked this transformation to the fate of women in Europe during this same period (roughly 1500–1800), and the fate of the environment. Although her analysis was centered on the early modern period, Merchant regularly pointed out the links between this crucial period and the crises of the twentieth century. Merchant did not shy away from the big picture—or from large-scale generalizations—and this is both a strength and weakness of the book. *Death of Nature* provided a provocative and challenging thesis that has forced subsequent scholars to at least consider its big questions, and explain why they do not agree with them, if they do not. Some historians find the book's argument frustrating, because it is ultimately unprovable (in the way that all good historical arguments centered on "why" are destined for ambiguity). A more cautious book might have drawn less criticism, but it would not have had the same impact upon gender studies, environmental history, and the history of science.[18]

Common to both Worster and Merchant is a strongly articulated moral vision. These books do not hesitate to pass judgments on the ways in which science has affected human societies and the non-human world for good or ill. Even Flader, who is ultimately more restrained and closely tied to her evidence than Merchant, still wrote from the perspective of an environmentalist. This is in keeping with the more applied focus of much environmental history, and could have been a factor in some historians of science dismissing environmental historians who write about science as activists or dilettantes.[19] Over the last forty years, a number of graduate students were trained in both the history of science and environmental history—for some, the job that they eventually landed helped to define them (whether they were hired as a historian of science, or an environmental historian, or even in an environmental studies program). In other cases, one or the other scholarly community was more receptive to the scholarship of a young scholar—many historians who work at the intersections of these two fields attended both the environmental history and history of science conferences, then regularly returned to whichever one seemed most amenable to their approach and worldview.

As it happened, Merchant's book was met with more acceptance in environmental history than in history of science, and this in part led to Merchant's greater involvement in the professional activities of the American Society for Environmental History (ASEH), including serving as that society's president, as did Flader and Worster.[20] A few scholars from that generation or the following one, such as the Wisconsin historian Gregg Mitman, have succeeded in maintaining high levels of involvement in both of these scholarly communities. Yet as Mitman admitted in a retrospective review article of Merchant's book, throughout the 1980s and 1990s it seemed that the two fields were moving in opposite directions. He suggested that environmental historians were involved in the swirling environmental politics of the period and thus constantly linking their scholarship to actual changing environments, while the sociology of scientific knowledge came to dominate the history of science and "any discussion of the nonhuman material world all but disappeared." For Mitman, the failure to concretely link environmental change (materiality) with the history of ecological change had thus far (as of 2006) divided these two fields.[21] This division had a personal cost for Mitman, as many environmental historians underutilized the history of ecology studies published by historians of science through the 1980s and 1990s, including his own 1992 history of ecology, *The State of Nature*.[22] In contrast, this same period, starting in the late 1980s and continuing to the present, was characterized by an ever increasing number of young environmental historians, often well trained in the sciences or history of science, that followed in what we might call the "Flader" vein. These scholars' work was centered on the ecological sciences and conservation concerns in the midst of changing natural and social histories, but paid more careful attention to the canonical history of science than was seen with Worster's *Nature's Economy*. Notable texts included Nancy Langston's *Forest Dreams, Forest Nightmares: The Paradox of Old Growth in the Inland West* (1995), James Pritchard's *Preserving Yellowstone's Natural Conditions: Science and the Perception of Nature* (1999), Thomas Dunlap's *Saving America's Wildlife* (1988), Stephen Bocking's *Ecologists and Environmental Politics: A History of Contemporary Ecology* (1997), Mark Barrow's *Nature's Ghosts: Confronting Extinction from the Age of Jefferson to the Age of Ecology* (2009), Jeremy Vetter's *Knowing Global Environments: New Historical Perspectives on the Field Sciences* (2010), Frederick Rowe Davis's *The Man Who Saved Sea Turtles: Archie Carr and the Origins of Conservation Biology* (2007), Benjamin Cohen's *Notes from the Ground: Science, Soil, and Society in the American Countryside* (2009), and Peter Alagona's *After the Grizzly: Endangered Species and the Politics of Place in California* (2013), among many others. All of these scholars focused upon the intersections between environmental management practices (often those of the government), cultural attitudes, landscapes or species, and the ecological sciences—and they got the science "right." More and more environmental historians are producing scholarship that, while it continues to share the central concerns of environmental history, is also credible history of science, and in fact, many of these scholars define themselves as members of both subfields.[23]

Given environmental historians' interest in situating the sciences—and particularly ecology—in the larger non-human world (both in terms of the impetus for how particular scientific conceptualizations emerged, and the impacts of those ideas and practices), it is perhaps not surprising that there have been a number of particularly effective studies that have examined the links between the environmental sciences and imperialism. This has been another area of particular strength within the interstices of environmental history and the history of science. Environmental historians are trained to think broadly, linking science, culture, and nature; thus an environmental historian writing about science in colonial India would typically make his or her work as much about modernism, development, and colonial politics as about science. Historians of science sometimes do the same thing, but often their narratives can be carefully (and narrowly) focused upon the developments of the science itself. [24] Peder Anker has argued specifically in *Imperial Ecology* that the science of ecology, globally, was in large part dependent for its growth and development upon British imperialism.[25] Much of environmental history as it has developed among scholars working in Africa and India includes analyses of the effects of imperialism upon landscapes, as well as people and politics.[26] Over the last decade a number of strong studies have illustrated that scientists—and the sciences that they practiced—were fully implicated in the imperial project, whether as rationalizing discourses, tools of control of people and landscapes, or extractive agents. Some studies have focused upon resistance to colonial-era science and others on complicity, and in almost every case they have shown that colonial-era science was never as monolithic or triumphant as memory would have it. But even further, some environmental historians working on the developing world have argued that the environmental sciences (and environmentalism) were fundamentally shaped by the periphery—not by the act of subduing the developing world, but more by the interchange of ideas, species (microbiotic and otherwise), and peoples that occurred during European expansion. Richard Groves' *Green Imperialism: Colonial Expansion, Tropical Island Edens, and the Origins of Environmentalism, 1600–1860* (1995) is one of the best of these studies at combining the history of the development of ecological science and the colonial experience.[27]

These African and Asian—particularly South Asian—environmental histories have been strongly influenced by the specific characteristics of environmentalism in their respective locations. Intriguingly, they have focused more upon conservation controversies than urbanization or pollution battles (though those are certainly present, in sheer numbers they have been less visible). In this, these studies and scholars perhaps mirror, to some extent, First World concerns with the fate of high biodiversity landscapes in the tropical world. Even so, social justice concerns are very much a part of this scholarship, particularly with regard to peoples that have been removed from forests and protected areas at the behest of scientists and managers seeking improved ecosystem health (often wrongly, as some of this scholarship suggests). Ramachandra Guha's *The Unquiet Woods: Ecological Change and Peasant Resistance in the Himalaya* (1989) is an exemplar of this tradition, and has proven influential in encouraging Americanists to reconsider conservation politics in the US.[28] Many of the scholars of India are inheritors of the subaltern school tradition, and the links between British institutions and universities and scholars working elsewhere in the Commonwealth have been strong, regardless of those scholars' nationalities. The British-based journal *Environment and History* has been, from its start in 1995, strongly tied to scholars working on Africa and Asia, and its contents have, overall, been more international and more focused upon science than the corresponding US-based journal, *Environmental History*. Yet another journal, *Conservation and Society*, begun in 2003 and based in India, offers an even more ambitious scholarly attempt to combine scholars from the sciences and the humanities in one publication. Environmental historians have been heavily involved from its conception, particularly those who have worked on Indian environmental history, but also science-oriented environmental historians who have studied other regions of the world.

Following the pattern of Richard Grove and his global scholarship, a number of environmental historians interested in science have attempted transnational studies that consider the global reach and production of science. Some of these, such as Tom Griffiths and Libby Robin's *Ecology and Empire: Environmental History of Settler Societies* (1997), or *Civilizing Nature: National Parks in Global Historical Perspective* (2012), edited by Patrick Kupper, Bernhard Gissibl, and Sabine Höhler, have achieved this through comparative edited collections that include scholars specializing in several different regions of the world. These books (including essays by several of the scholars mentioned here) succeeded in making connections between imperialism, the development of the science of ecology, and ecosystem changes throughout the world.

Others, like Jacob Darwin Hamblin's *Oceanographers and the Cold War* (2005) or my own *Inventing Global Ecology* (2004), have focused upon post-1945 transnational scientific collaborations by beginning in the United States and moving outward, implicitly treating the United States as at least a partial inheritor of the former British role of preeminent global scientific and political power. Still other scholars have chosen to focus upon conservation diplomacy, and look at the role of science as one factor in shaping international treaties to preserve wildlife, as with Kurk Dorsey's *The Dawn of Conservation Diplomacy: U.S.-Canadian Wildlife Protection Treaties in the Progressive Era* (1998). These books and others like them aspire to describe the connections between global environmental transformations, changing human attitudes toward the global environment, and transnational science. As a whole, these books demonstrate both the continuing relevance of nation-states in exerting power over the practices and uses of environmental sciences, as well as the need to illuminate the complex flow of scientific ideas, individuals, and ecological impacts across national borders. Few scholars even aspire to approach Grove's level of mastery of such a wide source-base or his sweeping argument, but these early attempts at writing transnational environmental histories centered around or including environmental sciences do suggest a potentially fruitful avenue for further research.[29]

POTENTIAL DIRECTIONS

In 2003, a US scholar, Edmund Russell, published "Evolutionary History: Prospectus for a New Field," an article which won the 2004 Leopold-Hidy prize for the best article published the previous year in *Environmental History*. In this article, Russell wrote that environmental history "has drawn on scientific ideas to understand" how humans have shaped nature. Russell challenged environmental historians to incorporate scientific insights even more thoroughly into their scholarship. Russell called for environmental historians to familiarize themselves with the insights of the modern synthesis that linked evolutionary theory and genetics within the biological sciences, and to apply those insights to his newly coined "evolutionary history." Russell followed this article with a monograph, *Evolutionary History: Uniting History and Biology to Understand Life on Earth* (2011), which more fully explicated, and demonstrated, how evolutionary history could work.[30]

Russell was not proposing that environmental historians wed themselves to a new idea—the modern synthesis was developed in the 1940s and 1950s, and is integral to all contemporary biology. In the attempt to study complex modern processes, "biology and history [are] leading the way along parallel, but too rarely intersecting, paths. Evolutionary history offers a way to link these endeavors." Russell argues that humans have shaped evolution itself, and the resulting evolutionary changes in turn shape the course of world history: "Evolutionary and social forces become entwined." This can be seen in many areas, whether speaking of agriculture, disease, domesticated animals, or even the ways that smoke in the air determines which color moths thrive in early industrial Britain. Russell's evolutionary history is an ambitious proposal, and would require that many environmental historians learn considerably more science than they currently know. As Russell writes in explaining the lack of evolutionary theory in current environmental history scholarship, "historians may have lacked familiarity with evolution in general and anthropogenic evolution in particular. Few graduate or undergraduate programs in history require courses in science, much less in evolutionary biology." It is still too early to see what effect Russell's call for an evolutionary history will have upon graduate programs and practicing historians, but he proposes a major reorientation of environmental history toward the sciences. In a review of his book in *Environmental History*, the reviewer claimed, "historians are frequently intimidated by science." This is certainly true of some historians—some others mentioned above, though, are quite comfortable with evolutionary theory. Thus, while evolutionary history might not ever become the mainstream for environmental history, there is certainly a critical mass of graduate students and faculty who could pursue this approach.[31]

It is important to note that Russell is quite aware of the dual nature of environmental historians' interaction with science. His call for greater attention to science (particularly evolutionary science) as a crucial tool is not a rejection of the intellectual legacy of the study of science as culturally mediated knowledge, and he applauds this complication of the notion of scientific objectivity. He does not believe that this should stop environmental historians from becoming more closely tied to evolutionary science: "we should not let skepticism necessarily lead to rejection." We can have our scientific cake, and eat it too, through careful and critical interrogation. In the 2010 *Companion to American Environmental History*, Russell had some supporters—Dan Flores and Don Worster both argued for more attention to evolution. These two senior scholars suggested that even human behavior was, in part, a product of human evolution—and while this is not radical within the context of evolutionary biology, it certainly runs contrary to the cultural turn of the social sciences. One scholar, commenting on these essays, claimed "my guess is that environmental historians will have to reckon with evolutionary ideas more closely in the coming years."[32]

A very different attempt to reorient the relationship between environmental history and the sciences was published in *Environmental History* in 2007. "The Problem of the Problem of Environmental History: A Re-reading of the Field" was a call for environmental historians to move away from their uncritical use of science, and into the welcome arms of social theory. The two authors, Sverker Sorlin and Paul Warde, work in Stockholm and Cambridge, respectively. Sorlin and Warde are concerned that "Environmental historians often have looked more to the natural than to the social sciences, both for data and at times for concepts." They suggest that insofar as environmental history "privilege[s] the natural sciences, it is hard to see environmental history as anything but an epiphenomenona in the study of nature, and it will be equally hard to convince the natural or social sciences of its importance." Sorlin and Warde see environmental historians as puzzlingly uninterested in attempting to "actually contribute to the vanguard of scholarship and to the theoretical progress of our profession." For these two European scholars, the problem with environmental history is not that its practitioners need to learn more science, but rather that they need to focus less upon science and more upon social theory. They provide several examples of how greater attention to social theory would help environmental historians to problematize and expand the concept of nature, and lead to a greater awareness of the role of science in rationalizing political relationships.[33]

These two calls to reform environmental history could not be more different. The authors' agendas, but also their experiences in the US and Europe, led them to view environmental history as if it were in fact two different disciplines. Within Europe, environmental historians are much more closely aligned with scientists, both collaborating more frequently, and more commonly involved in projects that might be described as historical ecology. North American environmental historians are more fully embedded in history departments, and have less contact with scientists. These differences are obvious at international conferences, where European and North American environmental historians often have predictably different topics, methodologies, and disciplinary orientations. Thus, it is perhaps not surprising that Sorlin and Warde see environmental history (from their vantage point in Europe) as nearly a subset of the environmental sciences, and the US-based Russell perceives environmental historians as largely disinclined to engage too closely with science, whether due to lack of training or skepticism born of social theory. It is a mistake to reduce these interventions to merely regional styles, however, for many historians, where ever located, operate somewhere between these extremes.

A third article published in *Environmental History*, "Opportunities in Marine Environmental History," by the US historian W. Jeffrey Bolster, was based on his experiences in an interdisciplinary research project. As he writes, "That collaboration's initial successes, tensions and criticisms illuminate not only the complicated marriage of interdisciplinary work, but fundamental conceptual disagreements about the relationship of the past to the present and future." He worked with ecologists as well as American and European environmental historians, and much of his article dealt directly with the difficulties involved in working across these different understandings of environmental history, and how they relate to science. Much as with the articles by Russell and Sorlin and Warde, Bolster hoped to persuade some readers to change their scholarly practices. Unlike the first two articles, however, Bolster was not trying to change scholars' orientation toward science, but rather trying to urge environmental historians to take more seriously the marine environment as a potential field for study. Since that article's publication, a number of other articles and books have come out of this larger research project—we focus here on this essay not because it is the final word on that particular research project, but because of how Bolster describes working at the intersection of environmental history and science on the History of Marine Animal Populations project (HMAP).[34]

HMAP, begun in 1999 with funds from the Alfred P. Sloan Foundation, mandated collaboration between historians and marine ecologists, and significantly, recognized a difference between historical ecology and marine environmental history.

Bolster suggested that marine environmental history is the only arena in which such collaboration between scientists and environmental historians has been "institutionalized and generously funded." In the seven years between the founding of HMAP and the publication of Bolster's article, though, not all collaborating historians were pleased: "Though relying on data from the past, the majority of HMAP projects have not been driven by the sort of questions that most American environmental historians ask; in fact, virtually all HMAP investigations have been quantitative, as if establishing benchmarks against which to measure loss is the *raison d'être* of marine environmental history." Bolster suggested that European environmental historians were more prepared to collaborate with scientists and to undertake quantitative investigations than their US counterparts. A Danish environmental historian, Poul Holm, was quoted as saying, "For historians to influence biologists...we need to present well-defined data and recurrent phenomena and hypotheses." Bolster, himself an American, was sympathetic to the concerns of US environmental historians, and their desire to use narrative and qualitative knowledge in studying the past. He argued that environmental historians "think about time in a radically different way from most scientists. They recognize that time (including ecological time) is not linear, and that long-gone contexts and now-invisible contingencies affected the past in such a way that it was qualitatively different." While Bolster described at least one prominent historian who quit HMAP due to dissatisfaction with working with scientists, he continued to see the collaboration as ultimately worthwhile, and encouraged other environmental historians to join him (and perhaps help change the culture of the collaboration away from exclusively quantitative history).[35]

These calls for potential directions describe three different ways to re-envision the relationship between environmental history and the sciences: one encouraging environmental historians to acquire more critical distance from science and participate more fully in the reflexive turn in the social sciences; another encouraging environmental history to integrate scientific theory (specifically evolutionary theory) into our central scholarly identity; and a third encouraging environmental historians to collaborate with scientists, but resist the perhaps inevitable pressure to join the more quantitative social sciences and turn the past into an ecological baseline. In obverse, their arguments suggest that environmental historians are relatively skeptical of social theory, quantification (at least in the US), collaboration, and theoretical science as a historical methodology. Stephen Pyne, in his 2006 presidential address for the ASEH, more positively expressed this same view, as he spoke against reductionism and in favor of narrative as essential to the environmental historian's craft. (This, of course, comes from a scholar who uses science constantly, and is even tenured in a School of Life Sciences.)[36] The flux and uncertainty surrounding science at the edges of this small but vibrant subfield speak to two things: a certain amount of disciplinary incoherence, but also a pragmatic confidence that the lack of a unifying plan of study provides a polyglot hybrid vigor. For environmental historians are nothing if not disciplinary mutts—we just hope that we are more the hardy academic sled-pullers of the Iditarod (hardly a purebred dog among them) and not intellectual mules—strong for a generation, a bit stubborn and set in our ways, and ultimately sterile.

Environmental historians continue to be fascinated by science both for what it can tell us about the material world, and for its place as one of the more significant and intriguing artifacts of human culture. To paraphrase Woody Allen's quip about nature, we are "two with science." This tension between embracing science and its discoveries, and holding it at arm's length for critical consideration, is shared by many contemporary non-scholars, but for different reasons; for science has been the primary epistemology for both appreciating the absolutely stunning complexity of the cosmos, and for irremediably changing—even destroying—some of those bits of the cosmos that we hold most dear. As environmental historians have shown time and again (and none more comprehensively than John McNeill in *Something New Under the Sun: An Environmental History of the Twentieth-Century World* (2000)), the scientific revolution, combined with gradual changes in economic and political organization, led to far-reaching environmental changes.[37] While some changes have been positive, especially from a human perspective, others have not. We live in a world that shows many signs of sweeping ecological and climatological transformation: the planet is warming; sea levels are rising; coral reefs are dying; and countless species of plants and insects— let alone microorganisms—are going extinct before they can even be catalogued. Other species are spreading rapidly: zebra mussels, cheat grass, and comb jellies, for example. Diverse forest ecosystems are being logged, and if replanted, often turned into mono-crop tree plantations. Wild fisheries are collapsing from over-harvest. Harmful artificial chemicals such as polychlorinated biphenyls (PCBs) can be found across the globe. Perhaps ironically, science has been as essential to the response to this crisis as it was to its creation. Although environmental historians might not need to be scientists in order to see which way the global wind blows, scientific research made possible all of the above claims about environmental degradation (and in so far as these claims are disputed, they are disputed with alternative scientific studies or explanations). It is no accident that many of the key figures in contemporary environmentalism have been scientists or were trained as scientists: Rachel Carson, Julian Huxley, Salim Ali, Barry Commoner, Paul Ehrlich, Wangari Maathai, and E. O. Wilson, to name just a few.[38]

In short, science, and the technological world that science has made possible, have given us both the environmental crisis and modern environmentalism, both endocrine disruptors and an understanding of the stunning diversity of life. So we can join Alexander Pope in celebrating Newton and his intellectual heritage, the scientific tradition that allows us to more fully understand nature in all of its glory, and to better understand the past as well. But we are still environmental historians, after all, all too aware of the material changes in the world that human actions and decisions, enabled by scientific knowledge, have achieved. And though none of us are irremediable declensionists, and we all hold onto the myriad hopes that knowledge and wisdom bring, late in the dystopian night we might not be blamed for wondering if it was only an apple that fell.

NOTES

* As per Alexander Pope: "Nature and Nature's laws lay cloaked in night, God said 'Let Newton Be!'—and all was light."

1. C. P. Snow, *The Two Cultures and the Scientific Revolution* (New York: Cambridge University Press, 1963). On environmental history and the two cultures, see Donald Worster, "The Two Cultures Revisited: Environmental History and the Environmental Sciences," *Environment and History* 2, no. 1 (1996): 3–14.

2. See, for example, Brian Black's argument that environmental history appeals to his environmental studies students in part due to its ability to synthesize many of their other disciplinary courses into one narrative. Brian Black, "Thickening our Stories: Models for Using Environmental History as Context," *The History Teacher* 37, no. 1 (November 2003): 57–66.

3. For example, see the soil and forest ecologist Daniel deB. Richter's article, "The Accrual of Land Use History in Utah's Forest Carbon Cycle," *Environmental History* 14 (July 2009): 527–542. For an overview of historical ecology, see Dave Egan and Evelyn Howell, eds., *Historical Ecology Handbook* (Washington, DC: Island Press, 2001).

4. History of science is a larger field than environmental history, and there are several shelves of books that share this perspective. For broad overviews of contemporary approaches to the history of science, see Peter Dear, "What Is the History of Science the History Of? Early Modern Roots of the Ideology of Modern Science," *Isis* 96 (2005): 390–406; Dear, "Focus: Global Histories of Science," *Isis* 101 (2010): 95–158; and for a summary of the combination of science, technology, and medicine often abbreviated as STM, see John Pickstone, "Sketching Together the Modern Histories of Science, Technology, and Medicine," *Isis* 102 (2011): 123–133.

5. Madhav Gadgil and V. D. Vartak, "Sacred Groves of India: A Plea for Continued Conservation," *Journal of the Bombay Natural History Society* 72, no.2 (1974): 313–320; Madhav Gadgil and Ramachandra Guha, *This Fissured Land: An Ecological History of India* (Delhi: Oxford University Press, 1992); M. D. Subash Chandran and Donald Hughes, "The Sacred Groves of South India: Ecology, Traditional Communities, and Religious Change," *Social Compass* 44, no.3 (1997), 413–427, among many other articles by these two authors.

6. For more on Gadgil and his science and its legacies, Michael Lewis, *Inventing Global Ecology: Tracking the Biodiversity Ideal in India, 1945–1997* (Athens: Ohio University Press, 2004).

7. Stephen Pyne has written several books on fire; see, for example, *Fire in America: A Cultural History of Wildland and Rural Fire* (Princeton, NJ: Princeton University Press, 1982); *Burning Bush: A Fire History of Australia* (New York: Henry Holt, 1991); *World Fire: The Culture of Fire on Earth* (Seattle: University of Washington Press, 1995); *Vestal Fire: An Environmental History, Told through Fire, of Europe an Europe's Encounter with the World* (Seattle: University of Washington Press, 1997); *Fire: A Brief History* (Seattle: University of Washington Press, 2001); *Awful Splendour: A Fire History of Canada* (Vancouver: University of British Columbia Press, 2008).

8. Alfred Crosby, *The Columbian Exchange: Biological and Cultural Consequences of 1492* (Greenwood: Greenwood Press, 1972); Crosby, *Ecological Imperialism: The Biological Expansion of Europe, 900–1900* (New York: Cambridge University Press, 1986); Peter Coates, *American Perceptions of Immigrant and Invasive Species: Strangers on the Land*

(Berkeley: University of California Press, 2006); Sarah Johnson, ed., *Bioinvaders: Themes in Environmental History* (Cambridge: White Horse Press, 2010); Harriet Ritvo, "Going Forth and Multiplying: Animal Acclimitization and Invasion," *Environmental History* 17, no. 2 (2012).

9. This claim was regularly leveled at Jared Diamond, *Guns, Germs, and Steel: The Fates of Human Societies* (New York City: W. W. Norton, 1999). Diamond is a professional scientist, but with a long history of varied interests and outstanding writing, and was invited to give a plenary address at the American Society of Environmental History's annual conference in 2002.

10. William Cronon, *Changes in the Land: Indians, Colonists, and the Ecology of New England* (New York: Hill and Wang, 1983); Richard White, *Land Use, Environment and Social Change: The Shaping of Island County, Washington* (Seattle: University of Washington Press, 1980).

11. "History of ASEH," American Society for Environmental History website, http://www.aseh.net/about-aseh/history-of-aseh.

12. Richard White, "Indian Peoples and the Natural World: Asking the Right Questions," in *Rethinking American Indian History*, ed. Donald Fixico (Albuquerque: University of New Mexico Press, 1997), 87–100. White and Cronon were not the only historians to struggle with this. For one additional example, see scientist Royce Ballinger, reviewing Ronald Tobey's *Saving the Prairies: the Life Cycle of the Founding School of American Plant Ecology, 1895–1955*, who was puzzled that Tobey had not given more attention to Henry Gleason, the famous ecologist who had challenged Clements ideas beginning in the 1920s: Royce Ballinger, "Review: Saving the Prairies," *Great Plains Quarterly* 1, no. 1 (1983): 54–55.

13. Brian Donahue, *The Great Meadow: Farmers and the Land in Colonial Concord* (New Haven, CT: Yale University Press, 2004).

14. Kenneth Sylvester and Eric Rupley, "Revising the Dust Bowl: High Above the Kansas Grasslands," *Environmental History* 17 (2012): 603–633.

15. Nicholas Robins, *Mercury, Mining, and Empire: The Human and Ecological Cost of Colonial Silver Mining in the Andes* (Bloomington: Indiana University Press, 2011).

16. Donald Worster, *Nature's Economy: A History of Ecological Ideas* (Cambridge: Cambridge University Press, 1977); Sharon Kingsland, *Modeling Nature; Episodes in the History of Population Ecology* (Chicago: University of Chicago Press, 1985).

17. Susan Flader, *Thinking Like a Mountain: Aldo Leopold and the Evolution of an Ecological Attitude toward Deer, Wolves, and Forests* (Columbia: The University of Missouri Press, 1974).

18. Carolyn Merchant, *The Death of Nature: Women, Ecology, and the Scientific Revolution* (New York: Harper and Row, 1980). Retrospectives include Gregg Mitman, "Where Ecology, Nature, and Politics Meet: Reclaiming *The Death of Nature*," *Isis* 97 (2006): 496–504; and Noel Sturgeon, Donald Worster, and Vera Norwood, "Retrospective Review," *Environmental History* 10 (2005): 805–815.

19. This happened less frequently in print than in conferences, lectures, and other informal events, where it happened frequently enough that it became, by the time that I was in graduate school in the late 1990s, a regular feature of graduate student life. Most environmental historians who worked at the intersection with the history of science at this time have at least one anecdote along these lines.

20. Mitman, "Where Ecology, Nature, and Politics Meet," 500.

21. Greg Mitman, "Where Ecology, Nature, and Politics Meet," 500.

22. Greg Mitman, *The State of Nature: Ecology, Community, and American Social Thought*, (Chicago: University of Chicago Press, 1992). Eugene Cittadino, *Nature as the Laboratory: Darwinian Plant Ecology in the German Empire, 1880–1900* (Cambridge: Cambridge University Press, 1990); Joel Hagen, *An Entangled Bank: The Origins of Ecosytems Ecology* (New Brunswick, NJ: Rutgers University Press, 1992); and Robert Croker, *Pioneer Ecologist: the Life and Work of Victor Ernest Shelford, 1877–1968* (Washington, DC: Smithsonian Institution Press, 1991), published in the same scholarly moment, never succeeded in crossing the late 1980s–1990s divide between history of science and environmental history.

23. Nancy Langston, *Forest Dreams, Forest Nightmares: The Paradox of Old Growth in the Inland West* (Seattle: University of Washington Press, 1995); James Pritchard, *Preserving Yellowstone's Natural Conditions: Science and the Perception of Nature.* (Lincoln: University of Nebraska Press, 1999); Thomas Dunlap, *Saving America's Wildlife* (Princeton, NJ: Princeton University Press, 1988); Stephen Bocking, *Ecologists and Environmental Politics: A History of Contemporary Ecology* (New Haven, CT: Yale University Press, 1997); Mark Barrow, *Nature's Ghosts: Confronting Extinction from the Age of Jefferson to the Age of Ecology* (Chicago: University of Chicago Press, 2009); Jeremy Vetter, ed., *Knowing Global Environments: New Historical Perspectives on the Field Sciences* (New Brunswick, NJ: Rutgers University Press, 2011); Frederick Rowe Davis, *The Man Who Saved Sea Turtles: Archie Carr and the Origins of Conservation Biology* (New York: Oxford University Press, 2007); Benjamin Cohen, *Notes from the Ground: Science, Soil and Society in the American Countryside* (New Haven, CT: Yale University Press, 2009); Peter Alagona, *After the Grizzly: Endangered Species and the Politics of Place in California* (Berkeley: University of California Press, 2013).

24. As seen, for example, in the difference between two books published in India within a year of each other: Deepak Kumar, *Science and the Raj, 1857–1905* (Delhi: Oxford University Press, 1995), which is a well received history of colonial Indian science, and Mahesh Rangarajan, *Fencing the Forest: Conservation and Ecological Change in India's Central Provinces, 1860–1914* (Delhi: Oxford University Press, 1996), a much more wide-ranging book which talks about forestry science within the context of a larger environmental history of central India.

25. Peder Anker, *Imperial Ecology: Environmental Order in the British Empire, 1895–1945* (Cambridge, MA: Harvard University Press, 2001), but also seen in works such as Richard Drayton, *Nature's Government: Science, Imperial Britain, and the "Improvement" of the World* (New Haven, CT: Yale University Press, 2000).

26. As summarized by Paul Sutter, "What can U.S. Environmental Historians Learn from non-U.S. Environmental Historiography?" *Environmental History* 8, no. 1 (January 2003): 109–129.

27. Richard Groves, *Green Imperialism: Colonial Expansion, Tropical Island Edens, and the Origins of Environmentalism, 1600–1860* (Cambridge: Cambridge University Press, 1996). Also see: Mahesh Rangarajan, *Fencing the Forest: Conservation and Ecological Change in India's Central Provinces, 1860–1914* (Delhi: Oxford University Press, India, 1999); R. Ravi Rajan, *Modernizing Nature:Forestry and Imperial Eco-Development, 1800–1950* (New York: Oxford University Press, 2006); Ramachandra Guha, *Unquiet Woods: Ecological Change and Peasant Resistance in the Himalayas* (Delhi: Oxford University Press, India, 1989); David Arnold, *The Problem of Nature: Environment,*

Culture, and European Expansion (London: Wiley-Blackwell, 1996); Ajay Skaria, *Hybrid Histories: Forests, Frontiers, and Wildness in Western India* (New York: Oxford University Press, 1999); Vasant Saberwal, *Pastoral Politics: Shepherds, Bureaucrats, and Conservation in the Western Himalaya* (New York: Oxford University Press, 1999); Paul Greenough and Anna Tsing, editors, *Nature in the Global South: Environmental Projects in South and Southeast Asia* (Durham, NC: Duke University Press, 2003); Helen Tilley, *Africa as a Living Laboratory: Empire, Development, and the Problem of Scientific Knowledge, 1870–1950* (Chicago: University of Chicago Press, 2011); Joseph Hodge, *Triumph of the Expert: Agrarian Doctrines of Development and the Legacies of British Colonialism* (Athens: Ohio University Press, 2007); Christopher Conte, *Highland Sanctuary: Environmental History in Tanzanias Usambara Mountains* (Athens: Ohio University Press, 2004); William Beinart, *The Rise of Conservation in South Africa: Settlers, Livestock, and the Environment, 1770–1950* (New York: Oxford University Press, 2008); William Beinart and Lotte Hughes, *Environment and Empire* (New York: Oxford University Press, 2009); Jane Carruthers, *The Kruger National Park: A Social and Political History* (Pietermaritzburg: Natal University Press, 1995); and James McCann, *Maize and Grace: Africa's Encounter with a New World Crop, 1500–2000* (Boston: Harvard University Press, 2007) among many others books, articles, and scholars which amply describe the intersections between environments, science, and colonialism.

28. Ramachandra Guha, *Unquiet Woods;* see also Ramachandra Guha, "Radical American Environmentalism and Wilderness Preservation: A Third World Perspective," *Environmental Ethics* 11 (1989): 71–83.

29. Tom Griffiths and Libby Robin, *Ecology and Empire: Environmental History of Settler Societies* (Seattle: University of Washington Press, 1997); Patrick Kupper, Bernhard Gissibl, and Sabine Höhler, *Civilizing Nature: National Parks in Global Historical Perspective* (New York: Berghahn Books, 2012); Jacob Darwin Hamblin, *Oceanographers and the Cold War: Disciples of Marine Science* (Seattle: University of Washington Press, 2005); Lewis, *Inventing Global Ecology;* Kurk Dorsey, *The Dawn of Conservation Diplomacy: U.S.-Canadian Wildlife Protection Treaties in the Progressive Era* (Seattle: University of Washington Press, 1998). Perhaps more appropriately located in the essay on technology, Etienne Benson, *Wired Wilderness: Technologies of Tracking and the Making of Modern Wildlife* (Baltimore: Johns Hopkins University Press, 2010) is another example of a history of science that aspires to a transnational scope.

30. The article Edmund Russell, "Evolutionary History: Prospectus for a New Field," *Environmental Histoy* 8, no. 2 (April 2003): 204–228 and the book *Evolutionary History: Uniting History and Biology to Understand Life on Earth* (Cambridge: Cambridge University Press, 2011).

31. Russell, "Evolutionary History: Prospectus for a New Field." John Herron, "review: Evolutionary History," *Environmental History* 17 (2012): 160–1.

32. Russell, "Evolutionary History: Prospectus for a New Field." Douglas Sackman, ed., *A Companion to American Environmental History* (Marden: Wiley-Blackwell, 2010); Robert Wilson, "Review: A Companion to American Environmental History," *Environmental History* 17, no. 1 (2012): 169.

33. Sverker Sorlin and Paul Warde, "The Problem of the Problem of Environmental History: A Re-reading of the Field," *Environmental History* 12, no. 1 (January 2007): 107–130. This critique is similar to that made by social historians and labor historians who called for environmental historians to pay more attention to the impact of unequal power

arrangements on their histories, as the environmental justice movement slowly influenced environmental history scholarship. See Gunther Peck, "The Nature of Labor: Fault Lines and Common Ground in Environment and Labor History," *Environmental History* 11, no. 2 (2006): 212–238.

34. W. Jeffrey Bolster, "Opportunities in Marine Environmental History," *Environmental History* 11, no. 3 (July 2006): 567–597. See also the wide-ranging "Marine Forum," *Environmental History* 18, no. 1 (2013): 2–126; and books such as W. Jeffrey Bolster, *The Mortal Sea: Fishing the Atlantic in the Age of Sail* (Boston: Belknap Press, 2012); and Bo Poulsen, *Dutch Herring: An Environmental History, c. 1600-1800* (Amsterdam: Aksant Academic Publishers, 2008).

35. Bolster, "Opportunities in Marine Environmental History," 567–597.

36. Stephen Pyne, "The End of the World," (2007 Presidential Address) *Environmental History* 12, no. 3 (July 2007): 649–653.

37. John McNeill, *Something New Under the Sun* (New York: W. W. Norton, 2000).

38. For more on each of these: Rachel Carson, *Silent Spring* (New York: Houghton Mifflin, 1962); Salim Ali, *The Fall of a Sparrow* (Delhi: Oxford University Press, 1986); Barry Commoner, *Science and Survival* (New York: Viking, 1967); Paul Erhlich, *The Population Bomb* (New York: Ballantine, 1969); Wangari Maathai, *Unbowed: A Memoir* (New York: Vintage, 2007); E. O. Wilson, *Naturalist* (Washington DC: Islands Press, 1994).

....

CHAPTER 9

TOWARD AN ENVIRONMENTAL HISTORY OF TECHNOLOGY

....

SARA B. PRITCHARD

THIS is an opportune moment to consider how environmental history has enriched—and perhaps can deepen even further—our understanding of technology, both as scholars and as planetary citizens. As the field has matured and professionalized over the past four decades, it has influenced not only the discipline of history, but also related fields. In addition, fifteen years' worth of sustained scholarship at the intersection of environmental history and the history of technology has emerged since the late 1990s. Yet there has been little historiographical synthesis of that work since Jeffrey Stine and Joel Tarr published their indispensable review essay in 1998, and even less explicit theoretical discussion of both the contributions and challenges of integrating the specialties.[1]

This opportunity also extends beyond academic circles, as environmental historians speak to pressing contemporary concerns. Current debates over genetically modified organisms, the biotech industry, global climate change, the Keystone XL pipeline, nuclear energy, the privatization of water, e-waste, and the Anthropocene, among other issues, demand historical reflection on the political stakes of technology and the environment in the modern world; on how and why historical actors and scholarly analysts have thought about nature and technology in the ways that they have; and on what it means, historically and analytically, to move to combine them, both physically and discursively. Scholars are now developing conceptual tools to wrestle with these urgent problems—problems that necessarily involve profound questions about nature, society, and technology, and their complex dynamics in both the past and the present. Environmental history, with its focus on the changing connections between people and nature, offers critical insights to this conversation.[2]

In light of these scholarly and contemporary issues, this chapter advances what I am calling an *environmental history of technology*.[3] The examination of technological artifacts and technical change through the lens of environmental history aims to bring vital contributions from this field to academic understandings of technology within the humanities and social sciences—and thus alter our understanding of technology—in at least three ways. First, environmental history questions the primacy of the sociotechnical within historical and contemporary studies of technology, and integrates nature prominently into their analyses. Second, it pushes us to consider how technology both reflects and mediates human-environment interactions. Finally, environmental historians have refined theories of technical development and change by stressing the ways that technology shapes, but is also shaped by, nonhuman nature. Needless to say, through such research, environmental history also serves to alter popular understandings of technology.

This essay is both historiographical and programmatic in that it synthesizes some of the major questions and approaches in environmental history, technology studies, and the eventual convergence of these specialties, while also calling attention to some productive ways forward thus far.[4] It begins with a broad discussion of how the fields have engaged with one another over the past generation of scholarship. I then examine several dominant ways in which environmental historians have analyzed technology in their stories, focusing specifically on technology and environmental change, as well as more recent work that has explored the ecological ambiguities of technology. In the final section, I highlight several concepts and approaches that have been especially generative for the writing of environmental histories of technology thus far—in particular, hybridity, multiplicity, evolutionary history, and envirotechnical analysis.

Environmental History and Technology Studies

A comprehensive discussion of how environmental history and the sociohistorical studies of technology have developed intellectually and professionally is beyond the scope of this chapter.[5] However, in advocating for an environmental history of technology, it is worth briefly considering how these fields have engaged with one another over the past forty years, including both the productive synergies and the tensions that have emerged from their dialogue. This section provides such an overview.

Technical objects and complex technological systems pervade not only the environment, but also the narrative landscape of environmental history.[6] Consider the axes, fires, and fences of colonial America, or the pesticides used in Central Valley agriculture. Recall the drill rigs and oil refineries of early twentieth-century Mexico and the extensive irrigation networks of the US West.[7] Although many early environmental historians lamented such technologies for the environmental problems, both past and present, as well as the general ecological degradation that they caused, subsequent environmental historians have sought more complicated, nuanced stories.[8] And until recently, rarely was technology a focal point of analysis.

Similarly, nonhuman nature is ever-present in historical and contemporary studies of technology. Late nineteenth- and early twentieth-century electricity networks in Europe and the United States depended on coal and rivers. Meanwhile, forests and streams surrounded early industrial paper mills, and uranium ore was essential to the dawn of the atomic age.[9] As these examples suggest, natural resources, and even entire ecosystems, have often been central to technology and thus its history, but they have been usually relegated to the background. Although it generally served as a stage for historical processes, rarely was the environment integral to explanations of technological development and change.[10]

There are striking parallels, though, between the two fields. Environmental historians and historians of technology have shared a tendency to each "black box" the cores of the other specialization: technology, in the case of environmental history; and nonhuman nature, in the case of technology studies. In doing so, environmental historians have tended to present technology—and historians of technology have likewise generally framed the environment—as outside their respective purview and basically irrelevant to their particular explanations of historical change. However, as I discuss below, these trends began to shift as the fields increasingly converged during the 1990s and ever since.

There are good reasons why scholars in both fields worked from these initial premises. In founding their specialty in the United States, environmental historians argued persuasively that not only do human interactions with nature merit historical inquiry; people's relationships with the environment can also fundamentally change our understandings of the past.[11] Thus, much of environmental historians' early work explored how the agency of the natural world shaped historical processes previously perceived as the exclusive domain of people. Technology was understandably a part, but not the centerpiece, of most narratives. In addition, as Helen Rozwadowski notes, environmental historians have been especially adept at using modern science (especially ecology) to help explicate the past, although the field has less consistently investigated and contextualized that knowledge.[12] Paul Sutter nicely encapsulates the theoretical conundrum: employing modern knowledge to interpret the past "has its merits, but it treats past and present knowledge inconsistently, using current science uncritically while assuming that past science (and the broader intellectual enterprise of thinking about nature) was the constructed political, social, and cultural practice that historians of science have rightly seen it as."[13] Overall, we can summarize this initial work as being predicated on a central binary of environment and society, with some studies emphasizing nature's agency, and others exploring more dialectical dynamics between them. Figure 9.1 offers a visual representation of this pattern.

Meanwhile, within technology studies in the 1980s and 1990s, a new generation of scholars responded critically to earlier notions of technological determinism by unpacking the social shaping of technology. Their critique became the heart of constructivist theories of technology. In the process, the environment essentially became a passive backdrop to stories of sociotechnical change. We can summarize this scholarship as working from a central binary of technology and society, with a notable shift in

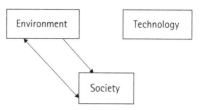

FIGURE 9.1 Schematic model of the theoretical relationship among environment, society, and technology within early environmental histories. The single-headed arrow represents a focus on nature's agency, while the double-headed arrow represents dialectical dynamics between natural and social processes.

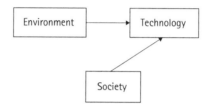

FIGURE 9.2 Schematic model of the theoretical relationship among environment, society, and technology within the sociohistorical studies of technology. The unidirectional arrow from environment to technology represents one form of technological determinism, what Wiebe Bijker calls a "theory of technology." The unidirectional arrow from society to technology represents later constructivist approaches that stress the social shaping of technology.

explanations of technical development and change from technological determinism to the social construction of technology. Figure 9.2 represents this scholarly trend.

Overall, the measured focus of each field enabled founding scholars to generate some of the distinctive contributions, and thus establish these new disciplines, as well as further distinguish them from their intellectual forebears. In particular, working from their initial assumptions enabled environmental historians to highlight the critical place and role of nature in history, thereby decentering people as the sole agents of historical change. Meanwhile, their own premises allowed scholars in technology studies to challenge technological determinism and develop rich analyses of the complex ways in which social processes shape technological objects and technical change, thus exposing the politics of technology and how people have agency. Each field's considered foregrounding of certain issues—which inevitably required placing other concerns in the background—spoke to notable gaps in previous literatures, and thus brought important new arguments to the academy and beyond.

Scholarship in environmental history and technology studies during the 1990s and 2000s subsequently complicated some of the fields' preliminary assumptions and began addressing arenas that had previously been part of the background. These shifts were due, in part, to the explicit engagement of one field with the other. The cultural turn, constructivism, and growth of science and technology studies (STS) together led later environmental historians to problematize nature and natural knowledge to a greater extent. Science was no longer a transparent tool that could open the black box of nature; instead, it too could—and should—be studied, historicized, and contextualized.[14] At the same time, the biological and field sciences became more prominent within the history of science and STS, thereby creating more topical overlap with environmental history, which has long had ties to the ecological sciences.[15] In addition, a renewed interest in materiality, and a growing attention to nonhuman entities within knowledge production and technological change by influential thinkers such as Angela Creager, Donna Haraway, Stefan Helmreich, Bruno Latour, Robert Kohler, Gregg Mitman, Michelle Murphy, Anna Tsing, and others, fostered more dialogue between science studies and environmental history.[16] Furthermore, as these and other leading scholars began training the next generation of academics—increasingly in *both* specialties—the research questions, methods, and analytic tools of the fields converged even more.

Over the past fifteen years, more environmental historians have framed three analytic categories—environment, society, and technology—as fundamentally entangled and intertwined. Consequently, these scholars have tended to consider all three concepts *together* in their analyses. We thus see a shift in conceptual models: from a central binary with a linear or reciprocal relationship between two factors, to studies that seek to flesh out more complicated, dialectical dynamics among all three factors simultaneously and on equal footing. As Matthew Evenden explains, one analytic category is no longer context for the other.[17] Of note, some historians of technology, especially those also engaged with environmental history, have made parallel moves in their own work.[18] This latest scholarly development is represented visually in Figure 9.3.

This recent work thus integrates chief insights from both environmental history and technology studies. The approach rejects multiple forms of determinism (environmental, social, technological) and thereby develops more complicated views of environment, technology, and their dynamics. In the process, environmental historians have begun to refine some of the central questions, foci, and theories of technical change within technology studies. This is most evident in overt technologies of environmental management such as dams, but several scholars have begun arguing for "hard"

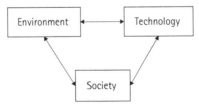

FIGURE 9.3 Schematic model of the theoretical relationship among environment, society, and technology within more recent environmental histories and some histories of technology engaged with environmental history. The double-headed arrows between all three possible pairs represent simultaneous reciprocal dynamics.

cases such as information and computing or consumer technologies, whose connections—or even relevance—to nature seem indirect at best.[19] Nonetheless, carrying out this approach certainly has its challenges, including opening (at least) two black boxes simultaneously and telling a coherent story that still captures the messiness of the world, both past and present.[20]

TECHNOLOGY IN ENVIRONMENTAL HISTORY

Several patterns emerge from examining self-defined environmental histories, focusing particularly on scholarship over the past three to four decades, as the field professionalized and became mainstream to the historical discipline.[21] The following discussion considers how environmental historians have framed technology within these studies. It analyzes some of the central tropes about technology and the technological within the field, and how they have shifted over time. In the process, I seek to highlight several of the wider implications, both theoretical and political, of these research questions, frameworks, and narratives.

"Technology" and "environmental" change: Moving away from dichotomies and declension

Technology as a powerful agent of environmental change is a pervasive theme within the field of environmental history. Carolyn Merchant's *The Death of Nature: Women, Ecology, and the Scientific Revolution* (1980) seamlessly links the scientific revolution, industrial technologies, and ecological changes that were not only dramatic, but in her view also detrimental. In *With Broadax and Firebrand: The Destruction of the Brazilian Atlantic Forest* (1995), Warren Dean traces the forest's transformation from the time of initial human settlement to the late twentieth century. Dean connects deforestation to a series of large-scale changes in Brazil's history—from Iberian colonization and plantation agriculture to recent efforts at "sustainable development." The title of his book suggests how technologies associated with these processes resulted in what he describes as relentless ecological "destruction." Mark Cioc's "eco-biography" of the Rhine follows that river's remaking since 1815. Through hydraulic engineering technologies such as dikes and dams, engineers and bureaucrats attempted to control the Rhine, at once seeking to adapt it to human use, while also juggling multiple, often competing demands. Cioc shows how these projects substantially altered the hydrology and ecology of the Rhine valley over the past two centuries. Although people were undoubtedly also affected in all of these instances, many environmental historians have particularly stressed the ecological consequences.[22]

Environmental historians often pay particular attention to the ecological changes associated with the implementation and use of new technologies. Some of these changes were planned. As Cioc and other historians of rivers have shown, engineers redesigned riparian ecosystems in an attempt to make them more suitable hydrologies for industrial mills, cities, hydroelectric plants, or the production of nuclear power.[23] During the twentieth century, chemical defoliants and pesticides killed vegetation and disease-carrying insects to protect the health of military troops in wartime, while open-pit mining leveled mountains at unprecedented scales.[24] Other studies have emphasized the unintended ecological consequences of new technologies. Edmund Russell has traced how the widespread adoption of insecticides in industrial agriculture actually ended up increasing the population of pesticide-resistant bugs. Meanwhile, supposedly healthy environments attracted allergy sufferers to the White Mountains and the Southwest of the United States. Yet expanding human populations, plus associated technological change and economic development, ended up radically remaking those places, ultimately and ironically creating unhealthy landscapes.[25]

Deforestation, dams, and development certainly had environmental "effects." However, a focus on various technologies' ecological consequences—a trend that likely reflects the field's strong materialist orientation—makes several theoretical assumptions.[26] Spotlighting the environmental effects of technology reproduces earlier views of technology as having impacts; but instead of the history of technology's prior emphasis on the social consequences of technology, environmental historians generally foreground its ecological repercussions. This scholarship provides a vital reminder to those in technology studies: technologies are always embedded within and reshape not only social worlds, but also natural ones.[27] At the same time, it defines the environment as fundamentally nonhuman and ultimately reifies a human/nonhuman nature divide that environmental history productively (and importantly) challenges in other ways. In addition, by concentrating on the environmental effects of technology, scholars can end up reinforcing the binary of technology and nature—a division which remains morally freighted within the field. Although many environmental historians have increasingly studied the built environment and shifted from discussing ecological "decline" or "degradation" to the more neutral "environmental change," the field's early political commitments (e.g., close ties to environmentalism) still often undergird these narratives.[28]

Scholarship over the past two decades has modified previous work in the field, including how scholars define "the environment" and how they have conceptualized the relationships among nature, technology, and society. For one, many environmental historians working during the 1990s and 2000s have reframed technology as less an agent of ecological than of *socio*-environmental change. This move challenges earlier conceptions of "nature" by expanding it to include humanity.[29] It thus speaks to a central critique of environmentalism (but also of environmental history) articulated by William Cronon in his seminal essay, "The Trouble with Wilderness."[30] It also highlights one of environmental historians' chief contributions: teasing out the porous boundaries and intricate dynamics between "nature" and "culture."

Scholars outside North America, or who write about issues outside that region, have been especially adept at integrating social and ecological analyses. Some of the earliest environmental histories of South Asia, such as *This Fissured Land: An Ecological History of India* (1992), brought together discussions of social and environmental change. A less stringent dichotomy also exists in Europe, where obviously managed landscapes have dominated for millennia. As Thomas Lekan and Thomas Zeller discuss in Chapter 13, Europe's environmental history (e.g., the history of human-natural interactions in Europe) has influenced the writing of European environmental history (e.g., the academic specialization that studies such interactions).[31] Scholarship focused on the colonial era, especially in Africa and Asia, has similarly shown the inextricable links between social and environmental processes. In particular, this work has explored how environmental management and technologies of "development" have served as means of sociopolitical control in imperial and now neoliberal contexts. Although diverse in many ways, much of this scholarship has been influenced by political ecology, subaltern studies, and colonial and postcolonial studies, which together have shaped studies of places outside the United States in numerous disciplines.[32] These literatures complement and ultimately deepen reconsiderations of American environmental history already undertaken by some leading scholars in the field, which have increasingly focused on sociopolitical control through environmental means and the uneven distribution of environmental resources and costs.[33]

Notably, it is not only analysts who have called attention to technologies and socio-environmental change. Historical actors have done so as well. Reformers in late nineteenth-century America and France, for example, complained about the detrimental influence of urban, industrial environments on human bodies. In the United States, they linked physical decline, particularly of middle- and upper-class white men, to industrialization, urbanization, and the loss of contact with "nature."[34] Roughly a century later, in the 1980s, the environmental justice movement forcefully exposed the unequal distribution of costs and benefits associated with industrial technologies on landscapes differentiated by race, class, and ethnicity. Thus, wealthier, predominately white communities have been more successful at mobilizing "not in my backyard" campaigns, while disadvantaged communities have had fewer financial and political resources to do so. Similarly, at times women have mobilized because they believed the built environment was having harmful effects on their own bodies.[35]

Other scholars have shown how socio-spatial differentiation shaped urbanization and thus the urban landscape itself. Integrating environmental history, urban history, and historical geography, Ari Kelman and Craig Colten traced the development of New Orleans, showing how that city's expansion depended upon ever-rising dikes to protect it from the Mississippi River. The resulting infrastructure ended up creating greater vulnerabilities, however, for the city's poorest and most marginalized residents who lived in the lowest areas, as Hurricane Katrina tragically demonstrated.[36] Since the late twentieth century, globalization has taken the socio-ecological disparities wrought by technological change to new levels on a worldwide scale. The processing of electronic waste such as computers and cell phones in the global South provides just one example.[37]

Some of this analysis, while maintaining an interest in the ecological changes associated with technological development, has pushed on the traditional definitions of technology, moving toward a sociotechnical understanding of technology that echoes views within technology studies that have emerged over the past three decades.[38] Several scholars of diverse historical and cultural contexts have shown, for example, how state-sponsored technological development involved dramatic environmental change. Explicit political imperatives—from monarchical consolidation and postwar reconstruction to international prestige and "development"—spurred the construction of large-scale projects. Old-regime France, nineteenth-century Germany, Stalinist Russia, Maoist China, and modernizing Brazil all undertook the construction of massive technological systems with vast stakes for the forests, waters, and landscapes (not to mention people) of the surrounding regions. Technological modernity, environmental transformation, and political order went hand-in-hand with ecological change manifesting state power over both humans and nonhumans.[39] Neil Maher and Sarah Phillips have demonstrated in their studies of New Deal America how conservation offered a way for the US federal government to expand its own authority. Both dramatically transformative technologies and conservation efforts served as effective means, then, by which to acquire or enhance political power.[40] Several environmental historians studying colonialism and capitalism have made similar moves, paying attention to the close relationship among different forms of political and economic power, technological change, and environmental regulation.[41] However, because they have brought perspectives from environmental history to their analysis, they simultaneously trace the importance of nature and environmental management to these processes.[42]

Technology's environmental ambiguities

As the previous subsection began to suggest, scholarship over the past twenty years has generally shifted toward more complex understandings of the place and role of technology in human-natural relations. These studies have proposed alternatives to both the progress story of technology's successful conquest of nature and the declensionist tale of technology's detrimental environmental effects. In part, such critiques emerge from questioning the presumed dichotomy of nature and technology that lies beneath both narratives. By examining the complex interplay of "environments" and "technologies" in specific case studies, scholars have begun challenging this binary, both in terms of its historical accuracy and its analytic value. Their work suggests that this dichotomy is more powerful as an ideal, or perhaps even an ideology, than as an effective description of actual historical relations.[43] As we will see in the following discussion, this scholarship has enriched, in fact, both environmental history and technology studies: environmental historians have helped recast the place of nature within not only narratives but also explanatory accounts of technical change within technology studies; at the same time, scholars in technology studies have offered more nuanced perspectives on technology and the technological within environmental history.

Several insightful threads have emerged from this scholarship thus far. In his influential essay, "Are You an Environmentalist or Do You Work for a Living?," Richard

White critiqued mainstream environmentalists for centering their political move-ment on a knowledge of nature based largely on leisure, recreation, and consumption. As White pointed out, white-collar professionals' ties to the environment were often masked in ways that blue-collar workers' were not. "Modern" technologies such as elec-tricity, cell phones, and the office building have, in fact, deepened humans' relationship with the natural world, including more distant environments, while simultaneously obscuring people's dependencies on those very places.[44] Even more recently, ubiquitous references to "the Cloud" mask the large-scale, resource-intensive technical systems on which the seemingly ethereal technology relies. Rather than simply condemning the ecological implications of mining or logging for their visible environmental "effects," White pushed both environmentalists and environmental historians to examine how workers have known nature through their various forms of labor. Subsequent environ-mental historians have heeded White's call with fruitful results.[45]

White's arguments can be extended to the scientists, engineers, technicians, users, and other groups associated with designing, developing, and operating technolo-gies. For example, French engineers conducted hydraulic research before construct-ing hydroelectric dams after World War II. Collecting reams of data, they determined average flow rates and flood levels, using these norms to guide their design of the dams. These transformative projects thus simultaneously yielded extensive knowledge about France's rivers and their hydrology. Environmental historians have been especially effective at underscoring the ecological costs of these projects and the frequent set-backs "experts" faced when they tried to realize their high-modernist visions.[46]

Other work has built on these insights, while also teasing out how these projects *pro-duced* environmental knowledge—knowledge that has sometimes led to a critique of those very projects.[47] At the same time, the context of these studies matters, even if technical publications, such as engineering journals and scientific papers, strip away "contextual" factors from "pure" data.[48] For instance, large-scale, state-sponsored river development in postwar France drove such hydraulic research. Ambitions to harness, control, and ultimately transform nonhuman nature spurred knowledge-making about the riparian environment. Environmental knowledge cannot, therefore, be iso-lated from environmental knowledge-*making*: exploring the specific processes and contexts by which knowledge is produced reveals the many ways in which social and political forces shape seemingly objective knowledge.[49] In addition, the knowledge produced through hydraulic engineering or other specialized fields competed with and frequently superceded other forms of knowledge and knowledge systems. For instance, local farmers and agricultural syndicates had together managed France's Rhône River long before hydraulic engineers sought its dramatic remaking after 1945. "Expert" knowledge was certainly not the first knowledge of those landscapes.[50]

Furthermore, some recent environmental historians, especially those also work-ing in the history of science and technology, have begun to investigate how numerous technical artifacts and systems *enabled* the production of environmental knowledge, whether by engineers, farmers, or others. Surveys, nails, drills, wells, radioactive trac-ers, traps, steamships, deep-sea dredging, and radio transmitters helped foresters calculate average tree size, engineers estimate rivers' flows, ecologists develop the notion of food webs, naturalists describe the deep ocean, and wildlife biologists track animals.[51] Knowledge of nature thus depended on a wide array of technical artifacts and practices. As such, new technologies, and new uses of established technologies, produced not only new forms of natural knowledge, but also new conceptions of "the environment" itself.

As Michelle Murphy puts it, these technologies helped materialize "the environment" in particular ways.[52] The US space program and widely distributed Apollo photographs of Earth from orbit have, for example, long been credited with helping foster a sense of the planet as a cohesive and interdependent, yet ultimately vulnerable, whole and thereby reinforcing the need for, in the words of a recent volume, "earthly politics."[53] Murphy has also demonstrated how surveys enabled women workers to materialize the late twentieth-century office building as an unhealthy environment, while Kim Fortun has shown how new computer and information technologies allowed activists to compile and process massive amounts of environmental data, yielding new natures and movements for environmental justice. In a final example, David Nye has explored how recent environmental movements have mobilized around growing concern over light pollution. The explosion in satellite and information technologies in the late twentieth century facilitated such conceptions of continental and even planetary landscapes defined by light; consider, for example, the image on the cover of this edited collection.[54] Although technology and nature are frequently opposed in the Western tradition, technologies such as these made certain kinds of environments—and ultimately certain forms of environmental politics—possible. Technological artifacts and technical change therefore had vast, ambiguous implications for nonhuman nature and humans' knowledge of it, far beyond simplistic characterizations as either powerful tools of progress or destructive agents of decline.[55]

Exploring the production of environmental knowledge, including the diverse technologies involved in environmental knowledge-making, enhances White's original insight. Nature was not simply there for workers to know; rather, workers produced nature in the very process of knowing it.[56] Their efforts even helped to define and possibly to redefine "the environment." At the same time, the process of studying and knowing nature frequently reshaped it. As studies by Matthew Klingle, Henry Lowood, and others have demonstrated, the boundaries between technoscientific "research" and environmental "management" were frequently murky. Models of forests, lakes, and rivers often ended up remaking those very environments—sometimes purposefully, at other times inadvertently.[57]

If some technologies helped constitute "the environment" and create new definitions and understandings of nature, others facilitated environmental conservation and protection. Paul Sutter established that the automobile played an important role in helping to catalyze the US wilderness movement during the first half of the twentieth century, as mass transportation technologies exposed many more Americans to national parks and other protected areas. More recently, David Louter traced significant changes in the desired relationship between cars, humans, and these parks in the western United States over the twentieth century, as they shifted from scenic land-scapes to roadless areas to ecological reserves.[58] Other scholars have identified ways in which modern technoscience laid essential groundwork for environmental science, politics, and policy. While the atomic age is justifiably seen as emblematic of death and destruction, it has also yielded valuable tools for knowing nature, including ways to recognize, comprehend, and explain ecological change. Radioactive tracers contrib-uted to the advancement of the ecological sciences during the postwar era by produc-ing new data, as well as contributing to new understandings of complex ecosystems. Environmental policymakers and activists marshaled these data and models to push for stricter regulations. Of course, the atomic age also spurred anti-nuclear protests and environmental movements in many countries around the globe, especially after the disasters at Three Mile Island, Chernobyl, and most recently Fukushima Daiichi, but these brief cases point to the complex ways in which ecology and the nuclear age were entangled.[59]

These examples suggest how technologies might serve as both motive and means of environmental knowledge-making and, in some cases, of environmental reform. If their benefits to the natural world are surprising, it suggests just how deeply their oppo-sition is presumed and engrained in some Western contexts. Together, such scholar-ship and non-American perspectives call into question not only the nature-technology binary, but also their presumed conflict.[60] Furthermore, they indicate how technologi-cal objects and systems can be at once technologies of environmental change *and* eco-logical awareness. The automobile contributed to the paving of vast tracts of land and global climate change over the twentieth century. Yet, as suggested above, Sutter and Louter showed how car tourism enabled many Americans to see, experience, and ulti-mately appreciate landscapes to which they might not otherwise have been exposed.[61] Similarly, complex technological systems of production, transportation, and distri-bution undergird modern consumerism. Yet, as Finn Arne Jørgensen has discussed, large-scale recycling networks have been developed in response to expanding con-sumer culture. Thus, technological systems can pose both environmental problems and solutions.[62] Ellen Stroud has studied how urbanization in New England cities actu-ally fostered reforestation in their hinterlands, thereby challenging the argument that cities are inherently antithetical to the natural world. Her work instead alludes to the uneven socio-spatial patterns of environmental use and non-use.[63]

Through such studies, environmental historians have effectively teased out technol-ogy's ecological ambiguities, offering frameworks that illuminate the complex inter-play of nature and technology in the past. The expansion of highways, the development of the atomic bomb, urbanization, and consumption may have indeed transformed specific ecosystems and contributed to global environmental change, at times dra-matically. The ecological, not to mention human, costs cannot—and should not—be overlooked. At the same time, technological development necessitated, and often gen-erated, valuable environmental knowledge. Moreover, some of these very same tech-nologies also fostered environmental concern, spurred reform, and ultimately enabled protection.

These frameworks allow, then, for the environmental complexities and ambiguities of diverse technologies. For this reason, neither the progressive nor the declensionist narrative is entirely satisfying. Moreover, although their stories are seemingly contrary, both narratives share a foundational dichotomy, and this binary results in a simple, linear story. In contrast, recent environmental histories that have explored hybrid landscapes and socio-spatial differentiation offer not only historically richer, but also more politically engaged, relevant, and ultimately useful narratives.

TOWARD AN ENVIRONMENTAL HISTORY OF TECHNOLOGY

Studies clustered in several areas have already begun enriching our thinking about technology *in*, and especially *as*, environmental history. In this section, I synthesize and extend work in three interrelated areas, which offer specific ways to write environmental histories of technology. Together, these conceptual tools and methods have enabled environmental historians to begin exploring technological artifacts and technical developments through central concerns in environmental history, thereby bringing new insights to technology studies. In particular, how do environmental factors and ecological processes constrain and shape (but not wholly determine) technical change? And how do technologies both reflect and mediate human-natural interactions? Furthermore, as twenty-first-century citizens increasingly contend with climate change, genetically modified organisms, and technologies that mimic natural entities, among other issues, the proliferation and intensification of new objects, relationships, and perceptions of nature and technology call for generative approaches and analytic tools such as those discussed here. I begin with a subsection on hybridity and multiplicity, turn to evolutionary history, and conclude with envirotechnical analysis.

Hybridity and multiplicity

Hybridity is now an influential concept in environmental history. While environmental historians have largely moved away from the notion of an untouched, pristine nature, epitomized by the idea of wilderness, Mark Fiege and Richard White explicitly theorized "hybrid landscapes."[64] Of note, hybridity has also become central to much recent STS scholarship. Perhaps of most relevance to environmental history, leading STS scholars Donna Haraway and Bruno Latour have developed the concepts of "natu-recultures" and "nature-culture," respectively; but other key analytic tools in STS are also predicated on hybridity—from the sociotechnical and technopolitics to technological culture.[65] Indeed, Brett Walker's "hybrid causation" exemplifies how some of this work in science studies has directly influenced environmental historians, including their engagement with and formulation of theories of hybridity.[66]

Fiege, White, Walker, and others have exemplified how the "hybrid turn" offers analytical power because it allows for greater complexity and captures the tight, mutually constitutive dynamics between what are often framed as dualisms—from nature and culture, and nature and technology, in environmental history, to society and technology in technology studies. For example, in *Irrigated Eden: The Making of an Agricultural Landscape in the American West* (1999), Mark Fiege studies the making of the agricultural heartland of southern Idaho. Irrigation networks and other technologies dramatically altered the area's ecology in the late nineteenth and early twentieth centuries. By explicitly calling it a hybrid landscape, Fiege emphasizes the persistent influence of ecological (and specifically hydrologic) processes within early industrial agriculture. As he demonstrates, those irrigation systems were often placed where creeks had formerly flowed. Similarly, forestry science and management may have remade trees and forests, while water management technologies substantially reshaped rivers and their geomorphology. Yet such organisms and ecosystems did not cease to have natural dimensions even when thinned or planted, dammed or diverted. Moreover, as Daniel Schneider shows, technological systems that may seem solely technical, such as modern metropolitan sewer networks, were arguably also ecosystems.[67]

In *Breathing Space: How Allergies Shape Our Lives and Landscapes* (2007), Gregg Mitman illustrates how environmental historians productively analyze the making of hybrids, tracing not only their specific configurations but also their larger stakes. In his study, Mitman traces how a single entity—allergy—was experienced differently by wealthy male Easterners, middle-class homemakers, and inner-city African Americans. In fact, he shows how the allergy itself may be entirely different depending on the context. However, multiple groups may also be exposed to the "same" allergen (e.g., ragweed or dust mites), but their allergic experiences could be radically different, based on each group's ability to flee risky landscapes at the height of allergy season, means of reducing exposure in the built environment, or access to health care. The specific historical relationships among these various social, economic, and ecological factors thus together mediated an individual's or group's exposure to and experience with the seemingly singular "allergy." What is especially important about Mitman's study is that, rather than simply concluding that allergies are hybrids of culture, political economy, and ecology, he instead unpacks exactly *how* allergies became such strikingly diverse objects and experiences for Americans in distinct contexts.[68] Thus, hybridity is not the culmination of Mitman's analysis, but rather the starting point for a rich exploration of the historically and culturally contingent contexts of its production.[69] It is precisely through this deeper reading that Mitman exposes some of the broader social and political stakes of allergies as hybrid entities in modern America: unhealthy environments, economic disparities, and unequal access to medical care, to name just three.[70]

Mitman's analysis is particularly instructive because it calls attention not only to the history but also the politics of hybrids. His study models how to unpack the power dynamics shaping the making and remaking of nature-culture in specific ways, thus revealing whose interests these hybrids serve. The biotech industry offers a pressing contemporary example of the enormous political and economic implications of blending "nature" and "technology," but historians have also explored these kinds of issues in the past.[71] Environmental historians also engaged with the history of technology have demonstrated, for instance, how Pacific Northwest forests, hemophiliac dogs, and modern chicken breeding all embody the strategic merging of "biological" organisms and "technical" objects, and therefore the ways these hybrids have served large corporations, big pharma, and industrial agriculture, respectively.[72]

Other environmental historians have developed an additional dimension to hybridity by exploring it as an actors' category. Thomas Zeller has shown, for example, how Nazi autobahn planners asserted that their freeways reconciled nature and technology. As he argues, such declarations were not unproblematic demonstrations of the National Socialist Party's "green" predisposition, but rather strategic proclamations. Such convenient assertions can also been seen in the claims advocates of river development in post–World War II France made when they emphasized the "natural" features of the dams and nuclear reactors they hoped to build. Some asserted such projects were extensions of the Rhône's existing hydrology.[73] By naturalizing technology and technologizing nature, they attempted to legitimate, even naturalize, large-scale development and ultimately efface politics.[74] Contextualizing actors' notions and invocations of hybridity in this way emphasizes its cultural and political purposes.

Both related to and distinct from the notion of hybridity is the concept of multiplicity. In *Sick Building Syndrome and the Problem of Uncertainty: Environmental Politics, Technoscience, and Women Workers* (2006), Michelle Murphy argues that a single object—say, bodies or buildings—"can concretely be many things at once."[75] Endangered fish might signify a loss of livelihood to fishermen, a crisis for environmental activists, declining habitat for birds dependent on that species, and an opportunity for non-native fish—*simultaneously*. A single object can thus have multiple meanings, but it can also be embedded within multiple worlds: the endangered fish is part of an aquatic ecosystem (ecology), the fishing industry (economy), and norms for healthy cuisine (culture). Murphy's study demonstrates how multiplicity can be an effective conceptual tool to other environmental historians. For one, it may help explain conflicts over environmental resources or debates because the "same" entity—fish, in this example—has different meanings and materialities for distinct constituencies.[76]

Overall, as these authors and their studies suggest, environmental historians have begun adopting these two conceptual tools—hybridity and multiplicity—to frame and analyze nature and technology in fruitful ways. By thinking with hybridity, environmental historians have also powerfully underscored the persistent nature of technology, thereby challenging scholars in technology studies not to remain saddled by what Wiebe Bijker calls the "haunting ghost" of technological determinism.[77] Acknowledging and embracing the environmental shaping of technical development and change instead pushes scholars in technology studies to consider society, technology, and the environment, and has thus contributed to the development of theories of co- or mutual production. Meanwhile, as Murphy shows, multiplicity complicates the linearity of both the progressive and the declensionist story because it moves beyond dualisms. Both hybridity and multiplicity therefore open up more complex interpretations that capture the complicated, ambiguous, and sometimes seemingly paradoxical dynamics of nature-technology in the past—and undoubtedly in the present as well.

Evolutionary history

Evolutionary history, "a new field" proposed and theorized by Edmund Russell, offers a specific approach to writing environmental histories of technology.[78] To paraphrase Russell, evolutionary history examines how humans have shaped variation, inheritance, and selection—usually seen as entirely "natural" phenomena—in the past. For instance, as Russell's own research showed, farmers' growing dependency on chemicals ended up selecting for insecticide-resistant bugs. Evolutionary history also explores how evolution may have, in turn, shaped historical processes, typically understood as deriving solely from humanity. A good example here is how Native Americans' isolation from Eurasian diseases meant that they were more susceptible to those diseases when eventually exposed to them, thereby enabling European colonization of the so-called New World.[79] As Russell argues, "uniting" the central insights of history and biology can enhance each field: understanding "historical" processes such as European imperialism elucidates biological contingencies such as what we now call invasive species, while "biological" processes, such as the selection of insecticide-resistant bugs, help account for historical particularities, such as the failing power of chemicals in industrial agriculture, over the course of the twentieth century.

Empirical studies by Russell and other scholars have revealed how technological development and use have had, at times, overt, intentional evolutionary consequences. Farmers selected plants with better disease resistance or bred horses, dogs, and pigs with characteristics people valued.[80] Other groups, such as scientists and the military, mediated the evolutionary trajectories of "undesirable" species such as mosquitos by attempting to reduce, if not destroy, their populations.[81] These examples indicate how technologies can serve as powerful agents of evolutionary change. Yet the development and use of technologies may have also had evolutionary implications, whether or not they were intended, and whether or not historical actors understood them in this way. Other environmental historians have traced, for instance, how hydroelectric dams and reservoirs ended up selecting for certain sizes and species of fish.[82]

Evolutionary history thus illustrates and bolsters two insights fundamental to environmental history. First, environmental phenomena, including evolutionary processes, may end up mediating historical change. Second, the complex dynamics of evolutionary history themselves exemplify the porous boundaries of nature and culture.

Two other conclusions also emerge from examining how different groups sought to direct evolutionary change through technology in the past. First, as scholars in technology studies remind us, values, politics, and power always mediate technical change.[83]

These insights underscore the fact that evolutionary change, like environmental change more broadly, often serves *certain* people's interests. As Deborah Fitzgerald has shown, hybrid corn benefitted corporations at the expense of small-scale farmers, while Tiago Saraiva has traced how sheep breeding enabled fascist imperialism in Europe and Africa.[84] These examples call attention to the larger contexts, histories, and politics of evolutionary change, including how and why these changes took place—and *for whom*. They therefore highlight the deep politics of evolutionary change and thus of "doing" evolutionary history, particularly in the early twenty-first century.[85]

Second, the idea of reworking "nature" for human benefit—for example, breeding dogs for medical research or planting GMO crops better suited to industrial agriculture—suggests that "nature" can essentially become technology, or at least technological. After all, if technologies are "people's ways of making or doing things," then "nature," undoubtedly already shaped by historic values and practices, can simultaneously be technology.[86] Evolutionary history offers, then, a compelling example of the problematic, porous boundaries of not only nature and culture, but also nature and technology.

It is worth noting that for Russell, evolutionary history is an analyst's project. He applies contemporary evolutionary theory to the past to help elucidate historical change. By doing so, we gain new understandings of how, for example, industrializing textile production in the eighteenth and nineteenth centuries involved not just using new machinery but also long cotton fibers from the New World whose evolution had been shaped by Amerindians for millennia, or why antibiotic resistance has emerged so quickly since World War II.[87] At the same time, using evolutionary theory without situating those ideas reifies current science. It may also end up obscuring historical interpretations of evolutionary phenomena: that is, how actors themselves understood and explained these events on their own terms. These brief examples suggest how environmental historians can harness vital lessons from Russell's approach, while remaining attentive to historical explanations.

Envirotechnical analysis

Scholars working at the intersection of environmental history and the history of technology, now known as "envirotech," have developed another approach to writing environmental histories of technology: envirotechnical analysis. Two clusters of work have emerged thus far. One group of studies, influenced by cultural and intellectual history, has focused on actors' understandings and representations of nature, technology, and their relationship. This work has generally examined the cultural specificity of these ideas and how they have changed over time. Particularly crucial is actors' strategic definition of these terms. In addition, scholars have traced how these ideas informed the physical (re)making of both "natural" entities and "technological" artifacts.[88] This second set of studies, which also reflects the materialist approaches to many environmental histories, has focused on the physical (re)production of "nature" and "technology."

These scholars have argued that "biological" organisms, "natural" environments, and "technological" objects are in fact envirotechnical entities: material hybrids of what is typically called nature and technology. This initial scholarship has thus begun developing envirotechnical analysis on two levels: the study of historical interactions between the environment and technology, both culturally and materially.

Through a number of empirical studies, envirotech scholars have exposed the historical blending of "natural" and "technological" entities. Dams may mediate rivers' hydrology, yet hydroelectric plants and nuclear reactors would not function without that flow. Envirotechnical analysis has thus problematized the categories, definitions, and boundaries of nature and technology, both historically and theoretically. It has both reflected and extended notions of hybridity, for instance, within environmental history. Some studies, however, have tended to focus on identifying objects as envirotechnical, while richer analyses have teased out the specific constitution and politics of blending "nature" and "technology" with an eye to their broader political, economic, social, and cultural stakes.[89] Still fewer scholars have theorized the integration of environmental history with the history of technology, particularly in terms of a central analytical tension: does incorporating the environment in theories of technical development, as most envirotechnical analysts do, mark a return to technological determinism?

In response, elsewhere I have sought to outline an overall mission for envirotechnical analysis and proposed three concepts to develop it in diverse contexts.[90] First, historians can examine how actors defined nature and technology, and thought about them in relation to one another. As with evolutionary history (discussed above), envirotechnical analysis is certainly an analyst's project.[91] Yet scholars can also ask questions about nature and technology in the past, while remaining sensitive to historical actors' own ideas and categorizations.

Second is the idea of envirotechnical systems. This concept builds on Thomas Hughes's idea of technological systems within the history of technology, which framed these systems as complex, large-scale networks of interrelated, interacting parts. Hughes, however, placed the environment explicitly outside his system.[92] Envirotechnical system instead emphasizes the ways in which the environment is always a part of technological systems. The term thus encapsulates this dynamic imbrication of natural and technological systems.

However, these specific reblendings of environmental and technological systems do not emerge out of nowhere. Rather, they arise in certain historical, cultural, and, importantly, political contexts. Thus, a related concept—envirotechnical regime—stresses the historical and political production of envirotechnical systems. As the political metaphor of "regime" suggests, this concept emphasizes the prescriptive formulation, production, and use of envirotechnical systems. In short, it highlights the specific, often strategic reblending of natural and technological systems to serve particular ends, although these configurations do not always develop exactly as the people and institutions promoting them intended.

Envirotechnical analysis ultimately seeks to integrate insights from environmental history and the history of technology. However, given this volume's focus, I will underscore one-half of the reciprocal relationship between the fields. In particular, each of these three concepts brings contributions from environmental history to technology studies. An examination of representations of nature and technology does not reject the current sociotechnical focus within technology studies, but instead calls attention to how ideas about the environment also matter in the design, production, and use of technologies. The concept of envirotechnical systems refines the influential concept of "technological" systems, which was originally theorized as entirely sociotechnical, to include the environment. Thus, technology shapes, but is also shaped by, nonhuman nature. Finally, envirotechnical regime stresses the strategic construction of envirotechnical systems by particular people and institutions, and therefore the ways in which contestation over the environment also mediates technological change. Moreover, envirotechnical analysis and these three concepts invite comparison and contrast across diverse historical, cultural, technological, and environmental settings.

Conclusion

Environmental history has both enriched and altered academic studies of technology since the 1990s in valuable ways. Environmental historians bring new perspectives and priorities to historical studies of technology. They are especially committed to exploring the place and role of nonhuman nature in all historical phenomena. As such, environmental history challenges the primacy of the sociotechnical within technology studies. More controversially, environmental historians push scholars in technology studies to make room for the environment in explanatory accounts of technological development—emphasizing the ways that ecological factors and processes constrain technical objects, systems, and change—without returning to technological determinism at its most extreme. Instead, combining these arguments with existing work in technology studies has led to more complex, nuanced models that allow for both the social and environmental shaping of technology. Consequently, environmental history has helped contribute to a reconsideration of major theories of technological development such as social construction, as well as prominent concepts such as technological systems. Most broadly, the field of environmental history has brought fresh questions and concerns to analyses of technology, encouraging scholars in technology studies to pay greater attention to the ways in which technologies reflect, mediate, and shape not only social worlds, but also human-natural interactions. These analyses deepen our understanding of the past, but they may also help us think about technology, society, and the environment in the present and the future.

NOTES

I would like to thank Drew Isenberg, Jay Turner, all of the *Oxford Handbook of Environmental History* workshop participants, and an anonymous reviewer, as well as Johanna Crane, María Fernández, Durba Ghosh, TJ Hinrichs, Stacey Langwick, Sherry Martin, Rachel Prentice, Jessica Ratcliff, Kathleen Vogel, Marina Welker, and Wendy Wolford for their insightful comments on previous versions of this chapter. I also thank Robert Kulik and Connie Hsu Swenson for their editorial assistance.

1. Jeffrey K. Stine and Joel A. Tarr, "At the Intersection of Histories: Technology and the Environment," *Technology and Culture* 39 (1998): 601–640. For updated perspectives, see Martin Reuss and Stephen H. Cutcliffe, eds., *The Illusory Boundary: Environment and Technology in History* (Charlottesville: University of Virginia Press, 2010), especially the Introduction, Afterword, and the chapter by Hugh S. Gorman and Betsy Mendelsohn, "Where Does Nature End and Culture Begin? Converging Themes in the History of Technology and Environmental History," 265–290; Thomas Zeller, *Driving Germany: The Landscape of the German Autobahn, 1930–1970* (New York: Berghahn Books, 2007), especially 6–8; Helen M. Rozwadowski, "Oceans: Fusing the History of Science and Technology with Environmental History," in *A Companion to American Environmental History*, ed. Douglas Cazaux Sackman (Malden, MA: Wiley-Blackwell, 2010), 442–461; Sara B. Pritchard, "Joining Environmental History with Science and Technology Studies: Promises, Challenges, and Contributions," in *New Natures: Joining Environmental History with Science and Technology Studies*, ed. Dolly Jørgensen, Finn Arne Jørgensen, and Sara B. Pritchard (Pittsburgh: University of Pittsburgh Press, 2013), 1–17. It is worth noting that Gorman and Mendelsohn reproduced the organizational structure of Stine and Tarr's essay from a decade earlier. This organization is useful at both moments in the fields' and subfield's history, but it does not always capture the emergence of new themes; for instance, the burgeoning literature on the body, environment, and technoscience (a concept from science and technology studies that emphasizes the ways in which the boundaries of science and technology have increasingly blurred since World War II).

2. For examples of environmental historians who have engaged explicitly with contemporary debates, see Ari Kelman, "Nature Bats Last: Some Recent Works on Technology and Urban Disaster," *Technology and Culture* 47 (2006): 391–402; Ari Kelman, "Boundary Issues: Clarifying New Orleans's Murky Edges," *Journal of American History* 94 (2007): 695–703; Joel A. Tarr, "There Will Be Gas," *Pittsburgh Post-Gazette*, August 2, 2009, http://www.post-gazette.com/pg/09214/987834-109.stm; Christopher Jones, "Defining the Problem," posted to H-Energy, June 27, 2010, http://www.h-net.org/~energy/roundtables/Jones_Gulf.html; Peter Shulman, "A Catastrophic Accident of Normal Proportions," posted to H-Energy, June 27, 2010, http://www.h-net.org/~energy/round-tables/Shulman_Gulf.html; Sara B. Pritchard, "An Envirotechnical Disaster: Nature, Technology, and Politics at Fukushima," *Environmental History* 17 (April 2012): 219–243, as well as the other essays in this special issue of the journal, "Japan Forum."

3. Maurits Ertsen uses the phrase in a recent book review. See his review of *Quagmire: Nation-Building and Nature in the Mekong Delta*, by David Biggs, *Technology and Culture* 53 (October 2012): 944–945.

4. Inevitably, it provides a partial overview in both senses of "partial": necessarily incomplete and reflecting my own research interests. I thank Wendy Wolford for the useful phrase.

5. For some classic works in environmental history that also provide historiographical and theoretical overviews at various points in the field's history, see Richard White, "American Environmental History: The Development of a New Historical Field," *Pacific Historical Review* 54 (1985): 297–335; Donald Worster, "Doing Environmental History," in *The Ends of the Earth: Perspectives on Modern Environmental History*, ed. Donald Worster (New York: Cambridge University Press, 1988), 289–307; Arthur F. McEvoy, "Toward an Interactive Theory of Nature and Culture: Ecology, Production, and Cognition in the California Fishing Industry," in *Ends of the Earth*, 211–229; "Environmental History: A Roundtable," *Journal of American History* (March 1990): 1087–1147; William Cronon, "A Place for Stories: Nature, History, and Narrative," *Journal of American History* 78 (March 1992): 1347–1376; Richard White, "Environmental History: Watching a Historical Field Mature," *Pacific Historical Review* 70 (2001): 103–111; J. R. McNeill, "Observations on the Nature and Culture of Environmental History," *History and Theory* 42 (December 2003): 5–43; Richard White, "From Wilderness to Hybrid Landscapes: The Cultural Turn in Environmental History," *Historian* 66 (2004): 557–564; J. Donald Hughes, *What is Environmental History?* (Cambridge: Polity, 2006); Sverker Sörlin and Paul Warde, "The Problem of the Problem of Environmental History," *Environmental History* 12 (January 2007): 107–130; Douglas Cazaux Sackman, "Introduction," in *A Companion to American Environmental History*, ed. Douglas Cazaux Sackman (Malden, MA: Wiley-Blackwell, 2010): xiii–xxi; and Frank Uekötter, *The Turning Points of Environmental History* (Pittsburgh: University of Pittsburgh Press, 2010). For some classic work in technology studies that also provide historiographical and theoretical overviews, see Thomas Parke Hughes, *Networks of Power: Electrification in Western Society, 1880–1930* (Baltimore: Johns Hopkins University Press, 1983); Thomas Parke Hughes, "The Seamless Web: Technology, Science, Etcetera, Etcetera," *Social Studies of Science* 16 (1986): 281–292; Wiebe Bijker, Thomas Hughes, and Trevor Pinch, eds., *The Social Construction of Technological Systems: New Directions in the Sociology and History of Technology* (Cambridge, MA: MIT Press, 1987); Wiebe E. Bijker and John Law, eds., *Shaping Technology/Building Society: Studies in Sociotechnical Change* (Cambridge, MA: MIT Press, 1992); Merritt Roe Smith and Leo Marx, eds., *Does Technology Drive History? The Dilemma of Technological Determinism* (Cambridge, MA: MIT Press, 1994), including the essay by Leo Marx, "The Idea of 'Technology' and Postmodern Pessimism," 237–257; Wiebe E. Bijker, "Sociohistorical Technology Studies," in *Handbook of Science and Technology Studies*, ed. Sheila Jasanoff et al. (Thousand Oaks, CA: Sage, 1995): 229–256; Ronald Kline, "Construing 'Technology' as 'Applied Science': Public Rhetoric of Scientists and Engineers in the United States, 1880–1945," *Isis* 86 (1995): 194–221; Wiebe E. Bijker, "Understanding Technological Culture through a Constructivist View of Science, Technology, and Society," in *Visions of STS: Counterpoints in Science, Technology, and Society Studies*, ed. Stephen H. Cutcliffe and Carl Mitcham (Albany: State University of New York, 2001): 19–34; R.R. Kline, "Technological Determinism," in *International Encyclopedia of the Social and Behavioral Sciences*, 3d ed., ed. Neil J. Smelsa and Paul B. Baltes (Amsterdam: Elsevier, 2001), 23, 15495–15498; Eric Schatzberg, "*Technik* Comes to America: Changing Meanings of Technology before 1930," *Technology and Culture* 47 (2006): 486–512; David E. Nye, *Technology Matters: Questions to Live With* (Cambridge, MA: MIT Press, 2006); Paul Forman, "The Primacy of Science in Modernity, of Technology in Postmodernity, and of Ideology in the History of Technology," *History and Technology* 23 (2007): 1–152; Leo Marx, "Technology: The Emergence of a Hazardous

Concept," *Technology and Culture* 51 (2010): 561–577. For overviews of STS, see Mario Biagioli, ed., *The Science Studies Reader* (New York: Routledge, 1999); Jasanoff et al., *Handbook of Science and Technology Studies*; Edward J. Hackett et al., eds., *Handbook of Science and Technology Studies* (Chicago: University of Chicago Press, 2009); Sergio Sismondo, *An Introduction to Science and Technology Studies* (Malden, MA: Blackwell, 2004).

6. Of course, the idea that technologies pervade the environment is predicated on a nature-technology dichotomy, which I discuss and problematize later in this chapter.

7. William Cronon, *Changes in the Land: Indians, Colonists, and the Ecology of New England* (New York: Hill and Wang, 1983); Linda Nash, *Inescapable Ecologies: A History of Environment, Disease, and Knowledge* (Berkeley: University of California Press, 2006); Myrna I. Santiago, *The Ecology of Oil: Environment, Labor, and the Mexican Revolution, 1900–1938* (New York: Cambridge University Press, 2006); Donald Worster, *Rivers of Empire: Water, Aridity, and the Growth of the American West* (New York: Oxford University Press, 1985).

8. For an overview of shifts in narrative within the field, see Sackman, "Introduction."

9. Hughes, *Networks of Power*; Judith A. McGaw, *Most Wonderful Machine: Mechanization and Social Change in Berkshire Paper Making, 1801–1885* (Princeton, NJ: Princeton University Press, 1987); Gabrielle Hecht, *The Radiance of France: Nuclear Power and National Identity after World War II*, 2d ed. (Cambridge, MA: MIT Press, 2009); Gabrielle Hecht, *Being Nuclear: Africans and the Global Uranium Trade* (Cambridge, MA: MIT Press, 2012).

10. This is due largely to the legacy of technological determinism, which I mention below. However, scholars working at the nexus of environmental history and the history of technology have begun challenging these arguments.

11. I set aside the question of predecessors such as the Annales School in France or other specializations within the historical discipline (such as the history of the US West) that influenced (American) environmental history. My argument here parallels Joan Scott's case for the transformative, not just additive, insights of new analytic categories. Her classic essay is Joan Wallach Scott, "Gender: A Useful Category of Historical Analysis," *American Historical Review* 91 (December 1986): 1053–1075.

12. Rozwadowski, "Oceans."

13. Paul S. Sutter, "Nature's Agents or Agents of Empire? Entomological Workers and Environmental Change during the Construction of the Panama Canal," *Isis* 98, no. 4 (2007): 724–753, quote from 728.

14. For useful commentaries on these points, see Sackman, "Introduction"; Rozwadowski, "Oceans"; and Aaron Sachs, "Cultures of Nature: Nineteenth Century," 246–265; all in *Companion to American Environmental History*. Sutter, "Nature's Agents or Agents of Empire?"

15. For a few examples of this trend, see Peder Anker, *Imperial Ecology: Environmental Order in the British Empire, 1895–1945* (Cambridge, MA: Harvard University Press, 2001); Angela Creager, *The Life of a Virus: Tobacco Mosaic Virus as an Experimental Model, 1930–1965* (Chicago: University of Chicago Press, 2002); Robert E. Kohler, *Lords of the Fly: Drosophila Genetics and the Experimental Life* (Chicago: University of Chicago Press, 1994); Robert E. Kohler, *Landscapes and Labscapes: Exploring the Lab-Field Border in Biology* (Chicago: University of Chicago Press, 2002); Robert E. Kohler, *All Creatures: Naturalists, Collectors, and Biodiversity, 1850–1950* (Princeton, NJ: Princeton

University Press, 2006); Erika Milam, *Looking for a Few Good Males: Female Choice in Evolutionary Biology* (Baltimore: Johns Hopkins University Press, 2010); Helen Tilley, *Africa as a Living Laboratory: Empire, Development, and the Problem of Scientific Knowledge, 1870–1950* (Chicago: University of Chicago Press, 2011); Jeremy Vetter, ed., *Knowing Global Environments: New Historical Perspectives in the Field Sciences* (New Brunswick, NJ: Rutgers University Press, 2010). On the place of ecology within environmental history, see Edmund Russell, *Evolutionary History: Uniting History and Biology to Understand Life on Earth* (New York: Cambridge University Press, 2011), especially 147.

16. For a few examples, see Creager, *Life of a Virus*; Donna Haraway, "A Cyborg Manifesto: Science, Technology, and Socialist-Feminism in the Late Twentieth Century," in *Simians, Cyborgs, and Women: The Reinvention of Nature* (New York: Routledge, 1991), 149–181; Donna Haraway, *The Companion Species Manifesto: Dogs, People, and Significant Otherness* (Chicago: Prickly Paradigm Press, 2003); Donna Haraway, *When Species Meet* (Minneapolis: University of Minnesota Press, 2008); Stefan Helmreich, *Alien Ocean: Anthropological Voyages in Microbial Seas* (Berkeley: University of California Press, 2009); Kohler, *Lords of the Fly*; Kohler, *Landscapes and Labscapes*; Gregg Mitman, *The State of Nature: Ecology, Community, and American Social Thought, 1900–1950* (Chicago: University of Chicago Press, 1992); Gregg Mitman, *Breathing Space: How Allergies Shape our Lives and Landscapes* (New Haven, CT: Yale University Press, 2007); Gregg Mitman, Michelle Murphy, and Christopher Sellers, eds., *Landscapes of Exposure: Knowledge and Illness in Modern Environments* (Chicago: University of Chicago Press, 2004); Michelle Murphy, *Sick Building Syndrome and the Problem of Uncertainty: Environmental Politics, Technoscience, and Women Workers* (Durham, NC: Duke University Press, 2006); Anna Tsing, "Unruly Edges: Mushrooms as Companion Species," *Environmental Humanities* 1 (2012): 141–154; see also the Matsutake Worlds website (www.matsutakeworlds.org).

17. I borrow and rework the phrase "equal analytic footing" from Matthew Evenden, "La mobilization des rivières et du fleuve pendant la Seconde Guerre mondiale: Québec et l'hydroélectricité, 1939–1945," *Revue d'histoire de l'Amérique française* 60: 1-2 (2006): 125–162, especially note 8.

18. For several examples, see Robert Gardner, "Constructing a Technological Forest: Nature, Culture, and Tree-Planting in the Nebraska Sand Hills," *Environmental History* 14 (2009): 275–297; Timothy J. LeCain, *Mass Destruction: The Men and Giant Mines that Wired America and Scarred the Planet* (New Brunswick, NJ: Rutgers University Press, 2009); Sara B. Pritchard, *Confluence: The Nature of Technology and the Remaking of the Rhône* (Cambridge, MA: Harvard University Press, 2011).

19. For the broader argument, see Pritchard, "Conclusion," *Confluence*, especially 245. On information and computing technologies specifically, see Djahane Salehabadi, "Second-Hand Technologies or E-Waste? The Composition of Germany's Progressive Environmental Legislation" (PhD diss., Cornell University, expected May 2014; Djahane Salehabadi, "Transboundary Movements of Discarded Electrical and Electronic Equipment," Solving the E-waste Problem Green Paper, United Nations University, April 11, 2013; Nathan Ensmenger, "Towards an Environmental History of Computing" (paper presented at the annual meeting for the Society for the History of Technology, Portland, Maine, October 10–12, 2013). For an example of consumer technologies, see Joy Parr, "What Makes Washday Less Blue? Gender, Nation, and Technology Choice in Postwar Canada," *Technology and Culture* 38 (January 1997): 153–186.

20. I am grateful to Paul Sutter for the pithy summary of the first challenge. My colleague Peter Dear has posed a version of it as a question: "What do you hold stable in your analysis?"

21. Markers of professionalization include the establishment of the American and European Societies for Environmental History and the International Consortium of Environmental History Organizations, including its associated World Congress of Environmental History; the journal associated with the ASEH (*Environmental History*) and a more European-based and globally oriented journal (*Environment and History*); and the creation of a separate category for "Environmental History" jobs on H-Net job listings. Sackman offers a periodization of environmental history scholarship in his Introduction to *A Companion to American Environmental History* pre-1977 (e.g., pre-ASEH), 1977–2000, and since 2000, which he describes as a period that has "not yet been subjected to the historians' test of time" (xix). I would suggest a slightly different periodization for North American scholarship: pre-1977; 1977–1995, with the publication of Cronon's "The Trouble with Wilderness" and the edited volume *Uncommon Ground*, which crystallized the cultural turn within environmental history; and since 1995, with the next break yet unclear.

22. Carolyn Merchant, *The Death of Nature: Women, Ecology, and the Scientific Revolution* (San Francisco: Harper and Row, 1980); Warren Dean, *With Broadax and Firebrand: The Destruction of the Brazilian Atlantic Forest* (Berkeley: University of California Press, 1995); Mark Cioc, *The Rhine: An Eco-Biography, 1815–2000* (Seattle: University of Washington Press, 2002). It is worth noting that Donna Haraway's now-famous "cyborg" was a response to Merchant's framing of science, technology, the environment, and gender. In brief, Haraway argued that feminists and environmentalists who idealize nature and critique (or simply ignore) modern science and technology do so at their peril. I thank Rachel Prentice for reminding me that Haraway was engaged with Merchant's earlier claims.

23. Cioc, *The Rhine*. See also Theodore Steinberg, *Nature Incorporated: Industrialization and the Waters of New England* (New York: Cambridge University Press, 1991); Richard White, *The Organic Machine: The Remaking of the Columbia River* (New York: Hill and Wang, 1995); Ari Kelman, *A River and Its City: The Nature of Landscape in New Orleans* (Berkeley: University of California Press, 2003); Jared Orsi, *Hazardous Metropolis: Flooding and Urban Ecology in Los Angeles* (Berkeley: University of California Press, 2004); Matthew D. Evenden, *Fish versus Power: An Environmental History of the Fraser River* (New York: Cambridge University Press, 2004); Craig E. Colten, *An Unnatural Metropolis: Wresting New Orleans from Nature* (Baton Rouge: Louisiana State University Press, 2005); Christof Mauch and Thomas Zeller, eds., *Rivers in History: Perspectives on Waterways in Europe and North America* (Pittsburgh: University of Pittsburgh Press, 2008); Pritchard, *Confluence*.

24. Rachel Carson, *Silent Spring* (Boston: Houghton Mifflin, 1987); Edmund Russell, *War and Nature: Fighting Humans and Insects with Chemicals from World War I to Silent Spring* (New York: Cambridge University Press, 2001); LeCain, *Mass Destruction*.

25. Edmund Russell, "Evolutionary History: Prospectus for a New Field," *Environmental History* 8 (2003): 204–228; Edmund Russell, "The Garden in the Machine: Toward an Evolutionary History of Technology," in *Industrializing Organisms: Introducing Evolutionary History*, ed. Philip Scranton and Susan R. Schrepfer (New York: Routledge, 2004): 1–16; Russell, *Evolutionary History*; Mitman, *Breathing Space*.

26. For several examples of materialist approaches, see Donald Worster, *Dust Bowl: The Southern Plains in the 1930s* (New York: Oxford University Press, 1979); Ellen Stroud, "Does Nature Always Matter? Following Dirt through History," *History and Theory* 42 (December 2003): 75–81; Russell, *Evolutionary History*. As Sackman succinctly notes, "a central analytical tension" has characterized the field of environmental history: namely, some environmental historians focus on the "so-called material dimensions of history," while others "have looked primarily at changing ideas about the natural world." As he puts it, "there is clearly a need for both kinds of histories, as well as for studies that show the interrelationship between the material and the ideological on all levels." See Sackman, "Introduction," xiv.

27. Of course, such statements reproduce binaries of society-technology and nature-technology.

28. For a useful overview of these shifts, see Sackman, "Introduction," especially xv–xvi.

29. Such arguments have been particularly developed by environmental historians interested in the nexus of the environment and the human body. See Conevery Bolton Valenčius, *The Health of the Country: How American Settlers Understood Themselves and Their Land* (New York: Basic Books, 2002); Mitman, Murphy, and Sellers, eds., *Landscapes of Exposure*; Nash, *Inescapable Ecologies*; Nancy Langston, *Toxic Bodies: Hormone Disrupters and the Legacy of DES* (New Haven, CT: Yale University Press, 2009); Joy Parr, *Sensing Changes: Technologies, Environments, and the Everyday, 1953–2003* (Vancouver: University of British Columbia Press, 2010); Brett Walker, *Toxic Archipelago: A History of Industrial Disease in Japan* (Seattle: University of Washington Press, 2011). See also Barbara Allen, *Uneasy Alchemy: Citizens and Experts in Louisiana's Chemical Corridor Disputes* (Cambridge, MA: MIT Press, 2003).

30. William Cronon, "The Trouble with Wilderness; or, Getting Back to the Wrong Nature," in *Uncommon Ground: Rethinking the Human Place in Nature*, ed. William Cronon (New York: W. W. Norton, 1995): 69–90.

31. Madhav Gadgil and Ramachandra Guha, *This Fissured Land: An Ecological History of India* (Berkeley: University of California Press, 1993). In addition to Lekan and Zeller's chapter in this volume, see also Michael D. Bess, *The Light-Green Society: Ecology and Technological Modernity in France, 1960–2000* (Chicago: University of Chicago Press, 2003).

32. Nancy Lee Peluso and Michael Watts, eds., *Violent Environments* (Ithaca, NY: Cornell University Press, 2001); S. Ravi Rajan, *Modernizing Nature: Forestry and Imperial Eco-Development, 1800–1950* (New York: Oxford University Press, 2006). For a novelist's beautiful portrayal of related issues, see Amitav Ghosh, *The Hungry Tide* (Boston: Houghton Mifflin, 2005). Environmental histories of colonialism are both rich and numerous. A few relevant works are now-classic overviews, such as Alfred W. Crosby, *Ecological Imperialism: The Biological Expansion of Europe, 900–1900* (New York: Cambridge University Press, 1986); Cronon, *Changes in the Land*. For Latin America, see Elinor G.K. Melville, *A Plague of Sheep: Environmental Consequences of the Conquest of Mexico* (New York: Cambridge University Press, 1997). For Africa, see Nancy J. Jacobs, *Environment, Power, and Injustice: A South African History* (New York: Cambridge University Press, 2003); Kate B. Showers, *Imperial Gullies: Soil Erosion and Conservation in Lesotho* (Athens: Ohio University Press, 2005); James C. McCann, *Maize and Grace: Africa's Encounter with a New World Crop, 1500–2000* (Cambridge, MA: Harvard University Press, 2005); Jacob A. Tropp, *Natures of Colonial*

Change: Environmental Relations in the Making of the Transkei (Athens: Ohio University Press, 2006); Diana K. Davis, *Resurrecting the Granary of Rome: Environmental History and French Colonial Expansion in North Africa* (Athens: Ohio University Press, 2007). Several scholars have looked at (postcolonial) "development," including its environmental dimensions. See Timothy Mitchell, *Rule of Experts: Egypt, Techno-Politics, Modernity* (Berkeley: University of California Press, 2002); Heather Hoag and May-Britt Öhman, "Turning Water into Power: Debates over the Development of Tanzania's Rufiji Basin, 1945–1985," *Technology and Culture* 49 (2008): 624–651.

33. Cronon, "The Trouble with Wilderness"; Louis Warren, *The Hunter's Game: Poachers and Conservationists in Twentieth-Century America* (New Haven, CT: Yale University Press, 1997); Karl Jacoby, *Crimes against Nature: Squatters, Poachers, Thieves, and the Hidden History of American Conservation* (Berkeley: University of California Press, 2001); Toby Craig Jones, *Desert Kingdom: How Oil and Water Forged Modern Saudi Arabia* (Cambridge, MA: Harvard University Press, 2010).

34. For discussion of these issues in the American case, see Peter J. Schmitt, *Back to Nature: The Arcadian Myth in Urban America* (Baltimore: Johns Hopkins University Press, 1990); Neil M. Maher, "A New Deal Body Politic: Landscape, Labor, and the Civilian Conservation Corps," *Environmental History* 7 (2002): 435–461; Neil M. Maher, "Body Counts: Tracking the Human Body through Environmental History," in *A Companion to American Environmental History*, ed. Douglas Cazaux Sackman (Malden, MA: Wiley-Blackwell, 2010), 163–180; Mitman, *Breathing Space*. A number of environmental historians have addressed the racial, ethnic, and class inequalities reproduced by such discourse. See Connie Chiang's chapter in this volume, as well as Andrew Hurley, *Environmental Inequalities: Class, Race, and Industrial Pollution in Gary, Indiana, 1945–1980* (Chapel Hill: University of North Carolina Press, 1995); Nash, *Inescapable Ecologies*. On parallel anxieties in France, although more from the perspective of the history of medicine, see David Barnes, *The Making of a Social Disease: Tuberculosis in Nineteenth-Century France* (Berkeley: University of California Press, 1995).

35. Studies by sociologist Robert Bullard particularly exposed environmental (in)justice. For two of his early works, see *Dumping in Dixie: Race, Class, and Environmental Quality* (Boulder, CO: Westview Press, 1990); *Confronting Environmental Racism: Voices from the Grassroots* (Boston: South End Press, 1993). Environmental historians such as Andrew Hurley have certainly put such issues in a deeper historical context; see his *Environmental Inequalities*. For a persuasive study of the gendered dimensions of environmental justice, see Murphy, *Sick Building Syndrome*.

36. Kelman, *A River and Its City*; Colten, *An Unnatural Metropolis*. On environmental inequities within urban landscapes, see also Matthew Klingle, *Emerald City: An Environmental History of Seattle* (New Haven, CT: Yale University Press, 2007), especially Chapter 6.

37. However, as Djahane Salehabadi notes in her dissertation and a recent publication, flows of electronic waste are far more complex than simply "from global North to global South." See Salehabadi, "Second-Hand Technologies or E-Waste?"; Salehabadi, "Transboundary Movements of Discarded Electrical and Electronic Equipment."

38. For two examples of these newer approaches (which are now classics in the field of technology studies), see Trevor J. Pinch and Wiebe E. Bijker, "The Social Construction of Facts and Artifacts: Or How the Sociology of Science and the Sociology of Technology Might Benefit Each Other," in *The Social Construction of Technological Systems: New Directions in the Sociology and History of Technology* (Cambridge, MA: MIT Press, 1987): 17–50; Bijker, "Sociohistorical Technology Studies."

39. David Blackbourn, *The Conquest of Nature: Water, Landscape, and the Making of Modern Germany* (New York: W. W. Norton, 2006); Jones, *Desert Kingdom*; Paul R. Josephson, *Industrialized Nature: Brute Force Technology and the Transformation of the Natural World* (Washington: Island Press, 2002); Chandra Mukerji, *Territorial Ambitions and the Gardens of Versailles* (New York: Cambridge University Press, 1997); Chandra Mukerji, *Impossible Engineering: Technology and Territoriality on the Canal du Midi* (Princeton, NJ: Princeton University Press, 2009); Pritchard, *Confluence*; Judith Shapiro, *Mao's War against Nature: Politics and the Environment in Revolutionary China* (New York: Cambridge University Press, 2001).

40. Sarah T. Phillips, *This Land, This Nation: Conservation, Rural America, and the New Deal* (New York: Cambridge University Press, 2007); Neil M. Maher, *Nature's New Deal: The Civilian Conservation Corps and the Roots of the American Environmental Movement* (New York: Oxford University Press, 2008).

41. On capitalism, see Worster, *Dust Bowl*; William Cronon, *Nature's Metropolis: Chicago and the Great West* (New York: W. W. Norton, 1991); Paul W. Hirt, *A Conspiracy of Optimism: Management of the National Forests since World War Two* (Lincoln: University of Nebraska Press, 1994). On socialism, see Josephson, *Industrialized Nature*; Shapiro, *Mao's War against Nature*.

42. For explicit discussion of these issues, see Jones, *Desert Kingdom*; Pritchard, *Confluence*.

43. On the ideology of the nature-technology binary, see Sara B. Pritchard and Thomas Zeller, "The Nature of Industrialization," in *The Illusory Boundary: Environment and Technology in History*, ed. Martin Reuss and Stephen H. Cutcliffe (Charlottesville: University of Virginia 2010): 69–100.

44. Richard White, "Are You an Environmentalist or Do You Work for a Living? Work and Nature," in *Uncommon Ground: Toward Reinventing Nature*, ed. William Cronon (New York: W. W. Norton, 1995): 171–185. Of course, these technologies are not limited to the workplace. Moreover, the blue-collar workers White references also (now) depend on many of these technologies, although I suspect environmentalists still associate loggers with axes and logging trucks, not GIS or modeling software. On the contemporary logging industry, see W. Scott Prudham, *Knock on Wood: Nature as Commodity in Douglas Fir Country* (New York: Routledge, 2005). For the high-tech worlds of the working class, see also the dissertation research of Hrönn Brynjarsdóttir (Information Science, Cornell University) on the extensive use of information technologies in the Icelandic fishing industry. On the hidden ecology of the digital age, see Salehabadi, "Second-Hand Technologies or E-Waste?"; Jane Anne Morris, "The Energy Nightmare of Web Server Farms," *Synthesis/Regeneration* 45 (Winter 2008). Of note, other historians have begun to explore how technologies of recreation can produce environmental knowledge. For instance, see Annie Gilbert Coleman, *Ski Style: Sport and Culture in the Rockies* (Lawrence: University of Kansas Press, 2004); Michael J. Yochim, *Yellowstone and the Snowmobile: Locking Horns over National Park Use* (Lawrence: University of Kansas Press, 2009); Joseph E. Taylor, III, *Pilgrims of the Vertical: Yosemite Rock Climbers and Nature at Risk* (Cambridge, MA: Harvard University Press, 2010); Michael W. Childers, *Colorado Powder Keg: Ski Resorts and the Environmental Movement* (Lawrence: University of Kansas Press, 2012).

45. For example, see Thomas G. Andrews, "'Made by Toile'? Tourism, Labor, and the Construction of the Colorado Landscape, 1858–1917," *Journal of American History* 92 (2005): 837–863; Thomas G. Andrews, *Killing for Coal: America's Deadliest Labor War* (Cambridge, MA: Harvard University Press, 2008); Maher, "A New Deal Body Politic."

46. James C. Scott, *Seeing Like a State: How Certain Schemes to Improve the Human Condition Have Failed* (New Haven, CT: Yale University Press, 1998).

47. Mara Goldman, Paul Nadasdy, and Matt Turner, eds., *Knowing Nature: Conversations at the Intersection of Political Ecology and Science Studies* (Chicago: University of Chicago Press, 2011).

48. The extensive use of the passive voice in scientific and technical publications illustrates this move. On the historicity of objectivity, see Lorraine J. Daston and Peter Galison, *Objectivity* (Cambridge, MA: MIT Press, 2007).

49. Pritchard, "Joining Environmental History with Science and Technology Studies."

50. On hydraulic research and competing knowledge systems, see Pritchard, *Confluence*, Chapters 3 and 4.

51. On forest calculations, see Henry Lowood, "The Calculating Forester: Quantification, Cameral Science, and the Emergence of Scientific Forestry Management in Germany," in *The Quantifying Spirit in the Eighteenth Century*, ed. Tore Frangsmyr, J. L. Heilbron, and Robin E. Rider (Berkeley: University of California Press, 1991), 315–342. On hydraulic data, see Pritchard, *Confluence*. On ecology, see Stephen Bocking, "Ecosystems, Ecologists, and the Atom: Environmental Research at Oak Ridge National Laboratory," *Journal of the History of Biology* 28 (1995): 1–47. On the history of the ocean, see Helen Rozwadowski, *Fathoming the Ocean: The Discovery and Exploration of the Deep Sea* (Cambridge, MA: Harvard University Press, 2005). On radio transmitters, see Etienne Benson, *Wired Wilderness: Technologies of Tracking and the Making of Modern Wildlife* (Baltimore: Johns Hopkins University Press, 2010). For an essay that dramatically reconfigures environmental historians' assumptions about "the environment" by showing how space scientists have begun to see Earth within a cosmic environment, see Valerie A. Olson, "NEOecology: The Solar System's Emerging Environmental History and Politics," in *New Natures: Joining Environmental History with Science and Technology Studies*, ed. Dolly Jørgensen, Finn Arne Jørgensen, and Sara B. Pritchard (Pittsburgh: University of Pittsburgh Press, 2013): 195–211.

52. Murphy, *Sick Building Syndrome*.

53. Sheila Jasanoff and Marybeth Long Martello, eds., *Earthly Politics: Local and Global in Environmental Governance* (Cambridge, MA: MIT Press, 2004).

54. Kim Fortun, "From Bhopal to the Informating of Environmentalism: Risk Communication in Historical Perspective," *Osiris* 19 (2004): 283–296; David E. Nye, *When the Lights Went Out: A History of Blackouts in America* (Cambridge, MA: MIT Press, 2010).

55. For two commentaries on the inseparability of nature and knowing nature, see Douglas R. Weiner, "A Death-Defying Attempt to Articulate a Coherent Definition of Environmental History," *Environmental History* 10 (2005): 404–420; Fortun, "From Bhopal to the Informating of Environmentalism." In addition, notions of "progress" or "destruction" beg the question: for whom? Environmental historians might add: for what (nonhumans)? Moreover, such assessments can be analyzed as actors' claims— that is, whether different actors believe such changes are positive (or not) and on what grounds.

56. Weiner, "A Death-Defying Attempt"; Fortun, "From Bhopal to the Informating of Environmentalism"; Rozwadowski, "Oceans."

57. Matthew Klingle, "Plying Atomic Waters: Lauren Donaldson and the 'Fern Lake Concept' of Fisheries Management," *Journal of the History of Biology* 31 (1998): 1–32; Lowood, "The Calculating Forester"; Pritchard, *Confluence*, Chapter 4.

58. Paul Sutter, *Driven Wild: How the Fight against Automobiles Launched the Modern Wilderness Movement* (Seattle: University of Washington Press, 2002); David Louter, *Windshield Wilderness: Cars, Roads, and Nature in Washington's National Parks* (Seattle: University of Washington Press, 2006). See also Zeller, *Driving Germany*; Christof Mauch and Thomas Zeller, eds., *The World Beyond the Windshield: Roads and Landscapes in the United States and Europe* (Athens: Ohio University Press, 2008).

59. Bocking, "Ecosystems, Ecologists, and the Atom"; Mark Fiege, "The Atomic Scientists, the Sense of Wonder, and the Bomb," *Environmental History* 12 (2007): 578–613; Mark Fiege, *The Republic of Nature: An Environmental History of the United States* (Seattle: University of Washington Press, 2012), especially Chapter 7; Richard Hindmarsh, ed., *Nuclear Disaster at Fukushima Daiichi: Social, Political and Environmental Issues* (New York: Routledge, 2013); Ian Jared Miller, Julia Adeney Thomas, and Brett L. Walker, eds., *Japan at Nature's Edge: The Environmental Context of a Global Power* (Honolulu: University of Hawaii Press, 2013).

60. See Thomas Lekan and Thomas Zeller's chapter on cultural landscapes in this volume for an extensive discussion of these themes in Europe. See also Bess, *Light-Green Society*. I am also influenced here by conversations with Julia Adeney Thomas. I have set aside important questions about "non-Western" traditions, as well as the problematic West/non-West divide. In addition, it seems likely that views of technology within environmental movements in the late twentieth century may have shaped the theoretical (and political) views of technology within environmental history as a field.

61. Sutter, *Driven Wild*; Louter, *Windshield Wilderness*.

62. Finn Arne Jørgensen, *Making a Green Machine: The Infrastructure of Beverage Container Recycling* (New Brunswick, NJ: Rutgers University Press, 2011).

63. Ellen Stroud, *Nature Next Door: Cities and Trees in the Northeastern United States* (Seattle: University of Washington Press, 2012).

64. Mark Fiege, *Irrigated Eden: The Making of an Agricultural Landscape in the American West* (Seattle: University of Washington Press, 1999); White, *Organic Machine*; White, "From Wilderness to Hybrid Landscapes." See also Cronon, "The Trouble with Wilderness"; Cronon, ed., *Uncommon Ground*; Aaron Sachs, "American Arcadia: Mount Auburn Cemetery and the Nineteenth-Century Landscape Tradition," *Environmental History* 15 (2010): 206–235; Aaron Sachs, *Arcadian America: The Death and Life of an Environmental Tradition* (New Haven, CT: Yale University Press, 2013). Exposing Native American environmental management practices certainly played a crucial role in destabilizing the idea of "natural nature." See Shepard Krech, III, *The Ecological Indian: Myth and History* (New York: Norton, 1999).

65. Haraway, *Companion Species Manifesto*, 1; Haraway, *When Species Meet*, 16; Bruno Latour, *We Have Never Been Modern*, trans. Catherine Porter (Cambridge, MA: Harvard University Press, 1993), 7. For two early articulations of the sociotechnical, see Bijker, Hughes, and Pinch, eds., *The Social Construction of Technological Systems*; Bijker and Law, eds., *Shaping Technology/Building Society*. On technopolitics, see Hecht, *Radiance of France*. On technological culture, see Bijker, "Understanding Technological Culture."

66. Walker, *Toxic Archipelago*.

67. Fiege, *Irrigated Eden*; Daniel Schneider, *Hybrid Nature: Sewage Treatment and the Contradictions of the Industrial Ecosystem* (Cambridge, MA: MIT Press, 2011).

68. Of note, hybridity complicates the idea that "nature's" agency can be completely disentangled from other factors. See Linda Nash, "The Agency of Nature and the Nature of Agency," *Environmental History* 10 (2005): 67–69; Mitchell, *Rule of Experts*; Paul Sutter, comment on J. R. McNeill's *Mosquito Empires: Ecology and War in the Greater Caribbean, 1620–1914*, annual meeting of the American Society for Environmental History, Tucson, AZ, April 13–15, 2011; Walker, *Toxic Archipelago*.

69. On "differential hybridity," see Suman Seth, "Putting Knowledge in its Place: Science, Colonialism, and the Postcolonial," *Postcolonial Studies* 12 (2009): 380.

70. Mitman, *Breathing Space*.

71. On contemporary biotech, see Robert H. Carlson, *Biology Is Technology: The Promise, Peril, and New Business of Engineering Life* (Cambridge, MA: Harvard University Press, 2010).

72. William Boyd and Scott Prudham, "Manufacturing Green Gold: Industrial Tree Improvement and the Power of Heredity in the Postwar United States," 107–139; Stephen Pemberton, "Canine Technologies, Model Patients: The Historical Production of Hemophiliac Dogs in American Biomedicine," 191–213; and Roger Horowitz, "Making the Chicken of Tomorrow: Reworking Poultry as Commodities and as Creatures, 1945–1990," 215–235, all in *Industrializing Organisms*. See also William Boyd, "Making Meat: Science, Technology, and American Poultry Production," *Technology and Culture* 42 (2001): 631–664.

73. Zeller, *Driving Germany*; Franz-Josef Brüggemeier, Mark Cioc, and Thomas Zeller, eds., *How Green Were the Nazis?: Nature, Environment, and Nation in the Third Reich* (Athens: Ohio University Press, 2005); Pritchard, *Confluence*, especially Chapters 2, 3, and 5.

74. Langdon Winner, *The Whale and the Reactor: A Search for Limits in an Age of High Technology* (Chicago: University of Chicago Press, 1986); James Ferguson, *The Anti-Politics Machine: "Development," Depoliticization, and Bureaucratic Power in Lesotho* (Minneapolis: University of Minnesota Press, 1994); Mitchell, *Rule of Experts*.

75. Murphy, *Sick Building Syndrome and the Problem of Uncertainty*, 12. See also her Introduction and especially the Epilogue.

76. See also Dolly Jørgensen, "Environmentalists on Both Sides: Enactments in the California Rigs-to-Reefs Debate," in *New Natures: Joining Environmental History with Science and Technology Studies*, ed. Dolly Jørgensen, Finn Arne Jørgensen, and Sara B. Pritchard (Pittsburgh: University of Pittsburgh Press, 2013): 51–68.

77. On the "nature of technology," see Andrew C. Isenberg, *Mining California: An Ecological History* (New York: Hill and Wang, 2005); Sara B. Pritchard, "Envirotech Methods: Looking Back, Looking Beyond?," paper presented at the conference of the Society for the History of Technology, October 2007; Gardner, "Constructing a Technological Forest"; Pritchard, *Confluence*; Wiebe E. Bijker, "Globalization and Vulnerability: Challenges and Opportunities for SHOT around Its Fiftieth Anniversary," *Technology and Culture* 50 (2009): 610.

78. Russell, "Evolutionary History"; Russell, "Garden in the Machine"; Russell, *Evolutionary History*.

79. Crosby, *Ecological Imperialism*; Cronon, *Changes in the Land*; David Igler, "Diseased Goods: Global Commodities in the Eastern Pacific Basin, 1770–1850," *American*

Historical Review 109 (2004): 693–719. See also William H. McNeill, *Plagues and Peoples* (New York: Anchor, 1998).

80. Ann Norton Greene, *Horses at Work: Harnessing Power in Industrial America* (Cambridge, MA: Harvard University Press, 2008); Haraway, *Companion Species Manifesto*; Haraway, *When Species Meet*; Sam White, "From Globalized Pig Breeds to Capitalist Pigs: A Study in Animal Cultures and Evolutionary History," *Environmental History* 16 (January 2011): 94–120. For examples pertaining to additional species, see essays in Philip Scranton and Susan R. Schrepfer, eds., *Industrializing Organisms: Introducing Evolutionary History* (New York: Routledge, 2004).

81. On the movement of "undesirable" organisms such as weeds and pests, see Crosby, *Ecological Imperialism*. On invasive species in the United States, see Peter A. Coates, *American Perceptions of Immigrant and Invasive Species: Strangers on the Land* (Berkeley: University of California Press, 2006). On attempts to eradicate insects on the home and war fronts, see Russell, *War and Nature*.

82. Russell, "Garden in the Machine"; Joseph E. Taylor, III, *Making Salmon: An Environmental History of the Northwest Fisheries Crisis* (Seattle: University of Washington Press, 1999).

83. Bijker, Hughes, and Pinch, eds., *The Social Construction of Technological Systems*; Bijker and Law, eds., *Shaping Technology/Building Society*; Hecht, *Radiance of France*. On the question, "Do artifacts have politics?" see Winner, *The Whale and the Reactor*.

84. Deborah Fitzgerald, *The Business of Breeding: Hybrid Corn in Illinois, 1890–1940* (Ithaca, NY: Cornell University Press, 1990); Deborah Fitzgerald, *Every Farm a Factory: The Industrial Ideal in American Agriculture* (New Haven, CT: Yale University Press, 2003). For a popular version of corn's history, see Michael Pollan, *The Omnivore's Dilemma* (New York: Penguin, 2007); Tiago Saraiva, "The Production and Circulation of Standardized Karakul Sheep and Frontier Settlement in the Empires of Hitler, Mussolini, and Salazar," in *New Natures: Joining Environmental History with Science and Technology Studies*, ed. Dolly Jørgensen, Finn Arne Jørgensen, and Sara B. Pritchard (Pittsburgh: University of Pittsburgh Press, 2013): 135–150.

85. This is a play on Worster, "Doing Environmental History."

86. Nina E. Lerman, Ruth Oldenziel, and Arwen P. Mohun, eds., *Gender and Technology: A Reader* (Baltimore: Johns Hopkins University Press, 2003), 2. Russell advocates using technological metaphors to frame and understand "natural" objects, processes, and phenomena in "technological" terms (and specifically industrial ones) in "Garden in the Machine." I develop the ideas of framing nature as technological and technology as natural in *Confluence*, Introduction.

87. Russell, *Evolutionary History*, especially Chapter 9.

88. For example, see Sara B. Pritchard, "Reconstructing the Rhône: The Cultural Politics of Nature and Nation in Contemporary France, 1945–1997," *French Historical Studies* 27 (2004): 766–799; Lissa Roberts, "An Arcadian Apparatus: The Introduction of the Steam Engine into the Dutch Landscape," *Technology and Culture* 45 (2004): 251–276.

89. A few relevant essays include Boyd, "Making Meat"; the essays in *Industrializing Organisms*; Gardner, "Constructing a Technological Forest."

90. Pritchard, "Envirotech Methods"; Pritchard, *Confluence*, Introduction; Pritchard, "An Envirotechnical Disaster."

91. It is probably no accident that interests in bridging nature and technology, both historically and historiographically, emerged in the late twentieth and early twenty-first centuries, as synthetic biology, engineering, nanotechnology, and other scientific and

technical fields increasingly challenge the borders of nature and culture, and nature and technology. I thank Djahane Salehabadi for reminding me of the historical context of our own work as scholars.

92. Hughes, *Networks of Power.*

CHAPTER 10

··

NEW CHEMICAL BODIES

Synthetic Chemicals, Regulation, and Human Health

··

NANCY LANGSTON

RECENT ecological thinking about chemicals challenges what the political scientists Steven Kroll-Smith and Worth Lancaster call the "Enlightenment-inspired idea that bodies and environments are genuinely discrete realities." They write, "It is customary to think of bodies and environments as if they are two separate and distinct entities. The pronoun 'my' in front of 'body' signals a possessive interest in human bodies that differs from the typical article 'the' that precedes 'environment.' 'My body,' and 'the environment' are lingual signals that two ontologically discrete things are being discussed. In a world enunciated with discrete categories, all of us simply know where our bodies end and the environment begins." Risk assessment, they argue, rest on these ontological assumptions of separateness. "By assuming a categorical distinction between bodies and environments, regulatory authorities can then issue a 'pollutant discharge permit' licensing the right to contaminate environments as 'long as the exposure is below the threshold at which' environmental toxins adversely affect bodies."[1] The implicit assumption is that bodies and environments are separate enough that one can contaminate the soil, water, or air, without contaminating people.

In contrast, ecological models of health envision the body as permeable to the environment, just as earlier generations of physicians did. Like most ecosystems, the body is affected and disturbed by natural toxins, parasites, solar radiation, and mutagens. In the 1950s, the medical researcher Rene Dubos argued that health can be viewed ecologically not as the simple absence of disease, but rather as "the ability to adapt to new or changing circumstances; compromised health may become apparent only when new sources of stresses are applied and the individual fails to adapt." Health consists of a "complex set of adaptations to stress, feedback loops, pathways of nutrients and energy, flows of energy and waste, and regulatory mechanisms that constrain these pathways."[2]

This chapter explores the intimate interconnections of synthetic chemicals, ecosystems, and bodies in the twentieth century. The chapter begins by examining the development of concern about industrial chemical exposures in the nineteenth and twentieth centuries. It then addresses the toxicological and regulatory frameworks that governments developed to address risks posed by chemicals, focusing on pesticide residues. Finally, I examine the emergence of an ecological perspective on health.

Since World War II, the production and use of synthetic chemicals have increased more than thirtyfold. The modern petrochemical industry, now a two-trillion-dollar-a-year global enterprise, is central to the world economy, generating millions of jobs and consuming vast quantities of energy and raw materials. Each year, more than 70,000 different industrial chemicals are synthesized and sold, resulting in billions of pounds a year of chemicals that make their way into our bodies and ecosystems.[3] Americans are now saturated in industrial chemicals, the products of a post-World-War-II boom in synthetic chemical manufacturing.

New technologies and methods for the detection of industrial chemicals, particularly endocrine disruptors, have drawn increasing attention to the pervasive and persistent presence of synthetic chemicals in our lives. Trace chemicals found in the air, water, and soil are now been being detected within our bodies. The chemical composition of our bodies is being altered in ways that reflect the transformations of our everyday environments. In July 2005, the US Centers for Disease Control released its *Third National Report on Human Exposure to Environmental Chemicals.*[4] Through the process of biomonitoring—that is, measuring the amount of a chemical in a blood or urine sample—the CDC aims to track the accumulation of synthetic chemicals into the human population through direct measurement of the populace. Endocrine disruptors present particular challenges to our current systems of monitoring and regulating synthetic chemicals in the environment. These chemicals have potential activity at orders of magnitude lower than current dose limits for other toxic chemicals. Perhaps more troubling, these chemicals leave no "smoking gun," with effects manifesting years, if not decades, later, and often in a body only indirectly exposed (such as a developing fetus).

INDUSTRIAL HYGIENE AND THE EMERGENCE OF THRESHOLD LIMIT VALUES

Synthetic chemicals, such as the persistent organic pollutants (POPs) and endocrine disruptors, are products of the World War II and Cold War era boom in chemical manufacturing, but the regulatory, conceptual, and political frameworks that enabled them to saturate our lives so quickly have deeper historic roots. As medical historian Christopher Sellers argues in *Hazards of the Job*, environmental health sciences developed out of industrial hygiene and its concerns with disease in the workplace.[5] Industrial hygiene was driven by environmental and public health concerns, particularly over the fate of workers in industries that produced and used chemicals.[6] Yet, powerful as it was in shaping a framework that challenged the right of industry to poison workers, the framework of occupational poisoning, which looked for acute poisoning effects in people exposed at high doses, eventually made it difficult to imagine other kinds of toxicity, such as low-level effects.

The Industrial Revolution was fueled by coal and other fossil fuels, leading to significant air and water pollution that affected workers and their families. The dye factories of England and Germany exposed workers to carcinogenic chemicals, while textile mills exposed workers to a host of natural and synthetic risks. Industrial hygienists challenged unrestricted pollution, but they also saw their role as enabling the continued expansion of industrial production. The goal was to protect worker and public health, not by banning chemicals but by finding thresholds beneath which workers could tolerate exposure.

As historian Linda Nash argues in *Inescapable Ecologies: A History of Environment, Disease, and Knowledge* (2006), before germ theories of health developed in the nineteenth century, earlier ecological ideas of health had envisioned the body as permeable to its environments. In contrast, emerging germ theory models viewed the body as separate from its environment, susceptible to being penetrated by a germ or a poison only when it crossed the threshold of the body's barrier to the world. The defeat of epidemic diseases had been a major public health victory in the late nineteenth century, and germ theories of health had been critical in helping to create these public health victories.[7] These models allowed medical scientists envision the effects of acute poisoning, but they made it more difficult to imagine the new chemical threats that were arising with the growth of industrial agriculture and industrial chemistry.[8]

By the 1920s, emerging toxicological opinion argued that toxic substances had threshold values, which were levels at which changes in normal physiological functioning could be detected. Below that threshold, the substance was believed to be harmless. Threshold values developed in a medical context which assumed bodies could be separated from environments. The origin of the word "threshold" hints at the implications of the threshold model of exposure. Threshold comes from the Old English *þrescold*, *þærscwold*, or *þerxold*, meaning "point of entering." A certain amount of chemical exposure was considered necessary to breach the bodily barrier, to cross the threshold of the body. Less than that, and the threshold would be inviolate, the separate home of the self unassailed.

Using the assumption that below a certain threshold a substance could not penetrate the body's defenses and become toxic, industrial hygienists developed threshold limit values (TLVs)—the concentrations of chemicals that were thought not to produce toxic effects in human workers. Nash describes how, in the 1920s, researchers moved from the workplace into the laboratory, where they could conduct experiments on laboratory animals, developing precise TVLs under carefully controlled conditions.[9] As Frederick Rowe Davis shows, laboratories such as E. M. K. Geiling's Tox Lab at the University of Chicago developed a series of tests to determine toxic thresholds for complex exposures to multiple compounds.[10]

By 1945, threshold limit values had been determined for 136 different compounds. Yet these laboratory values often had little relevance for worker or consumer exposure in broader environments such as agricultural fields, where precision was rarely possible and where confounding effects were numerous. Researchers came to see the workplace and the field as messy, inexact, and uncertain. The unpredictability of the field compared to the laboratory left TVLs developed in the laboratory unreliable for predicting the risks of any particular chemical exposure. Exposures that seemed safe in the lab might lead to poisoning in the agricultural environment. As Nash writes, "place mattered in multiple ways," yet experimental laboratory scientists attempted to remove or negate the effects of place.[11]

The conceptual frameworks of industrial hygiene that arose in response to threats posed by poisons such as lead worked reasonably well in addressing the effects of acute poisoning from natural toxins such as alcohol. But they proved to be inadequate tools for dealing with chronic threats posed by the synthetic chemicals of the postwar boom, particularly endocrine disruptors and other low-dose environmental exposures.

PESTICIDE RESIDUES

Concern about pesticide residues emerged well before Rachel Carson's work, James Whorton argues in *Before Silent Spring* (1974). Inorganic poisons, particularly arsenic, had been used in agriculture for two thousand years on a small scale, but grew in importance in the second half of the nineteenth century, in response to the ecological transformations shaped by the growth of monocultures, the loss of native ecosystems and native predators, and the increasing transportation of novel crops and pests around the globe.[12]

The movement of Colorado potato beetles from their native wild vegetation in the Rockies to cultivated fields threatened to devastate potato farming in the eastern United States. Paris green—an arsenic-containing green pigment—protected potatoes from the beetle, and despite early warnings that the poisonous effects might extend to human consumers, not just beetles, Paris green was soon being used enthusiastically on a wide variety of vine crops. The gypsy moth explosion in New England in the late nineteenth century brought lead arsenates into widespread use, for the lead mixture was gentler to foliage (if not to people) than arsenic. By the early 1900s, lead arsenate had become the most popular pesticide in use, and remained so until the introduction of DDT.

Few regulations limited the use of pesticides in the early twentieth century, and few agencies questioned their risks. The federal government, in fact, encouraged their use as part of the modernization of agriculture. Farmers could use pesticides in whatever quantities they wished, and economic entomologists encouraged them to do so. Some farmers were reluctant to spray, worried about the expense or their family's health, but by 1900 professional entomologists seemed to have set aside any doubts they might have had about pesticide safety. *Farmers' Bulletin*, a publication of the US Department of Agriculture, insisted that spaying arsenicals was "an operation virtually free from danger," and the department's chief entomologist, C. V. Riley, assured readers "how utterly groundless are any fears of injury."[13] By 1900 several states had acquired the regulatory power to force reluctant farmers to use pesticides.

Spray residues on crops became a concern in the early twentieth century, for arsenical residues had the potential to poison consumers as well as workers. A debate arose over the potential for harm from arsenic residues, and this debate created fault lines that still dominate current regulatory debates over toxic chemical exposure. Most agricultural experts were unconvinced about any risk from spray residues, for they were focused on acute poisoning: if a chemical didn't lead to immediate poisoning of applicators and farmers, surely it could not harm consumers, the reasoning went.

CHRONIC VERSUS ACUTE CONCERNS

While agricultural experts focused on the risks posed by acute poisoning, many health and consumer advocates grew increasingly concerned about low-dose, chronic exposure to arsenic in the early twentieth century. Consumers exposed themselves every day to substantial quantities of arsenic: in so-called "heroic" mineral medicines, in pigments, in paper, in labels, in toys, in soaps, in preserves and jams, in dental fillings, in money, and particularly in wallpapers, which slowly poisoned the air of the room as they released their toxins, creating low-level symptoms of arsenic poisoning. The symptoms were controversial, but they fit classic tests of environmental exposure: people got sick in their homes, and when they moved to other places, they felt better, but their symptoms recurred when they returned home. A group of doctors organized to appeal for legislation controlling arsenic in consumer products, until finally in 1900 such an act passed in Massachusetts. Yet even doctors could not agree on whether arsenic exposure really constituted a threat to public health, even as the possibility of slow intoxication—that is, a chronic health problem—from mercury, lead, and arsenic was being debated by twentieth-century toxicologists.[14]

By the 1920s, the medical community was starting to discuss the dangers presented by "the chemical artificiality of the modern industrial environment."[15] Medical scientists such as Dr. Karl Vogel began to note that "distribution of lead and arsenic was so complete that all members of industrialized populations carried at least traces of the metals in their tissues."[16] Laboratory research showed that chronic exposures to arsenic created tissue degeneration in rats, and physiologists such as Sister Mary McNicholas argued that the same was likely to be true in humans. Industry and agricultural scientists rejected these arguments, claiming that they were highly hypothetical—an argument that still resonates today with policymakers.

Debates among toxicologists and regulators grew over lead and arsenic tolerance levels in the 1930s: some wanted zero tolerance; others were convinced that would be impossible. In 1935 and 1937 the American Medical Association "took the position that the residues might become a public health hazard of the first order, and it urged further study and more stringent state regulation of the use of dangerous insecticides."[17]

The arsenical pesticides were developed at a time when the risks of chronic expo-sure were difficult to measure or understand. By the time these risks were understood, the pesticide industry and markets were too firmly established to yield easily—agricul-tural interests and chemical interest alike reacted passionately to the accusations that chemicals were neither necessary or safe. The use of synthetic pesticides emerged in a professional and regulatory context already dominated by the use of these arsenical compounds.[18]

Many Americans take for granted their right to be protected by government from poisons, but the government's power to regulate food and medicines was vigorously contested in the early twentieth century. Without legal compulsion, no manufacturers of chemicals were willing to undertake the slow, expensive studies needed to estab-lish the safety of their products, and this legal compulsion simply did not exist during the first decades of the Industrial Revolution. Repeated crises, with children poisoned by adulterated foods and contaminated medicines, shaped a consumer movement that pushed hard for food regulation. After years of battle over the right of the federal gov-ernment to regulate the safety of food and drugs, Congress passed the Pure Food and Drug Act in 1906, which gave limited power to the federal government to protect public safety from chemical harms.

The chemist and pioneering consumer advocate Harvey Washington Wiley had led the fight against impure foods for decades, and his efforts eventually resulted in the creation of the Food and Drug Administration.[19] Wiley, like other reformers in the Progressive Era, was no enemy to industry. He believed that progress was essential for prosperity and that business was a critical driver of progress. Progressives also believed in the lessons of history: namely, that the nineteenth-century excesses of the robber barons had proven that business could not police itself. Business could not protect citi-zens from injustice and injury while also seeking profits. It was therefore the responsi-bility of government to monitor industry and protect the public.

Wiley believed passionately in a form of the precautionary principle. Only precaution could protect the public from harm; waiting to see whether a particular chemical con-taminant in food might be harmful was unethical, for it would turn the United States into a nation of guinea pigs. Wiley argued that because the early effects of chronic arse-nic poisoning were almost undetectable, it would be impossible to regulate chemicals if proof of harm were required. Wiley's precautionary reasoning did not convince indus-try scientists, who demanded what the historian James Whorton calls "clear clinical signs of physiological damage" before they were willing to judge a substance harmful. These disagreements reflected both practical and ideological differences. Practically, the argument focused on where the burden of proof should lie. Wiley's precautionary principle held that a foreign substance should be presumed guilty until proven inno-cent; industry believed the opposite. As Whorton writes, "One approach imposes a risk on the public, the other a hardship on business."[20]

The differences between Wiley and industry were not just practical, however; they also had conceptual roots in the history of toxicology. Wiley had distrusted the pres-ence of any level, no matter how small, of a poison in food, but toxicologists were beginning to argue that substances generally had a threshold value below which the substance was unlikely to affect an individual's functioning or physiology. During the 1920s, industrial hygienists had developed techniques that, in Nash's words, enabled them "to quantify chemical exposures and to correlate those exposures with both physiologic variables and obvious signs of disease." The industrial hygienists envisioned chemicals "as akin to microbes, as singular agents that were capable of inducing a specific disease once they entered the body. What mattered was not the broader environment but the specific chemical exposure."[21] Researchers in industrial hygiene were well versed in physiology as well as germ theory, and, like physiologists, they believed in homeostasis, the concept that a body is a self-regulating system that can equilibrate itself, thus regaining balance after exposure to low levels of contamination. Wiley, trained as a chemist rather than a physiologist, grudgingly accepted the industrial hygienists' argument that chemical contamination might be inevitable in the modern world, yet he never agreed that bodies would equilibrate to this contamination.

Pesticide residue on fruit proved to be one of the first key challenges for regulators struggling to apply the 1906 Pure Foods Act in the face of concerted political opposition from the agricultural industry and its congressional allies. The central question was: how best could scientists establish the safety or harm of low levels of chemical residues on food? Where should the burden of proof lie? Given substantial scientific uncertainty over the potential effects of residues, how should uncertainty translate into policy?

One side, made up of farmers, their congressional allies, economic entomologists, the Public Health Service, and some medical scientists, believed that the criteria for safety should be clinical acute poisoning; doses that did not make someone immediately sick were assumed to be safe. The best way to test this would be to study people exposed to high levels of pesticides; if they were healthy, surely the public was safe. The burden of proof should be on those who would regulate.

The other group—including the FDA, consumer groups, and many medical scientists—took the opposite view. They believed that safety should not be defined by acute poisoning, but rather by chronic effects, and such data could best be gathered by extrapolating from lab studies on animals conducted over their entire lifetimes. The burden of proof should rest on those who used the chemicals, not on the regulators.[22]

Even though the acceptable residue levels were often quite generous, enforcing even those standards was difficult for federal agencies during the 1930s. The Bureau of Chemistry (which eventually became the FDA) could do little to enforce regulations, for its opponents campaigned vigorously against any federal regulation.[23] Manufacturers opposed strict interpretation of the Pure Food Act, which was intended to limit exposure to pesticide residues. Vigorous opposition to firm enforcement from within the United States Department of Agriculture (USDA) also made it difficult for the Bureau of Chemistry to act. The Bureau of Chemistry initially existed within the Department of Agriculture, which saw its mission as protecting American farmers. Every effort of the Bureau of Chemistry to regulate farmers was met with resistance from the greater USDA. By 1912, Wiley resigned in frustration, leaving to join the editorial staff of *Good Housekeeping*, where he could better advocate for consumers.[24]

After Wiley's resignation, the Bureau had a difficult time enforcing even limited seizures of adulterated food or food contaminated with residues, for orchardists took the Bureau to court when their products were seized. A habit of secrecy—keeping findings of illegal residues hidden from the public in order to give the industry a chance to clean up the problem first—led to a culture that promoted hiding information from consumers. In 1929, Walter Campbell, commissioner of the new Food and Drug Administration, warned the American Pomological Society that this strategy held grave risks: "What do you suppose would happen if the general public became acquainted with the fact that apples were likely to be contaminated with arsenic?...So far, we have not given the matter any publicity, and the public as a whole has no general knowledge on the subject...If they were to become curious today, we would probably have to admit that we are perhaps not going everything possible to remove excess arsenic from our fruit."[25]

Campbell was correct: when consumers learned how little was being done to protect public health from pesticide residue, outrage resulted. The consumer protection movement that emerged during the 1930s helped to transform federal policy with regard to pesticides and other chemical toxins, for consumer advocates destroyed the comfortable secrecy that had developed within government agencies. When the FDA responded to consumer pressure and lobbied for the Food, Drug, and Cosmetic Act in the late 1930s, the agency finally broke "with its tradition of sheltering the residue problem in secrecy." As a result, the agricultural industry was "made furious by what seemed the FDA's betrayal of trust."[26]

Fruit growers and chemical companies were able to fight off FDA attempts to restrict residues with help from their congressional representatives. For example, Representative Clarence Cannon of Missouri censored Walter Campbell's testimony about the health effects of residues in 1935. Two years later, when the FDA used experimental evidence from animal studies showing health problems of pesticides, Cannon made certain that the appropriations act for fiscal 1937 contained the provision "that no part of the funds appropriated by this act shall be used for laboratory investigations to determine the possible harmful effects on human beings of spray insecticides on fruits and vegetables."[27] This intense opposition to the use of animal experiments to demonstrate risks to humans was shaped by political concerns, and remained a persistent thread in regulatory debates for the next seven decades.

Historian Thomas Dunlap argues that the use of synthetic pesticides emerged in a professional and regulatory context already dominated by the use of arsenical compounds to control insects. Two decades before the introduction of DDT, federal and state regulatory agencies began setting standards for insecticide residues on food, so by the time DDT was available, experience with arsenic "had generated a set of working arrangements among the parties concerned with the problem—doctors, regulatory officials, and farmers—and a set of assumptions about the nature and extent of the residue problem that determined the response to DDT."[28] By the time synthetic pesticides such as DDT became available for civilian use after World War II, the FDA and the USDA had become accustomed to acting as "service agencies for interest groups."[29]

WAR AND CHEMICALS

Environmental historian Edmund Russell argues that understanding the explosive growth of modern chemistry requires that we attend to the close links between war and industry. The twentieth century marked the debut of modern chemical warfare against both pests and people. Twentieth-century wars increased demand for agricultural and industrial products, stimulating the chemical industry and transforming institutions of government. Russell notes that the "federal government did not wait for market forces to change the civilian economy; it intervened by creating and linking institutions, by expanding industry, agriculture, and science through direct funding and tax subsidies."[30] Disease control, and thus the chemical industry, was a key focus of these military and federal efforts.

The First World War changed the American chemical industry, because Americans supplied the Allies with industrial and agricultural products, thus increasing the size, expertise, profitability, and status of what had been a small, isolated industry. Chemical research before World War I had been dominated by Germans, but during the War, American companies began to develop research capacity, and American farmers began applying the new insecticides, in part to meet new market demands for agricultural products. The perceived need to wage gas warfare "forced the country to marry science and the military. This marriage increased commitment of military and civilian chemists to poison gas and created a new institution within the army, the Chemical Warfare Service. Gas research stimulated research on war gases as insecticides and ties between entomologists and chemical officers; it also increased the profile and activities of federal entomologists."[31]

World War I also helped foster a new set of ideologies, in which complete control of insects was conceptually desirable and victory over agricultural foes became as important as victory over military enemies. "Chemical companies capitalized on the capital and expertise gained in the war to grow and expand their work on insecticides, especially by searching for synthetic organic insecticides."[32] Entomologists promoted the idea that human beings were engaged in a war for survival with insects.

The Second World War had dramatic effects on the growth and power of the chemical industry. The War expanded markets for new chemicals and drugs, creating new institutional structures for scientific funding and research and new alliances between civilian university scientists, armed forces, industry. World War II strengthened a belief in the ideology of total victory, with agricultural experts soon calling for "total war against man's insect enemies, with the avowed object of total extermination instead of mere 'control.'"[33] The stunning practical power of science demonstrated during the War led leaders to develop ways to apply the lessons of World War II to the postwar world. The war led to intense advertising and marketing campaigns for new chemicals, linking victory in war with victory over insects, cementing synthetic chemicals' reputations as miracle workers, and submerging concerns over toxicity. Wartime publicity campaigns created civilian expectations that the new chemical wonders could solve all insect problems—on the farm, in the suburban landscape, and in the house.

During World War II, the Allies found that diseases killed more soldiers than weapons. Lice carried typhus, and before 1942, louse-control technology relied on pyrethrums from plants that needed to be imported from Africa. Wartime broke apart the international trade networks that supplied pyrethrums to the Allies, so a substitute that did not need to be imported was in high demand. DDT proved to be the answer. In November 1942, Geigy sent a sample of DDT to the Bureau of Entomology, preceded by some reports showing that it killed lice quite well while appearing to be relatively nontoxic to people. American researchers soon found that DDT killed mosquitoes, not just lice, even at extremely low does (1 part DDT to 100 million parts of water). The federal government ramped up production of DDT, and in late 1943, when typhus broke out in Naples, Allied health organizations dusted over a million civilians with DDT powder—the first time a typhus winter epidemic had been halted. The dramatic success "lifted hopes for 'total victory' against other insects—not just a relationship of uneasy balance, but total annihilation."[34]

Yet, while the spraying of DDT during World War II, and particularly spraying on islands in the Pacific theater to control malarial mosquitos, led to great health benefits for soldiers, not all scientists saw DDT as a wonder drug. Early tests raised concerns about DDT's safety for people, guinea pigs, rabbits, lab animals, and birds. Yet, because of the immediate threat of diseases carried by lice, the army was desperate, and "in spite of the earlier rather startling toxicity reports we had asked our people to start a limited manufacturing program" of DDT.[35] Further tests showed that skin absorption did not seem to lead to the same sorts of toxic reactions, and while researchers did not conclude that DDT was harmless, they did decide that the "hazards must be weighed against the great advantages of the material" for military use.[36]

While military researchers were concerned about the long-term safety of DDT, they did feel that the short-term risks of death from insect-borne diseases justified the long-term potential risks. Soldiers would be exposed for short durations, so the risks seemed reasonable. After the War, however, civilians would be exposed for very long, potentially interminable durations, leading to a completely different calculus of risk and benefit.

Pharmacologists joined entomologists in expressing concern, calling for increased testing before widespread use. Scientists at the FDA warned in 1944 that feeding experiments showed toxicity even in tiny amounts, and warned that "the safe chronic levels would be very low indeed."[37] Researchers were particularly concerned about the chronic effects, given that the chemical accumulated in the body fat. One scientist in 1944 warned that the very properties that made DDT so useful in war—persistence and a broad spectrum of activity—were the same traits that made scientists concerned about health effects when used in agriculture.

Ecologists also expressed concern about broader ecosystem-level effects of the chemical, for wartime DDT spraying had seemed to devastate some island ecologies. Entomologists feared that DDT might actually worsen pest problems by killing off competitors that kept pests in check. Field trials in Panama led to concern that DDT could create "biological deserts" and "areas devoid of life."[38] Fish and wildlife biologists shared this concern, and in May 1945, Clarence Cottam of the Fish and Wildlife Service urged that DDT not be released for civilian use until the service could better assess its ecological effects. Tests at Patuxent River Refuge in Maryland during the summer of 1945 found cause for grave concern for fish and bird populations. Other scientists urged that aerial spraying of DDT "be reserved for serious military emergencies."[39]

Even though most publicity and popular journalism in the late 1940s touted DDT as a wonder chemical, Russell shows that environmental concerns actually led to restrictions on DDT use immediately after the war, when the army was still able to control the drug. In April 1945, the army and the US Public Health Service announced restrictions on domestic spraying of DDT. These two agencies essentially controlled use of DDT in the US, because the War Production Board's priority system was able to regulate use and production of the chemical (but only during wartime). These two agencies briefly banned aerial spraying of DDT, urging that "much still must be learned about the effect of DDT on the balance of nature important to agriculture and wild life before general outdoor application of DDT can be safely employed in this country."[40]

Pressures for increased production after the war soon "eroded the authority of the Army and Public Health Service," for the War Production Board was concerned about a postwar Depression if war manufactures could not find new markets to replace military markets. The War Production Board decided to allow broader sales of DDT, and quickly encouraged wide civilian use to rebuild a postwar economy.[41] Ironically, after the war, no government agency actually had the authority to keep pesticides off the market, even if it was proved that they caused substantial harm. The USDA could enforce labeling requirements, and the FDA could seize food with pesticide residues, but neither agency could stop a company from selling a chemical.

Outside of expert circles, few members of the public were aware of medical, ecological, and scientific concerns with DDT or other pesticides. Intentional secrecy was one reason; research on DDT's potential for causing cancer had been classified by the army to "avoid disturbing rumors" during wartime.[42] Federal regulators were, by the late 1940s, frustrated that wide acceptance of these pesticides meant that the question of civilian safety was largely ignored. In 1949, the FDA Commissioner Paul B. Dunbar called for an assessment of DDT's safety in food, arguing that in "the postwar world people might take in small amounts over long periods," thus leading to greater risks than seen in wartime military use.[43] Federal scientists were particularly concerned about DDT in milk, finding that goats fed DDT produced milk that killed rats. When DDT showed up in cow milk destined for human consumption, federal entomologists recommended that dairy barn managers switch to other pesticides, but with little effect.

After the Second World War, aerial distribution of pesticides accelerated. Surplus planes, pilots, and DDT stimulated the crop-dusting industry, and the agriculture industry recovered from the Depression, providing new markets for sprays. By 1958, over 100 million acres of crops were sprayed with pesticides—one sixth of the cultivated lands in the US.[44] Yet as crop dusters sprayed more farm fields, they also sprayed the suburbanites who were moving out from the city, and this created a substantial backlash against the industry. Homeowners were happily using toxic chemicals in their houses, often at lethal doses, but at least they felt that they controlled what was sprayed on and around them. Aerial spraying threatened the very notion of home as a refuge, of control over the substances that entered the home and body. These suburbanites introduced what the philosopher Carl Cranor calls a sense of toxic trespass.[45] When workers, consumers, and even doctors challenged this trespass, the pesticide industry attacked them in court. As historian Pete Daniel describes in *Toxic Drift*, the USDA failed to protect public safety even when the medical evidence of harm was substantial, instead throwing its resources behind industry.[46]

The gypsy moth eradication program in the northeastern United States, which covered millions of acres of private land with a fine dust of chemicals, intensified the growing sense of trespass and powerlessness many people felt. Meanwhile, a fire ant control program took off in the South, complete with public relations campaigns vilifying the fire ant as a lethal threat to civilization. Plans to spray 20 to 30 million acres of the south in 1958 alone alarmed the public and state departments of conservation. Russell writes "confident of the technology they had gained during and after World War II, chemical officers and federal entomologists promoted eradication of human and insect enemies to the American public in the later 1950s. The strategy backfired as scientists and the public protested against chemical warfare and large pest eradication projects."[47] These efforts motivated two of the key critiques of Cold War power: President Eisenhower's criticism of the "military-industrial complex," and Carson's parallel argument about the dangers of chemical pest control in the hands of an increasingly authoritarian, technocratic state.

DDT became emblematic of the risks posed by organochlorines, and equally emblematic of the difficulties regulators faced in responding to those risks. By 1959, a fairly widespread public reaction against spraying campaigns began, and spread to reactions against chemical industry activities. These concerns helped to set the stage for Rachel Carson—and for the public response to her work. The Delaney hearings on pesticides in 1950 raised public awareness about potential toxicity. Acute poisoning incidents, particularly of farm workers, became more widely known, raising public concerns about citizens' personal risk of exposure.

As historian Linda Lear argues, Rachel Carson's *Silent Spring* gave voice not just to a growing fear of pesticides, but to growing challenges to the expertise models of professional knowledge, particularly their conceptualization of risk. After Carson's call to action, the 1960s and 1970s witnessed the emergence of consumer, environmental, and scientific forces that protested the misuse of synthetic chemicals.[48] Although Carson had questioned the technocratic, elite assumptions underlying the postwar boom in chemicals, the regulatory structures that emerged to control these chemicals relied on those same technocratic assumptions.

The quantitative science of risk assessment blossomed in the 1970s, leading to bans or restrictions on DDT, PCBs, and many persistent organic pollutants (POPs) such as the pesticides toxaphene, chlordane, dieldrin, endrin, and aldrin. Yet these regulatory actions soon proved incomplete and often ineffective. Many persistent organic pollutants remained in widespread use; new toxic chemicals such as polyvinylchloride became increasingly abundant, and banned chemicals continued to accumulate in the environment and in bodies.

ENDOCRINE DISRUPTORS

During the 1980s, concern developed about endocrine disruptors, chemicals that can mimic, block, or disrupt the actions of the body's own hormones, thereby altering reproduction and development, often with profound effects later in life. Their temporal effects, in other words, are often indirect: a tiny exposure to the fetus may have effects that are not obvious at birth, but that decades later lead to cancer. Moreover, their spatial effects can be equally puzzling, because they are rarely concentrated at the location of exposure. Many of them are quite volatile, evaporating from their points of use, moving with the earth's weather patterns, and migrating toward the poles. The most volatile of them are found at far higher concentrations in the Arctic Ocean than near the places they were used."[49]

Endocrine disruptors such as DDT have mechanisms of action that cannot be easily understood or contained within traditional paradigms of risk. Their effects rarely include acute poisoning, but instead a suite of puzzling, difficult to measure effects—effects that, as in the case of DDT, can lead to the near-extinction of entire species before any scientists manage to connect exposure with harm. In 1950, Howard Burlington and Verlus Frank Lindeman, two American biologists, showed that DDT could have estrogenic effects. Male chicks injected with a form of DDT had smaller testes (only 18 percent of normal size) and arrested development of secondary sexual characteristics compared to controls.[50] In effect, they were chemically castrated. Other researchers showed that DDT could alter the formation of enzymes in the liver, which would then alter the formation and regulation of estrogen, progesterone, and testosterone, affecting reproduction.[51] Yet regulatory agencies essentially ignored these findings, for they could not be quantified by risk assessments.

Meanwhile, signs of reproductive trouble in wildlife populations exposed to DDT and PCBs were emerging. An early sign that environmental chemicals might impair endocrine function was the discovery in the 1950s that DDT, a persistent organochlorine pesticide, caused bald eagles to lay eggs with thin shells by altering the actions of steroid hormones. In the 1960s, as Carson warned of the ecological effects of pesticides and the links between humans and wildlife, many wildlife biologists began asking why eagles, peregrine falcons, and similar birds were experiencing widespread reproductive failures. Carson singled out DDT as the likely culprit in eagle eradication, and noted that "the insecticidal poison affects a generation once removed from initial contact with it."[52]

In the 1980s, the researcher Theo Colborn observed that across the Great Lakes region, aquatic wildlife had developed reproductive problems, and she hypothesized that these problems might stem from the consequences of fetal events—namely, changes in the levels of steroid hormones such as estrogen and testosterone during fetal development.[53] Hundreds, perhaps thousands, of the synthetic chemicals added to the environment since the 1930s—agricultural chemicals such as organophosphate pesticides, industrial chemicals such as PCBs and dioxins, and many compounds in plastics—have the potential to magnify or reduce the action of steroid hormones, particularly on the developing fetus. Endocrine-disrupting chemicals are not rare; some of them are among the most common synthetic chemicals in production, particularly in the booming production of plastics.

In 1987, Ana Soto and Carlos Sonnenschein at Tufts University discovered the first hints that plastics might be leaching chemicals that could have estrogenic effects. Soto and Sonnenschein found that plastic test tubes were leaching chemicals into their cultures that stimulated the growth of breast cancer cells. This astonished Soto and Sonnenschein, because they knew of no other reports of estrogens leaching out of plastics. Everyone, including the manufacturers, assumed plastics were inert. The problem turned out to be nonylphenol, a chemical widely used in industry and domestic products such as paints, detergents, oils, toiletries, and agrochemicals. Nonylphenols are one in a larger class of related chemical compounds called alkylphenols, many of which turn out to weakly estrogenic, making breast cancer cells multiply in lab cultures. In Britain alone, 20,000 tons of these chemicals are used a year, and a third of these end up in our rivers and lakes at concentrations of fifty micrograms per liter—levels higher than those that induce cancer cell responses in the lab.[54]

Similar problems with other plastics emerged in the early 1990s. During the effort to create artificial estrogens in the 1930s, researchers had first synthesized bisphenol-A. Not as powerful as synthetic hormones such as diethylstilbestrol, bisphenol-A was largely ignored until researchers realized that, when polymerized, it formed a useful plastic known as polycarbonate, now extensively used for baby bottles, water bottles, and dental and can sealants. Bisphenol-A leaches out of those products, ending up in food, children's teeth, and in wildlife at concentrations higher than the levels that induce estrogenic responses in lab animals.

A third group of chemicals used in making plastics that leach estrogenic compounds are phthalates, perhaps the most abundant synthetic compounds in the environment. Phthalates are oily solvents that make plastics flexible but strong. Since phthalates need to be flexible, that means the molecules can't be too rigidly locked together, for flexibility requires molecules that slide over each other. But that lack of molecular rigidity also means they leach out easily. Phthalates keep your car dashboard from cracking; your nail polish from splintering; they allow plastic wrap to be shaped around food. They help chemicals absorb quickly into your skin, so they are added to shampoos, skin creams, sunscreens, and most cosmetics, without any testing required for adverse health effects.

In March of 2001, the US Centers for Disease Control released the results of its first study of the levels of twenty-seven chemicals found within American bodies. Researchers found phthalates in nearly every person they examined (out of 3,800 people drawn from healthy individuals around the country with no special exposure to toxic substances).[55] The highest concentrations came from certain phthalates (such as di-ethyl phthalate) used in toiletries like bar soaps, perfumes, and shampoos, perhaps because direct skin contact increases body burden. Some of the highest concentrations were in women of childbearing age—not the results anyone wanted to find, since fetal exposure is likely to be the riskiest.

Plastics, like pesticides, have become an intimate part of our everyday environments. Unlike pesticides (which are often perceived as threatening), most people think of plastic as inert, harmless materials. Yet, as Gerald Marcowitz and David Rosner show in *Deceit and Denial: The Deadly Politics of Industrial Pollution* (2002), plastics such as PVC (polyvinyl chloride) have the potential to kill. Rosner and Marcowitz examine the myriad ways PVC manufacturers have hidden evidence of grave harm to workers and exposed communities, using scientific uncertainty over mechanisms of action as a way of deflecting regulatory action. Devra Davis examines similar patterns of industrial deception and regulatory stalemates in *When Smoke Ran Like Water: Tales of Environmental Deception and the Battle Against Pollution* (2002).[56]

Controlling exposure to chemicals by setting tolerance levels and thresholds appeared to offer agencies a way to allow for beneficial uses of chemicals without exposing human populations to increased risks. Yet a threshold level-based approach made certain key assumptions: first, that scientific technologies allowed risks to be measured and monitored; second, that lower exposures to particular chemicals posed fewer risks than higher doses; third, that dilution could reduce chemical risks; and fourth, that the body's skin and cellular membranes provided a protective barrier against contamination from the outside world, while the placenta protected the developing fetus from chemicals ingested by the mother.

Synthetic endocrine-disrupting chemicals met none of these assumptions. Low doses of endocrine disruptors often have more problematic effects than high doses, overturning the basic precept of toxicology, namely, that the dose makes the poison, therefore risk can be regulated by reducing exposure rather than eliminating usage. The effects of endocrine disruptors often depend on *timing* of exposure rather than dosage, with fetal and early childhood exposures particularly problematic. Many endocrine disruptors resist the metabolic processes that bind and break down natural hormones, and their legacy effects can persist for generations. Many are lipophilic, meaning that they attach readily to fat (or lipid) molecules. This quality allows them to move across the membranes of skin and cells, penetrating the barriers that were thought to protect bodies from external contaminants. Many of these substances have pervaded environments on global scales, moving through atmospheric and oceanic currents to contaminate watersheds, wildlife, and people far from the initial point of production and consumption. On micro-scales, they have moved across cellular membranes to insert themselves into the most intimate spaces of reproduction.[57]

Endocrine-disrupting chemicals have posed novel challenges for scientists and regulatory agencies seeking to protect public health, because they do not easily fit within traditional risk paradigms. Although the threshold model may be useful for natural toxins such as aspirin, it is rarely relevant for endocrine disruptors. Even at extremely low levels, they can mimic, block, or disrupt the actions of the body's own hormones, thereby altering reproduction and development, often with profound effects later in life. In fact, endocrine-disrupting chemicals can actually have more powerful effects at low doses than at high doses. At low concentrations, hormones normally stimulate receptors, but at high concentrations hormones can saturate receptors, thus inhibiting their pathways. Low doses of endocrine disruptors might produce adverse impacts, even though higher doses might not. But the idea that a substance can have more powerful effects at low doses than at high doses fundamentally challenged toxicological paradigms.[58]

The effects of hormonally active chemicals such as DDT puzzled researchers and regulators because they differed dramatically among individuals, depending on the age of the individual and the timing of the exposure. These findings made little sense when interpreted through a standard toxicological paradigm, but they were less surprising when researchers considered how the endocrine system functions at different life stages. In adults, hormones mainly regulate ongoing physiological processes such as metabolism. Synthetic chemicals can lead to temporary endocrine changes, but adults are often able to recover from these disturbances.

Some of the most profound effects of endocrine disruptors come from their effects on fetal development. Tiny exposures during fetal development lead to surprising changes much later in life, yet few scientists, much less regulators, were initially able to envision that these changes could have been caused by fetal exposure. Conceptual models in developmental biology help explain why scientists and regulators discounted evidence of low-level chemical toxicity. Ecological and epigenetic concerns had played a major role in the development of embryology in the late nineteenth century, as investigators tried to understand how the environment shaped development of embryos. This changed in the early twentieth century. Scientists increasingly downplayed the role of the environment in development when social and technological forces made embryonic development easier to study internally than externally. By the 1940s, most developmental biologists had adopted a belief that the fetus was essentially determined by the genome, therefore invulnerable to influences from the environment. As the developmental biologist and historian of science Scott Gilbert writes, "genetics brought a new form of preformationism. Instead of a dynamically acting organism taking its cues from the environmental conditions and from the way that cells interact with each cell division, the 20th century brought a dominant and popular view that has often emphasized genes as programmed to carry the information of heredity, which was also the information necessary to construct an individual."[59]

An environmental focus in developmental biology reappeared in the 1990s, driven by new technologies, new environmental concerns, and a resurgent interest in epigenetics. Much of this research came from the work of amphibian biologists, who were jolted by the discoveries of amphibian limb deformities around the world. These biologists soon realized that, while the genes of a developing amphibian obviously played a role in its development, environmental effects, particularly the integrated effects of hormone mimics, parasites, and UVB radiation, were altering the patterns of development in ways that challenged strict reductionist paradigms.[60]

One of the central conceptual puzzles scientists, doctors, and regulators faced from the 1940s through the 1990s was imagining how something the fetus was exposed to during development could lead to cancer in puberty or adulthood. It was hard enough to conceive that a drug could cross the placental barrier and create immediate harm that was visible at birth, as thalidomide had done. Environmental influences on the fetus that lead to cancer or sexual changes in adulthood were much more difficult to imagine. Animal experiments in the 1930s and 1940s showed that fetal exposure to new chemicals could lead to reproductive problems that emerged only in adulthood, but researchers simply could not comprehend that this could occur in humans, because it violated a host of assumptions about genetically determined development.

The flourishing of epigenetics research since the 1990s has transformed conceptual models about gene-environment effects on the developing fetus. Every cell in the body contains the entire genetic code. But brain cells must use only the genes needed in the brain, while kidney cells should activate only the genes needed for renal function. Epigenetics explains how these different parts of the genome are activated or silenced during development. Cells commonly switch gene behavior on and off by attaching small molecules known as methyl groups to specific sections of DNA. The attachment and detachment of methyl groups is particularly important in the fetal development of the reproductive system.

Environmental scientists have long suspected that the environment somehow shapes the activities of genes, but they have had no way of explaining how. Most of the US cancer research establishment remains committed to the belief that gene mutations control cancer development, and so most of the funding goes to this type of research.[61] But some toxins do not cause mutations, even though exposures to those toxins may still be linked to increased cancer rates. Without a mechanism to explain how a toxin could cause cancer without causing a mutation, researchers tended to dismiss correlations as artifacts. Epigenetic changes are now emerging as the critical conceptual link between environment and genome.[62]

Changes that silence and unsilence genes, but leave the DNA sequence untouched, can set into motion environmental diseases that begin during development but appear only much later. A gene is silenced when something prevents its expression, often during transcription or translation. The effects of endocrine disruptors on the developing fetus provide an example of such epigenetic changes. Many cells have tumor-suppressor genes that keep tumors from becoming malignant. Chemical exposures can lead to epigenetic changes that silence these tumor-suppressor genes, even when their DNA sequence is unchanged. Likewise, cells also contain tumor-promoter genes which are normally not expressed. Exposure to toxins can turn on the expression of these genes, thereby allowing them to promote the growth of tumors. In animals bred to contain genes that make them particularly susceptible to fibroid tumors, those genes are normally silenced, but exposure to toxins will allow the expression of those genes in the fetus, and tumors will develop years later. Without the initial toxic exposure, a genetic susceptibility does not lead to cancer in adulthood. Exposure of the fetus to toxic chemicals can permanently reprogram tissue in a way that determines whether tumors will develop in adulthood.[63]

What do these low-level exposures mean for humans and wildlife? No one knows for certain, but there is an emerging consensus that toxicological models based on dose response models and thresholds—models at the core of risk analysis—fail to adequately address their potential for harm.

Environmental risk assessment has attempted to manage pollution by permitting chemical production, use, and release, "as long as discharges do not exceed an quantitative standard of 'acceptable' contamination. This approach assumes that ecosystems have an 'assimilative capacity' to absorb and degrade pollutants without harm. It also assumes that organisms can accommodate some degree of chemical exposure with no or negligible adverse effects, so long as the exposure is below the 'threshold' at which toxic effects become significant."[64] This, in essence, is the current "risk analysis model," which tries to calculate the mathematical likelihood that any individual chemical will harm the public. But as writer Michael Pollan notes, risk analysis "is very good at measuring what we can know—say, the weight a suspension bridge can bear—but it has trouble calculating subtler, less quantifiable risks. (The effect of certain neurotoxins on a child's neurological development, for example, appears to have more to do with the timing of exposure than with the amount.) Whatever can't be quantified falls out of the risk analyst's equations, and so in the absence of proven, measurable harms, technologies are simply allowed to go forward."[65]

In risk assessment, a lack of data is seen as evidence of safety, "so untested chemicals are allowed to be used without restriction. Since the vast majority of chemicals have not been subject to toxicity testing, ignorance becomes the dominant factor in environmental decisions."[66] The burden of proof of harm is placed on affected communities. Challenges to these assumptions have been present ever since Wiley called for a version of the precautionary principle in the early decades of the twentieth century. In recent years, they have coalesced around a call for an ecological approach to health that would place the burden of proof for safety on those who wish to profit from chemical exposures. As John Wargo argues in *Our Children's Toxic Legacy: How Science and Law Fail to Protect Us from Pesticides* (1998), many researchers are increasingly concerned that synthetic chemicals pose conceptual challenges to the risk assessment structures that technocratic governments have erected to control them.[67]

ECOLOGICAL PERSPECTIVES ON HEALTH

Traditional toxicological frameworks assumed a body, separate and individual, that might be protected from invasion by toxic substances. In contrast, an

ecological approach to health recognizes that the body is enmeshed in a web of ecological relationships, not an isolated castle whose threshold can only be breached
by a sustained attack from the outside. As historian of science Jody Roberts argues,
industrial chemicals occupy a position along the border between the natural and
synthetic worlds.[68] Such chemicals are abundant artifacts of an industrial society brought into being within a highly specific cultural infrastructure, against a
deeper historic backdrop of evolution that occurred without their presence. And
yet, increasingly they are a part of the natural world—and as persistent chemicals,
many of them will continue to be a part of our most intimate environments far into
the future.

Human bodies exist in dynamic ecological relationships with themselves, other
bodies, and the environments in which they are embedded. The body is an ecosystem
of its own, yet one linked to the world around it. Like all ecosystems, the body is constantly undergoing disturbances: natural toxins, parasites, mutagens. Health is not
the absence of stress, disturbance, or toxins; rather, it is the ability to respond to these
stresses. The immune response and mechanisms of cellular and DNA repair are all part
of a complicated ecosystem that regulates and repairs the human body.

Synthetic chemicals such as DDT are of concern not simply because they have potential to harm the body, for bodies are constantly negotiating exposures to substances
that have the potential to cause harm. Rather, certain synthetic chemicals are able to
transform the body's ecological repair mechanisms, often at the biochemical level. In
particular they alter the epigenetic processes that link environment and gene, leading to changes in gene expression, and in turn to changes in the numbers and types
of immune cells in the blood, and changes in hormone production and metabolism.
They alter ecological processes of human health, just as they alter broader ecosystem
processes.

It is easy to imagine that we are isolated individuals bombarded by synthetic
external disturbances that need to be fought off. But physiologists have moved away
from framing organisms as separate individuals, and they now sound like Donna
Haraway when they write about an organism as "an interaction between a complex,
self-regulating physiological system and the substances and conditions which we usually think of as the environment."[69] In other words, none of us are isolated individuals;
we are networks of self and non-self, of our own identities and DNA interwoven with
the colonies of parasites and bacteria and viruses that make up our bodies. Our deepest
sense of self reflects our personal, cultural, and evolutionary histories, including the
viruses that were once, millions of years ago, in immunological terms, "nonself invaders," but which eventually became incorporated into our DNA, and now modulate our
responses to the hormonal webs we exist within.

Chemical pollutants change the network of genetic, immunological, neurological, hormonal, and environmental interrelationships that control sex and reproduction in vertebrates, and this can kill an individual, eliminate a population, or drive a
species extinct. Understanding endocrine disruptors means reconsidering our bodies and our identities, seeing them not as separate isolated objects, but rather as what

Bruno Latour termed "hybrid networks." The material and the cultural, in Latour's terms, "weave our world together," yet these weavings have often become invisible to us.[70] Anne Fausto-Sterling writes that we have "forced the hybrid networks linking nature and culture underground... Although a strategy of ignoring hybrids worked in the beginning, it embodied a paradox. The better it worked, the more unacknowledged hybrids developed. The more we dominated nature, the more the proof of our domination poured into culture; the more culture dominated nature, and the more we created objects that were neither truly natural nor truly cultural."[71]

The hybrids we have created with endocrine disruptors resist our attempt to define clear boundaries between natural and synthetic, and between male and female. Synthetic hormone disrupting chemicals occupy a position along the border between the "natural" and the cultural or constructed world. Industrial chemicals are abundant artifacts of an industrial society brought into being within a highly specific cultural infrastructure, against a deeper historic backdrop of evolution that occurred without their presence. And yet, increasingly they are a part of the natural world—and as persistent chemicals, many of them will continue to be a part of our bodies and ecosystems far into the future.[72]

ACKNOWLEDGEMENTS

I thank the participants in the 2007 Temple University Workshop for their lively discussions and insights. Portions of this chapter appeared in a different form in *Toxic Bodies: Hormone Disruptors and the Legacy of DES* (New Haven, CT: Yale University Press, 2010). Yale University Press has been generous with its permission to allow use of this material. Research for this chapter was funded by a grant from the University of Wisconsin Graduate School.

NOTES

1. Steven Kroll-Smith and Worth Lancaster, "Review: Bodies, Environments, and a New Style of Reasoning," *Annals of the American Academy of Political and Social Science* 584 (2002): 203–212.
2. Rene Dubos, cited on page 117 in Joe Thornton, *Pandora's Poison: Chlorine, Health, and a New Environmental Strategy* (Cambridge, MA: MIT Press, 2000). For a longer discussion, see Rene Dubos, *Mirage of Health: Utopias, Progress, and Biological Change* (New York: Harper & Brothers, 1959).
3. Michael McCoy et al., "Facts & Figures of the Chemical Industry," *Chemical & Engineering News* 84, no. 29 (July 10, 2006): 35–72. The standard business history of the growth of the chemical industry is Alfred D. Chandler, *Shaping the Industrial Century: The Remarkable Story of the Evolution of the Modern Chemical and Pharmaceutical Industries* (Cambridge, MA: Harvard University Press, 2005).

4. Centers for Disease Control and Prevention, *Third National Report on Human Exposure to Environmental Chemicals* (Atlanta: CDC, 2005).

5. Christopher Sellers, *Hazards of the Job: From Industrial Disease to Environmental Health Science*, (Chapel Hill: University of North Carolina Press, 1997).

6. Sellers, *Hazards of the Job.*

7. For important essays on the various environmental and historical meanings of these shifts, see Gregg Mitman, Michelle Murphy, and Christopher Sellers, eds. "Landscapes of Exposure: Knowledge and Illness in Modern Environments," *Osiris* 19 (2004): 1–304.

8. Linda Nash, *Inescapable Ecologies: A History of Environment, Disease, and Knowledge* (Berkeley: University of California Press, 2007), and "Purity and Danger: Historical Reflections on the Regulation of Environmental Pollutants," *Environmental History* 13 (October 2008): 651–658.

9. Nash, *Inescapable Ecologies*, 142–144.

10. Frederick Rowe Davis, "Unraveling the complexities of joint toxicity of multiple chemicals at the Tox Lab and FDA," *Environmental History* 13 (October 2008): 674–683.

11. Nash, *Inescapable Ecologies*, 143

12. Thomas Dunlap, *DDT: Scientists, Citizens, and Public Policy* (Princeton, NJ: Princeton University Press, 1981), 19.

13. C. V. Riley, in *USDA Farmer's Bulletin 7*, cited in James Whorton, *Before Silent Spring: Pesticides and Public Health in Pre-DDT America* (Princeton, NJ: Princeton University Press, 1981), 71.

14. Whorton, *Before Silent Spring*, 21–34, 40–45.

15. Ibid., 176.

16. Ibid., 177.

17. Dunlap, *DDT*, 49.

18. Whorton, *Before Silent Spring*, 177; see also Dunlap, *DDT*, 5–6.

19. Whorton, *Before Silent Spring*, 99.

20. Ibid., 109.

21. Nash, "Purity and Danger."

22. Dunlap, *DDT*, 54.

23. Ibid., 49.

24. Whorton, *Before Silent Spring*, 111–117.

25. Ibid., 175.

26. Ibid., 201.

27. Ibid., 200ff.

28. Dunlap, *DDT*, 39.

29. Ibid., 55.

30. Edmund Russell, *War and Nature: Fighting Humans and Insects with Chemicals from World War 1 to Silent Spring* (Cambridge: Cambridge University Press, 2001), 11.

31. Ibid., 13.

32. Ibid., 14.

33. Russell, *War and Nature*, 156.

34. Ibid., 129.

35. Ibid., 125, note 8.

36. Ibid., 125.

37. Ibid., 157.

38. Ibid., 159.

39. Ibid., 160.
40. Ibid., 161.
41. Ibid., 162–163.
42. Ibid., 158.
43. Ibid., 175.
44. Ibid., 202.
45. Carl Cranor, "Some Legal Implications of the Precautionary Principle: Improving Information-Generation and Legal Protections," *Human and Ecological Risk Assessment* 11 (2005): 29–52.
46. Pete Daniel, *Toxic Drift: Pesticides and Health in the Post-World War II South* (Baton Rouge: University of Louisiana Press with the Smithsonian Museum of Natural History, 2005).
47. Russell, *War and Nature*, 213–4.
48. Linda Lear, *Rachel Carson: Witness for Nature* (New York: Henry Holt, 1997).
49. Mel Visser, *Clear, Cold and Deadly: Unraveling a Toxic Legacy* (Lansing: Michigan State University Press, 2007).
50. Endocrine disruptors and their histories are discussed in Nancy Langston, *Toxic Bodies: Hormone Disruptors and the Legacy of DES* (New Haven, CT: Yale University Press, 2010); "Gender Transformed: Endocrine Disruptors in the Environment," in *Seeing Nature through Gender,* ed. Virginia Scharff, (Lawrence: University of Kansas Press, 2003), 129–166; and Nancy Langston, "The Retreat from Precaution: Regulating Diethylstilbestrol (DES), Endocrine Disruptors, and Environmental Health." *Environmental History* 13 (2008): 41–65.

    See also Howard Burlington and Verlus F. Lindeman, "Effect of DOT on Testes and Secondary Sex Characteristics of White Leghorn Cockerels," *Society for Experimental Biology and Medicine Proceedings* 74 (1950): 48–51.
51. R. M. Welch, W. Levin, K. Kuntzman, M. N. Jacobson, and A. H. Conney, "Effect of Halogenated Hydrocarbon Insecticides on the Metabolism and Uterotropic Action of Estrogens in Rats and Mice," *Toxicol. Appl. Pharmacol.* 19 (1971): 234–246; A. H. Conney, R. M. Welch, R. Kuntzman, and J. J. Burns, "Effects of Pesticides on Drug and Steroid Metabolism," *Pharmacol. Therap* 8 (1966): 2–8.
52. Rachel Carson, *Silent Spring* (New York: Houghton Mifflin, 1962). For a fuller discussion of Carson, see Lear, *Rachel Carson*, and N. Langston, "Rachel Carson's Legacy: Gender Concerns and Endocrine Disrupting Chemicals," *GAIA* 21, no. 3 (2012): 225–229.
53. Theo Colborn, Dianne Dumanoski, and John Peterson Myers, *Our Stolen Future: Are We Threatening Our Fertility, Intelligence, and Survival?* (New York: Plume, 1997).
54. Soto describes this discovery in her untitled autobiographical essay on "in-cites," October 2001, http://www.in-cites.com/papers/dr-ana-soto.html. The research is reported in A. M. Soto, H. Justicia, J. W. Wray, and C. Sonnenschein, "p-Nonyl-phenol: an estrogenic xenobiotic released from "modified" polystyrene," *Environmental Health Perspectives* 92 (1991): 167–173.
55. Centers for Disease Control and Prevention, "National Report on Human Exposure to Environmental Chemicals," March 2001, Atlanta, Georgia.
56. Gerald Markowitz and David Rosner, *Deceit and Denial: The Deadly Politics of Industrial Pollution* (Berkeley: University of California Press/Milbank Books on Health and the Public, 2002); Devra Davis, *When Smoke Ran Like Water: Tales of Environmental Deception and the Battle Against Pollution* (New York: Basic Books, 2004).

57. Langston, *Toxic Bodies*.

58. Langston, *Toxic Bodies*, 5–12.

59. Scott Gilbert, "The Genome in its Ecological Context: Philosophical Perspectives on Interspecies Epigenesis," *Annals of the New York Academy of Sciences* 981 (2002): 202–218.

60. Marvalee H. Wake, "Integrative Biology: The Nexus of Development, Ecology and Evolution," *Biology International* (2004): 1–18.

61. Devra Davis, *The Secret History of the War on Cancer* (New York: Basic Books, 2009).

62. For an overview of epigenetics aimed at the general public, see Sharon Begley, "How a Second, Secret Genetic Code Turns Genes On and Off," *Wall Street Journal*, July 23, 2004.

63. Retha Newbold, "Perinatal Carcinogenesis: Growing a Node for Epidemiology, Risk Management, and Animal Studies," *Toxicology and Applied Pharmacology* 199 (2004): 142–150.

64. Joe Thornton, *Pandora's Poison: Chlorine, Health, and a New Environmental Strategy* (Cambridge, MA: MIT Press, 2000), 7.

65. Michael Pollan, "The Year in Ideas: A to Z; Precautionary Principle," *New York Times*, December 9, 2001.

66. Thornton, *Pandora's Poison*, 7–8.

67. John Wargo, *Our Children's Toxic Legacy: How Science and Law Fail to Protect Us from Pesticides* (New Haven, CT: Yale University Press, 1998).

68. Jody Roberts, draft report for the 2007 Gordon Cain Conference, Chemical Heritage Foundation (in possession of the author).

69. Glen Fox, "Tinkering with the Tinkerer: Pollution versus Evolution," *Environmental Health Perspectives* 103 (1995): Suppl 4, citing G. A. Bartholomew, "Interspecific Comparison as a Tool for Ecological Physiologists," in *New Directions in Ecological Physiology*, ed. M. E. Feder, A. F. Bennett, W. W. Burggren, and R. B. Huey (New York: Cambridge University Press, 1987), 11–37. Donna Haraway, "The Biopolitics of Postmodern Bodies: Constitutions of Self in Immune System Discourse," In *American Feminist Thought at Century's End: A Reader*, ed. Linda S. Kauffman (Cambridge: Blackwell, 1994), 199–233.

70. Bruno Latour, *We Have Never Been Modern*, (Cambridge, MA: Harvard University Press, 1993), 1–2, discussed in Anne Fausto-Sterling, "Science Matters, Culture Matters," *Perspectives in Biology and Medicine* 46 (2003): 109–124.

71. Fausto-Sterling, "Science Matters, Culture Matters."

72. Jody A. Roberts and Nancy Langston, "Toxic Bodies/Toxic Environments: An Interdisciplinary Forum." *Environmental History* 13 (2008): 629–635.

CHAPTER 11

....

RETHINKING AMERICAN EXCEPTIONALISM

Toward a Transnational History of National

Parks, Wilderness, and Protected Areas

....

JAMES MORTON TURNER

THE twentieth century has been described as the "prodigal century." The growth in population, agriculture, industry and, consequently, resource consumption, energy use, and pollution made it unlike any century that preceded it. Those changes contributed to a dramatic reordering of people and the environment. But amid so much change, the twentieth century was also distinguished by efforts to protect the natural world from the forces that were transforming it. Among the most sustained of such efforts were those to set aside protected areas for tourism, recreation, scenery, wildlife, and habitat conservation. By 2003, governments worldwide had designated more than 100,000 protected areas encompassing 18.8 million square kilometers of the earth's surface. If those protected areas were gathered together into a single landmass, it would rival South America in size.[1]

In 1970 Roderick Nash surveyed this growing parks movement and heralded the success of the United States in exporting "the national park idea around the world." He explained that the United States is "known and admired for it, fittingly, because the concept of a national park reflects some of the central values and experiences in American culture."[2] Indeed, the United States claimed the first national park, Yellowstone, in 1872. The US National Park Service, established in 1916, became a model for parks agencies worldwide. The United States helped organize and support the scientifically minded International Union for Conservation of Nature after World War II. And the United States hosted the first two World Conferences on National Parks in 1962 and 1972. In 2007 Donald Worster echoed Nash's celebratory tones, suggesting that "we have not fully appreciated how much the protection of wild nature owes to the spread of modern liberal, democratic ideals and to the support of millions of ordinary people around the world."[3]

But the rapid growth of protected areas since the 1970s, particularly in developing countries and often in service to conservation science, drew new attention to the social and political consequences of Western models of nature protection. In 1989, Indian scholar Ramachandra Guha warned that efforts to "transplant the American system of national parks onto Indian soil" trespassed on the rights of local peoples, catered to the interests of a conservation elite and foreign tourists, and privileged Western science and scientists.[4] In 1995, the historian William Cronon warned that idealizing wilderness as a pristine and untouched landscape masked the social consequences of protected areas.[5] In 2007, Christopher Conte, like other scholars and journalists, traced this "American wilderness model" back to the nineteenth-century United States, which "set the global example for the strict preservation of nature through the American national parks system."[6] These critiques played an essential role in drawing attention to theretofore under-addressed social consequences of protected areas, both in developing countries and developed countries, in the past and present.

This dualism between American cultural chauvinism (the national parks are America's best idea) and anti-imperial criticism (protected areas have been a form of imperial enclosure that romanticizes nature and alienates local peoples worldwide) has been central to the literature on parks, wilderness, and protected areas since the 1990s. But both sides in this debate have often shared one key assumption: that a coherent model of the American national park, predicated on wilderness protection, has been the definitive model for nature protection globally. But consider these figures: as of 2010, only 8 percent of protected areas (by number) and 47 percent of protected lands (by area) fall into categories of land protection and management that are national parks or wilderness areas—meaning most protected areas are otherwise classified and managed.[7] Many of the alternative categories of protected areas aim to support local communities, sustainable resource use, and cultural values; in recent decades, it is these protected-area categories that have been growing most rapidly. This is not to say that an American model of national parks has not been influential historically, but it is to suggest that there is considerable and under-appreciated diversity in the history of protected areas worldwide, which historians have only recently begun to explore.

This essay calls for an approach to national parks, wilderness, and protected areas that, instead of beginning with a transcendent American parks model, takes up protected areas as a matter of practice and politics that has been formed and contested in local, national, and international arenas. To advance this call, the essay is organized into five parts. First, it argues that the historical debates over protected areas in the United States challenge claims of a singular American model of parks protection. Second, it asks how an American wilderness park ideal came to occupy a mythic place in the national parks movement and related scholarship. Third, it considers recent scholarship that suggests that parks can more profitably be understood in the context of a transnational circulation of ideas, practices, organizations, and individuals which have informed the creation of a range of protected areas in the United States and abroad. Fourth, it examines how critical studies of protected areas sparked new debates over the social consequences of protected areas and strategies for managing them. It concludes that taking a transnational approach to the varieties of protected areas worldwide can inform efforts to advance more effective and socially just strategies for protected areas in the future.

DISENTANGLING WILDERNESS AND THE UNITED STATES NATIONAL PARK IDEAL

Champions of a United States tradition of nature protection have emphasized the significance and coherency of America's national park tradition. Roderick Nash championed the "American invention of national parks" in 1970.[8] Wallace Stegner popularized the idea that the national parks were "the best idea we ever had" in 1983.[9] And such sentiments inspired Ken Burns's twelve-hour documentary tribute to the national parks, titled *America's Best Idea,* which drew popular attention in the United States in 2009.[10] The usual assumption has been that the United States' spectacular wild lands and the nation's fascination with wilderness inspired a parks movement of national and, ultimately, international significance. Before considering how influential this parks model was globally, it is worth considering to what extent such a coherent and durable model of nature protection actually existed in the United States. In contrast to popular acclaim, a closer reading of the history of America's national parks and concerns for wilderness protection suggests a more contingent and contested history of nature protection that did not yield a singular model of global importance.

Few scholars have done more to draw attention to wilderness and the national park ideal in American history than Roderick Nash and Alfred Runte. Nash, in his classic 1967 study, *Wilderness and the American Mind,* advanced a sweeping history of the wilderness idea in American culture, explaining how a nation determined to harness its resources to a growing economy in the nineteenth century came to value wilderness as one of the nation's most distinct and formative influences in the twentieth century.[11] In 1979, Alfred Runte, one of Nash's students, offered the first synthetic history of the national park system, which emphasized how grand, scenic parks like Yellowstone and Yosemite became the inspiration for a national park ideal that hinged on monumental landscapes of little economic value that could stimulate an emerging tourist economy.[12] As Paul Sutter has aptly noted, Nash and Runte "welded wild nature and American exceptionalism together in a narrative particularly befitting a moment of national environmental awakening" in the 1960s and 1970s.[13] Although that celebratory narrative would rightly be subject to sharp criticism and revision in the 1990s and 2000s, these more recent critiques have often overlooked that in Nash's and Runte's analyses, the history of the wilderness ideal and American national parks, while related, were not congruent.

The United States Congress saddled the National Park Service with a broad and contradictory task in 1916, when it enacted an Organic Act for the national parks. The purpose of the new Park Service and the national parks "is to *conserve* the scenery and the natural and historic objects and the wild life therein and to provide for the enjoyment of the same in such manner and by such means as will leave them unimpaired for the enjoyment of future generations." [14] As both Nash and Runte explained, the initial motivation for protecting the early national parks was not the hope of protecting tracts of remote, impenetrable wilderness. Instead, authorizing legislation for the nation's grandest parks, such as Yellowstone (1872), Yosemite (1890), and Glacier (1910), highlighted their outstanding scenic values, emphasized protecting the parks in the public interest—rather than developing them for commercial gain—and made the parks available for the "enjoyment of the American people." This latter commitment reflected the partnership of park advocates and the railroad and automobile lobbies, which collectively urged Congress and the new Park Service to provide roads, rail stations, campgrounds, and hotels, which would make the national parks accessible to visitors and encourage Americans to "See America First!".[15] Although, as Runte and other scholars have noted, the Park Service did attempt to give more attention to the scientific and ecological values of parks (such as with the creation of the Everglades National Park in 1934 or Redwoods National Park in 1968), those goals have been in tension with the Park Service's enduring commitment to managing the parks for scenic tourism.[16]

When preservationists began to deliberately mobilize the concept of wilderness to justify nature protection in the United States in the 1920s and 1930s, the national parks figured prominently in such discussions, but for surprising reasons. In the view of wilderness champions, such as Aldo Leopold or Bob Marshall, both of whom worked for the Forest Service, the real threat to wilderness was not just loggers or miners; rather, it was roads, automobiles, and the culture of tourism and consumerism that came with them. Although national parks like Glacier or Mount Rainier might have included vast tracts of wilderness, the Park Service in the 1920s and 1930s seemed eager to open up the parks for tourists, not protect remote tracts of wild lands for scientific research or primitive recreation. Indeed, the Civilian Conservation Corps played a crucial role in expanding the tourist infrastructure in local, state, and national parks during the 1930s.[17] As Paul Sutter explains in his study *Driven Wild*, despite many, sometimes competing, ideas of what wilderness stood for, it gained significance among preservationists as a wild and primitive landscape devoid of cars and roads in the 1930s.[18] That made wilderness distinct from national parks, national forests, and other protected areas, where land managers often gave priority to opening up the landscape to tourists, resource extraction, or game management.

These separate, but overlapping, approaches to national parks and wilderness resulted in two distinct, but related, systems of nature protection in the United States. In 1964, Congress passed the Wilderness Act, which authorized the National Wilderness Preservation System, and designated "wilderness areas" on existing lands managed by the United States federal land agencies (including the Forest Service, Park Service, and Fish and Wildlife Service initially).[19] In contrast to the National Park

System more broadly, designated wilderness areas were meant to be primitive areas, off limits to roads, motorized vehicles, commercial activities (such as logging and mining), and other forms of development (including established campgrounds, lodges, or visitor centers).[20] The park system and wilderness system have been important strategies for protecting public lands in the United States. Since 1916, the park system has grown to 391 areas; it covered more than 84 million acres in 2013. Since 1964, the wilderness system has grown from 9.1 to 107.5 million acres, including 43.5 million acres administered by the National Park Service. [21] Thus, although the Park Service oversees extensive wilderness areas, much of the land it manages is not so protected.

The evolution of these two systems of land protection, and the debates that have defined them, suggests two important points regarding the history of protected areas. First, to conflate the history of national parks and wilderness in the United States— as later critics have often done—is to overlook the contingency of these distinct, but overlapping, strategies for establishing protected areas and the different ways in which people have tried to resolve tensions surrounding nature protection both in the United States and around the world. Second, emphasizing the national parks or wilderness as the primary strategies for protected areas in the United States has meant that other important approaches, such as federal wildlife refuges, state conservation programs, or systems of private land conservation in the United States, have drawn less scholarly attention, as have the ways in which those approaches have drawn upon and contributed to similar approaches elsewhere in the world.[22]

CONSTRUCTING AN EXPORT MODEL OF THE AMERICAN NATIONAL PARK

If there was no singular model of the national park, predicated on an American appreciation for wilderness, it is worth considering what explains the potency of an American model in histories of global nature protection. Astrid Swenson put this question eloquently in 2012: "Given the scale of global interaction, an interesting question remains…as to why the transnational history of the national park has been remembered as a primarily American one, and why the American side of the story has been reduced to a narrative about American love for wilderness?"[23] This myth took root after the 1950s, as the designation of protected areas began to expand rapidly worldwide. Before 1960, countries around the world had set aside a total of 11,708 protected areas. That figure surged to 61,871 protected areas by 1990.[24] For newly independent states seeking to establish their place in the international arena, designating national parks and related conservation programs was a relatively easy act of statecraft, required little commitment of financial resources, and often drew international recognition and support (including that of tourists, foundations, and foreign governments). Several factors contributed to the consolidation of an American model for parks development internationally.

The United States National Park Service began to formalize its international reach in the 1960s. Although the agency had provided information to foreign park services, been visited by park observers, and consulted from afar on foreign park projects prior to World War II, in the 1950s the United States expanded its international leadership. The agency began actively hosting visiting park officials and advocates (organizing and funding such visits), training foreign park managers and personnel (through regular programs), and providing technical assistance to park projects globally (including park design, construction, and scientific research and management). Such activities were consolidated under the National Park Service's Division of International Affairs in 1961. As Terence Young and Lary M. Dilsaver explain, with the creation of the new office, "the Park Service quickly experienced a marked increase in the frequency and scale in its interaction with foreign park agencies."[25] The chief partner for the Division of International Affairs was the United States Agency for International Development, which was a primary funder for the agency's international projects.[26] The scope and extent of this nexus between the leadership of the United States in parks policy and funding for parks projects is one that remains understudied, but essential to understanding both the history of the development of international parks policy and specific projects in countries ranging from Costa Rica to Kenya to Nepal.[27]

Although wilderness and national parks in the United States are distinct, these two approaches to protected areas did begin to align more closely in the 1960s and 1970s. Historically, the Park Service had actively manipulated the national parks to improve the visitor experience while resisting wilderness designations.[28] But the 1960s ushered in a shift in management goals in response to scientific reports, such as the Leopold Report of 1963, that focused renewed attention on managing the parks as primitive landscapes and for their scientific value.[29] These arguments resonated in the 1960s, as the science of ecology gained new authority, concerns about environmental degradation accelerated, and the government took on new responsibilities for environmental protection. As the historian Richard Sellars explains in *Preserving Nature in the National Parks: A History* (1997), the 1960s marked the advent of a renewed and lasting struggle "by scientists and others in the environmental movement to change the direction of national park management, particularly as it affects natural resources."[30] In the following decades, the park service would increasingly manage parks to protect and perpetuate native ecosystem elements and processes—which, to many observers, meant managing them more like wilderness. Thus, this convergence between parks and wilderness management in the United States contributed to a seemingly more coherent parks model worldwide.

One more factor helps explain the potency of the US parks model: the belief that the great American national parks were carved from a pristine and unpeopled wilderness. Such a selective remembrance of the history of land acquisition in the United States erased a history of conquest, enclosure, and alienation, which a new generation of historians brought to light in the 1990s and 2000s.[31] But that history was largely missing from narratives of American national parks and wilderness in the 1960s and 1970s—whether advanced from the podium at the 1962 or 1972 parks conferences or in the scholarship of Nash, Runte, and others. In erasing that history, it became possible to cast the national parks tradition not as a product of imperialism or colonialism, with consequences for indigenous peoples and other local groups—as it may well have appeared in many former European colonies in the 1960s and 1970s—but rather as an emblem of democracy, nationalism, and independence. As Swenson has suggested, "during decolonization and in the postcolonial world, the American model might have become even more attractive in order to rid the national park concept of some of its colonial legacies."[32]

The Second World Conference on National Parks is cited as playing a pivotal role in cementing the United States's role in the history of parks protection worldwide. The 1972 conference, sponsored by the International Union for the Conservation of Nature and the United Nations Environmental Program, was held in conjunction with the centennial of Yellowstone National Park.[33] Delegates to the conference visited the site, at the confluence of the Gibbon and Firehole Rivers, where the national park idea was supposedly broached around a campfire in 1869. And, as Nash recounts in the second edition of *Wilderness and the American Mind* (1973), afterwards "delegate after delegate from around the world rose to credit the United States with inventing the national park." Although some noted the shortcomings of the American park system, he explained, they heralded the "visionaries who anticipated human needs for nature and worked to institutionalize wilderness preservation."[34] It was a point that Runte echoed in his history of the national park system.[35]

By the mid-1970s, an American model of the national park, predicated on wilderness, had seemingly gained global significance. Indeed, for a generation, both champions and critics of protected areas would point to the significance of "America's Best Idea" or the "American wilderness model" for protected areas policy globally, even if they disagreed sharply on its significance and consequences.[36] Yet, even at the Second World Conference on National Parks in 1972, there was evidence to suggest that the history of national parks and protected areas was not so tidy or unidirectional. A cursory glance at the conference proceedings makes clear that for many participants, Yellowstone was a model in name only. As the conference's very first speaker, Belgian scientist Jean-Paul Harroy, noted, many places may be labeled a "national park," but "differ fundamentally" in purpose or management from Yellowstone. To understand the history of protected areas, he suggested, required an approach that considered the "diversities of outlook and biopolitical circumstances" of each nation. Instead of fêting the Yellowstone model, Harroy's address was both transnational and comparative in its consideration of the origins and challenges facing protected areas.[37]

TOWARD A TRANSNATIONAL HISTORY OF PROTECTED AREAS

The significance and originality of the policies put in place for protected areas and parks outside the United States has become an area of sustained and innovative historical research in the past decade.[38] Much of this scholarship questions the reach of the American model. For instance, Australia can rightly claim to have first used the term "national park" to designate Audley National Park in New South Wales in 1879—indeed, it turns out Yellowstone was deemed a "public park," not a "national park," upon its creation in 1872. Canada can take credit for establishing the first agency to administer a system of national parks when it established what is now known as Parks Canada in 1911.[39] Mexico can boast that its forty national parks were more numerous than the United States' thirty parks in 1940.[40] And many European countries pursued national parks and conservation projects domestically and in their colonies with goals and rationales that were distinct from those of the United States. As Ian Tyrell has noted, an earlier generation of scholars of national park and conservation history focused little attention on international efforts toward protected areas. When they did, it was "only to tell a story of American leadership and global dissemination of the American model of a national park idea."[41] Setting aside the export model of parks history not only reveals the variety of related ways that nations have pursued national parks and protected areas, it also suggests that in some instances, the United States' parks policy followed, rather than led, international trends.

Mexico's burst of parks development in the 1930s demonstrates the diversity of approaches to national park protection even in North America. In the 1920s and 1930s, the United States engaged Mexico in a joint effort to create an international park spanning the United States-Mexico border near present-day Big Bend National Park in Texas. That project ultimately failed, however. US officials saw the failure as the result of recalcitrance on the part of their Mexican counterparts. But the historian Emily Wakild makes clear that it was not so much that Mexican park officials were uninterested or unable to commit to an international park; rather, protecting such a remote, scenic, and wild landscape in northern Mexico did not align with Mexico's approach to national park protection. Mexico's national park program grew rapidly in the 1930s, as the Mexican leadership worked to achieve the social goals set forth in its Constitution of 1917, including advancing education, supporting unionization, and redistributing land to rural dwellers. This political context imparted a distinct set of social and geographical priorities to Mexico's national parks movement. In the 1930s, Mexico's leadership saw national parks as democratic, popular spaces available to "rural farmers, scientific bureaucrats, and urban workers."[42] To realize this goal, Mexico focused on protecting lands closer to urban areas, managing the lands for both their protection and resource use, and oftentimes forging compromises that kept landowners on their lands, rather than pushing them off (as was customary in the United States national parks). As Wakild notes, "overlooking the ways in which individuals and their

governments have conceptualized national parks runs the risk of reifying conservation as an easily transferable, universal concept rather than as a contested and pliable
notion that requires social significance and cultural understanding."[43]

National parks also proliferated in Europe during the early twentieth century. Some
countries adopted a parks model similar to that of the United States. Sweden, for example, created its first park in 1909, striking a balance between tourism and nature protection. But other countries, such as Germany, Italy, and Switzerland, pursued approaches
that were distinct from those in both the United States and Mexico. As one American
parks enthusiast noted after an investigative trip to Europe in 1928, the concept of the
national park had a different meaning in many European countries. Often, the creation
of parks in Europe was led by scientists, who prioritized scientific research and education. As Patrick Kupper has demonstrated, Switzerland was particularly influential in
advancing a science-based approach to parks protection. At its conception in 1909, the
vision for a Swiss park was that of the Swiss Natural History Society, which was a fore-
runner to the Swiss Academy of Sciences, and the Swiss League for Nature Protection,
which raised funds to purchase and lease land for the mountainous park in eastern
Switzerland. This park—conceived of by scientists and initially funded through private
donations—was then handed over to the Swiss federal state in 1914. Although the parks
movement appealed to Swiss nationalism (as did the American and Mexican parks
movements in the US and Mexico, respectively), the guiding rationale behind the new
Swiss park was a policy of complete preservation, or *Totalschutz*. That approach made
scientific research the park's main objective and gave priority to restoring previously
grazed, logged, and otherwise degraded lands to their natural state. Specifically, the
park's founders dismissed the goal of recreation—which they believed was overempha-
sized in the United States—and instead called for nonintervention in the park's natural processes: Swiss parks would be outdoor laboratories where nature could regain its
equilibrium. Carl Schröter, a Swiss professor of botany, envisioned a "grandiose experiment to create a wilderness."[44]

Europeans pursued similar projects, often on a grander scale, in their colonies.
Richard Grove argues in his seminal work, *Green Imperialism: Colonial Expansion,
Tropical Island Edens and the Origins of Environmentalism, 1600–1860* (1995), that the
development of modern environmentalism can only be fully understood by considering how it was manifest in the colonial state.[45] The example of the Parc National Albert
in the Belgian Congo suggests the already complex, transnational exchange of ideas
and practices important to protected areas in the interwar years. In 1919, American scientists John Merriam and Fairfield Osborn, the Belgian scientist Victor van Straelen,
and King Albert of Belgium discussed the possibility of "creating vast nature reserves
in Africa" which would be "protected from human influences and used for large-scale,
systematic scientific studies." King Albert, who had visited the American parks, acted
on this vision in 1925, when he created a gorilla sanctuary in present-day Rwanda and
Burundi, which van Straelen oversaw initially.[46] The park was distinguished by sharp
restrictions on access and an emphasis on scientific research, patterned after the Swiss
model. This approach to protected areas gained broader significance in 1933, when

European colonial powers and South Africa signed the 1933 London Convention Relative to the Preservation of Fauna and Flora. That agreement emphasized the importance of "preserving" nature through the creation of "national parks, strict natural reserves, and other reserves within which hunting, killing or capturing of fauna, and the collection or destruction of flora shall be limited or prohibited" throughout Africa.[47] By 1950, European powers had set aside protected areas in the Belgian Congo, South Africa, Malaysia, Sierra Leone, Madagascar, Uganda, Kenya, and Togo, among other colonies and countries. In most cases, the pastoralists, farmers, and others who lived on or relied upon these lands for their subsistence were displaced in the name of colonial conservation and nature protection.

Nature protection even became a focal point of political activism for scientists in the Soviet Union in the 1930s. Between 1928 and 1933, Stalin's revolution from above aimed to consolidate political control over the Soviet Union through sustained efforts toward collectivization, industrialization, and the elimination of civic opposition. It was a brutal period in Soviet history, when police terror permeated daily life, particularly for intellectuals perceived as hostile to the regime. In this context, Douglas Weiner explains in *A Little Corner of Freedom: Russian Nature Protection from Stalin to Gorbachëv* (1999), "nature protection emerged as a means of registering opposition to aspects of industrial and agricultural policy while remaining outwardly apolitical." While scientists gave voice to their concerns regarding Stalin's Five-Year Plan (at considerable personal peril and with little success), scientists found more security and influence in drawing on the language of science to position themselves as the dogged defenders of Russia's system of *zapovednikis* (nature reserves). Much like the scientifically minded Swiss approach to nature protection, *Totalschutz*, the *zapovednikis* were established as pristine, self-regulating, ecological communities. And even as they challenged the Stalinist state by couching their opposition in the apolitical language of science, Russian scientists succeeded both in guarding their autonomy and defending a system of nature reserves. In a repressive regime, Weiner explains, efforts toward nature protection became a small "archipelago of freedom." Although often threatened by development, by 1960 the system had grown to twenty-two reserves protecting some 4.2 million hectares—an expanse the Soviet scientists acknowledged fell short of similar systems in other countries, but which represented a significant victory in the face of the persistent threat of resource development and political persecution. [48]

These examples all suggest that to fully understand how different countries attempted to resolve some of the formative tensions in nature protection—such as balancing tourism and scientific research, local and national interests, democracy and authoritarianism, or preservation, conservation and restoration—it is necessary to set aside the export model of the American national park, and consider nature protection as a product of colonial power, national imperatives, and transnational exchanges. Such an approach not only challenges the singular importance of an export model of the American national park, it can also enrich understandings of American national park history as well. In the 1930s, for instance, the US National Park Service engaged in a halting and, ultimately, unsuccessful turn toward scientific management in the national parks. This effort can be attributed to John Merriam, who was keenly aware of the more scientifically minded focus of protected areas that had gained momentum in Europe and its colonies, in part through his work on Parc National Albert in the Belgian Congo. His efforts contributed to the development of educational programs in the US national parks, a study of the park fauna conducted by George Wright in the early 1930s, and a proposal to the National Park Service for the "establishment of research reserves in the national parks." Notably, when the agency's leadership took up the science reserves proposal, it was with reference to the Swiss model of *Totalschutz*. Despite these efforts, the US Park Service remained committed foremost to managing the parks for public enjoyment, and American scientists remained deeply divided over their role as political advocates.[49] But this debate in the United States over the scientific importance of national parks, and later wilderness areas, can only be fully understood in the context of more successful efforts to advance these goals in Europe and its colonies.

Even if these American scientists did not succeed in reorienting the National Park Service in the 1930s, in working with their counterparts in Europe and elsewhere, they did succeed in forging an international framework for nature protection that would have lasting significance. As Patrick Kupper has argued, the Swiss national park and related efforts in other European nations and colonies formed the "bedrock of early attempts to establish a global network of protected areas." In 1913, the Swiss hosted the first International Conference for the Protection of Nature, in Bern.[50] While the work of that group was cut short by World War I, the Swiss rekindled the project after World War II, hosting meetings in 1946 and 1947. These meetings led to the formation of what became the International Union for the Protection of Nature and Natural Resources in 1948, which has become the leading international organization for the stewardship of parks and protected areas globally. It was renamed the International Union for the Conservation of Nature and Natural Resources (IUCN) in 1956, bringing together government representatives, scientists, and non-governmental organizations and advocates in the shared goal of the "preservation of the entire world biotic community or man's natural environment." Toward that goal, it has promoted the transnational dialogue around protected areas through publications, sponsored exchanges, and a series of world conferences on the National Parks.[51] In its early decades, the IUCN was dominated by scientists, park managers, and government officials, who focused limited attention on the social dimensions of protected areas. In the early 1980s, however, as the creation of parks, wilderness, and other protected areas expanded globally, the IUCN became a focal point for a burgeoning set of concerns and frustrations over the social implications of protected areas, especially in developing countries.

THE CRITICAL TURN IN STUDIES OF THE NATIONAL PARKS AND WILDERNESS

With the publication of an essay titled, "The Trouble with Wilderness; or, Getting Back to the Wrong Nature," in 1995, the historian William Cronon publicized a critical turn in studies of parks and wilderness that had begun in the late 1980s. The article, which focused on the United States, aimed to provoke reflection and debate on the basic assumptions that underpinned the place of wilderness in American environmental thought and advocacy. In Cronon's view, Americans had come to idealize wilderness as pristine, unpeopled, and ecologically valuable—an antipode to modern civilization and its ills. But, as Cronon explained, wilderness was not simply a "pristine sanctuary" or a retreat from modernity; rather, it was "quite profoundly a human creation." In the United States, wilderness was a product of a culture that was beholden to a romantic view of sublime nature and entranced with the myth of the Western frontier. "The removal of Indians to create an 'uninhabited wilderness,'" explains Cronon, "reminds us of just how invented, how constructed, the American wilderness really is."[52] It was a construction that historically had held sway among urban tourists and wealthy sportsmen, and, more recently, among environmentalists and scientists, who did the most to champion parks and wilderness.[53] By idealizing pristine nature, minimizing the social consequences of nature protection, and drawing attention from other environmental issues, Cronon warned, "wilderness poses a serious threat to responsible environmentalism at the end of the twentieth century."[54]

Cronon's essay dumped a bucketful of water on the campfire around which Nash, Runte, and other celebrants of the American tradition of national parks and wilderness had circled for a generation. Since the 1960s, to the extent that scholars focused on wilderness or parks, they had largely celebrated their merits, rather than questioning their implications. That had begun to change in 1989, when Ramachandra Guha published an article arguing that wilderness protection was a uniquely American phenomenon with problematic implications for the developing world.[55] In quick succession, philosophers Baird Callicott and Holmes Rolston III, scientists Reed Noss, Arthuro Gómez-Pompa, and Andrea Kaus, geographer William Denevan, and activist Dave Foreman, among others, set out a series of affirmations and critiques of the wilderness idea.[56] By the mid-1990s, a new generation of American environmental historians, influenced by scholars such as E. P. Thompson and James Scott, began to reconsider the origins of national parks in the United States, giving new attention to the consequences for Native Americans and local communities.[57] When Cronon's essay was published, both as a scholarly article in an edited collection and as an abridged essay in the *New York Times Magazine*, such revisionist arguments began to draw wider attention.[58] The scientific, philosophical, and historical insights that emerged from this debate formed a powerful moment of reckoning for proponents of parks and wilderness, both among academics and environmentalists.

For all of its insights, however, "the great new wilderness debate" was framed fore-most as a reconsideration of the American wilderness ideal and, to a lesser extent, its consequences when exported abroad. Indeed, most of the excerpts and essays in the 1998 edited collection *The Great New Wilderness Debate* focused on North America. But many of the concerns that animated the debate in the United States had already become a staple of policy discussions at the IUCN, the Convention on Biological Diversity, and other international policy arenas and were emerging as central themes in a new generation of scholarship focused on the consequences of conservation policy in other parts of the world. The seeds of this critical turn in debates over protected areas were apparent in 1982 at the Third World Congress on National Parks in Bali, Indonesia. This was the first time the conference had been held in a developing coun-try, and the very first sentence of the Bali Declaration signaled a shift in the politics of protected areas. It read, "WE, the participants, in the World National Parks Congress, BELIEVE that: People are a part of nature." While emphasizing the threat to ecological processes and natural ecosystems, calling for the preservation of species diversity, and recognizing the importance of scientific research, the declaration also emphasized that protected areas had to be managed with attention to "economic, cultural, and political contexts" and must ensure "local support" through measures such as education, col-laborative decision-making, revenue sharing, and access to resources.[59]

Since the mid-1980s, the IUCN had become a hotbed for discussions regarding the social dimensions of protected areas. Although social scientists had been included in international discussions over national parks before then, their contributions most often focused on macro-level issues, such as the threat population growth posed to the establishment and management of protected area policy—not the micro-level issues important to the designation and management of individual parks and their social consequences.[60] Starting in the late 1980s, as social scientists—including geographers, anthropologists, sociologists, and historians—began to consider the assumptions and implications of protected area policy, they began to challenge the methods and goals advanced by the scientific community and international conservation organizations, such as the IUCN. Those efforts opened up a new set of questions that unsettled the assumptions and politics important to protected areas: What were the implications of conceptualizing nature as pristine wilderness? How had local people and communi-ties been affected by protected areas? How did local people rely upon resources in and around protected areas? Was the disregard for local peoples a product of the politi-cal authority claimed by ecologists, conservation biologists, and other scientists? What were the paradoxes of trying to protect and manage wild nature? It was these questions that shaped the critical turn in studies of parks, wilderness, and protected areas, in the United States, the international community, and other countries in the 1990s and 2000s. It is worth considering two of the most important lessons that emerged from these debates: the fallacy of pristine nature and the role of local communities in pro-tected areas.

One of the lessons of this critical turn in studies of protected areas is that even the wildest of landscapes have long histories of human use. As the geographer William

Denevan has explained, the "pristine view" is a cultural "invention" made possible in North America by the demise of Native American populations.[61] But what appears wild cannot be fully understood without considering activities such as burning, grazing, hunting, and early industrial activities, which have shaped the ecological compositions of the landscape for centuries.[62] The historian Mark Spence argues in *Dispossessing the Wilderness: Indian Removal and the Making of the National Parks* (1999), that "uninhabited wilderness had to be created before it could be preserved." Well-known parks such as Yellowstone, Yosemite, and Glacier were established while still inhabited by Indians who relied on these lands for their homes, their subsistence, and their culture. Yet, in each park, Indians were later forced out by the US government. "Generations of preservationists, government officials, and park visitors have accepted and defended the uninhabited wilderness preserved in national parks as remnants of a prior Nature," Spence explains. But to do so "forgets that native peoples have shaped these environments for millennia."[63]

The myth of pristine nature has been less pervasive in Europe. This was true in the case of the Swiss National Park, where scientists emphasized the importance of restoration. The historian Marcus Hall also makes this point in *Earth Repair: A Transatlantic History of Environmental Restoration* (2005), a comparative study of the culture of land restoration in the United States and Italy. As he explains, the Italians did not view the Alps as an unpeopled wilderness at risk of human destruction. Instead, Italians saw the mountain landscape as a product of layers of history and culture; they recognized that agriculture, grazing, and logging had helped maintain the stability of the mountain landscape. While "Americans see nature as the keeper of their Rockies," Hall argues, the "Italians see culture as the keeper of their Alps."[64] But Europeans were prone to fall prey to the pristine myth in their colonies. In the case of Guinea, for example, as James Fairhead and Melissa Leach explain in *Misreading the African Landscape: Society and Ecology in a Forest-Savanna Mosaic* (1996), colonial French conservation policies were founded on the assumption that the region's islands of forest surrounded by savannah represented "the legacy of a natural 'climax' vegetation remaining within an otherwise abused landscape." In fact, villagers, through seasonal agricultural practices and day-to-day disposal of waste and water, had created the conditions from which those forest patches sprung.[65] Seemingly wild forest patches were artifacts of local human habitation.

Such misconceptions played into domestic preservation initiatives in former colonial states too. In India, when Prime Minister Indira Gandhi announced Project Tiger, an extensive program of protected areas aimed at protecting India's iconic predator in the 1970s, she too appealed to the ideal of pristine nature. She urged Indians to recognize the "ecological value of totally undisturbed areas of wilderness."[66] In the 1960s, South African environmentalists—in this case, led by white women who were members of the National Council of Women of South Africa—protested the clearing of Dukuduku forest. They described the forest as a "true climax forest" and a rare example of undisturbed coastal forest left in the province. Yet, Dukuduku was hardly such as a pristine landscape. As Frode Sundnes explains, although the forest was little populated, it "had an important place in the rural economy and land use of Zululand in the nineteenth and early twentieth centuries."[67] Archeological and written evidence suggests that the forest was used for refuge, settlement, cultivation, and extensive grazing. But, between the 1930s and 1970s, forestry, conservation, and military policies forced locals from their ancestral homes. As a result of the displacement and subsequent forestry policies, the forest's ecological composition changed—prompting calls for its preservation.

The strategy of moving people from protected areas in the name of nature preservation is what Dan Brockington has described as "fortress conservation."[68] The assumption has often been that nature will best maintain values, such as stability or wildlife conservation, if insulated from people. But historians have demonstrated that this approach often resulted in local resistance that undermined goals for protected areas. This is the second key lesson of the critical turn in studies of protected areas. For example, at Mkomazi Game Reserve in northeast Tanzania, where thousands of villagers and their cattle were evicted to protect the park's wildlife in the 1980s, villagers continued to rely on park resources: they grazed cattle on park lands and hunted park wildlife.[69] The story is similar at Arusha National Park in Tanzania and Amboseli National Park in Kenya, where hunting of large mammals were persistent concerns. And at Bandhavgarh Tiger Reserve in India, villagers scheduled for relocation as part of reserve management, "simply lost the incentive to use the forest sustainably."[70] To park managers and scientists, such continued use of park resources for grazing, fuelwood gathering, and subsistence represented violations of park policies and threats to management goals. But, as Roderick Neumann explains in *Imposing Wilderness: Struggles over Livelihood and Nature Preservation in Africa* (1998), such illegal activities should instead be understood as representative of a determined, local peasant resistance. In his study of protected areas in Tanzania, he argues, "The 'crime' of poaching is not a crime at all, but a defense of subsistence, and the 'real crime' is that park animals are allowed to raid crops with impunity."[71] Although it is little recognized, such resistance was important even in Yellowstone in the early 1910s; Karl Jacoby's *Crimes Against Nature: Squatters, Poachers, Thieves, and the Hidden History of American Conservation* (2001) details how rural whites resisted federal regulations that banned hunting and subsistence practices in the park. In this respect, Yellowstone prefigured later conflicts over protected areas, but in ways that the usual celebratory narratives rarely acknowledged.[72]

That local resistance could undermine protected area goals would not have been a surprise to park officials in the 1960s and 1970s; the threats posed by pastoralists and poachers were active topics of debate at international park conferences. What would have surprised park officials were the instances in which the removal of local people undermined, rather than bolstered, ecological management goals—not because of local resistance, but because local subsistence activities contributed to ecological stability and wildlife diversity. A basic tenet of ecology during the twentieth century was the value of protecting habitat and guarding it from human disturbance: this was true in the 1920s and 1930s among wildlife scientists, ecologists in the 1950s and 1960s, and conservation biologists in the 1980s and 1990s.[73] In the case of Bharatpur in north-central India, which was prized for its migratory waterfowl, cattle and their owners were ousted in the early 1980s. In the decade after the ban was enacted, bird diversity dropped, rather than recovered. Without sustained grazing, weeds had taken over much of the park, reducing both food and habitat for migratory waterfowl. The historian Michael Lewis explains, "in the absence of intense use, a few 'weedy' plants were destroying the park as an open wetland habitat so well-suited for birds."[74] That example demonstrates a point the biologist Arturo Goméz-Pompa and anthropologist Andrea Kaus have made in the context of Latin America. In numerous instances, local peoples "have managed, conserved, and even created some of the biodiversity we value so highly." In some cases, achieving conservation goals, such as maintaining habitat or protecting species, depends on the presence of "human cultural traditions," not their absence.[75]

Without diminishing the injustices that have accompanied the establishment of protected areas, however, it is worth noting that many of those historical studies pinned the blame for such ill social consequences on a US model of national parks predicated on wilderness, both domestically and as an export to other countries. In the case of the United States, Carolyn Merchant has suggested that the Wilderness Act "reads Native Americans out of the wilderness and out of the homelands they had managed for centuries with fire, gathering, and hunting."[76] Mark Dowie, a well-regarded investigative journalist, did much to popularize such claims. Surveying protected areas globally, he argues that protected areas have resulted in the displacement of millions of "conservation refugees" worldwide. Indeed, he claims, the scale of such displacements is of the same magnitude as displacements attributed to more familiar activities, such as civil wars, famines, or large-scale development projects. But Dowie argues, much like Merchant, that in the case of conservation refugees, the root cause was an American "preference for 'virgin' wilderness." That viewpoint, he suggests, "value[s] all nature but human nature, and refused to recognize the positive wildness in human beings."[77] Further research will likely complicate such broad generalizations. With protected areas numbering in the tens of thousands, there is much work to be done to consider the scale, trend, and different manifestations of displacement. Some scholars, for instance, have suggested that such evictions have been less common in Latin America than elsewhere in the world.[78] It is also worth considering how such claims might change in the context of a more comparative, transnational approach to protected areas that gives more attention to historical context and more agency to local peoples.

For example, in most cases, there are numerous factors—apart from "wilderness"—that have contributed to or legitimated the displacement of local peoples from what are now protected areas. This is the lesson of Ted Binnema and Melanie Niemi's study of the aboriginal Stoney and the creation of Banff National Park in Alberta. As they explain, the Stoney were pushed out of the park in the late nineteenth and early twentieth centuries by sportsmen and conservationists, who feared that Stoney hunting activities threatened the long-term yield of big game in and around the park. In Binnema and Niemi's assessment, "aboriginal people were excluded from national parks in the interest of game (not wildlife) conservation, sport hunting, tourism, and aboriginal civilization, not to ensure that national parks became uninhabited wilderness."[79] The story is similar in the case of the Dukuduku forest in South Africa. As Sundnes argues, it was not a "preservationist vision" that explains the origins of the forest reserves, restrictions on local access, and displacement of native peoples. Rather, it was the British colonial government's practical and economic concerns for managing the forest and game for "sustainable yields," which was modeled on the Indian forestry tradition.[80] And, in the case of the United States in the 1970s and 1980s, debates over protecting both native interests and new protected areas in Alaska demonstrate how Native Americans actively deployed the concept of "wilderness" to ensure their continued access to homelands and subsistence resources they depended upon, particularly in the face of outside resource development.[81] Such examples offer two useful reminders. First, in many instances, displacement was driven by multiple factors, which may be related to, but cannot simply be attributed to, a wilderness ideal. Second, indigenous and local peoples have, at least in some cases, played an active role in defending their interests and shaping the creation of protected areas policies.

The lasting contribution of the critical turn in studies of protected areas in the 1990s was to draw attention to significant and understudied social consequences of protected areas. Considered against a transnational backdrop, it becomes clear that the great new wilderness debate centered in the United States in the 1990s formed a part of a broader, critical turn in the politics and discussions of protected areas internationally that began in the 1980s. This critical turn both contributed to and drew upon new discussions surrounding the creation of and goals for a broader array of protected areas that aimed to mediate between social and conservation goals in the 1990s and 2000s.

FROM PARKS AND WILDERNESS TO "NEW PARADIGM" PROTECTED AREAS

One of the most immediate consequences of the critical turn in studies of protected areas was to open up a breach between social scientists and conservation-minded scientists and activists. For example, Ramachandra Guha, who helped jumpstart the social critique of parks and wilderness in 1989, narrowed his critique to focus on conservation biologists in 2006. He described conservation biologists, who championed the protection of vast tracts of unpeopled wilderness, as scientific crusaders who had laid claim to "the same territory—uncultivated parts of the globe that are covered with what one group of scientists defines as 'forest,' the other as 'wild.'"[82] In advancing conservation in the name of biodiversity protection, Guha warned, the "biologist" had become a new force of imperialism disenfranchising local and indigenous peoples. Guha was not alone in highlighting the role of scientists, often working in concert with both governmental and non-governmental groups, in advancing conservation policies that ignored local environmental knowledge, alienated local peoples from lands and resources they had depended upon, and commandeered local lands for a global goal of protecting biodiversity.[83]

Some within the conservation biology community were quick to defend the essential role of strictly managed protected areas in protecting biodiversity. John Terborgh, a leading tropical ecologist and conservation biologist, published *Requiem for Nature* (1999), which argued forcefully against efforts to weaken protected areas to advance sustainable development.[84] John Oates, in *Myth and Reality in the Rain Forest: How Conservation Strategies are Failing in West Africa* (1999), warned conservationists to reject the myth that local, traditional, or indigenous peoples are conservationists. "On the contrary," he argues, "wherever people have had the tools, techniques, and opportunities to exploit natural systems they have done so."[85] And in 2006 two scientists, Harvey Locke and Philip Dearden, challenged the growing role of social scientists at the Fifth World Parks Congress. They reasserted the driving concerns of conservation biology—"Preservation of all components of biodiversity can only be attained if some areas are kept largely free of human alteration" and "[s]trictly protected areas where nature rules are needed"—as the fundamental goal of protected areas.[86] The advocacy of Terborgh, Oates, Locke, and Dearden was consistent with what some scholars had described as the resurgence of an authoritarian approach to conservation.[87]

But focusing on the stand-off between social scientists and conservation biologists draws attention away from the most dynamic arena of protected areas policy since the 1990s: efforts to designate areas and establish policies that can mediate between the social consequences and conservation goals of protected areas. Such attempts to foster more socially concerned and participatory approaches to protected areas policy have been a sustained topic of discussion at international conservation policy meetings since Bali in 1982. And as conservation scientists Ashish Kothari, Philip Camill, and Jessica Brown explain, "community-based conservation is now a central part of the prescriptions emanating from global institutions or forums such as the Convention on Biological Diversity and the IUCN."[88] These efforts have been advanced by the activism of indigenous and local peoples, the concerns of conservation practitioners and scholars, and the heightened attention to the rights of indigenous peoples internationally. In addition to efforts within conservation organizations, the United Nations adopted the United Nations Declaration on the Rights of Indigenous Peoples in 2007, which strongly affirmed the rights of indigenous peoples to their lands given their spiritual, subsistence, and conservation values. Some conservation officials are already envisioning a world where "displacement-based conservation will be consigned to the dustbin of history."[89] A wide array of protected areas has been developed over the past several decades that aim to advance such goals, including communal reserves in Peru, community conservation areas in Nepal, communal conservancies in Namibia, and indigenous protected areas in Australia.

The creation, management, and consequences of these "new paradigm" protected areas, created since the early 1980s, need to be studied further. Consider the Annapurna Conservation Area, set aside in Nepal in 1986, which is a magnet for trekkers to the Himalaya. The National Trust for Nature Conservation, which oversees the area in

Nepal, describes three goals for the area: conserve natural resources, foster sustainable social and economic development, and develop appropriate tourism. Notably, the trust describes Annapurna as a protected area rich in both biological and cultural diversity, citing not only the species it protects, but celebrating its 100,000 residents representing five different cultural and linguistic groups. Revenue from trekkers supports management of the area.[90] Or, consider the community conservancies in post-apartheid Namibia. The state government gave ownership of wildlife—including black rhinoceroses, elephants, and lions—to local community groups in the 1990s. Livestock farmers manage working landscapes that include cattle and predators. The success of these programs means Namibia's community conservancies (they now number sixty four) have expanded to exceed its national parks in size, creating new habitat for endangered species, lessening poaching, and rewarding locals with jobs and revenue from tourism (particularly fees from trophy hunting).[91]

While still new, these approaches have been evolving since the 1980s, and their origins, successes, and failures are ripe for historical study. But doing so means asking new questions. What historical antecedents to such approaches have been overshadowed by the historical focus on parks and wilderness? How have these new protected areas been shaped by local, national, and international contexts and priorities? What is the nexus between the priorities and policies developed at the international level and the practices in particular communities or regions? How effective are these protected areas in comparison to more traditional parks and wilderness areas? How are local peoples engaged in the creation and management of these areas: are they stake holders, rights holders, or property owners? Which models of participatory engagement have been most successful? It may seem that posing these questions represents a geographical shift toward the Global South and away from places like the United States in the study of protected areas. Although that may be true, this shift also offers an opportunity to reconsider the scope of protected areas studies in the United States as well. Just as scholarship on colonial conservation contributed to the critical turn in wilderness debates in the United States in the 1990s, US environmental historians have an opportunity to draw on these "new paradigm" approaches to better understand protected areas in the United States too. For example, collaborative conservation programs at the local level have gained momentum in places from Maine to Oregon in recent decades, but such programs remain understudied. As historians engage such approaches, examples such as the Namibian community conservancies may be even more illuminating than the scholarship on national parks, such as Yellowstone or Yosemite.

CONCLUSION: TRANSCENDING BOUNDARIES

Some scholars have argued that environmental historians have invested too much effort in studies of wilderness, national parks, and protected areas. In the case of wilderness, David Stradling has suggested, "what is truly ironic is that environmental historians have worn so many paths through the wilderness while there is so much work yet to be done in places where humans are not just visitors."[92] What such assessments miss is that wilderness, national parks, and other protected areas can only be understood in relation to the larger political, social, and ecological contexts that give them meaning. Examining the history of protected areas is not a flight from such realities or the issues that concern people. To take a transnational approach to protected areas, as I have suggested in this chapter, means grappling with issues such as the legacies of colonialism and postcolonialism, the evolution of international governance, the politics of science, the interface of international aid and conservation, contests over human rights, and the enfranchisement of indigenous peoples.

If the critical turn in studies of protected areas in the 1990s highlighted the social consequences of parks and wilderness, there is now an urgent need for historical studies that can contribute to our understanding of efforts to advance more inclusive, socially just, and ecologically resilient approaches to creating and managing a wide range of protected areas. As environmental historians turn to this challenge, there is one lesson they can draw on from conservation science: protected areas are not islands. Often, studies of protected areas have been hemmed in by boundaries—scholars have told stories about how protected areas come to be, rather than considering how protected areas have functioned.[93] But reconsidering how protected areas are products of and constituent parts of broader landscapes—ecological and social, local and transnational—and how those relationships affect the long-term success or failure of such protected areas can yield useful insights. Two recent studies, less concerned with the social consequences of protected areas, suggest how analyses that deliberately transcend the boundaries of protected areas can advance understandings of how protected areas function in a broader context.

This is one of the core contributions of *After the Grizzly: Endangered Species and the Politics of Place in California* (2013), Peter Alagona's study of protected areas and species conservation in California. By examining endangered species debates through the politics of place in California, Alagona makes clear, first, how contentious creating such protected areas can be and, second, that setting aside protected areas alone does little to guarantee the long-term viability of endangered species. In the case of the California condor, saving wilderness in the 1950s and 1960s failed to slow the bird's decline, for the threat was not simply habitat loss; rather, it was lead, cyanide, and DDT poisoning, which the animals suffered when feeding on carrion on surrounding lands. A controversial and intensive captive breeding program began in the 1980s and resulted in the capture of all surviving wild condors; by 2010, the number of animals returned to the wild was ten times greater than those captured. Or, consider the case of the kit fox, a species thought to be native to rural areas of California's Central Valley. The animal surprised scientists by thriving in a most unlikely place: the streets, parks, and backyards of Bakersfield, California. Although kit foxes may need better habitat, Alagona explains, scientists and conservationists also need a better understanding of "what constitutes habitat for species such as this and how different kinds of habitat areas—from wilderness to rangelands, farms, oil fields, and even cities—can contribute to the goal of recovering endangered species and integrating them into the broader cultural landscape."[94]

Robert Wilson's study of the Pacific Coast flyway, *Seeking Refuge: Birds and Landscapes of the Pacific Flyway* (2010), makes a compelling case for analyzing protected areas as hybrid landscapes, as much produced as preserved. Historically, wildlife refuges in the United States have been intensively managed to provide forage for migrating birds: in the 1930s and 1940s, for example, that meant planting, irrigating, and applying insecticides and pesticides (including DDT). The goal was to attract birds to refuges and away from agricultural fields, where they destroyed crops and frustrated farmers. What makes Wilson's study particularly illuminating is that instead of being organized around protected areas—their creation, management, and problems—the analysis follows the birds. In doing so, Wilson demonstrates how the analysis of protected areas can and must be woven into other landscapes, both physical and social, at local, regional, and transnational scales. Local debates over the consequences of waterfowl for agricultural lands drove the turn toward intensive management of the refuges. To undertake that management, refuge managers adopted agricultural techniques and technology from neighboring commercial farms. The Great Depression created an opportunity for both land acquisition and restoration, supported by a national constituency of sport hunters. But even as federal agencies coordinated such acquisitions, their efforts were informed by new research studies that revealed migratory patterns along long-distance flyways that reached from Asia to South America. Water for the new refuges hinged on competition with growers and cities. And, ultimately, it was the wastewater from an agricultural regime with an international market—California's Central Valley—that sustained the refuges and the migrations that depended upon them.[95]

Although the boundaries of protected areas may appear clearly marked on a map, as Wilson argues, "in reality, these were messy boundaries regularly transgressed by not only birds, but water, pesticides, weeds, insects, and other aspects of the nonhuman world."[96] Both of these studies demonstrate how following processes across boundaries can yield new insights into the consequences of protected areas, both good and bad. This approach offers an important model for historical studies of "new paradigm" protected areas as well. It is not just nonhuman nature that has been navigating the boundaries between protected areas and surrounding landscapes. As integrative approaches to protected areas have taken hold, humans have been crossing these boundaries more frequently, too. Historians need to consider how these different groups of people—indigenous peoples, local residents, land managers, international conservation officials, scientists, and tourists—have contributed to new forms of protected areas that reflect their varying commitments to cultural autonomy, biodiversity protection, economic growth, and respect for human rights. While both Alagona and Wilson spell out dire challenges for habitat and species protection—particularly in a world beset by climate change—their analyses also reveal unanticipated successes of wildlife thriving in intensively managed landscapes. We need more stories about how communities have both struggled and thrived in intensively conserved landscapes too. That means historical analyses not just of how protected areas came to be, but analyses that consider the extent to which protected areas have contributed to and drawn upon more sustainable landscapes that are both ecologically productive and socially just.

ACKNOWLEDGEMENTS

This essay has benefited from the exceptionally helpful comments and suggestions offered by Drew Isenberg, Matt Klingle, Michael Lewis, Tom Robertson, Paul Sutter, and an outside reader. I would also like to thank Leah Nugent for editorial assistance.

NOTES

1. On the environmental history of the twentieth century, see J. R. McNeill, *Something New Under the Sun: An Environmental History of the Twentieth-Century World* (New York: Cambridge University Press, 2000). On the expansion of protected areas (and the limitations of this data), see International Union for the Conservation of Nature and Natural Resources, *Benefits Beyond Boundaries: Proceedings of the Vth IUCN World Parks Congress* (Cambridge: International Union for the Conservation of Nature and Natural Resources, 2005), 289.

2. Roderick Nash, "The American Invention of National Parks" *American Quarterly* 22, no. 3 (1970): 726–735.

3. Donald Worster, "Nature, Liberty, and Equality," in *American Wilderness: A New History*, ed. Michael Lewis (New York: Oxford University Press, 2007), 263.

4. Ramachandra Guha, "Radical American Environmentalism and Wilderness Preservation: A Third World Critique," in *The Great New Wilderness Debate*, ed. J. Baird Callicott and Michael P. Nelson (Athens: University of Georgia Press, 1998), 235.

5. William Cronon, "The Trouble with Wilderness; or, Getting Back to the Wrong Nature," in *Uncommon Ground: Rethinking the Human Place in Nature,* ed. William Cronon (New York: W. W. Norton, 1996), 69–90.

6. Christopher Conte, "Creating Wild Places from Domesticated Landscapes," in *American Wilderness,* 226.

7. Calculations based on the International Union for Conservation of Nature, World Database of Protected Areas, 2009.

8. Nash, "The American Invention of National Parks."

9. For a more careful analysis of whether Stegner coined this phrase, see Alan MacEachern "Who Had 'America's Best Idea'?" NiCHE - Network in Canadian History & Environment, October 23, 2011. http://niche-canada.org/2011/10/23/who-had-americas-best-idea/.

10. Ken Burns, *The National Parks: America's Best Idea* (United States: PBS Distribution, 2009). For an insightful review of the documentary, see Paul S. Sutter, "Review: *The National Parks,*" *The Journal of American History* 97, no. 3 (December 2010): 892–897.

11. Roderick Nash, *Wilderness and the American Mind*, 4th ed. (New Haven, CT: Yale University Press, 2001).

12. Alfred Runte, *National Parks: The American Experience*, 3d ed. (Lincoln: University of Nebraska Press, 1997).

13. Paul S. Sutter, "The Trouble with 'America's National Parks'; or, Going Back to the Wrong Historiography: A Response to Ian Tyrrell," *Journal of American Studies* 46, no. 1 (March 2012): 29.

14. "An Act to Establish a National Park Service (1916)," in *America's National Park System: The Critical Documents*, ed. Lary M. Dilsaver (New York: Rowman & Littlefield, 1994), 46–47. On the history of the national parks and conservation issues more generally, see Robert B. Keiter, *To Conserve Unimpaired: The Evolution of the National Park Idea* (Washington, DC: Island Press, May 2013); Thomas R. Wellock, *Preserving the Nation: The Conservation and Environmental Movements* (Wheeling, IL: Harlan Davidson, 2007).

15. On the management of parks for public use and tourism, see Ethan Carr, *Wilderness by Design: Landscape Architecture and the National Park Service* (Lincoln: University of Nebraska Press, 1998); Stanford E. Demars, *The Tourist in Yosemite, 1855–1985* (Salt Lake City; University of Utah Press, 1991); Marguerite S. Shaffer, *See America First: Tourism and National Identity, 1880–1940* (Washington, DC: Smithsonian Institution Press 2001).

16. Runte, *National Parks: The American Experience*; Richard West Sellars, *Preserving Nature in the National Parks: A History* (New Haven, CT: Yale University Press, 1997).

17. Neil M. Maher, *Nature's New Deal: The Civilian Conservation Corps and the Roots of the American Environmental Movement* (New York: Oxford University Press, 2007); Sarah T. Phillips, *This Land, This Nation: Conservation, Rural America, and the New Deal* (New York: Cambridge University Press, 2007).

18. Paul Sutter, *Driven Wild: How the Fight against Automobiles Launched the Modern Wilderness Movement* (Seattle: University of Washington Press, 2002).

19. On the campaign for the Wilderness Act, see Mark Harvey, *Wilderness Forever: Howard Zahniser and the Path to the Wilderness Act* (Seattle: University of Washington Press, 2005).

20. On the National Park Service and wilderness, see John C. Miles, *Wilderness in National Parks: Playground or Preserve* (Seattle: University of Washington Press, 2009); James W. Feldman, *A Storied Wilderness: Rewilding the Apostle Islands* (Seattle: University of Washington Press, 2011).

21. Specifically on the expansion of the wilderness system, see James Morton Turner, *The Promise of Wilderness: American Environmental Politics since 1964* (Seattle: University of Washington Press, 2012). Other important books on the history of wilderness include Craig W. Allin, *The Politics of Wilderness Preservation* (Fairbanks: University of Alaska Press, 2008); Michael Frome, *Battle for the Wilderness* (Salt Lake City; University of Utah Press, October 1997); Michael Lewis, ed., *American Wilderness: A New History* (New York: Oxford University Press, 2007); Kevin R. Marsh, *Drawing Lines in the Forest: Creating Wilderness Areas in the Pacific Northwest* (Seattle: University of Washington Press, 2007); Nash, *Wilderness and the American Mind;* Doug Scott, *The Enduring Wilderness: Protecting our National Heritage through the Wilderness Act* (Golden, CO: Fulcrum Publishing, 2004). The best source of data on wilderness and protected areas on US federal lands is www.wilderness.net.

22. Peter Alagona makes this point effectively in "Homes on the Range: Cooperative Conservation and Environmental Change on California's Privately Owned Hardwood Rangelands," *Environmental History* 13, no. 2 (April 2008): 325–349. Also see Sally K. Fairfax, ed., *Buying Nature: The Limits of Land Acquisition as a Conservation Strategy, 1780–2004* (Cambridge, MA: MIT Press, 2005); Julie Ann Gustanski, Roderick Squires,

and Jean Hocker, *Protecting the Land: Conservation Easements Past, Present, and Future* (Washington, DC: Island Press, March 2000).

23. Astrid Swenson, "Response to Ian Tyrrell, 'America's National Parks,'" *Journal of American Studies* 46, no. 1 (March 2012): 42.

24. On the expansion of protected areas, see IUCN, *Benefits Beyond Boundaries*, 289.

25. Terence Young and Lary M. Dilsaver, "Collecting and Diffusing 'the World's Best Thought': International Cooperation by the National Park Service," *The George Wright Forum* 28, no. 3 (2011): 274.

26. Sterling Evans gives some attention to the role of USAID in funding activities in Costa Rica. Sterling Evans, *The Green Republic: A Conservation History of Costa Rica* (Austin: University of Texas Press, 1999).

27. On this point, also see Sutter, "The Trouble with 'America's National Parks'; or, Going Back to the Wrong Historiography," 28.

28. For examples of how the agency's management activities changed, see Stanford E. Demars, *The Tourist in Yosemite, 1855–1985*; James A. Pritchard, *Preserving Yellowstone's Natural Conditions: Science and the Perceptions of Nature* (Lincoln: University of Nebraska Press, 1999).

29. A. Starker Leopold et al., *Wildlife Management in the National Parks*. Prepared by the Advisory Board on Wildlife Management appointed by Secretary of the Interior Udall. March 4, 1963.

30. Sellars, *Preserving Nature in the National Parks*, 217. See also David M. Graber, "Resolute Biocentrism: The Dilemma of Wilderness in National Parks," in *Reinventing Nature?: Responses to Postmodern Deconstructionism*, ed. Michael E. Soulé and Gary Lease (Washington, DC: Island Press, 1995), 124. On the paradoxes of managing wildlife in national parks, see Etienne Benson, *Wired Wilderness: Technologies of Tracking and the Making of Modern Wildlife* (Baltimore, MD: The Johns Hopkins University Press, 2010).

31. Karl Jacoby, *Crimes against Nature: Squatters, Poachers, Thieves, and the Hidden History of American Conservation* (Berkeley: University of California Press, 2001); Rebecca Solnit, *Savage Dreams: A Journey into the Hidden Wars of the American West* (Berkeley: University of California Press 1994); Mark David Spence, *Dispossessing the Wilderness: Indian Removal and the Making of the National Parks* (New York: Oxford University Press, 1999); Louis S. Warren, *The Hunter's Game: Poachers and Conservationists in Twentieth-Century America* (New Haven, CT: Yale University Press, 1997).

32. Swenson, "Response to Ian Tyrrell," 43.

33. Hugh Elliott, ed., *Second World Conference on National Parks* (Morges, Switzerland: International Union for Conservation of Nature and Natural Resources, 1974).

34. Nash, *Wilderness and the American Mind*, 376.

35. Runte, *National Parks*, 166.

36. Nash, "The American Invention of National Parks"; Conte, "Creating Wild Places from Domesticated Landscapes."

37. Jean-Paul Harroy, "A Century in the Growth of the 'National Park' Concept Throughout the World," in *Second World Conference on National Parks*, 24–32.

38. Bernhard Gissibl, Sabine Höhler, and Patrick Kupper, *Civilizing Nature: National Parks in Global Historical Perspective* (New York: Berghahn Books, 2012).

39. Ian Tyrrell, "America's National Parks: the Transnational Creation of National Space in the Progressive Era," *Journal of American Studies* 46, no. 1 (March 2012): 1–21.

40. Emily Wakild, "Border Chasm: International Boundary Parks and Mexican Conservation, 1935–1945," *Environmental History* 14, no. 3 (July 2009): 455.

41. Tyrrell, "America's National Parks."

42. Wakild, "Border Chasm," 457.

43. Wakild, "Border Chasm," 455. For a fuller development of these arguments, see Emily Wakild, *Revolutionary Parks: Conservation, Social Justice, and Mexico's National Parks, 1910–1940* (Tucson: University of Arizona Press, 2011).

44. Patrick Kupper, "Science and the National Parks: A Transatlantic Perspective on the Interwar Years," *Environmental History* 14, no. 1 (January 2009): 58–81.

45. Richard H. Grove, *Green Imperialism: Colonial Expansion, Tropical Island Edens and the Origins of Environmentalism, 1600–1860* (New York: Cambridge University Press, 1995).

46. "Parks around the World," in *First World Conference on National Parks,* ed. Alexander B. Adams (Washington, DC: National Park Service, 1962), 408–409. See also Kupper, "Science and National Parks," 70–71.

47. *Convention Relative to the Preservation and Fauna and Flora in Their Natural State* (1933).

48. Douglas R. Weiner, *A Little Corner of Freedom: Russian Nature Protection from Stalin to Gorbachëv* (Berkeley: University of California Press, 1999).

49. On the debates over the role of scientists as advocates in the 1930s, see Sara Tjossem, "Preservation of Nature and Academic Respectability: Tensions in the Ecological Society of America, 1915–1979" (PhD diss., Cornell University, 1994).

50. Kupper, "Science and the National Parks: A Transatlantic Perspective on the Interwar Years," 64.

51. On these international efforts, see Timothy Farnham, *Saving Nature's Legacy: Origins of the Idea of Biological Diversity* (New Haven, CT: Yale University Press, 2007); Michael Lewis, *Inventing Global Ecology: Tracking the Biodiversity Ideal in India, 1947–1997* (Athens: Ohio University Press, 2004); Mark Cioc, *The Game of Conservation: International Treaties to Protect the World's Migratory Animals* (Athens: Ohio University Press, 2009).

52. Cronon, "The Trouble with Wilderness; or, Getting Back to the Wrong Nature," 69, 79.

53. On the role of rural Americans in the development of a conservation ethos, see Richard William Judd, *Common Lands, Common People: The Origins of Conservation in Northern New England* (Cambridge, MA: Harvard University Press, 1997).

54. Cronon, "The Trouble with Wilderness; or, Getting Back to the Wrong Nature," 81. See also a roundtable of responses to Cronon's essay in *Environmental History* 1, no. 1 (January 1996): 7–55.

55. Guha, "Radical American Environmentalism and Wilderness Preservation: A Third World Critique."

56. These responses and more are included in the collection, *The Great New Wilderness Debate* (1998).

57. On the cross-pollination between U.S. and non-U.S. environmental historiography, see Paul Sutter, "What Can U.S. Environmental Historians Learn from non-U.S. Environmental Historiography?" *Environmental History* 8, no. 1 (2003): 109–129.

58. Cronon, "The Trouble with Wilderness," *New York Times Magazine* (August 13, 1995), 42–43.

59. "Bali Declaration," in *National Parks, Conservation, and Development: The Role of Protected Areas in Sustaining Society*, ed. Jeffrey A. McNeely and Kenton R. Miller (Washington, DC: Smithsonian Institution Press, 1984), xi.

60. For instance, see Joseph L. Fisher, "Population and Economic Pressures on National Parks," in *Second World Conference on National Parks* (1972), 102–114.

61. William M. Denevan, "The Pristine Myth: The Landscape of the Americas in 1492," in *The Great New Wilderness Debate*, 414–442.

62. William Cronon, *Changes in the Land: Indians, Colonists, and the Ecology of New England* (New York: Hill and Wang, 1983).

63. Spence, *Dispossessing the Wilderness*, 5.

64. Marcus Hall, *Earth Repair: A Transatlantic History of Environmental Restoration* (Charlottesville: University of Virginia Press, 2005), 233. On French views of wilderness and nature, see Michael Bess, *The Light-Green Society: Ecology and Technological Modernity in France, 1960–2000* (Chicago: University of Chicago Press, 2003), 256–272.

65. James Fairhead and Melissa Leach, *Misreading the African Landscape: Society and Ecology in a Forest-Savanna Mosaic* (New York: Cambridge University Press, 1996), 6. James McCann, who cites the importance of this study, also warns against generalizing from it to explain African forest history. See McCann, *Green Land, Brown Land, Black Land: An Environmental History of Africa, 1800–1990* (Portsmouth, NH: Heinemann, 1999), 60–74.

66. As quoted in H. S. Panwar, "What to Do When You've Succeeded: Project Tiger, Ten Years Later," in *National Parks, Conservation, and Development* (1984), 186.

67. Frode Sundnes, "Scrubs and Squatters: The Coming of the Dukuduku Forest, an Indigenous Forest in KwaZulu-Natal, South Africa," *Environmental History* 18, no. 2 (April 2013), 277–308.

68. Dan Brockington, *Fortress Conservation: The Preservation of the Mkomazi Game Reserve, Tanzania* (Bloomington: Indiana University Press, 2002).

69. Brockington, *Fortress Conservation*.

70. Tiger Task Force, "The Report of the Tiger Task Force: Joining the Dots," (New Delhi, India: Union Ministry of Environment and Forests, 2005), 100.

71. Roderick P. Neumann, *Imposing Wilderness: Struggles over Livelihood and Nature Preservation in Africa* (Berkeley: University of California Press, 1998), 187. This line of analysis also draws on James Scott, *Weapons of the Weak: Everyday Forms of Peasant Resistance* (New Haven, CT: Yale University Press, 1985) and E. P. Thompson, *Whigs and Hunters: The Origins of the Black Act* (New York: Pantheon Books, 1975).

72. Karl Jacoby, *Crimes against Nature*.

73. Peter Alagona, *After the Grizzly: Endangered Species and the Politics of Place in California* (Berkeley: University of California Press, 2013), 6.

74. Lewis, *Inventing Global Ecology*, 212.

75. Goméz-Pompa and Kaus, "Taming the Wilderness Myth" in *The Great New Wilderness Debate*, 299.

76. Carolyn Merchant, "Shades of Darkness: Race and Environmental History," *Environmental History* 8, no. 3 (July 2003): 380–394.

77. Mark Dowie, "Conservation Refugees: When Protecting Nature Means Kicking People Out," *Orion* (November-December 2005); Mark Dowie, *Conservation Refugees: The Hundred-Year Conflict between Global Conservation and Native Peoples* (Cambridge, MA: MIT Press, 2009).

78. Avecita Chicchon, "Working with Indigenous Peoples to Conserve Nature," *Conservation and Society* 7, no. 1 (2009): 15; David Barton Bray and Alejandro Velazquez, "From Displacement-Based Conservation to Place-Based Conservation," *Conservation and Society* 7 (2009): 1, 11.

79. Theodore Binnema and Melanie Niemi, "'Let the Line Be Drawn Now': Wilderness, Conservation, and the Exclusion of Aboriginal People from Banff National Park in Canada," *Environmental History* 11, no. 4 (October 2006): 724–750.

80. Sundnes, "Scrubs and Squatters," 287.

81. Theodore Catton, *Inhabited Wilderness: Indians, Eskimos, and National Parks in Alaska* (Albuquerque: University of New Mexico Press, 1997); Turner, *The Promise of Wilderness*, chapter 5.

82. Guha, *How Much Should a Person Consume? Environmentalism in India and the United States* (Berkeley: University of California Press, 2006), 132–133.

83. For a summary of such critiques, see Arun Agrawal and Kent Redford, "Conservation and Displacement," *Conservation and Society* 7, no. 1 (2009): 1–10.

84. John Terborgh, *Requiem for Nature* (Washington, DC: Island Press, 1999).

85. John F. Oates, *Myth and Reality in the Rain Forest: How Conservation Strategies Are Failing in West Africa* (Berkeley: University of California Press, 1999), 55.

86. Locke and Dearden, "Rethinking Protected Area Categories and the New Paradigm," *Environmental Conservation*, 32, no. 1 (March 2005): 5.

87. Peter R. Wilshusen et al., "Reinventing a Square Wheel: Critique of a Resurgent 'Protection Paradigm' in International Biodiversity Conservation," *Society and Natural Resources* 15, no. 1 (2002): 17–40.

88. Ashish Kothari, Philip Camill, and Jessica Brown, "Conservation as if People Also Mattered: Policy and Practice of Community-Based Conservation," *Conservation and Society* 11, no. 1 (2013): 1–15.

89. David Barton Bray and Alejandro Velazquez, "From Displacement-Based Conservation to Place-Based Conservation," *Conservation and Society* 7, no. 1 (2009): 11.

90. For deeper analysis of Nepal, see Stan Stevens, "National Parks and ICCAs in the High Himalayan Region of Nepal: Challenges and Opportunities," *Conservation and Society* 11, no. 1 (2013): 29–45.

91. Richard Coniff, "An African Success: In Namibia, The People and Wildlife Coexist," *Yale Environment 360* (May 12, 2011). http://e360.yale.edu/feature/an_african_success_in_namibia_the_people_and_wildlife_coexist/2403/.

92. David Stradling, "Review of *The Promise of Wilderness: American Environmental Politics since 1964*, by James Morton Turner," *Environmental History* 18, no. 3 (July 2013): 637.

93. For instance, my history of wilderness politics in the United States focuses on the establishment of wilderness areas. Turner, *The Promise of Wilderness*.

94. Alagona, *After the Grizzly*, 196–197.

95. Robert M. Wilson, *Seeking Refuge: Birds and Landscapes of the Pacific Flyway* (Seattle: University of Washington Press, 2010).

96. Wilson, *Seeking Refuge*, 9.

CHAPTER 12

···

RESTORATION AND THE SEARCH FOR COUNTER-NARRATIVES

···

MARCUS HALL

IN *The World Without Us* (2007), Alan Weisman explores what would happen to the earth if humanity were suddenly and mysteriously removed from it. What would happen in our countrysides and in our cityscapes if every last human being abruptly boarded spaceships for the next galaxy or else succumbed to an instant plague? In this thought experiment, we realize that no longer would farmers grow crops; no longer would ranchers herd cattle or loggers cut down trees or miners excavate tunnels. No longer would buildings be built, roads maintained, petroleum combusted. Ceasing all human activities would have short- and long-term consequences: subway tunnels would flood, nuclear power plants would melt down, houses would fill with mold and rot and crumble. Some plants and animals would flourish, while others would disappear or revert to wild types. Atmospheric carbon-dioxide levels would diminish. Weisman's tale underscores how pervasively and dramatically the Earth has been modified by human action. Remove that human action, and an Eden "the way it must have gleamed and smelled the day before Adam" would eventually be restored.[1]

For the likes of Weisman and deep ecology enthusiasts, this earthly reversion to its natural state would be something to celebrate. Human achievements and high culture aside, *Homo sapiens* has become the most dangerous game, the definitive exotic species, the ultimate nest spoiler who, in the words of George Perkins Marsh, is "a power of a higher order than any of the other forms of animated life." My goal in this chapter is not to debate the extent of humanity's earthly imprint, or to judge that imprint as good or bad, or even to question Weisman's assumption that the earth will revert to a natural condition if humans are removed. Rather, my aim here is to question if—and to what extent—humans themselves can facilitate or accelerate the processes of environmental restoration, and to highlight how the restorative endeavor can shed light on other major issues in human history. Members of the Society for Ecological Restoration see themselves *assisting the recovery of an ecosystem that has been degraded, damaged, or destroyed*. With an earth riddled with degraded, damaged, and destroyed ecosystems, restorationists hope to repair some of that damage and bring back some of nature's former glory. A good number of environmentalists consider restoration to be the primary task of the twenty-first century. As we must assume that humanity is not departing for another planet any time soon, what is the promise of human-mediated restoration, and what has been our experience with it? How old is restoration, and how can the study of its history help us to understand and improve the restorative enterprise? Can restoration history clarify some of humankind's other challenges or dilemmas, environmental or otherwise?[2]

REWILDING, RENATURING, AND REGARDENING

Pleistocene Rewilding is what ecologist Josh Donlan and his colleagues call their proposal to bring back some of the land's pre-human splendor. Rather than imagining a world without us, in which nature would begin to recover if humans were removed, the Pleistocene rewilders propose that certain key organisms, such as high-trophic-level predators, be released in protected areas so as to begin restoring food webs that existed before *Homo sapiens* disrupted them, as in Pleistocene North America of 13,000 years ago. Donlan points out that when prehistoric humans crossed the Bering Land Bridge, creatures such as the saber-toothed cat and mastadon were quickly driven to extinction—by human hunting, climate change, or both—thereby erasing a variety of niche factors and other biological forces carried out by these animals. The argument goes that if modern-day proxies (or analogues) of these species in the form of African lions and elephants, for example, could be released in certain North American parks and preserves, their restored "ecological roles" would allow native ecosystems to continue once again on their evolutionary march, thereby promoting ever more complicated food webs while also fueling biodiversity. Pronghorn antelope gained their magnificent ten-meter leaps when pursued by once-abundant American cheetah, claim the Pleistocene rewilders, and introducing proxy predators will keep the pronghorns from losing that leap.[3]

Rewilding is generally considered an active process, planned and mediated by humans, as opposed to passive natural processes that might reinstate a more wild earth if humans and their activities were locally removed. Rewilders therefore hope to reactivate sufficient kinds and numbers of natural processes in order to begin creating newly wild systems—in effect, touching up the land so that it appears untouched. Pleistocene rewilding can be viewed as an extreme form of ecological restoration, whereas most other rewilding proposals focus on reinstating more recent pasts. Instead of 13,000 years ago, other rewilders see their highest goal as reestablishing, say, the ecosystems of 1492. These pre-Columbian rewilders assume that European settlement, even more than Native American settlement, produced truly serious degradation: this version of the noble savage myth posits that Clovis people, but not European settlers, could live lightly on the land. Still other rewilders pull their target date further forward, suggesting that recent pre-settled states are the best goals. Aldo Leopold's call to bring back pre-European settled prairies in a few old farm fields of Wisconsin placed the restorative goal somewhere in the early nineteenth century; Leopold called for a project that would "reconstruct...a sample of original Wisconsin—a sample of what Dane county looked like when our ancestors arrived here during the 1840s."[4]

Rewilding clearly raises a series of questions about the process of assisting an ecosystem to recover. Most obviously, one might ask which historical snapshot represents the ideally restored state, and how our visions of that state have changed over the decades. Are these snapshots better characterized by their former flora and fauna, or by their former qualities of, for example, diversity or productivity or healthiness? Or should restorative goals center on reinstating ecological potential rather than physical environmental states, as by reproducing ecological trajectories rather than discreet ecosystemic units? From a practical standpoint, should restorers concentrate their efforts on reintroducing plants (and animals), or perhaps on culling exotic species, rebuilding soils, or rekindling fire regimes, for example? And at what stage of recovery should the restorers be satisfied with their results? If we follow Marsh's nineteenth-century advice and become a "co-worker with nature in the reconstruction of the damaged fabric," then who decides on restoration's goals and who are its primary beneficiaries? Indeed, does restoration ultimately benefit nature or humankind?

We might enlarge these queries, and wonder how restoration fits into the history of conservation or the history of wilderness preservation. Can restoration's past offer insight into the nature of degradation or of natural disasters? How have other types of restoration—be they architectural, political, spiritual or religious—reflected challenges confronted by ecological restoration? Is an appreciation for history, and a sensitivity to the past, a prerequisite to doing good restoration? The following brief cases are ways to begin answering these questions.

RESTORATION IN EUROPE

Consider the project of rewilding outside of North America, such as in Europe, where this pursuit has become increasingly popular in the twenty-first century. In the Scottish Highlands, for example, land managers must be especially sensitive to the issue of identifying restoration's optimal former wild state. Lacking an undisputed date of discovery or time of human settlement, Scottish rewilders may point to the industrial revolution, or possibly the agricultural revolution, to mark the departure from wildness, which ranges from 200 to 10,000 years ago depending on one's conception of these revolutions. The challenges of selecting restorative goals is apparent at one site where rewilders are planting native Scotch pine, in an attempt to rebuild the "Great Wood of

Caledon" of Roman times, in locations that paleoecologists have concluded were, in fact, devoid of most pine pollen when ancient Romans marched across the British Isles. It seems that large tracts of Caledonia's great forests never were. Restoration in this case is recreating mythic instead of archival pasts; the optimal natural state exists in the ideals of the present rather than in documented historic states. When identifying goals, restorers use what information they can glean from the past, and sometimes that information is quite limited or even misconstrued. Given our necessarily incomplete record of former ecosystems, we must always rely to some degree on inference and intuition to fill in unknown gaps. Little surprise, then, that some rewilders seek other ways to gauge ideally restored conditions beyond identifying historical snapshots or dramatic changes in land use. Optimal ecosystem healthiness, stability, or biodiversity is their preferred goal, especially if these can be described and even measured. Or, it may be more fruitful to simply poll a site's user groups so as to identify and then restore their collective view of an ideally wild place.[5]

Unlike Scottish rewilders, German restorers commonly advocate "renaturing" (*Renaturierung*) at the same time that their Dutch colleagues practice "new naturing" (*natuurontwikkeling*). Weary of their dredged and straightened rivers, German activists and government engineers are working to reinstate more natural riparian nature, by removing dikes, reinserting river meanders, reintroducing endangered animals, and replanting rare species. Here, landscape design trumps landscape restoration: like landscape architects, these renaturers are mostly placing human designs on the land, and it may not matter much to their designs if the climate was quite different a hundred or a thousand years ago. Indeed, restoration enthusiasts must increasingly deal with the fact that former conditions may never be reproducible, not only because some degradation is irreversible, but because a site's rainfall or temperature may not be what it used to be; the biophysical novelty of the early twenty-first century is motivating more and more restoration ecologists to rename themselves "intervention ecologists." Yet any renaturer, new naturer, or intervener must remember that designing *ideal* natures, however one defines them, still requires one to ponder *past* natures, whether on the ground or in the mind. Our designs always stem from our accumulated historical understanding.[6]

The Dutch foster a slightly different restorative vision by promoting primitive environmental forces, as at the Oostvaardersplassen, a 6000-hectare park where Konik ponies, Heck cattle, and a few other primitive breeds have been recently introduced to serve as proxy creatures for extinct tarpan and auroch, progenitors of modern horses and cattle. The assumption here is that after several years of grazing by proxy breeds, the surrounding ecosystems will begin to morph into ones like those that thrived during the days of the ancient grazers, before human activities dominated. European and North American rewilders both advocate proxy species in their rewilding ambitions, except that the former rely on herbivores, while the latter emphasize (more controversially) the role of predators—which may possess big teeth and an ability to use them. Furthermore, it is not *wilderness* that concerns European restorationists so much as *self-willed* conditions: environments that allow natural systems to develop free from human constraints. To generalize broadly, Dutch rewilders promote the development of *wildness* in their projects; North American rewilders aim to bring *wilderness* back in theirs.[7]

Employing proxy species for extinct species creates additional complications for the rewilder. Most immediately, such proxy species can never be exact replicas of extinct species, so that slight variations in the kind and frequency of their grazing, for example, may result in significantly different grazing regimes, and so different vegetation compositions. Introduced grazers may also introduce novel parasites or other novel co-introductions. That is, proxy species inevitably introduce alien novelty into an ecosystem, and all good restorers work to keep novelty out. Alien species are enemy number-one in most restoration projects, and proxy species are undeniably alien at some level.[8]

In the case of Holland's new naturing project at Oostvaardersplassen, Konik ponies and Heck cattle promote still greater paradox, and not just because of their alien nature. It turns out that both breeds are not so much primitive breeds as human-manipulated breeds, for Konik ponies cohabited and coevolved with Polish peasants for centuries before being shipped across Europe, calling into question their ability to make nature anew free from human design. For their part, Heck cattle are an intensive breeding product of the Heck brothers, two German zookeepers of the Nazi era who sought to bring back the wild auroch of Central Europe, which, before becoming extinct in the eighteenth century, stood two meters at the shoulder and displaying impressive horns. This magnificent cow embodied Teutonic supremacy, and carried traits that the Heck brothers searched for in existing hardy cattle lines across Europe, but that they could only find in a few miscellaneous breeds. The brothers eventually concocted a new hybrid cow, utilizing various breeding stock, some of which came from as far away as Corsica—but with only partial genetic overlap with the auroch. In the end, the Hecks' painstakingly cross-bred result may be as artificial as the landscape it is supposed to rewild. In fact, it seems strangely harmonious to find out that a century ago, the Oostvaardersplassen itself once lay under the North Sea, with Holland's earlier diking and drainage schemes forming a prerequisite to its new naturing schemes by proxy grazers. A reclaimed landscape is now being rewilded. To the Dutch it makes sense that wildness must be ushered in by the human hand.[9]

Instead of renaturing or new naturing, the quintessential restorative style in Italy is gardening—or more properly, *regardening*. Although various forms of *rinaturazione* (or *restauro ambientale*) are practiced in Italy, Tuscany's or Lombardy's anciently manicured landscapes mean that Italians are used to seeing nature managed by the human hand and want to see that management continue: remove human stewardship and resource problems seem to arise. Restoring proper environmental conditions across this peninsula therefore means reinstating proper human care to the land and its biota. Forests are managed, soils are fertilized, terraces are built, vineyards and olive groves are pruned—all in the name of "restoration" to a former healthy (and humanized) condition. The gardens of Italy degenerate unless the nation's gardeners rehabilitate, resuscitate, and remake former gardens; Italy "is cultivated right up to the mountain tops" wrote Francesco Guicciardini in his 1568 history of the peninsula. Restoration as regardening means that when the Po River floods, blame is not directed at torrential downpours or nonabsorptive soils, but at farmers and foresters who inadequately managed the surrounding soils, shrubs, and trees that temper floods. This faith in humanized landscapes is also reflected by advocates of "naturalistic engineering" (*ingegneria naturalistica*) who at Circeo Park near Rome excavate ponds into the coastal littoral; the motives for these restorative projects center on creating waterfowl habitat more than reinstating primitive natures or rebuilding wild environments. Not surprisingly, architectural and environmental restoration are often inseparable in Italy. Even more than in northern Europe, the "built environment" in Italy includes nature. The few Italians enthralled with wilderness, or efforts to restore it, usually employ English terms to describe their activities, so foreign are the associated concepts.[10]

This brief sampling of the styles and varieties of contemporary restoration on both sides of the Atlantic suggests an extraordinarily rich human experience with repairing and renewing environmental conditions. Rewilding, renaturing, and regardening can all be considered ways of combating or reversing degradation, and so provide counter-narratives to the usual tale of environmental decline. Indeed calls for *improvement*, which rose during the Enlightenment, and were joined with calls for *progress* in the nineteenth century, continue in many circles up to the present, whether in the context of landscapes, river channels, animal breeds, or the human condition, and can be considered calls for restoration if one views primordial forms as having experienced falls from desirable states. "Wild nature is hideous and dying; it is I, I alone who can make it agreeable and living," declared Compte de Buffon in 1764, shying away from a declensionist environmental tale. Historian Carolyn Merchant explores this recovery narrative as promoted by Christian, modernist, feminist, and environmentalist traditions; she explains that the struggle for restoration in these narratives followed, respectively, original sin, political tyranny, female repression, and ecological degradation. David Lowenthal corroborates that "Yearning for restoration is age-old. So is faith it will come to pass." We realize that the "good old days" were often not so good—and we need to better explore how, in some cases, our land, sea, and air, and our lives and our lifestyles, have become better off than before.[11]

RESTORING THE IBEX

One of the great tales of environmental decline followed by recovery is that of the Alpine ibex, or European mountain goat, which was hunted almost to extinction across Europe during the nineteenth century. In 1821, less than a hundred ibex survived in a royal hunting reserve in Italy's northwestern Alps of Gran Paradiso. Then, following a dramatic program of wildlife management that included hunting prohibitions, armed game wardens, captive breeding, reintroducing, and monitoring, more than 40,000 ibex now wander across Alpine meadows in six countries. It is worth exploring how the

fall and rise of the ibex can enrich our understanding of the human ability to fix damaged natures.[12]

After glaciers retreated from the Alps during the last ice age, *Capra ibex* (otherwise
known as *stambeccho, steinboch, bouquetin,* and *kozorog* depending on region) colonized all suitable Alpine valleys, west to east. That ibex once roamed this entire mountain chain is shown by archaeologic and paleoarchaeologic evidence. Oetzi, the man
who fell in a crevasse on the Austrian-Italian border 5300 years ago and who was then
exhumed from the ice in 1991, holds shreds of ibex tissue in his gut. It is no surprise that
the Ibex's tasty meat, supple hide, and spectacular horns meant that this animal was
long prized by mountain peasants and lowland hunters alike. Moreover in the medieval
world, the powder of ibex horns bestowed magical cures, as did the occasional bezoar
stone extracted from the animal's gut, which was believed to counteract poison. The
ibex's many appealing qualities for humans, therefore, spelled almost certain doom for
the mountain goat, at the same time that these guaranteed intensive efforts to guard it
from extinction. As Michael Pollan would remind us, today's plentiful ibex numbers
demonstrate how successful certain creatures have been at appropriating human intelligence and human energy to protect and multiply their own kind (see Figure 12.1).[13]

Following royal hunting bans in 1821 and again in 1836, the turnaround in ibex numbers began in earnest in 1856, when King Vittorio Emanuele II established a hunting
reserve in the nearby massif of Gran Paradiso, employing a small army of guards to
protect his prize game animal. In the next decades, only the king and his associates
were allowed to kill ibex (although poachers managed to put occasional ibex steaks in
their frying pans.) Here again, one sees that charismatic megafauna are crucial to the
story of conservation when we note that the king's royal hunting reserve was converted
in the early 1920s into Gran Paradiso National Park, one of Italy's first two national
parks; it was inspired by Yellowstone's 1872 precedent, but has roots that stretch back
decades earlier (see Figure 12.2). One early park booster felt that Gran Paradiso would

FIGURE 12.1 Once isolated to the Gran Paradiso massif and numbering less than one hundred individuals in the 1820s, the Alpine ibex is now 40,000 strong across the Alps.

Photo by Marcus Hall.

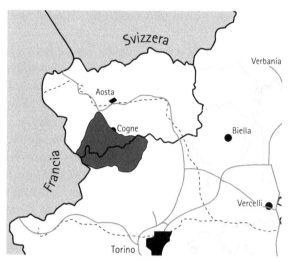

FIGURE 12.2 The current boundaries of Italy's Gran Paradiso National Park approximate those of the royal hunting reserve established in 1856. http://www.agraria.org/parchi/valleaosta/granparadiso.htm.

become "not a national park, but a monastery for the ibex." The Italian campaign to protect their wildlife icon thus indirectly conferred protection on a host of other park fauna, flora, and landscape elements within park borders. And as was the case for American national parks, efforts at *preserving* Italian parks also meant *restoring* some of the natural elements that formerly graced parklands. Nature's own regenerative forces served to replenish and revegetate when human activities were limited or excluded. Conservation has long invoked good measures of both preservation and restoration, and in many instances it is difficult to distinguish between the two.[14]

The naturally prolific ibex therefore contributed to its own recovery following the royal protective decrees, allowing the herd to grow to some 2600 individuals by the early twentieth century (even if a hundred males were harvested each year). This form of passive wildlife restoration was augmented by more active measures in 1906, when a Swiss wildlife zoo near St. Gallen managed to import and begin raising three ibex kids, doing the same with another hundred animals imported over the next few decades. The plot of this propagation project thickens when one realizes that the Italian king was long opposed to sharing his precious ibex, so that the Swiss wildlife enthusiasts resorted to bribing Gran Paradiso villagers to capture the animals and smuggle their live cargo over northern passes. Just as tropical countries today struggle to maintain control over their potentially lucrative species eyed by pharmaceutical companies, the international ibex struggle was an early episode in biological wealth management, sometimes referred to as *biopiracy*. In a New World conflict involving restoration and big game proprietorship, North America's Pleistocene rewilders have been criticized for proposing that lions and elephants be multiplied in the American West (as proxy species) because of the likelihood that tourist dollars will be diverted away from

Africa when big-game enthusiasts choose to take their "safaris" in the preserves of New Mexico and Arizona. Restoration decisions never depend on ecological science alone.[15]

The tale of ibex restoration also offers fresh insight into the wild. As "wildlife," ibex can survive harsh mountain weather and provide chase to the sportsman, but decades of contact with enclosures, selective breeding, reintroductions, and care-giving humans have bred domesticity into their veins. Conservation biologists talk of *bottlenecks* in plant and animal populations, whereby progeny descend from very small numbers of individuals, and such bottlenecks were narrow indeed in the case of the hundred-odd European mountain goats and their gene pool that grew to a population 400 times that size. One study clarifies that there were actually several bottlenecks in the expanding ibex population: the original endangered 1821 Italian herd, the 1906 Swiss relocation of a few individuals from this herd, the 1920 reintroduction of a few of these captively bred ibex to eastern Switzerland, and subsequent reintroductions from this region to other valleys in the Alps. Personal observation suggests that inbreeding is now manifested in morphology and behavior. Today it is a simple matter to leave the hiking trail and approach a group of basking ibex who return one's inquisitiveness with uninterested, blank stares. The fight-or-flight response is today largely absent in this magnificent creature. When inhabited by ibex, Alpine meadows becomes mere mountain pastures, rocky scree becomes so much zoo habitat. Europeans are fiercely proud of their heritage of *Capra* wildlife, but outsiders may wonder just how wild is this life.[16]

There is also the question of invasive alien (or exotic) species in ibex reintroduction projects. Subspecies of wild goat survive in mountain ranges from Spain to Bulgaria to Iran, and these goats have been useful for ibex restoration projects. For example, the mountains of Lebanon were scoured of their wild (Nubian) Capra populations a century ago, and it has been proposed that ibex specimens be brought there from southern Egypt to repopulate the area. In this way, southern Spanish ibex were introduced to the mountains of Portugal in 2006. Although exotic species are generally avoided in conscientious restoration projects—because the foreign element may further unbalance an already unbalanced ecosystem—perhaps one can justify importing and reproducing foreign-born ibex if doing so is the only way to multiply local ibex DNA, through hybrids. There have even been attempts to multiply the last individuals of the northern Spanish (Pyrenean) ibex through cloning. One must ask: should clones be considered quintessentially native or irredeemably exotic if they can survive laboratory conditions to multiply in alpine cirques? Is biodiversity enhanced if cloned individuals can be propagated?[17]

Complicating these wild matters are repeated attempts, since at least 1815, to breed rare ibex with local farm goats. The resulting hybrids sometimes sprouted the cherished horns, but such individuals did not survive long in reintroduction projects, which required animals to withstand the harsh alpine conditions of cold wind and deep snow. Still, such breeding experiments produced enough fertile progeny that trophy hunting clubs today warn their members against the taking of ibex specimens that may be more billy goat than *Capra ibex*. Indeed feral goats, that is, escaped domestic varieties, are sometimes mistaken for wild goats by hunters and tourists alike. From yet another perspective, some of these exotic and feral goats could serve crucial roles as proxy species if restorationists are primarily interested in reproducing grazing conditions instead of the native grazers themselves; proxy goats could theoretically help recreate foodwebs and niche factors that existed when large herds of native ibex grazed mountainsides across Europe and northern Africa. One therefore needs to question what exactly is being restored in the Alps' two-hundred-year ibex restoration project, which may be judged more successful at replenishing biomass than biodiversity. Perhaps European ibex enthusiasts have really been "renaturing" rather than "rewilding" their Alps—that is, reproducing idealized natures rather than prehistoric or untamed natures.[18]

But in last analysis, "regardening" may be the best the way to describe ibex recovery. Herbivores, carnivores, and scavengers; and grasses, shrubs, and trees, all exist in different numbers and proportions than they would if humans had never begun transforming the Alps centuries or millennia ago. Hunters since Oetzi the Iceman have altered its fauna; wood collectors since Hannibal have logged its trees; graziers since Charlemagne have herded ruminants across its high meadows. Evolving human visions of an ideal mountainous environment produced evolving cultural landscapes. Initially, clearing forests and exterminating wild beasts were the best ways to steward these mountains; today, building tourist villages and raising cows while repopulating wild (but harmless) animals is the most popular method. According to most sentiment, the great garden which is the Alps must be maintained, and ibex are now integral to that garden. Wolves and even bears are also making inroads in the Alps today, but local villagers are making sure that these carnivores will not enjoy the same success as the ibex. In 2006, farmers in Bavaria handedly disposed of the first brown bear spotted there in decades; in the French-Italian Alps, wolves now generate much fanfare along with much controversy. Only certain kinds of creatures receive generous restoration support. Plant-eaters, but not meat-eaters, are being nurtured and restored in Europe's Alpine rock garden.[19]

RESTORATION AND ENVIRONMENTAL DAMAGE

The ibex has become a symbol of the Alps. Its emblematic horns now grace park newsletters, websites, and t-shirts. Yet in some valleys, ibex reintroduction may be judged too successful. If one could "think like a mountain," wolves would receive a warmer welcome in the Alps if they could be seen as useful for controlling ibex herds, which may overgraze delicate meadows. Like all restorationists, ibex enthusiasts must consider which baseline, or target, is best when judging the outcome of their projects. The optimal baseline may be determined by identifying a historical snapshot, or by acknowledging that biological, meteorological, and especially, cultural conditions have changed over the last two centuries, so that any baseline must be updated accordingly. Ongoing global warming may well mean that ibex will need to migrate farther up the mountainsides in order to occupy the cooler, higher valleys. And restoration purists who aim to recreate baselines that incorporate natural changes while excluding cultural ones may be stymied when trying to distinguish the two.

One phenomenon that helps distinguish cultural from natural environmental damage is the thick vegetation sprouting in many areas of the Alps. Where cattle herders and wood users have withdrawn from formerly trammeled valleys, the meadows and woods are returning. Vast regions of the French and Italian Alps have lost more than half of their human residents during the last 150 years as more appealing lifestyles lure them to the plains. Those mountain villagers who hang on to their traditional livelihood consider such spontaneous regrowth deeply problematic, a manifestation of their disappearing way of life and directly responsible for such problems as flooding, burning, eroding, and insect-blighting. Environmental damage for these villagers is the unchecked shrubs and trees growing out of hillsides that are no longer mowed, grazed, and logged. No matter that a variety of forest animals—winged, scaled, clawed, and hoofed (including ibex)—are returning with the meadows and forests. From the villagers' perspective, there is every need to control that vegetation and assist, as restorationists do, "the recovery of an ecosystem that has been degraded, damaged or destroyed."[20]

But this spontaneous regrowth does not represent damage to everyone, now or in the past. Many city dwellers are quick to celebrate the mountain's increasing biomass and, by some measures, increasing biodiversity—even though exact species compositions may be quite different than in centuries past. From the comfort of their own living rooms, these urbanites consider the new forests to be providing not worse, but better protection from floods, fires, and erosion. Here again, one's gardening tradition influences how far one is willing to consider human-modified ecosystems as normal and good. To offer broad categorization, there is greater faith shown by rural than urban folk that humans can properly manage woods, waters, and wildlife. Mountain dwellers more than flatlanders feel that people can properly steward the alpine environment. Europeans more than Americans believe that their day-do-day activities can improve rather than damage nature. Former generations more than current generations assume that the land should be actively managed: Europe's leading ecologist between the world wars, Arthur Tansley, envisioned a harmonious "anthropogenic climax" that was a stable vegetative state formed by human activities, while Tansley's American counterpart, Frederic Clements, envisioned a stable "natural climax" that humans could only diminish or disrupt. Italian foresters in the mid-nineteenth century blamed their mountain problems on insufficient human management. American George Perkins Marsh, though, blamed such problems on recurring *mis*management, whereby humans carelessly or unknowingly disturbed the natural order: "we can never know how wide a circle of disturbance we produce in the harmonies of nature when we throw the smallest pebble into the ocean of organic life." Natural forces (spontaneous and entropic) were once considered the main source of environmental damage and the reason to restore; increasingly, cultural activities (deliberate or inadvertent) became the source of serious damage, and the reason why a handful of activists today propose that our twenty-first century be the "century of restoring the Earth."[21]

Certainly ibex have not occupied everyone's highest vision for the Alps, and so represented damage, especially to the shepherd or dairy farmer who viewed mountain meadows as fodder for their animals. After all, grass consumed by wild ungulates could not be consumed by domestic ones, so that greater ibex numbers meant fewer shepherd numbers, who did not harbor much love for their big-horned ouster. Gran Paradiso Park was enlarged several times in the decades after 1922, when adjoining valleys were added piecemeal to its core, sparking fiery protests from local villagers with each new enlargement. The expanding ibex herd, like the spreading shrubland, has been both cause and effect of the fluxes in the local alpine economy. At least at Gran Paradiso, the history of restoring ibex has also been a history of limiting shepherds and their flocks. Likewise at Yellowstone, the project of restoring wolves has represented a threat to certain cattle ranchers who run their animals across mountain meadows just beyond park boundaries. The Y2Y proposal for creating a mega-protected area connecting mountains and valleys from the Yukon to Yellowstone is deeply appealing to wildlife restorationists but utterly appalling to those who live nearby and may need to "work for a living." By refashioning ecosystems and rekindling natural processes, restoration has rarely been a win-win activity, whereby all parties benefit. Someone or some organization ultimately decides how and what to restore, and the corresponding political questions can be much more thorny than the ecological ones. Environmental restoration in its many forms has always extracted costs, directly or indirectly, expected or unexpected, and not all parties benefit equally. Helping nature heal, now and in the past, always entails making hard decisions over scarce resources, and political power influences the outcome of those decisions.[22]

There are some cases of restoration that seem successful by any measure, compromising the interests of no one. North American trout enthusiasts would point to the tale of the Pyramid Lake Lahontan cutthroat (*Oncorhynchus clarki henshawi*), originally found only in this large, brackish lake on the dry side of the Sierra Mountains in northwestern Nevada. The band of Paiute Indians whose ancestors (indigenous name, *Cuiyui Ticutta*, "fish eaters") subsisted on fish meat taken from Pyramid Lake, recount their great-grandfathers' stories of catching cutthroat as long as their arm and weighing up to sixty pounds, but then sadly watching their trophy fish disappear in the early twentieth century when hastily crafted reclamation projects diverted the Truckee River, the fish's main spawning grounds. Presumed extinct, the rare trout was miraculously discovered fifty years later swimming in an small creek on the other end of Nevada near Pilot Peak, transplanted there decades earlier by anglers probably motivated by the same desires as the reclamationists who envisioned the future of that desert as green, bountiful, and blossoming like a rose. After positive identification showed the Pilot Peak trout to be the same species as the treasured Pyramid Lake species, a methodical reintroduction project ensued, together with restorative measures for improving the flow and health of the Truckee River. In 2013, the *New York Times* ran an upbeat piece describing the amazing reappearance of the once-extinct Lahontan cutthroat and its fishery, with specimens now being hooked that reach twenty-four pounds and growing: "a rare win-win-win for native wildlife restoration, the tribe's economy and anglers."[23]

But beyond this unabashed restoration success story, one must still ponder the checkered legacies of how this fish disappeared and even how it was revived. For one, there were earlier restorative attempts that need recounting, such as various introductions of exotic fish into the Pyramid Lake drainage that included regional varieties of cutthroats as well as rainbow and lake trout. The resulting aquatic transformation and fish hybrids are now viewed as threatening a complete restoration, and may continue doing so for decades—despite today's popular tone of a triumphant, *fait accompli* fishery revival. There is also the celebratory listing of the Derby Dam on the National Register of Historic Places, a dam which (when combined with droughts) provided the last death blow to the struggling Lahontan fishery, by diverting the Truckee River to irrigate nearby farmland, and so destroyed the Pyramid Lake Paiutes' main livelihood. One learns that this modest dam is the (US) National Reclamation Act's first project, built in 1903 with federal money in the very district of Congressman Francis Newlands, the Act's main sponsor in Congress, who in other legislative matters advocated annexing Haiti and the Dominican Republic as repositories for African Americans, so as to cleanse the United States and preserve the nation "for all time for the white races." Lastly, one might also consider the question of who exactly discovered the surviving Lahontan cutthroat that turned this extinction story on its head. By some accounts, this rare fish was discovered by Robert Behnke, a renowned trout biologist working at Colorado State University; yet others claim that Don Duff, a passionate angler who knows Nevada like his backyard fishing hole, first found the rare cutthroat. But in the end, it was really Behnke's graduate student, Terry Hickman, who drove out to Pilot Peak to get his hands slimy in order to locate a few specimens of the "extinct" fish. We begin to see that a deeper inspection of the Lahontan cutthroat's recovery story suggests that restoration's celebrated counter narratives are always messy, offering details that reveal as much gloom as glory.[24]

THINKING HISTORICALLY ON
A DAMAGED EARTH

Integral to the restorative experience, whether of ibex or of trout, is a sensitivity to the past. Any conscientious act of environmental repair, rehabilitation, or remediation requires one to consider how the natural system once was before its present state. When setting out to put things right again, one might study an early description of a site, peer at an old picture, ask an elderly resident, or search an archives. Because the past holds information about a former, more desirable state, the practicing restorationist becomes an applied historian. Even when ecological process is judged more important than biophysical place, the restorationist is still keenly interested in how the past can serve as a model for the present. Restoration ecology becomes an historical science, whereby ecosystems and their managers depend on what came before to understand what exists now. The concerns for the European mountain goat and Pyramid Lake cutthroat arose from the knowledge that many more of them used to exist. Observations through time demonstrated that their numbers were dropping, and so efforts arose to begin protecting and multiplying the species. In the case of the ibex, the Italian king looked backward to envision his future herd; 40,000 *Capra ibex* grazing the Alps may now seem plentiful enough, but knowing this species' historical baseline, or baselines, would powerfully influence one's judgment of the proper size of the herd. Paiute Indians likewise queried their elders in order to estimate the extent and size of the proper fish population. Environmental restorers work within a longer time frame than environmental managers because the arc of recovery can take generations. Restoration both promotes, and is promoted by, the study of the past.

Other kinds of restorationists also foster the study and appreciation of the past. Art restorationists retouch and repaint a decaying masterwork so that its former luster will shine again. Architectural restorationists (sometimes called historic preservationists or heritage conservationists) renew and reinforce prized buildings and other built monuments, including whole landscapes, which were constructed during an earlier generation. Political or religious restorationists yearn for former times, when leaders still led and the clergy still prayed. Indeed, restoration appears to be a fundamentally conservative pursuit: one aims to recreate the past in the present and then defend it from reverting to a tarnished state. American environmentalist Dave Foreman, whose passion is to resuscitate wildness and heal ecological wounds, likewise advocates thinking historically. Foreman describes his book-long action plan, *Rewilding North America* (2004), as a work of science and of policy, but even more to the point, he explains that it is "first...a work of history." His book emphasizes a continent's former glory, and he reinforces his claim with excerpts from archives. In harking back to more desirable states, the several kinds of restorationists are all imploring us to read our history and to appreciate our past. "If you don't know how the present came to be," declares Foreman, "you stand in a nihilistic void and your words and actions lack coherence."[25]

Unlike the nostalgic, the restorer does not dwell on the past, but rather studies it in order to fashion a better present. The latter assumes that a pre-damaged past can serve as a model for the post-damaged future. There is certainly romance in bringing back better times and healthier processes, but restoration success is measured by how far the corrupted status quo has been converted to its former undamaged potential. Yet try as she might, the restorer does not bring back the whole past, but only desirable parts of it. Snakes and spiders were as integral to former foodwebs as were flowers and songbirds, but the latter creatures usually garner more human concern. Similarly, fires and blights, or droughts and downpours, may have been integral to these foodwebs, but only relatively recently were these events incorporated into restorative efforts. And spontaneous, natural recovery processes can bring back some, but not all, previous forms and functions. In their projects, today's restorationists undoubtedly omit other, seemingly unimportant—or undesirable—creatures and events that in time may be understood as crucial to optimal restoration. Like historians, restorationists necessarily filter and choose the most useful parts of the past for their projects, and their choices cannot escape biases and fashion. Good restorationists, like good historians, produce laudable results, but their creations are not facsimiles of what once was.

And so just as there is no such thing as total history, there is no such thing as total restoration. Only selections of former systems can be reinstated. An ambitious restorer will pay attention to animals and plants, to soils and waters, to prehistoric and pre-Pleistocene conditions, but she will ultimately need to favor one over the other, favor ibex over chamois over muflon—favor fish over waterfowl—favor one century over another, favor pre-industrial over pre-agricultural states. *Restoration* appeals because each person can envision remaking his own favorite element, even though any single project cannot hope to target everyone's favorite elements. Ecological baselines shift according to need and fashion. Marine biologist Daniel Pauly cautioned about shifting oceanic baselines, and about how researchers focus far too much attention on just one or a few accepted natural states that depend on their own limited temporal observations. Comb jellies (*Ctenophora*) picked up in a ship's ballast waters in Brazil's ports and then released two weeks later in eastern Mediterranean harbors are transforming the latter's ecosystems, so that resource managers must acknowledge these changed baselines, and manage accordingly.[26]

Laws also reflect the biases of baselines that favor certain kinds of natures. The US Pittman-Robertson Act of 1937, also called the Wildlife Restoration Act, helped multiply game animals in North America by taxing hunting supplies so that habitat could be purchased and preserved; however, it did not target non-game wildlife, much less all wildlife, but rather only the deer, elk, ducks, geese, and pheasants of interest to sport hunters. The massive restoration project currently underway in Florida's Everglades gains popular support through its promise to multiply spectacular shorebirds and manatees, yet town commissioners also quietly welcome new canals that will provide plentiful water for economic development. One might realize that the "Comprehensive Everglades Restoration Plan" is an oxymoron. With 9.5 billion dollars to be spent over thirty years, restoring the Everglades means different things to different people, but it can never reproduce a whole, brand new Everglades.[27]

There are always unknowns in formulating the goals of any restoration project, be they descriptions of past states or of projected future states. Some restoration practitioners prefer the terms *renewal* or *regeneration* to describe their pursuit because they allow greater flexibility in filling in the unknowns. If one is unsure which grass species once grew at a site, for example, or how dense such species were, then one must make educated guesses with available evidence and intuition before setting out to restore those grasses. Today in Chile's Patagonia, promoters of a large new national park feel that protected status is helping expansive grasslands "recuperate" following eighty years of intensive sheep grazing; beyond this hands-off method, their main hands-on restorative method is collecting and then sowing native grass seeds. In western North

America's vast, overgrazed grasslands, range managers in the 1950s saw themselves "rehabilitating" more than "restoring." From their perspective, it was unrealistic to try to reproduce conditions that existed before millions of sheep and cattle trampled the West. A leading restoration ecologist, Mohan Wali, entitled his textbook *Ecosystem Rehabilitation* (1992). A leading landscape architect, Robert France, notes that "restoration ecologists continue to refuse to admit that what they are doing is really a specialized form of landscape design and social revitalization." Some restorers therefore seek partial repair, others rely on their own hunches for generating goals, while still others utilize a variety of methods to do the best they can. These last enthusiasts may simply be *wilding, naturing,* or *gardening,* without invoking any *re-* counterparts.[28]

But even if Earth repairers are *storing* more than *restoring* nature, they still carry ideas and assumptions in their minds that derive from earlier generations. Notions of wildness swirling in their brains stem from an Eliot Porter photograph or a Sigurd Olson excerpt that are themselves reincarnations of works by William Henry Jackson or John Muir or J. J. Rousseau. American gardeners borrow from Andrew Jackson Downing whether they like it or not; British gardeners borrow from Joseph Paxton; Italians from the creators of Villa d'Este or Pratolino. Ideal natural states recapitulate the pastoral or the sublime (or Dante's paradise), and restorers bring back combinations of them all. Restorationist William Jordan III calls on fellow practitioners to maintain a "studied indifference to human interests" in their projects, in order to avoid inserting humanity's touch in a natural system as much as possible. But one must remember that humans can never be indifferent to their own interests, ignore their own perceptions, or deny their own understandings. If restoration is a human act, restoration will leave a human imprint. Just as Peter Novick cautions historians about pursuing that "noble dream" of objectivity, by pointing out that all historical tales are interpretations, restorationists must also be cautious about assuming that they can actually describe an historic ecosystem or disinterestedly identify the key components and processes to be brought back. Wilderness in the American mind may be different than wilderness in the European mind, and efforts at rewilding will produce very different results on the other side of the Atlantic. Restoration at Gran Paradiso meant ibex; at Yellowstone, it meant bison and elk, and then wolves; at Pyramid Lake it meant Lahontan cutthroats, and each of these animals embodies different kinds of wild natures. The Save China's Tigers project consists of removing neonatal tiger kittens from Chinese zoos, sending them to large preserves in South Africa for careful "rewilding" sessions, and finally reimporting the beasts fully grown and ready to hunt back to China's forests: such is how a bit of Africa red in tooth and claw is making its way to Asia.[29]

THE ROLE OF HISTORY IN RESTORATION

As both concerned citizens and active participants in restoration projects, what are we to do if we do not know the historical baseline, or cannot describe it accurately, or consider that baseline mostly irrelevant in light of overwhelming degradation or ongoing climate change? In attempting to manage a warming earth, how can we justify using the past as a model for our projects if we cannot expect to recreate that past? Asked more broadly, how can historians—and the historically minded—still be useful in a restoration project in a world surrounded by unprecedented ecological novelty?[30]

The best answer, and the key contribution of historians, is that they remind us that *time* is integral to the restorative endeavor, both on the land and in our thinking about it. Natural systems as well as human systems are dynamic entities that change slowly or rapidly, over the short- or long term, and rarely involve static states. It is only our relatively limited observations over brief periods that suggest there can be a single historic baseline that we should emulate. There are many former snapshots and many former processes that might be brought back. There are also numerous future states that could each represent acceptable goals of restoration. Evolution and random events (in the form of invasive species, novel microbes, or changing climates) guarantee that change to natural systems has occurred and will occur, and humans can only hope to mediate such changes for better rather than worse. Indeed, to *preserve* a national park, for example, we must continually *restore* it to a predetermined condition. Yellowstone's (or Gran Paradiso's) composition of flora and fauna, the structure of its foodwebs, even its cultural landscapes and historic buildings will all look rather different a century from now, unless restorative processes act to maintain the current state. Restoration is therefore a mechanism of preservation. A. Starker Leopold's 1963 report recommending that a US National Park be managed as a "vignette of primitive America" requires ongoing restorative maintenance to cull excessive elk numbers, say, or erase damage done by tourists. Restoration of a natural system can take thousands or millions of years, but it can also require just days or weeks if goals are adjusted to incorporate human values alongside natural processes, which they always must do.[31]

Historians can also teach that identifying primordial states—absolute beginnings—is as unrealistic in natural history as it is in human history. As W. G. Hoskins said, "everything is older than we think," and land managers should consider, for example, that today's degraded states may have once been even more degraded, so that historic snapshots—and even historic processes—cannot always serve as good models. As shown by Figures 12.3 and 12.4, landscapes depend on former landscapes, even if their echoes are not always detected. Restoration ecologists depend on former ecologists, even if their ideas are not always acknowledged.[32]

Restorationists therefore require not chronologies, but histories. Proper histories do not merely recount facts and list events, but rather set out to explain why a river was channeled or how a forest was logged, why a species became extinct or how a city was built. The reasons are in the details, and the historian tells stories with those details to distinguish between rivers, forests, species, and cities. Because each site is unique, each restoration project will require a different agenda. Even where there is convergence, so that two sites come to share similar biota or soils or landforms, a search in the archives (natural and cultural) can reveal the differences between two sites that will be so crucial for restoring each. Divergence, too, can also be accounted for by discrete

26898

FIGURES 12.3 and 12.4 Landscape vestiges. Even after the Limmat River near Zurich was canalled and straightened in the 1870s, sloughs and oxbows continued to signal where the river once flowed. Archives, Gemeinde Dietikon, Switzerland.

events through time, just as identical twins never grow up to be identical. One eco-logical study reveals that two nearby mountainsides in Idaho, with nearly identical sun exposures, soil stratas, and fire regimes, currently sprout very different forests because of miniscule but crucial historical events. Historians remind us that single reparative formulas do not exist.[33]

At the same time, historical conclusions can be cautiously generalized across cir-cumstances or sites. History does not repeat itself, but analogous circumstances mean that we strive to begin restoring with experience under our belts. When certain tech-niques show success, common sense dictates that under comparable conditions these should be used again. Analogies together with parables are the most relevant messages offered by the historian, and striving to learn about other restorative projects improves the chances of doing a better job. Learning about Europe's experience with ibex recov-ery improves our chances of recovering wildlife elsewhere. We are not condemned to repeat the past if we fail to remember it, but we are given a more powerful way to face our problems if we reflect on how others faced theirs.[34]

Not surprisingly, restoration histories are urgently needed on a range of subjects that cover more than just single species or ecosystem types. Spontaneous and assisted recoveries of lands and waters have been as common as episodes of damage and decline. Even though environmental historians have usually focused on pollution, extinction, and collapse, there are just as many counter tales waiting to be told about improve-ment, renewal, and revival—or attempts at doing so. Human relationships with the earth have not always been nasty, brutish, and short, and restoration historians can offer different interpretations. When it comes to understanding our earthly imprint, naïve optimism is not what is needed, but neither is naïve pessimism.

And unlike many environmental pursuits, restoration has a special relationship with history. There are histories *of* restoration, but there are also histories *in* restora-tion. As noted earlier, the very process of restoration requires one to think back, to consider what went previously so as to envision what might become. The past is not always better than the present but when we think it is, we can search for baselines and precedents that we want to emulate, in form and in process. Good restorationists foster historical thinking; good historians foster appreciation for the past and motivate us to rekindle and revive those parts of it that were better. Restoration and history are synergistic: practitioners of each promote the other. It should not be surprising that as more restorationists attend Thursday-night history clubs, more historians join Saturday-morning exotic weed pulls and native tree plantings.

Even as ecosystems in our twenty-first century step incrementally toward the poles or up mountain sides, seeking the cooler temperatures that they once knew, history still has a vital role in assisting the recovery of their damaged components. A turn to the past for documenting what went before is a glimpse into a living laboratory where one can search for wrongs as well as rights. The past can showcase more pris-tine conditions that can serve as models for the future's open spaces, but it can also reveal humanized places that functioned better than do their counterparts today. The past therefore furnishes the ingredients for auspicious counter-narratives that will be relevant in a changing tomorrow. Inspirational history reveals that humans can and have improved their world, as habitat for creatures domestic and wild. Historians raise a mirror to ourselves to see what we are capable of, and this includes living lightly on the land.

Notes

1. Alan Weisman, *The World Without Us* (New York: Picador, 2007), 5.
2. George Perkins Marsh, *Man and Nature; Or, Physical Geography as Modified by Human Action* (Cambridge, MA: Belknap, 1965 [1864]), preface. Definition of restoration, https://www.ser.org/resources/resources-detail-view/ser-international-primer-on-ecological-restoration.
3. Josh Donlan et al., "Re-wilding North America," *Nature* 436 (August 18, 2005): 913–914. On the causes of Pleistocene extinctions, see Paul S. Martin and H. E. Wright, Jr., eds., *Pleistocene Extinctions: The Search for a Cause* (New Haven, CT: Yale University Press, 1967), 75–200; Grover S. Krantz, "Human Activities and Megafaunal Extinctions," *American Scientist* 58 (March-April 1970): 164-170. Thanks to Drew Isenberg for pointing me toward the long debate over the causes of these extinctions.
4. Aldo Leopold quoted in William R. Jordan, III, "Looking Back: A Pioneering Restoration Project Turns Fifty," *Restoration and Management Notes* I, no. 3 (Winter 1983): 5.
5. Althea L. Davies, "Palaeoecology, Management, and Restoration in the Scottish Highlands," in *Restoration and History: The Search for a Usable Environmental Past*, ed. Marcus Hall (New York: Routledge, 2010): 74–86; Susan Tapsell, "River Restoration: What are we Restoring To? A Case Study of the Ravensbourne River, London," *Landscape Research* 20, no. 3 (1995): 98–111.
6. Stefan Zerbe and Gerhard Wiegleb, eds., *Renaturierung von Ökosystemen in Mitteleuropa* (Heidelberg: Springer/Spektrum Akademischer Verlag, 2009); Richard Hobbs et al., "Intervention Ecology: Applying Ecological Science in the Twenty-First Century," *Bioscience* 61 (2011): 442–450. There are many examples of how imagined and mythical pasts eventually replace more objective pasts; see for example Benedict Anderson, *Imagined Communities: Reflections on the Origin and Spread of Nationalism* (New York: Verso, 1983); Jared Farmer, *On Zion's Mount: Mormons, Indians, and the American Landscape* (Cambridge, MA: Harvard University Press, 2008). On intervention ecology, see Richard Hobbs, "Intervention Ecology: Applying Ecological Science in the Twenty-first Century," *BioScience* 61 (2011): 442–450.
7. Frans W. M. Vera, "Large-Scale Nature Development—the Oostvaardersplassen," *British Wildlife* 20, no. 5 (June 2009): 28–36; Marcus Hall, "Extracting Culture or Injecting Nature: Rewilding in Transatlantic Perspective," in *Old World and New World Perspectives in Environmental Philosophy: Transatlantic Conversations*, ed. M. Drenthen and J. Keulartz (New York: Springer, 2014).
8. Paul Robbins and Sarah A. Moore, "Ecological Anxiety Disorder: Diagnosing the Politics of the Anthropocene," *Cultural Geographies* 20, no. 1 (2013): 3–19.
9. Elizabeth Kolbert, "Recall of the Wild: The Quest to Engineer a World before Humans," *The New Yorker*, December 12, 2012, 50–60; Jamie Lorimer and Clemens Driessen, "Bovine Biopolitics and the Promise of Monsters in the Rewilding of Heck Cattle," *Geoforum* 48 (2013): 249–259.

10. Francesco Guicciardini quoted in Fernand Braudel, *The Mediterranean and the Mediterranean World in the Age of Philip II*, trans. Siân Reynolds, 2 vols. (New York: Harper & Row, 1966 [1949]), I, 66; Andrea Tosi, ed., *Degrado ambientale periurbano e restauro naturalistico* (Milano: FrancoAngeli, 1999); "Ingegneria Naturalistica," Regione Lazio, http://www.regione.lazio.it/web2/contents/ingegneria_naturalistica/. On the many linguistic conceptions of "wilderness" in several of the world's languages, see the descriptions of "Wilderness Babel," http://www.environmentandsociety.org/exhibitions/wilderness/overview.

11. Compte de Buffon, "De la Nature: Premiere Vué," *Histoire Naturelle, Générale et Particulière, Avec la Description du Cabinet du Roi* trans. Phillip R. Sloan, vol. 112 (1764), iii–iv, quoted in Clarence Glacken, *Traces on the Rhodian Shore: Nature and Culture in Western Thought from Ancient Times to the End of the Eighteenth Century* (Berkeley: University of California Press, 1967), 663; Carolyn Merchant, *Reinventing Eden: The Fate of Nature in Western Culture* (New York: Routledge, 2004), 11–38; David Lowenthal, "Reflections on Humpty-Dumpty Ecology," in *Restoration and History: The Search for a Usable Environmental Past*, ed. Marcus Hall (New York: Routledge, 2010), 13; Otto Bettmann, *The Good Old Days: They Were Terrible!* (New York: Random House, 1974).

12. Society Club International Online Record Book, http://www.scirecordbook.org/alpine-ibex-europe/.

13. Franco Rollo et al., "Oetzi's Last Meals: DNA Analysis of the Intestinal Content of the Neolithic Glacier Mummy from the Alps," *PNAS* 99, no. 20 (October 1, 2002): 12594–12599; Michael Pollan, *The Botany of Desire: A Plant's-Eye View of the World* (New York: Random House, 2001).

14. Achaz von Hardenberg and Bruno Bassano, "Long-term Ecological Research in Protected Areas: The Example of Alpine Ibex in the Gran Paradiso National Park," (2005). http://www.landesmuseum.at/pdf_frei_remote/NP-Hohe-Tauern-Conference_3_0073-0074.pdf; Georgio Anselmi quoted in James Sievert, *The Origins of Nature Conservation in Italy* (Bern: Peter Lang, 2000), 194; Marcus Hall, *Earth Repair: A Transatlantic History of Environmental Restoration* (Charlottesville: University of Virginia Press, 2005), 185.

15. Wildpark Peter and Paul, http://www.wildpark-peterundpaul.ch/html/geschichte.htm; Josh Donlan and Harry Greene "NLIMBY: No Lions in My Backyard," in *Restoration and History*, ed. Marcus Hall (New York: Routledge, 2010): 293–305.

16. Michael Stüwe and Kim T. Scribner, "Low Genetic Variability in Reintroduced Alpine Ibex (*Capra Ibex Ibex*) Populations," *Journal of Mammalogy* 70, no. 2 (May 1989): 370–373.

17. David M. Shackleton, ed., *Wild Sheep and Goats and their Relatives* (Gland: IUCN, 1997), 25–26, 66–67; Gisla Moço et al., "The Ibex Capra Pyrenaica Returns to its Former Portuguese Range," *Fauna and Flora International* 40 (2006): 351–354; Gavin Hudson, "Extinct Ibex Resurrected by Cloning ... then Goes Extinct Again," *Ecoworldly*, February 1, 2009, http://ecolocalizer.com/2009/02/01/extinct-ibex-resurrected-by-cloning-then-dies/. On the possibilities of bringing back extinct animals, see Stuart Brand, "The Dawn of De-Extinction. Are You Ready?" TED Talk, February 2013, http://www.ted.com/talks/stewart_brand_the_dawn_of_de_extinction_are_you_ready.html.

18. World Association of Zoos and Aquariums, "The Fall and Rise of the King of the Alps," http://www.waza.org/conservation/projects/projects.php?id=59; Safari Club International Online Record Book, http://www.scirecordbook.org/alpine-ibex-europe/;

Gila Kahila Bar-Gal et. al., "Genetic Evidence for the Origin of the Agrimi Goat (*Capra aegagrus cretica*)," *Journal of Zoology* 256, no. 3 (February 28, 2006): 369–377.

19. E.-D. Schulze et al., "Land-Use History and Succession Of *Larix Decidua* in the Southern Alps Of Italy–An Essay Based on a Cultural History Study of Roswitha Asche," *Flora* (2002): 705–713; *Spiegel Online International*, "Brown Bears Mysteriously Disappearing in the Alps," November 22, 2007, http://www.spiegel.de/international/europe/0,1518,518961,00.html; Nathalie Espuno et al., "Heterogeneous Response to Preventive Sheep Husbandry during Wolf Recolonization of the French Alps," *Wildlife Society Bulletin* 32 (2004): 1195–1208; Christian Glenz et al., "A Wolf Habitat Suitability Prediction Study in Valais (Switzerland)," *Landscape and Urban Planning* 55 (2001): 55–65.

20. Werner Bätzing, "La popolazione alpina: dall'urbanizzazione all'esodo dal territorio," in *Rapporto sullo Stato delle Alpi* (Torino: CIPRA, 1998): 90–98.

21. Hall, *Earth Repair*, 1167–1171; George Perkins Marsh, *Man and Nature; or, Physical Geography as Modified By Human Action* (New York: Scribner, 1864), 103; see also, Marcus Hall, "The Provincial Nature of George Perkins Marsh," *Environment and History* 10 (2004): 191–204; Trees for Life, "Proposal for the United Nations to Declare the 21st Century as 'The Century of Restoring the Earth,'" http://www.treesforlife.org.uk/tfl.intnl.html; Alan Featherstone, "Restoring the Earth," *Resurgence* 199 (March/April 2000), http://www.resurgence.org/magazine/issue199-You-are-therefore-I-am.html.

22. Remo Guerra, "I settanta anni del Gran Paradiso," *Parchi* 7 (November 1992): 26–28; "Parco Nazionale Gran Paradiso," *Correrenelverde*, http://www.correrenelverde.it/ambienteenatura/parchi/granparadiso.html; Richard White, "'Are You an Environmentalist or Do You Work for a Living?': Work and Nature," in *Uncommon Ground: Rethinking the Human Place in Nature,* ed. William Cronon (New York: Norton, 1995), 171–185; Yukon to Yellowstone Initiative, http://y2y.net/.

23. Joshua Zaffos, "Reinstating the Heir of the Truckee River," *High Country News*, July 7, 2003; Nate Schweber, "20 Pounds? Not Too Bad, for an Extinct Fish," *New York Times*, April 23, 2013.

24. Gordon Richards, "Demise of the Lahontan Cutthroat," Truckee Donner Historical Society, Inc., 2004, http://truckeehistory.owrg/historyArticles/history18.htm; Jonathan Stead, "Exploring Reintroduction of Lahontan Cutthroat Trout in a Headwater Stream," (Ms thesis, University of California, Davis, 2007), http://www.scientificjournals.org/journals2007/articles/1326.pdf; Francis Newlands quoted in Philip Perimutter, *Legacy of Hate: A short History of Ethnic, Religious and Racial Prejudice in America* (New York: M. E. Sharpe, 1999), 152; Samantha Carmichael, "Behnke Revisites [sic] Topic of Lahontan Cutthroat," *Trout Unlimited*, September 13, 2010), http://troutunlimitedblog.com/behnke-revisites-topic-of-lahontan-cutthroat/; Brett Prettyman, "Don Duff's Big Discoveries—Cuban Missiles and Extinct Cutthroat Trout populations," *Salt Lake Tribune*, October 29, 2012; Terry James Hickman, "Systematic Study of the Native Trout of the Bonneville Basin," (Ms thesis, Colorado State University, 1978). Dan Levin, "The Fish that Wouldn't Die," *Sports Illustrated*, March 17, 1980, http://sportsillustrated.cnn.com/vault/article/magazine/MAG1123274/2/index.htm.

25. Dave Foreman, *Rewilding North America: A Vision for Conservation in the 21st Century* (Washington, DC: Island Press, 2004), 6.

26. The classic statement on shifting ecological baselines is Daniel Pauly, "Anecdotes and the Shifting Baseline Syndrome of Fisheries," *Trends in Ecology and Evolution* 10, no. 10 (1995): 430. Bella Galil et al., "First Record of Mnemiopsis leidyi A. Agassiz, 1865

(Ctenophora; Lobata; Mnemiidae) off the Mediterranean Coast of Israel," *Aquatic Invasions* 4, no. 2 (2009): 356–362.

27. "The Journey to Restore America's Everglades," http://www.evergladesplan.org/about/about_cerp_brief.aspx.

28. "Restoring Biodiversity," Conservacion Patagonica, http://www.conservacionpatagonica.org/buildingthepark_biodiversity.htm; Hall, *Earth Repair*, 122–125; Mohan K. Wali, ed., *Ecosystem Rehabilitation—Preamble to Sustainable Development*, 2 vols. (The Hague: SPB Academic Publishing, 1992); Robert France, "Landscapes and Mindscapes of Restoration Design," in *Healing Natures, Repairing Relationships: New Perspectives on Restoring Ecological Spaces and Consciousness*, ed. Robert France (Sheffield, VT: Green Frigate Books, 2007).

29. William R. Jordan III, "Conclusion: Autonomy, Restoration, and the Law of Nature," in *Recognizing the Autonomy of Nature: Theory and Practice*, ed. Thomas Heyd (New York: Columbia University Press, 2005), 202; Peter Novak, *That Noble Dream: The "Objectivity Question" and the American Historical Profession* (New York: University of Cambridge Press, 1988); "Save China's Tigers," http://english.savechinastigers.org/rewilding; Hall, "Extracting Culture or Injecting Nature: Rewilding in Transatlantic Perspective," in *Old World and New World Perspectives in Environmental Philosophy: Transatlantic Conversations*, ed. M. Drenthen and J. Keulartz (New York: Fordham University Press, 2014).

30. Stephen Jackson, "Ecological Novelty is Not New," in *Novel Ecosystems: Intervening in the New Ecological World Order*, ed. R. Hobbs, E. Higgs, and C. Hall (New York: Wiley-Blackwell, 2013), chapter 7; Christoph Kueffer, "Understanding and Managing Ecological Novelty: Towards an Integrative Framework of the Socio-Ecological Risks of Novel Organisms," conference, Mt. Veritá, Switzerland, September 4–9, 2011.

31. A. Starker Leopold, et al., "Wildlife Management in the National Parks: The Leopold Report," (1963) http://www.nps.gov/history/history/online_books/leopold/leopold.htm.

32. W. C. Hoskins, *The Making of the English Landscape* (Harmondsworth, UK: Penguin, 1985 [1955]), 11.

33. Bruce McCune and T. H. F. Allen, "Will Similar Forests Develop on Similar Sites?," *Canadian Journal of Botany* 63 (1985): 367–376.

34. I borrow these insights from William Cronon, "The Uses of Environmental History," *Environmental History Review* 17, no. 3 (1993): 1–22.

CHAPTER 13

....

REGION, SCENERY, AND POWER

Cultural Landscapes in Environmental History

....

THOMAS LEKAN AND THOMAS ZELLER

ENVIRONMENTAL historians have been very successful in recent years in bringing the environment from the background to center stage in historical narratives. Their work has received considerable attention from other scholars who do not define themselves as environmental historians. In creating its own academic space, however, environmental history has not always acknowledged its intellectual debts to other fields. Geography, as the sometimes estranged great-uncle of environmental history, is chief among these neighboring fields. The concept of cultural landscape is and has been one of the strongest ligaments tying geography and environmental history together. Environmental historians continue to draw eclectically from geographic landscape studies, yet most environmental historians remain unaware of the longer conceptual pedigree of cultural landscape within Euro-American geographic thought, just as geographers remain sometimes unaware of environmental historians' contributions to the "telling of place stories."[1] Such lack of engagement with each other's research has prevented a deeper and potentially fruitful dialogue between these fields of inquiry.

Part of the reason for this lack of engagement between the two fields has been the shifting status of the cultural landscape within geographic studies. The cultural landscape was once a dominant concept in European human, cultural, and historical geography. It was also central to Carl Sauer's Berkeley School of landscape analysis from the 1920s to the 1950s. Yet the cultural landscape fell out of favor as a result of a shift toward quantification within geographic science in the 1960s and 1970s. Quantifiers such as Richard Hartshorne found the term "flaccid and unacceptable" and initiated an "assassination" of landscape that deepened existing divisions within geography. This division split scholars who embraced historicism, hermeneutics, and humanistic approaches from those who were allied more closely with empiricism, logical positivism, and quantifiable "space."[2] Similarly, the lingering association of landscape with scenery distinguished it from ecology, especially the concept of the ecosystem, which environmental historians such as Donald Worster have seen as an organic baseline for measuring human impacts in nature and, in turn, understanding nature's effect on human history.[3]

The new cultural geography of the 1990s brought both a renewed interest in and critique of the cultural landscape among geographers who were exploring postmodernist and poststructuralist social theories emphasizing discourse, identity, and the nexus of power and knowledge.[4] The outlines of this cultural turn in geography paralleled the epistemological and methodological debates within environmental history that occurred in that same decade. During this time, many environmental historians were forced to reassess nature's agency and the ontology of wilderness.[5] Many turned to landscape to convey the entangled and hybridized middle ground between nature and culture, society and space, and the organic and the machinic, but often with little reflection on the ontological status and political genealogy of the concept.[6]

This essay explores the ways that environmental historians can make landscape a useful category of environmental-historical analysis that examines the dynamic and reciprocal interplay between representations of nature, political and social development, and ecological change at varied scales and across different regions of the globe. After surveying briefly the history of cultural landscape within nineteenth-century and early twentieth-century geography and landscape preservation, this chapter analyzes the convergence of themes within the new cultural geography and recent environmental historiography on human-nature interaction and the agency of nature. It then examines the ways that recent work in cultural geography on landscape might invigorate environmental historians' own research, especially as the subdiscipline is becoming more transnational and takes seriously postcolonial critiques of Western nature conservation and nature perception that have revealed "untrammeled spaces" as homelands for local and indigenous peoples.

REGION OR SCENERY? HISTORICAL AMBIGUITIES OF THE TERM "CULTURAL LANDSCAPE"

The methodological controversy in geography over landscape stems in part from the term's dichotomous etymological roots. The first of these roots are the Old English and German notions of landscape (Dutch: *Landscap*; German: *Landschaft*) as a district owned by a lord or an area inhabited by a particular group of people; the second derives from late sixteenth-century Dutch landscape painters, who emphasized the visual or aesthetic traits of a place.[7] Historically, the first definition of landscape as an area or region prevailed in American geography until the 1930s, when geographers paid close attention to the cultural expression of human settlement that the researcher could detect using visual analysis of spatial patterns. This close association of the term with the *visual* appearance of the land and its inhabitants, however, was inspired initially by the emergence of landscape painting as a prominent genre of artistic production, which has led art historians to contribute numerous works on the definition and significance of landscape as a category of humanistic inquiry.

As Denis Cosgrove, Ann Bermingham, and others have argued, landscape paintings were not intended merely as pretty ornamentation in aristocratic or bourgeois households. In the seventeenth and eighteenth centuries, British and Dutch landowners, including bourgeois who bought their land for speculative purposes, often commissioned scenic paintings of their estates as symbolic weapons in the battle between feudal legal systems and capitalist land markets; Bermingham in particular examines the ways in which these paintings furthered the enclosure of common lands by justifying it visually.[8] Bird's-eye views using mathematically precise perspectives were important tools for the artistic representation of the land. As Cosgrove argues, "*landscape* privileges the sense of sight, and what started as a representation of space rapidly became a designation of material spaces themselves, which were referred to as landscapes and viewed with the same distanced and aesthetically discriminating eye that had been trained in the appreciation of pictures and maps."[9] For Cosgrove, in other words, the aesthetic conventions of landscape painting legitimized the development of capitalist land markets and elite social control.

Beginning in the 1920s, Carl Sauer and his acolytes tried to expunge these aesthetic and subjective dimensions of landscape from their analysis by narrowing their focus to the impact of human settlement over time on a particular region. They believed that landscape as "altered nature" could thereby be treated objectively as a byproduct or artifact of human habits and habitation.[10] But geographers inclined toward phenomenology and hermeneutics, as well as artists, novelists, and philosophers interested in the symbolism of nature, have also emphasized human *responses* to this apparently objective arrangement of material forms. For geographers who sought to transform geography into a mathematical "spatial science," however, this blending of region (which they assume can be analyzed objectively) with scenery irreparably contaminates landscape with subjective meanings and social complexities. For others, including many environmental historians and historical geographers, the methodological ambiguity of landscape provides a flexible tool that conveys the complex interweaving of space and society in a given territory.

Landscapes of the Nation

The blending of space and scenery in the concept of cultural landscape also made it an attractive medium for articulating regionalist and nationalist identities in nineteenth-century Europe, a legacy that has also fueled skepticism regarding the concept's methodological usefulness, particularly in Germany. There, the regionalist, nationalist, and ethno-racist meanings of landscape competed for academic recognition and political relevance. Even before the country was unified in 1871, the Romantic movement had invested certain regional landscapes and natural landmarks, such as the Black Forest, the Lüneburger Heath, the Bavarian Alps, and the Rhine River Valley, with spiritual and nationalist significance.[11] At the birth of German nationhood during the Napoleonic Wars, for example, Romantics such as Johann Gottfried Herder extolled untouched nature as a source of both divine presence and the Germanic character. Poets such as Friedrich Hölderlin and Novalis celebrated the untamed forest as a font of sublime sentiment, while nationalists such as Friedrich Jahn (the father of German gymnastics), the playwright Heinrich von Kleist, and the painter Caspar David Friedrich imagined the woodlands as the primeval home of the ancient Germanic tribes. Nationalists proposed that these sylvan clans had once rebuffed the advance of Roman civilization, setting a historical precedent for their contemporaries to cast off the yoke of Napoleon's "Latin" tyranny.[12]

Whereas these writers focused on the nationalist meaning of the forest, most of Germany's amateur geographers and naturalists looked to regionalist writers such as Wilhelm Heinrich Riehl to formulate their belief in cultural landscape diversity as the key to Germany's unique national character. A student of Herder, Riehl insisted in his multi-volume *Natural History of the German People* (1853–1869) that national character emerged organically from the topography and culture of a particular territory, rather than from abstract declarations of individual rights.[13] Riehl used the term cultural landscape, or *Kulturlandschaft*, to refer to the particular *physiognomy* of regions or townscapes produced by the interaction of various German "tribes" (*Stämme*) with the topographical particularities (climate, soil, waterways, etc.) of a given region. To his mind, one could quite literally visualize the distinctive features of a particular people, or *Volk*, by looking at ordinary features of the vernacular landscape, such as farms, peasant cottages, agricultural fields, or cemetery layouts, or by hearing them in the place names articulated in the various dialects of German.

In Riehl's view, this cultural landscape, which was the result of centuries of interaction between human communities and their natural surroundings, was the spatial and symbolic glue that anchored each German to a particular locality despite the dislocations of political and social modernization. Unlike American wilderness advocates such as John Muir, who valued spaces devoid of human influence, German landscape regional writers envisioned the ideal *Landschaft* as a pastoral scene that blended the natural, cultivated, and built environments in an aesthetically harmonious whole.[14] To be sure, cultural landscape traditions did develop in the United States, especially in the Northeast. While Muir and his disciples were extolling the supposedly wild Western territories of the United States, for example, local Vermonters in the late nineteenth century embraced the vision of a cultivated landscape, expressing the imposition of a moral order as the result of hard work. And for artists and those following them, the landscapes of the Hudson Valley in upstate New York achieved national prominence and meaning, leading to numerous comparisons between the Hudson and the Rhine in American and European travel accounts.[15] Yet by the late nineteenth century, anxieties about the "closing" of America's Western frontier, which famed historian Frederick Jackson Turner viewed as a threat to American individualism, ethnic assimilation, and democratic traditions, focused US federal policy on the expansion of national parks, not cultural landscape preservation.[16] For the disciples of Riehl in Germany, on the other hand, the two meanings of the landscape—the cultural and the natural—were often intertwined, so that the visual state of the *Landschaft* was thought to mirror the spiritual condition of the community. The landscape was the foundation of a harmonious homeland, or *Heimat*, signifying a uniquely and often regionally defined German sense of place that encompassed dialect, regional cuisine, customs, and distinctive landscapes.

While some scholars have interpreted Riehl's folkloric perspective as an ominous precursor of Nazi "blood and soil" rhetoric, his real foil in this mid-nineteenth-century context was Germany's nationalist rival France, not Jews or other "racial enemies." In that Gallic land, Riehl averred, "culture" emanated from the Parisian center outward into the provinces, whereas the German lands remained a collection of dozens of separate kingdoms, states, duchies, principalities, and free cities with territorial antecedents in the Holy Roman Empire and rival powers centers in Berlin and Vienna. This territory was thus a mosaic of culturally and geographically distinctive homelands that made up the whole without losing the integrity of the parts. Riehl's views created a new moral geography that charged ordinary features of the landscape—fields of grain, towpaths, meandering streams, even indigenous plant and animal species—with symbolic meaning. He argued that Germany's strength lay in the diversity of its regional landscapes, and elevated emotional identification with one's *Heimat* to a form of national patriotism. Even after Germany's unification in 1871, the country's status as a "nation of provincials," a country forged by Otto von Bismarck's agglomeration of once-independent provinces, cities, and kingdoms, led to a decentralized conception of homeland.[17]

This close association between the *Kulturlandschaft* and homeland sentiment ensured the concept a central role in Germany's early nature protection campaigns as well, a point often overlooked by environmental historians familiar only with the "Great Wilderness Debate" in North American or Australian contexts. As Riehl once remarked, "we must retain the forest not only to keep our stoves from growing cold in winter, but also to keep the pulse of our nation beating warmly and happily. We need it to keep Germany German."[18] Following Riehl, many nineteenth-century observers recast the aesthetic enjoyment of meadows and rock formations as a form of nationalist devotion, arguing that such landmarks needed to be revered and protected to sustain the German *Volk*. The members of Germany's early twentieth-century nature conservation (*Naturschutz*) and homeland protection (*Heimatschutz*) movements accepted that few "natural" areas still existed in Germany, but that ordinary landmarks, "natural monuments," and scenic areas deserved state protection.[19] They set aside historic oak trees, scenic agricultural fields, and indigenous species as symbols of a natural *Heimat* that was disappearing under the onslaught of new factories, mines, roads, power lines, and apartment buildings. In their eyes, the rationalized landscapes produced by modern industry also fueled social tensions and class divisions: how could factory workers and young people, they asked, feel loyal to a *Heimat* full of smoking factories, polluted slums, hydroelectric dams, and crass billboards? By protecting natural monuments and providing "healthy" recreation areas, politically and socially conservative preservationists hoped to reintegrate Germans into the imagined *Heimat* community and stave off the "red menace," rooting industrial workers into the national homeland to counter the attraction of socialist internationalism. Wilhelmine preservationists thus believed that environmental improvement and social stability went hand in hand, with homeland sentiments providing a bulwark against the leveling tendencies of modernity.[20]

German conservationists did admire the US government's decision to set aside large tracts of land as national parks, including Yellowstone in northwestern Wyoming (1872) and Yosemite in California (1890).[21] Organizations such as the Nature Protection Park Society, founded in Stuttgart in 1900, argued that Germany and the Austrian lands should follow America's lead by establishing such reserves in Central Europe, yet promotion of such large-scale reserves remained the exception rather than the rule in Imperial Germany.[22] The majority of nature conservationists and homeland protectionists prized the unspectacular, human-shaped vernacular landscapes of their regional and local homelands, not remote, sublime wilderness.[23] This humble perspective reflected not only the spatial limitations of Central Europe, but also longstanding institutional and cultural patterns of provincial self-determination and localist environmental perception. Hugo Conwentz, the head of Prussia's State Office for Natural Monument Preservation (*Staatliche Stelle für Naturdenkmalpflege*), recognized that large-scale reserves such as Yellowstone were inappropriate given Germany's economic conditions, size, and population density; this one park alone would have encompassed the entire Kingdom of Saxony. Unlike the National Forest Service or the National Park Service, the Prussian State Office served as an advisory agency for a decentralized array of regional and local efforts to secure individual natural objects or smaller conservation regions. In line with localist traditions in German environmental perception, Conwentz argued that a handful of large-scale reserves could never accomplish the broad-based environmental education necessary to secure the country's natural monuments. "It is much more appropriate and feasible," he suggested, "to preserve in their original state smaller areas with varied characteristics scattered throughout the entire region, preferable in every part of the country...here an erratic block, a piece of terminal moraine, or a group of cliffs, there a small moor, a heath, or a stretch of woodlands."[24] Conwentz also indicated that it was neither practical nor desirable to place all responsibility for environmental preservation in state hands. German conservationists relied instead on small-scale, public-private partnerships that united municipal governments, private organizations, and individual owners in common cause with the state to help local entities to identify and protect sensitive areas according to community priorities. European conservationists had thus learned to appreciate a middle ground between wilderness and civilization long before their counterparts in the US environmental movement.

CULTURAL LANDSCAPES IN EUROPEAN GEOGRAPHIC THOUGHT

While *Heimat* advocates explored the flora of nearby woods and relics of bygone settlements, Germany's academic geographers focused on the delineation and investigation of national spaces, as did their counterparts across Europe. The first university department of geography was founded in Germany in 1874, while the University of Oxford offered Britain's first undergraduate geography course in 1887. At this time, the discipline was dominated by a few individuals, including Friedrich Ratzel and Alfred Hettner in Germany; Paul Vidal de la Blache in France; A. J. Herbertson and Halford Mackinder in Great Britain; and W. M. Davis and Ellen Churchill Semple in the US. These early professional geographers operated with a narrowly determinist view of the relationship between environment and society; much of the task of early regional geography was to map these variations and to relate them to external, environmental factors.[25]

Conceptual and definitional work was necessary, and landscape was often both focus and locus. Scholars in Germany such as August Meitzen (1822–1910) and Siegfried Passarge (1867–1958) differentiated between a *Naturlandschaft* (natural landscape), the geographical entity untouched by man and rarely to be found in Europe, and the *Kulturlandschaft* (cultural landscape), the terrain resulting from the interplay between geography, climate, and a specific ethnic or national group.[26] According to Meitzen and Passarge, the visual appearance of an era could be dissected and examined scientifically, as an example of *Landschaftskunde*, the scientific study of landscape. For them, landscape was a palimpsest consisting of different layers, which could be analyzed and presented as evidence of the interplay between humans and their environment. Classifications of different types of villages, vernacular architecture, farming techniques, ground cover, and technological infrastructure would lead to landscape typologies. The goal was to categorize and analyze landscapes and to show the historical human-environmental interactions which had led to the current configuration of space and settlement.

At the same time, many German geographers' scientific efforts reflected a desire to find categorically distinct German landscapes. Well-known are the efforts by Friedrich Ratzel (1844–1904), who theorized that the state was an organic entity engaged in a Darwinian fight with other states over territories and "living space" (*Lebensraum*).[27] Ratzel was a fierce nationalist who fought in the Franco-Prussian War and strongly welcomed the unification of Germany under Prussian hegemony in 1871. Undoubtedly, his scholarly work was influenced by Herbert Spencer's popularizations of Charles Darwin's theories. Ratzel's particular concern was with the survival of states. Unlike previous geographers who looked for humankind's cultural footprints on earth, Ratzel sought to understand the effects of different physical features on historical developments. States, in his view, are never static; they must either grow or die, and a growing state shows its strength by usurping the living spaces of other states. Though states were driven by growing populations toward expansion, Ratzel also proposed that they were adapted to a particular homeland soil and climate; their borders were like membranes that protected the organism from outside invaders and processed raw materials for nourishment and industry. While Ratzel himself did not believe in the idea of superior and inferior nations and races, many of his later followers did. And while his interest in anthropogeography (the effect of geography on humankind) was never crudely deterministic, some of his followers, many of them outside of Germany, can be rightly accused of such reductionism. One of his students and a key representative of environmental determinism was the American Ellen Churchill Semple.[28] And when it was stripped of its lingering essentialism, this landscape morphology underpinned Carl Sauer's contribution to American geographical discourse from the 1920s onwards. Such environmental determinism also shaped the emerging science of ecology as well. The German botanist Oscar Drude produced *Deutschlands Pflanzengeographie* ("Germany's Geography of Plants") in 1896, which tied specific plants to specific environments. Drude's work was seminal for the adoption of the teleological climax community idea in the United States, formulated by Frederic Clements and Roscoe Pound, which achieved its apogee in the interwar era.[29]

During the 1920s, the era of Germany's first democratic regime, the Weimar Republic, landscapes studies reached a high point in both sophistication and politicization. Given the contentious and unstable history of this democratic interlude between a constitutional monarchy and a fascist dictatorship, geographic thought became enmeshed in battles over belonging, space, and home. In this context, the nationalist and Darwinist meanings of landscape became even more extreme, and environmental determinism figured prominently in these debates. While geographers were pondering landscapes in ever more intricate definitional debates, popular and semi-academic coffee-table books by Nikolaus Creutzburg and Eugen Diesel gained a wide audience.[30] Creutzburg aimed at capturing a totality of global landscapes by both displaying and describing them; Diesel's depictions and descriptions concerned what he perceived to be the Germanness of specific landscapes.

The Weimar era also witnessed a flourishing of geographical ideas that linked the historical evolution and political development of nations to their environmental conditions. Known as geopolitics, this new social-scientific perspective rediscovered and appropriated the pre-1914 work of Friedrich Ratzel, especially his notion that states, like organisms in the natural world, struggled against other states for natural resources and territory. In the geopolitical view, ordinary features of the landscape such as mountains, streams, and plains were not merely a passive setting for the unfolding of human agency, but worked actively as a "sculptor" of peoples. Geopoliticians in the 1920s envisioned this environmental determinism as a scientific field guide for statesmen. They believed that awareness of the state's natural conditions of existence would help officials to identify and realize the nation's destiny.[31]

Geopolitical theories were contradictory and, in the end, more programmatic than prescriptive. They gained acceptance in post-World War I Germany because they helped ordinary people to understand Germany's military defeat and its loss of colonies in the wake of the Versailles Treaty. Geopoliticians proposed that the nightmare of the two-front war and the British blockade of supplies to the homefront had reflected German military planners' inability to grasp the environmental limitations of Germany's vulnerable place in the center of Europe. Geopolitical ideas also fueled nationalists' sense of injustice with the Versailles Treaty; if borders were indeed membranes, then they could hardly be arbitrarily redrawn by politicians and diplomats. Geopolitics thus offered a scientific explanation for war and defeat beyond the confusing flux of recent historical events and a vague yet all-encompassing program of national renewal. If the climate, soil, rivers, and mountains were the true sculptors of peoples, then only a return to these biological foundations and a cultivation of extensive geographic knowledge could renew German society.[32] Not surprisingly, an essentialist, racialized version of the link between humans and their environment was one of the key tenets of German geographers' later ideological allegiance to the Third Reich's biopolitics.

During the Nazi dictatorship, Karl Haushofer provided the regime with the buzzword "Lebensraum," yet he prioritized space over race in his analysis of world-historical forces, whereas the Nazi "racial state" pursued biological racism toward expansionary and ultimately genocidal ends.[33] Rather than building upon Ratzel's and Haushofer's environmental determinism, Nazi ideologues, such as the Agricultural Minister Richard Walther Darré, made racial rejuvenation the regime's top priority and coined the phrase "blood and soil" to glorify the peasantry as the "racial stock" necessary for the rebuilding of Germany's bloodlines.[34] Such a worldview assumed that the "asphalt culture" found in cities such as Berlin was the result of alienation from the land caused by insidious "foreign races" or "racial degeneration," and that a typically German landscape had to be defended against the encroachments of un-German forces. If, in other words, landscapes became neglected and derelict, geographers postulated, this was the result of allowing Jews to play a dominant role in German society and economy. Indeed, four years before the Nazi takeover, Passarge had already published a book on "Judaism as a problem in landscape studies and ethnology" in which he suffused his geographical approaches with anti-Semitic beliefs.[35]

"Blood and soil" also provided Nazi practitioners with a justification for invading Poland and the Soviet Union, for it was only through encouraging rural settlement and homesteading in Eastern Europe that the Nordic race could stem the degenerative effects of city life.[36] The massive resettlement plans for Poland, based on the extermination of Jews and the enslavement of Slavs, showed the enthusiasm of landscape architects, geographers, and other planners to work on a racially cleansed *tabula rasa*. Landscape cleansing was seen as a necessary part of ethnic cleansing and mass murder; racial rejuvenation depended upon "restoring" and "Germanizing" the landscape.[37] Within Germany, landscape architects were also active in promoting a new kind of fascist industrial culture, one that overcame the "profligate" liberal era by harmoniously blending technology and nature. The Nazi autobahn was to be the showcase for such "German technology," and the propaganda for the roadways claimed that the concrete highways embellished rather than destroyed the rural landscape and touted the collaborative roles of landscape architects and civil engineers in building a new "crown" for the landscape. However, the results did not live up to the public claims, and regional conservationists, who had welcomed the rise of the Nazi dictatorship with open arms, were soon disappointed with the high-handedness of central planning.[38]

All in all, the Nazis employed the culturally charged term landscape to great effect in order to solidify their power in Germany, even though many Germans failed to realize the deeper racist meaning of their appeal to blood and soil. Rather than protecting the rural countryside, moreover, the regime pursued massive industrialization and preparation for war that led to increased urbanization and more intensive agriculture and forestry. Despite this gap between rhetoric and practice, the Nazi experience tainted landscape as a topic of geographic analysis and conservationist concern for decades to come, especially in the German-speaking countries. In Europe, French historians associated with the Annales School pioneered the integration of environment and history in a new form of *histoire totale*, while in the United States, Carl Sauer developed a new focus on human agency that played a critical role in many early works in American environmental history.

CARL SAUER, THE BERKELEY SCHOOL, AND AMERICAN GEOGRAPHIC THOUGHT

The subdiscipline of cultural geography, with its model of cultural landscape developed by Carl Sauer at the University of California at Berkeley during the 1920s and 1930s, was intended to break with the European emphasis on environmental determinism by reintroducing and elaborating upon the role of human agency in shaping the Earth's surface.[39] Sauer's work and that of his doctoral students dominated the middle decades of the twentieth century—so much so that historians of geography refer to this cohort as the "Berkeley School" to distinguish it from the "spatial science" that was being developed at Midwestern research universities from the 1940s onward. In addition to breaking with environmental determinism, the Berkeley School was known for specific research questions that reflected Sauer's personality and interests, most notably his life-long interest in ecological anthropology and the diffusion of premodern agricultural techniques in Meso-America.[40] Sauer laid out his model for studying the cultural landscape in the seminal 1925 essay "The Morphology of Landscape," though the most comprehensive expression of Berkeley school themes appeared in 1956 under the title *Man's Role in Changing the Face of the Earth*, a volume whose dire warnings about resource depletion, species extinction, soil erosion, and overpopulation signaled a renewed geographic interest in the limits of the earth to sustain human populations in the modern age.[41]

When Sauer began his career, Ratzelian environmental determinism tinged with social Darwinism still dominated the American scene. Ellen Churchill Semple of Clark

University, who was a follower of Ratzel's anthropogeography, reflected this in her classic *Influences of Geographic Environment* of 1911, in which she wrote:

> The mountain-dweller is essentially conservative. There is little in his environment to stimulate him to change, and little reaches him from the outside world...With this conservatism of the mountaineer is generally coupled suspicion towards strangers, extreme sensitiveness to criticism, strong religious feeling and an intense love of home and family. The bitter struggle for existence makes him industrious, frugal, provident; and, when the marauding stage has been outgrown, he is peculiarly honest as a rule...When the mountain-bred man comes down to the plains, he brings with him therefore certain qualities which make him a formidable competitor in the struggle for existence—the strong muscles, unjaded nerves, iron purpose and indifference to luxury bred in him by the hard conditions of his native environment.[42]

Sauer found such an account of human-environmental interactions inadequate for explaining the form, function, and use of landscapes and turned to the work of George Perkins Marsh, among others, to demonstrate how humans—working through the medium of culture—acted as active agents of environment transformation.[43] Like the early work of Meitzen and Passarge, Sauer believed that the geographer could study the forms of such interaction within a particular area and compare them to other landscapes, thereby developing a taxonomy of landscape forms that could be used to deduce underlying historical patterns of culture-nature interaction.

Sauer set forth these principles and marked his break with environmental determinism in his landmark essay "The Morphology of Landscape," in which he described the cultural landscape as a topography "fashioned from a natural landscape by a culture group. Culture is the agent, the natural area is the medium, the cultural landscape the result."[44] Sauer was especially interested in the anthropogenic shaping of natural vegetation by activities such as agriculture, fire, livestock grazing, mining, and gathering among indigenous peoples in the Americas before the era of European colonialism.[45] From these interests, he developed the concept of agricultural diffusionism, which proposed that agricultural systems in the Americas had emerged in the hilly, riverine terrain of the humid tropics, where hunting and gathering societies had had the leisure to experiment with plant domestication, and then spread to other environments, a process he traced from Central America to the US Southwest and beyond.[46] In separating agriculture's origins from its dispersal, Sauer and his students posited the physical environment—including topography, soils, watercourses, plants, and animals—as catalysts for human responses and adaptations, but rejected using them as universal causal factors in explaining the morphology and structure of human settlement patterns.[47] Despite its macro-scale and historicist precision, however, the Berkeley School geographers still relied primarily on the analysis of visual markers—agricultural fields, the dispersal of tools, and taxonomic classifications of domesticated plants, for example—to develop their arguments, a factor that later invited critique from advocates of spatial science.[48]

Sauer's visual-anthropological approach to landscape assumed a neat analytical distinction between nature and culture that allows the geographer to explore the effect of human activities upon an ecologically stable baseline.[49] For Sauer, culture acted in a direct, functionalist, and supra-organic form on the land to enable food procurement, human reproduction, and, in the modern era, profit-seeking: it is not mediated by contending socioeconomic, technological, or ideological forces. And despite his reliance on historicist methods and inductive interpretive strategies, Sauer gives virtually no weight to discrete cultural ideas, values, or perceptions as ecological agents in and of themselves.

Sauer's reliance on George Perkins Marsh to frame his analysis of human-nature interactions fueled his growing anxieties about the destructive potential of industrial technology on the environment. This shift in emphasis—from descriptive landscape morphology to environmentalist critique of ecological degradation—is already evident in his 1938 essay "Theme of Plant and Animal Destruction in Economic History," in which Sauer took social theorists to task for neglecting the "natural history of man," which, he argued, demonstrated "the revenge of an outraged nature on man" rather than a facile "mastery of man over his environment."[50] Sauer traced this "revenge of outraged nature" all the way back to a disturbance of the "symbiotic balance" during the Neolithic revolution, when settled agriculture and urban civilization denuded and desiccated wide swathes of the Mediterranean basin, the Middle East, and Central Asia. Given Sauer's interest in agricultural diffusionism in the New World, however, it is not surprising that he found his most appalling examples of wanton destruction in the Americas, where European colonialism had left a string of environmental catastrophes in its wake, such as the decimation of aboriginal peoples through epidemic diseases and the destruction of birds and mammals such as the passenger pigeon and the buffalo, which manifested the "suicidal quality" of mercantilist and capitalist extractive economies. "These are a few notes toward a history of the modern age," he wrote. "The modern world has been built on a progressive using up of its real capital."[51]

Anticipating Richard Grove's analysis of the relationship between colonial expansion and conservationism, Sauer believed that it was in such peripheries of Western civilization only recently subjected to mechanized agriculture, hydraulic engineering, and chemical fertilizers, such as the grasslands and deserts of the United States, that the worst excesses of technological hubris and capitalist expansionism were most manifest.[52] As the Dust Bowl catastrophe had demonstrated, Americans remained idealists whose "frontier attitude" contained a "recklessness of an optimism that has become habitual...We have not yet learned the difference between yield and loot."[53] By the 1950s, Sauer, along with other critics of modernity such as Pierre Teilhard de Chardin, Lewis Mumford, James Malin, and Paul Sears (all of whom appeared in the volume *Man's Role in Changing the Face of the Earth*), predicted that the spread of modern technologies and exploding populations in the developing world would soon endanger the globe's entire life support system by exhausting mineral, energy, soil, and water resources and destroying biodiversity at ever-accelerating rates. Given his ethnographic interest in Meso-American foodways, Sauer was particular worried about what modernization would mean for indigenous cultures. "Our programs of agricultural aid pay little attention to native ways and products," he lamented. "Instead of going out to learn what their experiences and preferences are, we go forth to introduce our ways and consider backward what is not according to our pattern...heedless that we may be destroying wise and durable native systems of living with the land."[54] Such hubris, he feared, reduced not only the adaptive possibilities of humankind, but limited the future range of possibilities for the organic evolution of the planet.

Despite the apocalyptic tone of many of the essays in *Man's Role in Changing the Face of the Earth*, not all of Sauer's inheritors focused on conservationist or environmentalist themes in their work. Sauer's emphasis on the visual and ethnographic interpretation of premodern agricultural fields, settlement formations, and material culture, for example, was echoed in the highly influential work of *Landscape* journal founder John Brinckerhoff Jackson, who championed the documentation of the prosaic, vernacular landscapes formed by ordinary people making a living in the natural world, much like regional homeland researchers in Central Europe. For Jackson, the individual dwelling is the elementary unit of the landscape and formed the nucleus of group identity formation that persisted across time in different regional environments.[55] Hostile to "environmentalism" as it emerged in the late 1960s, Jackson developed an eclectic, non-dogmatic approach that inspired numerous imitators, including the applied work of cultural landscape documentation and cultural resource assessment conducted by government agencies and historic landscape preservationists today.[56]

Sauer's concept of agricultural diffusionism, however, did underpin some of the best works of environmental history, including William Cronon's *Changes in the Land*, Timothy Silver's *A New Face on the Countryside*, and Mart Stewart's *What Nature Suffers to Groe*.[57] These works examined the transformation of various cultural landscapes over time—Native American, European settler, and African American—as European agricultural practices, forestry, commodity markets, and diseases were transported and adapted to a New World environment. Like Sauer, these works focus predominantly on premodern and preindustrial landscape change, although they reject his problematic tendency to delineate a teleological process of landscape evolution toward more homeostatic and complex forms. Instead, these authors emphasize the contingent historical factors that drove landscape change and the processes of hybridization that marked transitions from one anthropogenic terrain to the next. In effect, environmental historians were plowing the terrain that historical geographers had once tended by documenting the dialectical relationship between human and natural agency over time within a regional context.[58] In effect, environmental historians expanded the thematic scope and conceptual vocabulary of the Berkeley School even as their peers in geography largely abandoned cultural landscape in favor of quantitative, economic-materialist, and phenomenological approaches.

BEYOND THE BERKELEY
SCHOOL: QUANTIFICATION, MARXISM,
AND HUMANIST GEOGRAPHY

By the early 1960s the Berkeley School of cultural geography was at its apogee. Sauer's thirty-three doctoral students and, increasingly, his students' students, were established in almost all the major graduate schools in the United States and Canada. By this time, however, a quantitative turn was underway in the social sciences that would soon undermine cultural geography's central premises and methodological frameworks. This methodological shift caused a contentious reorientation and splintering of cultural geography into various camps, until its adherents began to regroup around the "new cultural geography" in the late 1980s. The quantification revolution brought increasingly sophisticated computing technologies and data sets and introduced into the discipline "central place" theories developed by German economic geographers such as Johann Heinrich von Thünen, August Lösch, and Walter Christaller. Such techniques led many geographers to steer the discipline away from its preoccupation with interpreting unique places, describing the society-nature interface, or charting the historical diffusion of agricultural practices and toward the articulation of universal spatial laws.[59] For economic geographers such as Hartshorne, the cultural landscape was an impediment to making geography a spatial science on par with economics or even physics; it was literally too superficial, dealing only with the outer epidermal layer of the earth's surface rather than deeper social laws that determined the pattern, distribution, and functions of human settlement.[60]

By the late 1960s, the quantitative revolution had elevated economic and social geography to a dominant place in the discipline, but its faith in the uncovering of universal spatial laws soon came under fire from a number of scholarly critics, including Marxists and human geographers, who pilloried what they deemed to be its simple, positivist faith.[61] Marxism's entrée into American geography was closely tied to the exodus of British academics to the United States from the late 1950s to the present, who brought with them a relatively flexible, non-dogmatic emphasis on historical materialism, class struggle, and bourgeois ideology heavily influenced by Antonio Gramsci's concept of cultural hegemony.[62] These authors also looked to the work of literary historian Raymond Williams, whose work demonstrated how moral and aesthetic values attached to the countryside in nineteenth-century Britain helped to reinforce repressive economic and political structures.[63]

Marxist-inspired scholars such as Denis Cosgrove and David Harvey used these insights to delve into social and economic questions neglected by the quantifiers, such as uneven regional development, urban inequality, and the impact of imperialist expansion in the Third World.[64] They also tackled questions of cultural hegemony; indeed, one of Harvey's best known essays explored how the Basilica of Sacré-Coeur in Paris became a powerful symbol of a resurgent Catholic monarchism on the skyline of revolutionary Paris in the wake of the Commune.[65] Harvey's semiotic approach later enabled culturally oriented Marxists to find common ground with poststructuralist and postcolonial approaches to geography. Still, Marxist geography was most illuminating when it examined the intersection of social and economic geography. It also tended to emphasize the exogenous role of dialectical materialism in shaping space, rather than the independent role of the landscape in actively shaping human society and culture.

Other human geographers rejected both spatial science and Marxist structuralism by exploring "landscapes of the mind" through existentialist, phenomenological, and psychological approaches.[66] This new generation of humanist geographers bemoaned the fact that both spatial science and Marxism took individuals out of the game in favor of collectivities, masses, and structures and created a false division between the observer and the outside world.[67] Humanists also argued that this division did violence to the subjective, emotional, and artistic bonds that human beings established with the outside world, which geographers such as Yi-Fu Tuan and David Lowenthal termed their unique "sense of place." Unlike the cultural geographers of Sauer's generation or the material cultural emphasis of Jackson and his admirers, human geographers' emphasis on sense of place explored how individuals made sense of and responded to their environments through a variety of media—artistic, literary, philosophical—leading to studies as diverse as the social values reflected in the bucolic English landscape and landscapes of fear.[68] Other geographers interested in behavioral science used interviews, surveys and questionnaires to produce generalizations about landscape cognition. This method underlay Kevin Lynch's *Image of the City* (1960), which explored how city dwellers use visual landmarks to organize their mental maps of place and enabled design professionals to promote specific planning solutions for enhancing urban quality.[69] Questions about the relative balance between Marxist structuration and humanist or cognitive agency remain an important fulcrum for studying landscape, one that was further complicated by feminist, postcolonial, and poststructuralist methods associated with the cultural turn of the 1990s, a period in which geography and environmental history developed parallel, but rarely mutually enriching, approaches to the textuality of landscape and the social and cultural construction of nature.[70]

THE CULTURAL TURN IN
HUMAN GEOGRAPHY

The cultural turn in human geography was part of a broader shift in the human sciences during the 1990s that challenged all three postwar developments in geography: quantitative geography's positivist faith in an orderly and rationally discernable universe; Marxism's assumptions about the relationship between material structures and ideology; and humanist geographers' phenomenological belief in a universal human subject. As both an epochal designation and an approach to knowledge, the "postmodern" in geography registered "incredulity towards metanarratives to grand plots that encompass the sweep of human existence in part because it supposes such metanarratives to be insensitive to the differences between peoples and places."[71] Whereas physical, economic, and social geographers had once dismissed "culture" as too fuzzy and subjective, the "cultural turn" implied that the "accumulations of ways of seeing, means of communicating, construction of value, and sense of identity should be taken as important in their own right, rather than just a by-product of economic formations... Suddenly 'culture' became intellectually fashionable as a starting point for interpretation, whereas it hitherto had been seen as lacking in rigor."[72] The quantitative turn had signaled geography's disciplinary desire to achieve the status of real science. The cultural turn marked many geographers' aversion to scientific pretension itself as a form of situated, value-laden knowledge produced in order to cement particular power relationships.[73]

The cultural turn rescued cultural landscape from analytical oblivion, but did so by jettisoning Sauer's anthropological concept of culture as a "total way of life." In fact, geographers such as Stephen Daniels and Peter Jackson argued that the very idea of culture as the possession of a bounded group of people is potentially divisive and dangerous because it reifies unnecessary boundaries between self and other.[74] They also maintained that existing landscape concepts were too atheoretical, politically conservative, and reliant on outmoded conceptualizations of culture and society.[75] Instead, they explicated how the material landscape is made meaningful through historically situated ways of seeing and social practices that generate meaning. As one cultural geographer describes it: "Cultural geography...becomes the field of study which concentrates upon the ways in which space, place and the environment participate in an unfolding dialogue of meaning."[76] Within this new definition of cultural geography, the cultural landscape is often depicted as a field of force that generates oppositional meanings without being reducible to any one representation or visual code.

The growing influence of literary studies and philosophy in postmodern thought found expression in James Duncan's analyses of the landscape as a text.[77] Such a model allowed geographers to examine which individuals or institutions "author" particular spaces in order to cement particular ideological positions or reify discriminatory notions of race, class, nationalism, or gender. As acute readers, they assert, we can deconstruct the meaning of such texts in ways that often depart from the author's intention in an open-ended process of interpretation, decoding, and destabilization.[78] Feminist geographers have found this emphasis on the textuality of landscape particularly fruitful in their explorations of the spatial dimensions of white, middle-class patriarchy, which called into question earlier phenomenological geography celebrating the place of "man" at the center of "his" world.[79] Postcolonial geographies influenced by Edward Said's *Orientalism* (1978) have also led geographers to critique the imperialist origins of their discipline and to interrogate how Western concepts of space and place have often assumed and subsumed an exotic other in ordinary techniques of mapping, journalism, or international development.[80]

Critics of the new cultural geography charge that the concentration on the semiotic qualities of landscape has led the discipline to abandon the material realm where individual and communal lifeworlds unfold. Yet most cultural geographers, much like their counterparts in environmental history, have viewed the symbolic realm as a window onto material relations, even if they no longer see regional ecology, topography, or climate as the starting point for analysis.[81] For Cosgrove and Daniels, for example, the "material" dimensions of landscape extend beyond its morphology or function and encompass the canvases, papers, photographs, and gardens in which individuals imagine the external world.[82] Don Mitchell has shown that arcadian depictions of California's irrigated fruit farms have tended to naturalize massive social and economic inequalities and obscure the social and cultural struggles over exploitation of surplus labor that went into their production.[83]

Yet the ecological thrust and conservationist urgency of Sauer's work rarely appeared in cultural-geographic works, and recent surveys of the field concede that environmental historians such as Cronon, Crosby, and Worster have more successfully engaged the interplay between cultural discourses, socioeconomic relations, and ecological change in recent decades than their peers in geography.[84] Michael Williams argues that geographers "relegated" environment in the widest sense from the 1950s to the 1970s due to the "spectre" of environmental determinism and the pressing epistemological search for new paradigms that have "diverted energies and interests away from the basic question of humans in nature, which many would regard as the nub of the subject."[85] The specificity of historical analysis can inform the level of generalization and structuration that geographers often aim for; environmental history, in this respect, has much to offer to cultural geography. Yet as environmental historians increasingly tackle questions about the social construction of nature and the class, race, and gender inequalities that result from this process, cultural geography's emphasis on power, knowledge, and space has much to offer environmental historians, as evidenced in more recent approaches.

CULTURAL LANDSCAPES AND ENVIRONMENTAL HISTORY: NEW OPPORTUNITIES

The intellectual ferment and epistemological controversies that animated the "new cultural geography" of the 1990s will be familiar to most environmental historians, who went through their own methodological debates during this same decade. William Cronon's 1995 landmark collection *Uncommon Ground* introduced readers to these new trends with contributions that engaged postmodern literary theory, cultural anthropology, and philosophy to challenge ideas about wilderness and critique the environmental movement for its white-collar, middle-class biases. In the wake of *Uncommon Ground*, environmental historians have shifted away from "Nature with a capital N" toward more complicated stories about the competing and contested "natures" of a host of different social actors and the non-human world they inhabit.[86] As Richard White noted in his 2004 essay "From Wilderness to Hybrid Landscapes: The Cultural Turn in Environmental History," this newer scholarship "emphasizes not just cultural but also hybrid landscapes rather than the wild, rural, and urban landscapes that were once treated as pure types; this often puts these scholars, even those most sympathetic to the political goals of environmentalism, at odds with sections of the modern environmental movement."[87] In this sense, environmental history in many ways has moved beyond the environmentalist moral and political impetus that drove the formation of the discipline in the 1970s—a factor that has created more common ground with the methodological and epistemological concerns of geographers and invited reflection on landscape as a category of environmental-historical analysis.

Seeing Beyond Wilderness in American Environmental History

Environmental history's cultural turn, and its growing distance from the mainstream nature conservation movement, found expression in Cronon's 1995 essay "The Trouble with Wilderness." In this essay, Cronon offered a provocative critique of the wilderness idea in American environmental thought. "The more one knows of its peculiar history, the more one realizes that wilderness is not quite what it seems," he argued. "Far from being the one place on earth that stands apart from humanity, it is quite profoundly a human creation—indeed the creation of very particular human cultures at very particular moments in human history...As we gaze into the mirror [that wilderness] holds up for us, we too easily imagine that what we behold is Nature when in fact we see the reflection of our own unexamined longings and desires."[88] The possibility of escaping into national or state parks, Cronon asserted, has exacerbated the dichotomy between nature and civilization, allowing us (one assumes he means North Americans especially) to shirk any responsibility for the more humble nature in our backyards. Cronon also argued that the aesthetic tropes that buttress our fascination with wilderness—the sublime and the frontier—reify elitist, colonialist, racist, and patriarchal assumptions about our place in the natural order. Wilderness thus denies modern societies a "middle ground" in which responsible human exploitation and natural processes might attain a balanced and socially just relationship where "city...[and]...wilderness, can somehow be encompassed in the word 'home.'"[89]

A number of environmental historians have discovered this "middle ground" in the cultural landscape, particularly as they have uncovered the tangled web of discourses, narratives, and social practices that have shaped human beings' relationship with areas once deemed "pristine" or "untrammeled," especially in the American West. Nancy Langston's *Where Land and Water Meet* (2003) and Joseph Taylor's *Making Salmon* (1999) begin with multiple and overlapping narratives, including Native American, rural Anglo-American, scientific, and urban-recreational accounts that have structured these communities' relationship to the Malheur Refuge and Columbia River salmon, respectively.[90] Though there are affinities here with Sauer's concept of culture acting as an agent on the environment, Langston and Taylor burst open culture into competing and often mutually exclusive narratives and images that do not add up to a coherent worldview. Even as they engage ideas about cultural or hybrid landscapes, however, environmental historians have for the most part maintained an epistemological commitment to narrating nature's agency in human affairs over time—a characteristic that has received short shrift in the new cultural geography. For example, Ari Kelman and the historical geographer Craig Colten have embraced landscape to describe the riverine environment of metropolitan New Orleans without losing sight of the Mississippi's power to destroy human lives, levees, and agricultural fields.[91] Nature's role in these societies is as complicated as society's role in nature. It would make little sense to separate natural from unnatural components in understanding the Mississippi and New Orleans. Rather, both are interlocked in a process of management and constant adaptation and readaptation.

Another manifestation of the cultural turn in environmental history is the increasing emphasis on the consumption of, rather than production in, the natural world.[92] Unlike earlier studies on the literal extraction of commodities through forestry or mining, consumption in these works deals especially with how landscapes and images of nature have been shaped in accordance with the phantasmagorical desires of drivers, backpackers, shopping mall visitors, and nature tourists. This shift toward the analysis of "landscapes of consumption" is well underway in geography, where scholars' engagement with the works of theorists Henri Lefebvre and Michel de Certeau has drawn their attention to the social production of space and power in capitalist-consumer societies.[93] As Gregg Ringer notes in his *Destinations: The Cultural Landscapes of Tourism* (1998), for example, there was nothing historically accurate or "natural" about the landscapes to which visitors flocked; these were highly modified, technology-dependent landscapes suited to the tastes and comfort of visitors.[94]

Such technological manipulation is especially evident in national parks, in which idyllic images of playgrounds for backpackers and naturalists have masked the conflicts that accompanied these landscapes' shift from spaces of work to places of leisure. Recent environmental-historical studies of North American national parks, such as Mark Spence's *Dispossessing the Wilderness* (1999) and Karl Jacoby's *Crimes Against Nature* (2001), for example, focus on the role of state authorities and capitalist powerbrokers in removing indigenous peoples or rural folk from their homelands and ending customary patterns of woodland use in the interest of creating nature reserves.[95] Influenced by James Scott's influential *Seeing Like a State* (1998), Jacoby's and Spence's works are attuned to the "moral ecology" of locals often portrayed as the enemies of conservation. Their works stand in sharp contrast to Roderick Nash's classic *Wilderness and the American Mind* (1967), which describes federal and state nature conservation and wilderness protection as the outcome of progressive American-nationalist, Transcendentalist, preservationist, and environmentalist beliefs.[96]

In their quest to experience untrammeled nature, tourists also helped to shape recreational hinterlands that linked "wilderness" and countryside to the city through the urbanized consumption of experience. William Cronon's *Nature's Metropolis* (1991), for example, showed that tourists' tendency to privilege the pastoral Midwest countryside over the corruption of Chicago masked the expanding material and cultural ties between the metropolis and its rural hinterland in the nineteenth and twentieth centuries.[97] According to Cronon, Chicago's development turned Frederick Jackson Turner on his head; whereas Turner believed that towns and cities emerged as a result of the growing density of homesteaders along the North American frontier, Chicago *created* the capitalist frontier by serving as the railway nexus for markets in grain, beef, and lumber. Eventually, this process transformed the second-growth forests of northern Wisconsin and Minnesota into a recreational hinterland for the city, replete with brochures and guidebooks touting the "pristine" quality of formerly clear-cut lands.

The Landscapes of European Environmental History

Environmental historians and historical geographers of Europe have been the most active in using and reformulating the landscape approach for their research. The 1955 publication of W. G. Hoskins' *The Making of the English Landscape* was a sweeping effort to study the changes in the physical landscape induced by humans in England over the last two thousand years using a variety of sources, including archival records, paintings, aerial photography, and archaeology. Accessibly written, it gained a considerable following inside and outside of academia; several regional studies followed. Hoskins was driven by the belief that preindustrial landscapes were more in harmony than industrial ones; in his eyes, industrialization marked a deep rupture in the history of landscapes. With the new upswing in cultural geography in the 1990s, historical geographers added considerations of power to their analyses of landscape. David Matless probed the ways in which landscape and a sense of English national identity have been tied together historically, for example, while Ian Whyte skillfully examined the interplay between landscape and history over the last five hundred years.[98]

On the Continent, the French *Annalistes* analyzed mountains, seas, climate, and soil as part of the *longue durée* that delimited the relative wealth and demographic growth of European agrarian societies before the industrial age.[99] In Germany, on the other hand, landscape studies did not establish themselves to the degree that they did in Britain because they suffered from the taint of Nazi "blood and soil," however inaccurate this categorization was. When environmental history appeared on the German academic scene in the 1980s, historical landscape change was conspicuous by its absence among the topics chosen by historians. Rather, they researched urban pollution and industrial ecology while their American colleagues were still debating whether such topics could properly be understood as environmental history. While the effects of industrial effluents and urbanization are undoubtedly critical environmental issues in the heartland of Europe's coal, steel, chemical, pharmaceutical, and automobile industries, scholars in the home country of the cultural landscape paid insufficient attention to studying changes in that landscape. The focus on air, water, soil, and noise pollution often relegated landscape to the realm of bourgeois aestheticism—at best an idealized alternative to industrial blight, at worst a bucolic veil that led middle-class Germans to turn a blind eye to the plight of manual laborers who suffered as a result of toxic exposure and unsafe working conditions.[100]

Only in the 1990s did a new generation of German-speaking environmental historians reappropriate landscape as an analytical category. While certainly being aware of the pitfalls of an essentialist human-landscape connection and often informed by the Anglo-Saxon reformulation of landscape studies, historians of Central Europe have been studying regions, rivers, and technological artifacts to the degree that they reflect landscape changes. The authors of this chapter are part of a group of scholars who embrace this trend. Thomas Lekan has argued that landscape preservationists in Germany before 1933 were by no means overwhelmingly proto-fascist in their political outlook and that landscape offers a fruitful lens for examining the contested meanings of German national and regional identities, as well as historicizing the possibilities and limits of environmental reform, between the 1880s and 1945.[101] Thomas Zeller has used the landscape of the German autobahn as both a site of contestation over professional interests and visions and as the outcome of those conflicts. Instead of being a refuge from society, as nineteenth-century Romantics had claimed, the landscaping of this major transportation network threw into sharp detail the social divisions and clashing aesthetic visions of landscape architects, civil engineers, and car drivers. What is more, the autobahn's landscape took on different meanings during the Nazi dictatorship and the Federal Republic.[102] Other scholars have contributed to the resurgence of landscape, including Gerhard Lenz, who has examined the "landscape of loss" in the heavily polluted industrial corridor in Saxony from the nineteenth century to the 1990s, as well as Daniel Speich, who has examined the modernization of riverine landscapes in Switzerland.[103] Regional studies understand landscape changes in conjunction with demographic and economic developments or seek to understand the relationship of humans and the North Sea through the medium of dams in the early modern period.[104]

The conspicuous absence of "wilderness" in the cultural landscapes of Central Europe is evident in much of the new work in this area. In line with current ecological theories of complexity and nonlinear systems, the ecologist Hansjörg Küster emphasized in a popular history of landscape in Central Europe that landscapes are in constant motion, with or without human interference.[105] Küster highlights that some of the oldest conservation areas in Germany, such as the Lüneburg Heath, are the result of human actions as much as of nature itself; indeed, the heath needs constant human and animal intervention in order to maintain its appearance. In a sweeping account of the role of water for and in German history, David Blackbourn has highlighted the hybridity of landscapes throughout the last three hundred years. He embraces the landscape approach emphatically and believes that designating it as a "Nazi" category does the Third Reich more honor than the regime deserves.[106] This dictatorship's ideological and intellectual assumptions consisted of an eclectic mix of preexisting and new ideas;

the fact that the landscape approach was used by the Nazis, Blackbourn argues, does not invalidate it from the outset. His analysis of water systems enables him to show how the history of Germany needs to be understood through an environmental angle in order to avoid precisely the kind of teleological assumptions that place the Nazi regime at the pinnacle of longer developments in Romantic or anti-modern thought.

Technologies of Landscape

The attraction of the landscape approach rests partially on the ease with which technological changes can be integrated into the historian's analysis. Rather than positing nature and technology as polar opposites, as in the "wilderness" idea, the cultural part of the landscape implies human action. By necessity, humans work the land in one way or other; they use tools ranging from sticks to backhoes. While the intensity of tool use varies greatly, the fact that technology is a necessary aspect of the human condition is invariable.[107] Thus, studying the environmental ramifications of technologies becomes one of the most important tasks of historians using the landscape approach. This is readily evident for the history of public works or of civil engineering, such as river management or road building—but for other technologies it is no less true. Landscapes are thus products of social contestation, not of technological determinism. David Nye's dictum that technology is not alien to nature, but integral to it, is therefore an important reminder for a student of cultural landscapes. It is important to emphasize that Nye embraces modern and industrial landscapes to a much higher degree than Sauer and early environmental historians, whose focus remained the transformation of the rural countryside.[108] Rather than delineating which elements of a landscape are natural or technological, studying the meeting zone between the two has been a heuristically helpful approach. Comparative studies which transcend one country or one continent are especially promising in this regard, as conservationists, engineers, and public policy makers have been far more transnational throughout the twentieth century than historians. For example, dam builders all over the world studied previous examples (and failures), while road planners met at international conventions and read foreign journals to form transnational communities of expertise. Forestry, while often national in origin, has been a global and often colonial instrument of control.[109]

POSTCOLONIAL NATURES: CULTURAL LANDSCAPES BEYOND EURO-AMERICA

Environmental history and cultural geography have begun to merge in their growing interest in the spatial and ecological dynamics of postcolonialism and globalization, which has stimulated innovative studies on the relationship between environmental justice and Third World movements, the movement and hybridization of people, commodities, and diseases on the colonial periphery, and the production and consumption of the "exotic." For example, landscapes have been important tableaux for studying the "tropicalization" of colonial parts of the world. Among environmental histories, Richard Grove's *Green Imperialism* (1995) is probably the best known work in this genre, since Grove argued that conservationist ideas first emerged in the eighteenth century, on the colonial periphery, where European land managers witnessed firsthand the devastating impact of deforestation and soil exhaustion caused by nascent capitalist expansion and sought ways to conserve forests and soils at home and abroad.[110] There is also an excellent and growing literature on South Asia. In a recent study, David Arnold has shown how the anticipation of India's tropicality through imagined landscapes in novels, poetry, and paintings preconditioned the way metropolitan Englishmen used botany as a way of appropriating India's plant "treasure."[111] Similarly, S. Ravi Rajan examined the connections between forestry and imperial development schemes for India by looking at how Prussian- and French-trained foresters erroneously transported European assumptions about scientific forestry to tropical landscapes.[112]

World-systems, Marxist, and postcolonial perspectives have also influenced the work of historians and political ecologists of Africa, who have taken European conservationists to task for their Eurocentric views of African wildlife and indigenous Africans in the creation of game reserves and national parks.[113] In his examination of the creation of Serengeti and Arusha National Parks in Tanzania, for example, historical geographer Roderick Neumann examined the way in which the European leisure classes' "pictorialization" of nature during the eighteenth and early nineteenth centuries erased the marks of human labor from "nature" and then legitimized a process of "recreational enclosure" in African colonies that alienated indigenous peoples from their land in order to create Edenic national parks between 1930 and 1960. He writes: "The idea of nature as a pristine, empty African wilderness was largely mythical and could only become a reality by relocating thousands of Africans whose agency had in fact shaped the landscape for millennia."[114] The spread of global tourism after the 1960s, and the desire to gain territorial control over rural peoples and resources, led many newly independent states, particularly Tanzania and Kenya, to continue these trends in conservation, thus perpetuating colonialist forms of spatial control in a postcolonial world.[115]

Neumann is one of the few historical geographers of Africa to engage the cultural landscape concept directly, though there is an expanding literature on how aboriginal forms of land and resource use contest colonial and neocolonial visions of the Continent that engage the new cultural geography's focus on symbolism and power. One prominent example is Melissa Leach and Robin Mearns' pioneering volume *The Lie of the Land: Challenging Received Wisdom on the African Environment* (1996).[116] In this work, the contributors challenge mass media and development agencies' self-evident assumptions about an overarching ecological crisis in Africa, which they argue has tended to justify external control by well-meaning but ill-advised Western non-governmental organizations and foreign affairs offices. The authors deconstruct overarching notions of ecological notions of climax, carrying capacity, and desertification, as well as macroeconomic analyses that ignore conditions on the ground, with an appeal to a nuanced, contextual, historical assessment of local conditions. The authors are particularly critical of a "received wisdom" that envisions ignorant, indiscriminate, or ill-adapted local land use as the cause of Africa's degradation while ignoring the logic and rationality of indigenous knowledge and practice. African hunters and herders are not merely agents or victims of ecological destruction; their voices and activities must be integrated into any development schemes that hope to shape fertile, resource-abundant landscapes. Leech and Mearns' perspective has numerous parallels with the recent historiography of North American national parks, such as Karl Jacoby's *Crimes Against Nature* and Mark Spence's *Dispossessing the Wilderness*, all of which seek to unravel the degradation narratives that underpin outside state control over rural people's hunting, herding, and farming. Yet explicit comparisons between the wilderness debate in North America and the social history of land use struggles in Africa, Asia, and Latin America remain few in number, despite recent calls for sharper comparative work. Here environmental historians have much to learn from historical and cultural geographers, whose work regularly engages fruitful mid-level points of comparison and generalization to understand historical patterns of land use, resource consumption, and landscape change at a global scale. [117]

CONCLUSION

For all its conceptual ambiguity and perhaps because of it, landscape offers one of the best tools for conceptualizing and moving beyond simple dichotomies between use and abuse, materialism and ideology, representation and reality. Today's landscape scholars avoid the essentializing tendencies of previous approaches, thus making room for open-ended, richer accounts. Looking at and historically understanding cultural landscapes as such not only more openly pays tribute to cultural and historical geography, the neighboring and overlapping sister disciplines of environmental history. It also helps to expand environmental history towards the current debate over the "spatial turn," the effort to understand historical changes not only in time, but also in space. Historically, many human actors in the modern West have thought about spaces as landscapes, and it is therefore incumbent upon environmental historians to unpack the category and use it for their analyses.

This is especially important for the current efforts to internationalize environmental history.[118] Put positively, the most active and largest community of environmental historians is currently found in the United States. Put negatively, these historians have often brought assumptions about US environments and historical approaches to their work that contribute little to understanding relationships between humans and their environment in non-American settings. The intellectual firestorm over William

Cronon's deserved attacks on the wilderness ideal might be seen as only a brushfire from a more transnational perspective. Embracing the landscape idea more fully would mean that historians of American and other environments could work with a tool that has a long transnational history, and can be used as such.

Secondly, taking the landscape concept seriously brings environmental historians into closer intellectual contact with their sibling discipline, historical geography. While we have tried in this chapter to demonstrate how much environmental historians can learn from the current productivity of historical and cultural geography, the latter can also benefit from being more closely aligned with the current boom of environmental history. Understanding the environment as cultural landscape is one of the most prominent ways to expand this mutually beneficial collaboration.

In conclusion, looking at landscape changes contributes to an enriched understanding of environmental history that includes a contemporaneous aesthetic understanding of nature.[119] In a sense, the concept of landscape was an ideal vessel for both material and aesthetic questions. Its earthen quality and its value as a cultural tool highlighted these comprehensive, integrative qualities. Landscape, in a sense, is the nexus between the material and the visual, between appropriation and appreciation.

NOTES

1. On this point, see Michael Williams's insightful essay "The Relations of Environmental History and Historical Geography," *Journal of Historical Geography* 20, no. 1 (1994): 3–21, here 3. A useful overview of the relationship between geography and history can be found in Alan R. H. Baker, *Geography and History: Bridging the Divide* (Cambridge: Cambridge University Press, 2003).

2. Lester B. Rowntree, "The Cultural Landscape Concept in American Human Geography," in *Concepts in Human Geography*, ed. Carville Earle, Kent Mathewson, and Martin S. Kenzer (Lanham, MD: Rowman and Littlefield, 1996), 127–160, here 127; Richard Hartshorne, *The Nature of Geography*, 2d ed. (Lancaster, PA: Science Press, 1946.) On the assassination of landscape, see Neil Smith, "Geography as Museum: Private History and Conservative Idealism in *The Nature of Geography*," in *Reflections on Richard Hartshorne's The Nature of Geography*, ed. J. N. Entikrin and S. D. Brunn (Washington, DC: Association of American Geographers, 1989), 89–120, here 107.

3. See Donald Worster, "Doing Environmental History," in *The Ends of the Earth*, ed. Donald Worster (Cambridge: Cambridge University Press, 1988), 289–306.

4. Rowntree, "The Cultural Landscape Concept," 139–147. Extensive literature can be found in this and other articles cited below. For a useful general history of geography, see David N. Livingstone, *The Geographical Tradition. Episodes in the History of a Contested Enterprise* (Oxford: Blackwell, 1992).

5. "A Roundtable: Environmental History," in *Journal of American History* 76, no. 4 (March 1990): 1087–1116; 1122–1131; 1142–1147.

6. William Cronon, "The Trouble with Wilderness, or Getting Back to the Wrong Nature," in *Uncommon Ground: Rethinking the Human Place in Nature*, ed. William Cronon (New York: W. W. Norton, 1995), 69–90; Richard White, "From Wilderness

to Hybrid Landscapes: The Cultural Turn in Environmental History," *Historian* 66, no. 3 (2004): 557–564; Mart Stewart, "If John Muir Had Been an Agrarian: American Environmental History West and South," *Environment and History* 11, no. 2 (May 2005): 139–162.

7. Rowntree, "The Cultural Landscape Concept," 128–130; Kenneth Olwig, *Landscape, Nature, and the Body Politic: From Britain's Renaissance to America's New World* (Madison: University of Wisconsin Press, 2002), 1–10.

8. Ann Bermingham, *Landscape and Ideology: The English Rustic Tradition, 1740–1850* (Berkeley: University of California Press, 1986).

9. Denis E. Cosgrove, "Landscape and *Landschaft*," *Bulletin of the German Historical Institute Washington D.C.* 35 (Fall 2004): 57–71. Reflections on imperial landscapes can be found in W. J. T. Mitchell, ed., *Landscape and Power*, 2d ed., (Chicago: University of Chicago Press, 2002).

10. Rowntree, "The Cultural Landscape Concept," 129–130.

11. For a recent example of the vast literature on German forests, see *Unter Bäumen. Die Deutschen und der Wald*, eds. Ursula Breymayer and Bernd Ulrich (Dresden: Sandstein, 2011). For examples of Romanticism, nature, and nationalism in German culture, the Rhine serves as a prime example; see, for example, Hans Boldt, "Deutschlands hochschlagende Pulsader," in *Der Rhein. Mythos und Realität eines europäischen Stromes*, ed. Hans Boldt (Cologne: Rheinland, 1988), 30–32; Helmut Mathy, "Der 'Heilige Strom.' Politische und geistesgeschichtliche Voraussetzungen der Rheinromantik," *Beiträge zur Rheinkunde* 36 (1984): 3–21; Hans M. Schmidt, Friedemann Malsch, and Frank van de Schoor, eds., *Der Rhein: ein europäischer Strom in Kunst und Kultur des zwanzigsten Jahrhunderts* (Cologne: Wienand, 1995). On the Lüneburger Heath, see Andrea Kiendl, *Die Lüneburger Heide: Fremdenverkehr und Literatur* (Berlin: Reimer, 1993).

12. On German forest Romanticism and national identity, see Michael Imort, "A Sylvan People: Wilhelmine Forestry and the Forest as a Symbol of Germandom," in *Germany's Nature: Cultural Landscapes and Environmental History*, ed. Thomas Lekan and Thomas Zeller (New Brunswick, NJ: Rutgers University Press, 2005), 55–80.

13. See Wilhelm Heinrich Riehl, *Die Naturgeschichte des Volkes als Grundlage einer deutschen Social-Politik*, vol. 1: *Land und Leute*, 5th ed. (Stuttgart, 1861). Riehl's work is available in translation; see *The Natural History of the German People*, ed. and trans. David Diephouse (Lewiston, NY: E. Mellen Press, 1990). On Riehl and regional identity, see Celia Applegate, *A Nation of Provincials: The German Idea of Heimat* (Berkeley, 1990), 34–41 and Jasper von Altenbockum, *Wilhelm Heinrich Riehl 1823–1897. Sozialwissenschaft zwischen Kulturgeschichte und Ethnographie* (Cologne: Böhlau, 1994). On the contrast between the French Enlightenment's constitutionalist and German Romantics' organicist visions of nationhood, see Jonathan Olsen, *Nature und Nationalism: Right-Wing Ecology and the Politics of Identity in Contemporary Germany* (New York: St. Martin's Press, 1999), 53–84.

14. The literature on *Landschaft* in German culture is quite extensive. See, for example, Alfred Barthelmeß, *Landschaft, Lebensraum des Menschen. Probleme von Landschaftschutz und Landschaftspflege geschichtlich dargestellt und dokumentiert* (Freiburg: Alber, 1988); Gerhard Böhme, *Natürlich Natur—Über Natur im Zeitalter ihrer technischen Reproduzierbarkeit* (Frankfurt: Suhrkamp, 1992); Norbert Fischer, "Der neue Blick auf die Landschaft: Die Geschichte der Landschaft im Schnittpunkt von Sozial-, Geistes- und Umweltgeschichte," *Archiv für Sozialgeschichte* 36 (1996): 434–442; Joachim

"Landschaft. Zur Funktion des Ästhetischen in der modernen Gesellschaft," in *Subjektivität: Sechs Aufsätze* (Frankfurt: Suhrkamp, 1974), 141–163.

15. Richard W. Judd, *Common Lands, Common People: The Origins of Conservation in Northern New England* (Cambridge, MA: Harvard University Press, 1997), 35–39; Angela L. Miller, *The Empire of the Eye: Landscape Representation and American Cultural Politics, 1825–1875* (Ithaca, NY: Cornell University Press, 1993).

16. Roderick Nash, *Wilderness and the American Mind*, 4th ed. (New Haven, CT: Yale University Press, 2001), 145–160; William Cronon, "Trouble with Wilderness," 76–77.

17. On German regional traditions, see Applegate, *Nation of Provincials* and Alon Confino, *The Nation as a Local Metaphor: Württemberg, Imperial Germany, and National Memory* (Chapel Hill: University of North Carolina Press, 1997). Recent environmental-historical research on France has also underscored the perdurability of regional identities in that country, but by the 1860s France had undergone over two hundred years of territorial and administrative rationalization under the absolutist, revolutionary, and Napoleonic regimes—a stark contrast to the separate kingdoms that still existed under the unified German Empire of 1871. On French regional identities, see Sara Pritchard, "'Paris et le desert français': Urban and Rural Environments in Post-World War II France," in *The Nature of Cities: Culture, Landscape, and Urban Space*, ed. Andrew Isenberg (Rochester, NY: University of Rochester Press, 2006), 175–192.

18. Riehl, *Natural History of the German People*, 49. Ronald C. Tobey, *Saving the Prairies: The Life Cycle of the Founding School of American Plant Ecology, 1895–1955* (Berkeley: University of California Press, 1981).

19. On the German environmental politics of home see William Rollins, *A Greener Vision of Home: Cultural Politics and Environmental Reform in the German* Heimatschutz Movement, *1904–1918* (Ann Arbor: University of Michigan Press, 1997) and Thomas Lekan, "The Nature of Home: Landscape Preservation and Local Identity," in *Localism, Landscape, and the Ambiguities of Place: German-Speaking Central Europe 1860–1933*, ed. David Blackbourn and James Retallack (Toronto: University of Toronto Press, 2007), 165–194.

20. On nature conservation and modernity, see Andreas Knaut, *Zurück zur Natur! Die Wurzeln der Ökologiebewegung* (Greven: Kilda-Verlag, 1993). Other works in this vein include Karl Ditt, *Raum und Volkstum: Die Kulturpolitik des Provinzialverbandes Westfalen, 1923–1945* (Münster: Aschendorff, 1988) and John Williams, "'The Chords of the German Soul are Tuned to Nature': The Movement to Preserve the Natural *Heimat* from the *Kaiserreich* to the Third Reich," *Central European History* 29, no. 3 (1996): 339–384 and Thomas Rohkrämer, *Eine andere Moderne? Zivilisationskritik, Natur und Technik in Deutschland, 1880–1933* (Paderborn: Schöningh, 1999).

21. On this point, see Lekan, "Nature of Home" and Theodor Ahrens, "Die Nationalparke der Vereinigten Staaten," in *Naturdenkmäler: Vorträge und Aufsätze*, ed. Staatliche Stelle für Naturdenkmalpflege, 3, no. 22 (1919): 4–46.

22. Wettengel, "Staat und Naturschutz, 1906–1945: Zur Geschichte der Staatlichen Stelle für Naturdenkmalpflege in Preussen und der Reichsstelle für Naturschutz," *Historische Zeitschrift* 257, no. 2 (October 1993): 355–399, here 362.

23. On this environmental ethic of home, see Rollins, *A Greener Vision of Home*

24. Hugo Conwentz, *Die Gefährdung der Naturdenkmäler und Vorschläge zu ihrer Erhaltung* (Berlin: Borntraeger, 1904), 82.

25. Paul Cloke, Chris Philo and David Sadler, eds., *Approaching Human Geography: An Introduction to Contemporary Theoretical Debates* (New York: Guilford Press, 1991), 4; Alfred Hettner, *Die Geographie. Ihre Geschichte, ihr Wesen und ihre Methoden* (Breslau: Hirt, 1927).

26. Cosgrove, "Landschaft," 64–66. For biographies and bibliographies of Meitzen and Passarge, see Robert C. West, ed., *Pioneers of Modern Geography. Translations Pertaining to German Geographers of the Late Nineteenth and Early Twentieth Centuries* (Baton Rouge: Department of Geography and Anthropology, Louisiana State University, 1990.)

27. Mark Bassin, "Imperialism and the Nation State in Friedrich Ratzel's Political Geography," *Progress in Human Geography* 11, no. 4 (1987): 473–495.

28. Geoffrey J. Martin, *All Possible Worlds. A History of Geographical Ideas*, 4th ed., (New York: Oxford University Press, 2005), 167–170; Hans-Dietrich Schultz, "Raumkonstrukte der klassischen deutschsprachigen Geographie des 19./20. Jahrhunderts im Kontext ihrer Zeit. Ein Überblick." *Geschichte und Gesellschaft* 28, no. 3 (2002): 404–434; idem, *Die deutschsprachige Geographie von 1800 bis 1970. Ein Beitrag zur Geschichte ihrer Methodologie* (Berlin: Geographisches Institut der Freien Universität, 1980).

29. Carl Ortwin Sauer, "The Morphology of Landscape," reprinted in *Land and Life: A Selection from the Writings of Carl Ortwin Sauer*, ed. John Leighley (Berkeley: University of California Press, 1963), 315–350.

30. Gerhard Hard, *Die "Landschaft" der Sprache und die "Landschaft" der Geographen: Semantische und forschungslogische Studien zu einigen zentralen Denkfiguren in der deutschen geographischen Literatur* (Bonn: Dümmler, 1970); Nikolaus Creutzburg, *Kultur im Spiegel der Landschaft. Das Bild der Erde in seiner Gestaltung durch den Menschen. Ein Bilderatlas* (Leipzig: Bibliographisches Institut, 1930); Eugen Diesel, *Das Land der Deutschen* (Leipzig: Bibliographisches Institut, 1931); Thomas Zeller, "Thomas Zeller on August Sander's Rhine Landscapes," *Environmental History* 12 (2007): 401–405.

31. David Thomas Murphy, *The Heroic Earth: Geopolitical Thought in Weimar Germany, 1918–1933* (Kent, OH: The Kent State University Press, 1997), 1–28. On the history and political significance of geopolitics in Germany, see Mark Bassin, "Imperialism and the Nation State in Friedrich Ratzel's Political Geography," *Progress in Human Geography* 11, no. 4 (1987): 473–495; "Race contra Space: The Conflict Between German *Geopolitik* and National Socialism," *Political Geography Quarterly* 6, no. 2 (April 1987): 115–134; Henning Heske, "Karl Haushofer: His Role in German Geopolitics and in Nazi Politics," *Political Geography Quarterly* 6, no. 2 (1987): 135–144. For an example of Haushofer's use of the concept in a domestic context, see Karl Haushofer, *Der Rhein: Sein Lebensraum, sein Schicksal*, 3 vols. (Berlin-Grunewald: Kurt Vowinckel, 1928).

32. Murphy, *Heroic Earth*, 1–60; Wilhelm Pessler, *Deutsche Volkstumsgeographie* (Braunschweig: Georg Westermann, 1931).

33. Mark Bassin, "Blood or Soil: The *Völkisch* Movement, the Nazis, and the Legacy of Geopolitik," in *How Green Were the Nazis? Nature, Environment, and Nation in the Third Reich*, eds. Franz-Josef Brüggemeier, Mark Cioc, and Thomas Zeller (Athens: Ohio University Press, 2005), 204–242. For a contemporaneous understanding of the territorial expansion of Germany, which essentializes its "ingrained habit of aggression," see

Derwent Whittlesey, *German Strategy of World Conquest* (New York: Farrar & Reinhart, 1942), especially the quote on p. 261.

34. Gesine Gerhard, "Breeding Pigs and People for the Third Reich," *How Green Were the Nazis?*, 129–146.

35. Siegfried Passarge, *Das Judentum als landschaftskundlich-ethnologisches Problem* (Munich: Lehmann, 1929). In 1936, Passarge published a thoroughly Nazified landscape primer: *Die deutsche Landschaft* (Berlin: Dietrich Reimer, 1936).

36. Gerhard, "Breeding Pigs and People for the Third Reich," *How Green Were the Nazis?*, 129–146.

37. Joachim Wolschke-Bulmahn, "Violence as the Basis of National Socialist Landscape Planning," *How Green Were the Nazis?*, 243–256.

38. Thomas Zeller, *Driving Germany: The Landscape of the German Autobahn, 1930–1970* (Oxford: Berghahn Books, 2007); Thomas M. Lekan, *Imagining the Nation in Nature: Landscape Preservation and German Identity, 1885–1945* (Cambridge, MA: Harvard University Press, 2004), chap. 5.

39. Rowntree, "The Cultural Landscape Concept," 129–130; Peter J. Hugill and Kenneth E. Foote, "Re-Reading Cultural Geography," in *Re-Reading Cultural Geography*, ed. Kenneth Foote et al. (Austin: University of Texas Press, 1994). The literature on the foundations and development of cultural geography and the cultural landscape concept is too voluminous to cite here in detail. Among the works used in the following sections: R. J. Johnston et al., eds., *The Dictionary of Human Geography*, 4th ed. (Oxford: Blackwell, 2000); Pamela Shurmer-Smith, ed., *Doing Cultural Geography* (London: Sage, 2002); Nigel Thrift and Sarah Whatmore, eds., *Cultural Geography: Critical Concepts in the Social Sciences* (London: Routledge, 2004); James Duncan, Nuala C. Johnson, and Richard H. Schein, eds., *A Companion to Cultural Geography* (London: Blackwell, 2004). On Sauer, see also Martin S. Kenzer, *Carl O. Sauer. A Tribute* (Corvallis: Oregon State University Press, 1987).

40. Hugill and Foote, "Re-Reading," 11–12.

41. Ibid. See Sauer, "Morphology" and "The Agency of Man on Earth," in *Man's Role in Changing the Face of the Earth*, ed. W.L. Thomas Jr. (Chicago: University of Chicago Press, 1956), 49–69.

42. Cited in Paul Cloke et al., *Approaching Human Geography*, 4–5.

43. On Sauer and the influence of Marsh's *Man and Nature* in the development of cultural-ecological approaches to human geography, see Karl S. Zimmerer, "Ecology as Cornerstone and Chimera in Human Geography," in *Concepts in Human Geography*, ed. Carville Earle, Kent Mathewson, and Martin S. Kenzer (Lanham, MD: Rowland and Littlefield, 1996), 161–188.

44. Sauer, "Morphology," 343; R. J. Johnston et al., *Dictionary of Human Geography*, 138.

45. Zimmerer, "Ecology as Cornerstone," 167–171.

46. Sauer, *Agricultural Origins and Dispersals* (New York: American Geographical Society, 1952); On the Latin America connection, see Sauer's books, *The Early Spanish Main* (Berkeley: University of California Press, 1966) and *Sixteenth-century North America: The Land and the People as seen by the Europeans* (Berkeley: University of California Press, 1971).

47. Hugill and Foote, "Re-Reading," 11–12.

48. Hugill and Foote, "Re-Reading," 19.

49. Johnston et al., *Dictionary of Human Geography*, 139.

50. Sauer, "Theme of Plant and Animal Destruction in Economic History," in *Land and Life*, 145–154, here 145.

51. Ibid., 148.

52. Richard Grove, *Green Imperialism: Colonial Expansion, Tropical Island Edens, and the Origins of Environmentalism, 1600–1860* (Cambridge: Cambridge University Press, 1995).

53. Sauer, "Theme of Plant and Animal Destruction," 154.

54. Sauer, "The Agency of Man on Earth," 69.

55. John Brinckerhoff Jackson, *Discovering the Vernacular Landscape* (New Haven, CT: Yale University Press, 1984); *The Interpretation of Ordinary Landscapes: Geographical Essays*, ed. D. W. Meinig (New York: Oxford University Press, 1979); *The Necessity for Ruins, and Other Topics* (Amherst: University of Massachusetts Press, 1980). For an evaluation of Jackson's legacy, see Chris Wilson and Paul Groth, eds., *Everyday America: Cultural Landscape Studies after J.B. Jackson* (Berkeley: University of California Press, 2003).

56. See, for example, the National Register Bulletins published by the US Department of the Interior, National Park Service, Cultural Resources Division, including *Guidelines for Evaluating and Documenting Rural Historic Landscapes* (1999) and *Guidelines for the Treatment of Cultural Landscapes*, http://www.nps.gov/tps/standards/four-treatments/landscape-guidelines/, accessed 31 March 2014.

57. Timothy Silver, *A New Face on the Countryside: Indians, Colonists and Slaves in the South Atlantic Forests, 1500–1800* (Cambridge: Cambridge University Press, 1998); Mart Stewart, *"What Nature Suffers to Groe": Life, Labor, and Landscape on the Georgia Coast, 1680–1920* (Athens: University of Georgia Press, 1996).

58. Michael Williams, "The Relations of Environmental History and Historical Geography."

59. Hugill and Foote, "Re-Reading"; Cloke et al., *Approaching Human Geography*, 9.

60. Rowntree, "The Cultural Landscape Concept," 132–133.

61. Ibid., 133–134.

62. Cloke et al., *Approaching Human Geography*, 16; Pamela Shurmer-Smith, "Marx and After" in *Doing Cultural Geography*, 29–41. Marxists have also drawn inspiration from Henri Lefebvre's structuralist Marxism as articulated in *The Production of Space* trans. Donald Nicholson-Smith (Cambridge, MA: Blackwell, 1991).

63. Raymond Williams, "Ideas of Nature" in *Problems of Materialism and Culture* (London: Verso, 1980), 67–85.

64. Shurmer-Smith, "Marx and After."

65. David Harvey, "Monument and Myth" in *Paris: Capital of Modernity* (London: Routledge, 2003), 311–340. Harvey's work best exemplifies the range of Marxist approaches to geographic space. See for example *The Urbanization of Capital: Studies in the History and Theory of Capitalist Urbanization* (Baltimore: Johns Hopkins University Press, 1985) and *The Condition of Postmodernity: An Enquiry into the Origins of Cultural Change* (Oxford: Blackwell, 1989).

66. Audrey Kobayashi and Suzanne Mackenzie, eds., *Remaking Human Geography* (Boston: Unwin Hyman 1989), 164–167.

67. Carol Ekinsmyth and Shurmer-Smith, "Humanistic and Behavioural Geography," in *Doing Cultural Geography*, 20.

68. Yi-Fu Tuan, *Topophilia: A Study of Environmental Perception, Attitudes, and Values* (Englewood Cliffs, NJ: Prentice-Hall, 1974); David Lowenthal, *The Past is a Foreign Country* (Cambridge: Cambridge University Press, 1985).

69. Rowntree, "The Cultural Landscape Concept," 135–136.

70. Vera Chouinard, "Structure and Agency: Contested Concepts in Human Geography," in *Concepts in Human Geography*, ed. Earle, Mathewson, and Kenzer, 383–410.

71. Cloke et al., *Human Geography*, 19.

72. Shurmer-Smith, *Doing Cultural Geography*, 1.

73. Ibid.

74. Stephen Daniels, "(Re)reading the Landscape," *Environment and Planning D: Society and Space* 6 (1988): 117–126; Peter Jackson, *Maps of Meaning* (London: Unwin Hyman, 1989); Denis Cosgrove and Stephen Daniels, eds., *The Iconography of Landscape: Essays on the Symbolic Representation, Design and Use of Past Environments* (Cambridge: Cambridge University Press, 1988). The best overview of the cultural turn can be found in Duncan et al., *Companion to Cultural Geography*.

75. Daniels, "(Re)reading the Landscape"; Jackson, *Maps of Meaning*.

76. Shurmer-Smith, *Doing Cultural Geography*, 3.

77. Duncan, "(Re)reading the Landscape"; Denis Cosgrove and Stephen Daniels, eds., *The Iconography of Landscape*.

78. Derek McCormack, "Introduction," in *Cultural Geography*, eds. Thrift and Whatmore, 4. Critical analyses of landscape and nationalism are particularly well-represented here; see Kenneth Olwig, *Nature's Ideological Landscape* (London: George Allen and Unwin, 1984) and Stephen Daniels, *Fields of Vision: Landscape Imagery and National Identity in England and the United States* (Cambridge: Polity Press, 1993).

79. The large literature on feminist geography cannot be cited here in detail. See, for example, Carol Ekinsmyth, "Feminist Cultural Geography" in *Doing Cultural Geography*, 53–66.

80. For an introduction to the larger literature on this theme, see Shurmer-Smith, "Postcolonial Geographies," in *Doing Cultural Geography*, 67–80.

81. Thrift and Whatmore, *Cultural Geography*; R. J. Johnston, *Dictionary of Human Geography*; Don Mitchell, "Cultural Landscapes: The Dialectical Landscape—Recent Landscape Research in Human Geography," *Progress in Human Geography* 26, no. 3 (2002): 381–389; idem., *Cultural Geography. An Introduction* (Malden, MA: Blackwell, 2000); Garth A. Myers, Patrick McGreevy, George O. Carney, and Judith Kenny, "Cultural Geography," *Geography in America at the Dawn of the 21st Century*, eds. Gary L. Gaile and Cort J. Willmott (Oxford: Oxford University Press, 2003), 81–96.

82. Cosgrove and Daniels, *The Iconography of Landscape*.

83. Don Mitchell, *The Lie of the Land: Migrant Workers and the California Landscape* (Minneapolis: University of Minnesota Press, 1996).

84. Rowntree, "The Cultural Landscape Concept," 140.

85. Williams, "The Relations of Environmental History and Historical Geography," 9–10.

86. Cronon, ed. *Uncommon Ground*; see also his important essay, "A Place for Stories: Nature, History, and Narrative," *Journal of American History* 78, no. 4 (March 1992): 1347–1376.

87. Richard White, "From Wilderness to Hybrid Landscapes: The Cultural Turn in Environmental History," *Historian* 66, no. 3 (2004): 557–564, here 562.

88. William Cronon, "Trouble with Wilderness," 73.

89. William Cronon, "Trouble with Wilderness," 89.

90. Nancy Langston, *Where Land and Water Meet: A Western Landscape Transformed* (Seattle: University of Washington Press, 2003) and Joseph Taylor, *Making Salmon: An Environmental History of the Northwest Fisheries Crisis* (Seattle: University of Washington Press, 1999).

91. Ari Kelman, *A River and its City: The Nature of Landscape in New Orleans* (Berkeley: University of California Press, 2003); Craig Colten, *An Unnatural Metropolis: Wresting New Orleans from Nature* (Baton Rouge: Lousiana State University Press, 2005).

92. White, "From Wilderness to Hybrid Landscapes," 559–562; Matthew W. Klingle, "Spaces of Consumption in Environmental History," *History and Theory* 42, no. 4 (December 2003): 94–110.

93. Henri Lefebvre, *The Production of Space*; Michel de Certeau, *The Practice of Everyday Life* (Berkeley: University of California Press, 1984).

94. Gregg Ringer, ed., "Introduction" to *Destinations: Cultural Landscapes of Tourism* (London: Routledge, 1998), 8.

95. See Mark David Spence, *Dispossessing the Wilderness: Indian Removal and the Making of the National Parks* (New York: Oxford University Press, 1999) and Karl Jacoby, *Crimes Against Nature: Squatters, Poachers, Thieves, and the Hidden History of American Conservation* (Cambridge: Cambridge University Press, 2001). For a recent analysis of the landscapes of race, see Richard H. Schein, ed., *Landscape and Race in the United States* (New York: Routledge, 2006).

96. See James Scott, *Seeing Like a State: How Certain Schemes to Improve the Human Condition Have Failed* (New Haven, CT: Yale University Press, 1998) and Nash, *Wilderness and the American Mind*.

97. William Cronon, *Nature's Metropolis: Chicago and the Great West* (New York: W.W. Norton, 1991).

98. David Matless, *Landscape and Englishness* (London: Reaktion Books, 1998); Richard Muir, *Approaches to Landscape* (Lanham, MD: Barnes and Noble, 1999); Ian D. Whyte, *Landscape and History since 1500* (London: Reaktion Books, 2002); Baker, *Geography and History*.

99. Classic examples of the *Annales* approach to historical geography include Emannuel Le Roy Ladurie, *The Peasants of Languedoc*, trans. John Day (Urbana: University of Illinois Press, 1974) and Fernand Braudel, *The Mediterranean and the Mediterranean World in the Age of Philip II*, trans. Siân Reynolds (London: Collins, 1972). The Annales approach tended toward seeing geography as a structural determinant of human activity, rather than as a cultural landscape of possibility.

100. Arne Anderson, "Heimatschutz: die bürgerliche Naturschutzbewegung," in *Besiegte Natur. Geschichte der Umwelt im 19. und 20. Jahrhundert*, eds. Franz-Josef Brüggemeier and Thomas Rommelspacher (Munich: C. H. Beck, 1987).

101. Lekan, *Imagining the Nation in Nature*; see also "Regionalism and the Politics of Landscape Preservation in the Third Reich," *Environmental History* 4 (July 1999): 384–404 and "'It Shall be the Whole Landscape!' The Reich Nature Protection Law and Regional Landscape Planning in the Third Reich," in *How Green Were the Nazis?: Nature, Environment, and Nation in the Third Reich*, cd. Mark Cioc, Franz-Josef Brüggemeier, and Thomas Zeller (Athens: Ohio University Press, 2005), 73–100.

102. Zeller, *Driving Germany*, 2007; "Molding the Landscape of Nazi Environmentalism: Alwin Seifert and the Third Reich," in *How Green Were the Nazis?: Nature, Environment, and Nation in the Third Reich*, 147–170, and "The Transformation of Nature under Hitler and Stalin" with Paul Josephson, in *Science and Ideology. A Comparative History*, ed. Mark Walker (London: Routledge, 2003), 124–155.

103. Gerhard Lenz, *Verlusterfahrung Landschaft. Über die Herstellung von Raum und Umwelt im mitteldeutschen Industriegebiet seit der Mitte des neunzehnten Jahrhunderts.*

Edition Bauhaus 4 (Frankfurt: Campus, 1999); Daniel Speich: *Helvetische Meliorationen. Die Neuordnung der gesellschaftlichen Naturverhältnisse an der Linth 1783–1823* (Zürich: Chronos, 2003). Also, see Stefan Kaufmann, ed., *Ordnungen der Landschaft. Natur und Raum technisch und symbolisch entwerfen* (Würzburg: Ergon, 2002) and David Gugerli, ed., *Vermessene Landschaften. Kulturgeschichte und technische Praxis im 19. und 20. Jahrhundert* (Zürich: Chronos, 1999).

104. Sabine Doering-Manteuffel, *Die Eifel. Geschichte einer Landschaft* (Frankfurt: Campus, 1995); Marie Luisa Allemeyer, *"Kein Land ohne Deich...!": Lebenswelten einer Küstengesellschaft in der Frühen Neuzeit* (Göttingen: Vandenhoeck & Ruprecht), 2006.

105. Hansjörg Küster, *Geschichte der Landschaft in Mitteleuropa. Von der Eiszeit bis zur Gegenwart* (Munich: Beck, 1995).

106. David Blackbourn, *Conquest of Nature: Water, Landscape and the Making of Modern Germany* (New York: W. W. Norton, 2006).

107. See Chapter 9 in this volume.

108. David Nye, ed., *Technologies of Landscape: From Reaping to Recycling* (Amherst: University of Massachusetts Press, 1999).

109. For transatlantic exchanges, see Daniel T. Rodgers, *Atlantic Crossings: Social Politics in a Progressive Age* (Cambridge, MA: Belknap Press, 1998); for global history on different levels, see Akira Iriye, *Global Community: The Role of International Organizations in the Making of the Contemporary World* (Berkeley: University of California Press, 2002); Helen M. Rozwadowski, *The Sea Knows No Boundaries: A Century of Marine Research under ICES* (Seattle: University of Washington Press, 2002). For an ongoing research project on automotive appropriations of landscape, see Thomas Zeller, "Staging the Driving Experience: Parkways in Germany and the United States," *Routes, Roads and Landscapes*, ed. Mari Hvattum, Brita Brenna, Beate Elvebakk, and Janike Kampevold Larsen (Farnham, UK: Ashgate, 2011).

110. Grove, *Green Imperialism*.

111. David Arnold, *The Tropics and the Traveling Gaze. India, Landscape, and Science, 1800–1856* (Seattle: University of Washington Press, 2006).

112. S. Ravi Rajan, *Modernizing Nature: Forestry and Imperial Eco-Development, 1800–1950* (Oxford: Oxford University Press, 2006).

113. The literature on the political ecology of Africa, especially national park development, is quite extensive and cannot be cited in detail. For an introduction see Jonathan Adams and Thomas O. McShane, eds., *The Myth of Wild Africa: Conservation Without Illusion* (Berkeley: University of California Press, 1992), esp. chapter 3; William M. Adams and Martin Mulligan, *Decolonizing Nature: Strategies for Conservation in a Post-Colonial Era* (London: Earthscan, 2003); James McCann, *Green Land, Brown Land, Black Land: An Environmental History of Africa, 1800–1990* (Portsmouth, NH: Heinemann, 1999). Gordon Matzke, *Wildlife in Tanzanian Settlement Policy: The Case of the Selous* (Syracuse, NY: Maxwell School of Citizenship and Public Affairs, 1977); Gregory Maddox, James L. Giblin, and Isaria N. Kimambo, eds., *Custodians of the Land: Ecology and Culture in the History of Tanzania* (Athens: Ohio University Press, 1996). On game reserves, see William Beinart, *The Rise of Conservation in South Africa: Settlers, Livestock, and the Environment, 1770–1950* (New York: Oxford, 2004) and his "Review Article: Empire, Hunting, and Ecological Change in Southern and Central Africa," *Past & Present* 128 (1990): 162–186; William Beinart and Peter Coates, *Environment and History: The Taming of Nature in the USA and South Africa* (New York: Routledge, 1995).

114. Roderick Neumann, "Ways of Seeing Africa: Colonial Recasting of African Society and Landscape in Serengeti National Park," *Ecumene* 2, no. 2 (1995): 149–169, here 150. Neumann is the most prolific and trenchant critic of national parks in Tanganyika/Tanzania and elaborates on this theme in numerous works, including *Imposing Wilderness: Struggles Over Livelihood and Nature Preservation in Africa* (Berkeley: University of California Press, 1998) and "Africa's 'Last Wilderness': Reordering Space for Political and Economic Control in Colonial Tanzania," *Africa* 71, no. 4 (2001): 641–665. On Tanganyika/Tanzania, see also Sunseri, "Forestry and the German Imperial Imagination and "Reinterpreting a Colonial Rebellion: Forestry and Social Control in German East Africa, *Environmental History* 8, no. 3 (2003): 430–451.

115. On this point, see Neumann, *Imposing Wilderness*; David Anderson and Richard Grove, eds., *Conservation in Africa: People, Policies, and Practice* (Cambridge: Cambridge University Press, 1987); Jim Igoe, *Conservation and Globalization: A Study of the National Parks and Indigenous Communities from East Africa to South Dakota* (Belmont, CA: Thomson/Wadsworth, 2004); Elizabeth Garland, "The Elephant in the Room: Confronting the Colonial Character of Wildlife Conservation in Africa," *African Studies Review* 51, no. 3 (December 17, 2008): 51–74. Jan Bender Shetler, *Imagining Serengeti: A History of Landscape Memory in Tanzania from Earliest Time to the Present* (Athens: Ohio University Press, 2007); Thomas Lekan, "'Serengeti Shall Not Die': Bernhard Grzimek, Wildlife Film, and the Making of a Tourist Landscape in East Africa," *German History* 29, no. 2 (June 2011): 224–264.

116. Melissa Leach and Robin Mearns, eds., *The Lie of the Land: Challenging Received Wisdom on the African Environment* (Portsmouth, NH: Heinemann, 1996).

117. On this point, see J. R. McNeill, "Observations on the Nature and Culture of Environmental History," *History and Theory*, Theme Issue 42 (December 2003): 5–43 and Thomas Lekan, "Globalizing American Environmental History," *Environmental History* 10, no. 1 (January 2005): 50–52.

118. Richard White, "The Nationalization of Nature," *The Journal of American History* 86, no. 3 (1999): 976–986.

119. For a landscape architect's perspective, see Anne Whiston Spirn, *The Language of Landscape* (New Haven, CT: Yale University Press, 1998).

PART III

WORKING AND OWNING

CHAPTER 14

A METABOLISM OF SOCIETY

Capitalism for Environmental Historians

STEVEN STOLL

If capitalism is thrown out of the door, it comes in through the window.
—Fernand Braudel, *The Wheels of Commerce*, 1979

ASKING FOR TROUBLE

CAPITALISM figures in almost every example of environmental history, though often
obliquely, as the unnamed social physics behind the clear cut, the dumpsite, the sooty
air that settles over the town, the eviction from the mountains, the tractor that slices
up the Plains, and the grain elevator that vacuums up wheat. It is rarely the primary
subject. Like the quake somewhere deep in the Pacific, it is the unmistakable cause but
one removed from the devastation of the tsunami, which is where historians usually set
their stories.

Anyone who chooses to avoid capitalism, preferring to assume its existence and
get on with other subjects, has my empathy. Imagine if historians disagreed about
who fought the Civil War and when, with some arguing that it began at Fort Sumter
and others with the founding of James Town. Still, the attempt is worth the trouble.
Environmental historians often reflect on the contradiction of unending economic
expansion on a finite planet, yet the ceaseless creation of surplus value is one of the
core assumptions of capitalism. If, as Theodore Steinberg observers, "It is hard to imagine...a system so good at producing wealth and so poor at distributing it, so steeped
in the commodity form and bent on bringing everything from land to water to air into
the world of exchange," then perhaps we should know the exact qualities of this system.
Documenting the effects of capital without confronting its architecture and foundation seems like an ineffective scholarly project.[1]

Teaching capitalism is important for another reason, having to do with our students. Historians know, along with Rousseau, that the social order does not come from nature, and they know that capitalism is around four hundred years old. Years of teaching, however, have convinced me that the most salient feature of commonplace knowledge is that capitalism *has no history*, or else it has a kind of *natural history* in which inherent capitalist qualities, shackled for centuries by savage privation and medieval backwardness, awaited the moment of their release. Most Americas would agree with this slop from Jonah Goldberg, a contributing editor to the *National Review*, who wrote in 2010, "Capitalism isn't some far-off theory about the allocation of capital; it is a commonsense description of what motivates pretty much all human beings everywhere." In this view, anyone who resisted—whether French peasants demanding lower bread prices or Massachusetts textile workers striking for higher wages—failed to behave like human beings everywhere, lacked commonsense, and thus can be dismissed.[2]

Environmental historians have a related problem. If in the minds of students capitalism is a basic human motivation, universal and inevitable, then what should they make of the havoc it causes? Those who teach a course in North American environmental history usually tell a hopeful story in which lethal smog and oil-befouled beaches ignite citizen action. Congress responds during a breathless decade of reform by requiring corporations to internalize the costs of their waste. Yet Earth Day gives way to reaction and eclipse during the 1980s. By the time we arrive at the North American Free Trade Agreement (NAFTA), we confront an insurgent and intellectually emboldened corporate establishment that succeeds in manipulating international law and the sovereignty of Mexico in order to recover the regulatory space it lost to the Environmental Protection Agency (EPA) and the Clean Air Act. Global warming comes next in our narrative: an epochal crisis arising in tandem with hundreds of billions of dollars in worldwide petroleum profits, the greatest accumulation of capital in human history. We might mention China's economic "miracle" and its staggering price as well.

There is good news, too—such as renewable energy and popular movements for environmental restoration and justice—but try as we might, such thin cloth cannot cover the naked, heaving beast standing next to us at the front of the classroom. At the very least, there is a problem with the end of the story and what students take away from it. They know (better than their professors do) that the same entities responsible for rivers that run black and accelerating carbon emissions also generate economic growth, future jobs, and enormous social power. Most students are optimists, which usually means that they seek a technological fix or a market-based solution. They pledge themselves to working within the brutal reality of the world, as though the same assumptions that unleashed the havoc will somehow bring about the sustainable and cooperative future they truly want. By failing to confront the social system, its actual history and false determinism, we limit their options for thinking about themselves. By tacitly confirming the inevitability of capitalism, we tell them that the havoc is inevitable too.[3]

We historians can combat the powerlessness of soggy hope by doing what we already do so well: historicize the social system. Environmental historians need to know how capitalism works in order to explain to students that it does not encompass the entire idea of society, does not contain unto itself freedom and the pursuit of happiness, is not identical to markets or entrepreneurship, and its invention was not historically inevitable. My view and point of departure is well stated by three historians, Joyce Appleby, Lynn Hunt, and Margaret Jacob, who have written, "One of the distinguishing features of a free-enterprise economy is that its coercion is veiled…Far from being natural, the cues for market participation are given through complicated social codes. Indeed, the illusion that compliance in the dominant economic system is voluntary is itself an amazing cultural artifact." Our task is to link this amazing cultural artifact to the stories we tell, so that environmental history offers a critique rather than an apology.[4]

In addition to Steinberg, a number of environmental historians have considered capitalism. One way or another, Andrew Hurley, Andrew Isenberg, Neil Maher, Carolyn Merchant, Colin Duncan, Donald Worster, John Richards, Mark Fiege, Michael Williams, Mike Davis, Robert Bullard, Richard White, William Cronon, and others have written works that link the control over environments with the imperatives of capital. Among these, Merchant and Duncan have been the most direct in attempting to grapple with early capitalism. I aim to suggest how we might integrate the work of many other scholars into our teaching and research, with reference to Karl Marx, Adam Smith, the French physiocrats, Karl Polanyi, Fernand Braudel, Immanuel Wallerstein, David Harvey, Joyce Appleby, Ellen Meiksins Wood, John Bellamy Foster, Robert Brenner, and others. In order to see factors of production like land, labor, and energy in a new light, we need a critical approach to capitalism and a familiarity with its most important interpretations. An ideal way to begin is with a definition that stresses the specificity of capitalism. At its center is a cycle or circuit of exchange.[5]

CIRCUITS AND IMPERATIVES

An understanding of capitalism begins with what it is not. Humans create things for use and exchange: commodities. Exchange often results in money that then purchases the items that people cannot grow or fashion themselves. Karl Marx expressed it like this: C→M→C, in which commodities sell for money, which is used to buy or make commodities. The circuit returns to where it begins, with useful and exchangeable things—cabbages to coffee, hogs to hoes. The circuit has subsistence production married to vigorous exchange, because no people in any place at any time did one without the other. This is not capitalism.[6]

Exchange does not define capitalism, and neither does the desire to better one's life by earning money; otherwise, the category would have to include ancient Athenian merchants as well as the fruit seller on the corner of 60th Street and Columbus Avenue in New York City. It is crucial and foundational to any historical demonstration of the characteristics of capitalism that exchange be distinguished from it and described as a universal practice. Michael Merrill makes this all-important point: Defining capitalism as exchange or the existence of markets, "as almost everyone now does, consigns the great struggles over the proper place of capital and capitalism in American society to a semantic no-man's land. If capitalism is little more than a synonym for a market economy, then any opposition to capitalism necessarily becomes an opposition to markets—in other words, an opposition so rarified and unreasonable to most people as scarcely to matter historically."[7]

Capitalism is a social system based on the creation of surplus value (what we can call, with some simplification, profit) from the purchase of labor-power. The investment of surplus value in order to realize additional surplus value is the circuit of capital, expressed by Marx as $M \rightarrow C \rightarrow MP/LP \rightarrow C' \rightarrow M + \Delta M$. Capitalists do not begin with useful or exchangeable things. They begin by entering the market to purchase commodities, including the means of production, raw materials extracted from environments, and labor-power. Production results in second-nature commodities, like gasoline and wool sweaters. These go back to market to bring the original sum advanced, plus surplus value. Capital is not a quantity and not coin in hand. It is not even the same thing as wealth. It is a social relationship between people who sell their labor and those who buy it, resulting in a sustained motion: surplus value that creates additional surplus value.

Yet notice that the two things that generate capital need to be generated. Labor is not really a commodity, but a capacity. It is also the absolute essence of capitalism because *it is the only factor of production that creates more value than it itself costs.* Buy someone's capacity to work for $1 a day, and you can use it to make $100 in products. Improve the machinery, and the difference between the wage and the output widens even more. Most people who work live on the commodity circuit: They sell their capacity to work (their one commodity) and exchange their wages for food and rent, continually recreating their capacity to work.[8]

The other part of capital is the one that most interests environmental historians. It consists of the things that come from forests and oceans or from under the Earth. Environmental historians are familiar with the notion that during the seventeenth or eighteenth centuries in England, and subsequently in the United States, lots of things that had not been commodities before took on that form. That is, production came to be directed toward surplus value. But commodities also have two qualities. Their outward form suggests their use (wool is good for weaving cloth) and points to the environmental and social conditions of production. Wool becomes useful only after a social and ecological process, requiring sunlight, water, meadow, shepherding, shearing, carding, and looming. But when wool takes on the generalized form of a commodity, the conditions of its creation are subsumed within the quantities and sums of trade. The processes and qualities that make a thing useful are necessary for it to *become* a commodity, but these no longer matter in its exchange *as a commodity.* Capitalism obscures every connection to the origins of things. It hides the politics, social relations, and environmental change inherent in how things turn into commodities.[9]

Environmental history is a powerful way of resisting the collective "forgetting" that is so much a part of consumption, but another field of study does this even better. Political ecology is a relatively new discipline that examines production within the critical definition of capitalism. It borrows from environmental history and also from anthropology, sociology, and geography. Paul Robbins defines it as "a field of critical research predicated on the assumption that any tug on the strands of the global web of human-environment linkages reverberates throughout the system as a whole." In a more pointed way than environmental history, political ecology assumes that every environmental change is the outcome of political conflict, of one group exercising power and asserting its conception of economy over others. I mention political ecology here because it encourages us to see the apparent singularity of the commodity as consisting of elements in conflict, and this is important in overcoming the mater-of-fact definition of capitalism as an "economic system" characterized by private ownership—or any other two-dimensional, textbook description. Marx said that every commodity contains within it "material relations between persons and social relations between things," and this two-part identity is useful for thinking about all kinds of things in the material world.

Definitions, however, cut across time. They do not have to do with history, even if they are derived from the study of history. This brief attempt at a definition begs the question: what is the history of the human and material relationships that generate capital? How might environmental historians tell this history, and what are their options? Generally, there are two dueling versions of the origin story, corresponding to two vast schools of historical thought. I will then suggest a path between them: Fernand Braudel.[10]

TRADE OR TORMENT?

The history of capitalism centers on a seemingly simple issue: whether or not it came from a long process of commercialization and accumulation, or from a more sudden transformation in the social relations of property. The first might be thought of as the ascent of calories and currency, the second as a legalized land grab.

By the 1750s, philosophers throughout Britain realized that something remarkable had happened in less than a century, and they tried to explain it. The Scots, in particular, including David Hume, Henry Home (Lord Kames), Adam Ferguson, and Adam Smith, favored the view that propensities within humans drove historical change. Theirs was a neat little tale of evolution from savage hunters to barbarian herders to semi-civilized farmers, and then on to the manufacturing and global shipping that defined their way of life as exceptional and entirely different from everything that had come before. They sometimes admitted that they had little evidence for the theory of stages. Instead it described what *must have* happened, which is another way of saying that it aligned with the commonly held assumption that the social order came from nature. Until the twentieth century almost no one in the Western world bothered to argue for or against the theory of stages or the role of human nature within it, because nothing could have been more obvious.[11]

This conjectural narrative did not remain conjectural. Rather, it gave birth to a more empirical view that argued for a gradual commercialization of European society. Hunters still turned into herders without much causation, but other things also happened. Money replaced rent in labor, trade resulted in accumulated wealth, gold and silver arrived from North and South America, banks issued mortgages, food proliferated, and so on. The "commercialization model" is not false, but depending on how scholars conceive of the forces driving it, it can easily turn into a just-so story. If capitalism is an expression of human nature and if this particular human nature has always existed, then capitalism has always existed in inchoate form. Like the theory of stages, no one really needed to prove how or when capitalism began by giving it a precise definition. It either existed in full flourishing or something prevented it from "takeoff." Karl Marx's version of that history has its own problems, but unlike any earlier one it begins with a material fact—the commodity—and not with subjective human qualities. In the meantime, the endlessly elastic and projectable human nature became the basis of some of the most influential tracts of the last two centuries, including Smith's *Inquiry into the Nature and Causes of the Wealth of Nations* (1776), Thomas Robert Malthus's *Essay on the Principle of Population* (1798), F. A. Hayek's *The Road to Serfdom* (1944), W. W. Rostow's *Stages of Economic Growth: A Non-Communist Manifesto* (1960), and countless other works.[12]

The most thoughtful recent rendition of commercialization comes from Joyce Appleby, in *Relentless Revolution: A History of Capitalism* (2010), which reviews and deepens the familiar themes, backed by a spectrum of scholarship. Her story runs something like this. Populations heaved and collapsed by turns in medieval Europe until the seventeenth century. The tension between agricultural output and famine explains much of the demographic and economic history of Europe before about 1650, when output in England began to increase. In the meantime, monarchs pursued trade and conquest, especially in the Americas. The gold and silver flowing into England, in particular, created inflation that put pressure on lords to do more with the land they controlled. It impelled them to rethink their aged leases and other arrangements with peasants.

Appleby does not launch her history from human nature and states at the outset, "there was nothing inexorable, inevitable, or destined about the emergence of capitalism." In her account capital did not come from a hardwired mentality of gain. Her earlier scholarship revealed *the invention* of human nature during early capitalism, culminating in Adam Smith's propensity to "truck and barter." She reprises *Economic Thought and Ideology in Seventeenth-Century England* (1978), with the argument that sixteenth-century advocates "depicted the incipient capitalist system as natural, liberating, progressive, and rewarding. Once they secured belief for this view, capitalists had the ideological punch to disrupt settled communities."

Yet *Relentless Revolution* mainly contends that capitalism came from an intensified and negotiated relationship between aristocratic landlords and their striving tenants— a relationship that had more to do with opportunity and mutual advantage, in her telling, than with the seizure and imperatives of private property. By deriving the system from a general upswing in human wellbeing and sufficiency, Appleby presents a progressive story, even as she details the contradictions between capitalism and democracy, along with capitalism's other "flaws." But where are the paupers—the English, Irish, and Scottish peasants evicted by acts of enclosure who wondered the highways starving? *Pauper* dates from the seventeenth century to describe someone with no means of subsistence, a kind of structural poverty never seen before.

In a sense, Appleby assumes the thing that needs to be explained. Her interpretation does not turn on human nature, but neither does it argue the underlying notion that people will behave like capitalists given certain circumstances—such as abundant food or opportunities for trade. Environments do not figure at all, and this is also part of the problem. Had Appleby asked how soils became available for production, as well as woods and waters, she might have sensed more acutely how enclosure dispossessed the people who once depended on land for livelihood.[13]

The environmental historian whose scholarship best represents commercialization is William Cronon. *Changes in the Land* (1983) and *Nature's Metropolis* (1991) present an evolutionary development from simple commodity exchange to industrial capital. Like Appleby, Cronon does not see a qualitative change in how people related to each other, but rather a quantitative intensification of practices that came ashore at Plymouth. "The market existed long before there was a Chicago," Cronon writes, "and although it attained new complexity in that city, it has since gone on to become a fact of life in most places." Yet this single market confuses capitalist accumulation with other exchanges. It says that all markets are the same, whether in the case of Englishmen and Wampanoag swapping fur, or that of speculators betting on the future price of wheat. In their social content, however, these exchanges could not have been more different. To watch a textile worker receive her wages might look like a simple payment for services rendered, but that description ignores every circumstance that brought the worker to the factory, the failure of the wage to meet her needs, and her lack of consent over the amount she is paid. An exchange is never merely an exchange.[14]

There is also a problem with the apparent dialectic between country and city. We learn that they form a single economy. The book's greatest achievement is to reveal that the apparent distinctions between city and country belie their deeper resemblance—how Iowa and Michigan, for example, would look different had the central market not organized their landscapes within a spatial division of labor, recreating them for its needs. The problem is in thinking of country and city as a single entity when (like exchange) that so-called unity did not extend to social relationships. *Nature's Metropolis* illustrates at terrific length how the Chicago Board of Trade and burgeoning corporations extracted every imaginable commodity from prairies, forests, and plains, creating a furiously expanding urban hinterland, without linking the extraction of value from human labor to the argument. First nature becomes second nature (use values become exchange values) through production, which requires human labor to be rendered available and then purchased. The violent enclosure of Indian homelands and the agrarian revolt against the domination of speculators and grain elevators over the wheat market receive no consideration. Coal miners never appear, though they dug the mineral that powered the engines and factories that changed first into second nature. A true dialectical argument would hold these conflicts at its center, not ignore them.

Cronon writes an ironic narrative about how progress brings waste and desolation to all the corners of the hinterland, yet the moral is curiously conciliatory. City and country "might come into conflict in a number of ways, but they also worked together as a system, joining to become the single most powerful environmental force...since the glaciers began their long retreat to the north." It is as though Cronon shares in the illusion "that compliance in the dominant economic system is voluntary." Finally, Chicago's hinterland is likened to an ice age in its transformative effect, lending to a human construction the immutability of the earth's orbital cycles. That is a peculiar lesson. *Nature's Metropolis* and *Relentless Revolution* are mirror images in the sense that they each project capitalism as a collective undertaking, and neglect half of what makes production possible.[15]

The person who did more than any other to think coherently about the two halves of production was Karl Marx. Marx developed a comprehensive critique of capitalism over five decades that integrated everything that had ever been written in four languages on the subject of political economy. Historians who borrow Marx's analytical categories have stressed various forms of compulsion as capitalism's spark point, involving the aristocracy, tenants, and peasants. As Karl Polanyi put it, buying and selling went from an expression of freedom, embedded in a moral economy that governed how people behaved in exchange, to the central organizing institution of social life, in which people—for the first time—depended on money in order to survive. Put in other terms, the new system (eventually, not at first) forced all classes, all land, and all labor to reproduce themselves through capital, whether in the form of wages, rents, or interest. Marxian historians do not deny that society became more commercialized or that commodity production spread for a variety of reasons or that merchant capital ended up as industrial capital; they simply deny that these developments explain capitalism, which they identify as a transformation in the relationship between humans and environments. This transformation first came about through the enclosure of the common lands in England.[16]

Ellen Meiksins Wood tells a version of the Marxian story in *The Origin of Capitalism* (2002). By the seventeenth century, British lords had lost their military function. They no long went around bashing in the heads of their French counterparts on battlefields, nor could they collect tribute by mugging peasants. By around the same time the aristocracy had also lent a great deal of their feudal authority to the monarchy, making the British state more centralized and more powerful than others in Europe. This had two implications. First, landlords could exercise power locally by economic means that did not threaten any of the functions of the state. Second, they could gain control of political institutions to promote their economic interests. This is how they set about to change the legal status of land.

Few people in sixteenth-century England owned land in a way that aligns with our understanding of ownership. Lords and peasants operated under a bundle of customary use rights that cascaded downward from the monarch and included the so-called common lands. The lords increasingly objected to this arrangement for a variety of reasons, one of which was a lucrative Atlantic wool trade, another of which was the high rate of wages some peasants won after the demographic implosion caused by the Black Death shifted the advantage in the labor market. Lords sought to extinguish customary use rights, and to replace peasants with sheep. They took this control by securing approximately four thousand Parliamentary Acts of enclosure, which seized perhaps six million acres, beginning in the fourteenth and fifteenth centuries, escalating in the sixteenth, seventeenth, and eighteenth, and slowing by the middle of the nineteenth. Adam Smith assumed without comment that "the appropriation of land and the accumulation of stock" replaced the "original state of things," in which every person owned the full value of their labor. Although forced appropriation went on right in front of him, Smith somehow assumed its necessity. In Marx's view, Smith came up with the idea that the circuit of capital required a "primitive accumulation" to start it churning—an assertion of rights intended to solve a long-standing problem. English lords of the seventeenth century could not purchase land and labor if these had no exchange value. Their solution was to seize them.[17]

Social scientists tend to think two-dimensionally, even about something as complicated as the shifting possession of land. If someone buys land according to law, what else is there to talk about? For economists the answer is nothing. But all private property came from non-private property, and the smoothness of legal exchange obscures the dislocation that happened in the past. Enclosure created two things at once. By transferring the means of production from one group to another, it provided the ecological basis for the accumulation of wealth and the necessary labor. In England, circa 1650, enclosure evicted peasants, forcing them into pauperism and thus the "choice" between wages or starvation. Lords might have paid wages and owned land before, but this new amalgam created something that had never existed before: capital.

Marx puts it this way in *Grundrisse*:

A mass of living labour powers was thereby thrown onto the *labour market*, a mass which was free in a double sense, free from the old relations of clientship, bondage and servitude, and secondly free of all belongings and possessions, and of every objective, material form of being, *free of all property*; dependent on the sale of its labour capacity or on begging, vagabondage and robbery as its only source of income. It is a matter of historic record that they tried the latter first, but were driven off this road by gallows, stocks and whippings, onto the narrow path to the labour market.

This dispossession, more than any other process or event, paved the way for the conversion of money into capital. It marked the separation of people from their longstanding ability to create their own subsistence, rendering them vulnerable and available to sell their labor to those with the money and motive to purchase it. Lords and tenants took land and labor and reassembled them in new ways.[18]

A document that illustrates the fallout of enclosure comes from Richard Price in *Observations on Reversionary Payment* (1812). People who once produced their own food on little farms, wrote the author, "will be converted into a body of men who earn their subsistence by working for others, and who will be under a necessity of going to market for all they want…There will, perhaps, be more labour, because there will be more compulsion to it." Price summed up the effect: "Upon the whole, the circumstances of the lower ranks of men are altered in almost every respect for the worse. From little occupiers of land, they are reduced to the state of day-labourers and hirelings." When workers starved on their wages the state reacted with Poor Laws, initiating another longstanding capitalist trend—private wealth with public costs. Lords externalized the subsistence of their workers and asked tax payers to make up the difference.[19]

Enclosure put lords in a novel situation. When land could be bought and sold, it needed to earn an income to justify the money invested in it. Enter the tenants. They paid rent on lordly fields with cash they earned by selling wheat and sheep. The rents on land went up and down, subjecting tenants to market pressures that shaped how they farmed, what they planted, and the productivity they demanded of labor. What happened next produced a kind of Big Bang. The tenants had an incentive. After paying rent, any money left over belonged to them. If they used their earned coin to purchase more efficient tools or pay more workers or rent additional fields, they would produce more commodities, which they could then exchange for more money. English tenants who deployed their gains to make greater gains entered the circuit of capital. They looked for new machines, new breeds of cattle, new methods of making and spreading manure.

Yet tenants acted as they did, not merely to seek advantage, but because capital created imperatives. As Wood puts it, "Capitalism can and must constantly expand in ways and degrees unlike any other social form. It can and must constantly accumulate, constantly search out new markets, constantly impose its imperatives on new territories and new spheres of life on all human beings and the natural environment." These imperatives meant that lord and tenant behaved in socially necessary ways, and workers tried to subvert them. Beyond that, the state itself took on a distinctive form and articulated rights and obligations consistent with the needs of accumulation. Counter movements formed, as workers called for relief from eighteen-hour days and starvation wages. Socialist experiments thrived, as intellectuals tried to marry the new scale of production with freedom from wages. The point is that capitalism became more than a set of commercial practices, but a complete social system, backed by all sorts of cultural authority and resisted from the moment it appeared.[20]

Authors in the Marxian tradition who have extended this narrative include Robert Benner, Mike Davis, David F. Noble, Charles Sellers, Michael Perelman, Eric Wolf, and Raymond Williams, whose *The Country and the City* (1973) is a literary history that depicts the ideological shift that took place in the seventeenth century when the countryside became the locus for accumulation. Williams and Brenner both hold that competition and pressure due to variable land rents forced tenants to accelerate the pace of enclosure. In other words, a tenant might have said something like this: I need those open fields so I can expand my farm, lower my marginal costs, and increase my profits because the lord charges such high rents. Williams argues for a long period of increasingly concentrated landholding that took a new form, in which "a capitalist social system was pushed through to a position of dominance, by a form of legalised seizure enacted by representatives of the beneficiary class."[21]

Enclosure, however, would seem to have little to do with events in North America, except that English peasants fled to Virginia as indentured servants to avoid pauperization. But did enclosure influence American capitalism? The question is too big to consider here, but there is at least one transitional figure between English lords and North American settlement. To a greater extent than any other English political philosopher of the seventeenth century, John Locke took a serious interest in North America. But as a primer on American capitalism his *Second Treatise of Government* (1690) is more of a riddle than a manual. The question of whether Locke advocated capitalism is a contentious subject within the study of his ideas. The basis of disagreement is this: Locke lived in a society in the midst of rapid and violent enclosure but never mentions it, even as he justifies the enclosure of American Indian territories.[22]

It is impossible to understate the importance of the "State of Nature" in Locke's formulation of North America. He did not acknowledge Cherokee, Iroquois, or any other Indian governance. The legal void he imagined in the New World opened the way for new political forms that would generate from the exercise of natural rights—without the illegitimate authority of a monarch. Yet while Locke might have hated the thuggish power of the British aristocracy over the lands they controlled, he tacitly recognized aristocratic legitimacy, calling private property "the law of the land, which is not to be violated." Odd that he should argue away any moral obligation to obey a king yet uphold the king's law, saying that no one could seize land as long as someone held a deed to it. How the primacy of property intersected with peasant eviction in Locke's mind is not clear, but one possibility must be that he said nothing about enclosure because he had given up on England as hopelessly corrupt. When he wrote, "In the beginning all the world was America," he expressed the fantasy of starting over.

Yet having grown up in English society, Locke knew that the only people with political independence owned land. They owned, in other words, the productive power behind subsistence, wealth, and the employment of others. Locke looked upon North America as a place where the material basis of social power could be diffused to the greatest extent in human history, where more people could own land, and thus become autonomous political actors, than anywhere else he knew of. A natural aristocracy could be created, in which common men could become peers. This idea was so radical that publishing it before the end of the English Civil War would have been an act of suicide.

The terrible paradox, however, is that the exercise of natural rights among Englishmen in America demanded the dispossession of people whom Locke and other Englishmen decided did not use land to socially beneficial purposes. Here is John Winthrop writing in 1628, sixty-two years before the *Second Treatise*:

> That which lies common, and has never been replenished or subdued, is free to any
> that possess and improve it; for God hath given to the sons of men a double right to
> the earth—there is a natural right and a civil right. The first right was natural when

men held the earth in common, every man sowing and feeding where he pleased. Then as men and their cattle increased, they appropriated certain parcels of ground by enclosing and peculiar cultivation, and this in time got them a civil right... As for the natives in New England, they enclose no land, neither have they any settled habitation, nor any tame cattle to improve the land by, and so have no other but a natural right to those countries. So if we leave them sufficient for their own use, we may lawfully take the rest, there being more than enough for them and for us.[23]

If we could spell out the underlying justification for this idea, it might go something like this. Agriculture creates surplus commodities that make possible dense populations and thus the expansion of civil society. Like the Scots, Locke saw farming as the material pivot of a complex society. This is the context for his assertion: "He who appropriates land to himself by his labour, does not lessen, but increase the common stock of mankind." Therefore, anyone living in a state of nature (and thus in the absence of civil authority and the legal title to land that governments conferred) who claimed a bit of earth but did not clear, plow, plant or otherwise "improve" it, had no right to keep it. They could be deprived of land and livelihood for the benefit of all humity. British lords used the same agument to justify enclosure.

Under Locke's argument, Native Americans could not be deprived of areas where they planted maize and squash. The rest could be enclosed, which meant in practice that that civil authority would seek continually to acquire it through treaties of cession. In the centuries that followed, white Americans justified moving onto Indian-controlled territory by asserting that Indians did not make profitable use of it. These cessions did not typically result in the "release" of labor as they did in England, but they made American capitalism possible nonetheless, in the sense that they created real estate and the ecological basis for commodity production. This land included the woods, animals, soils, and minerals that fed the proliferation of the white population and eventually industrial production.

The larger point about Locke, however, is that in his world, economic inequality came from political inequality, and he had every reason to believe that political equality would foster the economic kind. Yet his thinking took place within the context of his fantasy of starting over, and quite apart from facts on the ground. By the time that Locke wrote *The Fundamental Constitutions of Carolina* (1669), Englishmen had already begun to enslave Africans, forcibly assembling land and labor on plantations from Jamaica to Virginia. Locke did not comment on the seizing of another's body as a violation of natural rights (or as a kind of enclosure) nor did he seem to notice that political equality for some and the grossest degree of inequality for others went together hand-in-glove. Proponents of capitalism today who look to Locke as the founder of their thought fail to appreciate that he could not have imagined a situation in which legally constituted political equality could lead to the amassing of social power—and thus *de facto* political inequality—on a scale that makes the monarchy he hated look positively Lilliputian.[24]

Leaving theory aside, we turn to the environmental historian who best exemplifies the Marxian view, and whose work takes off from the enclosure of land in North

America. Carolyn Merchant's *The Death of Nature* (1980) and *Ecological Revolutions* (1989) together present a two-part intellectual and environmental history of the transformation of social and environmental relations in the Western world. The second book integrates elements of the first into its argument that New England underwent a colonial ecological revolution, followed by a capitalist one. Merchant shows that the colonial system stood on two contradictions. Taxation and other financial obligations requiring currency forced production beyond subsistence. Biological reproduction played a part in soil exhaustion and created a demand for land beyond the space available in towns, sending sons and daughters to new regions. Those who did not migrate took up intensive cultivation for the market in order to make up the difference between the lost opportunity of moving West and the degraded landscape of the old states. As Merchant puts it, "Inherent in 'extensive' farming in America was a fundamental contradiction between the requirements of production and those of reproduction."[25]

Merchant's analysis also touches on conservation, which she describes as "a utilitarian, homocentric ethic subservient to pragmatic means, not an ecocentric ethic that allowed a forest, lake, or swamp simply to exist." It deepened the psychological and social division between "ends and means, values and facts," as well as the alienation between men and women under a regime that reflected the division between production and reproduction in the larger economy. ("Within the household, as within the larger social whole, male and female, head and heart, calculation and emotion expressed the dualities of industrial capitalism.") Throughout, Merchant's aim is to reveal all the ways in which social change came about as cause and consequence of ecological change, and how the capitalist mode of production arose in contradiction to ecological processes and the reproduction of the household.[26]

Not every work, however, fits neatly into one or the other category. Colin Duncan's *The Centrality of Agriculture: Between Humankind and the Rest of Nature* (1996) presents a critique of capitalism in which the primary historical subject is the agriculture of *early English capitalism*. The idea is that, regardless of how tenants pressed on workers or how lords siphoned off profits, cultivation did not exceed ecological limits. According to Duncan, the major innovations of improved husbandry arrived before capitalism, in the sixteenth century. Tenants signed leases committing them to the "custom of the country," resulting in improved, not degraded, farmland. A similar capitalist agriculture flourished in the United States, and the American and English models lost their cultural currency for the same reason: an ecologically sustainable improved husbandry depended on close attention and physical labor, but colonization in the form of tropical plantations and the labor efficiencies of coal-burning steam engines invalidated this model of production.

Duncan sidesteps the social relations of the English case by arguing that convertible husbandry as a mode of production is detachable from the dispossession that made it operable. In other words, he tries to rescue restorative methods from the awkward fact that although they might have been worked out a century earlier, the system of lords, tenants, and laborers caused their difussion throughout the English countryside. Duncan's admiration of English farming in its non-industrial phase is similar to Marx's admiration of factories. For both authors, what might seem like a distinctly capitalist mode of productioin could become the basis for a socialist society. The difference is that Duncan (unlike Marx) rejects industrialism and looks forward to an ecologically benign socialism that would place a restorative agriculture at its center. *The Centrality of Agriculture* is one of the most intellectually fertile works of environmental history and one of the rare ones to address early capitalism.[27]

The Marxian interpretation offers a remarkable, indispensible critique, but historical nuance is not one of its strong points. At worst, everything falls into place so neatly, and every historical actor plays her part so inexorably, that it feels deterministic. Waving away the details of history results in sweeping theoretical pronouncements that often fail to set foot on the ground. This is why we need Fernand Braudel. No other historian of the last century combined theoretical mindfulness and scholarly achievement in an interpretation encompassing five centuries of European history, including the beginnings of capitalism. Braudel somehow presented a rigorous and specific historical method with conclusions about capitalism that transcend Europe.[28]

BRAUDELIAN STRUCTURES

Braudel's greatest achievement is the three-volume work, *Civilization and Capitalism, 15th to the 18th Centuries*, including *The Structures of Everyday Life*, *The Wheels of Commerce*, and *The Perspective of the World* (all three published in 1979). He attempted a grandly materialist history, in which structures served as the foundation above which arose the society and culture of Europe. Braudel and his mentor, Lucien Febvre, believed that the present could only be understood as deriving from climate, geography, and agrarian practices at the capillary level, which is to say that muddy peasant fields underlay politics and economy all the way up the social hierarchy, to the merchants and financiers at the top. Capital in its most transnational form rested on the labors of country people and their relationship to land.[29]

Braudel was not an environmental historian, but ecology figures in his writing, often in descriptions of changing landscapes and shifting uses. His sense of the interdepencence of life and landscape is especially apparent in is first work of history, *The Mediterranean and the Mediterranean World in the Age of Philip II* (1949). One scholar summarized Braudel's conception this way. For Braudel, "Human ecology begins with the axiomatic proposition that all living organisms need a source of sustenance. The source of this sustenance is to be found in the physical environment, and no matter how complex a social system becomes, it never overcomes its dependence upon the vital source. The physical environment is extended and modified through the social organization of social systems. In the process a social environment is created that has profound, if not equal, influence on the probabilities for human survival." Braudel offers environmental historians an overarching depiction of the grit and texture of material life constructed upon the practices that linked Europeans and their environments.[30]

Most of all Braudel is known as a historian of material life and the formation of capitalism. In this project he set himself against the explanations that he encountered among historians during the 1950s. "According to the text books," he wrote in *The Structures of Everyday Life*, a Europe isolated from the rest of the world underwent "gradual progress towards the rational world of the market, the firm, and capitalist investment." Instead, Braudel stressed the complexity of historical reality and attempted to pull patterns from the enormity of his data. There never was one market but many: local and national, merchant streets and rural fairs, and communities of buyers and sellers surrounding specific commodities.

Yet the boldest, most brilliant, and most useful distinction Braudel made is reflected in the three volumes themselves: they correspond to a three-tiered theory of economic life. Most people lived in households, producing food for direct consumption and exchange in small, localized regions. Beyond the household and village was the town, where commercial life thrived in a proliferating social division of labor, including vigorous trade with the countryside and competition among all makers and sellers. This is the realm of market relations—what Braudel refers to as a true realm of freedom. The third category, however, though it depended on the other two, departed from the rules that governed them. Braudel calls this "the zone of the anti-market, where the great predators roam and the law of the jungle operates. This—today as in the past, before and after the industrial revolution—is the real home of capitalism."[31]

Braudel's most important conclusion is that capitalists operate internationally, under conditions where they set the terms of exchange, sometimes under contract with the state, but always to their maximum gain, even to the point of monopoly. For Braudel, this new force in the world does not express human achievement, historical progress, or modernity. "The worst error of all," he writes in *The Perspective of the World*, "is to suppose that capitalism is simply an 'economic system', whereas in fact it lives off the social order, standing almost on a footing with the state, whether as adversary or accomplice: it is and always has been a massive force, filling the horizon." As Immanuel Wallerstein writes of Braudel on this point, "He turns the intellectual debate upside down. Rather than thinking of the free market as the key element in historical capitalism, he sees monopolies as being the key element. Monopolies dominating the market are the defining element of our system and it is this that distinguishes capitalism quite clearly from feudal society." The point is not that capitalist firms do not compete with each other, but rather that they have ways of avoiding competition by securing monopoly power, sometimes with the cooperation of the state. Braudel's division of economic life echoes the distinction between capitalism and exchange, in which the new social system took a practice that characterizes all human societies and recreated it as a tool for attaining unpredented social power over people and environments.[32]

All of this would situate Braudel within a Marxian frame, but he derives his entire story from a process that Joyce Appleby would applaud. Proliferating currency, improved and stabilized agriculture, regional and then national markets, oceanic trade and colonization, and the accretion of merchant capital: Braudel uses a commercialization narrative to arrive at Marxian conclusions. He stresses a slow and contradictory rise, stemming from town, not country. "Once a critical threshold had been reached, the proliferation of trading, of markets and merchants, occurred of its own accord. But this underlying market economy was only a *necessary*, not a *sufficient* condition for the formation of capitalist process." In one sense, Braudel reads much like Robert Brenner and Ellen Wood and most other Marxians, but he departs from them by placing little emphasis on changes in agrarian social relations. Braudel sees early capitalism as emanating from the *burghs* and as engaging very few individuals for centuries. He refuses to identify a moment of kick-start (like enclosure), writing at one point, "One thing seems to be beyond doubt: capitalism cannot have emerged from a single confined source."[33]

A related aspect of Braudel's work is his depiction of different kinds of markets existing at the same time and in the same places. In other words, no single form took over and transformed the others for a long time. In this way, Braudel might be one of the originators of *articulation*, the idea that peasant societies functioned within capitalist ones. The research of anthropologists like Michael Dove, Enrique Mayer, and Stephen Gudeman reveals instances of *campesinos* and other agrarians engaging with global markets while remaining in control of their subsistence. In the words of historian Elizabeth Dore, "Instead of viewing the sweep of history as the replacement of one social system by another, articulation theorists stressed the continuous interaction between different modes of production in one country. They argued that instead of free wage labor necessarily replacing unfree labor and capitalism replacing noncapitalist systems, the two often coexisted over the long term." The importance of this work for environmental history is manifold: it opens a realm of study in which there is no immediate and comprehensive capitalist "transformation."[34]

Not all of this is owed to Braudel, but Braudel fractured the unidirectional narrative of modernity, replacing it with something closer to articulation. He also suggested differentials in economic development between city and country, and especially between the market and the anti-market. Yet we are not finished with political economy. Two subjects, both directly relevant to environmental historians, sit at either ends of the history of economic thought and speak differently about the relationship between capitalism and nature: physiocracy and metabolism.

RULE OF NATURE

Capitalism overwhelmed medieval metaphors. Comparing the economy of a nation-state to a household or the human body no longer made sense. Economic society called for new interpreters, people who could describe the intersection of land, trade, currency, and taxation—factors that meant the difference between prosperity and poverty. The first identifiable philosophy of political economy outside of England attempted a synthesis of the state and agriculture. It did not really compute, but the virtues and flaws of the French physiocrats are worth considering.

While serving the French monarchy during the 1760s, Francois Quesnay and Victor Riqueti, Marquis de Mirabeau argued that the state was a "direct manifestation of the natural order," and that farming constituted the soul of its economy. Physiocracy means "rule of nature" and holds that only agriculture creates surplus product above costs and subsistence, because only agriculture actually *creates* anything. One way to think about this is productive versus unproductive labor. Unproductive labor is not the same thing inefficiency; it is simply labor that does not generate surplus value. Being a professor is a perfect example of non-productive labor.

The physiocrats beat Marx to this distinction by claiming that not only were farmers the only productive laborers, but that they did something peculiar: they reproduced themselves in the act of labor, consuming part of what they grew and leaving a surplus for everyone else. Other workers made shoes, iron, and all sorts of useful things, but according to the physiocrats (also known as the French Economists), other workers only moved matter around and consumed in order to work, resulting in a net loss to society.[35]

There was brilliance and folly in physiocracy. No one before about 1760 disputed that agriculture provided the basis of all wealth, that its intensification generated the calories that fed the population that performed the labor that turned out the commodities that earned the surplus value that served as capital. The French Economists made way for Smith and Marx by developing an overarching theory of political economy. They tried to locate the origin of value, a quest that occupied political economists for more than a century. But by rejecting manufacturing and commerce as sources of surplus value, they made themselves irrelevant during the eruption of factory production. The Physiocrats' more immediate problem had to do with their ties to France; aspects of their theory are so obscure and linked so closely to the French state that they do not translate very well into English, German, Italian, Russian, or Spanish. The Economists did have one adoring Scottish student, however, someone who took their insights and refashioned them into something entirely different: Adam Smith.

Adam Smith began writing his *Inquiry into the Nature and Causes of the Wealth of Nations* in 1766 during a sojourn in France, in which he consulted with the Quesnay. The meeting made a deep impression on Smith. A decade later he almost dedicated his book to Quesnay, and while by then physiocracy had fallen out of favor, Smith regarded it as "the nearest approximation to the truth that has yet been published upon the subject of political oeconomy." In fact, Smith did an intellectual flip with his own theory of value by substituting labor for agriculture, writing that though the Economists conceived of economy too narrowly, they nevertheless got it right by "representing the wealth of nations as consisting, not in the consumable riches of money, but in the consumable goods annually reproduced by the labour of the society."[36]

It would be tempting to declare the physiocrats the first environmental economists, but that would be a misreading. Production in their scheme remained ecologically benign only because it did not produce the externalities of industrial production.

Nothing in their thinking prevented or limited the full-throttle creation of wealth. Yet with their eclipse the biophysical environment faded from political economy altogether. It could then be re-assimilated as Nature, the infinite source of raw material, lacking independent processes that might clog up the humming "laws of motion," and within which capitalism functioned as a self-contained, ecologically isolated, ahistorical bubble. No political economist challenged this view or suggested otherwise until Marx posited that a social system is not well described as a mechanism, but as a metabolism.[37]

A METABOLISM OF SOCIETY

Ecosystems do not actually consist of resources; they consist of organisms and habitats. Any social system (whether socialism or capitalism) that views environments as consisting of commodities has no way of valuing ecological services or its own negative effects. An example of this thinking comes from Adam Smith, who observed in *Wealth of Nations* that consumption of fish would increase with the progress of society. Where would the extra fish come from? Smith's plan was for more and larger fishing boats to travel farther from shore. He assumed two things. First, abundance (and thus economic growth) was a technological problem. More energy and labor applied to the oceans *resulted* in more fish. Second, species existed like boxes on a supermarket shelf—they could be removed interchangeably without consequence.

Smith should not be blamed for not knowing about things that no one knew about in the 1770s, but something fundamental to capitalism operates in his fish problem. Finite environments mean finite growth, and that is the same thing as the stagnation of capital. Finding new environments to stave off that fatal condition has impelled the owners of capital to send fishing boats across ecological borders, to extract trees and coal from the Southern Mountains, to remove the Cherokee and Creek from Georgia in the persuit of more land for cotton plantations, and to manipulate the government of Nigeria in order to drill for oil. For people who manage capital, acknowledging limits is more frightening than the human and environmental fallout that comes from exceeding them.[38]

Yet for a long time socialists as well as capitalists looked to environments as infinite funds of raw material. In a sense, Marx did not transcend the theory of stages, with its presumption of growth; he merely added his own ending to the old story. Yet readers of Marx also know a different voice. In various places he wrote of the biophysical relationship between economic and ecological processes.[39]

Metabolism describes the cellular conversion of food into energy. Marx saw a parallel in the activity of labor, which converted first into second nature, creating use values. Yet all forms of production exist in tension with the capacity of environments to provide. Marx called labor "a process between man and nature, a process by which man, through his own actions, mediates, regulates and controls the metabolism between himself and nature." As we have already seen, however, the form of something can have different social meanings. The process between man and nature might sustain a community or generate surplus value. Marx, writing in the 1860s, suggested that a "rift" had opened between the constant demand for capital formation and the environments it required.[40]

Marx did not have data for proof of this rift—what he did have was widespread evidence for a decline in soil fertility. Marx had read the German chemist Justus von Liebig, who had read the American political economist Henry Carey, who knew the work of soil-improving American farmers from publications like *The Cultivator*. Carey's description of conventional agriculture as the "spoliation system" comes out of the vast literature of restorative husbandry. By the 1850s Liebig had emerged as not only the leading chemist of soil depletion, but also as an authoritative commentator on its causes. If ever-larger crops demanded practices "by which the ground must gradually lose the conditions of its fertility, by which it must be impoverished and exhausted," wrote Liebig in his *Letters on Modern Agriculture* (1859), "then such a system is not rational, though it enrich the individual who obtains these high returns." In criticizing spoliation, Liebig criticized capitalist farming.[41]

According to John Bellamy Foster, Marx developed his ecological critique while preparing to write the first volume of *Capital*. Since the eighteenth century the disjuncture between the fragility of soils on the one hand, and the imperative of capitalist agriculture to send the entire product of land (along with soil nutrients) down the turnpike to market on the other, had perplexed those who thought about it. (Former President James Madison hit upon the same metabolic rift in 1818, the year Marx was born.) But Marx made something out of the problem that no on else had, writing this in the third volume of *Capital*:

> Large landed property...creates conditions which cause an irreparable break in the coherence of social interchange prescribed by the natural laws of life. As a result, the vitality of the soil is squandered, and this prodigality is carried by commerce far beyond the borders of a particular state...Large-scale industry and large-scale mechanised agriculture work together. If originally distinguished by the fact that the former lays waste and destroys principally labour-power, hence the natural force of human beings, whereas the latter more directly exhausts the natural vitality of the soil, they join hands in the further course of development in that the industrial system in the countryside also enervates the labourers, and industry and commerce on their part supply agriculture with the means for exhausting the soil.[42]

Marx is the wellspring for writers today who explore the capitalist control over environments. These writers include geographers David Harvey, Don Mitchell, and George L. Henderson; literary theorists Ed White and Robert Pogue Harrison; political philosophers Neil Smith and John Bellamy Foster; political scientists James Scott, Michael Watts, and Karl Zimmerer; and historians Donald Worster, Jack Ralph Kloppenburg, Gunther Peck, Carolyn Merchant, and Colin Duncan. Harvey and Smith are interested in place making, or the imperative of capital to recreate itself geographically by gaining control of resources and people and reconfiguring them into systems that generate capital. Yet place making requires more than backloaders and bulldozers: it requires cultural meaning. Activities like logging and mining, though they destroyed an older mountain society in southern Appalachia, needed to be presented to a wider public as progress toward the shared goal of industrial expansion. Marx did not write about cultural hegemony, but he introduced the idea that environmental change is the sign of social power. Every social project has ecological implications and every environmental project contains the interests of one group over another.[43]

There is another essential aspect of metabolism: it requires fuel. This is a crucial part of the story left out until now: coal and oil did not just aid or enable capitalism, but should be understood as having created it. For although labor-power is the essence of the surplus value that becomes capital, coal jacked up the rate and magnitude of accumulation to such a degree that there would not have been industrial capitalism without it. No political economist—certainly not Smith or Marx—gave any attention to coal as the hydrocarbon infusion that took capitalism from the fiefdom of a few merchants and landlords into a social system that now dominates much of the world. The armies of workers, from Manchester in 1850, to Chicago and Moscow in 1950, to Guangdong in 2000, would never have existed without fossil fuels. And if the metabolism of society gyrates in a widening rift between ecological capacities and rates of production, fossil fuels are the real cause. The very thing that appears to define us as modern (more on that follows) is the thing that enables accumulation at the expense of every terrestrial environment. Historians are just beginning to assess the implications of this shift.[44]

TRANSFORMATIONS

There are a number of ways that historians might integrate the history of capitalism into their courses. One is to teach the definition of capitalism along with an illustrative example. I use the Caribbean sugar plantation. The plantation has a number of qualities that make it ideal for a lecture on slavery and capital. Its first flourishing coincides with and is related to capitalism in England. Colonizing organizations, like the Virginia Company, were among the first joint-stock corporations. As a business venture, sugar required investors, the enclosure of land from American Indians, and the forced appropriation of labor (Indians and then Africans)—all for the purpose of commodity production for a distant market. Sugar is one of the first instances of a technical division of labor (that is, a division of the manufacturing process) and the separation of production and consumption, points well made in *Sweetness and Power: The Place of Sugar in Modern History* (1986) by Sidney W. Mintz. The sugar plantation is not a perfect fit with capitalism—it says nothing about the importance of fuel and machinery to capitalism, and it overestimates the profitability of slaves. But it otherwise works. Dealing with capitalism early in a course has the benefit of allowing the instructor to return to the same points and make them differently with industrialization, and again with free trade and climate change.[45]

Another, related benefit of historicizing capitalism is that it helps students to unlearn the problematic notion that we are modern. Modernity might seem like a side issue to capitalism, but it is part of the self-justifying mechanism of the social system, a concept that combines the world-historical progress of capitalism with the separation between nature and culture. But if modernity turns out to be a false distinction between human beings, past and present, and different technologies—if, as Daniel Lord Smail and Andrew Shryock argue, there is no such thing as the "pre"—then the self-justification disintegrates. It finally becomes possible to talk about the relational differences between "pre" and "modern" technology. These differences provide a teachable example of how to look beyond the form of a thing and the received wisdom surrounding it to see the unevenness of its costs and benefits.[46]

What is the difference between a scythe and a John Deere harvester? The most obvious answer is that the harvester cuts more wheat in a given period of time with fewer workers, and thus at a lower cost per unit of wheat. In the prism of progress, the harvester's benefits could not be more obvious. Less labor expended and fewer people farming are better than more; gasoline is better than muscle; inexpensive food is better than expensive; large farms are better than small ones. But what are the differences between the tools? Let's assume that they do not differ because the inventors of the harvester were smarter than the inventors of the scythe. Humans have the same brains and abilities now as they had 200,000 years ago. *Homo sapien* behavior did not evolve between then and now; rather, it demonstrates variability. The harvester and the scythe differ because the owners of land adopted devices that lower the cost of labor, in order to increase relative surplus value (the difference between the value workers create and how much they are paid). The people who lost land or employment to harvesters joined a labor pool of the unskilled, forcing them to accept lower wages.[47]

The other part of the answer—that the harvester created more food through the added alchemy of chemicals and transportation—also needs qualification. There is more food than ever before, and it has made life easier for hundreds of millions of people in the developed nations. Inexpensive food makes it more possible for workers to live on low wages (thus "financing" relative surplus value across the entire industrial economy). The low cost of industrial food, however, does not reflect its social costs: poverty, pesticides, greenhouse gasses, and loss of genetic diversity, as well as the political power amassed by food-producing corporations. And the abundance of food is not evenly distributed; otherwise no one in the world would starve. Starvation might look like a problem that industrial agriculture can solve, but it is really a problem that industrial agriculture has caused. Throughout the twentieth century, as nations realized that farmland could be used to generate gross domestic product, they evicted peasants and turned land over to industrial firms. The food went for export, and the peasants went to slums. Note the contradiction: more food *and* more hungry people, unprecedented wealth *and* poverty. In contrast, people who practice subsistence cultivation rarely starve, but those who lose their land and move to slums often do.[48]

The differences between a scythe and a harvester are not merely technological, but also social and environmental. *They are tools that represent different assumptions about the purpose of production.* Furthermore, the two tools coexist. One is not past and the other present—rather, each is an articulated expression of needs and desires. *Modern* tells us nothing about the differences and similarities between them. It only tempts us to ascribe a sense of superiority and inevitability to the harvester, to see the surplus value it creates and the changes to landscapes it causes as part of a spirit of progress, muting its actual relations and costs. Our task as environmental historians is to bring historical truth to self-evident truth and open the deepest and most closely held assumptions to scrutiny. It is to provide students with the intellectual tools for thinking about the material world—and their place in it.

NOTES

1. I want to thank Andrew Isenberg and Colin Duncan for their comments on an early version of this chapter. Theodore Steinberg, "Can Capitalism Save the Planet?" *Radical History Review* 107 (2010): 7–24.
2. When the president of the United States Chamber of Commerce asserted in May of 1944, "We are all capitalists," he used that word to define Americans against the people they had just fought (Germans) and those they were about to (Russians), and he gave this elaboration: "Because we live in a capitalist society, all Americans are capitalists." *Milwaukee Sentinel*, May 11, 1944. Jonah Goldberg, "Capitalism vs. Capitalists," Townhall.com, April 23, 2010, http://www.aei.org/article/society-and-culture/free-enterprise/capitalism-vs-capitalists. Joyce Appleby, *Economic Thought and Ideology in Seventeenth-Century England* (Princeton, NJ: Princeton University Press, 1978). For context on this and other Atlantic universals, see Michel-Rolph Trouillot, "North Atlantic Universals: Analytical Fictions," *South Atlantic Quarterly* 101 (Fall 2002): 839–858.
3. I do not mean to suggest that environmental historians have ignored capitalism and its politics. Books like Robert Gottlieb's *Forcing the Spring: The Transformation of the American Environmental Movement*, 2d rev. ed. (Washington, DC: Island Press, 2005); Paul Charles Milazzo's *Unlikely Environmentalists: Congress and Clean Water, 1945–1972* (Lawrence: University Press of Kansas, 2006); and Adam Rome's *Bulldozer in the Countryside: Suburban Sprawl and the Rise of American Environmentalism* (New York: Cambridge University Press, 2001) demonstrate the political pressure that came from social activists and members of Congress, resulting in legislation that compelled corporations to absorb some of their most egregious externalities. Robert Bullard's *Dumping in Dixie: Race, Class, and Environmental Quality* (Boulder, CO: Westview Press, 2000) is about the uneven distribution of environmental pollution—toward restitution and restoration. Law professor Joel Bakan refers to corporations as "externalizing machines." See Mark Achbar and Jennifer Abbot (directors), Joel Bakan (writer), *The Corporation* (Zeitgeist Films, 2003). The North American Free Trade Agreement is a model for understanding how and why corporations seek to function away from any public oversight or environmental regulation. A number of nonprofit organizations have evaluated the agreement. See this example from the Sierra Club: "NAFTA's Impact on Mexico," http://www.sierraclub.org/trade/downloads/nafta-and-mexico.pdf.

4. Joyce Appleby, Lynn Hunt, and Margaret Jacob, *Telling the Truth About History* (New York: W. W. Norton, 1995), 121.

5. Andrew Hurley, *Environmental Inequalities: Class, Race, and Industrial Pollution in Gary, Indiana, 1945–1980* (Chapel Hill: University of North Carolina Press, 1995); Andrew Isenberg, *Mining California: An Ecological History* (New York: Hill and Wang, 2005); Neil Maher and David Kinkela, eds., "Editors' Introduction," in *Transnational Environments: Rethinking the Political Economy of Nature in a Global Age, Radical History Review* 2010 (Spring 2010). Carolyn Merchant, *Ecological Revolutions: Nature, Gender, and Science in New England* (Chapel Hill: University of North Carolina Press, 1989); Colin Duncan, *The Centrality of Agriculture: Between Humankind and the Rest of Nature* (Montreal: McGill-Queen's University Press, 1996); Donald Worster, *The Wealth of Nature: Environmental History and the Ecological Imagination* (New York: Oxford University Press, 1993); John Richards, *The Unending Frontier: An Environmental History of the Early Modern World* (Berkeley: University of California Press, 2003); Mark Fiege, *The Republic of Nature: An Environmental History of the United States* (Seattle: University of Washington Press, 2012); Michael Williams, *Deforesting the Earth: From Prehistory to Global Crisis* (Chicago: University of Chicago Press, 2003); Mike Davis, *Late Victorian Holocausts: El Niño Famines and the Making of the Third World* (New York: Verso, 2001); Bullard, *Dumping in Dixie: Race, Class, and Environmental Quality*; Richard White, *Railroaded: The Transcontinentals and the Making of Modern America* (New York: W. W. Norton, 2011); William Cronon, *Nature's Metropolis: Chicago and the Great West* (New York: W. W. Norton, 1992).

6. Notice that I did not mention barter. People do barter, but more often they do something that only *looks* like barter. Anthropologists have rejected the idea that a barter economy preceded money, which preceded credit. What really happened was that credit and money existed together, with debit and debt figured in terms of some currency. In almost every instance, parties that appear to barter convert boots and chickens into a *third* commodity that equalizes their value—*even if that third commodity is not present at the exchange.* See David Graber, *Debt: The First Five Thousand Years* (New York: Melville House, 2012).

7. Trade comes from environmental diversity; that is, people from different places trade their different things. The existence of ancient trade networks, between, for example, Chaco Canyon and Tenochtitlan, a distance of over 1,500 miles, supports this thesis. See Christina Dell'Amore, "Prehistoric Americans Traded Chocolate for Turquoise?" *National Geographic News*, http://news.nationalgeographic.com/news/2011/03/110329-chocol ate-turquoise-trade-prehistoric-peoples-archaeology. For a fuller discussion of what is not capitalism, see Ellen Meiksins Wood, *The Origin of Capitalism: A Longer View* (New York: Verso, 2002). Michael Merrill, "Putting 'Capitalism' in Its Place: A Review of Recent Literature," *William and Marry Quarterly* 52 (April 1995): 317–318.

8. Energy, too, creates more than it itself costs, and some raise it to the same importance as labor. We will consider coal at the end of the essay.

9. Karl Marx called this the "fetish" of commodities, in the sense that the use and consumption of commodities can be disconnected in our minds from the circumstances of their production. Karl Marx, *Capital: A Critique of Political Economy*, vol. I (New York: Penguin, 1990 [1867]), 165.

10. The best way to gain a strong working definition is to read the first volume of *Capital* along with one or more commentaries on capital. I suggest David Harvey, *A Companion*

to Marx's Capital (New York: Verso, 2010). Another good one is Ben Fine and Alfredo Saad-Filho, *Marx's Capital* (Sterling, VA: Pluto Press, 2004).

11. Kames wrote in "Sketch I," "The shepherd-state is friendly to population. Men by plenty of food multiply apace; and, in process of time, neighbouring tribes, straitened in their pasture, go to war for extension of territory, or mi-grate to land not yet occupied. Necessity, the mother of invention, suggested agriculture." Henry Home, Lord Kames, *Sketches on the History of Man* (1774). For Rousseau's counter view see A Discourse Upon the Origin and the Foundation of the Inequality Among Mankind (1755) and see Trouillot, "North Atlantic Universals: Analytical Fictions."

12. My review of Benjamin Friedman's *The Moral Consequences of Economic Growth* might be useful here: Steven Stoll, "Fear of Fallowing: The Specter of a No-Growth World," *Harper's Magazine*, March 2008.

13. Appleby's intellectual position and interpretation differ in her earlier and later books. *Relentless Revolution*, 21, 418, 436.

14. Cronon, *Nature's Metropolis*, 384.

15. The end of *Changes in the Land: Indians, Colonists, and the Ecology of New England* (New York: Hill and Wang, 1983) and the beginning of *Nature's Metropolis* leave a gap that more or less fits the period between the Revolution and the canal era, when agrarian capitalism in the United States took flight. For the best single book on agrarian capitalism in the United States, see Allan Kulikoff, *The Agrarian Origins of American Capitalism* (Charlottesville: University Press of Virginia, 1992).

16. Marx contributed to the early study of anthropology, sociology, political theory, and geography. As Eric Wolf notes, social science following Marx narrowed its purpose and its methods to exclude social conflict: "Contentless social relations, rather than economic, political, or ideological forces, thus become the prime movers of sociological theory," and of the other social sciences as well. Eric Wolf, *Europe and the People Without History* (Berkeley: University of California Press, 1982) 9, 21.

17. For primitive accumulation see Adam Smith, *The Wealth of Nations* (New York: Bantam, 2002 [first published as *Inquiry into the Nature and Causes of the Wealth of Nations* in 1776]), 91; Marx, *Capital*, vol. I, 873; Raymond Williams, *The Country and the City* (New York: Oxford University Press), 96.

18. Karl Marx, *Grundrisse* (New York: Penguin Books, 1993 [1939]), 505–507.

19. Richard Price, *Observations on Reversionary Payments* (London, 1812). Price is quoted by Marx in *Capital,* vol. I, 888. "Primitive accumulation cut through traditional lifeways like scissors. The first blade served to undermine the ability of people to provide for themselves. The other blade was a system of stern measures required to keep people from finding alternative survival strategies outside the system of wage labor." Michael Perelman's *The Invention of Capitalism: Classical Political Economy and the Secret History of Primitive Accumulation* (Durham, NC: Duke University Press, 2000), 14. On the Poor Laws and enclosure, see Keith Wrightson's *Earthly Necessities: Economic Lives in Early Modern Britain* (New Haven, CT: Yale University Press, 2000), 215–220. Wrightson explains that the Poor Laws came about for more than one reason, among them food crises that afflicted England before the last famine in 1624, but also to aid "labouring persons not able to live off their labour." Appleby says more, "What made the poor an issue in Restoration England was their tenuous position. They had been expelled from a traditional order and as yet were only conditionally and occasionally needed in the new economic structure growing up outside the old." Appleby, *Economic Thought*, 162–164.

Thomas Robert Malthus came up with his vicious proscription that the Poor Laws should be revoked and poor should die rather than sink the rest of the society (meaning the rich), without understanding poverty as the creation of the class he wanted to protect. The incoherence of the *Essay on the Principle of Population* (1798) is stunning considering its long career as a kind of natural history of society. My understanding of Malthus has benefited from John Bellamy Foster, *Marx's Ecology: Materialism and Nature* (New York: Monthly Review Press, 2000). The same pattern of externalized costs affects the working poor in the United States. See Barbara Ehrenreich, *Nickel and Dimed: On (Not) Getting By in America* (New York: Metropolitan Books, 2001).

20. Wood, *Origin of Capitalism*, 97. The power of capitalism to reproduce itself socially by taking away all alternatives to itself and by commanding the origin of all acceptable ideas is the subject of various works, two of the most important being John Gaventa, *Power and Powerlessness: Quiescence and Rebellion in an Appalachian Valley* (Urbana: University of Illinois Press, 1982) and Antonio Gramsci's prison notebooks. Gramsci says this of how a social practice (or praxis) shapes the way people see the world: "The philosophy of praxis itself is a superstructure, it is the terrain on which determinate social groups become conscious of their own social being, their own strength, their own tasks, their own becoming." David Forgacs, ed., *The Antonio Gramsci Reader* (New York: New York University Press, 2000), 196. For a history of the self-regulating market and responses to it see Karl Polanyi, *The Great Transformation: The Political and Economic Origins of Our Time* (Boston: Beacon Press, 2001 [1944]).

21. Williams, *The Country and the City*, 99. A work of history that flushes out all the dimensions of this process, neither squarely of one interpretation or the other, is Wrightson's *Earthly Necessities*. It includes a penetrating description of enclosure, as proceeding by fits and starts, over centuries not decades, sometimes by negotiation, and with plenty of protest and condemnation. It does not contradict the Marxian model but deepens and complicates it. Environmental Historians have authored a number of books and articles (not necessarily Marxian in interpretation) about land once held in common, squatted upon, or not owned that then enclosed, including Karl Jacoby, *Crimes Against Nature: Squatters, Poachers, Thieves, and the Hidden History of American Conservation* (Berkeley: University of California Press, 2003); Louis Warren, *The Hunter's Game: Poachers and Conservationists in Twentieth-Century America* (New Haven, CT: Yale University Press, 1999); Mark David Spence, *Dispossessing the Wilderness: Indian Removal and the Making of the National Parks* (New York: Oxford University Press, 2000); and Sara Gregg, *Managing the Mountains: Land Use Planning, The New Deal, and the Creation of a Federal Landscape in Appalachia* (New Haven, CT: Yale University Press, 2010). In these examples, either one of the states or the United States cordoned off what had been Indian country, fenced hunting grounds, and declared National Parks, resulting in eviction and the rendering of resources for industrial production.

22. Ellen Wood, *The Origin of Capitalism*, 110–115. Wood considers Locke a capitalist thinker. Neal Wood, *John Locke and Agrarian Capitalism* (Berkeley: University of California Press, 1984). Neal Wood (who is married to Ellen Wood) also sees Locke's interest in landed property and agriculture as evidence of a capitalist formulation. Neil J. Mitchell, "John Locke and the Rise of Capitalism," *History of Political Economy* 18 (Summer 1986): 291–305; John Henry, "John Locke, Property Rights and Economic Theory," *Journal of Economic Ideas* 33 (Fall 1999): 609.

23. John Winthrop, "General Considerations for the Plantations in New England, with an Answer to Several Objections," in *The John Winthrop Papers*, vol. II (Boston: Massachusetts Historical Society, 1931 [1629]), 120.

24. *The Fundamental Constitutions of Carolina* was never ratified.

25. Merchant, *Ecological Revolutions*, 185. For further evidence of this transition see Brian Donahue's *The Great Meadow: Farmers and the Land in Colonial Concord* (New Haven, CT: Yale University Press, 2004), which argues for a similar transition to intensive, market directed farming after the Revolution. My own *Larding the Lean Earth: Soil and Society in Nineteenth-Century America* (New York: Hill and Wang, 2002) benefited from *Ecological Revolutions*.

26. Merchant, *Ecological Revolutions*, 230, 233.

27. I confronted a similar problem in *Larding the Lean Earth* and borrowed Duncan's conception of improvement as ecologically benign, as an ethic of restraint and nascent conservation. I stand by my argument that Americans did not see improvement as a strategy for maximizing profit or leaning hard on land and labor, but I also underestimated the role of improvement as a transition to fulltime commodity production and the importance of that to capitalism. Duncan's book has the same problem. Colin Duncan, *The Centrality of Agriculture: Between Humankind the Rest of Nature*.

28. I am thinking of Karl Polanyi, whose *Great Transformation* is vulnerable to the criticism that his research did not match the scope of his conclusions. Braudel was not impressed, writing of Polanyi's classic, "Not the slightest effort has been made to tackle the concrete and diverse reality of history and use that as a starting-point... Sociologists and economists in the past and anthropologists today have unfortunately accustomed us to their almost total indifference to history. It does of course simplify their task." Fernand Braudel, *Wheels of Commerce, Volume II, Civilization and Capitalism, 15th–18th Century*, trans. Siân Reynolds (London: Phoenix Press, 2002 [1979]), 227. Yet one of the virtues of Karl Polanyi's *Great Transformation* is that it adds a much-needed corrective to Marx's own dismal view—that the movement to commodify everything and impose the radical notion of an unregulated market created a countermovement against it. Without that, the Marxian view has no way of evaluating American populism or the labor movement as responses to capitalism. The same criticism Braudel gave to Polanyi could be said of David Harvey, Neil Smith, and others.

29. Fernand Braudel, *The Structures of Everyday Life, Volume I, Civilization and Capitalism, 15th–18th Century*, trans. Siân Reynolds (New York: Harper and Row, 1981 [1979]); Fernand Braudel, *Wheels of Commerce, Volume II, Civilization and Capitalism, 15th–18th Century*; Fernand Braudel, *The Perspective of the World, Volume III, Civilization and Capitalism, 15th–18th Century*, trans. Siân Reynolds (London: Phoenix Press, 2002, [1979]). A version of the first volume appeared in French in 1967 and English in 1973.

30. I should point out that many of the studies Braudel depends upon are out of date. *The Mediterranean and the Mediterranean World in the Age of Philip II* (Berkeley: University of California Press, 1995, originally published in 1949) is probably the greatest dissertation of all time. For Braudel as human ecologist see James R. Hudson, "Braudel's Ecological Perspective," *Sociological Forum* 2 (Winter 1987): 146–165.

31. Braudel, *Wheels of Commerce*, 230.

32. Wallerstein himself, as one of Braudel's greatest proponents, extrapolated from the historical relationship between market city and countryside in his "world systems" model of capitalist expansion. Immanuel Wallerstein, *Unthinking Social Science: The Limits of*

Nineteenth-Century Paradigms (Philadelphia: Temple University Press, 2001), 197 and *The Capitalist World Economy: Essays by Immanuel Wallerstein* (Cambridge: Cambridge University Press, 1979); Braudel, *Perspectives of the World*, 623;

33. Braudel, *Wheels of Commerce*, 403, 600. Emmanuel Le Roy Ladurie takes a similar view, of slow accretion without a sense of culmination or the achievement of modernity in *The Peasants of Languedoc*, trans. John Day (Urbana: University of Illinois Press, 1974). Carlos Antonio Aguirre Rojas writes of Braudel, "Like Marx, Braudel adopts modern capitalist society as the center of his theoretical concern, dedicating to it an enormous and exhaustive labor of research. Though this research has a different focus than that of Marx, it also brings together enlightening and fundamental insights that permit a full understanding of modern bourgeois society and its genesis. This is what makes it possible for Braudel, not only to make brilliant assessments of the contemporary world, but also to ponder its possible destines." Carlos Antonio Aguirre Rojas, "Between Marx and Braudel: Making History, Knowing History," *Review* 15 (Spring 1992): 179.

34. Michael Dove, *The Banana Tree at the Gate: A History of Marginal Peoples and Global Markets in Borneo* (New Haven, CT: Yale University Press, 2011); Stephen Gudeman, *The Demise of a Rural Economy: From Subsistence to Capitalism in a Latin American Village* (London: Routledge, 1978); Enrique Mayer, *The Articulated Peasant: Household Economies In The Andes* (Boulder, CO: Westview, 2002); Elizabeth Dore, *Myths of Modernity: Peonage and Patriarchy in Nicaragua* (Durham, NC: Duke University Press, 2006), 20. In my own work I am interested in a similar phenomenon: instances in which subsistence gardens became integral to early coal mining. Not only did coal companies allow people from the mountains to keep gardens, they sometimes insisted on it as a subsidy to the low wages they paid. All over the world today, factory gardens and village rice paddies feed industrial workers.

35. Max Beer, *An Inquiry into Physiocracy* (London: George Allen and Unwin, 1939), passim; Elizabeth Fox-Genovese, *The Origins of Physiocracy: Economic Revolution and Social Order in Eighteenth-Century France* (Ithaca, NY: Cornell University Press, 1976), 9–11. There has been some attention paid to the physiocrats as originators of the idea of a steady-state economy, though it must be emphasized that they believed in economic growth. See W. A. Eltis, "Francois Quesnay: A Reinterpretation 2. The Theory of Economic Growth," *Oxford Economic Papers* 27 (November 1975): 327–351; Colin Duncan, *The Centrality of Agriculture: Between Humankind and the Rest of Nature* (Montreal: McGill-Queen's University Press, 1996); Ronald L. Meek, *The Economics of Physiocracy: Essays and Translations* (Cambridge, MA: Harvard University Press, 1963) 73.

36. This is the only time, except for a related footnote, that Smith uses the phrase "wealth of nations" in his book. Smith, *Wealth of Nations*, 862.

37. Samezo Kuruma, "Physiocracy," *History of Political Economy* (Tokyo: Iwanami Shoten, 1954), http://www.marxists.org/archive/kuruma/politicaleconomy-physiocracy.htm#n7.

38. Smith, *Wealth of Nations*, 318. "As population increases…there come to be more buyers of fish…A market which, from requiring only one thousand, comes to require annually ten thousand ton off fish, can seldom be supplied without employing more than ten times the quantity of labour which had before been sufficient to supply it. The fish must generally be sought for at a greater distance, larger vessels must be employed, and more expensive machinery of every kind made use of." I argue in *The Great Delusion* that growth and capitalism depend on the assumption of infinitude. Certain things

that are untrue need to be rendered as true in order for the expansion of capital (and thus capitalism itself) to make sense. Steven Stoll, *The Great Delusion: A Mad Inventor, Death in the Tropics, and the Utopian Origins of Economic Growth* (New York: Farrar, Strauss and Giroux, 2008). Herman Daly has done more than any other economist to bring conventional economics to biophysical reality. For a short piece on the discontinuity between economy and ecology see "Economics in a Full World," *Scientific American* (September 2005). For place-making as essential to capitalism and its allied view of nature, see Neil Smith, *Uneven Development: Nature, Capital, and Production in Space*, 3d ed. (Athens: University of Georgia Press, 1990).

39. For an analysis of socialist dreams of progress during the last century see Eric Hobsbawm, *How to Change the World: Reflections on Marx and Marxism* (New Haven, CT: Yale University Press, 2012).

40. "In all these forms—in which landed property and agriculture form the basis of the economic order...there is to be found...appropriation not through labour, but presupposed to labour; appropriation of the natural conditions of labour, of the *earth* as the original instrument of labour as well as its workshop and repository of raw materials." Karl Marx, *Grundrisse*, 485.

41. See Foster, *Marx's Ecology*, 155. Justus von Liebig, *Letters on Modern Agriculture*, ed. John Blyth (New York, 1859), 24. I had no idea that Marx had written about soils, nor had I read Foster when I published *Larding the Lean Earth: Soil and Society in Nineteenth-Century America* (New York: Hill and Wang, 2002). I knew about Carey's interest in soils but thought Carey inscrutable and largely unimportant. I continue to think that he was a dead end in American political economy. Marx gives a devastating critique of Carey in the last chapter of the *Grundrisse*.

42. Karl Marx, *Capital*, vol. III (New York: International Publishers, 1967), part VI, "Transformation of Surplus-Profit into Ground-Rent, Chapter 47: Genesis of Capitalist Ground-Rent," http://www.marxists.org/archive/marx/works/1894-c3/ch47.htm.

43. David Harvey, *Spaces of Hope* (Berkeley: University of California Press, 2000); Don Mitchell, *Lie of the Land: Migrant Workers and the California Landscape* (Minneapolis: University of Minnesota Press, 1996); George L. Henderson, *California and the Fictions of Capital* (Philadelphia: Temple University Press, 2003); Ed White, *The Backcountry and the City: Colonization and Conflict in Early America* (Minneapolis: University of Minnesota Press, 2005); Robert Pogue Harrison, *Forests: The Shadow of Civilization* (Chicago: University of Chicago Press, 1993); Smith, *Uneven Development: Nature*; John Bellamy Foster, *Marx's Ecology*; James Scott, *The Art of Not Being Governed: An Anarchist History of Upland Southeast Asia* (New Haven, CT: Yale University Press, 2009); R. J. Johnston, Peter. J. Taylor, and Michael J. Watts, *Geographies of Global Change: Remapping the World in the Late Twentieth Century* (Cambridge: Blackwell, 1995); Karl Zimmerer and Thomas J. Bassett, *Political Ecology: An Integrative Approach to Geography and Environment-Development Studies* (New York: Guilford Press, 2003); Donald Worster, *The Wealth of Nature*; Jack Ralph Kloppenburg, *First the Seed: The Political Economy of Plant Biotechnology, 1492-2000* (New York: Cambridge University Press, 1988); Gunther Peck, "The Nature of Labor: Fault Lines and Common Ground in Environmental and Labor History" *Environmental History* 11 (April 2006): 212–238.

44. Colin Duncan has thought about fuels and economies to a greater extent than any other environmental historian. See Duncan, "Adam Smith's Green Vision and the Future of

Global Socialism," in *New Socialisms: Futures Beyond Globalization*, ed. Robert Albritton, Shannon Bell, John R. Bell and Richard Westra (London: Routledge, 2004), 90–104.

45. Sidney W. Mintz, *Sweetness and Power: The Place of Sugar in Modern History* (New York: Penguin, 1986).

46. "The grammar of modernity functions as a temporally provincializing logic just as powerful, and as misguided, as the provincializing logic associated with Eurocentrism." Daniel Lord Smail and Andrew Shryock, "History and the 'Pre,'" *American Historical Review* 118 (June 2013): 709–737. Bruno Latour, *We Have Never Been Modern* (Cambridge, MA: Harvard University Press, 1993).

47. John J. Shea, "Refuting a Myth About Human Origins," *Science* 99 (March-April 2011): 128–135. The author demonstrates that tools nearly 200,000 years old and those much more recent do not differ. H. J. Habakkuk is still the authority on the circumstances leading to agricultural machinery. Habakkuk, *American and British Technology in the Nineteenth Century* (Cambridge: Cambridge University Press, 1962). For the social purpose of technology and capital as its driving force the very best book is David F. Noble, *Forces of Production: A Social History of Industrial Automation* (New York: Knopf, 1984).

48. The uneven distribution of food does not cut neatly between nation-states, but rather between classes. On the Green Revolution and the logic of industrial agriculture see Nick Cullather, *Hungry World: America's Cold War Battle Against Poverty in Asia* (Cambridge, MA: Harvard University Press, 2010). On the billion people (one-seventh of humanity) who now live in slums, see Mike Davis, *Planet of Slums* (New York: Verso, 2006). Critiques of industrial agriculture are many. See Deborah Fitzgerald, *Every Farm a Factory: The Industrial Ideal in American Agriculture* (New Haven, CT: Yale University Press, 2003).

CHAPTER 15

···

OWNING NATURE

Toward an Environmental History
of Private Property

···

LOUIS WARREN

> The first person who, having fenced off a plot of ground, took it into his
> head to say this is mine and found people simple enough to believe him,
> was the true founder of civil society.
>
> —Jean-Jacques Rousseau, 1755
>
> Thus in the beginning, all the world was America.
>
> —John Locke, 1690

PRIVATE property has been many things. According to some of its strongest advocates, it is the bedrock of freedom, democracy, personal security, capitalism, and wealth, as well as the surest defense against tyranny.[1] Friedrich Hayek, the leading economist of classical liberalism, extolled private property as the "moral basis" of "modern civilization." [2] Following in Hayek's footsteps, Milton Friedman, perhaps the most influential economist of the last fifty years, attributed virtually every aspect of any functioning government or economy to private property. "[I]f you want efficiency and effectiveness, if you want knowledge to be properly utilized, you have to do it through the means of private property."[3] It is not too much to say that for various prominent thinkers and politicians, private property is the central freedom, without which there can be no others.[4]

A full assessment of the merit of these claims is beyond the scope of this essay. But after decades of advocacy by property theorists who seem to believe that erecting boundaries around pieces of earth will raise a new Jerusalem, we might take this opportunity to explore what environmental history reveals about the meanings of property and its limitations. At its roots, any real property represents a political bounding of land. It is also a cultural inscription, an expression of how natural goods can be conceptually separated from the earth and tied to people. This essay will explore private property primarily as it has developed in the West and particularly in the United States, using more global examples to point out larger patterns that suggest useful directions for environmental historians of other regions and periods.

Private property, or freehold tenure, is usually defined as land owned by an individual who was more or less unfettered rights to sell, lease, or assign all rights to it.[5] But a close examination of the environmental meanings of property illuminates how limited conventional understandings of property have been. Theorists like Hayek and Friedman seem to equate private property with real estate—tidily bounded parcels swapped for set amounts of wealth. But modern real estate emerged out of a long history in which the characteristics of private property were hotly contested. Property and markets in it have not always facilitated freedom. In fact, at times they have obstructed personal liberties. Central to arguments over the right to sell property were fierce disputes about the rights and obligations of people who worked to make property from nature. As we shall see, some of the core rights of property holders in ages past actually impeded the emergence of property markets.

Moreover, as people have discovered continually throughout history, the workings of nature often undercut, erase, or confuse property lines and the very idea of ownership. Contrary to the claims of its loudest enthusiasts, private property has not always been obstructed by the existence of other forms of ownership, such as common or public property. In fact, private property has depended on these other kinds of property—*un*private kinds—to compensate for or contain natural forces that otherwise undermine it—or indeed, threaten its very existence.

The Origins Of Real Estate

Contemporary faith in private property is partly a product of its success as an institution, which is evident in the gigantic, centuries-long, ongoing process of privatization, or enclosure. Between 1520 and 1900, colonial powers put billions of acres in private hands. They achieved this mostly by wresting lands from indigenous owners at great cost in blood (most of it indigenous), and transferring titles to colonists who operated in a market economy.

After slowing over much of the twentieth century, the wave of privatization picked up again, this time through neoliberal economic policies rather than outright colonization. State properties, communal lands, and public resources from Russia to Java, the Amazon to Everest, found their way to the auction block. By the turn of the twenty-first century, in many parts of the world, the debate between left and right often was less about whether to privatize resources than how to do so. "Respect for private property has to be one of the foundations of the government's policies," said the new president of

Russia.[6] In China, in 2004, the National People's Congress amended the constitution of the People's Republic: "A citizen's lawful private property is inviolable."[7]

Given these developments, it is hardly surprising that some have cast the history of property as a tale of relentless, almost divinely ordained enclosure, in which land makes a steady march from the village common and other mostly inalienable holdings among premodern peoples to private, marketable parcels held by modern investors. The seeming triumph of privatization in places like China appears to bear out the most teleological of histories, such as Garrett Hardin's "Tragedy of the Commons," in which rising populations and market economies degrade natural resources until privatization becomes inevitable.[8]

Critics of such arguments point out that private property is no guarantee against environmental insult. From the windblown sands of the Dust Bowl in the southern Great Plains to the exploding rocks and toxic seepage at Hooker Chemical Company's Love Canal in upstate New York, private property has been no stranger to ecological decay. Around the world, privately held gold mines, plantations, and company towns earn few plaudits for environmental sustainability.

Moreover, for all its tragedies real and imagined, common property endures in many different forms. Indeed, if we define a commons as a resource claimed by a group of coequal users who restrict access, we can see not only that it exists, but even that in thrives in parks, fish and game, and other natural resources.[9]

But we might go further, to question the supposed incompatibility of private and common property, and to cast doubt on histories that narrate an inevitable eclipse of the latter by the former. Private holding and common estate do not only coexist, but they also reinforce and facilitate one another in ways that suggest that private property needs the commons in order to persist.

PROPERTY is not a static condition, but rather a constantly shifting bundle of use rights. Which sticks are in the bundle, and what they imply, depends on social acknowledgement of the rights of the user, and on political and environmental conditions that change frequently.[10]

Thus forms of property have varied tremendously. In some times and places, such as Mormon Utah in the 1850s, owning land conveyed the right to live on it and farm it, but not to sell it. In other places—modern metropolises from Hong Kong to today's downtown Salt Lake City—owning land usually implies the right to occupy land and to sell it, but usually not to farm it, which would violate numerous zoning regulations (and make little economic sense). At the same time, many crowded cities retain large community gardens.

These examples underscore that property rights are socially constructed, and the forms they take are particular to place and time. Derived from many economic, political, and cultural contingencies wherever it may be found, property seldom retains the same contours in any two places.

And yet a long-term homogenization of property has taken place. As much as the globe remains a gigantic patchwork of distinctive property holdings, in ages past it was even more variegated. For most of human history, nomadic hunting and gathering peoples projected a wide array of territorial claims to land, sometimes apportioning them to clans, sometimes to families, and sometimes extending them to certain products or animals. The slash-and-burn systems of horticulturalists, from the pre-contact Americas to Africa and Asia, assumed thousands of variations, in which land was only owned so long as it was used, but rights of owners differed from one people to the next, and a family's land, even if it could be bequeathed to heirs, could not be sold to outsiders.[11]

Even the development of agriculture did not lead directly to uniform codes of private property, for widely divergent methods of owning farmland existed simultaneously. For example, in Norway, in the nineteenth century, households on each large farm had no exclusive rights to any single parcel of land. Instead, each had a claim to a given portion of the value of the farm, which they took not only in the form of crops, but also as fish from streams, and firewood, lumber, resin, and game from the forest.

In Japan, by way of contrast, land was owned by village corporations, which allotted farm plots to each member of the village by lottery or on a rotating basis. Some plots might be marshy or infertile compared to others, but rotating-use rights meant the burden of bad land was potentially less onerous due to being widely shared.[12]

Complexes of usufruct expressed very particular use rights—property claims—to local land and peoples. They were tied to the peculiarities of local nature: the ripening of wild onions on the meadow; the maturing of honey in a hive; the pulse of a river and its herring. They also were bound to local cultures and their interior requirements. For these reasons, they often proved almost impossible to codify in legal codes of the state except in the most simplified manner. Faced with government requirements for the "legibility" of property arrangements, usufruct was often diminished or completely ignored, becoming an "outlaw" tradition even in some places where it had very ancient roots.[13]

And, as states began to regulate transactions in land, and as the market in private holdings developed, this remarkable variability in property holding gave way increasingly to a very different real estate system. In many countries, since the late nineteenth century, ever more has land has been located between surveyed boundaries, described on a legal title which was then traded by distant owners to buyers exchanging cash, sometimes involving banks on opposite sides of the world.

Historians usually trace the origins of this system to the rise of the nation-state in the early modern period. In post-Renaissance Europe, as kingdoms and principalities became nations, officials sought to project their power by mapping national boundaries. For this, they turned to cadastral maps—two-dimensional representations of property boundaries and land ownership. As rulers mapped national territories on paper (the term "cadastral" comes from Greek roots which mean "drawn by line"), they encouraged the identification of individual parcels with individual owners, partly out of a desire to tax those owners more effectively. Across Europe, this process abetted enclosure, or the transformation of common fields into private lands. Thus individual and national property grew, mutually constituting one another. To this day, nation-states remain the "ultimate owners" of all property within their boundaries, reminding us that private property in its modern form has never existed except under the umbrella of the state—and that Rousseau, in blaming private property for creating the state, had it exactly backwards.[14]

The appropriation and commodification of land constituted an economic revolution in Europe, and in colonial possessions it assumed truly spectacular dimensions. The British, Spanish, Portuguese, Dutch, and French projected maps and property lines as they sought to inscribe private landholdings over much of Africa, Asia, Australia, New Zealand, and the Americas. The most dramatic example, and since the time of John Locke the favorite of privatizers everywhere, was the United States. There, laws encouraged parceling out western lands, and thereby helped to create a vast, property-owning middle class.

For almost a century, scholars have debated the effects of enclosure, with some arguing that it liberated the countryside and others that it impoverished rural folk. [15] But because it contributed in many ways to the development of Western economies, the attachment of separate parcels to individual owners came to be seen as central to modern development. Consequently, after World War II, individual, transferable property rights became a core value of the "free world" in its contest with the Communist block.[16] Today, in the post-Soviet world, many theorists and world bankers set such rights as a baseline for the economic success of "post-communist" and developing countries.[17]

So it is that many scholars and theorists describe an arc of enclosure and allotment that ascends over the course of human history, culminating in modern capitalist economies. By these lights, the market in private land that began in Europe must sweep the world, upending the variegated property holding systems of antiquity. Any common property that remains is merely awaiting its conversion to real estate, for privatization has become the endpoint of world history.[18]

Indeed, the bounding of nature into private property has not only spread round the globe, increasingly it has also miniaturized onto ever smaller parts of it. The development of new technologies allows the parsing of nature in new ways, encouraging claimants to property rights in microorganisms and the DNA of animals, plants, and even indigenous peoples. Some foresee a new market in which exclusive rights to the resin of rare Amazon plants or to the genomes of Inuit children trade hands on world markets alongside land and other natural resources. If in the beginning all the world was America, then the coursing of every part of nature with property lines resembles the bounding of a new frontier, where instead of land turning to real estate, organisms and bodies are fragmented and abstracted into binary sequences commanding a price, and Marx's diagnosis of the capitalist condition—"all that is solid melts into air"—seems as astute as ever. [19]

But property is not one thing, and its history is far more complicated than a simple story of the rise of the real estate market. If today's advocates of privatization and the free market draw lessons primarily from the US example, a closer look at the history of property in the US suggests its dependence on the state, and its troubling relationship to some of America's most enduring social inequities.

KEYS TO THE KINGDOM

Broadly speaking, property has market value in proportion to the access it provides to natural resources and social goods. As any pioneer farmer could tell you, the worth of landed property derived not only from its potential for generating food and fuel, but also from the access it provided to critical institutions, especially markets. Proximity to canals, navigable rivers, and roads raised property values accordingly. Land in the United States privatized most quickly where there were markets nearby, such as those areas on the railroad near Chicago. Where there was little or no access to markets—as in much of Alaska—land often remained in public hands much longer.[20]

Indeed, much of the value of private property in today's United States derives less from its capacity for "productivity" in an agricultural or industrial sense than from other kinds of access. For example, many of today's most widely marketed properties are in suburbs, whose residents seldom trundle their produce to mart. Instead, in those curving streets and cul-de-sacs outside major cities, each home is a private point of access to public (or privately owned, publicly regulated) utility grids. The landowner does not own the drinking water, the sewer system, or the electrical grid, but her home provides access to them. If each suburbanite had to dig her own sewer line, generate her own electricity, and pave her own roads, we can be certain these properties would not even exist, or if they did, they would be of considerably less value.

Beyond these basic needs, the market value of the home is derived from what realtors euphemistically call "location." They could just as well call it access: to workplaces, shopping, and to public institutions like streets, highways, parks, schools, libraries, and swimming pools. In the end, the value of private property in the suburb is constituted by a network of public and private property holdings; what any home commands on the market is driven to a considerable extent by the webs of access in which it is entangled, and the value of that access is often dependent upon functioning public institutions, like the water district, the department of transportation, the board of education, and the police and fire departments.[21] Would Winnetka, Illinois be such a desirable location without its schools? Would home prices be what they are in Grosse Point, Michigan without its waterfront park? We can extend such arguments to non-suburban places around the globe. Would Hong Kong apartments be of such value on the market without the city's port connecting much of Asia to the rest of the world? And how much property of value could there be in the Netherlands without vast public dikes?

At the same time, the value of American suburban homes is also driven by their relative remoteness from other features of urban life such as busy streets, factory smokestacks, landfills, and junkyards. These absences in part express another layer of public context for the private home: zoning. Americans recognized the utility of zoning for sustaining the worth of suburban homes early in the twentieth century. Beginning with Los Angeles in 1908, alliances of developers, realtors, and city authorities have sought to control how private properties develop and what kinds of access and proximities they can have, with an eye toward encouraging investment by prospective homeowners in residential districts. [22]

Zoning pleased homebuyers because it provided not just remoteness from unsightly industry, but also from cheaper, less private dwellings like apartments and boarding-houses. The social consequences were plain to see: suburban bans on multi-family dwellings and on various kinds of rentals functioned in part as class barriers, separating suburban owners from downscale renters.

Moreover, in this way suburban property expresses and commodifies a newly self-conscious privacy, embedded in pastoral lawn and garden in ways that had not been possible in the metropolitan core. To this day, zoning remains central to suburban development, and the private property esteemed by most homeowners is thus constituted partly by governments that keep it remote from liquor stores, porn theaters, gun shops, and other less savory commercial establishments.

Just as an array of laws and government programs helped to segregate the suburbs by class, so too were they gendered male. Many of the same programs that inscribed disadvantages for poor city dwellers also weighted the scales against women. VA loans and the GI Bill, for instance, necessarily favored men because few women were allowed to serve in the military in the first place. In a society where women and minorities historically have earned lower wages than white men, the higher price of detached homes meant that most suburban owners would be male. [23]

Joining class and gender as markers of the suburban map, and makers of suburban real estate, was race. As early as 1890, when San Francisco barred Chinese residents from some neighborhoods, authorities sought to ban nonwhite peoples from owning lots in districts popular with white buyers. As suburbs grew in the early twentieth century, they remained racially restricted in ways that ensured the vast increase in real estate wealth of subsequent decades would flow disproportionately to white people. By the time of the Great Depression, exclusive racial covenants that began as local regulations were influencing federal policy on loan guarantees, and finding their way into federal law. The GI Bill and Veteran's Administration Home Loans, two of the programs most responsible for creating the suburban middle class that emerged after World War II, were both conceived and administered in ways that favored white veterans, helping to make suburban property a poor guarantor of freedom for aspiring buyers of color. The Supreme Court banned racially restrictive covenants in the early 1950s, but for decades after, realtors refused to show houses in "white" areas to prospective buyers of color. From the 1960s to the present, "white flight" from city centers to the suburbs has suggested the close connection between property values and race. [24]

Had suburban property been drawn differently, without such pronounced race and class exclusions, environmental politics in the US might well have taken a very different shape. Adam Rome has illuminated the many ways that the 1970s environmental movement emerged from suburban concerns about clean air, clean water, open space preservation, and wilderness protection. Impressive as it was, a suburban-generated movement was, necessarily, a movement that addressed concerns close to the hearts of white people. As is well known today, eventually environmentalists faced strong criticism (some of it from minority peoples who also happened to be suburban home-owners) for failing to address issues of environmental inequality.[25]

We might pause for a moment to consider one of the notable legacies of America's suburban development: that the landscapes of Levittown and Lakewood, which symbolized the allure of private property, were actually embedded in cultural and economic strictures that limited their fungibility on the open market. Private parcels in these districts transferred from developer to owner only insofar as they adhered to cultural rules about property owning that effectively made the suburban property market itself a kind of racial commons, one in which the market for homes was open only to white people. And thus the development of two parallel and sometimes contradictory strands of environmental reform politics, one called "environmentalism" and the other known as "environmental justice," has much to do with the shaping of the suburbs through patterns of exclusivity in property owning.

Suburban life is an age away from the travails of pioneers, but this brief consideration of access and zoning reminds us of two central characteristics of enclosure in US history, characteristics that are too easily obscured by abstract theories. First, parceling out the American landscape into private freeholds required public institutions, chief among them a landholding state. Even before the adoption of the US Constitution, Congress passed the Ordinance of 1785, which stipulated a system of surveying former Indian lands into a grid from which they would be sold to private owners. The unallotted lands became known as the public domain, and they represented a gigantic natural resource owned in common by the citizens of the US, and held in trust by the government, pending survey and public auction. According to the law, aspiring owners could buy it for no less than one dollar per acre. Adopted with the aim of shoring up democracy, the law would hopefully create a large class of investor freeholders on properties that were clearly demarcated and easily transferable. The grid pattern would help avoid title disputes of the sort that plagued many areas of the original colonies, where occupants pegged property boundaries to shifting landmarks. The entire package would guarantee that the public lands vanished into private holdings in orderly, democratic fashion, and causing a vast market in land to flourish.[26]

But in ways its creators could not have foreseen, governing the public domain required powerful institutions that revealed the effortless enclosure of the grid as a fantasy. The federal government's exclusive role as dispensor of western lands more or less ensured that it would play a major part in the settlement and society of the United States. By 1880, the United States Congress had passed nearly three thousand laws affecting the disposition of the public domain.[27] The American state may or may not have been weak relative to those of Europe, but what strength it had grew in no small measure from its role as the legitimating power behind most property transactions west of the Alleghenies.[28] That the parceling of the landscape fundamentally empowered the state was particularly ironic, because in no nation did the concept of private property, of a domain where the power of government is limited, become more highly resonant as a bulwark against the state. In other words, in the United States, the creation of a vast market in private property not only empowered some individuals, it also helped constitute the government authority that landholding ostensibly shielded those individuals from.

The second characteristic of the enclosure process, evident in suburbs and frontier alike, is its exclusivity along lines of race and gender and its attendant violence. For the first century of American history, enclosure enriched white owners at the expense of nonwhites, especially Native Americans, African slaves, and Mexicans. Arguably that was its purpose, since landownership was a fundamental expression of citizenship, a privilege by definition restricted to white people, with full rights exercised only by white men until after the Civil War. [29]

Indeed, surveying, allotting, and commodifying a landscape as dictated by the Ordinance of 1785 presupposed an absence of prior claimants, despite the ubiquity of such people in the very lands over which the ordinance extended. Making North American land into US property required extinguishing Indian claims, and that in turn required prolific bloodshed. Indian campaigns replete with massacres and atrocities expressed state power constituted by local or state militias (often supported with federal subsidies), or by the federal army itself. Environmental historians would do well to integrate into their histories the bloodshed at Tippecanoe, Palo Duro, and Wounded Knee. Only through the violent removal of Indian means to reproduce community and culture, only through destruction of their bodies, could the US turn Native American earth into American land.

Just as relevant to that history was the emergence of bureaucracies to contain and govern Indians, especially the US Bureau of Indian Affairs. Eventually, the US government ended warfare and separated Indians from the grid in reservations. Ultimately, with lawmakers passing laws to shrink even these holdings, Indian forms of communal and even individual property often proved untenable. Subsequent policy changes, such as the late nineteenth-century effort to break reservations into freehold parcels, had dramatic and often devastating effects on Indian communities and individuals. [30]

Thus behind every single property held by non-Indians lurks a preceding Indian claim. That claim is extinguished—if it is—only by the action of the state. Theorists who claim that modern landowners represent only a brake on the state ignore the centrality of governance in the making of modern property, and the degree to which landowners have been beneficiaries of violence and racism perpetrated on their behalf by state power. [31]

But even with a duly empowered national authority, the obstacles to making property from nature were many, not least because the abstraction of a two-dimensional grid was a poor fit for the three-dimensional contours of history and nature. One of the most persistent problems was that many of the lands that were swept into the public domain were already occupied, and not only by indigenous people. Even when laws allowed them to retain prior claims, Indians, Hispanos, French Americans in Old Louisiana, and sometimes America's own backcountry settlers could find themselves at odds with the rectilinear survey, which failed to account for forms of landholding that predated the United States.

Some of the most salient examples come from the former provinces of Mexico seized by the US in 1848. At that time, under Mexican law, a married woman could own, sell, lease, or bequeath property without consent of her husband. By contrast, in the US, husbands had to sign documents conveying authority for wives to exercise these rights. The US annexation of northern Mexico meant that married women across the region suddenly lost autonomy in the property market.[32]

Gender biases against women's ownership of property in US law remind us again that property ownership did nothing to empower individuals who could not actually own property. Although the Homestead Act of 1862 offered ownership prerogatives to independent women, it did so with the assumption that they would take up standard domestic roles. In general, other nineteenth-century property laws bestowed rights to own property on white women only in connection with their marital status.[33]

More public controversies erupted from other limitations of US law. Unlike the Ordinance of 1785, Spanish and Mexican law provided for large grants of land, often comprised of tens of thousands of acres, and sometimes much more. Recipients were either communities or individuals who were obligated to recruit communities to settle on their grants. In these settlements, which included towns from Texas to California, individual households had small freehold parcels for farming, but large areas, known as *ejidos*, were retained in common for grazing, wood gathering, and hunting.

Although the US promised to honor the property of former citizens of Mexico, US land law made no allowance for common property, and many of the old *ejidos* vanished into the public domain, or more directly into the hands of lawyers and speculators. This dispossession of Spanish and Mexican settlers has generated lawsuits that continue today, as well as a great deal of social friction, including an armed insurrection in northern New Mexico in 1967.[34]

As settlers of a prior colonial regime whose property rights were vacated by a successor state, New Mexico's Hispanos were not alone. Their counterparts stretched from Saskatchewan, Canada, where Metis settlers lost their common property holdings despite two rebellions against the British government, to southern Africa, where Boer farmers elected to escape the British grid that hemmed in their circular properties by trekking across the Vaal River and forming their own republic.[35] The enduring tension between gridded government and the ungridded governed suggests that creating the public domain out of preexisting holdings, and then converting it to private property in keeping with US law, called into question what property was, and whose customs and wealth it should protect. The bitter conflicts that resulted underscore that property making might look efficient and liberating in retrospect, but for many people, the transition from one property regime to another proved chaotic, bloody, and oppressive.[36]

THE CULTURE OF IMPROVEMENT AND THE NATURE OF THE GRID

For all the grid's failings, US authorities improvised it partly as a response to a crisis brought on not by foreign powers or Native Americans, but by American squatters and speculators (the categories were not mutually exclusive), who often spread westward in defiance of law and claimed territories large and small. Typically, those who staked a claim to western lands would then demand political leaders sanction their holdings with clear title to facilitate profitable sale.

Similar behavior obtained throughout the Anglophone world, in Canada, Australia, New Zealand, and southern Africa. In each of these places, colonists (squatters and speculators) pushed out ahead of state authorities, claiming land in advance of surveyors and then demanding surveys, laws, land offices, maps, and of course soldiers to uphold their claims. [37]

To U.S. settlers and authorities alike, unclaimed land was wilderness, a space that contained no property. If property were to exist there, it had to be made. Despite widespread agreement on these points, disputes over how best to turn nature into private holding energized political fights for decades after the Revolution.

The argument revolved around legal questions, but settlers were seldom in need of formal training in law. Throughout the English diaspora and in many other European colonies, colonists carried with them popular notions of property making through "improvement" of land.[38] Typically, to improve a parcel was to make a farm out of it so that it could be sold for higher value. (Indeed, the word "improve" comes directly from an Anglo-Norman root *emprouer*, "to turn to profit," which was itself derived from the Old French *prou*, "profit".)[39]

While improvement doctrine is commonly traced to John Locke, who viewed nature as a vast, unowned common that became private property when a man mixed his labor with it, such ideas were widely held before his time.[40] From the beginning of the English colonial effort, in the decade before Locke's birth, commentators claimed a right to American Indian lands (which they viewed as empty, undeveloped common) by virtue of their ability to make English farms on them. As the minister John Cotton put it, "In a vacant soyle, hee that taketh possession of it, and bestoweth culture and husbandry upon it, his Right it is."[41]

As environmental historians have shown repeatedly and to great effect, not only did Indians manage their lands assiduously—clearing, farming, selectively burning, hunting, and fishing—but settler enthusiasm for improvement led to tremendous simplification of natural systems. Improvement implied the clearing of forest, plowing meadows into fields and planting them in a monoculture, and eradicating predators and any other wildlife that competed with livestock. Species vanished, soil eroded, and fire and flood shifted in variability and intensity. Thus was improvement doctrine fundamental to massive biotic transformations from Maine to Hawaii.[42]

Some might see such destruction as the price of liberty in a market, but there are three frequently overlooked consequences of property making that point to the limits of enclosure for ensuring freedom and capitalism. The first concerns the limits of property as a guarantor of liberty. The second concerns the usefulness of improvement ideology for the propertyless and its incompatibility with capitalism. The third involves the dynamics of natural systems which, once improved, can actually make property harder to own. As it turned out, property owning guaranteed neither economic nor ecological wisdom.

Property theorists often esteem improvement ideology as a capitalist virtue. After all, settler willingness to risk their own labor in the creation of marketable commodities proved key to the making of the market in land.[43] In truth, on the frontier, settlers often did not risk their own labor but compelled the labor of others: African American slaves on the southeastern frontier in the early nineteenth century; Native Americans in the missions and estates of New Spain and Mexico. Such arrangements extended to US California, where the "Act for the Government and Protection of Indians" in 1850 established virtual slavery of native youths under the rubric of "indenture"; in Utah, too, Indian "adoption" served as a covert system of acquiring laborers for decades after the Mormons arrived in Salt Lake. Similarly in South Africa, the Boers who trekked to the Transvaal enslaved African children whose parents they had killed, pronouncing the children "apprentices." In northern Mexico, including New Mexico, colonists frequently enslaved Indians in various forms, while chattel slavery dominated the economies of Brazil and the Caribbean.[44]

These conjoined institutions—of property in land and property in people—thus expanded together. The process of enclosure entailed violence not just in the form of death for indigenes who resisted the theft of their land, but often slavery for indigenes and imported captives. Slavery was a key "infrastructure" for private property. Just as today's subdivisions would hold little value without access to electricity, southeastern plantations and many other New World landscapes were worthless without access to slaves.[45] The advancing liberties of western landowners thereby wrought a corresponding advance in misery for millions, sometimes an ocean away, through slave raids that entailed the complete loss of kin, property, and community and often lifelong captivity for victims. The violence of slavery, from war to starvation, physical abuse, and rape, hangs over the history of property, belying the sunny theories of free-market theorists. For environmental historians, a closer look at the history of human bodies and human bondage might open new windows on property and its role in world history.

To be sure, in the US and elsewhere there were explicit legal bans on slavery in many western states, beginning with the Northwest Ordinance of 1787. But Americans, like colonists elsewhere, shifted their ideals of liberty and landownership to accommodate economic needs and labor conditions in other regions, so that no one idea of freedom or property prevailed. If private property was a fundamental expression of liberal capitalism, the ubiquity of human chattel that accompanied it suggests that property ownership was something other than an unqualified bulwark of liberty.[46] Moreover, denying property rights to select outsiders remained a powerful way of constraining their political and economic power, and forcing them into the labor pool as something other than free labor. Thus, the Alien Land Act of 1913 denied Asian immigrants the right to own property in California, and a series of related actions aimed to keep "undesirable" social groups from acquiring freehold tenure, even as white landowners employed them as field hands.[47]

Beyond the question of who actually did the improving, the improvement doctrine, which so many credit with opening the land market, frequently impeded it. Long before the Revolution, many, perhaps most, Americans came to see improvement not just as *a* way to make property from nature, but as *the morally superior* way.[48] In this formulation, people who worked the land had more rights to it than somebody who simply bought title for cash. So, from the 1760s to the early 1800s, squatters in Maine, New Jersey, New York, South Carolina, North Carolina, Pennsylvania, Vermont, and Ohio rebelled against land companies and proprietors who claimed vast tracts of land and then offered them for rent or sale. To be sure, squatters were fervent capitalists: they shared a belief that improvement made land more valuable, and that improved land could be sold.

But squatters tempered their enthusiasm for profit with a moral economy of labor, in which they repeatedly and vehemently ranked personal improvement of land as a more honest and authentic claim than simple purchase. Powerful landowners often expressed frustration at the squatters who claimed portions of their tracts and thereby frightened away potential investors and buyers. Thus George Washington, one of the most successful land speculators of his age, tarred squatters as "banditti who will bid defiance to all authority while they are skimming and disposing of the cream of the country at the expense of many." In pegging property to the labor of settlers rather than to legal title, the ideology of improvement threatened the abstraction of land as unfettered commodity—an abstraction on which the wealth of George Washington and others often depended.[49]

The Ordinance of 1785 was intended partly to limit the potential for such conflicts by providing land rights only for buyers, not squatters. But it failed to change popular ideology, and conflicts between camps of squatter improvers and cash buyers roiled nineteenth century politics at least until the Civil War. In antebellum New York State, renters dismissed the exclusionary rights of wealthy proprietors, treating unworked acres as common property. Beginning in 1839, these renters rose up and demanded ownership of lands they farmed on the grounds that their labor entitled them. Some even envisioned a new legal order in which the cash market for land would vanish and property could only be acquired through labor.[50]

In the former Mexican provinces of California and New Mexico, on the other hand, squatters fought the speculators and corporations who acquired Spanish and Mexican land grants. Aspiring smallholders demanded the grants be returned to the public domain and thrown open to common settlers. In Sacramento, California, in the 1850s, riots pitting squatters against officials killed eight people, and subsequent battles on the streets and in courts traumatized the state for the better part of a decade.[51]

In Colorado and New Mexico, squatters and authorities faced off over the fate of the Maxwell Land Grant, which, at 1.7 million acres, was one of the largest properties vested by the Mexican government prior to US acquisition of the Southwest. Squatters refused to pay rent to a succession of grant owners, including Dutch investors who paid handsomely for title. In 1882, the antagonists finally shot it out in the town of Stonewall, Colorado, resulting in three squatter deaths. Only at great cost in money and blood did titleholders secure the grant.[52]

Elsewhere, on the public domain, squatters reshaped federal land sales to limit prices. One of the most common methods was to form a "claim association" of local squatters, who would attend a land auction en masse and bid a low, set price for every acre. Outsiders who attempted to bid higher prices (or any at all) were quickly intimidated into silence. Outside investors complained to authorities, and Congress was inevitably hard-pressed to find a compromise.[53]

So lawmakers split the difference, and inscribed limited squatter rights into federal land law. Beginning in the 1830s, "pre-emption" exceptions allowed squatters on the public domain to have the option of purchasing lands they improved at a minimal price before those lands could be auctioned to anybody else. In 1841, these exceptions became law with the Pre-Emption Act, which became a principal avenue to landed title for decades thereafter.[54]

The most famous example of the compromise between improvement and purchase occurred in 1862, in the Homestead Act, which stipulated that a homesteader could acquire title after five years' residence and completion of improvements, or after six months' residence and cash payment of $1.25 per acre. The endurance of the improvement doctrine was reflected in the term "proving up," which appears to have been invented by homesteaders to connote the process of improvement for title.[55]

After the original Homestead Act, the improvement doctrine influenced every revision of federal land policy through the First World War. But the Homestead Act was the beginning of the end for squatters, who had no rights under the law. Few homesteaders went the route of improvement; far more paid cash for title, and the Homestead Act itself proved more useful to speculators than to farmers.[56]

But the decline of squatters' rights should not blind us to the endurance of other ideas connecting improvement to property rights. Some have attributed such notions to universal attachments generated through labor. Geographer Jake Kosek posits that in working land, people come to feel it belongs to them, and in a sense, they to it, wherever legal title may actually lie. "The basis of ownership, and people's willingness to fight or die for rights of ownership, stem from a belief in the emotional bond of belonging," observes Kosek. That bond germinates and strengthens through physical work. Thus the Hispanic residents of Chimayo, New Mexico claim the forests surrounding their village in part because, through generations of cutting firewood and herding cattle through these mountain meadows, they feel their bodies have become part of the forests. These sentiments fuel ongoing conflicts between the US Forest Service—the forest's legal owners—and Chimayo residents, who nevertheless assert communal land rights. The confrontation has featured civil and incivil disobedience, numerous arrests, and lawsuits, preoccupying regional administrators, villagers, and environmentalists for decades.[57]

Kosek's analysis suggests potential for a more global and comparative scholarship of property making. In what ways do a given people invoke property rights through altering, or improving nature? What kinds of alterations are perceived as improvement, and what kind are seen as destructive? How have cultures exchanged ideologies of improvement and property making? Such questions could well inform new histories of property around the world.

Meanwhile, in the US, it seems likely that improvement doctrine will continue to resonate in environmental confrontations. In the 1990s, loggers in the Pacific Northwest fought new environmental restrictions, claiming a right to the forests as a place in which to labor. Ranchers across the West have claimed property rights in public lands by virtue of long tenancy and a history of family-centered work.[58]

Richard White warns environmentalists against elevating play and recreation over labor, because work is the primary means by which people have known nature.[59] But the enduring resonance of the improvement doctrine reminds us that labor has been a way not only of knowing nature, but of making it into property to be bought and sold. Indeed, it is striking how clashes between extractive industries like logging and ranching and environmentalists who seek to restrict those activities echo old fights between squatter-improvers and cash-buyers. Industrial advocates deploy powerful rhetoric in appealing to the public's enduring sense that people own the nature they (or their laborers) work. In turning against forms of labor and occupations that historically brought profit from nature, environmentalists have set themselves a formidable task.

BOUNDED PROPERTY AND NATURE UNBOUND

To be sure, even on private property, some resources remained public, or common. Rivers and streams could only be privatized to a limited degree, and in the arid West, water law developed in new ways to ensure that water could only be claimed when it was put to "beneficial use," that is, improvement of the land. Wildlife, too, remained a public resource, even on private holdings. State and federal authorities have increasingly managed deer, bears, and waterfowl as public goods, imposing rules of access that apply to both the landless and to landowners who own wildlife habitat.[60]

But improvement has entangled private and common property even more thoroughly. Fences, furrows, hoes, plows, and livestock—indeed, virtually every expression of an "improved" estate—could inadvertently subvert the very property it was meant to secure. When owners changed nature on one parcel of land, unintended consequences often undercut or threatened the boundaries separating it from another.

Thus farmers' plows and hoes, along with livestock grazing and trampling of soil, created ideal disturbance conditions for weeds.[61] As Mark Fiege explains, in Montana proliferating weeds changed property relations in ways that carried "enormous potential for spatial—and social—transformation."[62] Weeds dispersed seeds in the dung and on the coats of wild animals, on farm implements, and on the wind. They easily crossed the boundaries of the grid and created among individual owners a shared and unpleasant experience of a broader ecology.

No single landowner, no matter how diligent, could protect his property from the invasion of weeds via a neighbor's plot. This fact compelled neighbors to band together against this new environmental threat, ultimately enlisting the state in creating a new property regime to manage it. Thus Montana, which had become a state only in 1889, by

1895 adopted a law requiring landowners to destroy Canada thistle, Scotch bull thistle, and Russian thistle, and enabling county "weed supervisors" to enter private property and take action against weeds if the owner failed to do so. In imposing reciprocal obligations on private owners, the boundary-crossing weeds themselves came to represent a kind of shifting commons that continually erupted through the grid. The pesky plants became perfect examples of what Fiege calls "the incompatibility of human boundaries" with "forms of mobile nature—water, soil, and organisms—that these boundaries could not contain."

In other words, shifting ecology brought changes in use rights to private property, creating a new kind of common property overlaying the grid, a "weed commons," which all landowners were required to manage. Subsequent laws in the 1920s and 1930s mixed common and private more thoroughly, mandating quarantines of weed-infested farm products or equipment, and allowing public expenditures to control weeds on private property.[63] Private use rights to land thus carried new responsibilities in an expanding weed commons.

But such measures were never completely successful—far from it, judging by the proliferation of knapweed, dodder, white top, and morning glory—because too many obstacles got in the way. Making weed districts across patchworks of different kind of land—Indian allotments and reservations, private white-owned parcels, and public lands—proved difficult. But perhaps the most enduring obstacle was the grid itself, which inscribed individual ownership and reinforced an individualist ethos that militated against the mutualism that weed management required. "Consequently, the weed commons was always a qualified, contested, and contingent space."[64] In other words, the very intractability of weeds, and the perniciousness of their invasion, owed something to the private property that they so consistently undermined.

Weeds were hardly the only natural force to undercut private property after having been unleashed by it. The work of private owners in nature also brought on erosion, flood, and fire, erasing or confusing sacrosanct property boundaries. From the 1850s to the 1880s, California's hydraulic mining companies flushed billions of cubic yards of eroded granite from their private mountain claims into public rivers. The debris became a public menace as it flowed downriver to the rich farm country of the Central Valley, where seasonal floods washed it over private fields, severely compromising their fertility, and sometimes burying entire orchards. Mining companies claimed the right to treat their private property as they wished. But in the mid-1880s, the state finally decided to preserve private farms at the expense of the miners, and the legislature banned hydraulic mining to keep the mountainsides in place and out of the private fields of the Central Valley. Similar conflicts between upstream polluters and downstream owners have troubled riparian zones around the world.[65]

Other examples of private property mobilizing nature abound. Great Plains farmers removed buffalo grass and other plants to create wheat and cotton fields. But in the 1930s drought hammered the region, destroying crops and leaving the fields bare. Exposed to the wind, privately owned topsoil eroded into the air, massing it into thick, rolling clouds that choked infants to death, turned day into night, and at times rained

Oklahoma dirt on the Atlantic seaboard. Federal authorities intervened, repurchasing 11 million privately held acres and returning them to the public domain, planting "shelter belts" of trees to slow the wind, and encouraging contour plowing and other innovations to better secure the soil and the property of remaining farmers.[66]

Fire, too, has become a kind of common property. In many parts of the US, development of rural lands has led to invasions of foxtail, red brome grass, and brushy growth, creating a highly combustible plant complex. Year after year, dry seasons in the west bring ever more catastrophic fires. In 2007, fires in Southern California displaced a half million people, many of them from subdivisions whose construction had abetted the very plant growth that fueled the conflagration. Increasingly, landowners are legally required to control weeds, reduce brush, and incorporate fire-retardant materials into private structures. [67]

During the surge in foreclosures and property abandonment in the US after 2005, nature made hazards from improved real estate even in the most heavily paved suburbs. Mosquitoes proliferated in stagnant swimming pools, threatening West Nile virus and other maladies. Added to the routine tasks of code enforcement officers were the draining of untended pools and the removal of bobcats, rattlesnakes, and raccoons from empty homes. Managing the feral nature of suburbia thus became key to the welfare—and the home values—of remaining residents. [68]

The emergence of common property through the unstable grid reminds us how tentative and incomplete enclosure remains, precisely because property cannot conquer nature, but rather resides in it. No matter how individuated they may be, our freeholds are not free. They are entangled in natural systems that render them at the best of times tenuously controlled and imperfectly bounded. Nature remains dynamic, mostly unpredictable—intractable.

The Enduring Public Domain and the Failure of Private Property

That intractability explains why so much land in the US has never become private. At the end of the Great Depression, vast tracts of public domain persisted, as they have ever since. The US government today owns an acreage the size of India, and fully one-third of the American landscape consists of public lands. So much for the inevitable triumph of freehold and the market.[69]

Behind the persistence of the public lands lie natural causes and dramatic political shifts which have been explored at length elsewhere. Mostly, the steady march of private authority proved no match for real world aridity. Over and over again, homesteaders failed in trying to privatize the desert reaches of the public lands. Without sufficient water, farm improvements were impossible. Some time after 1920, the tide of privatization halted and reversed. The public domain began to grow.[70]

Beyond that, American ideas of property underwent dramatic transformations, and whole new categories of landowning emerged in the late nineteenth century. Partly out of nostalgia for the ungridded landscape they were yet in the process of bounding, Americans began demanding preservation of vast acreages as national parks at Yellowstone, Yosemite, Glacier, and elsewhere. Efforts to secure water supplies for lowland settlements led authorities to constrain settlement in high country forests, conjuring new categories of landholding that were decidedly public and inalienable. Reserving forests in this way became politically feasible in part because, as corporate trusts aggregated vast estates—Weyerhauser Corporation bought 900,000 acres in one deal in 1900—the democratic promise of dispensing land from the public domain rang increasingly hollow.[71] Thus, by 1907, the US government managed 151 million acres of national forest.[72]

But if the expansion of public lands, parks, and forests caused one form of enclosure to slow, it introduced another. Both national park and national forest restricted customary use rights of prior locals, particularly Native Americans, but also for Hispanos and other settlers. Their creation initiated long-term conflicts over what kind of property they were and to whom they belonged, which ever since have echoed in court rooms and other public forums from the Alaska to Florida and Maine to California.[73]

At the same moment they created vast new national holdings, Congress made various efforts to fend off monopoly control of western lands by creating new institutions to shore up smallholder democracy. One of the most influential of these was the Bureau of Reclamation, whose purpose was to irrigate desert lands to make small farms for freeholders. Beginning in 1902, the agency worked to create dams, reservoirs, and canals throughout the West. But in the end, smallholders were not the prime beneficiaries. The bureau succeeded best at delivering water to giant commercial farms, and to cities and their suburbs.[74] Thus, since its creation, the vast majority of private properties carved from US deserts have been in suburban oases like Los Angeles, Las Vegas, and Phoenix, where massive public investments in water delivery have allowed planned communities and subdivisions to proliferate over former public lands, for a time making this region the fastest growing in the United States. Without the development of water storage and distribution systems by the Bureau of Reclamation and by municipal counterparts such as the Los Angeles Department of Water and Power and urban water districts most of these private properties would not exist at all.

As Americans gravitated to metropolitan areas, the public domain continued to grow. With a few, relatively minor exceptions, the Great Depression brought the end of homesteading. In 1934, the Taylor Grazing Act barred private claims on the remaining public domain, and as we have seen, the US government bought up desolate farms and added them to public holdings. The vision of a fully private West would exist only in dreams.

Today, private properties survive in much of the rural West only as islands in a sea of public domain, without which many owners would fail. Where a rancher has a private holding, he or she has first option on leases of surrounding public acres. Leasing public land has often proved cheaper than leasing private land, and it can bring other advantages. The costs of fences, stock tanks, and other improvements are shared with federal owners (who also maintain the roads), and ranchers pay no property taxes on leaseholds. Many comparatively small freehold ranches would not be viable without leases to tens of thousands of surrounding acres of public land. Indeed, leasing land from the government has often proved cheaper than owning it.[75]

In the future, new technologies may enable landowners to better secure their properties in some ways, as the invention of the internal-combustion pump has allowed farmers on the Plains to draw precious water from deep below the earth's surface. But just as the depletion of groundwater could soon force such farmers into new common property arrangements to collectively manage the water table, the experience of the last half-century suggests it is just as likely that technology will force private and common property into new kinds of working relationships. This was a fact Montana farmers understood well by the 1960s, when technological solutions to the weed problem created new challenges to property lines. Weed controllers delighted in the invention of 2,4-D, a powerful herbicide that seemed to promise a weed-free Montana. But with its deleterious effects on wildlife, people, and livestock, its tendency to drift across property lines aroused vehement opposition. Legal controls on the use of 2,4-D represented, in effect, a greater ecological commons, one that encompassed private and public land alike.[76]

In similar fashion, environmental regulations have imposed various common property regimes on private and public landowners across the country.[77] The Endangered Species Act, the Clean Air Act, the ban on DDT, and other laws require certain kinds of behavior even toward private nature. The imposition of legal standards for air quality (and water quality) has limited the autonomy of private property owners. The results have often been politically explosive, testifying to the ongoing cultural and political work that goes into redrawing the boundaries of public and private property.

The endurance of the public domain and the emergence of new forms of common property in air and water suggest that whatever new technologies bring, common property will endure alongside private. Genetic technology, for example, has allowed both for new property rights in nature and the erosion of the lines around it. Monsanto, Novartis, and other corporate owners have created transgenic organisms (which have a gene from a different species inserted into their genome) for research purposes. In hopes that such creatures might be marketable, transgenic corn and lab rats are patented as exclusive corporate property.

But nature is porous and mutable, exactly the opposite of what property is meant to be. And a significant obstacle to this new property regime is that, like weeds, the newly patented organisms refuse to stay inside property boundaries. Through the dispersal of pollen, for example, genes from genetically modified corn have turned up in their supposedly organic counterparts, creating hybrid transgenic organisms which raise questions of ownership—and of culpability for altered genetic composition. In other words, new forms of private property in nature can threaten their own categories.[78] So private property and common property continue to overlap, comingling and complicating one another in a sign of how truly unfinished the task of bounding nature must always be.

Private property in the United States could not have evolved the way it did without a coevolution of common property holdings in game, fish, water, weeds, fire, and even land itself.[79] So long as private holdings reside in nature, it seems unlikely that the commons, for all its tragedies, will ever die. It persists across a vast range of resources in the US and the world, in water, forests, fish, air, pasture, and swampland. For all its remarkable consequences, the evolution of private property is only part of the story of owning nature in North America. Private property eroded common property regimes in many places, but it also resulted in a curious, complementary articulation and elaboration of other commons regimes, which paradoxically facilitated the maintenance of private property.

Few institutions have loomed as large in the making of economic liberalism as freehold land tenure, and few have been more poorly understood. Private property can be real estate, a commodity. But more than anything else it is an environmental relationship between people, nature, and the state. Only a history that acknowledges the full array of mutually supporting property regimes—private and public, individual and common, citizen and state—can illuminate the workings of property amid the work of nature.

NOTES

1. European Center for Economic Growth, http://e-growth.eu/?cat=6.
2. F. A. Hayek, "The Importance of Private Property," http://www.youtube.com/watch?v=wlCHkMKnV68.
3. "Free to Choose," 1980, http://miltonfriedman.blogspot.com/; Richard Pipes, *Property and Freedom* (New York: Knopf, 1999).
4. Thus, "Freedom only exists where there is complete respect for rights of property ownership." Ron Paul, "Liberty is the Answer," http://www.ronpaul2008.com/articles/240/respect-for-property-rights-necessary-for-freedom/.
5. John F. Richards, "Introduction," in *Land, Property, and Environment*, ed. John F. Richards (Oakland, CA: Institute for Contemporary Studies, 2002), 2.
6. Russia's Medvedev Pledges Freedom and Legal Reform," *International Herald Tribune*, February 15, 2008, http://www.iht.com/articles/reuters/2008/02/15/europe/OUKWD-UK-RUSSIA-MEDVEDEV-PRIORITIES.php.
7. The amendment came in 2004. "China Endorses Private Property," BBC News, http://news.bbc.co.uk/2/hi/asia-pacific/3509850.stm.
8. Garrett Hardin, "The Tragedy of the Commons," *Science* 162 (December 13, 1968): 1244–1248.
9. The definition is from Bonnie J. McCay and James M. Acheson, "Human Ecology of the Commons," in *The Question of the Commons: The Culture and Ecology of Communal Resources*, ed. McCay and Acheson (Tucson: University of Arizona Press, 1987), 8. For an introduction to the vast literature on common property, see also Garrett Hardin and John Baden, eds., *Managing the Commons* (San Francisco: W. H. Freeman, 1977); Elinor Ostrom et. al., eds., *The Drama of the Commons* (Washington, DC: National Academy Press, 2002); Bonnie J. McCay, Nives Dolak, and Elinor Ostrom, eds., *The Commons in the New Millennium: Challenges and Adaptations* (Cambridge, MA: MIT Press, 2003).

10. "Unless the people I live with recognize that I own something and so give me certain unique claims over it, I do not possess it in any meaningful sense." William Cronon, *Changes in the Land* (New York: Hill & Wang, 1983), 58. The image of property as a bundle of use rights appeared at least as early as World War I. Henry S. Maine, *Ancient Law* (New York: Dutton, 1917). In the words of Bonnie McCay and James Acheson, private property and common property, as expressions of conditional use, imply "a complex set of social duties, privileges, and mutualities." McCay and Acheson, "Human Ecology of the Commons," 8. For further discussion, see Huntington Cairns, *Law and the Social Sciences* (New York: A. M. Kelly, 1935), 59; A. Irving Hallowell, "The Nature and Function of Property as a Social Institution," *Journal of Legal and Political Sociology* 1 (1942–1943): 115–38; John C. Weaver, *The Great Land Rush and the Making of the Modern World, 1650–1900* (Montreal: McGill-Queen's University Press, 2003), 49.

11. Marshall Sahlins, *Stone Age Economics* (Chicago: Aldine-Atherton, 1972); Cronon, *Changes in the Land: Indians, Colonists, and the Ecology of New England*, 54–81; E. Adamson Hoebel, *The Law of Primitive Man* (Cambridge, MA: Harvard University Press, 1967), 46–63; Peter C. Perdue, "Property Rights on Imperial China's Frontiers," in Richards, ed. *Land, Property, and the Environment*, 71–93.

12. Richards, "Toward a Global System of Property Rights in Land," in *Land, Property, and Environment*, ed. John F. Richards (Oakland, CA: Institute for Contemporary Studies, 2002), 21; James C. Scott, *Seeing Like a State: How Certain Schemes to Improve the Human Condition Have Failed* (New Haven: Yale University Press, 1998), 38–9; Philip C. Brown, "Harvests of Chance: Corporate Control of Arable Land in Early Modern Japan," in Richards, ed., *Land, Property, and the Environment*, 38.

13. Scott, *Seeing Like a State*, 11–52.

14. Roger J. P. Kain and Elizabeth Baigent, *The Cadastral Map in the Service of the State: A History of Property Mapping* (Chicago: University of Chicago Press, 1992); John F. Richards, "Toward a Global System of Property Rights in Land," 31.

15. For a sample of the rich literature on English enclosure, see Edward C.K. Gonner's *Common Land and Inclosure* (New York: A. Kelley, 1966 [1912]); and John L. Hammond and Barbara Hammond, *The Village Labourer 1760–1832, a Study in the Government of England Before the Reforrm Bill* (New York: Longman's Green & Co., 1912).

16. This was partly because ruling elites who obstructed some privatization efforts in Europe had fewer objections to it elsewhere. John C. Weaver, *The Great Land Rush and the Making of the Modern World, 1650–1900* (Montreal: McGill-Queen's University Press, 2003), 46.

17. See for example Hernando de Soto, *The Mystery of Capital: Why Capitalism Triumphs in the West and Fails Everywhere Else* (New York: Basic Books, 2000). For correctives, see Robert Home and Hilary Lim, eds., *Demystifying the Mystery of Capital: Land Tenure and Poverty in Africa and the Caribbean* (Portland, OR: Cavendish Publishing, 2004) and Richards, ed., *Land, Property, and the Environment*.

18. Pipes, *Property and Freedom*; Hardin, "Tragedy of the Commons"; De Soto, *The Mystery of Capital*; Francis Fukuyama, *The End of History and the Last Man* (New York: Free Press, 1992).

19. Karl Marx and Friedrich Engels, *The Manifesto of the Communist Party* (Chicago: Charles H. Kerr, 1906), 17.

20. Such considerations underscore that property is merely one category of relationships people establish with land and with one another to secure access to goods. If property implies socially acknowledged rights to part of nature, there are many ways of gaining

access to nature without such rights. As Nancy Lee Peluso and Jesse C. Ribot have argued, poaching, stealing, squatting, and labor are just a few strategies for acquiring resources from nature without acquiring landed property rights. Jesse C. Ribot and Nancy Lee Peluso, "A Theory of Access," *Rural Sociology* 68, no. 2 (2003): 153–181.

21. To say nothing of the webs of credit and finance and regulation, from loans guaranteed by the Federal Housing Administration to standardized financial instruments generated by the Federal National Mortgage Association (Fannie Mae). See Kenneth T. Jackson, *Crabgrass Frontier: The Suburbanization of the United States* (New York: Oxford University Press, 1985), 190–218; Louis Hyman, *Debtor Nation: The History of America in Red Ink* (Princeton, NJ: Princeton University Press, 2011).

22. For the emergence of zoning, see Witold Rybczynski, *Last Harvest: From Cornfield to New Town* (New York: Scribner, 2007), 36; Marc A. Weiss, *The Rise of the Community Builders: The American Real Estate Industry and Urban Land Planning* (New York: Columbia University Press, 1987), 81.

23. Lizabeth Cohen, *A Consumers' Republic: The Politics of Mass Consumption in Postwar America* (New York: Knopf, 2003), 137–43.

24. For San Francisco and the Chinese, see George Lipsitz, *The Possessive Investment in Whiteness: How White People Profit From Identity Politics* (Philadelphia: Temple University Press, 2006), 25–33, and Nayan Shah, *Contagious Divides: Epidemics and Race in San Francisco's Chinatown* (Berkeley: University of California Press, 2001); for the racial context of the GI Bill and VA Loans, see Ira Katznelson, *When Affirmative Action Was White: An Untold Story of Racial Inequality in Twentieth-Century America* (New York: W. W. Norton, 2005), 113–129; for more on white flight, racial covenants, and the ways state subsidies and regulations segregated the metropolitan landscape, see Thomas J. Sugrue, *The Origins of the Urban Crisis: Race and Inequality in Postwar Detroit*, rev. ed. (Princeton, NJ: Princeton University Press, 2005); Kevin M. Kruse, *White Flight: Atlanta and the Making of Modern Conservatism* (Princeton, NJ: Princeton University Press, 2007); Matthew D. Lassiter, *The Silent Majority: Suburban Politics in the Sunbelt South* (Princeton, NJ: Princeton University Press, 2007).

25. Adam Rome, *The Bulldozer in the Countryside: Suburban Sprawl and the Rise of American Environmentalism* (New York: Cambridge University Press, 2001); Christopher Sellers, "Nature and Blackness in Suburban Passage," in *"To Love the Wind and the Rain": African Americans and Environmental History*, ed. Dianne D. Glave and Mark Stoll (Pittsburgh: University of Pittsburgh Press, 2006), 93–119; also see Christopher Sellers, *Crabgrass Crucible: Suburban Nature and the Rise of Environmentalism in Twentieth-Century America* (Chapel Hill: University of North Carolina Press, 2012), 63–73.

26. Peter S. Onuf, *Statehood and Union: A History of the Northwest Ordinance* (Bloomington: Indiana University Press, 1987), 21–43.

27. Weaver, *Great Land Rush*, 64.

28. For an introduction to the vigorous debate over the relative strength or weakness of the American state, see William J. Novak, "The Myth of the 'Weak' American State," *American Historical Review* 113, no. 3 (June 2008): 752–772; and Julia Adams, Gary Gerstle, William J. Novak, and Jonathan Fabian Witt, "*AHR* Exchange: On the 'Myth' of the Weak American State," *American Historical Review* 115, no. 3 (June 2010): 766–800.

29. Matthew Frye Jacobson, *Whiteness of a Different Color: European Immigrants and the Alchemy of Race* (Cambridge, MA: Harvard University Press, 1998), 22.

30. Frederick E. Hoxie, *A Final Promise: The Campaign to Assimilate the Indians, 1880–1920* (Lincoln: University of Nebraska, 1984); Emily Greenwald, *Reconfiguring the Reservation: The Nez Perces, Jicarilla Apaches, and the Dawes Act* (Albuquerque: University of New Mexico Press, 2002).

31. Christian W. McMillen, *Making Indian Law: The Hualapai Land Case and the Birth of Ethnohistory* (New Haven, CT: Yale University Press, 2007).

32. Maria E. Montoya, *Translating Property: The Maxwell Land Grant and the Conflict Over Land in the American West, 1840–1900* (Berkeley: University of California, 2002), 47–9.

33. Tonia M. Compton, "Proper Women/Propertied Women: Federal Land Laws and Gender Order(s) in the Nineteenth-Century Imperial American West" (PhD diss., University of Nebraska, 2009), http://digitalcommons.unl.edu/cgi/viewcontent.cgi?article=1018&context=historydiss.

34. The land grant issue has generated a sizeable body of scholarship. See Montoya, *Translating Property*; Jake Kosek, *Understories: The Political Life of Forests in Northern New Mexico* (Durham, NC: Duke University Press, 2007); William deBuys, *Enchantment and Exploitation: The Life and Hard Times of a New Mexico Mountain Range* (Albuquerque: University of New Mexico Press, 1985); Malcolm Ebright, *Land Grants and Lawsuits in Northern New Mexico* (Albuquerque: University of New Mexico Press, 1994); Peter Nabokov, *Tijerina and the Courthouse Raid* (Albuquerque: University of New Mexico Press, 1970); Charles L. Briggs and John R. Van Ness, eds., *Land, Water, and Culture: New Perspectives on Hispanic Land Grants* (Albuquerque: University of New Mexico Press, 1987).

35. For the Metis case, see Andrew Graybill, *Policing the Plains: Rangers, Mounties, and the North American Frontier, 1875–1910* (Lincoln: University of Nebraska Press, 2007), 86–8; for Boer farmers in South Africa see Weaver, *The Great Land Rush*, 118.

36. On these points, numerous helpful examples may be found among the following: Scott, *Seeing Like a State*; Hildegard Binder Johnson, *Order Upon the Land: The U.S. Rectangular Land Survey and the Upper Mississippi Country* (New Haven, CT: Yale University Press, 1976); Onuf, *Statehood and Union*; William L. Fox, *The Void, the Grid, and the Sign: Traversing the Great Basin* (Salt Lake City: University of Utah Press, 2000); John R. Stilgoe, *The Common Landscape of America: 1580–1945* (New Haven, CT: Yale University Press, 1982); John Opie, *The Law of the Land: Two Hundred Years of Farmland Policy* (Lincoln: University of Nebraska Press, 1994); William Pattison, *Beginnings of the American Rectangular Survey System, 1784–1800* (Chicago: University of Chicago Press, 1957); John Brinkerhoff Jackson, *Discovering the Vernacular Landscape* (New Haven, CT: Yale University Press, 1984); Kate Brown, "Gridded Lives: Why Kazakhstan and Montana Are Nearly the Same Place," *American Historical Review* 106, no. 1 (February 2001): 17–48.

37. Weaver, *The Great Land Rush*, 74–81.

38. Weaver, *The Great Land Rush*, 81. Thomas R. Dunlap, *Nature and the English Diaspora: Environment and History in the United States, Canada, Australia, and New Zealand* (Cambridge: Cambridge University Press, 1999).

39. "Improve," http://thefreedictionary.com/improve, also "Improve," *OED*, Compact Ed., V. 1, 1393. "To show that one has fulfilled the legal conditions for taking up (a grant of government land) so that a patent might be issued." This was the basis of the Americanism, to "prove up" a homestead, which meant to complete the proof of right to land. "Prove," *OED*, Compact E., V. 2, 2339.

40. "Excerpts from John Locke, Two Treatises of Government, The Second Treatise," http://www.users.muohio.edu/mandellc/locke.htm.

41. Cronon, *Changes in the Land*, 56–57.

42. There are too many books to cite here, but among the most influential are Cronon, *Changes in the Land*; Richard White, *Land Use, Environment, and Social Change: The Shaping of Island County, Washington* (Seattle: University of Washington Press, 1980); Timothy Silver, *A New Face on the Countryside: Indians, Colonists, and Slaves in South Atlantic Forests, 1500–1800* (New York: Cambridge University Press, 1990). The tradition of close examination of landscape-making and ecological change arguably began with George Perkins Marsh, *Man and Nature, or Physical Geography as Modified by Human Action* (New York: Scribner, 1864), and took its more recent impetus from Herbert Guthrie-Smith, *Tutira: The Story of a New Zealand Sheep Station* (Seattle: University of Washington Press, 1999 [1921]); William L. Thomas, Jr., ed., *Man's Role in Changing the Face of the Earth* (Chicago: University of Chicago Press, 1956).

43. Hernando de Soto, *The Mystery of Capital*.

44. James F. Brooks, *Captives and Cousins: Slavery, Kinship, and Community in the Southwest Borderlands* (Chapel Hill: University of North Carolina Press, 2002), 234–241; Michael Magliari, "Free State Slavery: Bound Indian Labor and Slave Trafficking in California's Sacramento Valley, 1850–1864," *Pacific Historical Review* 81 (May 2012): 155–192; Nigel Worden, *Slavery in Dutch South Africa* (New York: Cambridge, 1985); Herbert S. Klein and Francisco Vidal Luna, *Slavery in Brazil* (New York: Cambridge, 2009).

45. See for example Adam Rothman, *Slave Country: American Expansion and the Origins of the Deep South* (Harvard University Press, 2005).

46. James Oakes, *Slavery and Freedom: An Interpretation of the Old South* (New York: Knopf, 1990); Edmund S. Morgan, *American Slavery, American Freedom: The Ordeal of Colonial Virginia* (New York: W. W. Norton, 1975).

47. Michael J. Meloy, "Long Road to Manzanar: Politics, Land, and Race in the Japanese Exlcusion Movement, 1900–1942," (PhD diss., University of California at Davis, 2004); Mae Ngai, *Impossible Subjects: Illegal Aliens and the Making of Modern America* (Princeton, NJ: Princeton University Press, 2005), 46–50.

48. J. Hector St. John de Crevecoeur noted the centrality of improvement ideology in what he called "the true and only philosophy of an American farmer," which was as follows. "This *formerly rude soil has been converted by my father into a pleasant farm,* and in return it has established all our rights; on it is founded our rank, our freedom, our power as citizens, our importance as inhabitants of such a district." J. Hector St. John de Crevecoeur, *Letters from an American Farmer* (New York: E.P. Dutton, 1957 [1782]), 20–21, emphasis added.

49. George Washington, *The Writings of George Washington*, Vol. 8, ed. Jared Sparks (Boston: Russell, Odiorne, and Metcalf, and Hilliard, Gray, and Co., 1835), 481.

50. Reeve Huston, *Land and Freedom: Rural Society, Popular Protest and Party Politics in Antebellum New York* (New York: Oxford University Press, 2002), 28; Gunther Peck, "The Nature of Labor: Fault Lines and Common Ground in Environmental and Labor History," *Environmental History* (April 2006), http://www.historycooperative.org/journals/eh/11.2/peck.html.

51. Tamara Venit Shelton, *A Squatter's Republic: Land and the Politics of Monopoly in California, 1850 – 1900* (San Marino: Huntington Library Press, 2014); also Donald J. Pisani, "Squatter Law in California, 1850–1858," in *The Western Historical Quarterly* 25, no. 3 (Autumn 1994): 277–310.

52. Montoya, *Translating Property*, 120–136, 191–206.

53. Paul Wallace Gates, *History of Public Land Law Development* (Washington, DC: Public Land Law Review Commission, 1968), 152–59; Benjamin Hibbard, *A History of the Public Land Policies* (New York: Macmillan, 1928), 198–208; Allan G. Bogue, "The Iowa Claims Clubs: Symbol and Substance," in *The Public Lands: Studies in the History of the Public Domain*, ed. Vernon Carstensen (Madison: University of Wisconsin Press, 1963), 47–69.

54. Gates, *History of Public Land Law Development*, 219–247.

55. See note 37 above.

56. Benjamin Hibbard, *A History of the Public Land Policies* (Madison: University of Wisconsin Press, 1965 [1924]), 144–70, 347–455; for Homestead Act as the end of pre-emption, see Gates, *History of Public Land Law Development*, 246–7; for speculators benefitting from homesteads, see Gates, *History of Public Land Law Development*, see 387–434.

57. Jake Kosek, *Understories: The Political Life of Forests in Northern New Mexico*, 118–119.

58. Karen Merrill, *Public Lands and Political Meaning: Ranchers, the Government, and the Property Between Them* (Berkeley: University of California Press, 2002).

59. Richard White, "Are You and Environmentalist or Do You Work for a Living?", in *Uncommon Ground: Toward Reinventing Nature*, ed. William Cronon (New York: W. W. Norton, 1995?). See also, Cronon, *The Organic Machine: The Remaking of the Columbia River* (New York: Hill & Wang, 1995).

60. For water law see Donald Worster, *Rivers of Empire: Water, Aridity, and the Growth of the American West* (New York: Oxford University Press, 1985); Mark Reisner, *Cadillac Desert: The American West and Its Disappearing Water* (New York: Viking, 1986); for wildlife, see Michael Bean, *The Evolution of National Wildlife Law* (Council on Environmental Quality, 1977) 8–20; Richard Judd, *Common Lands, Common People: The Origins of Conservation in Northern New England* (Cambridge, MA: Harvard University Press, 1997); Louis S. Warren, *The Hunter's Game: Poachers and Conservationists in Twentieth-Century America* (New Haven, CT: Yale University Press, 1997); James C. Tober, *Who Owns The Wildlife? The Political Economy of Conservation in Nineteenth-Century America* (Westport, CT: Greenwood Press, 1981).

61. Mark Fiege, "The Weedy West: Mobile Nature, Boundaries, and Common Space in the Montana Landscape," *The Western Historical Quarterly* 36, no. 1 (Spring 2005): 24.

62. Fiege, "The Weedy West," 25.

63. Fiege, "The Weedy West," 31–33.

64. Fiege, "The Weedy West," 35–36.

65. Andrew Isenberg, *Mining California: An Ecological History* (New York: Hill & Wang, 2005), 23–52. David Hickman, "Landscapes of Green and Gold: The Environmental Vision of the California Horticulturalists, 1849–1900," (PhD diss., University of California at Davis, 2011); Mark Cioc, *The Rhine: An Eco-Biography, 1815–2000* (Seattle: University of Washington Press, 2002); Robert L. Kelly, *Gold vs. Grain: The Hydraulic Mining Controversy in California's Sacramento Valley* (Glendale, CA: Arthur H. Clark, 1959).

66. How much these measures helped (as opposed to the return of wetter conditions) remains a subject of debate. See Don Worster, *Dust Bowl: The Southern Plains in the 1930s* (New York: Oxford University Press, 1985), and for the opposing view see Paul Bonnifield, *The Dust Bowl: Men, Dirt, and Depression* (Albuquerque: University of New Mexico Press, 1979); also Joel Jason Orth, "The Conservation Landscape: Trees and Nature on the Great Plains," (PhD diss., Iowa State University, 2004).

67. "California Firest Force Thousands More From Their Homes," *USA Today*, Oct. 23, 2007, http://www.usatoday.com/weather/wildfires/2007-10-22-wildfires_N.htm; see also Steven J. Pyne, "Spark and Sprawl: A World Tour," *Forest History Today* (Fall 2008): 4–11.

68. See for example Lisa Ling, "Foreclosure Alley," KCET Los Angeles, September 23, 2008, http://www.kcet.org/shows/socal_connected/content/foreclosure-alley.html; "Honey, Time to Spray the Lawn," *Los Angeles Times*, May 2, 2009, http://www.latimes.com/news/local/la-me-spray-painted-grass2-2009may02,0,7272444,full.story; "Green Pools Sprout from Foreclosures," *Los Angeles Times*, May 3, 2009, http://articles.latimes.com/2009/may/03/nation/na-foreclose-pools3.

69. For comparison to India, see William K. Wyant, *Westward in Eden: The Public Lands and the Conservation Movement* (Berkeley: University of California Press, 1982), 27; for one-third of the landscape, see Richard N. L. Andrews, *Managing the Environment, Managing Ourselves: A History of American Environmental Policy* Second Ed. (New Haven, CT: Yale University Press, 2006), 55.

70. Richard Manning, *Rewilding the West: The Promise of Resurrection in a Prairie Landscape* (Berkeley: University of California Press, 2010) 109.

71. For Weyerhaeuser, see Richard White, *"It's Your Misfortune and None of My Own": A New History of the American West* (Norman: University of Oklahoma Press, 1991), 262.

72. White, *"It's Your Misfortune,"* 407.

73. Warren, *Hunter's Game*; Karl Jacoby, *Crimes Against Nature: Squatters, Poachers, Thieves, and the Hidden History of Conservation* (Berkeley: University of California, 2000); Mark David Spence, *Dispossessing the Wilderness: Indian Removal and the Making of the National Parks* (New York: Oxford University Press, 1999); Robert H. Keller, *American Indians and National Parks* (Tucson: University of Arizona Press, 1999); Philip Burnham, *Indian Country, God's Country: Native Americans and the National Parks* (Seattle: Island Press, 2000).

74. White, *"It's Your Misfortune,"* 524–525.

75. Manning, *Rewilding the West*, 125–135; Karen J. Merrill, *Public Lands and Political Meaning: Ranchers, the Government, and the Land Between Them* (Berkeley: University of California Press, 2002). The exceptions were relatively small-scale. Thus between 1946 and 1966, the federal government opened three thousand irrigated homestead for returning servicemen. See Brian Q. Cannon, *Reopening the Frontier: Homesteading in the Modern West* (Lawrence: University Press of Kansas, 2009).

76. Fiege, "The Weedy West," 15.

77. As Eric Freyfogle observes, since about 1900, "U.S. laws and regulations have increasingly embedded property rights in a communal order, aimed in important part at protecting the natural environment." Freyfogle, "Community and Market in Modern American Property Law," in *Land, Property, and the Environment*, 383. See also Eric Freyfogle, *The Land We Share: Private Property and the Common Good* (Seattle: Island Press, 2003); and by the same author, *On Private Property: Finding Common Ground on the Ownership of Land* (Boston: Beacon Press, 2007).

78. For an introduction to these issues, see Cori Hayden, *When Nature Goes Public: The Making and Unmaking of Bioprospecting in Mexico* (Princeton, NJ: Princeton University Press, 2003); Michael Flitner, "Biodiversity: Of Local Commons and Global Commodities," in *Privatizing Nature: Political Struggles for the Global Commons,* ed. Michael Goldman (New Brunswick, NJ: Rutgers University Press, 1998); for an example of how national

efforts to bound nature continue, see Larry Rohter, "As Brazil Defends Its Bounty, Rules Ensnare Scientists," *New York Times*, August 28, 2007, D1, 4.

79. See McCay and Acheson, *The Question of the Commons.*

CHAPTER 16

WORK, NATURE, AND HISTORY

A Single Question, that Once Moved Like Light

THOMAS G. ANDREWS

A tangle of submerged genealogies link labor history and environmental history into a larger whole. More of these lines of descent converge on the Welsh literary critic Raymond Williams than on any scholar before or since. Williams, a railway worker's son who rose from humble beginnings on the Welsh border to become a Cambridge don, joined E. P. Thompson and other British Marxists in charting a view of social transformation that was at once humanistic in the best sense of the term, and conscious of humankind's tenuous dependence on a fragile and increasingly burdened planet. Williams, especially in *The Country and the City* (1973) and the endlessly rich and suggestive essay, "Ideas of Nature" (1980), mounted a damning critique of the conjoined exploitation of working people and the natural world. Williams traced the origins of the plight of workers and nature to a single cause: dualistic conceptions of "nature" and "culture" that perpetuated a long-standing "separation of Nature from the facts of the labour that is creating it." *The Country and the City* offered a case study of one crucial phase in the development and solidification of the nature/culture binary: the capitalist transformation of the British countryside between the 1600s and 1800s.[1]

Williams, though, was never content simply to reveal that the conceptual frameworks that seemed to isolate nature from labor were at once historically constructed, and specious. The closing paragraphs of *The Country and the City* looked toward the future with particular optimism. Ideological divides created by a relatively small group of people via hotly contested historical processes, Williams reasoned, could be unmade if present and future generations could muster the will, strength, and organization needed to undertake struggles of their own. Williams looked upon the social, economic, and political upheavals that had erupted throughout the world in the 1960s with great hope. The old structures of capitalism—including the guiding ideas Williams sought to historicize, and thus destabilize in his influential *Keywords* (1976)—were beginning to crumble. "We are touching," Williams declared, "and know that we are touching, forms of a general crisis."[2] Out of the ashes of that "general crisis," Williams believed, new ways of thinking and acting could emerge.

Only a social-democratic revolution, Williams believed, could achieve a more just and equitable connection between the planet's diverse human populations and the various ecosystems they inhabited. Williams thus cast "resistance to capitalism" as "the decisive form of the necessary human defence." Williams saw ample evidence that the social movements of his time, whether green or pink, brown or red, had paved the way for such a "defence" by prompting a "change of basic ideas and questions."[3] Williams understood that the post-World War II order was beginning to enter what Daniel Rodgers would later call an "age of fracture."[4] Yet Williams also detected great promise in the fissures the "general crisis" had begun to inflict upon the intellectual moorings of western society. Indeed, a world-historical transformation now seemed within reach. As he drew *The Country and the City* to a close, Williams allowed himself to hope that the reductionistic, essentializing modes of thought that lay at the very heart of modernity might soon make way for a renewed and revitalized holism. He alluded to the "connection which I have been seeking for so long." The means to making this connection, Williams explained, was the pursuit of "the many questions that were a single question, that once moved like light."[5] Williams left it to others to articulate this "single question." His larger corpus of work, however, makes it clear that Williams's larger project was to dismantle the ideological "separation of Nature from the facts of the labour that is creating it."[6] Wherever else Williams's "single question" might eventually lead, it would begin by revolutionizing the long-obscured interactions of environment and work that together made the modern world.

"We can overcome division," Williams declared, "only by refusing to be divided"— such, at least, is the premise and the promise of the scholarship this chapter surveys.[7] In the wake of Williams's pathbreaking studies of the 1970s, historians began to rethink labor history and environmental history—two realms long consigned to distinct and sometimes hostile historical subfields—as related components of larger historical wholes. In the process, I trace the many questions scholars have posed about the long-overlooked relationships between work, workers, and the working classes, on the one hand, and nature in its ideological, material, social, and political dimensions, on the other. No commonly accepted term has emerged to describe this body of scholarship, but "workscape studies" serves as a defensible shorthand.[8] Workscapes constitute material environments made by the continuous confluence of human labor and natural processes. Both labor and nature can be fruitfully conceived of as streams that run together to shape particular types of places and spaces, bodies and processes, discourses and histories. The great strength of the workscape concept is its ability to convey the dialectic interrelationships through which work and nature have shaped each other over time.

Many factors have mediated and influenced this mutual shaping. First, workers labor in nature as part and parcel of the natural world—as living bodies, in other words. But they constitute their knowledge of their physical selves and the nature with which their bodies overlap and intersect through culture—which is to say, through epistemological and linguistic constructs; ideologies of race, class, gender; and other socially constructed processes of meaning-making ranging from the assessment of risk to the sensation of pain. Moreover, people labor in nature by wielding tools, machines, instruments of coercion, and other technologies upon and through human and animal subordinates. These technologies extend, magnify, and transcend the bodies of those who use them. To make matters more complicated still, people almost always labor as components of larger entities: exchange systems (such as markets), organizational forms (such as corporations), and technological networks (such as energy systems). Finally, state institutions serve to structure interactions between workers, their employers, and the natural world by defining and protecting property regimes, facilitating or foreclosing reforms, and executing or sanctioning some acts of violence while proscribing others. By combining multiple perspectives and planes of analysis, workscapes highlight the doubly constructed nature of a reality that is simultaneously independent of human thought and deed, *and* a product of artifice sensible only through limited human faculties of perception, comprehension, and conceptualization. Because workscapes stand at the intersection of internally differentiated human societies ordered by complicated, unequal relations of power and a still more diverse and differentiated natural world, such places, systems, and assemblages hold the promise of revealing fascinating and significant insights on past dynamics. Wherever people work, in short, the boundaries between nature and culture melt. We need to learn more about the alloys thus forged.

No single world region has monopolized the attentions of workscape scholars. At the same time, though, the sort of cosmopolitan holism that Williams invoked through his "single question" seems almost impossible to practice in today's academy. Ongoing disciplinary divides, the accelerating pace with which new studies appear in fast-growing fields like environmental history, and the continuing dominance of nation-states in the subdivision of historical knowledge continue to Balkanize our understandings of the past. In recent years, both labor history and environmental history have become increasingly transnational, international, and global—and yet the majority of workscape studies continue to focus on the US case. For this reason, and also because US historiography continues to exercise an outsized influence over historians of other places—a phenomenon especially true in environmental history—this essay focuses primarily on the US, with only occasional forays beyond American borders.

The cadre of unusually talented scholars who first brought wider prominence to environmental history used Marxian viewpoints both to challenge and to reinforce their precursors' insights. A raft of influential studies charting what one might call a "nature theory of value"—William Cronon's eloquent treatment of commodification in colonial New England, Caroline Merchant's incisive argument that the scientific revolution served to alienate women from nature, Richard White's innovative application of Latin American dependency theory to explain the waning power of Indian peoples vis-à-vis the United States government, Donald Worster's cutting examination of the interconnected exploitation of western waters and migrant farm workers—began to complement and complicate existing scholarly preoccupations with the "labor theory of value," a concept whose roots stretched past Karl Marx and the French physiocrats to the peasant revolts of the early modern world.[9]

The environmental histories of the 1980s and early 1990s showed just how much scholars had to gain by integrating environmental history and social history. In *Rivers of Empire* (1985), for instance, Donald Worster asserted that "the domination of nature can lead to the domination of some people over others." More important still, Richard White argued in a landmark 1995 essay, "'Are You an Environmentalist or Do You Work For a Living?,'" that environmentalists needed "to reexamine the connections between work and nature." White, concerned that environmentalism's callousness toward physical labor and working people sabotaged the movement's appeal to timber workers and other rural western Americans, posed two provocative questions: "How is it that environmentalism seems opposed to work? And how is it that work has come to play such a small role in American environmentalism?" White probed these quandaries in a wide-ranging, paradigm-bending exploration that built upon Raymond Williams's work of the 1970s.[10]

And yet despite the linkages exposed by Worster and White, labor history and environmental history subsequently developed mostly in isolation from each other, despite what Chad Montrie has rightly recognized as "common theory and shared subject matter." Labor historians and environmental historians tended to hail from different class backgrounds, to direct their work toward somewhat different political ends, and to study different kinds of places and social relationships (the city, the workshop, and the factory, on the one hand, and the countryside and the wilderness, on the other). Environmental historians moved away from Marxian approaches and toward eco-criticism, biography, central-place theory, and cultural history. Labor history, for its part, had entered a worrisome decline by the 2000s; fewer graduate students pursued doctorates in the field, and advertisements for academic positions in labor history all but vanished. A shared commitment to materialism and a common critique of alienation, it turned out, offered insufficient grounds for environmental historians and labor historians to forge common cause in the face of the vast amount of work the ruling and middling classes have expended over the centuries to vilify, erase, trivialize, or co-opt labor's connections to the land.[11]

Larger political and intellectual trends only served to reinforce divides between work and nature in historical scholarship. Nearly two decades after the spotted owl controversies of the early 1990s in the Pacific Northwest, many mainstream environmentalists remain ignorant or suspicious of people who perform physical labor and the collective movements that represent them. "Most greens," as *The Nation* explained of the recent Keystone Pipeline controversy, "rarely endeavor to take up an outright defense of workers, never mind their unions."[12] Since the mid-1970s, moreover, a ferocious and forceful neoconservative backlash has decimated union movements around the world—in few places with more success than in the US.[13] The same counteroffensive by business interests has also circumscribed environmental movements and the governmental institutions put in place during the upsurge of green activism that began in the late 1960s. Many scholars, meanwhile, remain enthralled with an orthodoxy esteeming race and gender above all other categories of analysis. In recent historical studies, class seems to matter—when it matters at all—only to the extent that it inflects or complicates race or gender. The environment, meanwhile, carries even less analytical power in mainstream historical studies.[14]

Larger global dynamics, such as the rise of neoliberal economic policies, have combined with the subordinate status of class and nature within contemporary humanistic research to insure that rich borderlands of labor and environmental history—the domains of experience, identity, and struggle in which the vast majority of humanity has spent much of its time dwelling and toiling—remain *terra incognita* to most scholars. It bears remembering, then, that work comprises much of what humans have done on, with, and to this planet.[15] As Raymond Williams once observed, "the degree to which the fact of labour is included, in observing a working country" has always been "historically conditioned."[16] Williams's dictum is well illustrated by the success with which scholars allied across history, archaeology, anthropology, ecology, and other fields have assailed two long-standing myths: the imperialist notion of North America as a virgin or lightly peopled land; and the primitivist fantasy that Indian peoples lived in simple, stable harmony with nature.[17] Well before sustained contact with European peoples began, many forms of productive and reproductive labor—family planning, plant-propagation practices that ranged from sporadic tending of essentially wild ecosystems to intensive agriculture, the development of short- and long-distance exchange networks, and selective burning, to name a few—had already transformed much of the continent into labored landscapes.

Scholars still need to devote more attention, however, to the social and political dynamics that shaped Native American workscapes. Gendered divisions of labor have preoccupied scholars since the first European accounts of the West Indies rolled off newly invented printing presses in the fifteenth century. By contrast, status hierarchies, coercive labor regimes, and other common modes by which Native American societies organized extraction, production, trade, and consumption—thus making indigenous workscapes—have only recently begun to receive much attention from historians.[18] "Historians in general," as Alexandra Harmon, Colleen O'Neil, and Paul Rosier recently argued, "have scarcely touched the subject of Indians and the American economy," let alone the topic of American Indian workscapes.[19]

Andrew Isenberg's *Destruction of the Bison* (2000), unlike many of its predecessors, dispenses with an exceptionally deceptive piece of noble-savage mythology: the notion that Native American societies comprised primitive, egalitarian polities composed of sovereign individuals. Isenberg filters accounts of trappers, traders, and explorers through a lens informed by feminist anthropology. The resulting analysis shows how women's status declined as native peoples on the northern plains became increasingly enmeshed in the bison trade during the early 1800s. No less important, inequalities of wealth widened between men possessing many horses and those possessing few or none. Men who had better access to equines could generally kill more buffalo than men who lacked ponies. Men who had better access to female labor (often via polygamy), meanwhile, could process more hides and robes for American and European markets. In the final reckoning, those Indian men who possessed the most horse wealth and the largest platoon of women to turn bison carcasses into commodities took a disproportionate toll on the buffalo populations of the Great Plains.[20]

Isenberg's later book, *Mining California* (2005), offers an even fuller application of Marxian political economy—this time to the Modoc War in northeastern California. Isenberg attributes this bloody 1860s conflict between native peoples and American invaders to ecological changes that "resulted in the decline of the Indians' resource base, their consequent shift into the wage labor market, and the decline of the population owing to epidemic disease and the migration of Modoc women into Euroamerican households as wives, prostitutes, or domestic laborers." In an all-too-common scenario, environmental changes undercut Native American lifeways. Native peoples subsequently struggled to carve out niches for themselves in Euroamerican-controlled labor systems, usually as captive, dependent, or wage workers. Isenberg convincingly contextualizes environmental and social dynamics among the Modoc as "remote extensions of the process known as enclosure that began in late medieval and early modern England and occurred elsewhere in northern Europe." On the Modoc Plateau, as in so many other parts of America, newcomers and natives encountered each other not as "two groups that lived in separate worlds," but as "employers and employees, the landed and the newly landless." Indians started performing old tasks under new circumstances; not infrequently, they turned to entirely new kinds of labor. In the process, they suffered from what Isenberg, following E. P. Thompson, calls the "imposition of social discipline by a new economic regime." Where Frederick Jackson Turner proposed in the 1890s a clear frontier demarcating "savagery" and "civilization," "nature" and "culture," Isenberg offers a twenty-first-century exposé of the messy, often violent ways in which peoples, places, and practices collided.[21] Imperialism and settler societies, in other words, restructured workscapes in far-reaching ways.

Isenberg's *Mining California* suggestively demonstrates how new ecological regimes spawned by the intrusion of settler economies alternately lured native peoples, and forced them to forge novel relationships with their own bodies as well as with non-human nature. The preoccupation of ethnographers, travelers, and subsequent historians with "traditional" Native American cultures means that we know all too little about these relationships.[22] Fortunately, a considerable body of recent studies on Native American workscapes—Kathy Morse on native guides and porters in Alaska and the Yukon, Gerald Ronning on Native American labor organizers, Susan Johnson on Miwok women gold miners, and Peter Iverson and others on Indian *vaqueros* in the California missions and Indian cowboys in the west more broadly—complicates the narratives of contact, decline, and false appropriation that mark the fence-lines of US environmental history's Indian ghetto.[23] In this regard, as in many others, students of American workscapes would do well to engage more fully with other historiographies, especially scholarship on indigenous laborers in Canada, Australia, and New Zealand.[24]

Even by drawing upon comparative and transnational approaches, though, students of Native American workscapes will still face the daunting task of developing a coherent analysis of the distinct and ever-changing ways in which hundreds of complex, internally segmented Native American societies worked to shape a seemingly infinite array of environmental relationships over many millennia. We know more than enough at this point, however, to urge our colleagues in other fields to acknowledge a crucial point: North America, like other "new worlds," was not just inhabited when "contact" began (and often quite densely), but also *worked*. It is high time for surveys, textbooks, and synthetic treatments of US labor history, economic history, the history of science and technology, and many other subdisciplines to abandon antiquated, exclusionary paradigms that continue to treat indigenous labor as a contradiction in terms.[25] To ignore the profound and ongoing importance of Native American workscapes, after all, is to unwittingly recapitulate a Turnerian vision of the American landscape in which white Americans penetrated a succession of virgin wildernesses—forests, swamplands, mountains, plains, prairies—each pristine and unmarred by the stigma of human labor.[26]

While scholars have proven slow to explore the intersections of labor and nature in Native American history, only the most rabid adherent to the "lost cause" mythology could deny the centrality of African American workers to plantation landscapes. Both before and after slavery, as Chad Montrie points out in *Making a Living* (2008), "race was at the center of a set of evolving social relations that determined to a great extent, or at least heavily influenced, how blacks as well as whites thought about and used the natural world around them."[27] Racialized labor systems and social hierarchies combined with African and African American environmental philosophies and practices, North American ecologies, and an array of organisms introduced from Europe, Africa, and beyond to indelibly shape southern landscapes, African American identities, and black experiences.[28]

In the Shadow of Slavery (2009), by Judith A. Carney and Richard Nicholas Rosomoff, reminds us that Africans created the planet's first workscapes tens of thousands of years ago. Carney and Rosomoff persuasively rescue African "pre"-history from imperialist condescension. "Perceptions of a continent populated by hapless farmers and herders in need of European instruction," they argue, "are inaccurate and fail to do justice to Africa's deep botanical legacy." By setting fires, domesticating animals and plants, and otherwise changing the land, Africans reshaped their continent's ecosystems over many millennia.

The driving purpose of *In the Shadow of Slavery,* though, is less a foray into deep history than an examination of the critical roles that African workscapes played in transatlantic slavery between the 1400s and the 1800s. Carney and Rosomoff show how African food staples literally rooted Atlantic slavery in African agricultural practices, exchange systems, and biotechnologies. As the transatlantic slave trade metastisized, enslaved laborers on large farms along Africa's coastal plains were coerced into producing the provisions upon which both the internal and external slave trades depended. Crops harvested by slaves bound to these farms fed raiding and war parties that struck deep into the African interior.[29] They also yielded a scant sustenance for the coffles of slaves these parties captured, the European posts (known as "factories") where most traffic in enslaved Africans took place, and the slave ships that subjected well over twelve million slaves to what Marcus Rediker has aptly described as a "pageant of cruelty, degradation, and death" on the Middle Passage to the New World's burgeoning plantation belts.[30] Provision farms grew a combination of African cultigens, such as millet and sorghum, as well as crops brought back from the Americas, particularly maize and manioc.[31] Many of these foods required laborious milling and preparation. "The provision trade," Carney and Rosomoff explain, "relied on the specialized skills of enslaved women," especially in the often arduous processes required to transform harvested plants into forms that were both edible and long-lasting; this, in turn, "increased the need for female labor" on slave-raiding expeditions. Because maize usually yielded more than sorghum, it largely replaced it, thus becoming "a crop increasingly emblematic of slave labor and sustenance." Cultivated "by agricultural slaves and those awaiting deportation," corn "fed the growing segment of the African population remanded to coastal port settlements and destined for American plantations. Surplus production was routinely sold to provision slave ships."[32]

Just as the introduction of New World cultigens changed African farming and inflicted devastation upon the coerced workers whose toil literally fed the slave trade, so, too, did the westward journeys of African seeds produce far-reaching impacts on the foodways, labor systems, and cultures of tropical America. Plantation societies relied heavily on "plant introductions." Carney and Rosomoff attribute the presence of African cultigens "in slave food fields to Africans themselves, who took the initiative in planting their dietary preferences from the leftover provisions that at times fortuitously remained from slave voyages."[33]

Carney's earlier *Black Rice* (2001) offers a classic example. Building upon previous work by historians Peter Wood and Daniel Littlefield, it examines "the African origins of rice cultivation in the Americas." Carney argues that the environmental expertise of enslaved Africans played an essential role in the emergence of the Carolina and Georgia Low Country as a highly productive and lucrative area for rice production. In an important corrective to Alfred Crosby, she argues that Columbian exchanges depended not just on the transfer of seeds across the Atlantic, but also upon the transplantation of a broad array of intricate cultural and environmental relationships from Africa to the New World. In a well-researched but still controversial "agrarian genealogy" that spans past and present, three continents, and several disciplines, Carney excavates the "indigenous knowledge systems" through which African peoples and their descendants cultivated, processed, and consumed rice, a crop now believed to have been independently domesticated by Africans and Asians. European planters desiring to establish rice along the coasts of Carolina, Georgia, and South America, Carney argues, relied upon slaves, especially West African girls and women whose considerable expertise in the grain's cultivation survived the Middle Passage intact. Significantly, Carney argues, "the share of slaves brought by British slavers into South Carolina from rice cultivation areas of Senegal, Gambia, and Sierra Leone" rose from 12 percent in the 1730s to 64 percent between 1769 and 1774. Though Carney's so-called "black rice thesis" has stimulated strident debate, her call to contextualize "American plantations in relation to African systems of production" has raised new questions about the interconnections of nature and labor in the making of the Atlantic World. No less significantly, *Black Rice* offers a fascinating spin on the metaphor David

Montgomery deployed so influentially in his landmark *Fall of the House of Labor* (1987). If in the factories of industrializing America, the manager's brain actually resided under the workman's cap, then no small measure of the rice planters' intelligence lay beneath the bondswoman's headscarf.[34]

For all of the contributions *Black Rice* makes, though, Mart Stewart's fascinating *"What Nature Suffers to Groe"* (1996) remains the most fully realized study of US plantation workscapes. "Early Americans," Stewart reminds us, "had a constant and intimate dialogue with nature in their everyday lives." While first-generation environmental historians tended to lament the progressive intensification of human mastery over nature, Stewart weaves a more nuanced story of the Georgia Coast from colonization through the early twentieth century. In his telling, planters, slaves, and various elements of nature engaged in an incessant process of "negotiation, exchange, manipulation, and transformation." Stewart documents the truly prodigious inputs of bonded African labor that made and maintained elaborate hydraulic works in low-country rice fields; in the process, he exposes both "the agency of slaves," and the "agency of nature in shaping the fate of plantation agriculture." Stewart asserts "that planters used the environment and appropriated knowledge about it to reinforce their own class interests." Slaves, in turn, "created counterstrategies to promote their own interests." They not only cultivated their own environmental knowledge, but also deployed it when and where they could. Stewart does not deny that slaveholders wielded disproportionate power in the resulting struggles: "The domination of nature and the domination of one group of humans by another," he argues, "evolved mutually." As slaves "became instruments of environmental manipulation," the environment "became an instrument to control the slaves."

Yet even the most formidable planters could maintain only a tenuous, laborious order: "Planters could shape, channel, and manipulate the energy of nature," Stewart concedes. Yet their ability to do so depended upon the success with which they could "coerce and manipulate the energy of slaves."[35] The productivity of rice workscapes was already wavering prior to the outbreak of the Civil War. Accumulating problems—declining soil fertility, hurricanes, a deteriorating disease environment, worsening ravages of bobolinks and other rice-eating birds, intensified flooding caused by the denudation of upcountry forests, tenacious resistance mounted by the slaves upon whose brawn and brains rice culture depended—all taxed yields and profits. Even the region's wealthiest planters struggled to maintain their fortunes.

Emancipation promised the slaves of the Georgia coast what Abraham Lincoln famously called "a new birth of freedom"; it delivered, at the very least, a literal unchaining of the deep familiarity African Americans had developed with the natural world, both on plantation workscapes and in the forests, swamps, and waters beyond.[36] After the federal government tragically failed to redistribute southern lands and outfit freedpeople with the tools and livestock they needed, whites "redeemed" and remade the Low Country. As Jim Crow's grip on the region tightened, however, Stewart argues that African Americans continued to deploy "keen awareness and precise knowledge of the environment" to cobble together a "patchwork" landscape of relative freedom, working to draw a range of resources from "small farms and open land, port towns and market gardens, and sea island hunting reserves and vacation retreats."[37]

Several other scholars have also traced enduring connections between African Americans' environmental knowledge and their struggles to create and preserve spaces of autonomy and resistance. Both before and after formal emancipation, people of African descent toiled to raise cash crops on cotton and sugar plantations throughout the Atlantic world. But African Americans also gardened and gathered, hunted and trapped, fished and marketed, told stories and traveled. Each of these ways of laboring the land helped blacks to carve out spaces of their own in the interstices of white-dominated landscapes. "On the margins" of what Chad Montrie calls "the more oppressive endeavor of plantation staple crop production," enslaved blacks managed to find "some relief from the intense alienation inherent in plantation cotton cultivation." With the end of slavery and the rise of sharecropping in much of the South, African Americans temporarily exercised greater control over their labor. Not surprisingly, they now found greater "opportunit[ies] than ever before," Montrie explains, "for gardening, hunting, and fishing."[38] These ways of interacting with the natural world seamlessly combined several functions, providing food, fibers, skins, and furs that blacks could use or exchange for other goods, creating and solidifying social ties and hierarchies, and offering respite and recreation.[39]

Given the vital roles that hunting, herding, gardening, and other kinds of work in nature had long played in the construction and preservation of zones of relative autonomy in the US South, it should come as no surprise that Jim Crow rule brought renewed assaults upon black natureways. In the late 1800s and early 1900s, white southern elites, as Steven Hahn, Nicolas Proctor, Scott Giltner, and others have argued, successfully enlisted the state to intensify their dominance over the land and its inhabitants. Through fence laws, game laws, and other policies, champions of white supremacy attempted to eliminate or circumscribe the islands of relative independence African Americans had worked so hard to establish.[40] Such campaigns jeopardized decades-old, even centuries-old, relationships between black folk and southern lands.

In areas such as the longleaf pine forests of the Red Hills along the Georgia-Florida border, the agricultural labors of black sharecroppers and tenants continued to play important roles in ecosystem maintenance. Bert Way's *Conserving Southern Longleaf* (2011) demonstrates the significance of African American cotton farmers to the ecology of these woodlands. Like the area's white landowners, blacks "continued the long tradition of burning in late winter and spring to rid the fields and forests of a year's worth of accumulated growth." Fire, local blacks and whites understood, "had several practical purposes." It prepared the ground for planting, controlled various insect pests, and encouraged regeneration of "succulent forbs and grasses for grazing livestock." The labor of fire-setting, Way shows, constituted a "cultural practice" that "allowed the region's longleaf pine woodlands to flourish" via an "unlikely confluence of ecological function and cultural preference."

In Way's Red Hills, as in Stewart's Sea Islands, the arrival of northerners seeking closer encounters with the natural world jeopardized longstanding social and ecological relationships. By single-mindedly pursuing bobwhite quail and other game, the wealthy owners of hunting preserves "did not realize," Way claims, "that to 'let alone' the longleaf pine-savannah woodlands was to change its very nature." Centuries of burning and other kinds of human work, after all, had indelibly shaped these ecosystems; misperceiving workscapes as leisure landscapes, well-heeled landowners unwittingly undercut some of the ecological relationships on which their sport depended. Elite sportsmen responded to a perceived decline in quail populations in the 1920s by seeking federal assistance; Herbert Stoddard, hired by the US Biological Survey to oversee bobwhite restoration in the area, advocated "vehemently for the continued use of fire in the southern coastal plain." Thus did one of the twentieth century's most influential ecologists become an ardent "defender not only of fire itself but also of a fire culture" rooted deeply in plantation workscapes forged by "old hierarchies of racial, political, and economic control."[41] As Stoddard recognized, ecological restoration not only required a lot of work; it also often depended upon the reintroduction of kinds of human labor that landowners had tried to eliminate in conservation's name.

The Deepest Wounds (2010), Thomas Rogers's study of landscape and labor in the sugar-producing countryside of northeastern Brazil, offers a useful point of comparison to environmental histories of slavery and emancipation in the American South. Drawing directly upon Raymond Williams, Rogers argues that planters saw their environment as "a *laboring landscape:* the landscape itself labored for planters, as they saw no distinction between land and labor. Land and people were bound together in their minds," Rogers explains, "by coercion, the sense of entitlement and control passed down across generations by those in power, and a history of slavery and plantation agriculture." From slavery through emancipation, cane workers contested the planters' tendency to treat them as extensions of plantation environments—"natural elements," as Rogers provocatively puts it, "alongside oxen." By the first half of the twentieth century, workers had become "intimately familiar with the many varieties of cane they worked with." They spoke of "black cane and purple cane, crystalline, ashen, red, bronzed, and 'little Cayenne' cane." Workers especially "liked 'hairless' cane because it lacked the spiky edges that abrade the skin, and they referred to some varieties as 'butter' canes because as they matured they slowly turned from light green to a creamy yellow." For the people of the cane fields, work helped "structur[e] their distinct view of the landscape. Besides giving them knowledge of and insight into natural processes, work also contributed to the accretion of meanings around the environment being worked, informing workers' sense of place." As Rogers demonstrates, "a history of workers is most effective when it is also a history of the environment they transform."[42]

Back in the US, meanwhile, the assault on black natureways by "progressive" conservationists worked in concert with endemic exploitation, racist violence, proliferating transportation systems, improved communications networks, and structural transformations in southern agriculture to loose African Americans from the countryside. The ensuing migrations brought African Americans to Harlem and Chicago's South Side, Pittsburgh and Detroit, the Colorado coalfields and other centers of industrial extraction and manufacturing. Brian McCammack, in one of the few works yet to examine Great Migration workscapes, contrasts the prior labors of African Americans in the South's fields and towns with their new jobs in Chicago's slaughterhouses and steel mills. Unlike the fieldwork into which many migrants had been born, the labor of "slaughtering took place indoors in a setting largely devoid of natural light, and the squeal and smell of a single hog's death was multiplied many times over as workers performed the same minute task hundreds of times per day." The steel industry, meanwhile, "deeply enmeshed workers in nature's raw materials and the elemental power of fire and metal." Black steelworkers in the South Chicago mills wrestled with the task of "harnessing and transforming" these elemental forces "with a series of tools and machines that seemed far from natural." When their shifts ended, Chicago's black migrants sought out "environments that more closely resembled the land they had left behind." In beaches, city parks, and rural resorts and camps, Chicago's black workers, like many of their white counterparts, "strove to come into closer connection with natural and landscaped environments in their new northern homes—but primarily through leisure, not labor."[43] In elucidating the connections between work, recreation, and nature in the lives of black participants in the Great Migration, McCammack and others have started to undermine what William Cronon once called "the arbitrary boundaries of labor and leisure that people impose on the natural world."[44]

Substantial numbers of blacks, meanwhile, migrated within the South, finding work in the mines of West Virginia, the orange orchards of Florida, and the pineries of Louisiana and Texas. Many of the tens of thousands of southern blacks pushed and pulled into wage labor in the turpentine industry undoubtedly developed what Cassandra Y. Johnson and Josh McDaniel call "an intimate lay knowledge of plants, animals, and a general environmental understanding." Not a few blacks, though, paid a heavy price to acquire such knowledge. "The exploitation, abuse, and racism that accompanied the industry," Johnson and McDaniel insightfully suggest, eventually "resulted in a dramatic decrease in the numbers of African Americans working in the forest." Johnson and McDaniel even speculate that enduring memories of toil in the naval stores industries may have "contributed to reluctance among African Americans" within the South "to engage in many forms of forest-based leisure time activities." This insight, provisional though it may be, nonetheless suggests how the divergent workscape experiences of northern and southern blacks could lead Chicago migrants to envision the woods of Michigan as a highly desirable retreat from industrial labor even as their counterparts in the yellow pine belt continued to perceive forests as haunting sites of ongoing exploitation.[45]

As all of these studies show, the historical intersections between labor and the environment can help historians of slavery, emancipation, and Reconstruction better understand the origins of plantation systems, the role progressive conservation played in imposing Jim Crow on the countryside, and the shifting strategies blacks mounted in the face of ongoing white resistance to the various forms of agency African Americans managed to wield in workscapes and landscapes alike.

In the northern and western United States, as in the South, agricultural work provided what Judith Carney aptly terms "the tissue linking culture to environment."

Unfortunately, studies of farming in New England, the Middle Colonies, and the Old Northwest have mostly focused on *either* work or nature. Historians of the trans-Mississippi West, by contrast, have more frequently and insightfully treated labor and the environment in concert.[46]

Mark Fiege's *Irrigated Eden* (1999), for instance, insightfully traces the transformation of southern Idaho's Snake River basin into a highly productive, heavily industrialized agricultural workscape. The region's pioneer farmers initially used the labor power of family members and neighbors to build and maintain dams, canals, ditches, and farms. With the rise of corporate farming and federal irrigation in the early twentieth century, however, "the work that people performed in the irrigated landscape" increasingly "reflected a convergence of the family farm ideal and industrialization." Fiege singles out the adoption of sugar beets as a major precipitant of this convergence. Growing beets required large inputs of hand labor, leading farmers and processors to recruit workers from Japan, the German Volga, the American Southwest, and Mexico. With the development of monogerm beet seeds in the 1960s, though, "the need for intensive labor" virtually vanished, abruptly reducing the beet industry's problematic dependence on migrant workers. "Labor altered the earth's surface on a massive scale" in the Snake Valley, Fiege concludes, "yet this was not a one-way process in which humans simply shaped the land. For the land" and the plants it supported "in turn affected the people who worked it." In Fiege's view, "land and work evolved together." Indeed, such mutualistic coevolution represents a core theme in recent studies on the laboring of North American lands.[47]

David Igler joins Fiege in viewing human labor as "the integral link between resource exploitation and large-scale production." Igler's *Industrial Cowboys* (2001) presents an incisive case study of Miller & Lux, a sprawling, California-based cattle company whose tentacles extended from the western range to slaughterhouses, irrigated farms, and other enterprises. The firm's power, Igler argues, "ultimately derived from the ability to tap both human and natural energy for its own ends." Wherever Miller & Lux did business, its activities "obscured the connections between production and consumption." Igler offers an important corrective by pointing out that "those who labored in the fields or slaughterhouses" (and, by extension, those who toiled in the mines, forests, factories, and other workplaces of industrializing America) often "held the best view on these linkages." Employees of Miller & Lux, in short, rarely enjoyed the privilege to imagine their lives as alienated from the natural world.[48]

Another work on California agriculture, Douglas Sackman's *Orange Empire* (2005), also uncovers the ongoing significance of productive labor in US environmental history. Sackman's signal contribution is to call into question broader historiographical trends in which consumerism and consumers have too often shoved producers and producerism to the margins of scholarly concern. Sackman effectively draws attention to the hidden but ongoing connections between consumption and production by urging his readers to question the orange-crate iconography that portrayed California citrus as a direct gift of the sun: "The neat trick of absenting the grower and other laborers not only heightened the consumer's sense of communing with nature, it masked the working conditions from which the fruit emerged. Erasing the workers who brought the fruit to the consumer's lips," Sackman argues, "made California an Eden in which fruits naturally materialized for the pleasure of people." To undercut the Edenic, anthropocentric narratives blazoned on orange crates, Sackman tracks oranges from grove to market. In so doing, he unveils the complicated ways in which race, gender, and ideas of nature structured life and labor in southern California's "orange empire." Work, Sackman demonstrates, "was critical to the transformation of oranges into artifacts that could be bought and sold." In *Orange Empire,* "the bodies of workers mediat[ed] the artificial zones of the market and the organic landscape of the groves." Laborers in the orange industry, like their counterparts across space, time, culture, and ecology, found their bodies constantly "shaped"—and reshaped—"by the physical work they performed."[49]

Though integrated histories of labor and nature have tended to focus on rural work, several studies have recently probed "the changing relationship between common people and the environment" in industrial centers.[50] Had an *Oxford Handbook of Environmental History* appeared around 2000, its chapter on labor and the environment probably would have focused largely on the august literature examining industrial health and safety in the so-called "dangerous trades" which fueled America's tremendous industrial expansion of the 1800s and 1900s. Barbara Allen, Claudia Clark, Alan Derickson, Gerald Markowitz, David Rosner, Christopher Sellers, and a host of other historians of public health built upon the pioneering work of Progressive investigators; like their predecessors, such scholars endeavored to expose the industrial economy's seemingly limitless woes through quantitative analyses and qualitative studies. In the US, the work of miners, factory hands, and other laboring people drove breakneck economic growth; historians of industrial health demonstrated that the workers' toil was often rewarded with sickness, injury, disability, and death.

Workers' bodies consequently became battlegrounds in the struggle to reconcile economic growth with democratic rights and human decency. Campaigns to ameliorate health and suffering in the nation's workplaces sometimes united reform professionals, labor aristocrats, and rank-and-file workers. The resulting coalitions sometimes succeeded at making industrial workscapes safer, healthier, and happier. In contrast to the aesthetic preservation and utilitarian conservation movements on which environmental historians have typically focused, movements dedicated to conserving workers' bodies and cultivating what contemporaries often called "industrial democracy" crafted a model of environmental reform in which economic and political elites played more limited roles. That efforts to improve industrial health and safety achieved only partial gains reveals less about the limits of the reformers' vision, and more about the ferocity with which capitalists and their allies often struck back against workers, unions, and state and federal agencies newly empowered to protect workers.[51] Historians of industrial health and safety have made undeniably significant contributions. Too often, though, they have tended to suggest that American workers interacted with nature only on the job, and only in highly negative ways. Nature figures in these works chiefly as the lead poisoning workers' blood or the silica scarring their lungs.[52]

In the wake of Richard White's call for historians to explore how labor has imparted knowledge of nature, historians have begun to pursue broader questions about the environmental history of industrial work.[53] Kathy Morse's *Nature of Gold* (2003), for example, explores both the bodily labor by which tens of thousands of Argonauts ventured north, and the arduous work these migrants subsequently devoted to coaxing wealth from the Yukon's buried depths. In the course of disassembling gold-bearing ecosystems, miners enlisted a succession of increasingly capital-intensive technologies, culminating in hydraulicking and dredging. The miners' labors soon "made it impossible for humans" along the region's waterways "to do any other kind of work, and made it difficult for nature itself to produce anything besides gold."[54]

By remaking the landscape "according to an industrial vision of a productive land," miners destroyed some components of nature. At the same time, though, Morse emphasizes that they also forged "a new and particular set of connections" with the natural world. Gold-seekers, Morse shows, came to "know nature" through their work. Grappling with diurnal and seasonal cycles, the "repetitive, grueling labor of thawing and digging through foot after foot of frozen muck and dirt," the ceaseless search for fuelwood to melt frozen earth and warm chilled bodies, and the accelerating impact of deforestation on the flows of water miners used to separate gold from less valuable materials, miners ultimately developed "an increasingly sophisticated understanding of how to take the earth apart to find gold." Only by integrating labor and environmental histories, Morse argues, can we grasp the real "nature of gold."[55]

Chad Montrie's study of female mill operatives in Lowell, Massachusetts in the mid-1800s offers another well-researched analysis of "the historical relationship between the work people do and the ways they think about nature." Montrie argues that "as the methods, organization, and purpose of labor changed for these workers, the way they used and understood the natural world also changed." Operatives drew important contrasts between textile mills and "their native rural homesteads." Nineteenth-century mill operatives, much like twentieth-century African American migrants to Chicago's steelmills and slaughterhouses, experienced profound on-the-job alienation from nature. Montrie demonstrates that textile workers often expressed this sense of alienation by invoking both remembered rural values, and romantic tropes borrowed from print culture. When strikes periodically erupted in and around Lowell, mill girls often represented their discontent by blaming wage work for separating them "from a factual as well as fictional rural landscape, one they believed more healthy, beautiful, spiritually meaningful, and conducive to the development of good morals."[56]

Few studies, alas, have examined the complex relationships between workscapes and industrial struggle in such rich detail. An exception is Thomas Andrews' *Killing for Coal* (2008), which explores the violent relationships between coal miners, coal companies, and the natural world in southern Colorado between the 1870s and the deadly coalfield war of 1913–1914. After examining the wide-ranging environmental and social impacts spurred by the advent of fossil-fueled industrialism in the Mountain West, *Killing for Coal* argues that the region's endemic labor violence was rooted in subterranean mine workscapes that cultivated common cause between inexperienced migrant workers and expert colliers known as "practical miners."[57]

Myrna Santiago's *The Ecology of Oil* (2006) pursues the story of labor, nature, and fossil fuels from coal to petroleum, from the US to Mexico, and from the early decades of the twentieth century into the 1940s. In so doing, it demonstrates just how much students of American workscapes have to learn from looking beyond US borders. Santiago traces the explosive growth of petroleum extraction in the Huasteca region of Mexico's Gulf Coast between the *Porfiriato* and the oil industry's 1938 nationalization. Throughout, Santiago focuses "on actors typically peripheral or neglected in the historiography of Mexican oil: indigenous people, nature, and oil workers." Together with entrepreneurs, professionals, and the state, each of these actors helped to create what Santiago calls "an entirely new ecology, the ecology of oil." Petroleum extraction devastated places shaped over the preceding centuries by indigenous peoples, hacienda owners, and *mestizos* who herded cattle and did other work for *hacendados*.

Oil discoveries stimulated labor migrations; both on the job or off, oil workers effected sweeping reconfigurations in the Huastecan landscape. Workers, whether foreign or Mexican, quickly learned that the labor hierarchies established by American, British, and Dutch oil companies had far-reaching consequences. "The top echelons" within the oilfield social order "reshaped the environment to fit not only productive needs, but also their own sense of aesthetics and pleasure." Santiago follows Richard White to argue that working-class men, by contrast, "knew nature through work." Santiago elaborates: "Even though bosses and workers changed the landscape together, their experience ... differed substantially depending on their position in the occupational ladder. That is, class and color deeply influenced the relationship between men and nature in the production process." Foreign drillers (highly skilled craftsmen who secured very high wages in Mexico) found that their "bodies literally stood between nature and wealth. Just as they were the agents that directly extracted the oil form the earth to transform it into material riches, theirs were the bodies liable to experience direct harm when nature released the sought-after chemicals. For that reason," Santiago explains, "it was the drillers and craftsmen who designed technologies to cap runaway wells and to prevent explosions." Foreign oil companies imposed neocolonial assumptions about race not just at work, but also on the camps built to house laborers and even the town of Tampico, which was fast becoming Mexico's major refinery center and oil port. Mexican oil workers in the Huasteca almost always performed the most dangerous, worst-paying jobs. When their shifts ended, they returned to home environments that suffered substantially higher risks of flooding, fire, disease, pollution, and other ills. "In the ecology of oil," Santiago laments, Mexicans "were the most vulnerable human beings."

Though oil company executives believed themselves to be masters of the Huasteca, the region's workers and workscapes together compromised the elites' commanding visions. Mexican workers, conceived in neocolonial frameworks as passive, obedient, primitive, and apolitical, adeptly resisted petro-capitalism. "There were tensions inherent in the network of relationships the oilmen built," Santiago notes, "and in the processes of change they set in motion. Neither men nor nature acted in accordance with the plan. Control proved difficult to achieve and maintain in both cases." The outbreak of the Mexican Revolution brought "a struggle for the ownership and control of nature itself." Revolutionary leaders "sought to wrestle control over nature from the oilmen as part of a nationalist development program based on a platform of conservation of natural resources." Petroleum companies faced even more strident opposition, however, from Mexican oil workers. Workscapes in the Huastecan oilfields, like those in the Colorado coalfields, catalyzed "a hyper masculine culture [that] bridged individual work experience and collective class consciousness and action."

Mexican workers, "consistently nationalist and to the left of the revolutionary leadership," repeatedly "challenged the social and environmental order the industry created." After decades of struggle and dozens of strikes, workers launched what Santiago calls a "daring bid to unravel the ecology of oil altogether." Uniting in a new industrial union, they pushed President Lázaro Cárdenas to nationalize the petroleum industry in 1938. Santiago considers Cárdenas's decree "the single most important moment for the forging of modern Mexico." By uncovering the previously hidden relationships linking environmental change, labor, and politics in a watershed moment in the history of Mexican nation-making, *The Ecology of Oil* suggests how workscape studies can help historians develop more persuasive interpretations of major national and transnational events.[58]

Taken as a whole, the large and growing body of scholarship on laboring the land has shown just how much scholars have to gain by examining historical relationships between working people and the natural world. Such an approach, as Montrie rightly emphasizes, "can and will alter the way we think about [laboring peoples'] experiences during industrialization, their changing identities, their varied and evolving culture and values, their efforts to create and maintain unions and other social organizations, as well as their role in politics." It should also change how we think about the character and causes of environmental change, the meaning of "nature," and the struggles that working people have waged over resources, hazards, space, and competing ways of "knowing nature."[59]

To build upon these important contributions, however, scholars interested in the laboring of the landscape would do well to devote greater attention to women. Environmental history, as Virginia Scharff argued almost a decade ago, "remains at present not a story of people and other things but is instead a story of man and nature." Drawing together evolutionary biology, etymology, and feminism to examine "why scholars in the field have found forest fires more fascinating than cooking fires," Scharff calls for "a broader, more nuanced, and far more gender-conscious understanding of work than the fundamentally Marxist formulation embraced" by environmental historians. Curiously, Scharff does not make an analogous case to women's historians: that studies of cooking fires and childcare, women wageworkers and female-dominated unions might benefit from incorporating non-human nature more fully into their analyses and narratives. Suellen Hoy's *Chasing Dirt* (1995), Claudia Clark's *Radium Girls* (1997), and a handful of other studies on domestic environments and industrial hygiene notwithstanding, the history of women's workscapes remains largely unwritten. Scholars committed to filling this considerable gap must overcome seemingly severe source constraints; they also need to develop new conceptual tools and theoretical frameworks capable of exploding the remarkably durable and insidious trope of "Man and Nature." The obstacles standing in the way of such a project are large, but the need for a more balanced view of work and nature in the American past is greater still.[60]

Equally necessary are studies to carry the story of the laboring of the American landscape into the twenty-first century. A question raised by Richard White's 1995 "Are You an Environmentalist?" still seeks an answer: how have everyday connections between nature and modern work shaped how Americans have thought about and acted toward the environment? A deepening alienation between working people and the natural world surely accompanied the rise of factory and office labor. But how best to characterize this alienation: absolute or partial, hegemonic or contested? What about the many kinds of work which have continued to bring people into intimate, sometimes deadly relationship with the environment? And how have ways of playing in nature not just "mimic[ked] work," as White presciently noted in his 1995 essay, but also contributed to the production of new workscapes through which apparel workers, equipment designers, river guides, and others have engaged various parts of the environment?[61] The twentieth century, as John McNeill argues in *Something New Under the Sun* (2000), witnessed the most rapid and far-reaching environmental transformations in many millennia; the emphasis on consumption and consumerism in recent US historiography should not obscure the extent to which the so-called "American century" was premised upon the most thorough reworking of the American landscape since the colonial era.[62]

Looking beyond the labor process and work politics, political ecologies of class have investigated the assorted environmental inequities workers have experienced from the nation's sparsely peopled wildlands to its densely packed neighborhoods. Environmental historians have used class analysis to complicate older narratives that celebrated Progressive-Era conservation as an enlightened, public-spirited, and unproblematic crusade against the wasteful destruction of natural resources. Karl Jacoby's *Crimes against Nature* (2001) applies the insights of Eric Hobsbawn, James Scott, and especially E. P. Thompson to controversies over poaching, timber theft, squatting, and other kinds of "local" land use. Jacoby's deeply researched and carefully argued book uncovers the workings of what Jacoby terms "moral ecologies" in the Adirondacks, Yellowstone, and Grand Canyon. Such local systems of "beliefs, practices, and traditions . . . governed how ordinary rural folk interacted with the environment." While earlier environmental historians tended to see past events through the eyes of conservation's elite proponents, Jacoby presents a pathbreaking "vision of nature 'from the bottom up.'" The story of American conservation, Jacoby convincingly claims, is in no small degree a story of how elite white sportsmen deployed state power to dismantle moral ecologies, then established in their stead "a formal code of law, created and administered by the bureaucratic state."[63]

Crimes against Nature, like Louis Warren's *The Hunter's Game* (1997) before it, places class struggle at the center of conservation history. To Warren, though, conflicts between elite Anglo sportsmen and rural hunters almost always pivoted upon gender, too. Sportsmen managed to shroud their own "ideas of masculinity" with "the mantle of state power." Native peoples, Hispanos, European immigrants, and other non-elite men all invested hunting with their own ideologies of manhood. As a result, state efforts to impose game regulations—and thus to regulate working-class masculinity—frequently led to violence. Local grandees bent on "reordering the land" dispatched game wardens, sheriffs, and other isolated representatives of state power to do the dirty work of enforcing conservation regulations. Agents of state and federal bureaucracies often drew first blood. The "recreational paradises" established through class conflicts on rural landscapes, Warren shows in an important corrective to the triumphalism of John Reiger's *American Sportsmen and the Origins of Conservation* (1975), "became to a significant degree the property—or the privilege—of a national, middle- and upper-class tourist public."[64] From the perspective of the rural working classes, the conservation movement seemed not "progressive," but deeply reactionary.

Earlier studies of American parklands emphasized the exceptionalism of the US case—perhaps the most crowd-pleasing argument environmental historians have ever managed to muster.[65] By recovering the "hidden histories" through which workscapes were enclosed and transformed into sacred wilderness playgrounds, Warren, Jacoby, and others have persuasively contextualized struggles over poaching in the late nineteenth and early twentieth-century United States within a long arc of international and transnational class conflict, stretching from the Black Laws of early eighteenth-century England to recent struggles between local land users and state power in present-day Africa, Asia, Oceania, and South America, as well as in Europe and North America.

Historians must be careful, though, not to lose sight of the contingencies that have always shaped relationships between local people and state conservation. Consider, for instance, Daniel W. Schneider's "Local Knowledge, Environmental Politics, and the Founding of Ecology in the United States" (2000). In a detailed and persuasive case study of pioneering ecologist Stephen Forbes's work for the Illinois Natural History Survey in the late 1800s and early 1900s, Schneider makes three important claims. First, he finds that Forbes and his fellow researchers drew heavily upon the tacit knowledge of experienced Illinois River fishermen. "Scientists new to a particular area," Schneider explains of field scientists more generally, "depended on local people to provide manual labor but also relied on their knowledge of its animals, plants, and habitats." Even "the equipment the survey scientists used to investigate the ecology of the river was modeled on that of the fishermen." Second, Schneider shows how efforts by local elites to enclose previously public places, such as backwater lakes, shaped the transfer of embodied understandings of the river workscape from fishermen to scientists. "Ecologists' reliance and local knowledge and involvement in local politics," Schneider argues, "jeopardized their assumptions about the objectivity of science and its demarcation from other kinds of knowledge, particularly that held by fishermen and hunters." Finally, Schneider contends that as Forbes became an advocate for the Illinois River fishermen whose understandings of the river he and his scientific colleagues synthesized and systematized, he began to perceive ecology as an avowedly practical and political discipline.

Forbes forcefully articulated this perspective in his 1921 presidential address to the influential Ecological Society of America, in which he defined ecology "not as 'an academic science merely' but, rather, 'that part of every other biological science which brings it into immediate relation to human kind.'" Forbes urged his fellow scientists to pursue what he called the "humanizing of ecology." Practitioners, he felt, must incorporate a concern for human welfare into their studies: "If people were part of nature—which Forbes thought self-evident—then ecology, as the science that examines the interactions of animals and plants with each other, was uniquely able, of all the biological sciences, to address practical problems."[66] Like other work on tacit knowledge, Schneider's study complicates simplistic understandings of science. Forbes and his colleagues, after all, not only relied upon the expertise of local fishermen; they also came to empathize with their informants, and even then to join with them in political and legal struggles against elite landowners along the river.

As the conflicts elucidated by Schneider, Warren, Jacoby, and others were unfolding across the forests and streams of the American countryside, critical stretches of the American coastline also witnessed mounting conflict between elite Anglo-Americans seeking refuge from industrial ills in wild nature and a polyglot population of workers toiling to make a living by harvesting and processing fish and other creatures extracted from the sea. Arthur McEvoy's *Fisherman's Problem* (1986) signaled the emergence of a rich body of scholarship on encounters between fisherfolk and government conservationists.[67] North American fisheries had long served as important crucibles for working-class identities, communities, and solidarities.[68] Subsequent scholarship by Jay Taylor, Lissa Wadewitz, and a host of others has compellingly analyzed fisheries workscapes of the nineteenth and twentieth centuries.[69] Such studies have revealed how capitalists, workers, and nation-states struggled in the context of declining fish populations. In her study of the marine borderlands of Washington State and British Columbia, for instance, Wadewitz enlists "social banditry" (another concept whose lineage runs from the great British social historians through Richard White) to make sense of clashes pitting the renegade fisherfolk of the Salish Sea against US and Canadian fisheries managers.[70] Wadewitz's work joins that of other fisheries historians to show how labor, identity, nature, and power became inextricably intertwined in marine workscapes.

Connie Chiang's masterful *Shaping the Shoreline* (2008) examines the simultaneous, often contentious emergence of an active tourist trade alongside the rise of industrial fisheries on California's Monterey Peninsula. "Because these enterprises sought to transform the same stretch of sand and sea," Chiang explains, "they often struggled for dominance. Even when they were not at odds, they were routinely defined in relation to one another." Over time, Chiang argues, "both industries—and the associated categories of labor and leisure, work and play—developed in reference to shifting notions of race, ethnicity, and class." The peoples of the shoreline associated different racial, ethnic, and national groups with distinct "activities in nature." In consequence, "the physical changes associated with the fisheries and tourism became markers of social difference." Competition over ocean resources intensified as Monterey's peoples became increasingly heterogeneous in the late 1800s and early 1900s. Italian, Portuguese, and other "white" fishermen, for instance, slashed the nets of Chinese fishermen while casting Asians "as destructive fishers" because they "use[d] trawl lines to take bottom fish." As Chiang demonstrates, "white" workers, like hotel and railroad owners and small businessmen, adroitly deployed "markers of social difference" "as tools to limit access to nature and assert their dominance over the coastline *and* other groups of people." Marine biologists, meanwhile, blamed the massive and luxurious Hotel Del Monte for dumping raw sewage straight into the bay, polluting the very beaches that drew tourists while destroying fish and shellfish populations.

One of the most interesting lessons of Chiang's story of Monterey, however, is that neither tourism nor fishing scored a total victory in the knock-down, drag-out battles between them. Tourism and fishing developed side-by-side, and they endured alongside each other for more than a century—from the peninsula's integration into American railroad networks through the post-World War II collapse of the sardine fishery. Even after many fishermen left for more prolific waters or grappled with unemployment, "city planners and developers" determined that Monterey's "fading industrial ambience could attract tourists." The old fishing port's "decayed built environment" served to accentuate "Monterey's nature even as it stood for its exploitation." Focusing on the transformation of the old Hovden fish plant into the Monterey Bay Aquarium, a world-class tourist attraction that opened in 1984, Chiang notes a crucial irony: on California's Central Coast (and presumably many other places, too), "Working landscapes were often tourist attractions; tourist attractions were also working landscapes. Those who played in nature," then, "were not separate from the world of labor; work, in fact, made their leisure possible."[71]

Class exclusion, as Raymond Williams well understood, has also shaped representations of American nature—not just on the nation's coastal margins, but in its agricultural heartlands. In *Lie of the Land* (1996), geographer Donald Mitchell argues: "California has not so much been discovered as *made*—and not only in the imagination." Integrating the theoretical insights of such neo-Marxians as Williams, Henri Léfebvre, and David Harvey with selective historical case studies spanning from the Wheatland Riot of 1913 through the Great Depression, Mitchell emphasizes the importance of material production—work—in the representation and misrepresentation of America's most productive farmlands. "By examining the labor geography of a place—how it is that labor is spatially and socially organized (through struggle) and how it quite literally makes places," Mitchell contends that scholars "can better describe and analyze the labor history." Ultimately, Mitchell posits a "labor theory of landscape" that opens up new perspectives on American workers' "unceasing history of struggle."

Demonstrating how the central role of human labor in constructing landscapes came to be obscured, even forgotten, Mitchell's analysis of the Central Valley, like Sackman's analysis of orange marketing, my essay on the erasure of work from Colorado's leisure landscapes during the late 1800s, and several other studies, problematizes visual and textual representations of American places. The landscape gaze, this scholarship demonstrates, has included working people only rarely, and then almost never as real human beings, but instead as symbols or types: the prospector, the hunter, and so forth. Not surprisingly, American labor movements repeatedly allied with sympathetic artists such as Lewis Hine, John Steinbeck, Dorothea Lange, and the California Labor School in the quest to reclaim a place for working people in an American landscape "made," in Mitchell's terms, through labor.[72]

Union campaigns engrossed workplaces and workscapes; they also sometimes involved campaigns by working people to secure more time and space for recreation. The class politics of leisure became particularly contentious in public places of amusement, especially urban parks built to bring city-dwellers of various backgrounds into closer contact with domesticated nature. Colin Fisher, in a fascinating chapter on May Day in late nineteenth- and early twentieth-century Chicago, shows how a traditional European holiday in which rural folk celebrated the advent of spring became an occasion for massive protests in Chicago's parks. The city's workers had long celebrated May Day by picnicking, drinking, and otherwise enjoying the outdoors. After many employers refused to honor a state law implementing the eight-hour workday in 1886, though, more than 80,000 unionists, radicals, and anarchists paraded through the city's streets. A campaign best remembered for its tragic culmination at the Haymarket actually began with song.

Like Montrie's mill girls, Chicago's workers expressed longing for a closer connection to the natural world: "We want to feel the sunshine," they sang. "We want to smell the flowers." The May-Day marchers' lyrics seemed to harmonize with the sentiments of many leading preservationists; in 1898, to cite just the most famous instance, John Muir would proclaim, "Thousands of tired, nerve-shaken, over-civilized people are beginning to find out that going to the mountains is going home." Fisher shows that the well-known workers' mantra— "Eight hours for work, eight hours for rest, eight hours for what we will"—was premised upon the desire of laboring people for "greater leisure-time contact with nature." Even after the execution of the alleged Haymarket conspirators transformed May Day into an international day of celebration and militancy for socialist and anarchist workers, members of Chicago's working classes continued to draw upon "pagan traditions of nature worship" in their efforts to unite "a radical critique of capitalism with a pre-modern or anti-modern sensitivity to the agency of nature."[73]

In the country as in the city, race joined class, region, and consumerism to fragment the US labor movement. Brian McCammack's pathbreaking study of Idlewild, the epicenter of a resort and camp region in lower Michigan that catered largely to African American Chicagoans, reveals a similar class dynamic. "Reformers in the black elite," McCammack argues, "believed that 'better than all the medicine in the world' was a retreat to pastoral environments outside the city." As Idlewild and other rural retreats became more popular in the 1920s and 1930s, self-appointed "race leaders" changed course. Instead of maximizing the number of blacks who would receive the "medicine" to be found in contact with pastoral nature, respectable blacks attempted to render "the more pristine, aesthetically pleasing, desirable, and exclusive environments largely inaccessible to the working classes."[74]

Lawrence Lipin's *Workers and the Wild* (2007) complements the work of Fisher, McCammack, and others on the evolving class politics of recreation in nature. Lipin documents how Oregon's Portland-based union leaders initially used "producerist language" drawn from such sources as Henry George and the Bible to oppose the use of state funds for scenic highways and wildlife conservation. Such sweetheart deals, unionists objected, benefited elite northwesterners instead of the region's working people. By the 1920s, though, many skilled workers had managed to secure shorter hours, longer vacations, and new automobiles. Henceforth, Portland's craft unionists began to pour out of the city. Weekend tourism, Lipin claims, led "labor activists and union members to reassess what they thought about the proper uses of nature." The natural world of forests, rivers, beaches, and mountain ranges became much more than a "source for productive labor"; it also became a site for workers to consume recreational experience. Workscapes thus became playscapes. Yet whether Portland workers experienced the natural world by reeling in salmon, taking down deer, or summiting the high Cascades, they could not leave social tension and class conflict behind. "As soon as workers turned to the great outdoors," Lipin notes, "voices were raised to prevent them from ruining it for others."[75] The increasingly affluent trade unionists of Portland and other urban areas also found themselves ever more at odds in the post-World War II era with some of the same rural workers whose causes they had previously advocated in controversies over natural resource use.[76]

Neil Maher's *Nature's New Deal* (2008), the first environmental history of Franklin Roosevelt's Civilian Conservation Corps (CCC), adapts an approach that is geographically more expansive but chronologically more focused than Lipin's *Workers in the Wild*. In a sweeping and compelling argument, Maher shows how one of the most popular of all New Deal programs helped to transform the American landscape while remolding the bodies and minds of some three million enrollees. For Maher, the New Deal marks an important but underappreciated bridge between progressive conservation and modern environmentalism. Enrollees, drawn primarily from the white working classes of the nation's cities, worked in forests and parks, as well as on farms. Through their labor, as well as through night classes on forestry and other conservation-related topics, the young men of the CCC "gained an understanding of conservationist philosophy and the various techniques for implementing it." The CCC, Maher argues, thus "broadened the conservation movement's composition" and "democratized this movement by continually introducing the theory and practice of conservation to the CCC's working-class enrollees, to residents of local communities situated near Corps camps, and finally to the country as a whole through national media coverage." The CCC became especially popular among European immigrants;

for Poles, Italians, Jews, and others, "labor in nature" for federal work crews offered an opportunity to "becom[e] more 'white,'" garnering support for the New Deal from across the US political spectrum.

By the mid-1930s, though, the CCC faced mounting criticism, especially from wildlife ecologists and wilderness preservationists. Groups like the Emergency Conservation Committee and the Wilderness Society blamed the work Corps enrollees performed for destroying habitat, enabling automobiles to penetrate previously roadless areas, and otherwise harming the natural world. The resulting debates foreshadowed the emerging themes of post-World War II environmentalism—encapsulated by Bernard DeVoto in a 1950 essay as "the balances of Nature, the web of life, the interrelationships of species."[77] Maher has less to say, though, about what the story of the CCC can tell us about the origins of anti-environmentalism among the working classes. Future students of New Deal workscapes would also do well to examine to what extent criticism leveled against the Corps by scientists and activists shaped the class dimensions of postwar conflicts between conservationists, on the one hand, and those dissenting voices that would later coalesce to forge the modern environmental movement.

A final strand of scholarship in the political ecology of class examines what Andrew Hurley calls *Environmental Inequalities* (1995). Historians, Hurley explains, "can attempt to determine who benefited and who suffered when a particular society altered its relationship with the surrounding natural and built environment." Peter Thorsheim's *Inventing Pollution* (2006), in the course of tracing evolving discussions of smoke and pollution in London from the nineteenth through the twentieth centuries, shows how "those in power used their privileged positions to shift the burden of pollution to others." Some even employed "hereditary arguments" to justify the exposure of "poor and working-class people...to higher levels of pollution than those experienced other sections of the population." As one London physician argued in 1905, "Adults engaged in manual labour...do not require pure air such as is found in the country," while "people engaged in mental work, and those whose occupation preclude [*sic*] a sufficient amount of exercise, fail in health unless from time to time they make excursions into the country."[78]

Hurley himself evaluates with admirable nuance the distribution of gains and losses that characterized the rise and decline of Gary, Indiana, a steel-making center stratified by class and race. He finds that "despite the strength of organized labor, the rise of a civil rights movement, and a liberal political order that pledged itself to uplifting the nation's underprivileged, the political process and the dynamics of the marketplace gave industrial capitalists and wealthy property holders a decisive advantage in molding the contours of environmental change." By contrast, "those groups who failed to set the terms—African Americans and poor whites—found themselves at a severe disadvantage, consistently bearing the brunt of industrial pollution in virtually all of its forms: dirty air, foul water, and toxic solid wastes." The last part of Hurley's book shows how working people and the movements that represented them sought to address these environmental inequalities through reforms that extended from the workplace to the workers' communities beyond. In the process, *Environmental Inequalities* reveals the ongoing connections between political ecologies of class and the class politics of environmentalism. As Hurley astutely points out, "it was no coincidence that the age of ecology" unfolded in lockstep with "the rise of environmental inequality."[79]

In October of 1991, an African American activist and organizer from Harlem delivered a rousing rebuke to mainstream US environmental organizations at the inaugural People of Color Environmental Leadership Summit. "For us," Dana Alston explained, "the issues of the environment do not stand alone by themselves. They are not narrowly defined. Our vision of the environment is woven into an overall framework of social, racial, and economic justice." To Alston and her audience, "the environment…is where we live, where we work, and where we play." Alston's purpose was to galvanize the nascent environmental justice movement by reminding the Sierra Club, the National Resources Defense Council, and other mainstream environmental organizations that for working people, the politics of the environment involved much more than wilderness preservation.[80]

Almost four years later, the noted environmental historian William Cronon published an article in the *New York Times Sunday Magazine* that would spawn what some have called "the great new wilderness debate." Cronon's thesis in "The Trouble with Wilderness" overlapped with Alston's. The environmental movement, Cronon declared, suffered from a debilitating preoccupation with an idea of wilderness tainted by romantic ideologies that left humans no rightful place in nature. Environmentalists sometimes succeeded at protecting relatively small, often remote stretches of land; Cronon worried, though, that efforts to preserve these "wilderness" tracts actually led to the sacrifice of lands which failed to meet the exacting statutory requirements of the 1964 Wilderness Act. The latter parts of Cronon's essay revealed the scholar's fundamental support for a vision of environmental health premised upon the achievement of a just and equitable society. Alas, Cronon broached his support for environmental justice only after having spent much of his essay characterizing "the environmental movement" as a singular entity defined chiefly by its obsession with wilderness. This may have been good strategy: mainstream environmentalists surely merited much of the criticism Cronon directed toward them. Yet the trouble with "The Trouble with Wilderness" was that the article's rhetorical strategy unwittingly wrote Dana Alston and the entire environmental justice movement out of the history of American environmentalism.[81] This tack elided the long-running engagement of such wilderness thinkers as Robert Marshall and Benton MacKaye with a range of labor questions; Paul Sutter's *Driven Wild* (2002), for example, portrays MacKaye's conservation activism as initially motivated by his growing concern over "the relation between labor unrest and unsustainable patterns of resource extraction."[82] Even more important, Cronon's reduction of environmentalism to wilderness advocacy obscured the many forms of grassroots activism working-class Americans and their unions had mounted in the postwar era.

Fortunately, other scholars were already striving to rectify the long-standing omission of working people from histories of American environmental politics. Some, as we have already seen, revisited political ecologies of class in the late nineteenth and early twentieth centuries to examine how the imposition of conservation and preservation policies hurt poor working people in the nation's hinterlands. Others sought to identify historical precursors to contemporary environmental justice movements. And still others attempted to explore unfolding relationships from the New Deal through the Reagan Revolution between conservationists and environmentalists, on the one hand, and working people and their unions, on the other. Taken together, this body of work offers an important rejoinder to orthodox understandings of environmentalism, labor politics, and anti-environmentalism.

Robert Gottlieb's *Forcing the Spring* (1993) ambitiously treats environmentalism not as a monolith, but rather as an array of overlapping and "complex movements" which sprouted from "diverse roots." Gottlieb places cities, industrial workplaces, women, people of color, and workers at the center of his story. The result is nothing less than a "people's history" of US environmental politics. Throughtout, Gottlieb unearths previously overlooked historical actors and gives them their due, from the chemist and industrial-health reformer Alice Hamilton to consumer advocate Florence Kelley to environmental justice crusader Dana Alston. *Forcing the Spring,* like the scholarship that has followed in its wake, draws particularly promising connections between labor and environmental history in the course of examining progressive conservation and the rise of modern environmentalism in the 1960s and 1970s.[83]

Chad Montrie has usefully divided the political history of work and nature in the postwar era into two broad themes: labor environmentalism and environmental justice. Montrie, elaborating upon Scott Dewey's earlier research, argues that working-class sportsmen's associations and the United Auto Workers (UAW) leadership together played significant roles in shaping Michigan's policies of game conservation and pollution control. Together, UAW president Walter Reuther and rank-and-file workers turned the union into "the leading proponent among organized labor for improvements in environmental quality." By the 1960s, Reuther was tirelessly championing what he called "a decent, wholesome living environment for everyone." Toward this end, the UAW established a conservation department, which lent critical financial support to organizers of the first Earth Day (much as the Carpenters Union, at Stewart Udall's request, had lent its weight to support the Wilderness Act of 1964). Montrie persuasively demonstrates that the UAW constituted "one of labor's foremost advocates for local, state, and national pollution controls." Alas, the faltering position of American automobile manufacturers and Reuther's tragic death (his plane crashed in 1970 while en route to the UAW recreation and conservation center) insured that the UAW would subsequently part ways with environmentalism. The union had done much to inaugurate the movement, yet by the early 1970s, environmentalism no longer seemed to serve the interests of the UAW or its members.[84]

Laurie Mercier's study of the copper smelting center of Anaconda, Montana reinforces Montrie's interpretation of working-class environmental advocacy. Anaconda unionists, concerned about dangerous workplaces and toxic smelter smoke, pressured the United Steel Workers (USW) from the 1950s onward to give environmental concerns a more prominent role in the union's bargaining agenda with management.[85] This push by Anaconda workers dovetailed with similar initiatives undertaken by American unions over the previous decades. The United Steelworkers implored the Public Health Service to investigate the deadly Donora, Pennsylvania smog incident of 1948; the UAW had orchestrated an unsuccessful campaign to stop Detroit Edison from building the Fermi nuclear generating station on Lake Erie; and Oil, Chemical and Atomic Workers (OCAW) from Long Island "collected thousands of baby teeth for the campaign which made strontium 90 contamination a national issue and mobilized public support for the 1963 limited Nuclear Test Ban Treaty." Building upon this foundation, OCAW, the United Mine Workers of America, and the International Association of Machinists began in the late 1960s to forge alliances with the Sierra Club, Friends of the Earth, Environmental Action, and other environmental groups. This coalition between environmentalism and labor brought a raft of legislative triumphs, including the Coal Mine Safety and Health Act of 1969, the Occupational Safety and Health Act of 1970, and the National Environmental Policy Act of 1970.[86] Unfortunately, such workscape reform legislation seems to have fallen into a blank space in the historiographical map, largely overlooked by environmental and labor historians alike.[87]

While scholarship on labor environmentalism has uncovered a forgotten history of advocacy by a handful of powerful industrial unions, the historical literature on environmental justice has also uncovered the abiding importance of workscape health and safety to California farm workers. Linda Nash's *Inescapable Ecologies* (2006), to cite the most prominent example, reconstructs unfolding ideas about disease, landscapes, and human bodies in the Central Valley. Nash demonstrates how vernacular understandings of the intimate relationship between bodily health and the external environment persisted well into the twentieth century, offering an alternative to the deeply reductionist theories of disease favored by scientists and physicians.

Nash's findings regarding pesticides are especially noteworthy. Migrant workers, she shows, had become intimately familiar with various forms of chemical poisoning by the mid-1950s. Rachel Carson's epochal *Silent Spring* (1962) helped to translate the local knowledge of California farm laborers into a broader indictment of pesticides. United Farm Workers (UFW) leaders, though, largely ignored both the complaints of rank-and-file members about pesticides, and the national outrage generated by *Silent Spring*. Finally, in 1969, the UFW decided to make "worker health and safety its primary issue, granting it precedence over long-standing demands for collective bargaining." "We have to come to realize," Cesar Chavez explained, "that the issue of pesticide poisoning is more important today even than wages." By "point[ing] out that the polluted agricultural landscape posed risks not only to farmworkers but also to everyone who passed through it or consumed its products," the UFW, Nash argues, "foreshadowed the environmental justice movement of the 1980s." Environmental justice advocates, Nash demonstrates, used commonplace understandings of diseased environments to contest medical discourses that located the sources of illness and health not in the landscape, but in individual human bodies. "It is not simply that what we think of as 'nature' is really a complex mixture of nature and culture," Nash concludes. "What we call 'human' is similarly mixed. Not only have humans mixed their labor with nature to create hybrid landscapes; nature—already a mixture of human and nonhuman elements—has intermixed with human bodies, without anyone's consent or control, and often without anyone's knowledge."[88] In this sense, the very flesh of California farm workers, like that of Anaconda smeltermen, came to bear indelible imprints of workscape relationships.

Agricultural laborers, of course, were hardly the only group of workers grappling with permeable bodies and hybrid workscapes. Yet even as Chavez was declaring "the unity which the union movement can have with the environmentalists" to be "crucial to our survival," prospects for an enduring alliance between environmentalists and the labor movement were beginning to waiver.[89] The Sierra Club's refusal to support the UFW's call for stronger pesticide regulations in California in the early 1970s revealed the two organizations' differing worldviews and clashing priorities. In subsequent years, the energy crisis and the ensuing economic downturn combined with the divisive politics of nuclear power to expand these cracks into gaping fissures.[90]

No synthesis of labor environmentalism's nadir has yet appeared, nor have any studies attempted to explain the relationship between the eclipse of labor environmentalism and the emergence of environmental justice movements. Piecing together insights from a range of studies, however, helps to illuminate some of the broader factors at play. For starters, large industrial corporations and their conservative allies managed with much success to pin blame for the nation's economic woes on the new environmental regulations implemented in the late 1960s and early 1970s. In truth, the nation's manufacturing heartlands were already suffering from deindustrialization by the 1950s. The auto industry began to abandon Detroit in that decade, while textile mills and other factories had long since shuttered plants in New England and the Midatlantic region to take advantage of lower labor costs and a friendlier political environment in the so-called "right-to-work" states of the US Sunbelt, as well as in northern Mexico and Asia. Though environmental regulations played little if any role in propelling capital flight, corporations and their political allies nonetheless blamed these laws for the plant closures and layoffs that decimated the American Rust Belt in the 1970s.[91] Leaders of embattled unions generally reacted to "jobs vs. the environment" propaganda by removing issues of environmental health and safety from their bargaining agendas.

Deepening economic woes led organized labor to abandon broader issues like air and water pollution. In the summer of 1972, Nixon's secretary of labor, George Schultz, asked AFL-CIO president George Meany about the unions' stance on a clean water bill; Meany reportedly replied that his members "don't want to be put out of jobs by environmental kooks."[92] Meany's remark is all the more notable for its timing: the most powerful man in the American labor movement was already perceiving environmentalism and unionism as competing interests in a zero-sum game.

Then the real trouble started. Historians continue to disagree vociferously about when and why the New Deal coalition and the liberal ideals it embodied began to break apart. This larger historiographical debate has enormous implications for how we make sense of the complex relationships between post-Earth Day environmental movements and an American labor movement that was disintegrating into what one observer accurately called "a movement of movements." The existing literature already outlines some important intersections between labor history and environmental history during the pivotal 1970s. First, organized labor's reputation for corruption was further sullied when profiteering unions such as the Teamsters undercut the painstaking workscape reforms secured in previous years by the UFW and other democratic unions. Second, the OPEC oil crises that began in 1973 revealed just how profoundly American industries and consumers—including, of course, working-class consumers—had come to depend on foreign petroleum. As gas supplies faltered and prices soared, three decades of robust, broadly shared economic gains came to an abrupt halt. Jefferson Cowie and Nick Salvatore argue that 1972 had been "the most egalitarian year in US history, the point on the graph where society's largess was shared most equitably; unemployment was at historic lows, [and] earnings were at their all time high for male wage earners, having climbed an astonishing forty percent since 1960."[93] All that changed in 1973–1974, as "stagflation"—stagnant growth combined with inflation—"sapped the nation's economic strength," as Cowie explains, thus "mark[ing] the end of the postwar boom."[94] For the first time in nearly two generations, the American pie stopped getting bigger.

As workers, unions, and corporations battled to survive an increasingly dire situation, the fiscal strains inaugurated by the energy crises began to drag down the American welfare state. Labor environmentalism, an agenda that had seemed marginal during the 1950s and 1960s, now began to strike the embattled leaders of the AFL-CIO and most of its largest member unions as downright utopian. It should come as no surprise, then, that when self-conscious environmental justice movements began to emerge in such disparate locales as Love Canal, New York, Louisiana's Cancer Alley, and Warren County, North Carolina, such campaigns would stand apart from and sometimes in opposition to organized labor.[95]

Unfortunately, we still know all too little about the roles workscapes played in another critical trend: the shift among many working-class whites from Franklin Roosevelt's liberalism to Ronald Reagan's neoconservatism. Neither labor historians nor students of American political economy have done enough to integrate environmentalism into their interpretations of the all-important 1970s.[96] Most environmental historians, for their part, have tended to skip from heroic narratives of environmentalism's rise in the 1950s and 1960s to the anti-environmentalist backlash of the late 1970s and 1980s, without devoting much thought to the intervening years—perhaps the most critical stretch of a decade that historians have increasingly cast as a key turning point in US and world history. Two important sets of questions seem especially worth exploring. First, what roles did environmentalism play in the cultural politics of the 1970s? To what extent did environmentalism's associations with college students, experts, the ruling classes, protestors, and other undesirable "sixties" types help to erode its support among working people? To what extent did Nixon, Reagan, and other neoconservatives manipulate these associations via superficially populist rhetoric whose real aim was the destruction of the New Deal order? Second, how did neoconservatives, who generally embraced extreme free-market ideologies that cast scorn on virtually any form of environmental regulation by the state, eventually prevail in internal power struggles within the Republican Party against moderate and even liberal GOP stalwarts who self-consciously built upon foundations laid by Teddy Roosevelt, Herbert Hoover, and Richard Nixon?

Whatever the answers to these questions, Reagan rode to victory in 1980, then began a systematic effort to turn back the clock on "liberal" reforms dating to the Great Society, the New Deal, and even the Progressive Era. The labor and environmental movements joined the urban poor and the welfare state as prime targets of Reagan's disingenuous brand of "conservatism." Reagan and his subordinates—most notably James Watt in the Department of the Interior and Ann Burford Gorsuch at the Environmental Protection Agency—decisively reversed a longstanding record of Republican Party support for conservation and environmentalism.

In the course of knocking mainstream environmental organizations onto their heels, the GOP-led backlash against environmentalism leveraged "jobs vs. the environment" into a devastating rallying cry against regulation. A 1982 critique aptly lamented the impact of such "environmental job blackmail." Despite widespread cooperation between labor and environmental groups over the preceding years, a series of "highly publicized conflicts" fostered a "perception that environmentalism and unionism do not mix." Together with the deepening crisis of the American labor movement, these realignments rendered it increasingly difficult during the 1980s for unions and environmentalists to engineer effective alliances on the national level.[97]

It is still too early to look upon the post-1980s era with much clarity. A few themes, though, seem like promising candidates for future histories of work and nature at the end of the twentieth century: the bitter conflicts that erupted between environmental organizations and overwhelmingly white (and male) loggers, ranchers, and fishermen; a countervailing trend in which working people, particularly women and people of color, built grassroots movements that wedded concern for environmental health and social justice; and the cresting of a "green"-"red" coalition in American politics that made international headlines during the World Trade Organization protests of 1999. As historians turn their attention to these and other recent events, they would do well to avoid exceptionalist treatments of American environmental politics; powerful labor-environmentalist coalitions, for example, have emerged in Germany, India, and many other nations, providing some tantalizing possibilities for comparison. Scholars should also endeavor to balance an honest confrontation with the limitations and shortcomings of the environmental and labor movements with the imperative not to provide further succor to the triumphant forces of reaction, whose incessant attacks on the regulatory welfare state have demonstrably injured both the well-being of working-class Americans, and the health of American ecosystems.[98]

The scholarship on work, nature, and history is even more expansive than this chapter has suggested, reaching beyond the US to encompass many parts of the world—a logical and welcome development for a body of literature that has drawn inspiration from Karl Marx, Raymond Williams, and E. P. Thompson, as well as from social and environmental activists across the globe. Though admittedly selective, the portrait I have sketched here is nonetheless complete enough for the reader to judge the extent to which environmental historians of labor have both added to and transformed our understandings of the past. It is difficult to tell whether anyone else in academia or beyond is listening, but there are ample reasons that they should.

By paying more attention to nature, for instance, labor historians can better understand the structure and character of work, as well as the ways in which relationships between working people and the environment shaped conflict with managers, employers, and others. They can ask how inequities of class have been mapped onto lands and bodies. They can explore the strategies by which members of the laboring classes have worked to mitigate, contest, or evade the environmental dimensions of their subordination. And they can investigate the extent to which working people have formulated identities and collective movements not simply in relationship to people, place, and work, but also to other living things, toxins, and various other elements of their environment.

Environmental historians, for their part, have just as much to gain by more fully incorporating labor into the stories they tell. It has taken work to change the land; we can see these changes in a fuller light by examining them from the perspectives of laboring people. Ideas of nature, for instance, have always been varied, complex, and contradictory. Nature has meant many more things to many more people than conventional meta-narratives have generally allowed. So why do some environmental historians persist in treating the capacity to observe, appreciate, and experience a connection to nature as an amenity that only Americans of certain classes or races have been sufficiently privileged to enjoy?

In other ways, too, restoring labor to the land reveals new insights about the politics of conservation and environmentalism. Working-class peoples related to these movements in complicated and contingent ways. In many cases, they strategically allied with activists from more elite backgrounds to secure safer, healthier workplaces and communities. But working people also tended to steadfastly oppose measures such as game laws that jeopardized their livelihoods, not to mention parkland preservation and transportation projects that threatened to divert limited public money toward pet projects of the ruling classes. Environmental historians must be careful not to turn complicated, still-unfolding dynamics into a simplistic cartoon that parrots the questionable conventional wisdom of a "jobs vs. the environment" trade-off.

Broader audiences also have much to learn from histories of work and nature. Environmentalists need to know that men and women of many classes and colors have been fighting for environmental justice for more than a century, that California farmworkers spoke out about the hazards of pesticides before Rachel Carson wrote *Silent Spring*, that major American unions championed game conservation and anti-pollution campaigns in the decades before Earth Day, and that the discomfort of many working-class Americans with environmental regulation draws upon a long history in which well-heeled sportsmen and their allies employed state power to sever the material and symbolic connections that subsistence hunters, freedpeople, and others had formed with the land. Union leaders and organizers need to view environmentalism not simply as an elitist movement bent on destroying jobs and sequestering resources, but also as a potential commitment of rank-and-file workers, as well as a fruitful ally against the broader neoconservative campaign to destroy the protections against unbridled free-marketeering secured at such immense cost by the great American social movements of the nineteenth and twentieth centuries. Historians who have heretofore privileged race and gender as categories of analysis, meanwhile, might want to open their eyes more fully to the world in which we actually live—a world whose inequalities and inequities have grown out of more than two root causes, and developed in more than two dimensions.

Yet for all this, the impulse that might finally thrust interest in histories of work and nature more fully into the public eye seems destined to come, if it comes at all, not from the halls of academe, but instead from the workscapes and homes, the bodies and minds of those who possess neither the luxury nor the curse of single-mindedly pursuing scholarly vogues. As we historians read and write and dither, the creative destruction of the world unfolds, terrible and grandiose. With each passing moment, it nudges us a little more urgently, daring us to notice where we are while demanding that we remember how we really got here. The work we have done, the landscapes we have changed—the histories of struggle many of us feel in our bones and hear on the wind—some day these stories must cry out in full voice. Only then will Raymond Williams' "single question" again move like light.

NOTES

1. Raymond Williams, *The Country and the City* (New York: Oxford University Press, 1973); Raymond Williams, "Ideas of Nature," in *Problems in Materialism and Culture: Selected Essays* (London: NLB, 1980), 67–85.
2. Williams, *Country and the City,* 292.
3. Ibid., 302.
4. Daniel Rodgers, *Age of Fracture* (Cambridge, MA: Belknap Press of Harvard University, 2011).
5. Williams, *The Country and the City* (New York: Oxford University Press, 1973), 305.
6. Ibid., 141.
7. Ibid., 306.
8. Dissatisfied with the alternatives then available, I coined the term "workscape" while working on the dissertation that became *Killing for Coal* (2008). Thomas G. Andrews, *Killing for Coal: America's Deadliest Labor War* (Cambridge, MA: Harvard University Press, 2008). Scholarship informing the workscape idea includes: Richard White, "Are You an Environmentalist or Do You Work for a Living? Work and Nature," in *Uncommon Ground: Toward Reinventing Nature,* ed. William Cronon (New York: W. W. Norton, 1995); Elaine Scarry, *The Body in Pain: The Making and Unmaking of the World* (New York: Oxford University Press, 1985); Nancy Quam-Wickham, "Rereading Man's Conquest of Nature: Skill, Myths, and the Historical Construction of Masculinity

in Western Extractive Industries," in *Boys and Their Toys: Masculinity, Class, and Technology in America*, ed. Roger Horowitz (New York: Routledge, 2001), 91–108; Ava Baron, ed., *Work Engendered: Toward a New History of American Labor* (Ithaca, NY: Cornell University Press, 1991); Linda Nash, "Finishing Nature: Harmonizing Bodies and Environments in Late-Nineteenth-Century California," *Environmental History* 8 (2003): 25–52; Conevery Bolton Valencius, *The Health of the Country: How American Settlers Understood Themselves and Their Land* (New York: Basic Books, 2002); Christopher Sellers, "Thoreau's Body: Towards an Embodied Environmental History," *Environmental History* 4 (1999): 486–515; Rosalind Williams, "Nature Out of Control: Cultural Origins and Environmental Implications of Large Technological Systems," in *Cultures of Control*, ed. Miriam R. Levin (Amsterdam: Harwood Academic Publishers, 2000), 41–68; Arthur McEvoy, "The Triangle Shirtwaist Factory Fire of 1911: Social Change, Industrial Accidents, and the Evolution of Common-Sense Causality," American Bar Foundation Working Paper 9315 (Chicago: American Bar Foundation, 1993); Tim Ingold, "Tools, Minds and Machines: An Excursion in the Philosophy of Technology," in *The Perception of the Environment: Essays on Livelihood, Dwelling and Skill* (London and New York: Routledge, 2000), 294–311.

9. Donald Worster, *Rivers of Empire: Water, Aridity, and the Growth of the American West* (New York: Oxford University Press, 1992); Donald Worster, *Dust Bowl: The Southern Plains in the 1930s* (New York: Oxford University Press, 1979); William Cronon, *Changes in the Land: Indians, Colonists, and the Ecology of New England* (New York: Hill & Wang, 1983); Richard White, *Roots of Dependency: Subsistence, Environment, and Social Change Among the Choctaws, Pawnees, and Navajos* (Lincoln: University of Nebraska Press, 1983); Carolyn Merchant, *Biological Revolutions: Nature, Gender, and Science in New England* (Chapel Hill: University of North Carolina Press, 1989). See also the forum on environmental history in *Journal of American History* 76 (1990), which included contributions from Cronon, Merchant, and Worster. Worster elsewhere asserted that "Nature as a real, intrinsically significant autonomous entity gets obliterated, by workers and owners alike, in Marx's onward march of social progress." Worster, *Rivers of Empire*, 26.

10. White, "Are You an Environmentalist or Do You Work for a Living?", 171.

11. Worster, *Rivers of Empire*, 50; Chad Montrie, *Making a Living: Work and Nature in the United States* (Chapel Hill: University of North Carolina Press, 2008), 8. The erasure of labor from the landscape will be explored later in this essay.

12. Jane McAlevey, "Unions and Environmentalists: Get It Together!," *The Nation*, May 7, 2012, http://www.thenation.com/article/167460/unions-and-environmentalists-get-it-together#.

13. The number of union members in the US workforce peaked in 1979, at 21 million; the percentage of American workers who belonged to a union, meanwhile, peaked at 28.3 percent in 1954. Gerald Meyer, "Union Membership Trends in the United States," *Federal Publications*, Paper 174, http://digitalcommons.ilr.cornell.edu/key_workplace/174. In 2010, these figures stood at 14.7 million and 11.9 percent, respectively. AFL-CIO, "Trends in Union Membership," http://www.aflcio.org/joinaunion/why/uniondifference/uniondiff11.cfm. An important exception to the broader waning of the American labor movement involves public-sector unions, which have flourished. See Joseph A. McCartin, "Bringing the State's Workers In: Time to Rectify an Imbalanced US Labor Historiography," *Labor History* 2006 (47): 73–94.

14. Claims of this magnitude are, of course, difficult to substantiate. One sign, though, of the diminution of class as a category of analysis—and of labor history as a field of

study—is the omission of any essay explicitly devoted to labor history in the most recent stab at a synthetic treatment of US historiography, in contrast to its predecessor volume. Leon Fink, "American Labor History," *The New American History,* ed. Eric Foner (Philadelphia: Temple University Press, 1997), 333–52; *American History Now,* ed. Eric Foner and Lisa McGirr (Philadelphia: Temple University Press, 2011).

15. It also arguably encompasses many things that animals have done, since other beings have long served as humankind's tools and allies in transforming nature. For a provocative call to expand labor history to include non-human animals, see Jason Hribal, "'Animals Are Part of the Working Class': A Challenge to Labor History," *Labor History* 44 (2003): 435–53.

16. Williams, *Country and the City,* 290.

17. On modes of representing native men and their work, the large literature on contact and exchange is most instructive. A succinct introduction can be found in Cronon, *Changes in the Land,* 34–53.

18. For a critique of Indian ecological innocence, see Shepard Krech III, *The Ecological Indian: Myth and History* (New York: W. W. Norton, 2000). For a review of the literature on native interactions with the environment, see Louis S. Warren, "The Nature of Conquest: Indians, Americans, and Environmental History," in *Companion to American Indian History,* ed. Philip J. Deloria and Neal Salisbury (Malden, MA: Blackwell, 2002), 287–306. For a recent approach that presents a more complicated range of perspectives, see *Native Americans and the Environment: Perspectives on the Ecological Indian,* ed. Michael E. Harkin and David Rich Lewis (Lincoln: University of Nebraska Press, 2007).

19. Alexandra Harmon, Colleen O'Neill, and Paul C. Rosier, "Interwoven Economic Histories: American Indians in a Capitalist America," *Journal of American History* 98 (2011): 699.

20. Andrew C. Isenberg, *The Destruction of the Bison: An Environmental History, 1750–1920* (New York: Cambridge University Press, 2000).

21. Andrew Isenberg, *Mining California: An Ecological History* (New York: Hill & Wang, 2005), 137, 154, 162. Isenberg devotes virtually no attention to the ecological underpinnings of widespread hostility between Chinese miners and native peoples. Robert F. Heizer, ed., *Destruction of the California Indians: A Collection of Documents from the Period 1847 to 1865 in Which Are Described Some of the Things That Happened to Some of the Indians of California* (Lincoln, Neb.: Bison Books, 1993 [1974]), 287–92.

22. For an extended exposition and critique of this "history of academic segregation," see Harmon, O'Neill, and Rosier, "Interwoven Economic Histories," especially 700–705.

23. Peter Iverson, *When Indians Became Cowboys: Native Peoples and Cattle Ranching in the American West* (Norman: University of Oklahoma Press, 1994); Kathryn Morse, *The Nature of Gold: An Environmental History of the Klondike Gold Rush* (Seattle: University of Washington Press, 2003); Gerald F. W. Ronning, "I Belong in This World; Native Americanisms and the Western Industrial Workers of the World, 1905–1917" (PhD diss., University of Colorado at Boulder, 2002); and Susan Lee Johnson, *Roaring Camp: The Social World of the California Gold Rush* (New York: W. W. Norton, 2000). See also Eric V. Meeks, "The Tohono O'odham, Wage Labor, and Resistant Adaptation, 1900–1940," *Western Historical Quarterly* 34 (2003): 469–489; Colleen O'Neill, *Working the Navajo Way: Labor and Culture in the Twentieth Century* (Lawrence: University Press of Kansas, 2005); William J. Bauer, Jr., "'We Were All Migrant Workers Here': Round Valley Indian Labor in Northern California, 1850–1929," *Western Historical Quarterly* 37

(2006): 43–64; Coll Peter Thrush, *Native Seattle: Histories from the Crossing-Over Place* (Seattle: University of Washington Press, 2007).

24. To give just a few relevant Canadian examples, see John Sutton Lutz, *Makúk: A New History of Aboriginal-White Relations* (Vancouver: University of British Columbia Press, 2008), Mary-Ellen Kelm, *A Wilder West: Rodeo in Western Canada* (Vancouver: University of British Columbia Press, 2011), 81–86. For a thorough if now somewhat dated overview of the Australian literature, see Ann Curthoys and Clive Moore, "Working for the White People: An Historiographical Essay on Aboriginal and Torres Strait Islander Labour," *Labour History* 69 (1995): 1–30. For a more recent synopsis, see Lucy Taska, "What's in a Name? Labouring Antipodean History in Oceania," in *Global Labour History: A State of the Art*, ed. Jan Lucassen (Bern: Peter Lang, 2006), 343–4, 349–50.

25. Consider the temporal scope of an all-too typical recent work: *Rethinking U.S. Labor History: Essays on the Working-Class Experience, 1756–2009*, eds. Donna Haverty-Stacke and Daniel J. Walkowitz (New York: Continuum, 2010). My point here builds upon a critique leveled by Clara Sue Kidwell, "Native American Systems of Knowledge," in *Blackwell Companion to American Indian History*, ed. Deloria and Salisbury, 85–102.

26. For a survey of economic history, see Harmon, O'Neill, and Roser, "Interwoven Economic Histories," 705–711. These authors note that although American Indians figure centrally in recent syntheses and overviews of US economic history during the colonial era, they become increasingly peripheral in works addressing later time periods. In the field of labor history, by contrast, native peoples rarely make a significant appearance even in early chapters.

27. Montrie, *Making a Living*, 10.

28. Slavery continued in much of the North into the 1800s, and white and Indian southerners persistently endeavored to extend the "peculiar institution" to California and other parts of the West, so workscapes of African and African-American slavery extended well beyond the Mason-Dixon line and the compromise lines of the 1800s. See, for instance, Paul Finkelman, "Evading the Ordinance: The Persistence of Bondage in Indiana and Illinois," *Journal of the Early Republic* 9 (1989): 21–51; Stacey L. Smith, *Freedom's Frontier: California and the Struggle over Unfree Labor, Emancipation, and Reconstruction* (Chapel Hill: University of North Carolina Press, 2013).

29. Marcus Rediker rightly notes that in the context of the African slave trade, "war was a euphemism for the organized theft of human beings." *The Slave Ship: A Human History* (New York: Viking, 2007), 99.

30. Ibid., 120.

31. As Rediker explains, the morning meal aboard ship "usually consisted of African food according to the region of origin of the enslaved: rice for those from Senegambia and the Windward Coast, corn for those from the Gold Coast, yams for those from the Bights of Benin and Biafra." Ibid., 237. Rediker's use of the term "African food" suggests how Africans effectively indigenized plants of New World origin. See especially James A. McCann, *Maize and Grace: Africa's Encounter with a New World Crop, 1500–2000* (Cambridge, MA: Harvard University Press, 2007).

32. Judith A. Carney and Richard Nicholas Rosomoff, *In the Shadow of Slavery: Africa's Botanical Legacy in the Atlantic World* (Berkeley: University of California Press, 2009), 4, 60, 56–7, 66.

33. As Rediker points out, "much of the wood for the slavers was hewn by slaves, many of whom had crossed the Atlantic on slave ships. Liverpool shipbuilders even imported pine

from the slave-based colonies of Virginia and Carolina with which to build Guineamen in their own yards. This suggests one of the ways in which the slave trade helped to reproduce itself on an international scale. The ships brought the laborers and the laborers cut the wood to make more ships." *Slave Ship*, 53.

34. Judith Carney, *Black Rice: The African Origins of Rice Cultivation in the Americas* (Cambridge, MA: Harvard University Press, 2001), 2, 77, 85, 89–90, 100, 140. For rejoinders to Carney, See David Eltis, Philip Morgan, and David Richardson, "Agency and Diaspora in Atlantic History: Reassessing the African Contribution to Rice Cultivation in the Americas," *American Historical Review* 112 (2007): 1329–1358; "*AHR* Exchange: The Question of 'Black Rice,'" *American Historical Review* 115 (2010): especially 125–171. For Montogomery, see; David Montgomery, *The Fall of the House of Labor: The Workplace, the State, and American Labor Activism* (New York: Cambridge University Press, 1987), chap. 1.

35. Mart A. Stewart, Jr., "*What Nature Suffers to Groe*": *Life, Labor, and Landscape on the Georgia Coast, 1680–1920*, paperback ed. with a new preface by the author (Athens: University of Georgia Press, 2002 [1996]), xii, 9, 148, 174, 135. Notable works on the environmental history of slavery in other parts of the Americas include Warren Dean, *With Broadax and Firebrand* (Berkeley: University of California Press, 1995) and the many citations to be found online at: http://www.stanford.edu/group/LAEH/html/caribbean.htm.

36. On watermen, see David S. Cecelski, *The Waterman's Song: Slavery and Freedom in Maritime North Carolina* (Chapel Hill: University of North Carolina Press, 2001).

37. Stewart, "*What Nature Suffers to Groe*" 194. For an accessible survey of slavery, reconstruction, and the environment, see Theodore Steinberg, *Down to Earth: Nature's Role in American History* (New York: Oxford University Press, 2002), 71–155.

38. Montrie, *Making a Living*, 34–52 (quotes from 51 and 52); Mart A. Stewart, "Slavery and African American Environmentalism," in "*To Love the Wind and the Rain*": *African Americans and Environmental History*, ed. Dianne D. Glave and Mark Stoll (Pittsburgh: University of Pittsburgh Press, 2006), 9–20; Scott Giltner, "Slave Hunting and Fishing in the Antebellum South," in ibid., 21–36; Joseph P. Reidy, "Obligation and Right: Patterns of Labor, Subsistence, and Exchange in the Cotton Belt of Georgia, 1790-1860," in *Cultivation and Culture: Labor and the Shaping of Slave Life in the Americas*, ed. Ira Berlin and Philip D. Morgan, 138–154; John Campbell, "As 'A Kind of Freeman'? Slaves' Market-Related Activities in the South Carolina Up Country, 1800–1860," in ibid., 243–274; Roderick A. McDonald, "Independent Economic Production by Slaves on Antebellum Louisiana Sugar Plantations," in ibid., 275–302.

39. On animal property and social ties, for instance, see Dylan C. Penningroth, *The Claims of Kinfolk: African American Property and Community in the Nineteenth-Century South* (Chapel Hill: University of North Carolina, 2003).

40. Steven Hahn, "Hunting, Fishing, and Foraging: Common Rights and Class Relations in the Post-Bellum South," *Radical History Review* 26 (1982): 37–64; Stuart A. Marks, *Southern Hunting in Black and White: Nature, History, and Ritual in a Carolina Community* (Princeton, NJ: Princeton University Press, 1991); the spirited debate between Steven Hahn, on the one hand, and Shawn Everett Kantor and J. Morgan Kousser, on the other, in *Journal of Southern History* 59 (1993): 201–266; Nicolas W. Proctor, *Bathed in Blood: Hunting and Mastery in the Old South* (Charlottesville: University of Virginia Press, 2002); and Scott E. Giltner, *Hunting and Fishing in the New South: Black Labor and White Leisure after the Civil War* (Baltimore: Johns Hopkins University Press, 2008).

41. Albert G. Way, *Conserving Southern Longleaf: Herbert Stoddard and the Rise of Ecological Land Management* (Athens: University of Georgia Press, 2011), 13–16, 27, 53.

42. Thomas Rogers, *The Deepest Wounds* (Chapel Hill: University of North Carolina Press, 2010), 8, 72, 81, 89, 12–13.

43. Brian McCammack, "Recovering Green in Bronzeville: An Environmental and Cultural History of the African American Great Migration to Chicago, 1915–1940" (PhD diss., Harvard University, 2012), 68–75.

44. William Cronon, "On the Shore between Work and Play," foreword to Connie Chiang, *Shaping the Shoreline: Fisheries and Tourism on the Monterey Coast* (Seattle: University of Washington Press, 2008), xi.

45. Cassandra Y. Johnson and Josh McDaniel, "Turpentine Negro," in *"To Love the Wind and the Rain,"* ed. Glave and Stoll, 51–52.

46. Carney, *Black Rice,* 136. The revisionist literature on New England farmers arguably focuses on work, yet labor processes, class dynamics in the countryside, and similar themes are mostly absent from Richard Judd, *Common Lands, Common People: The Origins of Conservation in New England* (Cambridge, MA: Harvard University Press, 1997); Brian Donahue, *The Great Meadow: Farmers and the Land in Colonial Concord* (New Haven, CT: Yale University Press, 2004).

47. Mark Fiege, *Irrigated Eden: The Making of an Agricultural Landscape in the American West* (Seattle: University of Washington Press, 1999), 117–142.

48. David Igler, *Industrial Cowboys: Miller & Lux and the Transformation of the Far West, 1850–1920* (Berkeley: University of California Press, 2001), 124, 144.

49. Douglas Cazaux Sackman, *Orange Empire: California and the Fruits of Eden* (Berkeley: University of California Press, 2005), 9, 89, 121–122, 126. California orange crates represented just one part of a larger set of designs on boxes, posters, advertisements, and so forth that "vividly linked nature, work, and trade" in the British Empire and other economic entities. William Beinart and Lotte Hughes, *Environment and Empire,* Oxford History of the British Empire Companion Series (New York: Oxford University Press, 2007), 228.

50. Montrie, *Making a Living,* 135 fn. 3.

51. Arthur F. McEvoy, "Working Environments: An Ecological Approach to Industrial Health and Safety," *Technology and Culture* 36 (supplement 1995): S145-S172; Alan Derickson, *Workers' Health, Workers' Democracy: The Western Miners' Struggle, 1891–1925* (Ithaca, NY: Cornell University Press, 1988); Alan Derickson, *Black Lung: Anatomy of a Public Health Disaster* (Ithaca, NY: Cornell University Press, 1998); Jacqueline Karnell Corn, *Response to Occupational Health Hazards: A Historical Perspective* (New York: Van Nostrand Reinhold, 1992); David Rosner and Gerald Markowitz, *Deadly Dust: Silicosis and the Politics of Occupational Disease in Twentieth Century America* (Princeton, NJ: Princeton University Press, 1991); Claudia Clark, *Radium Girls: Women and Industrial Health Reform, 1910–1935* (Chapel Hill: University of North Carolina Press, 1997); Christopher C. Sellers, *Hazards of the Job: From Industrial Disease to Environmental Health Science* (Chapel Hill: University of North Carolina Press, 1997); Mark Aldrich, *Safety First: Technology, Labor, and Business in the Building of American Work Safety, 1870–1939* (Baltimore: Johns Hopkins University Press, 1997); Robert Gottlieb, *Forcing the Spring: The Transformation of American Environmentalism* (Washington, DC: Island Press, 1993); Angela Gugliotta, "Class, Gender, and Coal Smoke: Gender Ideology and Environmental Injustice in Pittsburgh, 1868–1914,"

Environmental History 5 (2000): 165–193; Barbara Allen, *Uneasy Alchemy: Citizens and Experts in Louisiana's Chemical Corridor Disputes* (Cambridge, MA: MIT Press, 2003); and Steinberg, *Down to Earth,* 138–157. In a shadow of things to come, workers such as locomotive firemen sometimes found their work practices and even their livelihoods threatened by measures such as anti-smoke ordinances; David Stradling, "Dirty Work and Clean Air: Locomotive Firemen, Environmental Activism, and Stories of Conflict," *Journal of Urban History* 28 (2001): 35–54. Studies of industrial safety do not always take environmental history seriously; see, for instance, John Fabian Witt, *The Accidental Republic: Crippled Workingmen, Destitute Widows, and the Remaking of American Law* (Cambridge, MA: Harvard University Press, 2006).

52. The same fault applies even to the most recent scholarship on industrial health and safety, i.e. Christopher Sellers and Joseph Melling, eds., *Dangerous Trade: Histories of Industrial Hazard across a Globalizing World* (Philadelphia: Temple University Press, 2012).

53. A regrettable consequence of the choices I have made in organizing this essay is that I have too cleanly severed agricultural and industrial work. The growing use of machines, chemicals, geographic information systems, and so forth made many parts of the American countryside as "industrial" as the most modern factory, depending on how one defines that most elusive of historical concepts, "industrialization." See, for instance, Deborah Fitzgerald, *Every Farm a Factory: The Industrial Ideal in American Agriculture* (New Haven, CT: Yale University Press, 2003).

54. Kathryn Morse, *The Nature of Gold: An Environmental History of the Yukon Gold Rush* (Seattle: University of Washington Press, 2003), 92.

55. Ibid., 88, 92, 94.

56. Chad Montrie, "'I Think Less of the Factory: Than of My Native Dell': Labor, Nature, and the Lowell 'Mill Girls,'" *Environmental History* 9 (2004), http://www.historycoop-erative.org/journals/eh/9.2/montrie.html. A version of this article appears as chapter 1 in Montrie, *Making a Living.*

57. Andrews, *Killing for Coal.*

58. Myrna Santiago, *The Ecology of Oil: Environment, Labor, and the Mexican Revolution, 1900–1938* (New York: Cambridge University Press, 2006), 1–8, 148, 185, 187, 230. Santiago notes that outsiders in the nineteenth and early twentieth centuries typically viewed the Huasteca "as devoid of 'real men'—men who worked, sweated, raised crops, and produced culture" [36]. Like their counterparts to the north, in other words, newcomers to the region wrote native peoples and native labor out of the landscape.

59. Montrie, *Making a Living,* 6.

60. Virginia J. Scharff, "Man and Nature! Sex Secrets of Environmental History," in *Seeing Nature through Gender,* ed. Virginia J. Scharff (Lawrence: University Press of Kansas, 2003), 3, 9, 12; Claudia Clark, *Radium Girls: Women and Industrial Health Reform, 1910-1935* (Chapel Hill: University of North Carolina, 1997); Suellen M. Hoy, *Chasing Dirt: The American Pursuit of Cleanliness* (New York: Oxford University Press, 1995). Recent steps in the right direction include Jennifer L. Morgan, *Laboring the Land: Reproduction and Gender and New World Slavery* (Philadelphia: University of Pennsylvania Press, 2004), especially ch. 5; and Nancy C. Unger's new *Nature's Housekeepers: American Women in Environmental History* (New York: Oxford University Press, 2012).

61. White, "'Are You an Environmentalist Or Do You Work for a Living?,'" 174. In addition to references in this paper on sport hunting, sport fishing, and parks, see Annie Gilbert Coleman, *Ski Style: Sport and Culture in the Rockies* (Lawrence: University Press of

Kansas, 2004); Michael J. Yochim, *Yellowstone and the Snowmobile: Locking Horns over National Park Use* (Lawrence: University Press of Kansas, 2009); and Joseph E. Taylor III, *Pilgrims of the Vertical: Yosemite Rock Climbers and Nature at Risk* (Cambridge, MA: Harvard University Press, 2010). But Taylor frequently diverges from White in the greater attention he devotes to the contexts which shaped climbers' "sporting labors" (6). He notes, for instance, that Alaska's Mount McKinley was first summited in 1910 by a party of miners; emphasizes how English and American gentlemen climbers erased the work guides performed; and argues that Yosemite's Beat climbers comprised "the first generation of outdoor athletes to make a living from play"; 15, 24, 29, 175, 188. At the same time, though, Taylor argues that though "climbing was exceptional in that lives really could hang in the balance...it was still a form of risk that was both totally voluntary and had no purpose beyond self-satisfaction"—a conclusion that could hardly be applied to most forms of labor (105).

62. J. R. McNeill, *Something New Under the Sun: An Environmental History of the Twentieth-Century World* (New York: W. W. Norton, 2000). Chad Montrie's *Making a Living* begins to tackle the environmental history of modern work, yet in his twentieth-century case studies he mostly shifts away from the labor process and workers' experiences, instead focusing on their leisure and political activism.

63. Karl Jacoby, *Crimes against Nature: Squatters, Poachers, Thieves, and the Hidden History of American Conservation* (Berkeley: University of California Press, 2001), 3, 195.

64. Louis S. Warren, *The Hunter's Game: Poachers and Conservationists in Twentieth-Century America* (New Haven, CT: Yale University Press, 1997), 14, 177; John F. Reiger, *American Sportsmen and the Origins of Conservation* (New York: Winchester Press, 1975). Other important studies of American hunting and conservation include Benjamin Heber Johnson, "Subsistence, Class, and Conservation at the Birth of Superior National Forest," *Environmental History* 4 (1999); Mark David Spence, *Dispossessing the Wilderness: Indian Removal and the Making of the National Parks* (New York: Oxford University Press, 1999).

65. Witness, for instance, the feel-good public response to documentarian Ken Burns's recent "The National Parks: America's Best Idea," which overwhelmingly hewed to the storylines first imported into the academy from the conservation and environmental movements by scholars such as Roderick Nash and Alfred Runte.

66. Daniel W. Schneider, "Local Knowledge, Environmental Politics, and the Founding of Ecology in the United States," *ISIS: Journal of the History of Science in Society* 91 (2000): 684, 691, 702.

67. Arthur McEvoy, *The Fisherman's Problem: Ecology and Law in the California Fisheries, 1850–1980* (New York: Cambridge University Press, 1986).

68. Daniel Vickers, *Farmers & Fishermen: Two Centuries of Work in Essex County, Massachusetts, 1630–1850* (Chapel Hill: University of North Carolina Press, 1994).

69. Wayne M. O'Leary, *Maine Sea Fisheries: The Rise and Fall of a Native Industry, 1830–1890* (Boston: Northeastern University Press, 1996); Joseph E. Taylor III, *Making Salmon: An Environmental History of the Northwest Fisheries Crisis* (Seattle: University of Washington Press, 1999); David F. Arnold, *Fishermen's Frontier: People and Salmon in Southeast Alaska* (Seattle, University of Washington Press, 2008); Matthew G. McKenzie, *Clearing the Coastline: The Nineteenth-Century Ecological and Cultural Transformation of Cape Cod* (Hanover, NH: University Press of New England, 2010); Brian J. Payne, *Fishing a Borderless Sea: Environmental Territorialism in the North Atlantic, 1818–1910* (East Lansing, MI: Michigan State University Press, 2010); and Lissa K. Wadewitz, *The*

Nature of Borders: Salmon, Boundaries, and Bandits on the Salish Sea (Seattle: University of Washington Press, 2012).

70. Wadewitz, *Nature of Borders,* especially chs. 4 and 5. On social banditry, see Eric Hobsbawm, *Primitive Rebels: Studies in Archaic Forms of Social Movement in the 19th and 20th Centuries* (Manchester: Manchester University Press, 1959), published in the US as *Social Bandits and Primitive Rebels* (Glencoe, IL: Free Press, 1960); *Bandits* (London: Liedenfeld and Nicholson, 1969); Richard White, "Outlaw Gangs of the Middle Border: American Social Bandits," *Western Historical Quarterly* 12 (1981): 387–408.

71. Chiang, *Shaping the Shoreline,* 6–7, 9, 15, 133. For more on this theme, see above, especially Taylor, *Pilgrims of the Vertical.*

72. Don Mitchell, *The Lie of the Land: Migrant Workers and the California Landscape* (Minneapolis: University of Minnesota Press, 1996), 1–9; Thomas G. Andrews, "'Made By Toile'? Tourism, Labor, and the Construction of the Colorado Landscape, 1858–1917," *Journal of American History* 92 (2005): 837–863; *At Work: The Art of California Labor,* ed. Mark Dean Johnson (Berkeley: Heyday, 2003). A key to understanding the role nature plays in the thought of Mitchell and other critical geographers is Mitchell's emphasis on "landscape morphology," a definition that implies a sort of static materiality quite different from the more nuanced and dialectical approach of environmental historians such as Mark Fiege and Katherine Morse. Note also Mitchell's preoccupation with "space," which connotes a sort of field, stage, or set of coordinates rather than a dynamic and living world capable of exerting agency within human history.

73. Colin Fisher quotes from unpublished draft supplied via personal communication from Fisher's book project, *Urban Green: Working-Class Nature Tourism in Industrial Chicago* (Chapel Hill: University of North Carolina Press, under contract); Muir, "The Wild Parks and Forest Reservations of the West," *Atlantic Monthly* 81 (January 1898), 15, later republished in slightly revised form in *Our National Parks* (Boston: Houghton, Mifflin, 1901), 1; Roy Rosenzweig, *Eight Hours for What We Will: Workers and Leisure in an Industrial City, 1870–1920* (New York: Cambridge University Press, 1983). The classic work on these topics is Roy Rosenzweig and Elizabeth Blackmar, *The Park and the People: A History of Central Park* (Ithaca, NY: Cornell University Press, 1998).

74. McCammack, "Recovering Green in Bronzeville," 311–322.

75. Lawrence M. Lipin, *Workers and the Wild: Conservation, Consumerism, and Labor in Oregon, 1910–1930* (Urbana: University of Illinois Press, 2007), 89, 112. Ending his story roughly where Sutter's *Driven Wild* begins, Lipin reinforces Sutter's path-breaking exploration of an important tension in American environmental politics: a see-saw struggle between popularization and exclusion whereby different groups of people have sought to reconcile the nation's democratic promise with the desire to minimize human impact on wilderness preserves.

76. Taylor's *Making Salmon,* ch. 6, reinforces Lipin's portrait of a growing urban-rural divide among Oregon workers on fisheries issues.

77. Neil M. Maher, *Nature's New Deal: The Civilian Conservation Corps and the Roots of the American Environmental Movement* (New York: Oxford University Press, 2008), 91, 10–11, 109; DeVoto, "Shall We Let Them Ruin Our National Parks, *Saturday Evening Post,* 1950, quoted in ibid., 224.

78. Peter Thorsheim, *Inventing Pollution: Coal, Smoke, and Culture in Britain since 1800* (Athens: Ohio University Press, 2006), 75.

79. Andrew Hurley, *Environmental Inequalities: Class, Race, and Industrial Pollution in Gary, Indiana, 1945–1980* (Chapel Hill: University of North Carolina Press, 1995), xiv, xiii, 172.

80. Quoted in Robert Gottlieb, *Forcing the Spring: The Transformation of the American Environmental Movement* (Washington, DC: Island Press, 1993), 5.

81. William Cronon, "The Trouble with Wilderness, or, Getting Back to the Wrong Nature," *New York Times Sunday Magazine*, August 13, 1995, 42–43. The fuller version of Cronon's essay was published under the same title with comments by other scholars and a response by Cronon in *Environmental History* 1 (1996): 7–55. On the ensuing debates, see *The Great New Wilderness Debate*, ed. J. Baird Callicott and Michael P. Nelson (Athens: University of Georgia Press, 1998); *The Wilderness Debate Rages On: Continuing the Great New Wilderness Debate*, ed. Michael P. Nelson and J. Baird Callicott (Athens: University of Georgia Press, 2008). As Andrew Hurley pointed out, "Simply measuring commitment to environmental reform against a middle-class standard is inadequate" compared to an understanding of "the full range of social responses to pollution." *Environmental Inequalities*, 12.

82. Paul Sutter, *Driven Wild* (Seattle: University of Washington Press, 2002), 142–153;

83. Gottlieb, *Forcing the Spring*. One anecdotal piece of evidence that suggests the success of Gottlieb and others at recasting the history of American environmentalism was the American Society for Environmental History's establishment of the Alice Hamilton Prize for the best essay on environmental history published outside of the ASEH's journal, *Environmental History*, first awarded in 1997.

84. Montrie, *Making a Living*, 106, 93, 103; Scott Dewey, "Working for the Environment: Organized Labor and the Origins of Environmentalism in the United States, 1948–1970," *Environmental History* 3 (1998): 45–63. On UAW support for Earth Day, see Nelson Institute for Environmental Studies, University of Wisconsin-Madison, "Gaylord Nelson and Earth Day: The Making of the Modern Environmental Movement," http://www.nelsonearthday.net/earth-day/coalition.htm. The AFL-CIO joined the UAW as the first organizations to pledge funding for Earth Day. On Udall, the Carpenters Union, and the Wilderness Act, see Taylor, *Pilgrims of the Vertical*, 274.

85. Laurie Mercier, *Anaconda: Labor, Community, and Culture in Montana's Smelter City* (Urbana: University of Illinois Press, 2001), 193–201.

86. For a useful if incomplete overview, see Brian K. Obach, *Labor and the Environmental Movement* (Cambridge, MA: MIT Press, 2004), ch. 1. See also Richard Kazis and Richard L. Grossman, *Fear at Work: Job Blackmail, Labor and the Environment* (New York: Pilgrim Press, 1982), 243; Robert Gordon, "Working Class Environmentalism and the Labor-Environmental Alliance" (PhD diss., Wayne State University, 2003); Hurley, *Environmental Inequalities*, 10–12. Revealingly, industrial health and safety joined other environmental issues as key planks of several of the insurgent democratic union movements of the early 1970s; on Black Lung activism among West Virginia's Miners for Democracy and South-Chicago-based USW dissident leader Ed Sadlowski, see Jefferson Cowie, *Stayin' Alive: The 1970s and the Last Days of the Working Class* (New York: New Press, 2010), 31–35, 41.

87. Marco Armiero and Marcus Hall note that "unlike the 'silent spring' of some environmentalist traditions, ecological concerns were hardly represented in Italian social movements of the sixties." Italian "environmentalism had to negotiate a complicated path between a powerful communist party and an elite protectionist culture"; labor unions sometimes

sought a middle path by "experiment[ing] with new discourses and new concerns, such as occupational health and environmental justice." "*Il Bel Paese:* An Introduction," in Armiero and Hall, eds., *Nature and History in Modern Italy* (Athens: Ohio University Press, 2010), 4.

88. Linda Nash, *Inescapable Ecologies: A History of Environment, Disease, and Knowledge* (Berkeley: University of California Press, 2006), 129–165, 209.

89. Chavez quoted in Robert Gordon, "Poisons in the Fields: The United Farm Workers, Pesticides, and Environmental Politics," *Pacific Historical Review* 68 (1999): 51.

90. Ibid.; Montrie, *Making a Living*, 126–8.

91. See, for starters, Jefferson R. Cowie, *Capital Moves: RCA's Seventy-Year Quest for Cheap Labor* (Ithaca, NY: Cornell University Press, 1999).

92. Memo from Schultz to Nixon, quoted in Cowie, *Stayin' Alive*, 158.

93. Jefferson Cowie and Nick Salvatore, "The Long Exception: Rethinking the Place of the New Deal in American History," *International Journal of Labor and Working-Class History* 74 (2008): 20.

94. Cowie, *Stayin' Alive*, 12.

95. Kazis and Grossman, *Fear at Work*, 248; Thomas Sugrue, *The Origins of the Urban Crisis: Race and Inequality in Postwar Detroit* (Princeton, NJ: Princeton University Press, 1996); Montrie, *Making a Living*, 125–127; Gordon, "Poisons in the Fields," 54; Robert Gordon, "Shell No!: OCAW and the Labor-Environmental Alliance, 1968-1984," *Environmental History* 3 (1998): 459–486. Andrew Battista explains that national labor politics in the late 1970s reflected a divide between the AFL-CIO leadership and the construction and building trades unions, on the one hand, and a "bloc of liberal and progressive labor leaders drawn from unions in the manufacturing, public, and private service sectors." "Labor and Coalition Politics: The Progressive Alliance," *Labor History* 32 (1991): 404; Hurley, *Environmental Inequalities*, chapter 4.

96. Cowie's otherwise excellent *Stayin' Alive* constitutes a prominent and illustrative example: "environmentalism" and related terms do not appear in the index, and environmentalism plays virtually no role in Cowie's analysis.

97. Battista, "Labor and Coalition Politics"; Kazis and Grossman, *Fear at Work*, x.

98. The notion of "cresting" seems particularly appropriate given sociologist Obach's characterization of relationships between labor and environmental groups as buffeted by "waves of cooperation and conflict." *Labor and the Environmental Movement*, 22. In a compelling overview of the green-labor alliance in international politics, Victor Silverman provocatively claims that "Labor thinking about the environment today grows from a profound, if halting, reorientation in trade union ideology about human beings' place in the natural world." Unfortunately, I have found no studies in which scholars situate the US case in this broader context, but one point of coincidence between Silverman and Americanists such as Mercier bears mentioning: labor involvement in environmentalism began less with a concern for occupational health then with "a labor commitment to community and society as a whole," though in practice such a distinction may not have been so evident. Victor Silverman, "Sustainable Alliances: The Origins of International Labor Environmentalism," *International Labor and Working-Class History* 66 (2004): 119, 122.

CHAPTER 17

THE NATURE OF DESIRE

Consumption in Environmental History

MATTHEW KLINGLE

INTRODUCTION: EMPTYING THE BACKPACK

"I'VE been a businessman for almost fifty years. It is as difficult for me to say those words as it is for someone to admit being an alcoholic or a lawyer." With this confession, Yvon Chouinard opened his 2005 memoir, *Let My People Go Surfing: The Education of a Reluctant Businessman.* Founder and owner of Patagonia', the planet's best known and most emulated outdoor clothing manufacturer, Chouinard had long regarded businessmen as "greaseballs." He struck this pose at a young age. After graduating from high school in 1956, Chouinard became an itinerant surfer and mountaineer. He made his reputation in Yosemite in the early 1960s, where he and others ascended the world's most difficult routes and styled themselves "the Valley Cong," dodging rangers, scarfing leftovers, and defying the park's two-week camping limit at Camp 4. "We were," he recalled, "rebels from the consumer culture."[1]

But, naturally, Chouinard's relationship to consumerism was more complicated than one of simply rejecting it and dropping out. That is the reason why his career is one departure point to explore the environmental history of consumption. Beginning with the epitome of outdoor chic seems to play to stereotype. Yet by unpacking Chouinard's career, we can see how completely this greenest of companies is enmeshed in the modern history of consumption.

Before Chouinard became an outdoor gear producer, he embraced his inner consumer. In the late 1950s, disappointed by the quality and cost of available climbing gear, he forged his own equipment. His Lost Arrow pitons, named after a famous Yosemite feature and branded to emulate a fellow noted gear maker, were soon the most coveted in the sport. His product line expanded from hardware to software, and in 1973 Chouinard named his new spinoff company for a treasured place, at once "far-off, interesting, not quite on the map." "Patagonia" fit because it conjured "romantic visions of glaciers tumbling into fjords, jagged windswept peaks, gauchos, and condors." It evoked Chouinard's wish to make clothing durable enough to survive the worst weather on earth. The name also reflected a marketer's savvy: Patagonia could be "pronounced in every language."[2]

Chouinard's aspirations meshed with the times. Concerns over a diminished planet, its plentitude exhausted by ravenous consumers, had begun to mount. So had the interest in demanding outdoor activities like rock climbing, which were seen as opportunities to reconnect with an imperiled nature. Here, too, Chouinard was a pioneer of sorts. With the publication of his 1972 catalogue for Chouinard Equipment (the predecessor to Patagonia), he helped to launch the clean climbing movement in North America by eschewing rock-eating pitons and bolts—basically spikes or screws hammered or drilled into the rock—in favor of artificial chocks and nuts that used natural features like cracks or holes to protect climbers and aid ascent. As Chouinard and fellow climber, Tom Frost, wrote in the catalog, reliance upon old technologies was deteriorating both "the physical aspect of the mountains and the moral integrity of the climbers."[3]

Chouinard's allegiance to sustainability initially was limited to protecting his mountain playgrounds. Others were taking the broader view. By the early 1990s, the World Commission on Environment and Development, better known as the Brundtland Commission, had defined "sustainable development." Protests over disappearing tropical rainforests, vanishing whales, and an eroding ozone layer led to international political action. Eventually, Chouinard expanded his loyalties. For him, the fulcrum for personal and professional transformation was a period of poor sales that almost destroyed his company. In July 1991, during a national recession, Patagonia laid off 20 percent of its workforce. In the aftermath, Chouinard rebuilt the firm by limiting annual growth, critiquing supply chains, switching to organically grown and recycled materials, encouraging safe and equitable work environments, and promoting corporate donations to grassroots environmental organizations. His wife, Malinda Pennoyer Chouinard, was central in rebuilding the company's social and environmental bottom lines. Patagonia became a pioneer in sustainable business initiatives, and when its profits rebounded—the company had a reported $540 million in sales in 2012 alone—imitators soon followed. By the early twenty-first century, it was a global company with retail fronts from London to Tokyo, plus thriving mail order and online businesses.[4]

Chouinard's standing as both a dirtbag climber and corporate chief opens a neglected storyline in the history of environmentalism: how concern for the earth has grown from the consumption of nature. As he explained in his memoir, he applied his core values of "living a life close to nature" through "what some people would call risky sports" to his business.[5] Superficially, Chouinard can seem like a colossal hypocrite. Read cynically, *Let My People Go Surfing* is an extended advertisement for Patagonia, just as Chouinard Equipment was the quintessence of Chouinard's alpinist ego. The book's tone is a curious hybrid: part *The Lorax* and part Clayton Christensen, the Harvard Business School professor who coined the buzzword "disruptive innovation." Chouinard may not see himself as an environmental prophet, but his fanatical customer base believes otherwise. The label more often roams the Lower East Side of Manhattan than the Haute

Route, yet those urban-bound, fleece-wearing consumers have also ponied up for many far-flung environmental causes championed by the company.[6]

In all its contradictions, *Let My People Go Surfing* is a remarkably fitting medium for thinking about the environmental history of consumption. Before we can explore that history, however, we need to revisit how earlier historians have defined and researched consumption and consumerism. Until recently, most historians have focused on cultural and social explanations. This is not surprising. As David Steigerwald notes, consumer history began with postwar intellectuals decrying the corrosive effects of "mass culture" on North American social, cultural, and political institutions. From David Riesman's "lonely crowd" to William H. Whyte's "organization man" to Marshall McLuhan's "the medium is the message," cultural critics portrayed the public, to quote Dwight Macdonald, as "passive consumers" duped into a Hobson's choice "between buying and not buying."[7] Paradoxically, while Main Street obediently aped the ad men of New York and Toronto, consumer critics also redefined culture as an arena in which "power was manifested, imposed, and challenged." Movies, toys, clothing, or television became subjects worthy of serious inquiry.[8]

By the 1970s, a new generation of scholars reared on Walt Disney, Coca-Cola, and television reruns adopted a more sanguine view. Embracing the cultural turn in the humanities and social sciences, they "debunked mass-culture theory as warmed-over puritanism in pseudo-Marxist garb." Boomer scholars on both sides of the forty-ninth parallel exchanged the term "mass culture" for "popular culture" and imbued consumption with democratic energy. This new generation of scholars peopled their studies with actors who "defended their ethnic, racial, or sexual integrity" through "active engagement" with consumer society. A redefined consumerism became a way to resist, perhaps even overthrow an older order. This emphasis upon human agency and cultural plasticity soon reshaped disciplines as diverse as anthropology, literary theory, communications and media studies, and psychology.[9]

The scholarly attraction was obvious. Consumers could be cast as proto-revolutionaries, what one historian called "the vanguard of history."[10] Liberated from the constraints of mass-culture theory, North American historians debated when and where people first identified as consumers. Some located the origins in the early Enlightenment; others fixed its birth in the late nineteenth century. Jan De Vries suggested the Industrial Revolution was an "industrious revolution" spurred by innovations within households.[11] Other historians projected contemporary concerns backwards, finding "no shortage of cultural nags and scolds." Figures divergent as Adam Smith and Thorstein Veblen turned into "foils for defending popular culture."[12]

In the years since the consumerist turn, scholars have followed several paths of inquiry to unmask the "image making of advertising" and "consumers' social identity and desires."[13] One probes the mass market, advertising, and corporate manipulations of consumer behavior, or reactions against consumerism.[14] Another track meanders through women's history, labor history, immigration history and ethnic studies, discovering how diverse peoples have defined their lives through consumables.[15] A third route merges these concerns to ask how consumer behavior shaped regional and national ideology and political culture.[16] Historians have learned that North Americans have always embraced consumerism with ambivalence, like Yvon Chouinard, relishing its abundance while feeling uneasy, even hostile about goods and services as therapeutic agents.[17]

Environmental historians engaged this consumer history with their own biases. The first generation emphasized the material over the cultural in a critique of capitalism run amok. One salient work was Donald Worster's *Dust Bowl* (1979), which obliquely critiqued consumption as part of a larger system where nature, "seen as capital," became "desanctified and demystified." Using Marx and Weber as analytical frameworks, Worster dammed corporate "suitcase farmers" for swindling smaller landowners, soil conservation scientists for failing to thwart erosion, and federal bureaucrats for mismanaging the land and impoverishing the people who dependent on it. Worster's conclusions meshed well with other early works in environmental history, but the focus remained on primary commodity production—grain farming on the Southern Plains—rather than how wheat got to market or who ate it.[18]

The cultural turn in environmental history sparked a reassessment of capitalism and the salience of consumerism. In his groundbreaking book, *Nature's Metropolis* (1991), William Cronon drew upon Karl Marx's insight that raw resources—trees, grasses, or animals—what he called "first nature," took on economic and cultural exchange value when they became commodities—lumber, grain, or meat—as "second nature." Chicago's explosive growth in the nineteenth century rested upon the city's ability to extract wealth from pine forests in Michigan or wheat fields in the Dakotas. City and hinterland were an interdependent whole. While Cronon extended Worster's analysis of capitalism as a motive force behind consumerism, something he had done previously in his 1983 book, *Changes in the Land*, the individuals that did the consuming were even more invisible than in his study of colonial New England. As with Worster, some critics at the time saw Cronon's work, perhaps unfairly and confusingly, as "(Birken)-stock condemnations of accumulation, and Club of Rome-like denunciations of economic growth."[19]

In contrast, Jennifer Price, author of *Flight Maps* (1999), used irony instead of earnestness to study actual consumers. Price argued that as Americans turned passenger pigeons, snowy egrets, and pink flamingos into meat pies, feathered hats, and plastic ornaments, they also defined which Americans had the right to hunt pigeons, patronize upscale milliners, or decorate lawns. Consumerism thus enabled distinctions based on class and gender by attenuating some bonds to nature. Yet Price's scrutiny sometimes obscured the material consequences of that transference. In emphasizing the pleasures and ironies of consumption, the ecological costs became afterthoughts. If Worster demonized capitalism as negative or Cronon lamented tallying the market's true costs, Price domesticated the beast by chastising her readers to "get real."[20]

Environmental historians today still employ both materialist and cultural approaches in their work. This is important because no historical subject, perhaps, demands integrating materialism with culture more than consumption. It is the crux of our many environmental calamities today: decimated fisheries, denuded forests, diminished biodiversity, scarce or unevenly distributed energy, urban sprawl, pollution, and, of course, climate change. It is the motivation behind environmentalism, just as it was for the earlier conservation movement. And it is a primary force within modern economies, affecting labor practices to commodity prices to equity markets across the globe. The challenges facing environmental historians willing to study consumption are many, especially if they range beyond North American shores.

Four specific challenges face scholars intrepid enough to begin such journeys. First, political economies driven by consumer demand are neither uniquely American nor recent. In the modern age, consumerism was born transnational. By the 1700s, consumers from London to Amsterdam were being schooled in the rules and rewards of consuming. So, too, were their colonial subjects in Bombay or Jakarta, who often crafted their own forms of consumerism during and after European rule.[21] Simply going abroad might not be sufficient, however. As previous histories of consumption have shown, unpacking the global and transnational dimensions of consumer economies that became ever more integrated, rapid, and complex in the twentieth century is no small challenge. Framing such projects transnationally can provide opportunities to break free from the nation-state, provided other boundaries are clearly defined.[22]

Therein lies the second challenge. Capitalism is spatially and temporally complex, hardly an unchangeable motive force. Comparing British mercantile policies with American colonial reactions in the eighteenth century to the Americanization of immigrant workers in Depression-era Chicago may be fruitful, but it can also yield generalizations that say little about the lives of consumers in either period or place. As with any historical subject, context matters.[23]

This leads to a third caveat specific to environmental history: matter also matters. Consumption transforms and is transformed by an unstable physical environment across wide ranges of space and scale: from individual bodies and households to factory or field to supermarket or big box store shelf. Tracing cause and effect can be very difficult. Even environmental historians, who are more attentive to the spatial, sometimes neglect the multidimensional world of human relations with nature.[24] Another problem is nature itself. Animals and plants move, evolve, diminish or go extinct. Pathogens and parasites ravage crops or sicken consumers. Weather and climate play havoc with primary production, affecting supply chains and buying habits. History is filled with examples of consumers' tastes or entire markets changing thanks, in part, to the contingencies of an unstable nature.[25]

Finally, just as historians need to engage with nature, they also need to engage with the basic material processes that link particular products to particular markets and consumers. Technology matters because technology is matter—and politics, too. Tracing the production and consumption of pharmaceuticals, personal computers, and genetically modified organisms requires different methods and concepts than older histories of commodity goods and staple crops. Writing environmental histories of consumption means building still more bridges to science and technology studies. Thankfully, scholars working on this emerging synthesis—commonly called "envirotech"—have already started this important undertaking.[26]

The recent "hybrid turn" within environmental history is at the core of these four challenges.[27] Environmental historians of consumption, like others in the larger field, now try not to fix lines between nature and culture too rigidly. Indeed, the rise of the so-called "ecological consumer" demonstrates how quickly lines have dissolved. Once seen as a selfish vice, consumerism has morphed into an altruistic virtue with whole websites and book series devoted to living well through enlightened environmental consumption. Purchasing organic food, wearing Patagonia clothing, and recycling household waste are statements of cultural identity as much as political declarations. They are also as equally fraught with ambiguities and contradictions. When Woody Allen quipped that he was "at two with nature," he captured our dilemma of venerating nature even as we devour it.[28]

Yvon Chouinard might appreciate such wit: he is a hybrid himself. As a child, he wanted to be a fur trapper because "no young kid growing up ever dreams someday of becoming a businessman." Instead, he grew up to market youthful dreams of outdoor play to desk-bound, urban-dwelling, well-meaning adults.

Chouinard's company and its products thus capture the three themes of this essay: political economies and cultures of consumption, commodity chains and flows, and the embodiment of consumerism as health. The explosive growth of the postwar North American economy and rising interest in outdoor recreation made Chouinard's dual careers as climber and clothier possible. Before the 1950s, neither the technologies to manufacture lightweight, durable equipment, nor the green-minded consumers hungry to buy it, existed. Once Chouinard began worrying over the social and environmental costs of his merchandise, he asked his employees to follow his company's commodity chains and document where and how Patagonia sourced raw materials and assembled its goods. Beginning in 2005, the company launched its Common Threads Initiative to make as much of its clothing recyclable as possible. After discovering that some processes and products were harmful to human or ecological health, like pesticide-grown cotton and waterproof polyvinyl chloride fabric, he pushed to find alternatives, like organic cotton, or eliminate entire product lines altogether. At the foundation of Patagonia's entire existence is an environmental history of consumerism full of contradictions and ironies.

Chouinard has admitted that Patagonia falls short. It will neither be "completely socially responsible" nor ever make "a totally sustainable non-damaging product" in its lifetime. "But it is committed to trying," he concluded.[29] Like Chouinard's corporate manifesto, this essay is a selective survey. It is not an endpoint, but a beginning to a longer trek.

SHADES OF GREEN: POLITICAL ECONOMIES AND CULTURES OF CONSUMPTION

As Raymond Williams observed, the "intellectual separation between economics and ecology" is a hallmark of the modern industrial world.[30] Until the invention of the term "ecology" by Ernst Haeckel in 1866, naturalists used economy to describe the inter-related relationships between living beings and their environment.[31] Reconnecting the two fields has come back into vogue among economists and businesspeople. The divorce often persists among historians, however. Untangling commodity chains back across time and space is one way to mend the rift, but such histories often overlook another hallmark of modernity: how the political economies and cultures of consumption render consumerism a natural social process. Consumption morphs into a way of life, if not living itself.

Not surprisingly, consumption is a key tension within the modern environmental movement, especially in North America. For every prophet of green profits, others rail against the marketplace. Jeremiads against consumption have become rhetorical staples of environmentalism: have one child, cut addiction to fossil fuels, eat low on the food chain, buy local or, better still, buy nothing at all. The rise of "No Impact Man" and his gospel of green consumption, however, has a deeper history that historians are only now exploring.[32] For example, did the emergence of consumer rights alongside environmentalism blunt or redirect efforts to change North American consumerism wholesale? Individual anxieties over cleaner water, organic foods, and safer products today often trump collective hazards to social and ecological well-being. The resulting effects have helped to change the marketplace for safer, greener goods without changing underlying patterns of North American consumption.[33]

Unmasking how the natural becomes social or the social becomes natural thus requires addressing the political economies *and* political cultures of consumption. Historians need to see past the modern state while also seeing like a state because, after all, there is nothing natural about nations.[34] In contemporary social theory about state formation and function, nature is all but absent, while the "massive appropriation of natural resources upon which the modern world depends" unfolds daily, often out of sight.[35]

This blindness begins with consumption. Individually, some consumers can and do try to act in informed ways, but their actions alone cannot overcome the structural barriers against change. Bicycling to work can reduce individual petroleum consumption provided one lives in a bicycle-friendly community. Live in the suburbs or the countryside, and throwing away your car keys can be next to impossible. For other well-intentioned consumers, trying to do their part does not insure that their neighbors or fellow taxpayers will do theirs. Paying extra for municipal curbside recycling or wind-generated electrical power is less attractive if free riders benefit without covering their share. While some historians have started to investigate such debates, we still need more studies of how consumer activism has influenced larger-scale environmental reform.

Historical studies of economic emergencies, when social and environmental book-keeping can sometimes change, are more commonplace in the literature. The Great Depression was one such crisis. Franklin Roosevelt's New Dealers wanted to rescue the nation by reforming its political economy. As during the earlier Progressive Era, New Deal conservation, with its own gospel of efficiency plus equity, was central to state reformation and power.[36] "National welfare," writes Sarah Phillips, "required that rural welfare underlay both New Deal and postwar liberalism."[37] But the New Dealers' ambitions were greater than their Progressive Era ancestors' goals because the problems they faced were larger, more complex, and immediate. Inspired by the economic philosophies of John Maynard Keynes, their reforms spanned geographical and political scales: the contested Agricultural Adjustment Administration, which regulated prices and provided subsidies; the massive hydroelectric and irrigation projects of the Tennessee Valley Authority and Bonneville Power Administration, which generated inexpensive power and regulated unruly waterways; or the Rural Electrification Administration, which brought electricity to impoverished farmlands. Damming rivers, building turbines, electrifying homes, retiring exhausted acreage, and controlling flood-prone rivers stimulated both conservation and wealth redistribution. They were also methods of resource redistribution.

The unanticipated consequence of this "New Conservation," however, was an industrial agricultural system that maximized production to sustain high wages and cheap food, often at the later expense of rural communities and landscapes. Big agribusiness predated the New Deal, as did government support for farmers, ranchers, and other rural producers. From subsidizing irrigated agriculture to sponsoring scientific research into increased crop yields, big agriculture and big government had worked together since the late nineteenth century to harvest the fat of the land for American consumers.[38] After the Second World War, thousands of government experts shipped American agricultural and technical know-how abroad to spur development, boost production, expand consumers' purchasing power, and thwart Communism. The results of Cold War state-building overseas were usually as mixed as the New Deal-era results at home.[39]

The effects of the Great Depression ramified long after economic upheaval subsided. Within industrializing nations, the quest for natural resources was often concurrent with state efforts to incorporate and develop peripheral regions. Earlier networks of resource extraction and consumer markets provided frameworks for later systems to extract other commodities. In the Canadian subarctic, the region's countless lakes and rivers were aquatic highways for generations of *coureur du bois*, beginning in the 1700s, who shuttled furs from Rupert's Land to Montreal and returned with textiles, tobacco, and worked metal to trade with Native brokers. Twentieth-century demand for forest products, fish, precious metals, and uranium ore made the Canadian North valuable again. Beginning in the 1930s, the Canadian government, working closely with private firms, used the resource boom to disassemble the region's raw resources, process them into salable commodities, dispatch the products south, and import industrial technology and consumer goods to the North. This exchange was far from equal and the effects were often catastrophic. Infectious diseases buffeted once-isolated Native communities; heavy metals and toxics polluted waterways; fisheries in Great Slave and Great Bear Lakes almost collapsed; and the replacement of traditional diets with processed foods helped to create new epidemics of chronic diseases like diabetes. As Liza Piper concludes, the "final divorcing of end products from local nature" reflected the choices of state agencies and private firms, rather than the "inevitable consequences" of industrialization and consumption.[40]

Political economies of industrial resource extraction and consumption intimately connect with political ideology as well. Take Venezuela. Following the discovery of oil along Lake Maracaibo in 1922, the survival of the Venezuelan state relied upon the ability of both dictators and democratically elected leaders to wed the nation's "two bodies": the political body of its citizens and the natural body of its rich natural resources. As Venezuela devolved into a modern petrostate, its leaders monopolized political violence as well as natural wealth. "Democracy and dictatorship," anthropologist Fernando Coronil writes, "became two sides of the same oil coin" in what he calls the "magical state."[41]

This magical state could not have emerged without its magicians: the wealthy, powerful Venezuelan elite who collaborated with British and North American petroleum companies. Nor could it have sustained itself without the cravings of modern consumers, who were driving the modern petroleum economy. As historian Miguel Tinker Salas argues, the nationalization of Venezuela's oil industry in 1976 only compounded the contradictions by creating a state within a state. Petróleos de Venezuela was effectively an autonomous holding company for foreign concerns that benefitted foreign consumers. Few benefits from the oil economy before or after nationalization went to Venezuela's poor, and the environmental consequences were dreadful: oil-slicked beaches, contaminated groundwater, and skies smudged by refinery gas flares. By the late twentieth century, Salas concludes, the ability of foreign oil companies to "appropriate a Venezuelan persona" had helped to create a new political economy and culture of entitlement where petroleum financed "the growth of the state, private development and commerce, and the urban middle class."[42]

In this context, it is easy to suggest that North American consumers helped to make a new Venezuela possible. Their consumption, however, did not take place without some significant help. The rise of the Organization of Petroleum Exporting Countries, or OPEC, as a political force in the 1970s helped Venezuela, a member state, to gain power as part of a powerful oligopoly. Petroleum pushed Caracas to align with Washington during the Cold War, attracting lucrative American foreign aid as a result. The particular material properties of oil itself—trapped beneath the earth, volatile and flammable, unevenly distributed across space—meant that foreign investment was especially critical for building Venezuela's production capacity as well as insuring the state's monopoly on political power and violence.[43]

The election of Hugo Chávez in 1999 redirected the wealth of the Venezuelan petrostate without changing the fundamental dynamic between oil and power. Invoking discontent over political corruption tied to oil money, Chávez seized control of Petróleos de Venezuela, transferred more funds into domestic social programs, and redirected foreign policy. Previously, the Venezuelan petrostate existed independently of civil society as a "para-state." Chávez reconstituted the "para-state" as a "meta-state" to bankroll his "Bolivarian" revolution and free the country from United States control. As a result, foreign consumers' endless thirst for petroleum stokes the equally obsessive impulse in Caracas to control the domestic petroleum industry.[44]

At the core of the Venezuelan example is an important point: state linkages to nature have never been natural. Whether they seek to monopolize the production and distribution of natural resources and commodities, or they arrange the political landscape to benefit private enterprise, states structure social relations to nature. As owner or as referee, state power derives from trying to dictate those relationships. Economic and political difficulties reveal how and when political figures reconfigure these relationships. The state can make consumption and production seem like natural processes, but they are not.

Technocratic state-building was never a peculiarly North American phenomenon, nor was the United States the only version of what Lizabeth Cohen calls the "Consumer's Republic." As historians of modern Europe note, many nations depend upon "the standardization of leisure habits and on the creation of iconic vistas" to standardize "nature's cultural meanings." The lines between exploiting social and natural resources blur, becoming "inextricably linked projects" to promote industrial production as well as citizen consumption.[45] Since the Bismarck period, successive German regimes have tethered nation building to natural conquest.[46] The apotheosis of central planning came with the Nazis and their 1935 Reich Nature Protection Law, the most comprehensive policy of its age. The early history of the *autobahn* illustrates the tensions within fascist Germany. National highway engineers outflanked resistance from local preservationists by stressing how roads would restore the land to its "natural" condition while promoting both car sales and automobile-driven leisure. When opposition grew too strident, the state simply seized property and pushed aside opposition. Consuming nature through leisure was seen as a means to cement anti-modern mysticism about unsullied Nature and Nazi racial ideals of a pure German homeland peopled by a pure German *Volk*. Highways and mass-produced automobiles, like the eponymous Volkswagen, were designed to make Germans more German by getting them outdoors.[47]

The ambiguities of Nazi views persist into the present. Postwar currency reform and the Marshall Plan propped up the fragile western half of a divided Germany and distinguished it from the Soviet-occupied eastern zone by promoting consumption. In a 1948 election poster for Karl Adenauer and his Christian Democratic Union (CDU) party, a mother with an overflowing shopping bag pulls a handful of banknotes from her purse as her rosy-cheeked daughter eats an apple. The poster reads *Endlich wieder kaufen können*, or, "Finally we can buy again," and urges *Wahlt CDU*, or, "Vote CDU." As the first Chancellor of West Germany from 1949 to 1963, Adenauer presided over the postwar *Wirtschaftswunder*, or "economic miracle," that remade German culture and society. Federal policies encouraged consumer spending. Employers provided generous vacation benefits. Companies like Volkswagen, owned in part by the state of Lower Saxony, promoted cars and roads to help speed Germans into the countryside. Even when Germans began questioning the dogma of mass consumption, they never fully rejected it. The rise of the Greens in the 1970s and 1980s gave political force to critics of nuclear power and industrial pollution, compelling traditional parties such as the CDU to champion environmental policies. Yet the Greens did not challenge the assumption that Germans could become more efficient consumers by consuming ecologically. Many of the strictest environmental regulations enacted by the European Union following the 1993 Maastricht Treaty had their origins in German domestic politics.[48]

The paired processes of "the nationalization of nature" and "the naturalization of the nation" helped to create postwar mass-consumer society in France as well.[49] The upshot was an ambiguous hybrid, what Michael Bess calls the "light-green society," where French greens and technological enthusiasts together produced "the partial greening of the mainstream, in which neither side emerged wholly satisfied nor utterly dismayed."[50] During the early postwar decades, state planners and engineers extolled the importance of technological and scientific independence to insure French security and prosperity. They promoted civilian nuclear power, the high-speed *Train à Grande Vitesse* or TGV, the supersonic Concorde jet, and enormous public works projects such as the Donzère-Mondragon hydroelectric complex on the Rhône River, calling it a "modern-day Notre-Dame or Chartres." These and other projects sanctioned economic expansion, mass consumption, and postwar Gallic radiance.[51]

By the early 1970s, a nascent French environmental movement was attacking the priests of high modernism and their religion of endless growth. Eventually, the French Greens joined the church, so to speak, while altering the faith only slightly. Unlike their North American counterparts, French Greens venerated the countryside, not wilderness; theirs was a nature long touched by human labor. It was also something they consumed, often ravenously, on weekend jaunts to rural inns and at daily meals filled with fresh cheese and produce. As a result, the French state and its subsidized industries adeptly incorporated the Greens and their rhetoric to appeal to green-minded voters and their appetites. Ultimately, the hard-green critique faded in favor of the light-green compromise. Consumerism sealed the union. Atomic reactors and fast trains were part of the same landscape as homeopathic remedies and artisanal foods. The light-green French decided that they could live well and not go without.[52]

This is a French story, but it has variants throughout the developed world. When faced "with a tough choice between a technological modernity and a green vision," North Americans and French alike "hedged, and in effect chose both."[53] In the United States, the desire to blend modern convenience with back-to-nature primitivism began with the woodcraft and nature study crazes of the Progressive Era. At the dawn of the twentieth century, going into the woods equipped only with a hatchet and know-how, building fires and sleeping in a handmade lean-to, was a way to reject temporarily the trappings of modernity. A century later, being a good steward of the earth meant choosing carefully the right consumer products so you would have no effect whatsoever on the land: aluminum backpack frames, nylon tents, Gore-Tex jackets, and freeze-dried food. As James Morton Turner concludes, in rejecting industrial and consumer society, many Americans today run headlong into its embrace, simultaneously preaching the ethic of "Leave No Trace" when traipsing the trails.[54]

Of course, there were critics of even this more mindful consumerism. During the first Earth Day celebration in 1970, Denis Hayes said pointedly one "can't live an ecologically sound life in America" because the marketplace made "the individual almost powerless."[55] Changing individual shopping habits was not the route to reform, he warned, whereas collective political activism could promote long-term change. Finding an organizing principle, however, proved elusive. Some warned that over-population, particularly in the developing world, would strip the Earth bare. These neo-Malthusians, led by Paul Ehrlich, author of *The Population Bomb* (1968), also railed against overconsumption in developed nations by calling upon everyone to limit their family size. Others, like Barry Commoner, blamed instead an unsustainable economic system and the uneven distribution of wealth for the environmental crisis. Commoner's famous "four laws of ecology," especially his last aphorism—"there is no such thing as a free lunch"—was also a reproach of consumption.[56]

Translating these concerns into political action proved difficult. The same subur-ban women and college students who bolstered the environmental movement also supported consumer rights and benefitted from postwar affluence. The Sixties coun-terculture that helped to spawn the wilderness craze also birthed the *Whole Earth Catalog*, first published in 1968, a hippie stepchild of Sears, Roebuck. The *Catalog's* father, Stewart Brand, a San Franciscan entranced by the futurist Buckminster Fuller, used the catalog to espouse technological optimism, personal development, and social entrepreneurship. According to Andrew Kirk, the *Catalog* was as much political text as shopping list. It was an ink-and-paper ancestor of the Internet, an ever-changing emporium of goods and Californian entrepreneurialism that eventually bred Silicon Valley. By the end of the twentieth century, many Americans, like many Europeans, believed they could have their environment and consume it, too.[57]

Yet as in France and Germany, there was nothing natural or preordained about the American light-green society. Many would contest Kirk's overly sanguine (if prescient) assessment that Brand and his band of merry pranksters pioneered a more pragmatic environmentalism "that lies somewhere between conservative and liberal."[58] Brand and his acolytes imagined a green America that transcended class conflict, but their dreams failed to speak to anyone unable to partake in their communion of personal affluence. Like other Americans of means, his band of happy profiteers gleefully skirted the social divides rooted in class privilege and political ideology. Brand was more Ayn Rand than Henry David Thoreau.

The clash between labor and leisure endemic to industrial and postindustrial society suggests another approach to this history. The origins of this clash lay in the trans-formation of the early conservation movement following the decline of Progressivism. Herbert Hoover, the energetic Quaker reared in Oregon and schooled at Stanford, merged conservation and mass consumerism in the 1920s during his eight years as Secretary of Commerce. He negotiated the Colorado River Compact to provide irriga-tion and power for Southwestern states, expanded the National Park System, reformed the scientific management of the nation's commercial and sports fisheries, and fostered private-state partnerships to promote outdoor recreation. In 1924, discussing plans to build dams on the Colorado River, he suggested that "some waterfalls...could be con-structed with a view to their public availability as scenery." "Human intelligence added to the resources of nature," he said without sarcasm, "was an unstoppable force for good and beauty and order. [59] Hoover, the little engineer that could, believed that tech-nocratic elites could and should merge efficiency and splendor to smooth economic inefficiencies and eliminate class conflict through consumption.

An older concept of producers' rights, tempered in the often-bloody labor fights during the Gilded Age and Progressive Era, still resonated with many workers even as they, too, embraced the gospel of consumerism. The rise of tourist economies through-out the industrializing West and declining northern New England often pitted rural working residents against weekend or seasonal sojourners. [60] In some cases, as Thomas Andrews found in Colorado, the very workers who dug the state's mines also built its tourist infrastructure. Over time, they were pushed to the margins as aspirant fron-tiersmen and mining engineers remade the West into a tourist site. [61] In other cases, workers preserved the primacy of extractive industry over tourism, but only when the economic climate was favorable. During the 1890s, Connie Chiang notes, the propri-etors of the posh Hotel Del Monte expelled the smelly Chinese squid fishery to protect guests' health and happiness, but during the Great Depression, organized labor joined capital to repulse efforts to restrain fish reduction in the name of odor abatement. [62]

Yet extractive industrial workers were also consumers of outdoor recreation. By the 1920s they began to amend their allegiance to producer conservation—setting aside lands and waters for productive industrial use to benefit everyone—to incor-porate consumerist conservation—opening some public lands for recreational use. Blue-collar conservationists advocated for more roads, more camping and hunting grounds, and more hatcheries to rear more fish for more sportsmen. Theirs was a con-servation philosophy situated at a right angle to the emerging wilderness crusade; they wanted protection, but they did not want an exclusive preservation that left little room for workers in the wild. [63] Alliances between rural and urban union workers created some common political ground for working-class conservationism. On the eve of the Great Depression, however, workers in Oregon questioned what Lawrence Lipin called the "producers' republic." Old alliances shifted. The wedge issue was fishing. Thanks to cars, urban Oregon workers now identified with fellow weekend sport anglers and "looked to the state to help reduce the scarcity of both producer and consumer goods." The Oregon landscape, transformed by decades of resource extraction, was now more than "a source of plenty and productivity . . . but also a source of leisure and recreational fulfillment." [64]

But it was middle-class Americans who were the bulwark of the environmental con-sumers' republic. Many had flocked to the exploding postwar suburbs, lured by the promise of spacious homes close to the outdoors. Once settled there, they resisted the city's further advance into the countryside. [65] Consumer entitlements as homeowners translated into political entitlements as environmentalists. The story of Wisconsin's Fox River Valley is an exemplar of this national trend. Paper and pulp mills built the city of Appleton and its environs. Generations of residents could see, smell, and hear the source of prosperity, but some began questioning the cost of progress. As Appleton repositioned itself as a commercial center after World War II, clean electricity and new roads filtered "out the majority of nature's unpleasant qualities while creating new windows of access to its numerous amenities."[66] Fox River Valley residents began seeing themselves as consumers, then environmentalists. They championed efforts to protect new amenities, calling for the reduction or elimination of pulp and paper mill pollution. Consumer society offered "a tenuous foundation for environmentalism," Greg Summers concluded, because while citizens "objected to pollution," they did not decry the "basic direction of industrial and commercial development" that had made outdoor play so convenient and valuable.[67]

Contradictions within the environmental politics of consumption eventually split Americans along class lines, but not as cleanly or as simply as historians have sometimes argued. When middle-class environmentalists pressed for pollution controls or open space protections, they could find blue-collar allies. Howard Zahniser called upon organized labor to help pass the 1964 Wilderness Act and early wilderness legislation; the union leadership delivered. These alliances were often fragile and subject to economic and political pressures. Industrial interests and state agencies exploited workers' anxieties by threatening to ship imperiled jobs overseas or to cut production in response to increased regulation, but even tenuous alliances offered alternatives to the intractable choice between jobs and environment.[68]

In most instances, however, the story of Appleton became the story of North America, if not the world. By the 1980s, a sour and unpredictable economy, plus globalization, helped to undercut what remained of the blue-green alliance. Hard-hatted union members linked arms with sea-turtle-costumed protesters at the 1999 World Trade Organization meeting in Seattle. It was more theater than reality. For some protestors, the Battle of Seattle was a rally against consumerism, not a march for labor rights or endangered species. Protestors targeted storefronts of Nike, The Gap, McDonald's and other corporate franchises during the riots that followed the marches. Their ire could be seen as misplaced, however, since just a short walk from the downtown business district was one of Seattle's iconic global brands: Recreational Equipment, Incorporated, or REI, the outdoor equipment retailer. REI embodied the shift toward purchasing premium green products, just as it embodied the dream of the Emerald City itself—a place where pristine nature and modern conveniences coexisted side by side.[69]

Even Seattle, however, was too built up for some green consumers. Throughout the North American West, an influx of vacation homebuyers to Missoula, Montana, Bend, Oregon, Prescott, Arizona, or the Okanagan Valley of British Columbia pitted "neo-locals" against displaced extractive industrial workers. Former mining and ranching towns became destination resorts, bucolic refuges, and authentic Western retreats for wannabe cowboys.[70] Tourism seemed to promise an environmental ethic that balanced production and consumption. Instead, tourism was a "devil's bargain," as Hal Rothman called it, in which tourist communities relied increasingly upon cheap immigrant labor from Latin American and Africa, as home and retail developments devoured nearby wildlife habitat or pushed into hazardous fire-prone terrain. As conservation ecologist William Jordan aptly put it, "ten thousand Thoreaus" and their holiday homes were doing as much damage as the most rapacious logging or mining company.[71] By the end of the twentieth century, those fed up with the high price of out-door living in North America were lured still further afield by the promise of affordable second homes in Mexico, Costa Rica, Chile, or Argentina.

Ultimately, the origins of this green consumerism lay in the modern political economy of mass consumerism. This paradox has not gone unnoticed. Surveying the landscape of postwar prosperity in 1958, economist John Kenneth Galbraith asked, "How much should a country consume?" Wondering why Americans would not restrain their appetites, the "appetite itself," he confirmed, was "the ultimate source of the problem."[72] North American conservationists have sometimes, but not always, been reluctant to probe the source of their endless affluence. As Ramachandra Guha notes, Mahatma Gandhi asked the same question as Galbraith, only more pointedly: how much should a person consume? As Guha concludes, both questions underscore how the personal and the national blur in the "selective environmentalism" typical of the developed world.[73] In Guha's estimation, modern consumption enabled Western nations to set aside wilderness for biodiversity and recreation by displacing their appetites on to ever more distant lands and waters.[74] The separation of economics from ecology that made modern consumerism possible was also responsible, in part, for the birth of modern environmentalism in all of its complexity.[75]

TANGLED CHAINS: COMMODITIES, ENERGY, AND WASTE

The earliest environmental histories grew from the fertile soil of agricultural and natural resource histories. The famed *Annales* School provided the intellectual seedbed. Fernand Braudel's *longue durée* illustrated the production and flow of commodities across the early modern Mediterranean World. Alfred Crosby extended the analysis by tracing the deployment of what he famously called the "shock troops" of Western imperialism: introduced flora and fauna, plus pathogens, that tempered native resistance. After contact, Native peoples shaped and extracted wealth from changed landscapes, often to purchase manufactured goods from European invaders. None of this was possible without the expansion of a commodity-hungry Europe.[76] The environmental histories that followed Crosby, detailing the battles and casualties of invasion from the Americas to Oceania, were not explicitly consumer-minded, but consumption was at the foundation of these histories. Clearing forests, cutting trees, importing livestock, and producing crops were a function of an expanding global economy that incorporated far-flung ecosystems with distant consumers. Some historians realized this, notably William Cronon, who concluded that colonial New England became a *new* England because settlers believed that "the wilderness should turn a mart" to provision voracious English and American consumers.[77]

Sidney Mintz's *Sweetness and Power* (1985) was one of the first commodity histories to look holistically at what circulated through the market: in this case, sugar, which became a vehicle for linking metropole to colony without losing the connection to one place or the other. The bitter outcomes of sugar production, from slavery to soil exhaustion, were inextricably bound up with sugar consumption by English tea drinkers and candy-craving American children, underscoring that "in understanding the relationship between commodity and person we unearth anew the history of ourselves." Mintz established the framework for what historians now call transnational or global history.[78] The linkages he traced were commodity chains, also called commodity flows.

The notion of commodity chains grew out of Latin American history. Immanuel Wallerstein created the term as part of his "world systems theory," which in turn was based on Braudelian history and dependency theory, itself a Latin American product.[79] Other scholars, notably Latin American historians, soon followed Mintz's lead. At first some studies concentrated on the export sector as colonial exploitation. Other historians concentrated on the relationship between land tenure and labor markets, ignoring the metropole altogether. In both instances, Latin American nations and peoples were at the mercy of the "commodity lottery" and the global market set the odds, a view that confirmed the most rigid parameters of world systems theory. Few scholars explicitly considered how commodity production and trade shaped and were shaped by environmental change over time. [80]

Environmental historians now can do more to help expand commodity chains as analytical tools. Drawing from other disciplines can help. Geographers and other social scientists have long recognized the value of commodity chains while noting where their utility can fall short. They offer many useful suggestions for environmental historians to follow. First, many commodity chain studies tend to ignore the origins of chains in local environments and markets. Lengthening chains can ground them specific modes of extraction, labor, and material conditions. Second, paying attention to space as well as time can reveal how specific commodity chains are tied to particular geographies and societies. Petroleum and precious metals may be extracted from many places and homogenized as commodities in the global market place. Yet not all oil and mining industries are equal, nor are the locales from which the raw products flow. Third, by deepening social and material analyses together across many industries or sectors, scholars can better follow historical changes in the world economy and the environment. Commodity chains are always plural, akin to networks as they span "the very largest and very smallest geographical scales."[81]

Several environmental historians have already started down this path and Latin American historians have been in the vanguard. Warren Dean was one notable trailblazer. The development of Brazil's rubber economy, he argued, was a biological and an industrial process that depended as much upon the health of rubber trees as the price fluctuations in São Paulo or New York City. Demand by American and European companies, like Goodyear, drove Brazilians to create monoculture plantations that put domesticated rubber plants at risk of highly contagious pathogens. Eventually, firms had to relocate plantations to Southeast Asia to outrun infections and insure supply.

Dean challenged prevailing ideas about dependency theory by concentrating primarily on the production of rubber, and not how consumers used auto tires or latex gloves.[82] Subsequent historians emulated Dean's approach to scrutinize an array of trade goods: silver; indigo; cochineal; cacao; guano; petroleum; rubber; henequen; cocaine; coffee; and the old standard, sugar.[83] The rise of commodity chain theory and its analogs offered ever more powerful tools for environmental historians reexamining trade as part of mass consumerism.[84]

Other environmental historians focused on the denudation of tropical landscapes to supply world markets. North American and Western European demand for beef, coffee, and sugar came with substantial, even irreversible environmental damage to coastal wetlands in Cuba, mountainsides in Columbia, and forests Brazil, Indonesia, and Liberia. By the end of the twentieth century, Richard Tucker concludes, the "race to stabilize production systems in the tropics" was as futile as "the effort to preserve a remnant of tropical nature."[85]

Tucker's consumers, like most environmental histories of the era, were an undifferentiated mass that ate, drank, and bought their way into oblivion. More recent studies have put a human face on commodity chains without neglecting the consequences of consumer appetites. Bananas make for a compelling story, as it took producers and consumers working together to make it the omnipresent North American fruit. For most of the nineteenth century, bananas were luxuries. Aggressive vertical integration by foreign companies in Central America drove down prices. Faster steamships and railroads and air transport made the fruit more accessible. When sovereign governments resisted companies such as the United Fruit Company, known as *El Pulpo* or "the octopus," the United States sometimes sent troops to enforce corporate dominance in places like Nicaragua. Over time, companies purchased compliance by offsetting budget deficits or paying for internal improvements, as in Honduras. Either way, gunboat diplomacy or corporate filibustering helped to build the region's pejorative "banana republics." Prices for bananas fell and consumers responded. Distance and politics made such machinations almost invisible to consumers back home, giving sellers an opportunity to generate more demand. Housewives began adding bananas to shopping lists, enticed by grocers and wholesalers who marketed the fruit as a healthful and digestible food for children. A new character created by the United Fruit Company in 1944, Miss Chiquita, modeled on the popular singer and film star Carmen Miranda with her fruit-festooned hat, helped to promote bananas as an exotic but affordable pleasure.

North American demand for bananas had concrete effects on Latin American lives and landscapes. As with rubber cultivation, fruit corporations found that planting the favored Gros Michel cultivar rendered banana trees vulnerable to fungal pathogens that thrived in plantation agro-ecosystems. Growers first tried to outpace the disease by moving plantations, flooding infected plots, or spraying toxic chemicals. Eventually, growers turned to another cultivar, the Cavendish, which resisted fungal attacks but was easily bruised when shipped. To make the new variety palatable to consumers, growers harvested and shipped the still-green and bruise-resistant fruits, then used basic botany to prepare them for sale once they arrived at their destination: enclosing the unripe fruits in an airtight space and adding additional gaseous ethylene, a natural plant hormone, to finish ripening them. In postwar North America, consumers expected plentiful bananas in all seasons, decorating their breakfast tables year round. What they did not see in their perfectly yellowed bananas were the Honduran or Ecuadorian planation laborers complaining about exposure to toxic chemicals and fighting for safer working conditions, or the many acres of forest felled to expand production and thwart pathogens.[86]

While historians have traced the circuits of crops from field to market, fewer have chased the trails of animals from forest or feedlot to table. This is beginning to change. As with agricultural histories of consumption, these tales were first told by historians in other historical fields. American and Canadian historians have written many sophisticated studies of the early fur trade, and Western United States and business historians have examined the cattle and meat packing industries. All these enterprises were part of transcontinental markets that linked voyageurs in Rupert's Land and cowboys in Laredo to milliners in London and meatpackers in Chicago. Between the destruction of fur-bearing animals and the slaughter of cattle came the near-eradication of the bison, killed in such numbers for the hide and tallow trade that the American zoologist William Hornaday described the southern plains as "one vast charnel-house."[87] Yet bison were not merely provender for butchers or raw materials for tanneries. Bison were saved from extinction, in part, because consumers demanded frontier spectacle and amusement at private ranches, zoos, Wild West shows, and public reserves. Sometimes, those same consumers also ate what they ogled. During the 1920s, railroad passengers traveling to the National Bison Range in Montana could dine on bison steak while scanning the horizon for their meal on the hoof. Twenty years later, in the 1940s, biologists and wardens in Wood Buffalo National Park backed round-ups and operated abattoirs to cull diseased bison herds and promote tourism. Native hunters, long barred from hunting in Canadian and American national parks, were excluded from the harvest.[88]

Domesticated animals have received even less attention, but this is changing, too. Several recent works explore the propagating and harnessing of industrialized animals for mass consumption. Animal husbandry depends upon selective breeding. Since the advent of genetic science in the late nineteenth century, scientists and farmers have molded animals in ever more intricate ways to meet to changing consumer demands. High-protein beef, low-fat chicken, and disease-resistant hogs render animals as consumer-driven technologies. The promise and pitfalls of this biological tinkering remain uncertain.[89]

Aquatic fauna, too, have been relentlessly exploited as food, although the planet's watery reaches are even more mysterious than its terrestrial realms.[90] Arthur McEvoy's revaluation of the so-called "fisherman's problem" explained the collapse of California's fisheries as a dynamic process. Fluctuating international markets, ethnic and racial conflicts to control valuable stocks, and unstable river and oceanic ecologies yielded a legal and industrial framework that encouraged over-harvesting. Joseph

Taylor's subsequent examination of the Northwestern salmon fisheries complicated this picture still further. Artificial propagation emerged to circumvent habitat destruction and over-fishing, but hatcheries were never the promised panacea. Scientists, regulators, and commercial and sport anglers have slowly grasped the necessity, as Taylor argues, of moving away from "the fallacy of trying to control nature and toward the more realistic goal of trying to govern ourselves."[91] Others have explored the exploitation of sea life, but few with the same theoretical sophistication. The burgeoning aquaculture industry, an enterprise as complex and global as hog farms and cattle abattoirs, is but one example of a topic awaiting environmental historians.[92]

Even more technologically complex products and processes await further investigation. Work by historians on mining and metal production point in the right direction. Computers, smart phones, batteries, and light bulbs—all entwine with multiple commodity chains. Metals and rare earths compose just one set of links: coltan from sub-Saharan Africa (for tantalum capacitors), lithium from South America and Australia (for batteries), and europium and yttrium oxides from China (for fluorescent bulbs and liquid-crystal displays, or LCDs). Primary metal production is only part of these chains. Several popular books and numerous investigative articles have documented labor abuses, unsafe working conditions, and widespread pollution at production facilities or following disposal of discarded products.[93] Even the Internet, invisible in its ubiquity, has an environmental history waiting to be searched. Companies like Google are expanding their infrastructure and looking for accessible energy and affordable labor to meet consumer demand for web services. Not coincidentally, Google has also become a major energy consumer. According to its website, Google obtains almost one-third of its energy from renewable sources—wind, solar, or hydropower—and offsets the rest to claim carbon neutrality. Consumers can feel better about their computing habits because, as Google claims, it provides "efficiency at scale" to reduce energy costs and carbon output.[94]

Energy chains are commodity chains, too, even if they are concealed. Despite its importance for industrial economies, environmental historians have only recently paid attention to energy beyond the environmental consequences of industrialization or political action against particular types of production. The problem lies in how to interpret the technology so central to energy production, transmission, and consumption. We all too often assume that technology simply explains itself, or is devoid from politics. The road from the invention of the internal combustion engine to the family car, as Warren Susman argued, was never preordained. In the United States, it took the genius of Henry Ford to turn invention into cultural innovation.[95] Energy is no exception. Consumers were as salient in the development of our modern energy ecology as they were in the creation of modern agricultural ecology. One of the most prolific energy historians, David Nye, rejects the "simple equivalence of rising energy and cultural advance," arguing that only a historical approach explains Americans' appetite for energy.[96]

Electricity has been the most examined energy system in terms of consumption. From the late nineteenth century, producers needed to nurture demand to make electricity profitable and affordable. Building generation and delivery systems required massive investments by the state and industry. Getting electrical power to businesses, factories, and homes was only half of the struggle. There also had to be machinery and appliances to consume the electricity. Producers encouraged consumers to purchase new-fangled appliances; consumers then used these and other goods to reshape domestic and public spaces. Affordable and plentiful electricity alleviated household labor while adding new chores and expenses. It lit dark nights for work and leisure, opened new opportunities for art and commerce, and remade American public and private institutions from Ford automotive assembly lines to city public utilities.[97]

Less noticeable at first was how electrical production shaped and was shaped by nature. Richard White's imaginative history of the Columbia River explores how "the river was an organic machine, an energy system which, although modified by human interventions, maintains its natural, its 'unmade' qualities."[98] Native peoples harvested the energy of the river as salmon they consumed. American colonists who followed used the river's currents to ship commerce. By the New Deal, engineers put the river to work powering turbines to generate electricity. The massive hydroelectric systems now spanning the Columbia were only the latest of many attempts to harness nature's energy for human ends. Subsequent scholars have applied White's felicitous metaphor to assess the environmental and social consequences of energy production. Inexpensive electricity was only inexpensive because some social and ecological costs were displaced onto communities at the margins of regional or national politics. Building hydroelectric dams on the upper Columbia or Fraser in British Columbia or coal-fired power plants in Arizona's Four Corners region only seemed harmless when urban consumers ignored displacement of native peoples, eradication of salmon runs, and pollution of lands and waters beyond the city lights.[99]

Fossil fuels have received less attention as consumer products until recently. Petroleum is of special importance because it has, literally and figuratively, greased the rails of industrialism. The modern hydrocarbon economy today grew from institutional structures created as much by national governments as private enterprise, including stable property regimes, regulatory controls, tax policies, infrastructure, and public investment. As with electricity, there was never a truly free petroleum market. And as with other energy regimes, petroleum consumption eventually heightened public concern over a degraded natural world.[100]

California's petroleum history is an instructive example. At first, in the late nineteenth century, political struggles over the allocation of oil lands and the regulatory rules governing production yielded wild price swings, scaring off investors and ruining companies. By the early twentieth century, state intervention had calmed the markets, resulting in an oil glut just as new markets opened with automobiles and asphalt. Tax and investment policies to spur consumer spending also funded highways and roads. Californians enjoyed energy convenience even as their state slipped from oil exporter to oil importer. More worrisome was the maze of highways clogged with smog-belching cars that had, as Paul Sabin concludes, trapped the Golden State "in an automobile landscape of its own making."[101] California was merely the harbinger of things to come. The creation of the 1956 Highway Trust Fund, fueled by federal gas-tax revenue, elevated state policies for road construction to the federal level. State and federal gas tax policies, as Christopher Wells points out, were the concealed lubricants behind "the growth of a vast automotive infrastructure" designed to stimulate incessant "American demand for gasoline." Recent efforts to escape this endless cycle are behind attempts to tax the latest and perhaps most dire externality of our petroleum dependency: carbon emissions.[102]

It is easy to vilify cars and all they represent. In the demonology of modern environmentalism, automobiles are Satan's chariots, but at the dawn of the motor age, not all consumers were convinced that cars and nature were unholy enemies. Initially, cars were sanitary angels. Automobiles cleaned up American cities by ejecting horses, which consumed enormous amounts of feed, produced tons of manure, and left large, unsightly bodies when they died.[103] Automobiles also opened the scenic countryside to city-weary tourists. The subsequent fights over roads and cars reshaped ideas about nature and nation.[104] Seen at first as a badge of indulgent luxury, automobiles soon became necessities. Manufacturers such as Henry Ford or financiers such as John Raskob, creator of General Motors Acceptance Corporation, or GMAC, one of the first finance companies, put automobiles within reach of middle-class Americans. As car culture took off, Detroit's Big Three rushed to grow and promote demand. At least one big automaker—General Motors—built cars with planned obsolescence in mind. When consumers drifted from Detroit's offerings in the 1960s, the Big Three criticized Washington for insufficient support and remonstrated buyers that imports were unpatriotic.[105]

The advent of postwar environmentalism and the energy crisis of the early 1970s finally shocked lawmakers into action. Their responses were temporary and partial. A flurry of legislation by Republicans and Democrats forced the Big Three to make concessions on fuel economy and emissions. More Americans purchased smaller cars, often from Japanese and European importers, looking for better mileage and cleaner emissions. Consumers' ardor for big cars never faded completely. Once the 1980s ushered lower gas prices back in, American and foreign manufacturers began marketing sport utility vehicles and light trucks as safe family vehicles or outdoorsy fashion statements. Given the choice between protecting the environment and stopping sprawl or preserving individual choice and freedom, which was often equated with escaping into nature, most Americans opted for the latter. Cars were demonic destroyers of nature; they were also the fastest and most liberating way to get out of town.[106]

Environmentalists revile automobiles because they produce waste: choking clouds of exhaust, leaking gas, dripping oil pans, and carcasses of steel and plastic at the end of their life. Consumption of all kinds produces waste, so garbage also occupies a special spot in environmentalist iconography. And trash has a history, too, one bound to a mature consumer culture. At the dawn of the American republic, reuse and repair were commonplace, as were the jobs that made them possible. Cobblers resoled shoes. Seamstresses patched trousers and darned socks. Rag pickers sorted scraps to make paper and felt. The industrial revolutions that followed made it both cheaper and more fashionable to replace rather than repair. What was old and used up now often stayed that way. It became trash.

By the end of the nineteenth century, garbage itself had become an industrial problem best solved by trained experts: engineers, home economists, and sanitarians. Those who continued to pick and sort and hoard were deemed unsanitary. Most of these laborers were immigrants, children, or women at the margins of society. As well-meaning reformers put an end to the trash trades, manufacturers touted the efficiency and convenience of disposable products. "New products raised standards," says Susan Strasser, "and consumption itself became a new kind of work" that unfolded, predictably, along gender lines.[107] Postwar prosperity brought still more trash into American homes: TV dinners in aluminum foil, frozen vegetables in plastic pouches, breakfast cereal in cardboard boxes. When the environmental movement began to advocate the recycling of this mushrooming household waste, a new industry erupted to meet the moral and economic demand of responsible disposal. Seen historically, Strasser concludes, nothing is "inherently trash."[108]

Like other businesses, the trash and recycling business is a global enterprise. The globalization of trash arguably began in the summer of 1987, when the tragicomic story of a vagabond barge, the *Mobro 4000*, loaded with more than six million tons of trash from New York City and Long Island, captured the world's attention. The entrepreneur who chartered the barge had planned to ship refuse from New York to North Carolina, where it would be burned to produce methane. When inspectors found possible evidence of hospital waste, they prevented the barge from docking, a pattern that repeated itself over next five months in five other states and three countries. At last, the *Mobro 4000* returned to New York City, anchored offshore until a judge ordered its load incinerated in Brooklyn and the ash buried in the barge's port of origin, Islip, Long Island. In the aftermath, city and state governments began implementing recycling programs to reduce pressure on crowded landfills. Meanwhile, enterprising businessmen, learning from the fate of the *Mobro 4000*, started a now-thriving global trade in waste and recycling.[109]

In this new planetary marketplace of garbage, something old can be new again. Well-meaning Westerners donate used clothing to charity organizations that sell t-shirts and blue jeans and out-of-fashion suits by the ton to distributors who then ship it to the developing world. Known in the Bemba language as *salaula*—"to select from a pile in the manner of rummaging"—second-hand clothing was big business by the mid-1980s across sub-Saharan Africa. Like so many Scarlett O'Haras, tailors made new clothing from discarded textiles, wholesalers distributed *salaula*, and retailers sold the items in small and big bazaars. Critics decried *salaula* as undercutting domestic garment industries, reopening painful debates over development and postcolonial dependency. Overlooked in the debates, according to anthropologist Karen Tranberg Hansen, was how "the donors are complicit in this as consumers in a global world where Third World products rarely earn a fair price."[110]

That discarded clothing could be a valued consumer good as well as the detritus of consumerism is another absurdity environmental historians have yet to explain. The line between waste and want is as thin as ever in our interconnected world. Several years ago, a container ship carrying plastic toys from China lost part of its cargo in the North Pacific. Yellow ducks and green frogs washed ashore from Alaska to California. They also joined the great gyres of floating trash in the central North Pacific. Made with imported oil in Chinese factories for export markets, then shipped overseas on ships powered by petroleum, the toys never reached their intended consumers. Yet they still tell a story as trash. Their oceanic resting places are the newest links in the tangled chains of consumption that bind us all. [111]

EMBODIED CONSUMERISM:
HEALTH AND DISEASE

Along the saguaro-dotted rim rock and arroyos of south-central Arizona are the homelands of the Akimel O'Odham, commonly known as the Pima, whose ancestors have lived there for at least 1,500 years, possibly longer, hugging the life-giving Gila River, the center of their world. Akimel O'Odham means the "River People." The Gila yielded waterfowl and fish, grasses for weaving baskets and shelter, and water for irrigating crops of squash and melons and beans. In the late 1600s, the Spanish introduced new crops, like wheat, which the Pima adopted with great success, selling the surplus to Spanish and Mexican towns to the south. [112]

The Akimel O'Odham mastered their fate until the advent of American rule. After the Mexican-American War, permanent colonists began to arrive—a trickle at first, then a steady stream. A reservation was established for the Pima in 1859, encircling the Gila River from its union with the Salt River southward to Casa Grande, allowing the community to continue its lucrative agricultural trade. When the widening stream of Americans became a torrent after the Civil War, new arrivals hijacked the Gila and Salt Rivers for farming and urban development in Phoenix, just forty-odd miles downriver. [113] Beginning in 1892, the Gila disappeared altogether during the dry summer months. Crops failed, groundwater proved insufficient, and starvation stalked the reservation. Over time, the losses mounted as thirsty and hungry speculators purchased tribal land and water rights. In the course of four generations, the Akimel O'Odham watched their once-vibrant agricultural economy evaporate.

As the landscape changed, the Pimas' bodies changed as well. The US Department of Agriculture, in conjunction with the Bureau of Indian Affairs, redirected the surfeit of subsidized postwar industrial agriculture to help Indian wards—bacon, lard, cheese, beans, corn syrup, and pasta. Cut off from traditional foods, often lacking electricity and running water, becoming increasingly sedentary and impoverished, the Pima saw their health decline. In 1965 National Institutes of Health researchers reported that the prevalence of adult-onset (now Type 2) diabetes among the Pima was fifteen times the national rate. The resulting complications—renal disease, retinopathy, neuropathy, and cardiovascular disease—were equally prevalent and horrific. By the 1990s, the Akimel O'Odham had become the face of the world's mounting diabetes epidemic. A leading diabetes expert, Peter H. Bennett, suggested that diabetes among the Pima was "a clear example of genetic-environment interaction." Despite evidence that the Pima and other Native groups may be genetically predisposed toward diseases like diabetes, federal policies and environmental changes likely increased their susceptibility.[114]

Diabetes has been described as a "disease of civilization," but it is perhaps more accurate to call it a disease of consumerism. To physicians and social critics in Enlightenment-era England, the Pima's plight would have come as no surprise. They mapped the "body politics upon the body human" only to find that "health for the human body was itself highly problematic" in the modern era. [115] The consumer revolution of eighteenth-century England posed significant challenges for traditional religious and political injunctions against greed and envy. Previously, gluttony was smart preventative medicine adapted for an era of frequent famine. By the 1730s, however, improvements in agriculture and transportation "made produce abundant and gin dead cheap," resulting in an excess of indulgence and the spread of so-called "wasting diseases": scurvy, cancer, catarrhs, asthma and consumption, later known as tuberculosis. Just as in the body politic, "wealth easily mutated into waste," compelling some physicians, such as George Cheyne, to ask whether "the wealth of nations secured the health of nations."[116]

The English were uncertain about how best to answer Cheyne's question. For his part, Cheyne blamed consumer excess. Englishmen ate and drank too much and suffered from the coal- and dust-choked airs of their industrializing cities. As he explained in *The English Malady*, the "nervous disorders" afflicting his countrymen were symptomatic of their separation from "simple, plain honest and frugal" living thanks to the now widely available "whole Stock of Materials for Riot, Luxury, and to provoke Excess."[117] Cheyne's solution was dietary austerity and abstinence.[118] Other physicians, like the radical Scot Thomas Beddoes, identified the same causes but believed the substitution of traditional physical hard labor for "the almost feminine occupations" of weaving and glazing and painting was turning men into invalids.[119] "In this age of self-made men," Roy Porter notes, "people made their own illnesses.[120]

Beddoes and Cheyne became durable oracles. In the young United States, the links between health and consumption became manifest in the landscape of the expanding nation. Humid, hot climates yielded disease and lassitude; temperate climates fostered vigor and enterprise. When the landscape did not conform to ideals of health, Americans tried to change the land to promote well-being: draining swamps, redirecting rivers, and opening lands to cultivation. The advent of modern medicine and germ theory did not fully uproot environmental explanations of disease grounded in older traditions of medical geography. Mass consumer culture masked older and enduring links between body, health, and place even as it intensified or created new ones.[121]

Place had been central to the early history of asthma and allergies, and changing places was often seen as the way to banish ill health. Physicians instructed sufferers to take hay fever holidays in arid or cool locations. At first, only the most affluent could flee to the White Mountains of New Hampshire or the upper peninsula of Michigan. Expanding railroad, and then highway, networks put rural health resorts in the reach of middling Americans by the end of the nineteenth century. As physicians discovered the vectors for asthma and allergy attacks—pollen, pets, and pollution—the quest for more distant landscapes of health boosted the rise of resort communities in the arid Southwest.[122]

It was "upon hope and past historical landscapes," Gregg Mitman writes, "that today's lucrative allergy industry is built."[123] The origins of that industry lay in the city as well as the countryside. A century ago, physicians and public health officers started observing that asthma and allergy rates were highest in the nation's poorest, minority neighborhoods. The postwar advent of chemical prophylactics (first antihistamines, then steroids) and technological fixes (filters, vacuum cleaners, allergy-proof bedding) opened another front in the war on asthma and allergies, even as many allergists pushed for environmental solutions (better housing, pollution abatement, or weed removal). Despite the widespread use of prescription inhalers and over-the-counter pills, asthma and allergy rates continued to climb. We now live in an era where asthma and allergies are "enmeshed in a world of billion-dollar drugs." The lowest common denominator in public health is not social equity but "economics and marketing," as health care itself has become a consumer good. "Allergy," Mitman concludes, "is not a thing but a relation" that connects Americans' bodies to their landscapes of consumption.[124]

These landscapes of consumption can also be landscapes of exposure and contamination. In the mid-nineteenth century, industrial workers' bodies became proving grounds for new theories of industrial disease. The introduction of new materials and new processes endangered laborers' health as manufacturers pushed to produce consumer goods: brilliant white paints made with lead, strike-anywhere matches with phosphorous tips, or glow-in-the-dark watches tinged with radioactive radium. Pioneering scientists and public health advocates, such as Alice Hamilton, sounded alarms that the hazards of the job were crippling and killing workers, plus endangering consumers' health as well. Hamilton and her colleagues helped to create the modern field of toxicology and pushed for legislation to protect workers' safety.[125]

Other reformers unmasked the horrific consequences of the emerging mass food and drug industries. Upton Sinclair, the famous muckraking author who critiqued Chicago's meatpacking industry in his 1906 novel *The Jungle*, was more concerned with workers' rights than consumers' well-being. He famously quipped the he had "aimed at the public's heart, and by accident I hit it in the stomach."[126] Sinclair's poor aim unintentionally but firmly fixed public ire on contaminated food, poisonous and untested drugs or patent medicines, and unsafe drinking water instead of brutal labor conditions. The passage of the Pure Food and Drug Act and the Federal Meat Inspection Act of 1906 started a longer, if uneven, trend of federal oversight to protect consumers' health, but not always workers' safety.[127]

The landscapes of consumers' health changed again in the postwar era, as chemical exposure became the newest addition to the litany of anxiety. As with the furor over Sinclair's exposé, it was worries over food that prompted change. Like bananas, oranges and other citrus fruit were not commonplace until horticulturists marketed them to uneducated consumers. Citrus boosters also marketed California as a healthful place—abundant sunshine, fertile soil, proximity to Pacific Rim markets—while securing public subsidies for water projects, favorable labor laws, and efficient rail and highway networks to get their crops to market. Citrus crate labels and magazine ads portrayed a bucolic landscape of happy workers and verdant groves. As with bananas, however, growers could only sustain these illusions by exploitative labor practices, increased applications of pesticides and herbicides, and immense diversions of water. Within every orange was "an economic and moral failure" that embodied how California's agribusiness withheld nature's bounty "from deserving citizens."[128]

Some of those deserving citizens suffered most from the partitioning of nature into production and consumption. In the 1940s, scientists and physicians in California's Central Valley were recording many farm worker poisonings and deaths from herbicide and pesticide application. Researchers first tried "to contain the dangers of industrialization, not challenge it" through prophylactic measures such as safer protective suits, working practices, and improved housing.[129] As contamination spread, farm workers and adjacent residents challenged the prevailing premises of epidemiology and occupational medicine. Claiming that decades of chemical applications had poisoned groundwater supplies and injured harvesters, these advocates adopted an ecological approach toward public health and the political economy of consumption. The initial organizing drive to create the United Farm Workers in 1962 coincided with the publication of Rachel Carson's *Silent Spring,* and the international boycott against non-union grape growers that soon followed appealed to both worker safety and consumer health. The timing was not coincidental. According to Linda Nash, farm labor activists realized that as humans industrialized the land, the land had "industrialized them."[130]

By the 1980s, the politics of environmental health and safety had moved from field and factory to office space and the family home. Office workers complaining of rashes, headaches, blurred vision, and inexplicable immune disorders began to blame exposure to cleaning solvents, paint, synthetic carpets, ink and copier toner, adhesives and other workplace products for their maladies. Scientists and physicians quarreled over how to measure health effects of "multiple chemical sensitivity," or if the dangers existed at all. Unable to determine specific mechanisms for exposure, researchers initially created a new category to describe the problem: "sick building syndrome."[131] The term also highlighted how, as Janet Ore argues, building materials from pressure-treated lumber to formaldehyde-soaked drywall created new "biotic realms in where flesh and blood interact with chemicals in unpredictable, even wild, ways."[132] Activism by chemically sensitive consumers has changed construction codes, created demand for greener building materials and cleaning products, and reshaped ideas about environment and health.

These intimate geographies of health and industrialization extend into the minutest of scales within the human body. Endocrine disruptors, the group of chemicals that either mimic or distort animal hormonal systems, efface boundaries between bodies and landscape completely. The first synthetic estrogen, diethylstilbestrol or DES, once held great promise to regulate female hormone levels. That promise came under early scrutiny. Studies in the 1930s suggested that DES caused cancers and triggered reproductive and sexual developmental problems in laboratory animals, but drug companies dissuaded the US Food and Drug Administration from its precautionary stance in 1941. They instead emphasized DES's potential to reduce pregnancy complications, miscarriages, and fetal weight gain, and to treat menopause. Soon thereafter, DES became a widely prescribed drug, and was also approved as an additive in industrial animal feeds to boost livestock and poultry weight gain. The human consequences were catastrophic: 95 percent of the two to five million children exposed to pharmaceutical DES in the womb developed reproductive problems from cancer to infertility. Other costs were as significant if not as easily quantified. At its peak industrial use in the 1960s, more than 90 percent of feedlot cattle in America were fed DES. What livestock bodies could not consume or absorb made its way "into aquatic ecosystems, with unknown effects" at the time.[133]

Until recently, toxicologists adhered to the old dictum that "the dose makes the poison," developed during the first era of industrial contamination. They only slowly recognized that chemicals like the plasticizer bisphenol-A (BPA), which is used in a variety of consumer goods from water bottles to canned food liners, disrupts endocrine systems even at very low levels of concentration when leached into foods and liquids. Scientific awareness, in turn, has driven product regulation and consumer demand; stainless steel water bottles and BPA-free baby products now crowd store shelves.[134] But BPA is only one of thousands of compounds of concern. Scientists have discovered, according to Nancy Langston, that "our pee is doused with poisons—metabolites from the breakdown of birth control pills, caffeine from all the coffee and Mountain Dew we're quaffing, remnants from the aspirin and Tylenol and anticholesterol drugs we use to stanch the pain of our modern aliments."[135] Although the science as of this writing is still in dispute, in 2012 the world's largest professional group of endocrinologists called for further study and restrictions on the most questionable endocrine-disrupting chemicals. Nicholas Kristof, the noted *New York Times* columnist, has helped to publicize this concern as well.[136] Meanwhile, because of the ubiquity of older products and their long environmental fates, some chemicals and heavy metals will "continue to be part of the world far into the future" beyond the point of "remembering their origins."[137] Rising rates of hermaphrodism, cancers, and reproductive problems in wildlife and humans alike may be yet another consequence of mass consumption.

We may have, perhaps, even changed human and animal evolution through consumption. Selective breeding and biotechnology has crafted organisms to meet our every desire: allergy-friendly designer dogs, enormous hogs, richly marbled beef, plants that kill potential insect or fungal attackers, and patented laboratory mice that help researchers study and treat disease.[138] Demand for organic and locally grown food has been one reaction by consumers to escape their tainted lives. This may seem little more than individual deliverance. Big organic agriculture still produces environmental and social costs, from petroleum dependence to unfair labor practices, most of which are not passed onto consumers.[139]

But if the focus shifts from Earthbound Farm Organics or Whole Foods, the picture becomes more complicated. Buying big organic at a national franchise is an individual action based on consumer preferences for health, taste, aesthetics, or politics. Going to a farmers' market or purchasing a community supported agriculture (or CSA) share can be a means of finding more absolution in the checkout line. It can be a commitment to community, to a particular place, to civic environmentalism. Farmers' markets and CSAs are curious hybrids, appealing to individual consumers' appetites but also encouraging people to act as citizens.[140]

Farmers' markets and locavorism can only provide so much salvation. Try buying bananas for a hungry toddler in Maine at the local co-op in January, or coffee any time of the year in most of the Northern Hemisphere without selling out to the global agricultural system. And North Americans are not alone in their implication. As industrial agricultural techniques and Western-style diets have circled the globe, together with rising affluence, consumers from Beijing to Manila and Mexico City to Warsaw have come to expect meat and processed food on more than special occasions. Rising demand overseas can only explain part of this shift. Environmental opposition in North America and Western Europe is pushing industrial manufacturing and agribusiness to nations with laxer environmental and labor laws. Scientists and physicians are now documenting the costs: polluted waterways, tainted food, rising obesity rates, and mounting cases of non-communicable diseases once uncommon outside of the industrialized West, including diabetes, hypertension, cardiovascular disease, and many cancers. New variants of infectious diseases are another consequence. Industrial feedlots and genetically modified and chemically enhanced livestock make nutritional abundance possible, but they can also harbor antibiotic resistant super strains of bacteria or newly-virulent viruses. Scares over SARS, avian flu, and swine flu underscore two clichés: we are what we eat, and there may be no such thing as a free lunch after all. What we consume may consume us in the end as well.[141]

CONCLUSION: TOWARD AN ALL-CONSUMING HISTORY

Consumption has blossomed into a vital historical field, if one with sometimes poorly marked boundaries covered in weeds. Almost two decades of scholarship on consumption—by anthropologists, sociologists, geographers, and historians—stress how "commodities, like persons, have social lives."[142] We know much about how commodities travel from producer to consumer, take on symbolic value, and forge connections between people and places. Meanwhile, nature itself often remains unseen yet ubiquitous.

Environmental history offers the promise of a more grounded history of consumption. Until recently, we have all too often replicated the split between material and ideological approaches that launched our field and defines it still. Studies on consumption are no exception. Many notable works trace how representations of nature—in film, at amusement parks, as art or literature—circulate in the marketplace of ideas and commerce.[143] Other scholars unravel how capitalism demystifies and disenchants nature, often with profound consequences for those places and people that provide resources.[144] Studying consumption can blend the material and the ideological across wide spatial scales. It is a return to the vital heart of what makes environmental history unique: the relationship between the human and the world beyond us.[145]

Environmental history is prepared to advance significantly how historians theorize and study consumption. It can remind us that consumerism is where our affection for nature begins. The capitalist system that many environmentalists and historians criticize as the source of our modern problems with nature is often the source of our love for it. Consumption is both the wellspring of purity as well as an enumerated sin in the secular religion of environmentalism.[146] When critics such as Paul and Anne Ehrlich declaim consumption and call "into question the sustainability of the human enterprise," they launch an uncomfortable crusade without entirely recognizing how it implicates them as well. They also point to another awkward truth: by hiding behind the label of consumer choice, many North Americans absolve themselves of collective political action.[147]

Environmental historians are well poised to analyze consumption as something more complex than nature rendered into dead objects and political discourse. They can provide a powerful critique of why, despite jeremiads against consumerism, it remains materially and culturally intrinsic to environmentalism, including the recent surge in green-tinted commerce and politics. Advertising campaigns by multinational petroleum companies such as British Petroleum, which proclaimed that BP stood for "Beyond Petroleum" (at least until the disastrous 2010 Deepwater Horizon oil spill in the Gulf of Mexico), or Starbucks, which sells organic free-trade coffee blends to assuage caffeinated consumers' guilt, or Paul Hawken's so-called "natural capitalism" based on a green industrial revolution, are part of a longer and more interesting story. By reconnecting producers to consumers, such research may also help people today. Seen historically, the "distancing effect" of consumption has been twofold: consumers have been distanced from the environmental effects of natural resource exploitation and from the tangible connection to nature through work.[148] The historical consequences of this distancing effect have been profound for producers and consumers, human beings, and the natural world.

Already, some businesses and consumers understand the value of minding the gap. Beginning in 2008, Patagonia put its environmental assessment program on line. Called "The Footprint Chronicles," the interactive website allowed customers to follow their garments from organic cotton harvested in Texas or polyester fibers manufactured in India to textile mills in North Carolina or China. An evolving reference library allows customers to conduct further research. Patagonia's decision to reveal its supply chains was more than idealism. As Jill Dumain, director for environmental analysis, explained in an interview with business magazine *Fast Company*, "the green marketplace has become crowded," so Patagonia had to learn "to communicate in circles that are very different than they were 10 to 15 years ago."[149]

Meanwhile, Chouinard has remained comfortable with the contradictions of doing good deeds while living well. He practices his MBA style of leadership—"management by absence"—by playing outdoors as often as possible.[150] Despite his pessimism about the state of the planet, he had good reason to keep on playing. In 2007 *Fortune* named Patagonia the planet's coolest company. He had become a corporate rock star, a recipient of numerous honorary degrees and awards, fêted by heads of state, and lauded by fellow business leaders. Chouinard's gospel of green consumerism had become part of Patagonia's product line. He began selling it relentlessly. In his latest book, *The Responsible Company*, Chouinard explained how "everyone in business—at every level—has to deal with unintended consequences of a 200-year old industrial model that can no longer be sustained ecologically, socially, or financially." Talking about Patagonia's footprint is a loaded term, filled with assumptions about consumerism, politics, and ecology. For a company specializing in human-powered sports and promoting environmental virtues, it is an apt measure of success.[151]

It seems strange that today corporations are defining environmental virtues and the common good. The model of modern industrial capitalism appears broken. Government and other civic institutions seem to have failed. Only enlightened business practice and direct citizen action can save the planet while saving our souls. Smart consumers can be the mediators by making ethical purchasing the centerpiece of their lives. This is the memo at the heart of Chouinard's corporate *cri de coeur*—and it is a powerful message. But it is not a novel one. Humans have sought solace and salvation in the comforts of the marketplace since the dawn of capitalism. We have always invested more in our possessions than we can ever hope to withdraw. What has changed, perhaps, is our realization that our habits as consumers cost more than we have known—and that the avenues for change are different in today's interconnected world..

"We patrol a strict boundary between Nature and not-Nature," writes Jennifer Price, "with a hefty set of desires."[152] The environmental history of consumption will only come of age when scholars prove willing to ask tough questions both of consumerism and those who critique it. It will need to embrace the hybridity that has defined the field recently without becoming politically or analytically trapped by it.[153] Writing the environmental history of consumption in the twentieth and twenty-first centuries is also more complicated, in several key ways, than doing that history in the nineteenth century or before. Changes in financing, corporate structures and governance, political economy, and transportation and communication, plus the scale and scope of environmental change, pose different challenges for future scholars. Even the qualities of desire itself are historically specific. The consumer dream world of a seventeenth-century factor in the Dutch East Indian Company buying spices in Java cannot easily be compared to a software engineer in Vancouver buying free-trade coffee online today.

What unites past and present is desire, the thrill of something getting something new, the joy of a life made easier thanks to material abundance. It is easy to gainsay consumption as an environmental sin. But consumption is not always evil. It reflects one of our most basic yearnings: to connect with something real. In getting real, however, historians always should ask who sets the terms and who or what bears the expenses. Ultimately, we are all bound to the ever-changing nature of our desires, even if some wishes never deliver dreams of endless green.

NOTES

Thanks to Connie Chiang, David Gordon, Drew Isenberg, Jennifer Scanlon, Rachel Sturman, Joseph "Jay" Taylor, Jay Turner, and Allen Wells for their comments and criticisms. I developed some of these ideas previously in Matthew Klingle, "Spaces of Consumption in Environmental History," *History and Theory* 42 (December 2003): 94–110.

1. Yvon Chouinard, *Let My People Go Surfing: The Education of a Reluctant Businessman* (New York: Penguin, 2005), 3, 18.
2. Chouinard, *Let My People Go Surfing*, 38.
3. Chouinard and Tom Frost, *Chouinard 1972 Catalog*, http://www.frostworksclimbing.com/gpiw72.html. For context on Chouinard's second career as a businessman, see Joseph E. Taylor III, *Pilgrims of the Vertical: Yosemite Rock Climbers and Nature at Risk* (Cambridge, MA: Harvard University Press, 2010), 175–278.
4. Hugo Martin, "Outdoor Retailer Patagonia Puts Environment Ahead of Sales Growth," *Los Angeles Times*, May 24, 2012, http://articles.latimes.com/2012/may/24/business/la-fi-patagonia-20120525. Sales figures were based on the previous twelve months ending in April 2012.
5. Chouinard, *Let My People Go Surfing*, 3–4.
6. For past and present environmental grants and political action, see http://www.patagonia.com/us/environmentalism.
7. Dwight Macdonald, "The Theory of Mass Culture," in *Mass Culture: The Popular Arts in America*, ed. Bernard Rosenberg and David Manning White (Glencoe, IL: The Free Press, 1957), 60, quoted in David Steigerwald, "All Hail the Republic of Choice: Consumer History as Contemporary Thought," *Journal of American History* 93 (September 2006): 387–88.
8. Steigerwald, "All Hail the Republic of Choice," 388.
9. Steigerwald, "All Hail the Republic of Choice," 388.
10. Daniel Miller, "Consumption as the Vanguard of History: A Polemic by Way of an Introduction," in *Acknowledging Consumption: A Review of New Studies*, ed. Daniel Miller (London: Routledge, 1995), esp. 8–9, 41–42, quoted in Steigerwald, "All Hail the Republic of Choice," 388.
11. The best summary of this debate remains Jean-Christophe Agnew, "Coming Up For Air: Consumer Culture in Historical Perspective," in *Consumption and the World of Goods*, ed. John Brewer and Roy Porter (London: Routledge, 1993), 19–39. Another good synthesis is Peter N. Stearns, *Consumerism in World History: The Global Transformation of Desire*, 2d ed. (New York: Routledge, 2006). For other perspectives, focused largely on the Anglo-American world, see T. H. Breen, *The Marketplace of Revolution: How*

Consumer Politics Shaped American Independence (New York: Oxford University Press, 2004); Joyce Appleby, "Consumption in Early Modern Social Thought," in *Consumer Society in American History,* ed. Lawrence B. Glickman (Ithaca, NY: Cornell University Press, 1999), 130–46; James Axtell, "The First Consumer Revolution," ibid., 85–99; and Neil McKendrick, John Brewer, and J. H. Plumb, eds., *The Birth of a Consumer Society: The Commercialization of Eighteenth-Century England* (Bloomington: Indiana University Press, 1982). For a thoughtful etymology of "consumption," see Raymond Williams, *Keywords: A Vocabulary of Culture and Society* rev. ed. (New York: Oxford University Press, 1976), 78–79. For "industrious revolution," see Jan De Vries, *The Industrious Revolution: Consumer Behavior and the Household Economy, 1650 to the Present* (New York: Cambridge University Press, 2008).

12. Steigerwald, "All Hail the Republic of Choice," 389–90.

13. Lizabeth Cohen, "Is There an Urban History of Consumption?," *Journal of Urban History* 29 (January 2003): 87; see also Joy Parr, "Reinventing Consumption," *Beaver* 80 (February/March 2000): 66–73; and Donica Belisle, "Toward a Canadian Consumer History," *Labour/ Le Travail* 52 (Fall 2003): 181–206.

14. For a limited set of examples, see Stuart Ewen, *Captains of Consciousness: Advertising and the Social Roots of American Culture* (New York: McGraw-Hill, 1976); T. J. Jackson Lears, *No Place of Grace: Antimodernism and The Transformation of American Culture, 1880–1920* (New York: Pantheon, 1981); Roland Marchand, *Advertising the American Dream: Making Way for Modernity, 1920–1940* (Berkeley: University of California Press, 1985); Susan Strasser, *Satisfaction Guaranteed: The Making of the American Mass Market* (New York: Pantheon Books, 1989); Lears, *Fables of Abundance: A Cultural History of Advertising in America* (New York: Basic Books, 1994); Ian McKay, *The Quest of the Folk: Antimodernism and Cultural Selection in Twentieth-Century Nova Scotia* (Montreal: McGill-Queen's University Press, 1994); David Monod, *Store Wars: Shopkeepers and the Culture of Mass Marketing, 1890–1939* (Toronto: University of Toronto Press, 1996); and Russell Johnston, *Selling Themselves: The Emergence of Canadian Advertising* (Toronto: University of Toronto Press, 2001).

15. Selected works include Susan Porter Benson, *Counter Cultures: Saleswomen, Managers, and Customers in American Department Stores, 1890–1940* (Urbana: University of Illinois Press, 1988); Lizabeth Cohen, *Making A New Deal: Industrial Workers in Chicago, 1919–1939* (New York: Cambridge University Press, 1990); Joy Parr, *The Gender of Breadwinners: Women, Men, and Change in Two Industrial Towns, 1850–1950* (Toronto: University of Toronto Press, 1990); Dana Frank, *Purchasing Power: Consumer Organizing, Gender, and the Seattle Labor Movement, 1919–1929* (New York: Cambridge University Press, 1994); Kathy Peiss, *Hope in a Jar: The Making of America's Beauty Culture* (New York: Metropolitan Books, 1998); Jennifer Scanlon, *Inarticulate Longings: The Ladies' Home Journal, Gender, and the Promise of Consumer Culture* (New York: Routledge, 1995); Suzanne Morton, *Ideal Surroundings: Domestic Life in a Working Class Suburb in the 1920s* (Toronto: University of Toronto Press, 1995); Robert E. Weems, *Desegregating the Dollar: African American Consumerism in the Twentieth Century* (New York: New York University Press, 1998); Karen Dubinsky, *The Second Greatest Disappointment: Honeymooners, Heterosexuality, and the Tourism Industry at Niagara Falls* (New Brunswick, NJ: Rutgers University Press, 1999); Jefferson Cowie, *Capitol Moves: RCA's Seventy-Year Quest for Cheap Labor* (Ithaca, NY: Cornell University Press, 1999); Valerie J. Korinek, *Roughing It in the Suburbs: Reading Chatelaine Magazine*

in the Fifties and Sixties (Toronto: University of Toronto Press, 2000); and Steve Penfold, *The Donut: A Canadian History* (Toronto: University of Toronto Press, 2008).

16. Selected titles include Doug Owram, *Born at the Right Time: A History of the Baby Boom Generation* (Toronto: University of Toronto Press, 1996); Dana Frank, *Buy American: The Untold Story of Economic Nationalism* (Boston: Beacon Press, 1999); Joy Parr, *Domestic Goods: The Material, the Moral, and the Economic in the Postwar Years* (Toronto: University of Toronto Press, 1999); Gary Cross, *An All-Consuming Century* (New York: Columbia University Press, 2001); Meg Jacobs, *Pocketbook Politics: Economic Citizenship in Twentieth-century America* (Princeton, NJ: Princeton University Press, 2004); Kristin L. Hoganson, *Consumers' Imperium: The Global Production of American Domesticity* (Chapel Hill: University of North Carolina Press, 2007); Cohen, *A Consumer's Republic*; and Béatrice Craig, *Backwoods Consumers and Homespun Capitalists: The Rise of a Market Culture in Eastern Canada* (Toronto: University of Toronto Press, 2009). For a recent survey of this literature in US history, focusing on politics, see Meg Jacobs, "State of the Field: The Politics of Consumption," *Reviews in American History* 39 (September 2011): 561–73.

17. Warren I. Susman, "The People's Fair: Cultural Contradictions of a Consumer Society," in *Culture as History: The Transformation of American Society in the Twentieth Century*, rev. ed. (New York: Pantheon Books, 1984), 211–229.

18. Donald Worster, *Dust Bowl: The Southern Plains in the 1930s* (New York: Oxford University Press, 1979), 6.

19. William Cronon, *Nature's Metropolis: Chicago and the Great West* (New York: W. W. Norton, 1991); Peter A. Colcanis, "Urbs in Horto," *Reviews in American History* 20 (March 1992): 14–20.

20. Jennifer Price, *Flight Maps: Adventures with Nature in Modern America* (New York: Basic Books, 1999), 256.

21. Stearns, *Consumerism in World History*, 151. For other analyses of the global dimensions of consumerism, see Craig Clunas, "Modernity Local and Global: Consumption and the Rise of the West," *American Historical Review* 104 (December 1999): 1497–1511; and Jeremy Prestholdt, "On the Global Repercussions of East African Consumerism," *American Historical Review* 109 (June 2004): 755–781.

22. Joseph E. Taylor III, "Boundary Terminology," *Environmental History* 13 (July 2008): 454–481.

23. T.H. Breen, "Will American Consumers Buy a Second American Revolution?," *Journal of American History* 93 (September 2006): 405–406.

24. For the challenges of writing spatial environmental history, see Richard White, "The Nationalization of Nature," *Journal of American History* 86 (December 1999): 976–986; and Taylor, "Boundary Terminology." For this critique applied to consumption, see Klingle, "Spaces of Consumption in Environmental History."

25. For an insightful analysis of this problem for environmental historians, see Mark Fiege, "The Weedy West: Mobile Nature, Boundaries, and Common Space in the Montana Landscape," *Western Historical Quarterly* 36 (Spring 2005): 22–48.

26. For representative studies, see Paul R. Josepheson, *Resources Under Regimes: Technology, Environment, and the State* (Cambridge, MA: Harvard University Press, 2005); Tom McCarthy, *Auto Mania: Cars, Consumers, and the Environment* (New Haven, CT: Yale University Press, 2007); Timothy J. LeCain, *Mass Destruction: The Men and the Giant Mines that Wired America and Scarred the Planet* (New Brunswick, NJ: Rutgers

University Press, 2009); and Matthew Evenden, "Aluminum, Commodity Chains, and the Environmental History of the Second World War," *Environmental History* 16 (January 2011): 69–93. For a recent summary of "envirotech," see Dolly Jørgensen, Finn Arne Jørgensen, and Sara Pritchard, eds., *New Natures: Joining Environmental History with Science and Technology Studies* (Pittsburgh: University of Pittsburgh Press, 2013). For technology as politics, see Langdon Winner, "Do Artifacts Have Politics?," *Daedalus*, 109 (Winter 1980): 121–136, and *The Whale and the Reactor: A Search for Limits in an Age of High Technology* (Chicago: University of Chicago Press, 1986).

27. Paul S. Sutter, "The World With Us: The State of Environmental History," *Journal of American History* 100 (June 2013): 94–120.

28. Quoted in Price, *Flight Maps*, xix.

29. Chouinard, *Let My People Go Surfing*, 258.

30. Raymond Williams, "Ideas of Nature," in *Problems in Materialism and Culture: Selected Essays* (London: Verso, 1980), 67–85. Donald Worster makes a similar point, but instead of going back to Karl Marx, as Williams did, he referred to the Frankfurt School thinkers, Max Horkheimer and Theodor Adorno. See *Rivers of Empire: Water, Aridity, and the Growth of the American West* (New York: Pantheon, 1985), 22–60.

31. For entwined and divergent trajectories of the two terms, see Donald Worster, *Nature's Economy: A History of Ecological Ideas* (San Francisco: Sierra Club Books, 1977), 1–56; and Robert P. McIntosh, *The Background of Ecology: Concept and Theory* (New York: Cambridge University Press, 1985), 1–27.

32. For a very small sampling of these critiques in popular culture, see Alan Thein During and John C. Ryan, *Stuff: The Secret Lives of Everyday Things* (Seattle: Northwest Environment Watch, 1977); Bill McKibben, *Maybe One: The Case for Smaller Families* (New York: Plume, 1999); Colin Beavan, *No Impact Man: The Adventures of a Guilty Liberal Who Attempts to Save the Planet, and the Discoveries He Makes About Himself and Our Way of Life in the Process* (New York: Farrar, Strauss, and Giroux, 2009); and Annie Leonard, *The Story of Stuff: The Impact of Overconsumption on the Planet, Our Communities, and How We Can Make it Better* (New York: Free Press, 2010)

33. Andrew Szasz, *Shopping our Way to Safety: How We Changed from Protecting the Environment to Protecting Ourselves* (Minneapolis: University of Minnesota Press, 2007); see also the collected essays in *Confronting Consumption*, ed. Thomas Princen, Michael Maniates, and Ken Conca (Cambridge, MA: The MIT Press, 2002).

34. James Scott, *Seeing Like a State: How Certain Schemes to Improve the Human Condition Have Failed* (New Haven, CT: Yale University Press, 1998).

35. Fernando Coronil, *The Magical State: Nature, Money, and Modernity in Venezuela* (Chicago: University of Chicago Press, 1997), 21.

36. Samuel P. Hays, *Conservation and the Gospel of Efficiency* (Cambridge, MA: Harvard University Press, 1959), remains the classic, if now widely contested, summary of this era.

37. Sarah T. Phillips, *This Land, This Nation: Conservation, Rural America, and the New Deal* (New York: Cambridge University, 2007), 4. See also Neil M. Maher, *Nature's New Deal: The Civilian Conservation Corps and the Roots of the American Environmental Movement* (New York: Oxford University Press, 2007), 115–210.

38. Perhaps nowhere was the linkage between agribusiness and federal support stronger than in the trans-Mississippi West. For example, see Donald J. Pisani, *From Family Farm to Agribusiness: The Irrigation Crusade in California and the West, 1850–1931* (Berkeley: University of California Press, 1984); Steven Stoll, *The Fruits of Natural Advantage: Making the Industrial Countryside in California* (Berkeley: University

of California Press, 1998); David Igler, *Industrial Cowboys: Miller & Lux and the Transformation of the Far West, 1850–1920* (Berkeley: University of California Press, 2001); and Douglas Cazaux Sackman, *Orange Empire: California and the Fruits of Eden* (Berkeley: University of California Press, 2005).

39. For the New Deal, see Phillips, *This Land, This Nation*, 238–83; and "Lessons from the Dust Bowl: Dryland Agriculture and Soil Erosion in the Untied States and South Africa, 1900–1950," *Environmental History* 4 (April 1999): 245–266. For Cold War foreign policy, see Nick Cullather, *The Hungry World: America's Cold War Battle against Poverty in Asia* (Cambridge, MA: Harvard University Press, 2010).

40. Liza Piper, *The Industrial Transformation of Subarctic Canada* (Vancouver: University of British Columbia Press, 2009), 288. For later health effects, see Piper, "Chronic Disease in the Yukon River Basin, 1890–1960," in *Locating Health: Historical and Anthropological Investigations of Health and Place*, ed. Erika Dyck and Christopher Fletcher (London: Pickering and Chatto, 2011), 129–49; and "Nutritional Science, Health, and Changing Northern Environments," in *Big Country, Big Issues: Canada's Environment, Culture, and History*, ed. Nadine Klopfer and Christof Mauch (Munich: Rachel Carson Center Perspectives Series, 2011/4): 60–85.

41. Coronil, *The Magical State*, 9.

42. Miguel Tinker Salas, *The Enduring Legacy: Oil, Culture, and Society in Venezuela* (Durham, NC: Duke University Press, 2009), 239.

43. For a thoughtful comparison to another petrostate, see Michael Watts, "Petro-Violence: Community, Extraction, and Political Ecology of a Mythic Commodity," in *Violent Environments*, ed. Nancy Peluso and Michael Watts (Ithaca, NY: Cornell University Press, 2001), 189–212; and Watts and Ed Kashi, *Curse of the Black Gold: 50 Years of Oil in the Niger Delta* (New York: PowerHouse Books, 2008).

44. Salas, *The Enduring Legacy*, 237–250.

45. Deborah R. Coen, "The Greening of German History," *Isis* 99 (2008): 143.

46. David Blackbourn, *The Conquest of Nature: Water, Landscape, and the Making of Modern Germany* (New York: W. W. Norton, 2006), 77–250; Mark Cioc, *The Rhine: An Eco-Biography, 1815–2000* (Seattle: University of Washington Press, 2002).

47. Thomas M. Lekan, *Imagining the Nation in Nature: Landscape Preservation and German Identity, 1885–1945* (Cambridge, MA: Harvard University Press, 2004); Thomas Zeller, *Driving Germany: The Landscape of the German Autobahn, 1930–1970* (New York: Berghahn Books, 2007); and *How Green Were the Nazis?: Nature, Environment, and Nation in the Third Reich*, ed. Franz-Josef Brüggemeier, Mark Cioc, and Thomas Zeller (Athens: Ohio University Press, 2005).

48. For postwar Germany, East and West, see Blackbourn, *The Conquest of Nature*, 311–346; Sandra Chaney, "For Nation and Prosperity, Health and a Green Environment: Protecting Nature in West Germany, 1945–1970," in *Nature in German History*, ed. Christof Mauch (New York: Berghahn Books, 2004), 93–118; and Sandra Chaney, "Protecting Nature in a Divided Nation: Conservation in the Two Germanys, 1945–1972," in *Germany's Nature: Cultural Landscapes and Environmental History* (New Brunswick, NJ: Rutgers University Press, 2005), 207–243. For the CDU campaign poster, see http://www.monheim.de/stadtprofil/historisches/lexikon/einzel-handel.html. Thanks to Drew Isenberg for helping me to navigate modern German history.

49. Sara B. Pritchard, "Reconstructing the Rhône: The Cultural Politics of Nature and Nation in Contemporary France, 1945–1997," *French Historical Studies* 27 (Fall 2004): 766–767.

50. Michael Bess, *The Light-Green Society: Ecology and Technological Modernity in France, 1960–2000* (Chicago: University of Chicago Press, 2003), 4; see also Pritchard, *Confluence: The Nature of Technology and the Remaking of the Rhône* (Cambridge, MA: Harvard University Press, 2011).

51. Bess, *The Light-Green Society*, 11–114; Pritchard, "Reconstructing the Rhône," 781.

52. Bess, *The Light-Green Society*, 161–217.

53. Bess, *The Light-Green Society*, 241.

54. James Morton Turner, "From Woodcraft to 'Leave No Trace': Wilderness, Consumerism, and Environmentalism in Twentieth-Century America," *Environmental History* 7 (July 2002): 462–484. For another article that traces these relationships with an attention to gender as well, see Phoebe S. Kropp, "Wilderness Wives and Dishwashing Husbands: Comfort and the Domestic Arts of Camping in America, 1880–1930," *Journal of Social History* 43 (Fall 2009): 5–30. For a reassessment of Leave No Trace, see Gregory L. Simon and Peter S. Alagona, "Beyond Leave No Trace," *Ethics, Place, and Environment* 12 (March 2009): 17–34.

55. Denis Hayes, "The Beginning," from *Earth Day: The Beginning—A Guide for Survival*, ed. Environment Action (New York: Arno Press and *The New York Times*, 1970), ii. Hayes originally gave his remarks in a speech at the Sylvan Theater in Washington, DC, on April 22, 1970. For more on Earth Day, see Adam Rome, *The Genius of Earth Day: How a 1970 Teach-In Unexpectedly Made the First Green Generation* (New York: Hill & Wang, 2012).

56. Thomas Robertson, *The Malthusian Moment: Global Population Growth and the Birth of American Environmentalism* (New Brunswick, NJ: Rutgers University Press, 2012); Paul R. Ehrlich, *The Population Bomb* (New York: Ballantine Books, 1968); Paul Sabin, *The Bet: Paul Ehrlich, Julian Simon, and the Our Gamble over Earth's Future* (New Haven, CT: Yale University Press, 2013); and Michael Egan, *Barry Commoner and the Science of Survival: The Remaking of American Environmentalism* (Cambridge, MA: The MIT Press, 2007).

57. Adam Rome, "'Give Earth a Chance': The Environmental Movement and the Sixties," *Journal of American History* 90 (September 2003): 535–554; Andrew G. Kirk, *Counterculture Green: The Whole Earth Catalog and American Environmentalism* (Lawrence: University Press of Kansas, 2007); see also Jeffrey Craig Sanders, *Seattle and the Roots of Urban Sustainability: Inventing Ecotopia* (Pittsburgh: University of Pittsburgh Press, 2010), for counterculture environmentalism at work in an urban context.

58. Kirk, *Counterculture Green*, 217.

59. William Hard, "Giant Negotiations for Giant Power: An Interview with Herbert Hoover," *Survey Graphic* 5 (March 1924): 577; also quoted in Kendrick A. Clements, *Hoover, Conservation, and Consumerism: Engineering the Good Life* (Lawrence: University Press of Kansas, 2000), 78.

60. For example, see Richard W. Judd, *Common Lands, Common People: The Origins of Conservation in Northern New England* (Cambridge, MA: Harvard University Press, 1997), 123–145, 197–228.

61. Thomas G. Andrews, "'Made by Toile'? Tourism, Labor, and the Construction of the Colorado Landscape, 1858–1917," *Journal of American History* 92 (December 2005): 837–863.

62. Connie Y. Chiang, *Shaping the Shoreline: Fisheries and Tourism on the Monterey Coast* (Seattle: University of Washington Press, 2008), 12–100.

63. Lawrence M. Lipin, *Workers in the Wild: Conservation, Consumerism, and Labor in Oregon, 1910–30* (Urbana: University of Illinois Press, 2007).

64. Lipin, *Workers in the Wild*, 156, 158.

65. Adam Rome, *The Bulldozer in the Countryside: Suburban Sprawl and the Making of American Environmentalism* (New York: Cambridge University Press, 2001); Peter Siskind, "Suburban Growth and Its Discontents: The Logic and Limits of Reform on the Postwar Northeast Corridor," in *The New Suburban History*, ed. Kevin Kruse and Thomas Sugrue (Chicago: University of Chicago Press): 161–182.

66. Gregory Summers, *Consuming Nature: Environmentalism in the Fox River Valley, 1850–1950* (Lawrence: University Press of Kansas, 2006), 113.

67. Summers, *Consuming Nature*, 200.

68. Chad Montrie, "A Decent, Wholesome Living Environment for Everyone: Michigan Autoworkers and the Origins of Modern Environmentalism," in *Making a Living: Work and Environment in the United States* (Chapel Hill: University of North Carolina Press, 2008), 91–112; Andrew Hurley, *Environmental Inequalities: Race, Class, and Pollution in Gary, Indiana, 1945–1980* (Chapel Hill: University of North Carolina Press, 1995), 135–153; Mark Harvey, *Wilderness Forever: Howard Zahniser and the Path to the Wilderness Act* (Seattle: University of Washington Press, 2005), 96–128.

69. For this context on the 1999 WTO protests and riots, see Matthew Klingle, *Emerald City: An Environmental History of Seattle* (New Haven, CT: Yale University Press, 2007), 250–253; see also the WTO History Project, http://depts.washington.edu/wtohist/.

70. Bonnie Christensen, *Red Lodge and the Mythic West: Coal Miners to Cowboys* (Lawrence: University Press of Kansas, 2002); Annie Gilbert Coleman, *Ski Style: Sport and Culture in the Rockies* (Lawrence: University Press of Kansas, 2004), 117–214; and William Philpott, *Vacationland: Tourism and Environment in the Colorado High Country* (Seattle: University of Washington Press, 2013).

71. Hal Rothman, *Devil's Bargains: Tourism in the Twentieth-Century West* (Lawrence: University Press of Kansas, 1998); William Jordan, "Ten Thousand Thoreaus," *Ecological Restoration* 18 (Winter 2000): 215. For a variation on Jordan's argument, focusing on how the modern sport of rock climbing has yielded both a new industry of leisure and attendant environmental damages in pursuit of outdoor fun, see Taylor, *Pilgrims of the Vertical*, 233–278.

72. John Kenneth Galbraith, "How Much Should a Country Consume?," in *Perspectives on Conservation*, ed. Henry Jarret (Baltimore: The Johns Hopkins University Press, 1958), 91–92, as quoted in Ramachandra Guha, *How Much Should a Person Consume?: Environmentalism in India and the United States* (Berkeley: University of California Press and Delhi: Permanent Black, 2006), 221–222.

73. Guha, *How Much Should a Person Consume?*, 223.

74. For a damning indictment of how California's environmental policies have saved places at home while encouraging the exploitation of pelagic fisheries and boreal forests in Canada and petroleum reserves in Amazonian Ecuador to feed Californians' demand for food, housing, and energy, see Tom Knudson, "State of Denial" and "Moving Beyond Denial," *Sacramento Bee*, April 27, 2002, *www.sacbee.com/static/live/news/projects/denial/*.

75. Guha's critique of Western environmentalism extends to the practice of environmental history itself. As he writes, North American environmental historians practice "a studied insularity" in method and approach. Few have "seriously studied the global consequences of consumerism, the impact on land, soil, forests, and climate, of the American way of life," an irony, he writes, made all the more poignant in the wake of the Gulf War and the quest to secure what the British newspaper *The Guardian* called "The American Way of Driving." As Guha concludes, "no American historian has to my knowledge taken

to heart the wisdom in that throwaway remark, to reveal in all its starkness the ecological imperialism of the world's sole superpower." Guha overstates the case, but only slightly. See Guha, *How Much Should a Person Consume?*, 227–228.

76. Alfred W. Crosby, *Ecological Imperialism: The Biological Expansion of Europe, 900–1900*, rev. ed. (New York: Cambridge University Press, 1984), 272–280; see also *The Columbian Exchange: Biological and Cultural Consequences of 1492* (Westport, CT: Greenwood Press, 1972).

77. William Cronon, *Changes in the Land: Indians, Ecologists, and the Ecology of New England* 20th anniversary ed. (New York: Hill and Wang, 1983); 159–170; for similar analyses of other locations, see also Richard White, *Land Use, Environment, and Social Change: The Shaping of Island County, Washington* (Seattle: University of Washington, 1980); and Timothy Silver, *A New Face on the Countryside: Indians, Colonists, and Slaves in South Atlantic Forests, 1500–1800* (New York: Cambridge University Press, 1990).

78. Sidney W. Mintz, *Sweetness and Power: The Place of Sugar in Modern History* (New York: Penguin Books, 1985), 214.

79. Immanuel Wallerstein, *The Modern World-System*, vol. I: *Capitalist Agriculture and the Origins of the European World-Economy in the Sixteenth Century* (New York: Academic Press, 1974); *The Modern World-System*, vol. II: *Mercantilism and the Consolidation of the European World-Economy, 1600–1750* (New York: Academic Press, 1980); *The Modern World-System*, vol. III: *The Second Great Expansion of the Capitalist World-Economy, 1730–1840's* (San Diego: Academic Press, 1989). For a small selection of dependency theory, see André Gunder Frank, *On Capitalist Underdevelopment* (New York: Oxford University Press, 1975); and Fernando Enrique Cardoso and Enzo Faletto, *Dependency and Development in Latin America*, trans. Majory Mattingly Urquidi (Berkeley: University of California Press, 1979). For an overview of commodity chains and rival (if complimentary) approaches, see Phlip Raikes, Michael Friis Jensen, and Stefano Ponte, "Global Commodity Chain Analysis and the French *filière* Approach: Comparison and Critique," *Economy and Society* 29 (August 2000): 390–417. A fuller comparison of these two approaches is beyond the scope of this essay. For an early application of commodity chains, see Gary Gereffi and Miguel Korzeniewicz, eds., *Commodity Chains and Global Capitalism* (Westport, CT: Greenwood Press, 1994). For a more recent overview, see Jennifer Bair, ed. *Frontiers of Commodity Chain Research* (Stanford, CA: Stanford University Press, 2009).

80. For one example of these theories applied to Latin America, see Kenneth Duncan and Ian Rutledge, eds., *Land and Labor in Latin America* (New York: Cambridge University Press, 1977). For a summary of this scholarship, coupled with a thoughtful rereading of world-systems theory as environmental history, see Jason W. Moore, "The Modern World-System as Environmental History?: Ecology and the Rise of Capitalism," *Theory and Society* 32 (June 2003): 307–377.

81. For two useful syntheses by social scientists that critique commodity chains, see Paul Ciccantell and David A. Smith, "Rethinking Global Commodity Chains: Integrating Extraction, Transport, and Manufacturing," *International Journal of Comparative Sociology* 50 (June/August 2009): 361–384 and Moore, "The Modern World-System as Environmental History?," 359.

82. Warren Dean, *Brazil and the Struggle for Rubber: A Study in Environmental History* (New York: Cambridge University Press, 1987). Dean's last and posthumous book further explored environmental history but paid little attention to global trade or consumption.

See *With Broadax and Firebrand: The Destruction of the Brazilian Atlantic Forest* (Berkeley: University of California Press, 1995), 347.

83. Salient studies include Allen Wells, *Yucatán's Gilded Age: Haciendas, Henequen, and International Harvester* (Albuquerque: University of New Mexico Press, 1985); Carl E. Solberg, *The Prairies and the Pampas: Agrarian Policy in Canada and Argentina, 1880–1930* (Stanford, CA: Stanford University Press, 1987); Paul Gootenberg, *Between Silver and Guano: Commercial Policy and the State in Post-Independence Peru* (Princeton, NJ: Princeton University Press, 1989); Jeremy Adelman, *Frontier Development: Land, Labour, and Capital on the Wheatlands of Argentina and Canada, 1890–1914* (Oxford: Clarendon, 1994); Jimmy M. Skaggs, *The Great Guano Rush: Entrepreneurs and American Overseas Expansion* (New York: St. Martin's Press, 1994); William Roseberry, Lowell Gudmundson, and Mario Samper Kutschbach, eds., *Coffee, Society, and Power in Latin America* (Baltimore: The Johns Hopkins University Press, 1995); Fernando Coronil, *The Magical State: Nature, Money, and Modernity in Venezuela* (Chicago: University of Chicago Press, 1997); Steve Striffler, *In the Shadows of State and Capital: The United Fruit Company, Popular Struggle, and Agrarian Restructuring in Ecuador, 1900–1995* (Durham, NC: Duke University Press, 2002); Sven Beckert, "Reconstructing the Worldwide Web of Cotton in the Age of the American Civil War," *American Historical Review* 109 (December 2004): 1405–1438; Gregory T. Cushman, "'The Most Valuable Birds in the World': International Conservation Science and the Revival of Peru's Guano Industry, 1909–1965," *Environmental History* 10 (July 2005): 477–509; Sterling Evans, *Bound in Twine: The History and Ecology of the Henequen-Wheat Complex for Mexico and the American and Canadian Plains, 1880–1950* (College Station: Texas A&M University Press, 2007); Paul Gootenberg, *Andean Cocaine: The Making of a Global Drug* (Chapel Hill: University of North Carolina Press, 2008); and Cushman, *Guano and the Opening of the Pacific World* (New York: Cambridge University Press, 2013).

84. For commodity chain theory and its analogs, see Terence K. Hopkins and Immanuel Wallerstein, "Commodity Chains in the World Economy Prior to 1800," *Review* 10, no. 1 (1986); 151–170; Gary Gereffi, "The Organization of Buyer-Driven Global Commodity Chains: How U.S. Retailers Shape Overseas Production Networks," in *Commodity Chains and Global Capitalism*, ed. Gary Gereffi and Miguel Korzeniewitz (Westport, CT: Praeger, 1994), 95–112; John Talbot, "The Struggle for Control of a Commodity Chain: Instant Coffee from Latin America," *Latin American Research Review* 32 (1997): 117–135; and Philip Raikes, Michael Fris Jensen, and Stefano Ponte, "Global Commodity Chain Analysis and the French *Filière* Approach: Comparison and Critique," *Economy and Society* 29 (August 2000): 380–417. For an excellent example of these theories applied to specific commodities, see *From Silver to Cocaine: Latin American Commodity Chains and the Building of the World Economy, 1500–2000*, ed. Steven Topik, Carlos Marichal, and Zephyr Frank (Durham, NC: Duke University Press, 2006). The best and most cogent survey of this scholarship is Steven C. Topik and Allen Wells, "Commodity Chains in a Global Economy," in *A World Connecting, 1870–1945*, ed. Emily S. Rosenberg (Cambridge, MA: The Belknap Press of Harvard University Press, 2012), 593–812.

85. Richard Tucker, *Insatiable Appetite: The United States and the Ecological Degradation of the Tropical World* (Berkeley: University of California Press, 2000), 11. For a less scholarly summation of this relationship, see John Vandermeer and Ivette Perfecto, *Breakfast of Biodiversity: The Truth about Rainforest Destruction*, foreword by Vandana Shiva

(Oakland, CA: Food First, 1995). The exploitation of the tropical world for luxury goods began well before the twentieth century. For one compelling study on the tropical hardwood trade in mahogany, see Jennifer L. Anderson, *Mahogany: The Costs of Luxury in Early America* (Cambridge, MA: Harvard University Press, 2012).

86. John Soluri, *Banana Cultures: Agriculture, Consumption, and Environmental Change in Honduras and the United States* (Austin: University of Texas Press, 2005); and Steve Marquardt, "'Green Havoc': Panama Disease, Environmental Change, and Labor Process in the Central American Banana Industry," *American Historical Review* 106, no. 1 (February 2001): 49–80.

87. For an insightful overview of this history, see Richard White, "Animals and Enterprise," in *The Oxford History of the American West*, ed. Clyde A. Milner II, Carol A. O'Connor, and Martha A. Sandweiss (New York: Oxford University Press, 1994), 237–273 (quotation at 237). For bison, see Dan Flores, "Bison Ecology and Bison Diplomacy: The Southern Plains from 1800 to 1850," *Journal of American History* 78 (September 1992): 465–485; Andrew C. Isenberg, *The Destruction of the Bison* (New York: Cambridge University Press, 2000); and Theodore Binnema, *Common and Contested Ground: A Human and Environmental History of the Northwestern Plains* (Norman: University of Oklahoma Press, 2001).

88. Isenberg, *The Destruction of the Bison*, 164–192; John Sandlos, *Hunters at the Margins: Native Peoples and Wildlife Conservation in the Northwest Territories* (Vancouver: University of British Columbia Press, 2007), 87–105.

89. Eric Schlosser, *Fast Food Nation: The Dark Side of the All-American Meal* (New York: Houghton Mifflin Harcourt, 2001); *Industrializing Organisms: Introducing Evolutionary History*, ed. Susan R. Schrepfer and Philip Scranton (New York: Routledge, 2004); Steve Striffler, *Chicken: The Dangerous Transformation of America's Favorite Food* (New Haven, CT: Yale University Press, 2005); Roger Horowitz, *Putting Meat on the American Table: Taste, Technology, Transformation* (Baltimore: The Johns Hopkins University Press, 2006); and Sam White, "From Globalized Pig Breeds to Capitalist Pigs: A Study in Animal Cultures and Evolutionary History," *Environmental History* 16 (January 2011): 94–120.

90. For a useful survey of this continuing oversight within the field, see W. Jeffrey Bolster, "Opportunities in Marine Environmental History," *Environmental History* 11 (July 2006): 567–597; see also Joseph E. Taylor III, "Knowing the Black Box: Methodological Challenges in Marine Environmental History," *Environmental History* 18 (January 2013): 60–75. One study seeking to redress this gap is Nancy Shoemaker, "Whale Meat in American History," *Environmental History* 10 (April 2005): 269–294.

91. McEvoy, *The Fisherman's Problem*; Joseph E. Taylor III, *Making Salmon: An Environmental History of the Northwest Fisheries Crisis* (Seattle: University of Washington Press, 1999), 257. Another recent work explores the fisheries decline in the North Atlantic is W. Jeffrey Bolster, *The Mortal Sea: Fishing the Atlantic in the Age of Sail* (Cambridge, MA: The Belknap Press of Harvard University Press, 2012).

92. For example, see David F. Arnold, *The Fishermen's Frontier: People and Salmon in Southeast Alaska* (Seattle: University of Washington Press, 2008); Brian J. Payne, *Fishing a Borderless Sea: Environmental Territorialism in the North Atlantic, 1818–1910* (East Lansing: Michigan State University Press, 2010); Matthew McKenzie, *Clearing the Coastline: The Nineteenth-Century Ecological and Cultural Transformations of Cape Cod* (Lebanon, NH: University Press of New England, 2011); and Lissa Wadewitz, *The*

Nature of Borders: Salmon, Boundaries, and Bandits on the Salish Sea (Seattle: University of Washington Press, 2012). Other salient and best-selling examples include Mark Kurlansky, *Cod: A Biography of a Fish that Changed the World* (New York: Penguin, 1998); and Paul Greenberg, *Four Fish: The Future of the Last Wild Food* (New York: Penguin, 2010). For one recent popular work on aquaculture, see Paul Molyneaux, *Swimming in Circles: Aquaculture and the End of Wild Oceans* (New York: Basic Books, 2007). For a thoughtful analysis of aquaculture by an environmental historian, see John Soluri, "Something Fishy: Chile's Blue Revolution, Commodity Diseases, and the Problem of Sustainability," *Latin American Research Review* 46, Special Issue (2011): 55–81.

93. One notable popular history is Elizabeth Grossman, *High Tech Trash: Digital Devices, Hidden Toxics, and Human Health* (Washington, DC: Island Press and Shearwater Books, 2006). For another suggestive piece, see Aaron Sachs, "Virtual Ecology: A Brief Environmental History of Silicon Valley," *World Watch* 12 (January/February 1999): 12–21. For the Pulitzer Prize-winning investigation into Apple's production practices in China, see the 2012 special series by Charles Duhigg and Keith Bradsher, "The iEconomy," *New York Times*, http://www.nytimes.com/interactive/business/ieconomy.html. For liquid-crystal displays, or LCDs, see Nicholson Baker, "A Fourth State of Matter: Inside South Korea's LCD Revolution," *The New Yorker*, July 8 and 15, 2013, 63–73. For representative histories of mining, see Kathryn T. Morse, *The Nature of Gold: An Environmental History of the Klondike Gold Rush* (Seattle: University of Washington Press, 2003); LeCain, *Mass Destruction*; Evenden, "Aluminum, Commodity Chains, and the Environmental History of the Second World War"; and Kent Curtis, *Gambling on Ore: The Nature of Metal Mining in the United States, 1860–1910* (Boulder: University Press of Colorado, 2013).

94. For Google's energy use, see James Glanz, "Google Details, and Defends, its Use of Electricity," *New York Times*, September 9, 2011, B1. For Google's explanation, see http://www.google.com/about/datacenters/, http://www.google.com/green/bigpicture/ and "Google's Green Computing: Efficiency at Scale," http://static.googleusercontent.com/external_content/untrusted_dlcp/www.google.com/en/us/green/pdfs/google-green-computing.pdf.

95. Susman, "Culture and Communications," in *Culture as History*, 252–270.

96. David E. Nye, *Consuming Power: A Social History of American Energies* (Cambridge, MA: The MIT Press, 1998), 1.

97. The book that established electricity as a social process was Thomas P. Hughes, *Electrification in Western Society, 1880–1930* (Baltimore: The Johns Hopkins University Press, 1983); see also David E. Nye, *Electrifying America: Social Meanings of a New Technology* (Cambridge, MA: The MIT Press, 1990). For electricity and domestic work, see Ruth Schwartz Cowan, *More Work for Mother: The Ironies of Household Technology from the Open Hearth to the Microwave* (New York: Basic Books, 1983), 69–101.

98. Richard White, *The Organic Machine: The Remaking of the Columbia River* (New York: Hill & Wang, 1995), ix.

99. Bruce Stadfeld, "Electric Space: Social and Natural Transformations in British Columbia's Hydroelectric Industry to World War II" (PhD diss., University of Manitoba, 2002); Matthew D. Evenden, *Fish versus Power: An Environmental History of the Fraser River* (New York: Cambridge University Press, 2007); Sarah S. Elkind, *How Local Politics Shapes Federal Policy: Business, Power, and the Environment in Twentieth-Century Los Angeles* (Chapel Hill: University of North Carolina Press, 2011); Pritchard, *Confluence*;

and Andrew Needham, *Power Lines: Energy and the Making of the Modern Southwest* (Princeton, NJ: Princeton University Press, 2014).

100. This is not to say that environmental historians and others have not examined the political economy or environmental consequences of petroleum and other hydrocarbons. For selected examples, see Paul Sabin, "Voices from the Hydrocarbon Frontier: Canada's Mackenzie Valley Pipeline Inquiry, 1974–1977," *Environmental History Review* 19 (Spring 1995): 17–48 and "Searching for Middle Ground: Native Communities and Oil Extraction in the Northern and Central Ecuadorian Amazon, 1967–1993," *Environmental History* 3 (April 1998): 144–168; Brian Black, *Petrolia: The Landscape of America's First Oil Boom* (Baltimore: The Johns Hopkins University Press, 2000); and Myrna I. Santiago, *The Ecology of Oil: Environment, Labor, and the Mexican Revolution, 1900–1983* (New York: Cambridge University Press, 2006); Thomas Andrews, *Killing for Coal: America's Deadliest Labor War* (Cambridge, MA: Harvard University Press, 2008); and "Oil in American History: A Special Issue," *Journal of American History* 99 (June 2012): 18–255, http://www.journalofamericanhistory.org/projects/oil/. One of the most prolific scholars of petroleum is Joseph A. Pratt. See, for example, "Letting the Grandchildren Do It: Environmental Planning During the Ascent of Oil as a Major Energy Source," *The Public Historian* 2 (Summer 1980): 28–61, and "Warts and All?: An Elusive Balance in Contracted Corporate Histories about Energy and Environment," *The Public Historian* 26 (Winter 2004): 21–39.

101. Paul Sabin, *Crude Politics: The California Oil Market, 1900–1940* (Berkeley: University of California Press, 2005), 6. For a survey of California's energy economies, see James C. Williams, *Energy and the Making of Modern California* (Akron, OH: University of Akron Press, 1997).

102. Christopher W. Wells, "Fueling the Boom: Gasoline Taxes, Invisibility, and the Growth of the American Highway Infrastructure, 1919–1956," *Journal of American History* 99 (June 2013): 73.

103. Clay McShane and Joel A. Tarr, *The Horse in the City: Living Machines in the Nineteenth Century* (Baltimore: The Johns Hopkins University Press, 2007).

104. Paul S. Sutter, *Driven Wild: How the Fight Against Automobiles Launched the Modern Wilderness Movement* (Seattle: University of Washington Press, 2002); David Louter, *Windshield Wilderness: Cars, Roads, and Nature in Washington's National Parks* (Seattle: University of Washington Press, 2006); Rudy Koshar, "Organic Machines: Cars, Drivers, and Nature from Imperial to Nazi Germany," in *Germany's Nature: Cultural Landscapes and Environmental History*, ed. Thomas M. Lekan and Thomas Zeller (New Brunswick, NJ: Rutgers University Press, 2005), 111–139.

105. For GMAC, see Louis Hyman, *Debtor Nation: The History of America in Red Ink* (Princeton, NJ: Princeton University Press, 2011), 20–27; for Henry Ford and early environmental history of automobiles in the United States, see McCarthy, *Auto Mania*, 1–147 and "Henry Ford, Industrial Ecologist or Industrial Conservationist?: Waste Reduction and Recycling at the Rouge," *Michigan Historical Review* 27 (2001): 52–88. For planned obsolescence, see McCarthy and Ted Steinberg, *Down to Earth: Nature's Role in American History* (New York: Oxford University Press, 2002, 2008), 206–228.

106. For the environmental history of automobiles and their effects on the United States landscape, see Christopher W. Wells, *Car Country: An Environmental History* (Seattle: University of Washington Press, 2012); for SUVs and postwar debates, see McCarthy, *Auto Mania*, 148–266.

107. Susan Strasser, *Waste and Want: A Social History of Trash* (New York: Henry Holt, 1999), 200–201; see also Carl A. Zimring, *Cash for Your Trash: Scrap Recycling in America* (New Brunswick, NJ: Rutgers University Press, 2005).

108. Strasser, *Waste and Want*, 5.

109. For the tale of the garbage barge, see Steinberg, *Down to Earth*, 235–236, and Michael Winerip, "Voyage of the *Mobro 4000*: Retro Report—The Big Stories Then in the Light of Now," *New York Times* (May 6, 2013), http://www.nytimes.com/2013/05/06/booming/new-video-series-re-examines-garbage-barge-fiasco.html. For the globalization of the garbage trade, see Emily Brownell, "Negotiating the New Economic Order of Waste," *Environmental History* 16 (April 2011): 226–261.

110. Karen Tranberg Hansen, *Salaula: The World of Secondhand Clothing and Zambia* (Chicago: University of Chicago Press, 2000), 256.

111. Donovan Hohn, *Moby-Duck: The True Story of 28,800 Bath Toys Lost at Sea and of the Beachcombers, Oceanographers, Environmentalists, and Fools, Including the Author, Who Went in Search of Them* (New York: Penguin, 2011); Richard C. Thompson, Ylva Olsen, Richard P. Mitchell, et al., "Lost at Sea: Where Is All the Plastic?" *Science* 304 (May 7, 2004): 838. See also Curtis Ebbesmyer and Eric Scigliano, *Flotsametrics and the Floating World: How One Man's Obsession with Runaway Sneakers and Rubber Ducks Revolutionized Ocean Science* (Washington, DC: Smithsonian, 2009).

112. There is a rich and extensive historical and anthropological literature on the O'Odham that is too large to cite here. The general contours of the details in this paragraph come from Robert Rossell, "History of the Pima," Gary Witherspoon, "Pima and Papago Ecological Adaptations," and Leyland C. Wyman, "Pima and Papago Social Organization," from *Handbook of North American Indians: Volume 10: Southwest*, ed. Alfonzo Ortiz (Washington, DC: Smithsonian Institution, 1983): 149–192.

113. For the history of the Gila seizure, see David H. DeJong, *Stealing the Gila: The Pima Agricultural Economy and Water Deprivation, 1848–1921* (Tucson: University of Arizona Press, 2009).

114. For diabetes prevalence, see Max Miller, Thomas A. Burch, Peter H. Bennett, and Arthur G. Steinberg, "Prevalence of Diabetes Mellitus in American Indians: Results of Glucose Tolerance Tests in the Pima Indians of Arizona" (Abstract), *Diabetes* 14 (July 1965): 439–440, and Maurice L. Sievers, "Disease Patterns Among Southwestern Indians," *Public Health Reports* 81 (December 1966): 1078. For one survey of this history, see M. Yvonne Jackson, "Diet, Culture, and Diabetes" and Cynthia J. Smith, Elaine M. Mahahan, and Sally G. Pablo, "Food Habit and Cultural Changes among the Pima Indians," in *Diabetes as a Disease of Civilization: The Impact of Culture Change on Indigenous Peoples*, ed. Jennie Rose Joe and Robert S. Young (Berlin: Mouton de Gruyter, 1994), 381–434; see also Carolyn Smith-Morris, *Diabetes Among the Pima: Stories of Survival* (Tucson: University of Arizona Press, 2006). For the final quotation, see Peter H. Bennett, "Type 2 Diabetes Among the Pima Indians of Arizona: An Epidemic Attributable to Environmental Change?," *Nutrition Reviews* 57 (May 1999): S53.

115. Roy Porter, "Consumption: Disease of the Consumer Society?," in *Consumption and the World of Goods*, 59.

116. Porter, "Consumption," 59, 63.

117. George Cheyne, *The English Malady; or, A Treatise of Nervous Diseases* (London: G. Strahan, 1733), 174, 49, as quoted in Porter, 63–64.

118. Porter, "Consumption," 64.

119. Thomas Beddoes, *Essay on the Causes, Early Signs, and Prevention of Pulmonary Consumption for the Use of Parents and Preceptors* (Bristol, UK: Biggs and Cottle, 1799), 85, 87–88, as quoted in Porter, 66.

120. Porter, "Consumption," 70–71.

121. Conevery Bolton Valenčius, *The Health of the Country: How Americans Understood Themselves and Their Land* (New York: Basic Books, 2002); Linda Nash, *Inescapable Ecologies: A History of Environment, Disease, and Knowledge* (Berkeley: University of California Press, 2006).

122. Gregg Mitman, *Breathing Space: How Allergies Shape our Lives and Landscapes* (New Haven, CT: Yale University Press, 2007), 10–129.

123. Mitman, *Breathing Space,* 129.

124. Mitman, *Breathing Space,* 248, 252. For health care as consumer good, see Nancy Tomes, "Merchants of Health: Medicine and Consumer Culture in the United States, 1900–1940," *Journal of American History* 88 (September 2001): 519–547.

125. Gregg Mitman, Michelle Murphy, and Christopher Sellers, eds., *Landscapes of Exposure: Knowledge and Illness in Modern Environments, Osiris* 19 (Chicago: University of Chicago Press, 2004). Salient works include Claudia Clark, *Radium Girls: Women and Industrial Health Reform, 1910–1935* (Chapel Hill: University of North Carolina Press, 1997); Christopher Sellers, *Hazards of the Job: From Industrial Disease to Environmental Health Science* (Chapel Hill: University of North Carolina Press, 1999); Gerald Markowitz and David Rosner, *Deceit and Denial: The Deadly Politics of Industrial Pollution* (Berkeley: University of California Press, 2000); and Christian Warren, *Brush with Death: A Social History of Lead Poisoning* (Baltimore: The Johns Hopkins University Press, 2000).

126. Upton Sinclair, "What Life Means to Me," *The Cosmopolitan: A Monthly Illustrated Magazine* 41 (October 1906): 594.

127. James Harvey Young, *Pure Food: Securing the Federal Food and Drugs Act of 1906* (Princeton, NJ: Princeton University Press, 1989). Surprisingly, there are only a few environmental histories of food as a consumer good, although that is quickly changing. For examples, see Cindy Ott, *Pumpkin: The Curious History of an American Icon* (Seattle: University of Washington Press, 2012); and Kendra Howard-Smith, *Pure and Modern Milk: An Environmental History since 1900* (New York: Oxford University Press, 2013).

128. Sackman, *Orange Empire,* 11; see also "Putting Nature on the Table: Food and the Family Life of Nature," in *Seeing Nature Through Gender,* ed. Virginia Scharff (Lawrence: University Press of Kansas, 2003), 169–193; and Stoll, *The Fruits of Natural Advantage.* For an insightful study of another commodity fruit and the creation of an agricultural-labor system, see W. Thomas Okie, "'Everything is Peaches Down in Georgia': Culture and Agriculture in the American South" (PhD diss., University of Georgia, 2012); and "Under the Trees: The Georgia Peach and the Quest for Labor in the American South," *Agricultural History* 85 (Winter 2011): 72–101.

129. Nash *Inescapable Ecologies,* 151.

130. Nash, *Inescapable Ecologies,* 210.

131. Michelle Murphy, *Sick Building Syndrome and the Problem of Uncertainty: Environmental Politics, Technoscience, and Women Workers* (Durham, NC: Duke University Press, 2006).

132. Janet Ore, "Mobile Home Syndrome: Engineered Woods and the Making of a New Domestic Ecology in the Post–World War II Era," *Technology and Culture* 52 (April 2011): 286. For one popular and fictional example of this new environmental awareness, see *Safe,* directed by Todd Haynes (1995; Los Angeles, CA: Sony Home Video, 2001), DVD.

133. Nancy Langston, "The Retreat from Precaution: Regulating Diethylstilbestrol (DES), Endocrine Disruptors, and Environmental Health," *Environmental History* 13 (January 2008), 51; see also "Gender Transformed: Endocrine Disruptors in the Environment," in *Seeing Nature Through Gender*, 129–166 and *Toxic Bodies: Hormone Disruptors and the Legacy of DES* (New Haven, CT: Yale University Press, 2010).

134. Sarah Vogel, "Forum: From 'The Dose Makes the Poison' to 'The Timing Makes the Poison': Conceptualizing Risk in the Synthetic Age," *Environmental History* 13 (October 2008) 667–673; see also *Is it Safe?: BPA and the Struggle to Define the Safety of Chemicals* (Berkeley: University of California Press, 2012).

135. Langston, "Gender Transformed," 131. For one popular study, see Sandra Steingraber, *Living Downstream: A Scientist's Personal Investigation of Cancer and the Environment* (New York: Addison-Wesley, 1997). For one recent scientific study, see WuQiang Fan, Toshihiko Yanase, Hidetaka Morinaga, et al., "Atrazine-induced aromatase expression is SF-1 dependent: implications for endocrine disruption in wildlife and reproductive cancers in humans," *Environmental Health Perspectives* 115 (May 2007): 720–727.

136. R. Thomas Zoeller, T. R. Brown, L. L. Doan, et al., "Endocrine-Disrupting Chemicals and Public Health Protection: A Statement of Principles from The Endocrine Society," *Endocrinology* 153, no. 9 (September 2012): 4097–4110; and Nicholas D. Kristof, "Big Chem, Big Harm," *New York Times*, August 25, 2012, SR11; see also "How Chemicals Change Us," *New York Times*, May 3, 2012, A31, and "Warnings from a Flabby Mouse," *New York Times*, January 19, 2013, SR11.

137. Jody A. Roberts and Nancy Langston, "Forum: Toxic Bodies/Toxic Environments: An Interdisciplinary Forum," *Environmental History* 13 (October 2008): 629.

138. For summaries of evolutionary environmental history, see Susan R. Schrepfer and Philip Scranton, eds., *Industrializing Organisms: Introducing Evolutionary History* (New York: Routledge, 2004); Karen A. Rader, *Making Mice: Standardizing Animals for American Biomedical Research, 1900–1955* (Princeton, NJ: Princeton University Press, 2004); Edmund Russell, "Evolutionary History: Prospectus for a New Field" *Environmental History* 8 (April 2003): 204–228; and *Evolutionary History: Uniting History and Biology to Understand Life on Earth* (New York: Cambridge University Press, 2011).

139. Julie Guthman, *Agrarian Dreams: The Paradox of Organic Farming in California* (Berkeley: University of California Press, 2004); Michael Pollan, *The Omnivore's Dilemma: A Natural History of Four Meals* (New York: Penguin, 2006), 134–184.

140. For civic environmentalism, a capacious term, see William Shutkin, *The Land that Could Be: Environmentalism and Democracy in the Twenty-First Century* (Cambridge, MA: The MIT Press, 2000) and DeWitt John, *Civic Environmentalism: Alternatives to Regulation in States and Communities* (Washington, DC: Congressional Quarterly Press, 1994). For one of the few extended historical studies of community-supported agriculture and its effect upon people and place, see Brian Donahue, *Reclaiming the Commons: Community Farms and Forests in a New England Town* (New Haven, CT: Yale University Press, 1999).

141. "A Hog Giant Transforms Eastern Europe," *New York Times*, May 5, 2009; "Swine Flu Outbreak Could be Linked to Smithfield Factory Farms," *Grist Magazine*, April 25, 2009, http://www.grist.org/article/2009-04-25-swine-flu-smithfield/. For non-communicable diseases, see Derek Yach, Corinna Hawkes, C. Linn Gould, et al., "The Global Burden of Chronic Diseases: Overcoming Impediments to Prevention and Control," *Journal of the American Medical Association* 291 (June 2, 2004): 2616–2622.

142. Arjun Appadurai, ed., *The Social Life of Things: Commodities in Cultural Perspective* (New York: Cambridge University Press, 1986), 3.

143. For examples in this vein, see Susan Davis, *Spectacular Nature: Corporate Culture and the Sea World Experience* (Berkeley: University of California Press, 1997); Gregg Mitman, *Reel Nature: America's Romance with Wildlife on Film* (Cambridge, MA: Harvard University Press, 1999); Thomas R. Dunlap, "Tom Dunlap on Early Bird Guides," *Environmental History* 10 (January 2005): 110–118; and *In the Field, Among the Feathered: A History of Birders and Their Guides* (New York: Oxford University Press, 2011).

144. A notable example is Richard Tucker, *Insatiable Appetite*.

145. For critiques of environmental history in this vein, see William Cronon, "Modes of Prophecy and Production: Placing Nature in History," *Journal of American History* 76 (March 1990); 1123; Klingle, "Spaces of Consumption"; and Douglas Cazaux Sackman, "Consumption and the Angel of History" in "Anniversary Forum," *Environmental History* 10 (January 2005): 86–88.

146. For purity in environmentalism, see Richard White, "The Problem with Purity," Tanner Lecture on Human Values, delivered at the University of California at Davis, May 10, 1999, available at http://www.tannerlectures.utah.edu/. For environmentalism as secular faith, see Thomas Dunlap, *Faith in Nature: Environmentalism as Religious Quest* (Seattle: University of Washington Press, 2004).

147. Paul and Anne Ehrlich, *One with Nineveh: Politics, Consumption, and the Human Future* (Washington, DC: Island Press, 2004), 215.

148. Richard White, "'Are You an Environmentalist or Do You Work for a Living?': Work and Nature," in *Uncommon Ground: Toward Reinventing Nature*, ed. William Cronon (New York: W. W. Norton, 1995), 171–185.

149. Alissa Walker, "Measuring Footprints," *Fast Company* 124 (April 2008): 59–60, http://www.fastcompany.com/756443/measuring-footprints. For the "Footprint Chronicles" website, see http://www.patagonia.com/us/footprint/suppliers-map/.

150. Chouinard, *Let My People Go Surfing*, 58, 180.

151. Chouinard and Vincent Stanley, *The Responsible Company: What We've Learned from Patagonia's First 40 Years* (Malibu, CA: Patagonia Books, 2012), 2. For "coolest company," see Susan Casey, "Éminence Green," *Fortune* 155 (April 2, 2007): 62–70. Patagonia was a key example in a recent documentary chronicling "responsible companies" produced by an advertising and web-based marketing firm. See *The Naked Brand*, directed by Jeff Rosenblum and Jeff Sherng-Lee (San Francisco, CA: Questus, 2013), DVD and http://thenakedbrand.com/. Full disclosure: in 2008, this essay's author wrote the honorary degree citation for Chouinard at Bowdoin College. For details, see http://library.bowdoin.edu/arch/collections/college-archives-and-records-management/honors/chouinard08.pdf. The concept of an "ecological footprint" was first used by William E. Rees (and his student Mathis Wackernagel): see Rees, "Ecological Footprints and Appropriated Carrying Capacity: What Urban Economics Leaves Out," *Environment and Urbanization* 4 (October 1992): 121–130. The concept of the "footprint" remains understudied by historians. For one notable exception, see James Morton Turner, "Counting Carbon: The Politics of Carbon Footprints and Climate Governance from the Individual to the Global," *Global Environmental Politics* 14 (February 2014): 58–79.

152. Price, *Flight Maps*, xix.

153. For the tradeoffs of the "hybrid turn," see Sutter, "The World With Us," 96–100, 118–119.

CHAPTER 18

··

LAW AND THE ENVIRONMENT

··

KATHLEEN A. BROSNAN

EARTH's human population grew exponentially as agriculture and pastoralism developed separately in southwest Asia, China, and Mesoamerica and spread to other regions over several thousand years (c. 5,000 BCE). Complex, hierarchical, settled societies emerged, placing more intensive pressures on natural resources.[1] Questions arose about who controlled what resources, and to what ends those resources could be used. Possible resolutions ranged from violence to conciliation, but over time, most communities chose a middle path in which customs determined how people interacted with each other vis-à-vis nature. Accepted practices became more formalized in legal systems. Law constitutes the rules of conduct and of relations prescribed by a controlling authority with binding force that must be obeyed by community members; failure to comply results in a civil or criminal penalty. Societies with a predominant legal culture, such as the United States, generally resolve conflicts through courts and other institutions.[2] Laws represent a society's normative values and reflect the goals and policies of those in authority. Like agricultural systems, legal ideas migrated with people around the world. The law that emerged in the United States, for example, drew upon legal traditions in England and, to a lesser extent, other European nations, but also evolved in America's social, economic, and physical environments. And concepts from US environmental law circulated around the globe at the end of the twentieth century.

Legal subfields emerged over time to guide distinct social, economic, and political behaviors, reflecting the ever-increasing complexity of human societies. Environmental law developed in the United States and other nations after World War II to adjudicate the use of government authority to protect the natural environment and human health from the impacts of pollution and development.[3] Long before this subfield existed, however, law was a profound cultural influence on the environment. A new activism produced a spate of environmental protection laws in the 1960s and 1970s, but there were legal antecedents. Americans historically employed property law and nuisance law to define human relations with nature and to mediate between humans in disputes about resources. "Forms of property and other legal and social institutions are not immutable," as Arthur McEvoy observes, "rather, they are creatures of history, evolving in response to their social and natural environments even as they mediate the interaction between the two."[4]

Past actors rarely looked at nature holistically, and law reflected and reinforced this perspective. English common law treated nature as divisible and its components as severable. Contemporary concerns about ecological health were not part of historical legal debates. Politics, economic liberalism, and the diversity of jurisdictions within the United States thwarted coherent resource management. Meanwhile, changes in the principles that defined the relationship between law and the environment were evolutionary. Shifts in legal thinking that once seemed radical are revealed, from the perspective of history, to maintain continuity with the past. The centrality of land to American identity remained a constant from the colonial era, but shifting normative values altered interpretations of property rights and their limits, concepts of nuisance, the role of economics and science in evaluating legal obligations, and acceptable exercises of state power.

Environmental historians, legal historians, law professors, and other scholars grapple with the law-environment relationship, in the process revealing a persistent, unresolved tension. Historically jurists and others debated how much weight to give the liberty of individuals to pursue private interests in comparison to the effect of their actions on the general welfare of society. Lawmakers, in deciding whether to protect traditional economic activities or foster new ones, did so with an eye to the impact on the larger community. For example, legal historian J. Willard Hurst argues that nineteenth-century judges promoted the release of creative entrepreneurial energy through the reallocation of resources, but William Novak finds evidence of efforts to protect the people's welfare in a range of venues.[5] Like other legal concepts, laws affecting the environment embodied society's complex, changing culture. They evolved over time and were never static. Law and the actors who made the law constantly negotiated and re-negotiated the balance between individual freedom and general welfare as they contemplated the human-nature relationship. With the more intensive resource usage under industrialization, courts seemed to shift toward protecting and promoting the dynamically propertied, the profitable, and the politically powerful. Subsequent conservation laws tried to curb some excesses of this exploitation, but generally still embodied an unquestioned assumption that nature existed for human utility. Post-World War II laws approached nature more holistically, accorded intrinsic value to nonhuman actors, gave greater weight to scientific evidence, and in doing so perhaps redefined the community good, but litigation under these laws still frequently pitted vibrant economic interests against this good.

COMMON LAW AND THE COLONIAL ERA

Before Europeans arrived in the New World, Native Americans exercised their own resource management systems.[6] In the colonial era, Europeans brought laws from their homes and supplanted indigenous customs. Elements of Spanish law persist in parts of the western United States, as do, to a lesser extent, minor aspects of Dutch and French systems in other regions, but English common law formed the foundation of most American law.[7] Colonists used those aspects of English common law that served their needs in a new environment and rejected those that did not. Common law is distinguished from a civil law tradition that dates back to ancient Egyptian codes (c. 3000 BCE). Codification positively outlined acceptable behavior and relied on judicial administration rather than interpretation. There were periodic attempts since European settlement to impose codes, but this civil law tradition made only minor impacts on American jurisprudence.[8]

"Common law" refers to principles and rules of action for the security of people, property, and government, which derive authority from customs and folkways, or from judicial decisions enforcing such customs. It involved actual cases where judges applied, refined, or transformed past principles in the face of changing societal, economic, political, or ecological pressures. Following precedents handed down over generations, judges supposedly provided a sense of stability and predictability. The word "common" referred to geographic diffusion, rather than the law's application to a broad swath of people. The common law primarily served English aristocrats. Different local practices created a varied, dynamic foundation, but there was one clear constant: land remained a priority of the legal process, and the same was true in the Americas.[9] Most colonists lived in rural environs where success or failure depended on what they took from the land.

In their obsession with land, common-law courts in England and its colonies adjudicated competing rights in natural resources. Property law and nuisance law provided the main tools. Issues of resource management since the colonial period were "interwoven with questions of property rights: who had what rights to use or transform the environment and its resources, what responsibilities went with those rights, and what restrictions might governments impose on the exercise of them."[10] The law assumed the physical world could be divided, allocated, and owned. Real property consisted of land and whatever was erected on, attached to, and exercisable within the land.[11] Issues of law and the environment often revolved around the question of what property ownership meant. Many Americans today wrongly assume that property ownership is absolute and always has been so. Historically real property consisted of a bundle of rights subject to limitations on its usage, and the courts constantly have reinterpreted those rights.

British colonists did not receive the common law as a whole from England. Most carried little knowledge of the law, and the few who possessed more information brought distinct permutations based on their place of origin.[12] Temporal variations occurred depending upon the date a colony was founded and the stage of economic and social growth at the time of a decision. Geography and religion also produced discontinuities.[13] Finally, the king remained the sovereign in theory. Absentee landlords assumed that English law protected their properties, but colonial actors, isolated on the edge of empire, adapted traditions to suit their own needs and rejected those of no value.[14] In theory, for example, colonial courts should have applied the English principle of tenure, in which no lands were owned absolutely because all were held in subordination to the crown. The idea that property constituted a bundle of rights emerged from this principle. Landholders' possessory titles only accorded them designated rights in land. In England, land was the basis for social, political, and economic influence, in part because of its scarcity. In the British colonies, by contrast, there was a shortage of people and an abundance of land, particularly with the rejection of native titles. The king's interest in colonial lands required quit-rents, or the annual payment of money or goods, but colonists found the principle antiquated.[15] The Body of Liberties in 1641 forbade all feudal incidents, effectively barring quit-rents in Massachusetts. In the middle colonies, openly hostile juries refused to enforce rents. Southern colonies reduced the amount the colonists owed and allowed them to pay with commodities instead of cash. Colonists often sent trash tobacco; its poor quality meant it was not sellable.[16]

Colonial courts generally supported free tenure of land. "All in all, it was easier to use land as a cheap, convenient subsidy than as a means to fetter a restless population. Hence, in general, land policy and land law tended to discard many traits of the past." In a world of landed plenty, the king gave holdings to friends. Colonial governors offered land for political favors. Indentured servants, upon completing their terms, claimed as many as fifty acres. Distributions occurred profligately. More people and types of people owned land than in England. The transfer of land rights moved from the realm of family and tradition to the domain of contracts.[17] Most New England colonies abandoned primogeniture (the superior right of the eldest son to succeed to his father's estate) by the end of the sixteenth century. It persisted in cases of intestacy (the condition of dying without a will) in other British colonies through the American Revolution, but ceased to be of great significance. Given an abundance of land, it was less necessary to pass land within the family to protect social position. New systems of inheritance contributed to the ever-increasing number of holders exploiting the land for profit.[18]

Legal historians recognize distinct legal adaptations to the American setting, but focus on economic and social consequences of common-law interpretations of property, rather than their environmental implications. William Cronon's *Changes in the Land* (1983), by contrast, explores how the introduction of European and English concepts of property, among other factors, brought interrelated ecological and cultural changes to colonial New England. Law is an integral part of culture. In the Indians' dynamic system of land "ownership," he explains, property was expressed in the sovereignty of the sachem and defined by a village's political and ecological territory. Colonists' laws also changed. Communitarianism, in the form of mutually owned common and pasture lands, gave way to individual holdings. Initial transfers specified uses for particular parcels, but as property rights became more entrenched, such limitations vanished, facilitating the division, enclosure, transformation, and sale of individual plots as commodities. Cronon contends that the alienation of land as an abstract commodity, and the accompanying transition toward capitalism, contributed to the reckless extraction of resources, but he grapples with law on a philosophical rather than practical level, drawing on theories by John Locke and John Winthrop instead of court decisions.[19]

The environmental historian Brian Donahue alternatively argues that colonial agriculture, which combined mixed husbandry and a common field system, was ecologically sustainable. Over several decades, men in Concord, Massachusetts acquired hundreds of acres, albeit in dozens of discontinuous parcels. The irregular divisions incorporated a desire to achieve equity but also mirrored the area's natural diversity. Each man possessed several kinds of natural resources.[20] Donahue implicitly incorporates George Haskins's contention that colonists developed a legal system adapted to their new environment and committed to expanding the number of landowners, but there is no sense of how the evolving common law facilitated these transitions. The absence of the law is all the more striking when Donahue discusses the demographic pressures on agrarian resources in Concord at the end of the eighteenth century, although the scarcity was less significant than what England had encountered in the sixteenth century. The crisis initially abated through lower fertility and outmigration, but the yeoman world of mixed husbandry disappeared by 1850.[21] These very issues of abundance and scarcity were at the core of colonial law and thus shaped resource usage in North America.

American interpretations of English common law created a tension in the rules governing property rights. One theory of property limited owners to the "natural use" of their land; courts defined agrarian uses as both natural and morally superior. However, courts also recognized the priority of property rights, which in theory allowed those who were first in time to pursue any desired activity. This second rule was potentially more compatible with economic development, but in the eighteenth century, with an economy characterized by slow, measured activity, courts balanced these tensions, and these theories acted as limitations on property.

Nuisance law provided another limitation on freehold ownership and became an early source of environmental control. It prevented individuals from using property in ways that interfered with the rights of others. Courts defined private nuisances as wrongful acts, even if legal and non-trespassory, which destroyed or diminished another's property or interfered with his lawful use or enjoyment of the land. Such a nuisance vexed William Aldred of Norfolk, England in 1610. His neighbor, Thomas Benton, converted an orchard into a pig sty, a legal activity. Aldred claimed the fetid odors of the sty interrupted enjoyment of his own property. The King's Bench agreed that the enclosure constituted a nuisance and famously declared *sic utere tuo ut alienum non lædas*: "use your own property in such a manner as not to injure that of another."[22]

Colonial courts adopted the *Aldred's Case* precedent. A mainstay of US jurisprudence, it became a foundation of modern environmental law. Courts barred nuisances such as loud noises, thick smoke, or noxious smells,[23] and *sic utere tuo* became a guiding principle in the negotiation of competing property rights. It encouraged people to move foul activities to locales that limited their harm, but its application was difficult.[24] It did not specify what level of injury merited action or what remedies should correct or eliminate harm. Colonial courts protected traditional agrarian uses by estopping (halting) others, but over time US judges hesitated to issue injunctions to halt the offensive but highly profitable activities. Courts increasingly ruled that nuisances should be limited to cases with manifest and serious injuries.[25] Cases involving multiple sources of pollution were problematic because, as in all torts, a plaintiff must prove the proximate cause. In 1906, for example, the State of Missouri hoped to stop Chicago from dumping raw sewage into Lake Michigan because it found its way via waterways to St. Louis and other Missouri towns that experienced typhoid epidemics. The US Supreme Court rejected the claim because Missouri could not prove the causal connection given the large number of other Missouri and Illinois communities that disposed of their waste in those streams.[26]

In other North American colonies, different legal traditions emerged. The Spanish, for example, used metes and bounds to set boundaries according to the natural contours of the land. Easily adaptable to local environments, the system supported the subsistence agriculture that defined most settlements in New Mexico, particularly given their isolation from markets in the Americas and Europe. Following the Pueblo Revolt of 1680, the Spanish imposed on the upper Rio Grande Valley a system of individual and communal land grants that prompted small settlements along the river and its tributaries. Individuals received house and irrigation allotments, and shared common lands for pastures, animal watering, and wood harvesting. These communities frequently employed "long-lots or narrow strips of land emanating from the waterways. They were a necessary response to the physical environment and represented a practical and equitable method of partitioning land." Over time, given the region's aridity and cultural practices, which divided these strips for subsequent generations, more densely populated communities intensified resource usage. The commons lost grasses and trees; settlers pushed grazing and cutting higher into the hills. Overused, heavily salinized irrigation plots were unproductive.[27]

A New Republic and its Public Domain

Natural law provided essential ideological foundations for the American Revolution. Enunciated by many but most closely associated with Locke, natural law shaped the human condition, from the individual person to society as a whole, according to its proponents. Under theories of natural law, historian Mark Fiege explains, "Each person owned his body and the labor it could perform. By cutting trees, plowing soil, hunting animals, or mining gold, a person in effect joined his labor to nature and so made the modified nature his property."[28] American colonists of all classes embraced natural law, whether elites who read the philosophers or others who learned of it through sermons, pamphlets, or workingmen's clubs. They believed that their labor and their predecessors' labor transformed unimproved nature into farms, ports, roads and forms of property. Land, and unfettered access to land, remained a central part of the Revolution.[29]

Common legal trends emerged, but the diversity of jurisprudence that defined the colonial era continued after the Revolution. Each new state underwent its own political and legal changes. Some Americans questioned whether they should maintain the common law inherited from England, and instead suggested civil law as an alternative offering clarity and a supposed absence of class bias. Many revolutionaries argued that the flaw had not been in the common law but in the king's perversion of it. Common law, they contended, embodied the fundamental norms of natural law. In the end, lawyers and jurists, who lacked the education and exposure to contemplate alternatives, insured the common law's persistence, and over the next century, Americanized it to meet the needs of the greater numbers of landowners and others who claimed rights in nature.[30]

The new nation's challenges quickly became apparent. Kentucky was organized as a county of Virginia, which made grants there, often without verifying locations. The Kentucky legislature also distributed lands haphazardly. White settlers arrived, seeking farms, and speculators were eager to exploit the market. Inadequate surveys contributed to the chaos. In the early 1780s, Kentucky utilized metes and bounds without referent baselines; surveyors defined boundaries with landscape features. Many surveyors lacked adequate training even in the crude methods of the time. They obtained their positions through political connections or because of rudimentary literacy. Described as an honest and average surveyor, Daniel Boone used prominent trees as markers, a practice that caused confusion when others attempted to retrace his lines on the ground.[31] With such surveys, multiple claims existed for individual parcels, and fraud and inequities ensued. The wealthiest one-tenth of the taxpayers, including many nonresidents, held a third of the land, while more than half of white adult males possessed none. Claim duplications led to litigation. Baffled by the legal intricacies of securing title and frustrated with the cost and pace of litigation, many Kentuckians questioned the fairness of the courts as richer litigants hired attorneys and waited through long appeals. In 1797, five years after it became a state, Kentucky finally passed an occupying-claimant law as a compromise. Under the act, the Kentucky Court of Appeals refused to protect squatters who had not perfected titles, but required nonresidents to reimburse squatters for improvements.[32]

The federal government also entered the realm of property. When the Articles of Confederation proved unsatisfactory, a newly ratified US Constitution in 1787 provided for a stronger centralized government.[33] Between 1781 and 1802, the individual states ceded their western land claims to the federal government; the latter acquired other lands through acquisition, treaty, and conquest. In disposing of its public domain, the United States expanded the number of property owners and reinforced the notion of land as the cornerstone of American economy and law. The chaos in Kentucky confirmed the need for a rational and uniform survey system.[34] The Land Ordinance of 1785, perhaps the most important law passed by the Confederation Congress and the basis for future land policies, provided mechanisms for the division, sale, and settlement of public lands. Through its surveys, the federal government implicated itself in the chain of title and molded most landscapes west of the Appalachian Mountains (and north of the Ohio River east of the Mississippi River) into a patchwork grid readily observable today.[35] The initial debate captured tensions about what type of nation the United States would be. Paul Gates, the leading scholar of public land law, explains that Thomas Jefferson envisioned a country of yeoman farmers whose land ownership and self-sufficiency inured noble qualities of citizenship in a democratic society. He wanted the government to sell land directly to farmers in small lots at nominal prices and proposed the US Rectangular Land Survey, known as the grid, to facilitate such sales.[36]

A different school of thought, represented by Alexander Hamilton, favored policies that exploited the nation's natural resources, aided rapid industrialization, and lessened US reliance on foreign trade. The sale of large land blocks at high prices to moneyed individuals, who in turn divided and resold them, would insure revenues and bolster a burgeoning manufacturing sector. The Land Ordinance of 1785, though written in part by Jefferson, reflected this position. Reformulated land laws between 1800 and 1841 became more favorable to individual farmers by reducing the minimum acreage. New sales policies, along with veterans' bonuses, created a landscape where small freeholds mixed with speculators' numerous and extensive properties. Despite more favorable land laws, however, many settlers still could not afford even the minimum purchase price.[37] Finally, the Homestead Act (1862) gave a freehold title to 160 acres to settlers who occupied and improved the land over five years. Settlers, who otherwise lacked capital, became landowners with all property rights under common law, but 160 acres proved inadequate for agriculture in the arid West and many failed. Moreover, large property consolidations persisted. Railroads received land grants, and others sometimes engaged in fraud to exploit the Homestead Act.[38]

Disposal of the public domain involved every national issue in the nineteenth century: relations between the states; relations between the states and the federal government; westward expansion and treaties with Native Americans; the extension of slavery; transportation; foreign policy; veterans' affairs; and the limits of federal power, particularly in the acquisition of new territories, such as Louisiana, Texas, and California. The 1785 ordinance and subsequent public land laws stabilized the young republic. They prepared western regions for settlement by white Americans and provided systems for establishing territories and their later admission as states.[39] According to Fiege, "By providing a systematic, standardized method of measuring and mapping property boundaries, the rectangular survey fostered a dependable market in land."[40]

In other situations, the government followed its citizens and land laws were reactive. In principle, no one could claim title until federal surveyors completed their work and the land office made a parcel available for sale. In practice, some settlers moved into new territories before surveyors. They lacked legal standing, but believed their efforts to improve the land—cutting trees, filling wetlands, planting crops—gave them a moral claim. After a decade of piecemeal provisions, Congress passed a general preemption law in 1841 that gave heads of households who occupied and improved a parcel the first option to buy it at the government's minimum price, once it was surveyed. Congress included safeguards to prevent abuse. Nonetheless, settlers used collusion and sometimes violence to insure outsiders did not bid up the price, and speculators hired dummy entrymen who held legal title in name only.[41] The Oregon Territory presented a different challenge. Americans migrated there in the 1840s as the United States and Great Britain disputed the region's ownership. Settlers petitioned Congress to form a territorial government and sanction their parcels. When the nations agreed to a border at the Forty-ninth Parallel in 1848, Congress organized Oregon Territory, but to the settlers' dismay, nullified prior land titles. Two years later, perhaps out of a sense of obligation to those who risked safety and wealth to stake a US claim to Oregon, Congress reversed its decision. The Oregon Donation Act of 1850 provided 320 acres to single white male citizens and 640 acres to married couples who had cultivated land for four years.[42]

The historian Ted Steinberg contends that federal land laws did more than generate a stable land market; they helped to redistribute the nation's natural wealth. The concept of the commodity was essential to dispersing resources. Efficient management and reallocation of natural assets necessitated their pricing.[43] "The grid was the outward expression of a culture wedded not simply to democracy, but to markets and exchange as well." Commodification transformed the land. Farmers, particularly after rail transport improved, replaced native vegetation with crops that held value in distant markets. In the world of exchange, monocultural production enhanced economies of scale and profits, but also created a series of ecological crises, from pest infestations to drought and erosion.[44] Farmers pushed into marginal lands on the plains, contributing over time to the Dust Bowls of the twentieth century.[45]

In theory the dispersal of public lands shored up the family farm, the dominant American form of land tenure. In practice, inefficiencies in the system and sometimes outright fraud allowed wealthy property holders to dominate the landscape. West of the Mississippi River, land developers, ranchers, irrigation companies, and other corporations often frustrated the democratizing impulse of the Homestead Act and other federal laws designed to deal with the region's aridity, such as the Desert Land Act of 1877 or the Timber and Stone Act of 1878. California rancher and land developer James Ben Ali Haggin lobbied California's senator to secure the passage of the 1877 law, received advance word of its passage, and paid hundreds of San Francisco residents to file dummy claims in the Kern River Valley. To reclaim the desert there, he bought water rights and gained control of irrigation channels, leaving the small farmers the law intended to aid dependent upon him for water and often for temporary employment with his Kern County Land Company. The Miller & Lux Company previously purchased twenty miles along the Buena Vista slough of the same river, and its control of a limited water supply allowed the company to otherwise control almost 80,000 acres where it grazed cattle. While these San Francisco-based corporations later litigated who had the superior water rights, both assumed that despite the stated intention of these policies, the reclamation of these arid lands required the efforts of large, capitalized private ventures.[46]

Unrestrained capitalism disrupted the democratic dispersal of the public domain. Such developments led Gates to conclude that private ownership was the wrong primary policy for the public domain. He views the gradual twentieth-century shift away from sales to permanent federal land management as an incomplete recommitment to democratic principles.[47]

AN INDUSTRIALIZING NATION

Migrations across the continent, whether first occasioned or later sanctioned by federal land laws, were part of a revitalized, fast-paced economy that defined the nineteenth-century United States. Early in the century, Americans moved toward industrialization and more market-oriented agriculture, although the nation faced a persistent shortage of capital that threatened such ventures. They turned to the law to resolve this problem, and in doing so, reinforced and then expanded earlier legal concepts. In the United States, land was a commodity, an interchangeable good subject to the forces of supply and demand, but it also became a source of capital for the production of greater wealth.

Inspired by Aldo Leopold, Hurst studied Wisconsin's lumber industry in an effort to push the boundaries of legal history beyond questions on the transformation of the common law and the acceleration of economic change to discover the interrelations between nature and human society. Hurst argues that state courts stressed property as an institution of growth rather than the key to financial security, a symbol of social status, or a familial bulwark. "[O]n the whole, the nineteenth-century United States valued change more than stability and valued stability most often where it helped create a framework for change...the main current ran to the protection of property in action."[48] Expressed in results more than formal language, law promoted the release of creative entrepreneurial energy through the reallocation of resources. Hurst tackles the environmental effects of this release of energy in his study of Wisconsin's timberlands. Nineteenth-century Americans did not always agree on how best to use resources, but they did share a common assumption that people should materially control the environment through technology and legal mechanisms. At the state and federal level, government actively participated, developing policies through the public land distributions, powers to spend and tax, and extensions of eminent domain that mobilized capital in areas where it had its greatest multiplier effect. The courts effectively defined a new priority that promoted present wealth over future social or ecological concerns. "Lumbermen mined the forests for the most immediately marketable timber, leaving uneconomic islands of residue and cutting without regard to new growth or control of fire hazards." In applying the common law, judges did not simply respond to a changing economy. They promoted certain property owners over others by reinterpreting precedents and restructuring remedies.[49]

"Over the course of the nineteenth century," Eric Freyfogle notes, "the landowners' right to use the land intensively expanded, while the landowner's right to halt interferences [under nuisance law, for example] contracted."[50] Theories of natural use and priority that emerged in eighteenth-century law increasingly came into conflict with the new economic ethos. In the early nineteenth century, courts began to resolve these conflicts by adopting a new balancing test of efficiency and assigning priority in land usage to people willing to accept the risks of investment. Claims founded on natural use receded. Courts defined property in terms of autonomy for private decisions, even if the uses of the land did not comport with traditional agrarian activities.[51] Courts adopted a reasonableness standard in which they measured competing property uses and rejected activities, even if first in time, which inefficiently employed natural resources. Morton Horwitz writes: "The idea of property underwent a fundamental transformation—from a static agrarian conception entitling an owner to undisturbed enjoyment, to a dynamic, instrumental, and more abstract view of property that emphasized the newly paramount virtues of productive use and development."[52]

For Horwitz, property is an evolving legal concept, not an ecological subject. The instrumentalism proffered by Hurst and Horwitz responded to scholarly theories of liberal constitutionalism which argued that the law drew political authority when the Supreme Court, primarily, defined and protected the border between public powers and private rights. Their more realistic socioeconomic interpretation stresses the proactive role of private law and state courts in the release of capitalist energy. Novak offers a third view from the perspective of nuisance law: "Public regulation—the power of the state to restrict individual liberty and property for the common welfare—colored all facets of early American development."[53] Nuisance law remained central to the economy and Americans' relations with their environment, and according to Novak, constituted the most important legal doctrine in ordering a well-regulated society—a theory and practice of governance, including law, committed to the *salus populi,* that is the people's welfare. Judges favored commerce and dynamic property, but recognized the public's superior rights. Offensive trades provide the clearest example of a regulatory common law. Neighbors sued to halt thick smoke, pungent smells, and contaminated groundwater. Successful plaintiffs obtained injunctions. Other cases brought monetary relief. A string of decisions reinforced the fact that property rights remained subject to the principle of *sic utere tuo.*[54]

Christine Rosen joins legal history and environmental history to reconcile the differing interpretations from Horwitz and Novak. In the first wave of industrial pollution litigation before the Civil War, Rosen contends, contradictions in nuisance case law reflected the challenges of evaluating the environmental effects of dynamic new industries. "The courts adjudicated cases involving pollution from the traditional, stench-emitting industries that had long been considered per se (inherent) 'nuisance' industries very differently than those involving pollution from new manufacturing and railroad industries. The conflict reveals the trouble judges had applying a precedent-based common law system to the unprecedented environmental impacts of the industrial revolution."[55] Juries, through experience, and judges, through precedents, understood traditional nuisances, but deeply embedded ideas about what constituted a material nuisance made it more difficult for courts to assess and restrain dynamic new industries such as smelters and steam-powered textile factories. Despite environmental degradation, plaintiffs lacked credible evidence to prove proximate cause and convince juries that pollution from new industries constituted a threat to health or property values. Judges resisted applying the *sic utere tuo* doctrine to new forms of pollution.[56]

Many business historians argue that the post-Civil War success of modern industrial firms was economically rational and inevitable and thus protected by courts. Most notably, Alfred Chandler contends that new production systems utilizing steam power and electricity resulted in more capital-intensive industries. These large-scale businesses supposedly drove economic growth because of greater productivity, lower costs, and higher profits. Their managers mobilized and applied capital and visibly shaped the economy more than the invisible market forces defined by economic philosopher Adam Smith.[57] Despite technological innovations and economies of scale and scope, Rosen advises, American society tried to curb environmental damage from activities, such as industrial meatpacking, as it became more familiar with their practices. Sanitary regulations, nuisance suits, and protests forced meatpackers, for example, to address pollution. The search for more innovative methods of waste disposal propelled them into the by-product manufacturing that insured profit maximization.[58] The regulation of industry and common law of nuisance neither disappeared nor stood as obstacles to principles of economic liberalism.

In *Railroaded* (2011), Richard White challenges Chandler, shattering the myth that nineteenth-century railroad corporations were independent models of organization, efficiency, and rationality.[59] "There is no such thing as a market set apart from particular state policies, institutions, and social and cultural practices." White argues, "The question is not whether governments shape markets; it is how they shape markets."[60] Construction of the transcontinental lines was chaotic, haphazard, corrupt and unnecessary. It resulted not from industry leaders' managerial expertise, but from their success in wresting vast amounts of public funding in loans and grants from the public domain. Other federal support involved the appropriation of Indian lands. Legal privileges, such as the treatment of corporations as persons, corporate laws that permitted overcapitalization, and injunctions against worker organizations, contributed to the "creative destruction" that followed. "The advantages of being near a railway line and obtaining public land prompted settlers to take greater chances with aridity. The combination of cheap or free land with railroad connections was a powerful lure." The costs included the near extermination of the bison, the emergence and later collapse of the open-range cattle industry, and an influx of farmers ill-prepared for the arid plains. [61]

Historian Brian Balogh adds, "Americans sought active governance at the federal level time and again over the course of the nineteenth century. Their efforts produced a variety of mechanisms, from debt assumption to Supreme Court decisions that undercut health and safety prerogatives of states in the name of interstate commerce."[62] In *Swift v. Tyson* (1842), a case involving a negotiable instrument, the question was whether the Judiciary Act of 1789 required federal courts to follow state law in cases brought under diversity jurisdiction.[63] The Supreme Court held that federal courts could follow a "general" law of commerce even if its principles diverged from states' precedents. The Court sought to unify the nation's commercial law, believing commerce to be national and international in character and something that should not be subject to parochial local interests and unnecessary regulations. Common law principles diverged from state to state, but commonalities emerged in commercial cases.[64] The *Swift* decision allowed federal courts to thwart local laws that unduly disrupted commercial activity in the courts' opinion.[65] Minnesota, for example, passed a law with environmental and public health implications, requiring state inspectors to examine animals twenty-four hours before slaughter if their meat was to be sold there. Despite constitutional provisions and precedents that upheld inspection rights, and despite scientific evidence about the benefits of inspections, the Supreme Court in 1890 struck down the law as an inappropriate exercise of the state's police powers and an unconstitutional prohibition on the interstate sale of meat.[66]

WATER LAW

The common law traditionally treated water resources as common rather than private property. Riparian owners abutted streams. They were entitled to what they needed for "natural uses" assuming that they did not materially interrupt other riparian holders' reasonable enjoyment, but life-sustaining running water could not be claimed as private property nor diverted for private profit. English and colonial common law developed the doctrine of natural flow; it limited the amount of water that a riparian holder could take for artificial uses, such as mining, manufacturing, or other non-agrarian activities. In the nineteenth century, Americans challenged this law based on economic and environmental conditions in different regions of the country.

In the humid eastern states with abundant water resources, judges balanced conflicting interests to determine the greater economic efficacy and in doing so, reduced restrictions on artificial uses. In *Mulligan v. Palmer* (1805), the New York Supreme Court recognized the right of an upstream owner to construct a dam for milling purposes, despite the fact that he possessed no priority and impaired downstream users' ability to enjoy their rights.[67] By the Civil War, eastern states witnessed a near-universal judicial acceptance of the idea that owners had the right to develop water for business purposes.[68] In *Nature Incorporated* (1991), Steinberg explains that New England relied on a booming textiles industry which required larger dams than the ones grain mills built to service the agricultural economy. However, the larger dams blocked fish migrations. Pollution from textile factories diminished oxygen levels in and human health along the Merrimack River, but judges abandoned the natural flow rule and balanced the competing interests. When courts provided relief to injured riparian owners, they awarded monetary damages that allowed dynamic economic activities to continue, rather than issuing injunctions that abated the nuisance. Steinberg concludes that a viable ecological solution was not possible in the nineteenth-century universe of instrumental property law, industrial development, and unreflective beliefs in anthropocentric control of resources.[69]

West of the Mississippi River, for nearly three centuries prior to 1848, Spanish water law dominated much of the region. Given the aridity in many areas, cooperative irrigation and water management were essential to community persistence.[70] Under Spanish colonial rule, the crown retained ownership, but a flexible system gave courts great latitude in responding to local needs. Judges first considered *derecho*, a word with no specific English equivalent which translates most closely to "justice." Local customs and a sense of fairness were as important as legal precedents. The core principle was that surface water be shared as a resource in common.[71] All community members contributed to water system maintenance, and people could not build mills or canals to further their own interests at the community's expense. The powerful local authority that managed the community system also marketed its own water rights, which set a precedent for other holders to rent or sell theirs.[72] In 1747 in San Antonio, for example, Antonio Rodriguez Medoros, the *mayordomo* or overseer, sold two *suertes* of land to Miguel de Castro and one day of water every twenty days. Other transactions separated water rights from the land. In 1782, in the same region, Juan Joseph Montes de Oca traded a twenty-three-hour water right to Juachin Flores de Sandeja for a sixteen-hour water right.[73] Environmental and legal historian Donald Pisani emphasizes that such transfers conveyed a right of use, not title. "The main purpose of irrigation was to create a permanent community, not to encourage individual enterprise. And since formal ownership of water remained with the state, water rights were far more adaptable than under Anglo-American law.... When conflict occurred, judges tried to promote the greatest good for the greatest number."[74]

Spanish legal systems persisted with few modifications in Texas, New Mexico, and California after the formation of the Republic of Mexico in 1821. Mexico ruled Texas until 1836, and California and New Mexico until 1848. As Americans poured into California and other parts of the larger trans-Mississippi West, including its more arid regions, settlers and courts in newly formed states and territories debated what law should apply to water use. When sovereignty passed to the United States after the war with Mexico, the Treaty of Guadalupe Hidalgo promised that community water systems would survive. In 1850, the California legislature passed an act incorporating the city of Los Angeles and guaranteed that it succeeded in the water rights of the Pueblo de Los Angeles, a community irrigation system of 18,000 acres, including 1,000 irrigated acres. By 1854, the state legislature extended principles of community water use to other pastoral communities in Southern California with large Hispanic populations and no significant mining or other claims to water. Townships established special boards to oversee water use, but over the next three decades these systems of public control eroded, as the influx of settlers and new economic activities placed greater demands on limited resources. "Large landowners regarded the (community) ditches as relics of primitive subsistence economy and as a limitation on speculative land profits." Californians advocated for water laws that recognized more dynamic uses. Community water systems in New Mexico and Texas similarly gave way by the end of the nineteenth century.[75]

In northern California, newcomers also rejected the common law principle of natural flow and fashioned an alternative doctrine of prior appropriation. It granted a permanent right to the first stakeholder who put water to a beneficial, continuous use, regardless of the proximity of his land to the stream or the impact on later priorities. Prior appropriation took hold in the mountains of California and Colorado, where miners diverted often plentiful waters away from stream beds in search of gold. Through laws and constitutions, many, but not all, western states and territories extended the system to irrigation. Appropriative rights became a class of property that could be sold in whole or part. Like their eastern counterparts, western courts were pro-development, and as the legal historian Gordon Bakken writes, "encouraged entrepreneurs to scramble for water, quickly construct works, and apply the asset to the industry of the region."[76] Eager to show that the rule of law guided new territories, jurists justified departures from the common law of property and water. Moses Hallett, chief US Territorial Judge in Colorado, opined in *Yunker v. Nichols* (1872) that the arid environment created appropriative rights: "Rules respecting the tenure of property must yield to the physical laws of nature, whenever such laws exert a controlling influence."[77]

Some legal historians accept Hallett's logic, contending that climate and the economic challenges necessitated the creation of appropriative rights.[78] The explorer John Wesley Powell had believed that the aridity defined much of the West. He warned against widespread settlement and industrialization in fragile environments.[79] In practice, however, aridity became a rationale for certain legal decisions, rather than a limitation on development. Prior appropriation developed in response to local efforts to balance tensions between community welfare and unrestricted capitalism, not as a monolithic entity predestined by nature. In 1882, in *Coffin v. Left Hand Ditch Co.*, the Colorado Supreme Court, while recognizing the difficulties of farming on the arid plains, emphasized the need to promote development and legitimated the state's support and regulation of priorities under its police powers.[80]

Pisani observes, "There are many ways to allocate scarce resources, and the triumph of prior appropriation was not inevitable." The doctrine was not an environmentally driven necessity, but rather a legal innovation designed to make water a commodity.[81] Pisani does not suggest that aridity is irrelevant, but rather that the rugged western environments exacerbated an American predilection toward localism inherent in fifty jurisdictions. Fragmentation and the absence of a coherent national water policy for the management of regional rivers that crossed multiple state borders intensified competition and insured private exploitation of natural resources. Western courts' ambiguous support of prior appropriation in the nineteenth century (some states still utilized riparian law or a mixture of the two) reflected concerns about its monopolistic potential. Where it took hold, prior appropriation left the welfare of communities to the whims of multiple discrete decisions made in the pursuit of individual profit. The law made no value judgment on what constituted a beneficial use. "First in time, first in right" seemed to reflect American principles of liberty by suggesting everyone potentially had access, but appropriative rights became a form of property favoring landowners and speculators with greater capital, even if courts attempted to protect local interests.[82]

The doctrine of prior appropriation spread to other Anglo settler societies. In 1872, Canada passed the Dominion Lands Act (DLA), a law often compared to the US Homestead Act, in an effort to encourage settlement and development in Manitoba and the Northern Territories. The DLA authors adopted prior appropriation, or prior allocation as it was known in Canada, because the prairies presented new challenges. "The inability to remove large quantities of water for irrigation, problems with some landowners not directly connected to a watercourse, and the instability of common law water rights resulted in the dominion government's acquiring more regulatory control over the resource." Wary of investing in large water projects, settlers, local governments and the Canadian Pacific Railway Company believed that prior allocation offered clarity and stability, particularly for long-term usages.[83] Critics later questioned this water regime. "Prior allocation locks into place past use patterns. It does not protect ecosystem services or reflect interactions between ground and surface water. Prior allocation does not easily accommodate new users or uses, and is not flexible enough to address emerging challenges such as increased urbanization, new priorities, or climate change."[84]

Occupying Earth's driest continent, Australia sought water management systems to insure settlement and development. Attention focused on the semi-arid Murray-Darling River basin straddling four states in southeastern Australia. In 1886, after investigations of western US water law and irrigation systems in Egypt, India, and Italy, the state of Victoria passed an Irrigation Act abolishing riparian rights and nationalizing the state's water. A licensing system reserved to the state all authority to approve diversions on a first-come, first-serve basis.[85] Neighboring New South Wales soon offered its first comprehensive law. The Water Act of 1912 also rejected common law riparianism in favor of a state-controlled process that granted rights on the same basis. Farm production soared, including the introduction of thirsty cotton crops, but environmental problems followed: administrators overallocated limited water, salinity levels rose, and indigenous flora and fauna diminished. Facing drought, New South Wales passed the Water Management Act of 2000, which prioritized rights during shortages, adopted principles of adaptive management (constant re-evaluation and, when needed, readjustment of goals), and encouraged public participation to reach consensus among stakeholders. Unlike the United States, the maintenance of public ownership allowed Australia to adapt its form of prior appropriation to changing environmental and social conditions.[86]

MINERALS AND WILDLIFE

Prior appropriation in the western United States was tied to a paralysis of federal authority with respect to mining. The government leased mineral lands in the public domain, but failed to prevent trespassers and returned little revenue. With mid-nineteenth-century gold and silver discoveries, miners arrived in large numbers before officials appeared or territories formed. In this vacuum, miners adopted codes that bound actors within a district to customs for staking and recording claims, resolving disputes, and enforcing the decisions of miners' courts. Miners took ores for free and assumed that the federal government would grant them preemptive rights and extinguish Indian titles.[87] Federal officials continued open access to minerals, which some scholars contend represented an enduring commitment to economic liberalism.[88] Market values pervaded the pursuit of mineral wealth, but this early governance remained local, cooperative, and participatory. Self-imposed regulation obligated and liberated.[89]

Mining was essential to California's economy. Consequently, Andrew Isenberg argues, the state's jurists, like others across the nation, turned away from common law precedents which treated farming as the highest and best use of the land. California law allowed miners to enter public lands already occupied by farmers and ranchers, hoping to create for ordinary workers the opportunity to earn greater wealth. In 1855, the California Supreme Court extended to hydraulic miners the right to divert streams, limited only by the principles of prior appropriation. They were responsible only for damages caused by the negligent construction or maintenance of flumes.[90] As mining became an increasingly capital-intensive activity, the laws that were intended to spread the mineral riches to the many supported the corporate interests of the few.

The federal Mining Act of 1872 still allowed prospectors to claim land without paying for the minerals they exploited, but Congress hoped to halt the endless litigation that defined mining by replacing diverse district codes with uniform standards like the apex doctrine. Under this principle, a claimant who located his claim at the top of an exposed vein, or its apex, possessed the right to follow the continuous vein, even under others' surface claims. In Leadville, Colorado, juries hostile to its monopolistic potential rejected the apex principle. Colorado districts clung to the sideline principle that allowed miners to take ore from veins underneath their surface claims. In theory, the sideline seemed more equitable; no individual stakeholder unfairly disadvantaged others. No court followed the apex doctrine until a verdict in an 1886 trial in Aspen, Colorado. If this law promoted economic liberalism by privatizing resources and redistributing wealth, its erratic enforcement revealed an unwillingness to abandon the people's welfare and the idea of more equitable access to mineral wealth.[91] Bakken argues that the law was not an isolated effort to provide a subsidy worth billions to the mining industry, but instead was part of the larger federal policy designed to place public lands in individual hands and promote economic growth. In practice, the law regularized large-scale exploitation of minerals by capital-rich predatory conglomerates and contributed to equally endless litigation over who controlled the apex. Tailings, slickens, slimes, and cyanide residues fouled local waters and toxic fumes filled the air, but these environmental consequences were left for inadequate resolution by local communities under the common law.[92]

Treating property as a bundle of rights reinforced the idea that minerals could be severed from the land, regardless of the impact.[93] Hydraulic mining, an activity that focused on removing a single aspect of nature at any cost, was particularly devastating.

Its debris raised riverbeds, flooded farms, drove salmon from waterways, and smoth-ered oyster beds. By 1880, public opinion on hydraulic mining began to shift. Farmers, anti-monopoly groups, urban workingmen, and irrigation advocates organized an Anti-Debris convention in California to insure the passage of a new state constitu-tion more favorable to agricultural and urban interests. In 1882, Marysville landowner Edwards Woodruff sued mining companies along the Yuba River. Justice Lorenzo Sawyer of the Ninth US Circuit Court, a former miner, traveled to the region to observe changing conditions. In *Woodruff v. North Bloomfield Gravel Mining Co.* (1884), Sawyer issued an injunction to prevent further dumping. Grounded in the precedents of nui-sance law and riparian doctrines, the case focused on the people's common law right to untrammeled access to navigable waterways, but also balanced the efficacies of competing economic interests. By the 1880s, mining was a declining industry, and the state's wheat industry was booming. Finally, the decision involved a nascent recogni-tion of the interdependencies of nature, which the law rarely considered.[94]

Nevertheless, ecological damage associated with mining continued into the new millennium. Post-World War II environmentalism and new regulatory schemes worked to curtailed some activities of mining companies in the United States, but efforts to repeal the Mining Act of 1872 and replace it with a stricter law did not suc-ceed. More intensive technologies introduced and enhanced more destructive extrac-tion methods, such as those involved in open-pit mining and mountaintop removal. The environmental damage and associated social dislocation that accompanies the use of such methods is particularly severe in nations lacking strong regulatory structures.[95]

McEvoy addresses the ecological consequences of hydraulic mining in *The Fisherman's Problem* (1986), but his larger concern is the often conflicting actions of groups exploiting California's fisheries. A short-term bias toward capitalism, he con-tends, ignored long-term social and ecological costs. The myopic behavior of fisher-men partly resulted from the operation of property law. Even as overfishing depleted fisheries, courts resisted overturning customs that permitted open access. It became clearer that the fisheries constituted "a commonly owned resource whose preserva-tion required government intervention in the economy."[96] By the twentieth century, California had in place a body of law for fishery management, reinforced by federal laws such as the Lacey Act (1900), which prohibited interstate trade in certain plants and wildlife. In *Making Salmon* (1999), Joseph Taylor similarly shows how "culture, nature, and technology both encouraged and moderated exploitation" of the salmon population in the Pacific Northwest and led many participants there to conclude that government oversight was necessary. Nonetheless, politics sometimes trumped law and the commercial fishing industry often lobbied to overcome the regulatory scheme and keep fisheries open.[97]

Traditionally courts defined human relations with wild animals by the common law rule of capture. In *Pierson v. Post* (1805), the New York Supreme Court answered the question of which hunter owned a fox. Post gave chase to a fox across unowned land. Pierson knew someone else was in pursuit, but he killed the fox and took it for his own. Post sued for damages and a trial court ruled that the chase gave him possession. The

New York Supreme Court reversed the decision, concluding that pursuit did not grant a legal right. Regardless of his "uncourteous" conduct, Pierson gained ownership when he killed the fox.[98] *Pierson* is a well-known precedent, but there were few such cases under British or US law. Hunting was an elite activity in England, reserved, in theory, for those with rights from the Crown who owned much of the land. In the United States, hunting was more ubiquitous, but hunters rarely sued over a single animal.

Thus, nineteenth-century cases involving the rule of capture often focus on the challenging circumstances of commercial whaling. Courts acknowledged the realities of ocean hunting.[99] A wounded whale might not die immediately, and when it did die, it might sink and not re-surface for days after a ship had moved on. Who owned it—the whaler who killed it or a person who found it? In 1881, a federal court in Boston turned to local usage and ruled that title belonged to the whaler who committed the only possible act of capture in the situation, i.e. embedding an identifiable bomb-lance in the whale.[100] Freyfogle and Dale Goble conclude, "The rule that emerges... [is] an animal that is lawfully available for capture (an important 'if,' of course) belongs to the first person to reduce it to actual possession. Mortal wounding is enough if the hunter continues to chase. Short of that, courts consider the characteristics of the species, the specific challenges of hunting it, and customs that apply widely among those who pursue the particular type of animal."[101]

Other courts applied the rule of capture to subsurface minerals on private lands. Traditionally, landowners possessed exclusive, severable, and transferable rights to such resources. However, the Pennsylvania Supreme Court concluded, in *Westmoreland & Cambria Natural Gas Co. v. DeWitt* (1889), that when natural gas, oil, or groundwater "escape, and go into other land, or come under another's control, the title of the former owner is gone. Possession of the land, therefore, is not possession of the gas."[102] Texas courts emphasized that while oil and gas are subject to capture, a landowner must be afforded a reasonable opportunity to recover resources beneath his land.[103] These decisions spurred landowners in resource-rich regions to drill as many wells as possible before neighbors sank their own. Dense drilling often caused a dissipation of pressure, resulting in incomplete extractions, or subsidences, a problem also associated with coal mining and hard rock mining.[104] Various states and the federal government passed conservation laws to limit the waste of oil and gas resources, but these had limited effectiveness in precluding more destructive practices.[105]

CONSERVATION AND THE LAW

The Lacey Act, mentioned above, represented a new development in federal law, as Congress sought to preserve game and wild birds by making it a federal crime to poach them in one state and sell them in another. It was one of many Progressive-Era laws reflecting a stronger national conservation ethos that questioned and sought to regulate inefficiencies and the outright waste of resources under the market principles that increasingly dominated the common law. Samuel Hays's *Conservation and the Gospel of Efficiency* (1959) became a foundational book in environmental history. For Hays, conservation began in the 1890s as a scientific movement, rather than a grassroots initiative. It emerged from Washington, DC in struggles over the separation of powers and the creation of new executive agencies, such as the Forest Service and the Bureau of Reclamation, which enhanced presidential authority. Agencies hired experts who, conservationists hoped, would rise above normal politics and make decisions based on science, rationality, and efficiency.[106] The Progressive Era regulatory schemes analyzed by Hays embraced greater state power over urban spaces, waterways, public lands, and natural resources, but other historians identify earlier origins of conservation. John Reiger, for example, uses a top-down analysis but emphasizes upper-class sports hunters who grew concerned about vanishing game species in the 1870s. They formed hunting clubs and launched national publications. These efforts formed the first wildlife refuges and provided training for later conservation leaders such as Theodore Roosevelt.[107]

Richard Judd challenges top-down, expert-driven notions of conservation, giving a central role to New England farmers and fisherman. They believed that forest removal violated laws of nature, and grew concerned in the 1840s about the implications of deforestation for their democratic agrarian communities. By the 1880s, they formed organizations, such as the Massachusetts Forestry Association, and lobbied state governments for protective legislation. New forestry boards lacked authority to dictate land uses but launched education campaigns.[108] New England towns maintained local fisheries in systems dating from the colonial era. When commercial fishermen, who followed the railroad north, constructed weirs to harvest fish for an export market, they threatened local food sources and river traffic. Villages demanded state laws to protect a way of life previously governed through custom. By the 1860s, Maine had some 400 fishing laws to protect local actors against outside commercial interests.[109]

New Englanders influenced and were influenced by Vermonter George Perkins Marsh. In his first conservation speech in 1848, he observed: "loss of the primeval forest left a void in the spiritual as well as the economic scheme of rural life."[110] In *Man and Nature* (1864), Marsh argued that ancient Mediterranean civilizations failed because deforestation eroded soils and diminished productivity. Marsh gave literary voice to a long-present moral impulse that found expression in new laws.[111] Isolated federal efforts to regulate the hunting of Pacific seals (1870), investigate the causes of declining fisheries (1871), and establish a national park at Yellowstone (1872) led to more comprehensive management schemes under the Forest Reserve Act (1891) and the Newlands Reclamation Act (1902). In some states, fears that a timber famine might cause desertification prompted new laws and bureaucracies to enforce them by the 1870s.[112]

Pisani questions whether the more powerful, federal bureaucracies that emerged, such as the US Forest Service and the Reclamation Service, achieved the scientific purity or efficiency that Hays proposes. Agencies competed for funding and political power, often pursuing contradictory agendas, which undermined unified resource management.[113] Gates argues that the Reclamation Act, which established greater federal control over western rivers, represented a break from nineteenth-century land policies that theoretically emphasized individualism. Pisani counters that the Reclamation Act "pandered to home rule and institutionalized fragmentation." It ratified preexisting state and territorial statutes and constitutions, guaranteeing that coordinated planning for the protection of watersheds or other ecological and social goals remained a challenge.[114]

The twentieth century witnessed continuities between these earlier conservation efforts and New Deal programs. "Other policies that dated to the opening decades of the twentieth century, such as multiple-use water planning and sustained yield forestry came into their own in the 1930s, and new policies such as soil conservation and outdoor recreation became a part of conservation for the first time." A more powerful nation-state evolved as many Americans questioned the propriety of unrestrained economic activities and hoped to protect resources. The national government increasingly encroached on the authority of the states and their common law.[115]

President Franklin Roosevelt's New Deal launched agencies such as the Soil Conservation Service (SCS), the Tennessee Valley Authority (TVA), and the Civilian Conservation Corps (CCC) to address economic and environmental crises through conservation.[116] Created under the Soil Conservation Act (1935), the SCS responded to a legacy of public land law. With an agricultural boom between 1909 and 1914, farmers claimed more homesteads than in the nineteenth century. When the market bottomed out in the 1920s, they tilled more and more marginal lands in an effort to overcome debts and eke out a living the Great Plains. In the 1930s, winds swept away unprotected topsoil, and floods washed away the underlying landforms. According to Donald Worster, the Dust Bowl "came about because the expansionary energy of the United States had finally encountered a volatile, marginal land, destroying the delicate ecological balance that had evolved there. . . . What brought them to the region was a social system, a set of values, an economic order." The courts as well as federal land policy had helped release this energy in the past, and Worster argues, New Deal policies often reinforced the worst trends. Going forward, laws such as the Agricultural Adjustment Act, the Flood Control Act, and others tended to subsidize the wealthiest landowners rather than assist poor farmers or tenants, failed to alter long-term agricultural practices on the Plains, and ignored the fundamental capitalist relationship at the core of the problem.[117]

URBAN ENVIRONMENTS

As the federal government entered new regulatory areas, local and state governments became proactive, limiting property rights by exercising newly realized police powers to protect human health, direct the economy, and shape the physical landscape. Urban policymakers and progressive reformers challenged the free-market approach in land-use law that proliferated in the nineteenth century when Americans faced ubiquitous, more intense exposure to air and water pollution.[118] As Joel Tarr explains, society's use of water, air, and land as sinks for industrial wastes first prompted cities to use technology to reduce pollution, but they did so haphazardly. Nearly six hundred cities installed waterworks by 1875, but no city simultaneously constructed a sewer system. Once municipalities built sewers, local sanitary conditions improved, but morbidity and mortality rates soared downstream. Sewers simply shifted the problems to another locale. Over time, cities instead adopted policies with cost-benefit analyses based on health and nuisance concerns.[119] Private nuisance law and zoning had different origins in the common law and municipal law, respectively, but both became important land-use controls.

Mid-nineteenth-century law offered city residents few solutions beyond costly, protracted nuisance suits. Courts showed a growing inclination to support dynamic enterprises at the expense of other landowners. Successful parties might receive monetary awards, but courts rarely abated nuisances. St. Louis formed a Board of Health in 1867 to address the waves of cholera and seemingly related industrial odors, particularly those associated with slaughterhouses. The Board possessed the legal authority to halt offensive activities, but its members hesitated to penalize profitable businesses and feared tight regulation might drive industry outside the city, where they lacked jurisdiction. Instead, the Board targeted industries near affluent neighborhoods and exercised laxity in enforcing regulations in lower-class ones. These practices encouraged the formation of nuisance zones in poorer neighborhoods.[120]

Limits of jurisdictional authority also frustrated municipalities that tried to control pollution at their edges. For example, New York City became the nation's leader in kerosene manufacturing in the mid-1800s. When growing operations required specialized districts, refiners looked beyond the city, in part, because its Board of Health vigorously prosecuted manufacturers who violated municipal nuisance ordinances. Refiners settled along Newtown Creek on Long Island and a few other locales, but the stench wafted to Manhattan and oil slicks and other wastes impaired oyster beds. The municipalities that housed the refineries and enjoyed their economic benefits showed little inclination to regulate them. The New York City Board of Health challenged the refiners but lacked jurisdiction to take action in Queens or Brooklyn (which were not yet part of the city). The governor ordered the abatement of Newtown Creek nuisances in 1881, but left enforcement in the hands of Brooklyn prosecutors who still chose not to act. Political fragmentation proved a substantial impediment to controlling these nuisances.[121]

Twentieth-century reformers viewed the industrial city as a failing organism, its uncontrolled sprawl comparable to a cancer. Its only remedy lay in integrated planning and government interventions into private property.[122] Local governments are creations of, and derive their authority from, their states under municipal law. State constitutions grant general powers to local governments, and from the late nineteenth century, courts liberally interpreted municipal authority.[123] In New York City, a group of reformers, merchants, and realtors pushed for a law to regulate specific environmental conditions: surging population growth, the spread of factories with undesirable workers into elite neighborhoods, and the traffic congestion and blockage of light and air associated with skyscrapers. Policymakers responded with the comprehensive Zoning Resolution of 1916. It outlined boundaries of exclusively residential districts, prohibited incompatible, usually noxious property uses in those districts, and designated height and setback controls that limited skyscrapers to the business districts. The ordinance became a model for other cities.[124] In St. Louis, Board of Health practices helped create industrial districts, but noxious businesses still crept into affluent neighborhoods, leading the city to pass a new ordinance in 1918 with zones for single-family homes, commercial activities, and industry. The law banned factories from residential areas, but allowed housing in industrial zones. Workers often had no choice but to live near factories, while residents in other neighborhoods found their property values enhanced. "Begun under the auspices of protecting the general public from disease, pollution control increasingly became a tool for protecting private property and thus, distinguishing the environmental experiences of rich and poor, powerful and powerless."[125]

Planners believed that zoning ordinances must comport with carefully researched plans to withstand constitutional challenge as uncompensated takings. In *Village of Euclid v. Ambler Realty Co.* (1926), the US Supreme Court held that a zoning statute was a reasonable extension of the Ohio town's police powers. In finding a valid government interest in regulating the character of a neighborhood, the court implicitly supported a vision of urban environments in which the highest land use was a neighborhood of single-family homes—a vision that still shapes sprawling metropolises. The *Euclid* court found a precedent for its reasonableness test in private nuisance law: "In solving doubts, the maxim *sic utere tuo ut alienum non lædas*, which lies at the foundation of so much of the common law of nuisances, ordinarily will furnish a fairly helpful (clue). And the law of nuisances, likewise, may be consulted, not for the purpose of controlling, but for the helpful aid of its analogies in the process of ascertaining the scope of the power."[126] Some commentators called for a strict liability standard for any substantial loss, but courts continued to balance the social utility of the defendant's offensive activity against the gravity of the harm to the neighbor.[127] Perceptions of what constituted a nuisance changed, reflecting ecological knowledge, the influence of federal regulations, and a modern emphasis on the quality of life.

POSTWAR ENVIRONMENTAL LAW

Many Americans trace the rise of environmentalism, and the regulations it spurred, to the publication of Rachel Carson's *Silent Spring* in 1962. The book prompted new nongovernmental organizations, such as the Environmental Defense Fund and the Natural Resources Defense Council, to challenge federal and state agencies to ban DDT.[128] An onslaught of other environmental statures followed in the 1960s and 1970s, but scholars emphasize continuity with the past in both policy and law.[129] Building on

New Deal precedents, for example, Congress passed the Water Pollution Control Act of 1948, which authorized the Surgeon General to work with federal, state, and local agencies on comprehensive programs to reduce pollution in interstate waters and provided federal funds to construct sewage treatment plants.[130] Other federal laws included the Fish and Wildlife Coordination Act (1946) and the Air Pollution Control Act (1955).

Karl Brooks traces the basic processes of environmental law, although not its content, to the Administrative Procedure Act (APA) of 1946. The New Deal and mobilization for World War II accelerated environmental problems, as the government harnessed the nation's natural resources to resolve economic and military crises, but they also helped solidify the administrative state. Under the APA, coalitions of hunters, anglers, recreationalists, conservationists, and their lawyers crafted procedures allowing them to intervene in administrative decision making that affected their aesthetic, recreational, or conservation interests long before legal precedents such as the *Scenic Hudson* case (1965) gave such citizens legal standing in the courtroom. These coalitions pragmatically selected which agency decisions to contest and developed media campaigns to woo public opinion. Environmental lawmaking began locally but transformed federal programs and national nongovernmental organizations.[131]

With this administrative structure in place, a stronger environmental ethos gripped the nation by the 1960s, and Congress responded with dozens of laws to protect wildlife and wilderness and to limit harmful exposure of humans and the environment to toxins.[132] These statutes, and similar ones at the state level, addressed "fundamental questions about how people drew sustenance from the larger community of life." They contemplated ecological interactions, biodiversity, sustainability, and interdependence.[133] Over time, the nation, its courts, and its citizens struggled to understand these new ecological concerns in the legal context of federalism and the common law of property and nuisance.[134] In adjudicating the use of government authority to protect the natural environment and human health, federal courts changed the procedural landscape. In *Sierra Club v. Morton* (1972), the US Supreme Court heard the appeal of the Sierra Club in its effort to halt development in an area adjacent to Sequoia National Park. The Court concluded that the organization as a corporate entity had not sustained a sufficient injury to give it standing, but it might file suit on behalf of individual members whose aesthetic or recreational interests were affected. As the complaint had not mentioned such interests, the Court denied the appeal, but the decision expanded the pool of future litigants. Like economic interests, environmental concerns merited protection.[135]

Signed into law on January 1, 1970, the National Environmental Policy Act (NEPA) attempted to provide an integrated environmental policy. Influenced by advancing knowledge in ecology, the law gave scientific evidence priority. The fragmentation of power among federal agencies and resistance from political forces hindered coherent planning, but NEPA added an important element to the legal mix by requiring federal agencies to submit an environmental impact statement (EIS) for any actions affecting the quality of the human environment.[136] Gathering information on the negative environmental impacts did not automatically bar damaging activities, but the Supreme

Court ruled that agencies had an affirmative duty under NEPA to develop and include plans outlining how they will mitigate environmental damage.[137]

Environmentalists recognized that the Endangered Species Act (ESA) of 1973 offered another effective tool. The federal government had managed game species since the Progressive Era. Congress designed this new law to protect any threatened or endangered species from the pressure of economic development on private and public land based solely on the best available scientific evidence. "The ESA was an idealistic and perhaps naïve attempt to preserve humanity by preserving other species in the ecological support system that makes life possible."[138] Because habitat loss was the main reason for species decline, the law allowed for the estoppel of activities that threatened critical habitat. Like many new environmental statutes, this controversial law was redistributive.[139] The nation benefited as a whole from the preservation of plant and animal species, but the burden of compliance fell on an isolated few. The TVA had nearly completed construction of the Tellico Dam, for example, when it was discovered that operation of the dam would harm the snail darter, an endangered fish species. Because Congress intended endangered species to receive the highest protection, the Supreme Court concluded, in *Tennessee Valley Authority v. Hill* (1978), that the only viable choice was to permanently halt the dam.[140] Congress later exempted Tellico Dam from the ESA so that construction could be completed (snail darter populations were moved to nearby streams), and amended the law by complicating the procedures to list a species, imposing new requirements for local hearings, and limiting the time frame during which the listings could be made. The ESA became a "tool of last resort that can slow but not prevent the accelerating loss of biodiversity from the American landscape."[141]

State and local governments enacted statutes to reduce pollution, but also pursued more aggressive planning schemes that dictated private land usage. Gates, an advocate for coherent planning of the public domain, adds, "In fact, the old distinction between public and private is losing its sharpness, or is being eroded away, and for the sake of later generations it should be. Has a man a right to destroy good, irreplaceable agricultural land by covering it up with cement or stripmining it?"[142] In the 1960s, states like California and New York enforced new assessment programs to encourage farming. In 1968, Napa County, California, in a backward-looking measure designed to conserve an agrarian lifestyle, prevented dense urban activities by imposing the nation's first agricultural preserve and barring construction on lots smaller than forty acres.[143]

Critics claimed that rigid, inefficient environmental laws addressed narrow risks, but imposed high economic costs. In the 1980s, a property rights movement "reified" property as a cultural ideal and "timeless institution" now threatened by regulation and land-use controls. The movement ignored common law traditions that limited property rights.[144] A more conservative judiciary at the state and federal level still supported land-use controls, but constrained the scope of laws curtailing pollution and protecting habitats. The Supreme Court, for example, ruled in 1992 that under South Carolina's Beachfront Management Act, which deprived a landowner of any beneficial use of his property, the state must make just compensation because the prohibited activity—construction of vacation homes—was not restricted under the common law of nuisance.[145] The Supreme Court also retracted regulatory authority. The Environmental Protection Agency relied on individual states to enforce its regulatory schemes. In a 1992 case involving the Low-Level Radioactive Waste Policy Amendments Act (1985), the Court ruled that the agency cannot dictate a state's manner of administration without violating the Tenth Amendment, although it can block funds to states that do not meet the guidelines.[146] Environmental lawmaking lost some of its vigor; according to Brooks, "Reaganism" and Americans' failure to develop sustainable plans led to "crisis-mode law enforcement" and an environmental law with "more administrative form than legal substance."[147]

The policies of other federal agencies, some of which maintained close relationships with the industries they regulated, had profound environmental impacts. Historian Nancy Langston relates that Diethyl stilbestrol (DES), a synthetic estrogen proposed to treat the symptoms of menopause, presented one of the first tests of the Food and Drug Administration's expanded authority under the Food, Drug, and Cosmetic Act of 1938. The FDA's first commissioner advocated review procedures for new drugs that paralleled the precautionary principle environmentalists promote for toxic chemicals: "if an action might cause severe or irreversible harm to complex systems, the burden of proof should be on the industry to show that it is safe, rather than the affected communities to show that it is harmful."[148] With respect to DES, the drug companies denied the applicability of animal studies and argued that regulation in the absence of evidence of harm undermined industry innovation and placed unreasonable costs on companies and their customers. By 1941, the FDA abandoned the precautionary principle and approved DES. The FDA moved into a partnership with the pharmaceutical industry. It later approved the widespread use of DES to prevent miscarriages. DES also entered the food supply through implants in chicken and cattle for rapid growth. The environmental and health effects of DES, while not immediately apparent, were devastating: reproductive disruptions in wildlife, birth defects in humans, prostate cancer, infertility, and early sexual maturity. Langston reminds us that the regulators who administer laws often seek pragmatic solutions in a world of scientific uncertainty and partisan politics.[149]

Environmental activists lobbied legislators to integrate the precautionary principle into environmental statutes, and argued that courts should consider it in interpreting those acts or applying the common law. At times, some agencies made administrative decisions and Congress passed statutes, such as the Endangered Species Act and the Clean Water Act, among others, that incorporated precautionary approaches.[150] To date, however, the United States has not adopted the principle universally. A 2006 case involving a question about the authority of the Army Corps of Engineers to administer wetlands under the Clean Water Act, for example, revealed that the Supreme Court justices continued to assign different weights to competing values in determining how environmental risks should be managed, with more conservative justices, such as Anton Scalia and John Roberts, emphasizing property first and foremost.[151]

International Environmental Law

The Rio Declaration of United Nations Conference on Environment and Development in 1992 recommended for international environmental policy a precautionary principle which stipulates that where there are "threats of serious or irreversible damage, lack of full scientific certainty shall not be used as a reason for postponing cost-effective measures to prevent environmental degradation."[152] The declaration is not a rigid obligation; the nations that signed it only agreed to the principle's general applicability, not its specific use in every case.[153] The 1992 Maastricht Treaty creating the European Union (EU) explicitly adopted a precautionary principle that requires proportional responses and stresses the use of science in evaluating risks. This principle provided the justification for prohibiting the use of hormones in livestock in member nations, for example, although in practice, decisions also have reflected economic and cultural concerns such as the preservation of family farms. Despite the EU's position, the Rio Declaration has never enjoyed universal application around the world.[154]

While the United States has not taken the lead on the precautionary principle, it played a foundational role in developing the field of international environmental law. With NEPA, the United States was the first nation to adopt an environmental impact assessment (EIA) process as part of its law. Other nations recognized that EIA potentially facilitated sound economic development, allowed for citizen participation, required mitigation plans for damages, and provided the means to hopefully avoid government-induced environmental crises. By 2013, perhaps as many as 125 nations had adopted some permutation of NEPA.[155] NEPA provides but one link between the United States and the larger world of international environmental law.

Since its inception, the United States has negotiated treaties with other nations; many of these dictated the terms of trade which, in turn, influenced the use of natural resources in a competitive global economy. Fisheries were often the scenes of conflicts and the subject of treaties.[156] The Great Lakes fisheries long constituted a point of tension between the United States and Canada. A joint commission in 1891 proposed that the United States require the same kind of licenses that Canada did and recommended fish propagation efforts and fishing net specifications. Favorably received in Canada, the report met resistance in the United States. President Theodore Roosevelt later tried to resurrect fishery regulations, but US commercial fishermen fiercely lobbied Congress to prevent a new agreement, claiming it would bring economic hardship. While a treaty with Great Britain on the Great Lakes and Puget Sound fisheries was ratified in 1908, Congress did not pass the necessary laws to implement its provisions. The British and Canadian governments later withdrew from the treaty.[157]

A few years later, the United States entered into a convention with Great Britain to protect birds that migrated between the United States and Canada from the commercial trade for feathers and pets. Backed by a coalition of scientists, women's clubs, hunters, and birdwatchers, these negotiations confirmed the importance of popular support, scientific evidence, and even economic incentives for such pacts.[158] The Migratory Bird Treaty Act of 1918 implemented the convention, making it unlawful to hunt, kill, or sell listed birds. Traditionally, states regulated hunting through statutes or the common law. Courts held earlier federal attempts to protect birds that migrated among the states unconstitutional because such authority was not among the enumerated powers. Missouri challenged the Migratory Bird Treaty Act on the same basis, but in 1920, the Supreme Court upheld the statute because Article VI of the Constitution made treaties the "supreme law of the land," superseding any state concern. The United States later signed similar agreements with Mexico, Canada, Japan, and Russia.[159] This treaty was one of the first international efforts designed to constrain trade in the name of nature protection. Partner nations later banned substances that contributed to ozone depletion (the Montreal Protocol), reduced acid rain (U.S.-Canada Air Quality Agreement), or worked to restrict greenhouse emissions (Kyoto Protocol). In recent decades, almost all international agreements on any subject, such as the North American Free Trade Agreement between the United States, Canada, and Mexico, contained specific environmental provisions to protect shared waterways, to regulate transboundary pollution, or to prevent the introduction of nonindigenous species.[160]

International environmental law (IEL), a branch of international law that mediates problems between nations, differs from the broader field. First, the policies of nation-states, particularly industrialized nations, contributed to the creation and growth of this subfield. The momentum from domestic environmental polices led nations to address problems that spilled across geopolitical borders. Nongovernmental entities from labor unions to professional associations participate. Finally, the foundational norm of IEL is "sustainable development," a political concept that calls for the balancing economic growth, social growth, and environmental protection.[161] Coined by the United Nations World Commission on Environment and Development in 1987, sustainable development "meets the needs of the present without compromising the ability of future generations to meet their own needs."[162] It has worked better as a goal than as a guideline for nations for implementing environmental policies. Nations do not agree about what constitutes sustainable development. The refusal of the United States to ratify the Kyoto Protocol (1997), which calls for reduced emissions of greenhouse gases, reflects this tension. The United States objected to different standards for nations based on levels of development and the consequent exclusion of China and India, two of the world's largest polluters, from more rigid provisions. Right or wrong, the US position undermined the nation's credibility in the global community, belying a longer history of US leadership in IEL and in the promulgation of regulatory models for improved environmental protection.

CONCLUSION

International environmental law added new layers to the complex legal relationship between humans and nature. To a large extent, it relies on the goodwill of nations and, just as earlier law often balanced individual liberty against larger community needs, it asks those nations to balance short-term economic self-interest against the long-term ecological health of the planet. Countries have met such burdens with varying levels of acceptance. The United States rejected certain pacts on the basis that they favored less developed nations which recklessly exploited resources and produced excessive pollution in an effort to participate more fully in the global economy. Ironically, in the nineteenth century, as the United States pushed into the industrial age, its courts frequently transformed the common law to favor economically powerful interests who dynamically employed capital, even when their activities badly damaged the environment and interfered with more traditional resource usages enshrined for centuries in the common law. With few exceptions, it was only in the post-World War II era that science emerged as a larger part of the legal agenda and new statutes gave aspects of nature value beyond anthropocentric utility. Nonetheless, as lawmakers continue to balance individual rights against a more broadly defined general welfare that incorporates ecological health, environmental law remains subject to shifting morality, economic pressures, political stresses, and international tensions.

NOTES

I would like to thank Andrew Isenberg and the anonymous reader of this handbook for their fruitful suggestions for improving this chapter. I also would like to thank Jacob Blackwell of the University of Oklahoma for his research assistance.

1. Clive Ponting, *A Green History of the World: The Environment and the Collapse of Great Civilizations* (New York: Penguin, 1991), chapter 4 *passim*.
2. James Willard Hurst, "Legal Elements in United States History," *Perspectives in American History* 5 (1971): 3–5.
3. Richard J. Lazarus, *The Making of Environmental Law* (Chicago: University of Chicago Press, 2004), xi–xiv, 251–252.
4. Arthur McEvoy, *The Fisherman's Problem: Ecology and Law in the California Fisheries, 1850–1980* (New York: Cambridge University Press, 1986), 13.
5. Hurst, *Law and the Conditions of Freedom*, 24–29; and William J. Novak, *The People's Welfare: Law & Regulation in Nineteenth-Century America* (Chapel Hill: University of North Carolina Press, 1996), 2.
6. By the nineteenth century, the federal government had displaced most Native Americans pursuant to the supposed demands of the globally oriented market economy, and had imposed new property regimes across the continent, although the last three decades of the twentieth century witnessed a resurgence of indigenous rights to land, water, and hunting and fishing rights. The complexities of this federal Indian and tribal law, however, are too great to allow for adequate discussion in this chapter. See Richard N. L. Andrews, *Managing the Environment, Managing Ourselves: A History of American Environmental Policy*, 2d ed. (New Haven: Yale University Press, 2006), 77–79; and Paul Wallace Gates, *History of Public Land Law Development*, prepared for the U.S. Public Land Law Review Commission (Washington, DC: Government Printing Office, 1968), 369–370, 453–455.

7. Lawrence M. Friedman, *A History of American Law*, 3d ed. (New York: Touchstone, 2005), xiii.

8. Kermit L. Hall, *The Magic Mirror: Law in American History* (Oxford: Oxford University Press, 1989), 10.

9. Friedman, *A History of American Law*, xvii; and Gary Nash, *The Unknown Revolution: The Unruly Birth of Democracy and the Struggle to Create America* (New York: Viking, 2005), 215.

10. Andrews, *Managing the Environment, Managing Ourselves*, 49.

11. Eric T. Freyfogle, *Agrarianism and the Good Society: Land, Culture, Conflict, and Hope* (Lexington: University of Kentucky Press, 2007), 95; and Herbert Hovenkamp and Sheldon F. Kurtz, *Principles of Property Law*, 6th ed. (St. Paul, Minnesota: Thomson West, 2005), 56.

12. Peter Charles Hoffer, *Law and People in Colonial America* (Baltimore: Johns Hopkins University Press, 1998), 2–6; and Friedman, *A History of American Law*, 5.

13. Hall, *The Magic Mirror: Law in American History*, 13.

14. George L. Haskins, *Law and Authority in Early Massachusetts: A Study in Tradition and Design* (Lanham, Md.: University Press of America, 1960), 4–6; David T. Konig, *Law and Society in Puritan Massachusetts: Essex County, 1629–1692* (Chapel Hill: University of North Carolina Press, 1979), 20–23; and William E. Nelson, *Americanization of the Common Law: The Impact of Legal Change on Massachusetts Society, 1760–1830* (Athens: University of Georgia Press, 1994), 13–16.

15. American dissatisfaction with and frequent unwillingness to pay such obligations contributed to the subsequent dissolution of ties to the Mother Country. Haskins, *Law and Authority in Early Massachusetts*, 69–70; and David G. Allen, *In English Ways: The Movement of Societies and the Transferral of English Local Law and Custom to Colonial Massachusetts Bay in the Seventeenth Century* (New York: W. W. Norton, 1983), 35–41.

16. Beverley W. Bond, Jr., "The Quit-Rent System in the American Colonies," *American Historical Review* 17 (April 2012): 496–516.

17. Friedman, *A History of American Law*, 26. Also see Hoffer, *Law and People in Colonial America*, 101; and David T. Konig, *Law and Society in Puritan Massachusetts*, 39–45. Lacking men skilled in the language of conveyances, particularly in the 1600s, colonists adopted less encumbered land forms to facilitate numerous and frequent property transfers. Treating land as a commodity, colonists created systems to record deeds and protect titles in a volatile market.

18. Hall, *The Magic Mirror*, 44.

19. William Cronon, *Changes in the Land: Indians, Colonists, and the Ecology of New England* (New York: Hill and Wang, 1983), chapter 4 *passim*. Also see Hall, *The Magic Mirror*, 42; and regarding the transfer of larger cultural values toward nature in the colonial era, see Thomas Dunlap, *Nature and the English Diaspora: Environment and History in the United States, Canada, Australia, and New Zealand* (New York: Cambridge University Press, 1999).

20. Brian Donahue, *The Great Meadow: Farmers and the Land in Colonial Concord* (New Haven, CT: Yale University Press, 2004), 107.

21. Ibid., 219–225.

22. *Aldred v. Benton*, IX.57b, *The Reports of Sir Edward Coke* (London: Joseph Butterworth and Son, 1826), 103–108.

23. *Camfield v. United States*, 167 U.S. 518, 523 (1897).

24. William Blackstone, *Commentaries on the Laws of England*, 12th ed., (London: 1793), 305; and J.G. Lammers, *Pollution of International Watercourses* (The Hague: Martinus Nijhoff Publishers, 1984), 570.

25. *Palmer v. Mulligan*, 3 Cai. R. 307 (1805).

26. *Missouri v. Illinois*, 200 U.S. 496 (1906).

27. The *alcade mayor* provided a possible check on overexploitation. He determined whether local resources were sufficient to support new settlers under proposed grants. While designed to insure the colony's economic survival, this mechanism helped conserve water and land resources. Robert MacCameron, "Environmental Change in Colonial New Mexico," in *A Sense of the American West*, ed. James E. Sherow (Albuquerque: University of New Mexico Press, 1998), 49–50, 52. Also see D. W. Meinig, *The Shaping of America: A Geographical Perspective on 500 Years of History*, vol. I (New Haven, CT: Yale University Press, 1986), 240; and more generally Malcolm Ebright, *Land Grants and Lawsuits in Northern New Mexico* (Albuquerque: University of New Mexico Press, 1993).

28. Mark Fiege, *The Republic of Nature: An Environmental History of the United States* (Seattle: University of Washington Press, 2012), 73.

29. Ibid., 74–75; and Nash, *The Unknown Revolution*, 96, 215–216.

30. Friedman, *A History of American Law*, 65–70.

31. John Mack Faragher, *Daniel Boone: The Life and Legend of an American Pioneer* (New York: Henry Holt and Company, 1992), 238–240.

32. Ibid.; and Stephen Aron, *How the West Was Lost: The Transformation of the West from Daniel Boone to Henry Clay* (Baltimore: Johns Hopkins University Press, 1996), Chapter 4 *passim*.

33. Friedman, *A History of American Law*, 71.

34. Faragher, *Daniel Boone*, 240.

35. Article IV, Section 3, of the US Constitution asserted this federal power by granting Congress the authority "to dispose of and make all necessary Rules and Regulations governing the Territory or other Property belonging to the United States." Benjamin H. Hibbard, *A History of Public Land Policies* (New York: Macmillan, 1924), 17–47; Gates, *History of Public Land Law Development*, 62–72, and "An Overview of American Land Policy," *Agricultural History* 50 (1976): 213–229.

36. Gates, *History of Public Land Law Development*, 62. Also see Ted Steinberg, *Down to Earth: Nature's Role in American History* (Oxford: Oxford University Press, 2002), 59–60.

37. Gates, *History of Public Land Law Development*, 62–72; Richard White, *"It's Your Misfortune and None of My Own": A New History of the American West* (Norman: University of Oklahoma, 1991), 143–145; Paul Wallace Gates, "Land Policy and Tenancy in the Prairie States," *Journal of Economic History* 1 (May 1941): 60–82, and "Tenants of the Log Cabin," *Mississippi Valley Historical Review* 49 (June 1962): 3–31. The 1800 law, for example, reduced the minimum acreage for purchase from 640 to 320 acres, but it also doubled the purchase price per acre from $1 to $2. The 1820 law reduced the acreage to 160 and the purchase price to $1.25, but public land remained beyond the means of many Americans.

38. Gates, *History of Public Land Law Development*, 50–65, and "An Overview of American Land Policy," 213–229; and Leo Marx, *The Machine in the Garden: Technology and the Pastoral Ideal in America*, 35th anniversary ed. (New York: Oxford University Press, 2000 [1964]), chapter III *passim*.

39. Fiege, *The Republic of Nature*, 95. Also see John Opie, *The Law of the Land: Two Hundred Years of American Farmland Policy* (Lincoln: University of Nebraska Press, 1987), chapter 1 *passim* and Hildegard Binder Johnson, *Order Upon the Land: The U.S. Rectangular Land Survey and the Upper Mississippi Country*. (New York: Oxford University Press, 1976).

40. Fiege, *The Republic of Nature*, 170.

41. Gates, *History of Public Land Law Development*, 238–247; and Friedman, *A History of American Law*, 171.

42. The law also allowed patents for white men who intended to become citizens. Nearly 7500 patents were issued under the Oregon law. Congress passed similar measures in 1842 and in 1853 to assist early American settlers in Florida and New Mexico, respectively, although the acreage was smaller. Gates, *History of Public Land Law Development*, 118–119.

43. Steinberg, *Down to Earth*, 55.

44. Ibid., 61. See also William Cronon, *Nature's Metropolis: Chicago and the Great West* (New York: W.W. Norton, 1991), chapter 3 *passim*.

45. Donald Worster, *Dust Bowl: The Southern Plains in the 1930s* (New York: Oxford University Press, 1979), 83–90.

46. David Igler, *Industrial Cowboys: Miller & Lux and the Transformation of the American West, 1850–1920* (Berkeley: University of California Press, 2001), 97–101. In *Lux v. Haggin*, 69 Cal. 255, 10 P. 674 (1886), Charles Lux, co-owner of the Lux-Miller cattle company, sued James Haggin. Organized upstream, Haggin's Kern Valley Land and Water Company diverted large amounts of river water to irrigate crops and thereby reduced the amount that reached the Lux-Miller ranch. When some 10,000 head of cattle died, Lux-Miller sued Haggin over who had the best claim to the water. The California Supreme Court decided that newcomers were not permitted to divert water from the landowners who previously established rights to the water. In other words, the court concluded that "appropriation" rights were secondary to "riparian" rights. This decision formed the basis of California's subsequent approach to water rights.

47. Gates, *History of Public Land Law Development*, 50–65, *The Illinois Centralization and its Colonization Work* (Johnson Reprint Group, 1968 [1934]), *Fifty Million Acres: Conflicts over Kansas Land Policy, 1854–1890* (Norman: University of Oklahoma Press, 1997 [1954]), and *The Jeffersonian Dream: Studies in the History of American Land Policy and Development*, ed. Allan G. and Margaret Beattle Bouge (Albuquerque: University of New Mexico Press, 1996), 10–25; Harry N. Schreiber, "The Economic Historian as Realist and Keeper of Democratic Ideals: Paul Wallace Gates's Studies of American Land Policy," *Journal of Economic History* 40 (Sept. 1980): 585–593; Allan Bogue, *From Prairie to Corn Belt: Farming in the Illinois and Iowa Prairies in the Nineteenth Century* (Chicago: University of Chicago Press, 1963) ; and Friedman, *A History of American Law*, 168–169.

48. Hurst, *Law and the Conditions of Freedom*, 24–25.

49. Ibid., 70 (quotation), and chapter II *passim*. Also see James Willard Hurst, *Law and Economic Growth: the Legal History of the Lumber Industry in Wisconsin, 1836–1915* (Cambridge, MA: Belknap Press of Harvard University Press, 1964).

50. Eric T. Freyfogle, *On Private Property: Finding Common Ground on the Ownership of Land* (Boston: Beacon Press, 2007), 95.

51. Hurst, *Law and Economic Growth* generally; and Hurst, *Law and the Conditions of Freedom*, chapter I *passim*.

52. Morton J. Horwitz, *The Transformation of American Law, 1780–1860* (Cambridge, MA: Harvard University Press, 1977), 31.

53. Novak, *The People's Welfare*, 22–24.

54. Ibid., 60–62, 217–233.

55. Christine Meisner Rosen, "'Knowing' Industrial Pollution: Nuisance Law and the Power of Tradition in a Time of Rapid Economic Change, 1840–1854," *Environmental History* 8 (October 2003): 567.

56. Ibid., 574–584.

57. Alfred D. Chandler, Jr., *The Visible Hand: The Managerial Revolution in American Business* (Cambridge, MA: Harvard University Press, 1977), 5–20, 288–300.Also see Robert Wiebe, *The Search for Order, 1877–1920* (New York: Hill & Wang, 1967), 180–186.

58. Christine Meisner Rosen, "The Role of Pollution Regulation and Litigation in the Development of the U.S. Meatpacking Industry, 1865–1880," *Enterprise and Society* 8 (June 2007): 297–347.

59. Chandler, *The Visible Hand*, chapters 3–5 *passim*.

60. Richard White, *Railroaded: The Transcontinentals and the Making of Modern America* (New York: W. W. Norton, 2012), xxv.

61. Ibid. 111, 170, 189, 327, 419–29, 462–81, 485 (quotation). In *Santa Clara County v. Southern Pacific Railroad Company* 118 U.S. 394 (1886), for example, the court concluded that corporations were persons in litigation under the 14th Amendment because corporate shareholders were persons meriting protection under the Amendment. "Subsequent Supreme Court decisions moved far beyond this tangential corporate status. The Court soon treated corporations as integrated wholes, readily distinguishable from those individuals who happened to won share. Brian Balogh, *A Government Out of Sight: The Mystery of National Autority in Nineteenth-Century America* (New York: Cambridge University Press, 2009), 334.

62. Balogh, *A Government Out of Sight*, 19.

63. Pursuant to Article III, Section 2 of the US Constitution, federal courts maintained jurisdiction in cases in which parties to a lawsuit are citizens of different states. Federal question jurisdiction alternatively involves issues arising under federal law.

64. *Swift v. Tyson*, 41 U.S. 1 (1842); Balogh, *A Government Out of Sight*, 235; and Friedman, *A History of American Law*, 192. In 1938, the US Supreme Court overturned *Swift* and ruled that there was no general federal law. Justices Louis Brandeis and Felix Frankfurter and others argued that the *Swift* decision had given federal courts undue power and worked to the benefit of large corporations who exploited diversity jurisdiction to avoid local regulations. *Erie Railroad v. Tompkins*, 305 U.S. 63 (1938); and Michael E. Parrish, "The Great Depression, the New Deal, and the American Legal Order," in *American Law and the Constitutional Order: Historical Perspectives*, 2d ed., ed. Lawrence M. Friedman and Harry N. Scheiber (Cambridge, MA: Harvard University Press, 1988), 388.

65. Balogh, *A Government Out of Sight*, 276.

66. The Minnesota law also partly reflected the lobbying efforts of the Butchers Proctective Association as it fought against the concentrated power of the big four, nationally active meatpackers. *Minnesota v. Barber*, 136 U.S. 313 (1890); and Balogh, *A Government Out of Sight*, 342–343. Police powers refer to the authority conferred by the Tenth Amendment upon the individual states through which they may adopt laws and regulations that secure the safety, health, morals, and general welfare of their citizens.

67. *Mulligan v. Palmer*, 3 Cai. R. 307, 2 A.D. 270 (1805), discussed in Horwitz, *The Transformation of American Law*, 37, and in Hall, *The Magic Mirror*, 115.

68. Horwitz, *The Transformation of American Law*, 38.

69. Theodore Steinberg, *Nature Incorporated: Industrialization and the Waters of New England* (Amherst: University of Massachusetts Press, 1994), 13, 114–116, 142–143.

70. Charles R. Porter, "The Hydraulic West: The History of Irrigation," in *The World of the American West*, ed. Gordon Morris Bakken (New York: Routledge, 2011), 312; and David J. Weber, *The Mexican Frontier, 1821–1846: The American West under Mexico* (Albuquerque: University of New Mexico Press, 1982), 35–40.

71. Michael C. Meyer, *Water in the Hispanic Southwest* (Tucson: University of Arizona Press, 1984), 178 and Charles R. Cutter, "Community and Law in Northern New Spain," *The Americas* 50 (1994): 467, cited in Porter, "The Hydraulic West," 319.

72. Porter, "The Hydraulic West," 320–324; and Meyer, *Water in the Hispanic South*, 167–180.

73. Porter, "The Hydraulic West," 324.

74. Donald J. Pisani, *To Reclaim a Divided West: Water, Law, and Public Policy, 1848–1902* (Albuquerque: University of New Mexico Press, 1992), 39.

75. Ibid., 40–46, quotation, 45.

76. Gordon M. Bakken, *The Development of Law on the Rocky Mountain Frontier: Civil Law and Society, 1850–1912* (Westport, CT: Greenwood Press, 1983), 71.

77. The plaintiff, Yunker, brought an action in trespass against the defendant, Nichols, for diverting all the water from a jointly constructed irrigation canal from a creek adjacent to the defendant's farm. The canal passed through the defendant's farm to the plaintiff's farm. Under the traditional common law, Nichols as the riparian landowner arguably possessed superior rights. An 1861 territorial statute extended a right of way, for the purpose of irrigation, over adjacent lands to property owners "on the bank, margin or neighborhood of any stream of water." *Yunker v. Nichols*, 1 Colo. 551, 564 (1872); and 1861 Colorado Session Laws 67, Section 1. Also see Kathleen A. Brosnan, *Uniting Mountain and Plain: Urbanization, Law, and Environmental Change along the Front Range* (Albuquerque: University of New Mexico Press, 2002), 78–79; and Ralph H. Hess, "The Colorado Water Right," *Columbia Law Review* XVI (1916): 649–664. Hallett became the US District Court judge following Colorado's statehood in 1876.

78. Norris Hundley, *Water and the West: The Colorado River Compact and the Politics of Water in the American West* (Berkeley: University of California Press, 1966), 66–73; John D. W. Guice, *The Rocky Mountain Bench: The Territorial Supreme Courts of Colorado, Montana, and Wyoming, 1861–1890* (New Haven, CT: Yale University Press, 1972), 124; Robert G. Dunbar, *Forging New Rights in Western Waters* (Lincoln: University of Nebraska Press, 1983), 78, and "The Adaptability of Water Law to the Aridity of the West," *Journal of the West* 24 (January 1985): 57; and Gordon Bakken, "The Influence of the West on the Development of Law," *Journal of the West* 24 (January 1985): 67.

79. J. W. Powell, *Report on the Lands of the Arid Regions of the United States*, 2d ed. (Washington, DC: Government Printing Office, 1979). Powell argued for the use of watersheds to define the boundaries of irrigation systems.

80. In this case, the Colorado Supreme Court held that Article XVI of the state constitution ("The right to divert the unappropriated waters of any natural stream to beneficial use shall never be denied") simply ratified that the doctrine of prior appropriation that had been in place since the earliest uses of water by white settlers in Colorado. *Coffin v. Left Hand Ditch Co.*, 6 Colo. 443, 446 (1882); and Brosnan, *Uniting Mountain and Plain*, 79.

81. Pisani, *To Reclaim a Divided West*, 31. Gordon Bakken similarly argues that lawmakers in the Rocky Mountains borrowed from the English common law, but also adapted this law to meet

the needs of the frontier. Bakken returns to Earl Pomeroy's earlier analysis which emphasized a juridical balance between continuity in the transmission of law from the eastern
United States and innovations to the conditions in the western territories in the nineteenth
century. See Gordon Bakken, "Contract Law in the Rockies, 1830–1912," *American Journal
of Legal History* 18 (1974): 34; Earl Pomeroy, *The Territories and the United States*, 52–61; and
Aaron Steven Wilson, "Government and Law in the American West," in *The World of the
American West*, ed. Gordon Morris Bakken (New York: Routledge, 2011), 508–510.

82. Pisani, *To Reclaim a Divided West*, xv-xvii, 26–31. Pisani covers water issues in California,
    Colorado, Nebraska, and Wyoming.
83. Tristan M. Goodman, "The Development of Prairie Canada's Water Law, 1870–1940," in
    Law and Societies in the Canadian Prairie West, ed. Louis A. Knafla and Jonathan Scott
    Swainger (Vancouver: University of British Columbia Press, 2006), 269–277.
84. Oliver M. Brandes, Linda Nowlan, and Katie Paris, *Going With the Flow? Evolving Water
    Allocations and the Potential and Limits of Water Markets in Canada* (Ottawa: The
    Conference Board of Canada, 2008), ii., http://poliswaterproject.org/sites/default/
    files/09_going_w_flow_1.pdf.
85. Peter N. Davis, "Australian and American Water Allocation Systems Compared," *Boston
    College Industrial and Commercial Law Review* 9 (Spring 1968): 650–659.
86. Yee Huang, "A Tale of Two Countries: Lessons from Australia for Water Law in the United
    States?," Center for Progressive Reform, http://www.progressivereform.org/cprblog.
    cfm/%3Chttp:/fixtheclimate.com/fileadmin/rss/CPRBlog.cfm?idBlog=860CB207-0
    2D4-BB7A-B891B40E9ECFF220; and Brian Haisman, "Impacts of Water Rights Reform
    in Australia," *Water Rights Reform: Lessons for Institutional Design*, ed. Bryan Randolph
    Bruns, Claudia Ringler, and Ruth Suseela Meinzen-Dick (Washington, DC: International
    Food Policy Research Institute, 2005), 113–152, http://www.ifpri.org/sites/default/files/
    publications/oc49.pdf.
87. Among other rules, district codes limited the claim size. By custom, mining claims had
    smaller borders than agricultural lands because the depth of the lode, not the amount of
    surface, determined the potential wealth. Smaller borders also provided greater oppor-
    tunities for the first wave of prospectors. Brosnan, *Uniting Mountain and Plain*, 14–16;
    Bakken, "The Influence of the West on the Development of Law," 66–67; and Robert
    W. Swenson, "Legal Aspects of Mineral Resources Exploitation," in Gates, *History of
    Public Land Law Development*, 716–723.
88. See for example Christopher McGrory Klyza, *Who Controls the Public Lands? Mining,
    Forestry, and Grazing Policies, 1870–1990* (Chapel Hill: University of North Carolina
    Press, 1996), 31.
89. On this alternative perspective, see Brosnan, *Uniting Mountain and Plain*, 16; Donald
    J. Pisani, "Promotion and Regulation: Constitutionalism and the American Economy,"
    Journal of American History 74 (December 1987): 769; and Novak, *The People's Welfare*,
    10–11.
90. Andrew C. Isenberg, *Mining California: An Ecological History* (New York: Hill & Wang,
    2005), 31–34.
91. The appellate court in Denver later upheld the verdict although the parties had reached
    a settlement in the interim. Brosnan, *Uniting Mountain and Plain*, 29–33; Malcolm
    Rohrbough, *Aspen: The History of a Silver-Mining Town, 1879–1893* (New York: Oxford
    University Press, 1986), 94–104; and "The Romance of a Mining Venture," unpublished
    manuscript, David M. Hyman Collection, Colorado Historical Society.

92. Gordon Morris Bakken, *The Mining Law of 1872: Past, Politics, and Prospects* (Albuquerque: University of New Mexico Press, 2008), 6, 81, 119–25, 157–63.

93. Eric T. Freyfogle, *Natural Resources Law: Private Rights and Collective Governance* (St. Paul, MN: Thomson West, 2007), 47–51.

94. *Woodruff v. North Bloomfield Gravel Mining Co.*, 18 F. 753 (C.C.C. Cal. 1884) discussed in Robert L. Kelley, *Gold v. Grain: The Hydraulic Mining Controversy in California's Sacramento Valley* (Glendale, CA: Arthur H. Clarke Co., 1959), 228–39; Isenberg, *Mining California*, 171–175; and McEvoy, *The Fisherman's Problem*, 114.

95. See for example Timothy LeCain, *Mass Destruction: The Men and Giant Mines that Wired America and Scarred the Planet* (New Brunswick, NJ: Rutgers University Press, 2009), 209.

96. McEvoy, *The Fisherman's Problem*, 94, citing Hurst, *Law and Economic Growth*, 440.

97. McEvoy, *The Fisherman's Problem*, 118–119. Other advocates hoped that artificial propagation within hatcheries provided a better solution than the regulation of fishermen, but it failed as panacea to compensate for losses associated with the pressures humans placed on natural habits. Joseph E. Taylor, III, *Making Salmon: An Environmental History of the Northwest Fisheries Crisis* (Seattle: University of Washington Press, 1999), 38, 70–75, 125–130.

98. *Pierson v. Post*, 3 Cai. R. 175 (1805).

99. Eric T. Freyfogle and Dale D. Goble, *Wildlife Law: A Primer* (Washington, DC: Island Press, 2009), 39–40.

100. *Ghen v. Rich*, 8 F. 159 (1881).

101. Freyfogle and Goble, *Wildlife Law*, 40–41.

102. *Westmoreland & Cambria Natural Gas Co. v. De Witt*, 130 Pa. 235 (1889).

103. *Elliff v. Texon Drilling Co.*, 210 S.W.2d 558 (1948).

104. Terry Lee Anderson and Fred S. McChesney, *Property Rights: Cooperation, Conflict and Law* (Princeton, NJ: Princeton University Press, 2003), 216–225; Christopher Gordon Down and John Stacks, *Environmental Impacts of Mining* (Applied Science Publisher, Ltd., 1977), 316; and Fred G. Bell and Lawrence J. Donnelly, *Mining and its Impact on the Environment*, chapters 3–5 *passim*.

105. Louisiana passed the first such law in 1906. The US Congress ratified the Interstate Compact to Conserve Oil and Gas of 1935, with the endorsement of six states. Now consisting of thirty member states, it regulates on a voluntary basis and possesses no enforcement powers to control production. Interstate Compact to Conserve Oil and Gas of 1935, United States in *The Encyclopedia of the Earth*, http://www.eoearth.org/article/Interstate_Compact_to_Conserve_Oil_and_Gas_of_1935,_United_States.

106. Samuel P. Hays, *Conservation and the Gospel of Efficiency: The Progressive Conservation Movement, 1890–1920* (Cambridge, MA: Harvard University Press, 1959), 2–4, 219–231.

107. John F. Reiger, *American Sportsmen and the Origins of Conservation*, rev. ed. (Norman: University of Oklahoma, 1986), 21–23.

108. Richard W. Judd, *Common Lands, Common People: The Origins of Conservation in Northern New England* (Cambridge, MA: Harvard University Press, 1997), 25–34, 91–98.

109. Ibid., 124–45. Judd subsequently argued that pre-Darwinian writings of early American naturalists influenced perceptions of nature and laid the groundwork for the later conservation movement by investing the natural world with value. The naturalists emphasized the utility of resources, but observed nature's interconnectedness and inculcated a Romantic attachment to the natural world. The knowledge accumulated throughout the

nineteenth century aided in the passage of the Forest Reserve Act of 1891 and partly laid the foundation for the Progressive Era conservation movement. Richard W. Judd, *The Untilled Garden: Natural History and the Spirit of Conservation* (New York: Cambridge University Press, 2009), 8–9, 279–305.

110. Quoted in Ibid., 94. Marsh gave his speech at 1848 at the Rutland County (Vermont) Agricultural Society.

111. George Perkins Marsh, *Man and Nature; or, Physical Geography as Modified by Human Action*, ed. David Lowenthal (Cambridge, MA: Belknap Press of Harvard University Press, 1965 [1865]).

112. Brosnan, *Uniting Mountain and Plain*, 166–67; and Donald J. Pisani, *Water, Land, and Law in the West: The Limits of Public Policy, 1850-1920* (Lawrence: University Press of Kansas, 1996), 119–140.

113. Pisani, *Water, Land, and Law in the West*, 141–158.

114. Pisani, *To Reclaim a Divided West*, 324–325; and Gates, *The Jeffersonian Dream*, 113–114. Klyza alternatively suggests that different philosophies informed the land-management policies of various federal agencies. In mining, theories of economic liberalism called for the transfer of public lands to private hands. Technocratic utilitarianism in forestry led to government ownership and management. Finally, interest-group liberalism allowed private parties interested in grazing to set government policy on public lands. Klyza, *Who Controls the Public Lands?*, 11–24.

115. Pisani, *Water, Land, and Law in the West*, 197; and Robert Gottlieb, *Forcing the Spring: the Transformation of the American Environmental Movement* (Washington, DC: Island Press, 1993), 115.

116. Neil M. Maher, *Nature's New Deal: the Civilian Conservation Corps and the Roots of the American Environmental Movement* (New York: Oxford University Press, 2007), 17–15, 46–56, 211–214.

117. Worster, *Dust Bowl*, 5, 158–160, 229–230. Also see Sarah T. Phillips, *This Land, This Nation: Conservation, Rural America, and the New Deal* (New York: Cambridge University Press, 2007), who argues that attempts to raise rural standards of living were directly tied to conservation programs, although such efforts were not always successful.

118. Freyfogle, *On Private Property*, 96–97.

119. Joel A. Tarr, *The Search for the Ultimate Sink: Urban Pollution in Historical Perspective* (Akron, OH: Akron University Press, 1996), 11–13, 29, 179–211.

120. Andrew Hurley, "Busby's Stink Boat and the Regulation of Nuisance Trades, 1865–1918," in *Common Fields: An Environmental History of St. Louis* (St. Louis: Missouri Historical Society Press, 1997), 151–158.

121. Andrew Hurley, "Creating Ecological Wastelands: Oil Pollution in New York City, 1870–1900" *Journal of Urban History* 20 (1994): 345–356.

122. Lewis Mumford, *The City of History: Its Origins, its Transformations, and its Prospects* (New York: Harcourt, Brace, & World, 1961), 34, 110–111.

123. Gerald E. Frug, "The City as a Legal Concept," *Harvard Law Review* 93 (April 1980): 1057–1154.

124. Other cities previously implemented limited-purpose controls on building bulks, such as height strictures to prevent fires in Boston or preserve aesthetic lines in Washington, DC. Other controls banned particular land uses in certain districts, including offensive trades such as tanneries, slaughterhouses, and garbage dumps. Robert C. Ellickson and A. Dan Tarlock, *Land-Use Controls: Cases and Materials* (Boston: Little, Brown and Company,

1981), 39–40; and "About NYC Zoning," New York City Department of City Planning, http://www.nyc.gov/html/dcp/html/zone/zonehis.shtml. After many amendments, New York City replaced its original zoning ordinance with a new resolution in 1961.

125. Hurley, "Busby's Stink Boat," *Common Fields*, 162.

126. *Village of Euclid, Ohio v. Ambler Realty Co.*, 272 U.S. 365, 387 (1926). Zoning ordinances and their underlying plans have grown more complex and accommodated or limited land uses unanticipated in 1916, but since the *Euclid* decision, no state supreme court has challenged the basic constitutionality of the general practice of zoning.

127. Robert Ellickson, "Alternatives to Zoning: Covenants, Nuisance Rules, and Fines as Land Use Controls,: *University of Chicago Law Review* 40 (1973): 728–731.

128. For example, see Oliver A. Houck and Richard J. Lazarus, "The Story of Environmental Law," in *Environmental Law Stories*, ed. Lazarus and Houck (New York: Foundation Press, 2005), 2–6.

129. Andrews, *Managing the Environment, Managing Ourselves*, points to a long tradition of federal resource management dating to the early republic. Gottlieb, in *Forcing the Spring*, identifies continuities between Progressive Era conservation and postwar environmentalism.

130. Federal Water Pollution Control Act (1948), Digest of Federal Resource Laws of Interest to the US Fish and Wildlife Service, http://www.fws.gov/laws/lawsdigest/fwatrpo.html; and Robert V. Percival, "Environmental Law," in *The Oxford Companion to American Law*, ed. Kermit L. Hall (New York: Oxford University Press, 2002), 260.

131. Brooks, *Before Earth Day*, 40–92. Also see *Scenic Hudson Preservation Conference v. Federal Power Commission*, 354 F.2d 608 (2d. Cir. 1965).

132. A short list of federal statutes includes the Clean Air Act (1963); the Wilderness Act (1964); the Wild and Scenic Rivers Act (1968); the Clean Air Act (1970); Coastal Zone Management Act (1972); Endangered Species Act (1973); Safe Drinking Water Act (1974); Resource Conservation and Recovery Act (1976); and the Comprehensive Environmental Response, Compensation, and Liability Act, known as Superfund (1980).

133. Dale D. Goble and Eric T. Freyfogle, *Wildlife Law: Cases and Materials* (New York: Foundation Press, 2002), v.

134. Richard O. Brooks, Ross Jones, and Ross A. Virginia, *Law and Ecology: The Rise of the Ecosystem Regime* (Burlington, VT: Ashgate, 2002), 26; and Eric T. Freyfogle, *The Land We Share: Private Property and the Common Good* (Washington, DC: Island Press, 2003), 206–207.

135. *Sierra Club v. Morton*, 405 U.S. 727 (1972). *Sierra Club v. Morton* is perhaps best remembered, however, for Justice William O. Douglas's dissent. A noted outdoorsman, Douglas argued that environmental issues should be litigable in the name of the despoiled natural resources where damage is the subject of public outrage.

136. Ray Clark and Larry Canter, eds., *Environmental Policy and NEPA: Past, Present, and Future* (Boca Raton, FL: St. Lucie Press, 1997).

137. *Robertson v. Methow Valley Citizens Council*, 490 U.S. 332 (1989).

138. J. Michael Scott, Dale D. Goble, and Frank W. Davis, "Introduction," in Goble, Scott, and Davis, eds., *The Endangered Species Act at Thirty: Renewing the Conservation Promise*, vol. I (Washington, DC: Island Press, 2006), 3.

139. Lazarus, *The Making of Environmental Law*, 29–47.

140. *Tennessee Valley Authority v. Hill*, 437 U.S. 153 (1978). Also see Jason E. Shogren, ed., *Private Property and the Endangered Species Act: Saving Habitats, Protecting Homes* (Austin: University of Texas Press, 1998).

141. Scott, Goble, and Davis, "Introduction," 15; and Holly Doremus, "The Story of *TVA v. Hill*: A Narrow Escape for a Broad New Law," in *Environmental Law Stories*, ed. Lazarus and Houck, 109–140.

142. Gates, "An Overview of American Land Policy," 227–228. Also see Tim Lehman, *Public Values, Private Lands: Farmland Preservation Policy, 1933–1985* (Chapel Hill: University of North Carolina Press, 1995), 98–101; and Freyfogle, *Agrarianism and the Good Society*, 83–90.

143. The Napa County supervisors later extended the agricultural preserve, which applied only to unincorporated agricultural lands, to the adjacent hillsides and increased the minimum acreage. Kathleen A. Brosnan,"Crabgrass or Grapes: Urban Sprawl, Agricultural Persistence and the Fight for the Napa Valley," in Char Miller, ed., *Cities and Nature in the American West* (Reno: University of Nevada Press, 2010), 34–56.

144. Freyfogle, *On Private Property*, 74–83.

145. *Lucas v. South Carolina Coastal Council*, 505 U.S. 1003 (1992). Courts, however, rarely recognized regulatory takings.

146. *New York v. United States*, 505 U.S. 144 (1992). Also see Percival, "Environmental Law," 263.

147. Brooks, *Before Earth Day*, 203.

148. Nancy Langston, *Toxic Bodies: Hormone Disruptors and the Legacy of DES* (New Haven, CT: Yale University Press, 2010), ix.

149. Ibid., chapter 3 *passim*, 61–64, 75–77, 155–56. New urgency for the approval of DES emerged, in part, because the medical profession began to treat menopause as a medical crisis requiring a beneficial therapy.

150. Phillip M. Kannan, "The Precautionary Principle: More than a Cameo Appearance in United States Environmental Law?" *William & Mary Environmental Law and Policy Review* 31, no. 2 (2007): 435–456.

151. *Rapanos v. U.S. Army Corps of Engineers*, 547 U.S, 715 (2006). The court failed to reach a majority opinion as to the appropriate balance and remanded the case for further deliberation.

152. Rio Declaration on Environment and Development, UN Conference on Environment and Development, UN Doc. A/CONF.151/5/Rev.1 (1992).

153. Kamman, "The Precautionary Principle," 420. Some six years later the Wingspread Conference, which included treaty negotiators, scientists, activists, and scholars, broadened the principle by removing words such as "serious" or "irreversible." "Wingspread Statement on the Precautionary Principle," Science and Environmental Health Network, http://www.sehn.org/wing.html.

154. See Beanal v. Freeport-McMoran, Inc., 197 F.3d 161, 167 (5th Cir. 1999); and Jonathan Weiner and Michael Rogers, "Comparing Precaution in the United States and Europe," Journal of Risk Research 5 (2002): 317–349.

155. Charles H. Eccleston, *NEPA and Environmental Planning: Tools, Techniques and Approaches for Practicioners* (Boca Raton, FL: Taylor & Francis, 2008), 299–300; C. Wood, "What has NEPA Wrought Abroad?" in *Environmental Policy and NEPA: Past Present, and Future*, ed. Ray Clarke and Larry W. Canter (Boca Raton, FL: CRC Press, 1997), 99–111; and John Cronin and Robert F. Kennedy, Jr., *The Riverkeeper: Two Activists Fight to Reclaim Our Environment as a Basic Human Right* (New York: Simon & Schuster, 1999).

156. See chapter 24 in this volume.

157. Margaret Beattie Bogue, *Fishing the Great Lakes: An Environmental History, 1783–1933* (Madison: University of Wisconsin Press, 2000), 192–193, 238–248; and Kurkpatrick

Dorsey, *The Dawn of Conservation Diplomacy: U.S.-Canadian Wildlife Treaties in the Progressive Era* (Seattle: University of Washington Press, 2010), 76–105.

158. Dorsey, *The Dawn of Conservation Diplomacy*, 3–18, 215–237.

159. *Missouri v. Holland*, 252 U.S. 416 (1920).

160. Alan E. Boyle, "International Law and the Protection of Global Atmosphere: Concepts, Categories and Principles," in *International Law and Global Climate Change*, ed. Robin Churchill and David Freestone (London: Graham & Trotman, 1991), 7–19.

161. Lakshman Guruswamy, *International Environmental Law in a Nutshell* (St. Paul, MN: Thomson West, 2007), 1–4, 34–50; and "Environmental Law: International Environmental Law in the United States," in *The Oxford Companion to American Law*, ed. Kermit L. Hall (New York: Oxford University Press, 2002), 264–265.

162. *Report of the World Commission on Environment and Development*, United Nations, December 11, 1987.

CHAPTER 19

....

CONFLUENCES OF NATURE
AND CULTURE

Cities in Environmental History

....

LAWRENCE CULVER

MANY environmentalists in North America and Europe do not like cities. Perceiving them as the antithesis of nature, these environmentalists have perpetuated a long-standing distrust of cities that goes back to Thomas Jefferson, and then even farther back to British suspicions of urbanism, which stood in contrast to a preference for pastoral landscapes and the idealized country life. Better to be a hale and hearty farmer, so the thinking went, than an enervated and effete urbanite. Since the nineteenth century, European Romantics and Americans from John Muir to Theodore Roosevelt have likewise embraced wilderness and wild nature as a place of purity and escape—not merely the opposite of urban life, but its antidote.

Environmental historians, perhaps unsurprisingly, initially also neglected urban history as a component of environmental history. Histories of national parks, forests, or farming? Those were environmental history. There was no nature in the city to interest them. Urban historians likewise eschewed the environment as a factor in their analysis. Politics, crime, culture, infrastructure, and other topics composed urban history.

Yet in recent decades environmental and urban histories have begun to intersect. Historians have begun to see connections between natural and built environments, between humans and ecosystems, and between cities and surrounding landscapes and regions. Beyond these recent developments, there is also a much older literature connecting nature to the city, largely produced not by historians, but by urban reformers including Frederick Law Olmsted, and turn-of-the-century reformers including Ebenezer Howard and the members of the British Garden Cities Movement, all of whom wanted to bring nature into the city. By examining the place of nature in the city, and the place of the city in nature, environmental historians have discovered a powerful lens through which to view the world entire, and to perceive cities as central to environmental history as confluences of nature and culture. The historians Joel Tarr and

Martin Melosi forced other historians to think about waste and energy systems within cities.[1] William Cronon brought to light the intricate connections between cities and their hinterlands.[2] Andrew Hurley examined urban environmental inequalities.[3] Taken together, in the work of urban environmental historians such as Ari Kelman and Matthew Klingle, urban environmental history suggests that urbanites organize social and environmental systems both within and around cities.

URBAN ORIGINS AND NATURAL SETTINGS

The original forces that drove humans to gather in cities are not known. The factors may have been connected to climate change, to the rise of agriculture, to the emergence of religion, or other causes. In North America before 1492, religion seems to have been a driving force, from the Maya city-states in present-day Yucatán, to Cahokia near the confluence of the Missouri and Mississippi Rivers. In fact, some sites long assumed to be urban centers, such as Chaco Canyon in New Mexico, are now seen primarily as religious centers, as well as destinations for periodic religious pilgrimage, but not home to large permanent populations, akin to medieval European pilgrimage cities such as Canterbury. At the same time, other sites with a religious purpose were quite clearly cities in the sense that we would understand. The Mexica or Aztec capital of Tenochtitlán, for example, was an imperial capital, with a stratified and complex society, temples, palaces, markets, elaborate aqueducts, and infrastructure that rivaled ancient Rome.[4]

The ancient European world played a role in city-making in North America as well: towns founded by the Spanish, or Native American cities remade by them, contained plazas and gridded street plans that were a direct inheritance of Roman urban forms. London, in contrast, might have been founded by Romans, but its streets were a notorious sprawl even in the early modern period. Boston's streets were likewise a chaotic, unplanned maze; the city resembled an oversized English country village transplanted to Massachusetts. Anglo-American cities ultimately adopted the grid as well, though often less due to Roman models than to the fact that the Land Survey Ordinance of 1785 had spread a Cartesian surveyor's grid across North America, creating square parcels that were easy to sell. In the nineteenth century, the Frenchman Emile Boutmy would succinctly characterize the new nation's character, and its relationship to North America: "Their one primary and predominant object is to cultivate and settle these prairies, forests, and vast waste lands. The striking and peculiar characteristic of American society is, that it is not so much a democracy as a huge commercial company for the discovery, cultivation, and capitalization of its enormous territory. ... The United States are primarily a commercial society ... and only secondarily a nation."[5]

Whatever their urban forms, North American colonial cities created by settlers all shared one key characteristic: they were advantageously located, as best as their founders could surmise, to profit from local natural resources or geography. For example, Quebec was founded on a high promontory above the St. Lawrence River, New

Amsterdam was located at the mouth of the Hudson, New Orleans was founded on the last relatively high, solid ground before the mouth of the Mississippi, or St. Louis, settled near the confluence of the Mississippi, Missouri, and Ohio Rivers, just like Cahokia before it. Geographers call such urban places, where rivers converge or where rivers meet the sea, "breaks in transport."[6] Even Plymouth and Boston, cities founded for religious reasons by Pilgrims and Puritans, were nevertheless located at good harbors, close by Cape Cod, the long, sheltering peninsula named for the fish that had first drawn the English to these waters.

Of course, settlers could also massively misread local landscapes. Jamestown, the first English colony to survive, was initially an utter disaster of starvation and disease. The investors who had selected the location chose it in part because it seemed defensible, an understandable consideration. Yet they also assumed that the surrounding countryside would abound in gold and Mesoamerican-style Native American cities full of potential laborers, and that an easy Northwest Passage to the Pacific could be found nearby. They and other early settlers were also baffled by the climate, assuming that New England and the Atlantic Coast would have a climate similar to the Mediterranean weather of Spain, on the same latitude across the Atlantic. The Vikings might have named eastern North America Vinland, but Mediterranean wine grapes were not going to prosper in its colder and sometimes more capricious climate. For neither the first nor last time, settlers would misread an unfamiliar environment, with disastrous consequences.[7]

Transforming Nature into Capital and Cities

Over time, as settler populations grew, these tenuous North American colonies stabilized and grew, profiting from the natural abundance of a continent that had lost as much as 95 or 98 percent of its indigenous population to diseases brought by Europeans, the most horrific consequence of the Columbian Exchange initiated in 1492. Colonists transformed resources into commodities that were shipped throughout the Atlantic World. The forests of New England would supply the British with timber for ships, especially the tall tree trunks needed for masts, long since harvested in Europe. Cod and other fish were salted and shipped across the Atlantic. Grain and other nonperishable crops could be shipped as well. Finished goods, from furniture to fine porcelain, made the return trip. As a result, the communities on the Northeast coast—Boston, New York, and Philadelphia—emerged as true port cities, and centers inhabited by mercantile, consumer-oriented, literate populations that would serve as crucibles of the American Revolution.[8]

Farther south, however, a very different economic model was emerging. Tobacco, the crop that had saved Jamestown, was highly labor intensive. Without the large, urbanized Indian populations they had expected to dominate, English settlers turned to a new labor market: African slaves.[9] Slaves would power an agricultural revolution from Virginia south to the Carolinas and Georgia. By the beginning of the nineteenth century, in the lower South, a new crop—cotton—would transform Southern agriculture into a highly profitable monoculture dominated by the largest slave owners. Aside from one large port city—New Orleans—which acted as the entrepôt for the entire Mississippi River Valley and its tributaries, other southern cities would remain smaller port towns, such as Savannah, Charleston, and Mobile. The interior would remain rural, and large cities would not appear there until after the Civil War.[10]

Yet this rural region's chief crop would play a central role in the industrialization and urbanization of the North. Cotton bales were transformed into thread and cloth at Lowell, Massachusetts, and at other New England mill towns located on the fall lines of rivers, where drops in elevation gave flowing water the power to turn wheels, turbines, and the machinery of spinning and weaving machines. King Cotton might dominate the South through the nineteenth century, but it transformed the North as well. Steamboats, trains, and canals would increase trade and transport, accelerating urban and industrial growth. The completion of the Erie Canal in 1825 would make New York the dominant urban center of the northeast, leaving Boston and Philadelphia behind. It would also open up an east-west axis connecting the Hudson River Valley with the Great Lakes region, which together composed the industrial heartland of the United States.[11]

It was here that a truly industrial city arose, yet one where industrial development was inextricably connected to the surrounding landscape and its exploitation through agriculture: Chicago. Famously called "Nature's Metropolis" by William Cronon, Chicago and its grain silos, stockyards, slaughterhouses, and railroads grew alongside agriculture in the Midwest, Mississippi River Valley, and the Great Plains, making up together a vast hinterland that Chicago boosters called the "Great West." Natural abundance created economic abundance, and by the later nineteenth century, the world's first skyscrapers constructed inside the Loop attested to Chicago's meteoric rise.[12]

Farther west, however, aridity, deserts, and difficult terrain made travel, let alone settlement, far more difficult. Through irrigation, and communal ownership of water, small communities rose in the Spanish and then Mexican North, from Santa Fe to Tucson to Los Angeles. The first Anglo Americans to pursue a concentrated irrigation strategy were the members of the Church of Jesus Christ of Latter-day Saints, better known as the Mormons. After the death of their prophet Joseph Smith, the Mormons trekked west, settling in the valley of the Great Salt Lake. Here, under the leadership of their new prophet Brigham Young, the Mormons created a theocratic city-state, one that quickly harnessed local streams to irrigate crops. They brought with them their church founder's "Plat of Zion," a distinctive gridded city planning system that incorporated farmlands into the community, but remained dense enough to be defended or evacuated in the event of an attack, whether from Ute Indians or the federal government. This model of irrigation and distinctive urban planning would spread in communities across the Great Basin, and ultimately to Mexico and Canada.[13]

It was riches, however, not religion, that opened the floodgates of urban western settlement. James Marshall's discovery of gold at Coloma in 1848 transformed California and the nation. In the foothills of the Sierra Nevada, ambitious argonauts founded cities to serve as commercial centers of the gold country. Some, such as Sacramento and Marysville, outlived the gold rush; others, such as Vernon, established at the juncture of the Feather and Sacramento Rivers, or Sutterville, below the juncture of the American and the Sacramento, lasted only a few years. The gold rush also transformed the sleepy port of San Francisco into the largest city in a now US-controlled West. As a "new Boston" arose overnight on the West Coast, and the world rushed in to dig for gold, San Francisco would prosper, and its capital would be used to fund mining elsewhere, from the Comstock to Colorado and the Klondike, creating dozens of chaotic mining boomtowns. Some, such as Denver, survived, while others withered when the veins of ore emptied out.[14]

One of those ores in particular—iron—would create a new industrial city, Pittsburgh, dwarfing anything ever imagined at a river mill town such as Lowell. A geographical logic located Pittsburgh and other cities of the "steelbelt"—Chicago, Gary, Detroit, and Cleveland—just north of a huge concentration of coal stretching from Pennsylvania to Illinois, and within easy shipping distance via the Great Lakes from the Mesabi Range in northern Minnesota. By 1900, the United States had surpassed Britain's industrial production. All that steel and iron would also allow the construction of steel frame skyscrapers, elevated commuter lines, bridges, and even pipes for new water and sewer systems. It also allowed a network of railroads to spread across the nation. For the first time, heavy industry also appeared outside the Northeast and upper Midwest. In 1870, Pueblo, Colorado, was founded. Tapping into Colorado's coal, it serving as a center of steel and iron production for the next century. A year later Birmingham, the first industrial city in the South, was founded in Alabama at the heel of the Red Mountain, so named for its iron ore. Its founders, a mix of northern and southern investors, followed a plan first imagined before the Civil War to create an "iron plantation" in the iron and coal-rich region using slave labor. In reality, Birmingham's labor system, highly dependent on forced black convict labor, was terrible enough. Meanwhile, in Georgia, Henry Grady boosted the "New South" and the rebuilt and expanding city of Atlanta, one of a number of inland or highland cities that would grow in the decades following the Civil War. All of these new industrial cities would also witness the construction of streetcar suburbs, affluent communities removed from the pollution of industry, but connected to it by public transportation.[15] Whether in California, on the southern shores of the Great Lakes, or in Alabama, cities served as collection points for capital, labor, and natural resources—the three "factors of production" in industrial economies.

In the same era, another city was growing in a very different "Southland"—Southern California. Bypassed by the gold rush, Los Angeles in 1870 was a small pueblo of five thousand people. It lacked water, any obvious natural resources, or a good harbor. What it did have, however, was climate—sunny days, warm winters, and, especially near the coast, mild summers. It was, and remains, one of the rarest environments on earth: a humid desert. After first promoting it as "tropical," which brought forth unfortunate associations with malaria, boosters settled on a Mediterranean comparison, as "Our Italy." In an era when tuberculosis was a terrifying prospect, and many diseases were thought to spread by "miasmas," or unhealthy air emanating from swamps and other supposedly unclean places, moderate temperatures, mild humidity, and sunshine were hugely appealing. Perhaps as much as a quarter of all the people who came to the region prior to 1900 came for Southern California's supposed healthfulness.[16]

Of course, cheap land and labor provided by Indians, Mexicans, Chinese, and others also proved appealing. Henry Huntington, nephew of Collis Huntington (one of the four major partners in the Central Pacific Railroad) and inheritor of his uncle's vast fortune, made an entirely new fortune as the largest landowner in Southern California. His interurban train system connected a dispersed network of communities and housing developments, meaning that Southern California sprawled well before the automobile. The car proved hugely popular in this region, and Los Angeles County soon had the highest rates of car ownership in the world. New Yorkers and Londoners were baffled when they visited the city, finding a middling downtown surrounded by a vast and diverse landscape of housing developments, agricultural land, retail and business centers, and open space. Mass suburbia had been born.[17]

All those cars needed gasoline, and the large petroleum reserves under the city would help provide it. One of the many products made from petroleum was celluloid, and by 1910 the film industry had arrived, forever associated with a housing development where many studios initially settled: Hollywood. The city without a reason to exist finally had reason enough to prosper, and outgrew San Francisco by 1920. Yet all that growth came about not *because* of natural advantages, but rather *despite* natural disadvantages. With federal funds, Los Angeles constructed an artificial harbor to compete with San Francisco and San Diego, which had two of the finest natural harbors in the world. Also with federal help, it appropriated the Owens River, which flowed east of the Sierras. Los Angeles voters approved the construction of the longest aqueduct ever constructed, bringing the river to the San Fernando Valley. When the waters cascaded down to the valley floor, an assembled throng rushed toward it, tin cups in hand. William Mulholland, the self-educated engineer who had designed the aqueduct, was asked to give a speech marking the occasion. He gestured towards the water, and said, "There it is. Take it." His words could be the motto, not only for the city, but the nation, and its view of natural resources in the 150 years following the Revolution.[18]

ACCELERATING URBANIZATION AND THE RISE OF THE SUNBELT

While the South began to urbanize after the Civil War, it remained the nation's most rural region, with the largest proportion of rural residents anywhere in the country. By 1910, however, that population was on the move. Spurred first by new industrial jobs in Southern cities, the devastation of cotton crops by the boll weevil, and the tight-ening grip of segregation and the Klan, black southerners began the Great Migration north to Harlem, Chicago's South Side, and other northeastern and Midwestern cities. A smaller number headed west as well. Following World War I, an agricultural depres-sion sent many white southerners on a similar journey. Even before the Dust Bowl and the "Okie" exodus to California, Southerners were moving north and west, transform-ing the nation's demography and urban culture in the process.[19]

World War II made this migration even more pronounced. As America mobilized for war, military personnel, support staff, and industrial workers poured into port cit-ies from Norfolk to Seattle. West Coast cities grew the most in sheer numbers, but it was the cities of the interior West, such as Phoenix and Las Vegas, which grew by the largest percentages, sometimes doubling in size. It was only after World War II, how-ever, that it became clear that this was a decisive shift in American urban history.[20]

In the decades to come, what had been among the most rural, and in some cases the most forbidding, regions of the US would become the destination for the largest migra-tion in the nation's history. In the Southwest, specifically the Sunbelt, the new dams and aqueducts that reengineered the hydrology of the West made large-scale growth pos-sible in places like Phoenix. Los Angeles at least had that balmy climate. Now, for the first time, cities grew in places that not only did not have obvious natural resources—their environments, from Arizona to Florida, could be actively hostile, and deadly without climate control. In the Southeast, the Civil Rights Movement, integration, and air conditioning made the region attractive to investors and migrants alike, including, by the later twentieth century, the children and grandchildren of those who had taken part in the Great Migration. In both regions, the Cold War would provide well-paying jobs, and the GI Bill and the Federal Housing Authority loan system would make home ownership affordable. Factories shifted from war production to consumer goods, from cars and televisions to Formica, particle board, concrete block, and other materials that made home construction cheaper, faster, and feasible on a mass scale for the first time. While the Northeast as a whole lost residents, Sunbelt-style suburbs spread outward from New York, Chicago, Philadelphia, and other northern cities. From Levittown on Long Island, to Lakewood, south of Los Angeles, mass suburbia had become a national phenomenon. So, however, had mass segregation; FHA loans, disbursed by local banks, were not available to prospective African American homebuyers. Nor could they be used to renovate existing housing. Increasingly, black Americans would be marooned in decaying central cities, while white Americans fled to the suburbs. When it first opened, Levittown was the whitest community in the US: it did not have a single black resident. Some conservative baby boomers in the early twenty-first century look back on the 1950s as the "real" America, not realizing just how alien it was to the era before it.[21]

Alien or no, the Sunbelt boomed from 1945 to the early twenty-first century. By 2010, California possessed 38 million residents, Texas 25 million, and Florida—nearly half of which was malarial swamp avoided by all but the Seminoles until the late nine-teenth century—had 21 million residents. A ranking of the largest cities in the United

States showed a similar shift. In the census of 2010, Phoenix supplanted Philadelphia as the fifth-largest city in the nation, and San Jose, center of Silicon Valley, had pushed de-industrializing and depopulating Detroit out of the top ten.

Yet this unprecedented boom has been followed by an epic bust. Housing prices peaked in 2006, and then began falling. The worst economic downturn since the Great Depression hobbled the world economy beginning in 2008. Oil prices also spiked, as more and more consumers in China, India, and other rising economic powers began to buy cars. The former boomtowns of the Sunbelt, particularly ones dependent on tourism or real estate, suffered severe economic contractions, from California and Nevada to Florida. Perhaps this will someday be remembered as the end of sprawl, of edge cities and epic commutes, and perhaps it signals the end of the single-family home itself. Yet the US weathered the OPEC oil embargo of the 1970s, and suburban development continued unabated. For that matter, American-style suburban developments have now been constructed around some Chinese cities. American urbanism, for better and worse, has gone global. This, then, is a question pitting culture against economics and environment. Will rising economic and environmental costs force Americans to surrender the suburban good life? Or will they persist in constructing dispersed communities of single-family homes that provide family space, but also force reliance on oil and the automobile? Or, is it possible that the United States will find a third way, one that allows a more dispersed urbanism, but one that does not have the same environmental and economic costs? Only time will tell, and that is a question for the future, while this chapter is concerned with the past.

What—If Anything—Makes American Cities Distinctive?

Unlike European cities, or Spanish colonial cities in central Mexico or Andean South America, cities in the colonial or national US were new. They were not grafted onto indigenous foundations, as Mexico City grew from Tenochtitlán. Nor, outside of the Southwest, could they draw on existing patterns of indigenous labor or tribute, though Native Americans lived and labored in many American cities, from Minneapolis to Seattle. As a result, American cities, which might at first have been founded at strategic locations, almost always grew at places assumed to be economically advantageous, near good agricultural land, valuable ores, a confluence of rivers, or a pass that settlers or rains could traverse. These cities were founded and inhabited by people who had no long-term knowledge of their local environments, and this could—and did—lead to disaster.

American cities, because of their newness, were also especially likely to be transformed by technology. Subways were added into the existing urban matrix of London or Paris. The same was true of cars—they entered an existing urban landscape. That landscape might well be modified by new technology, but it continued otherwise intact. In the US, a city such as Los Angeles—or Houston, or Phoenix, or Atlanta—literally grew up with mass transit and then the automobile. New technologies shaped the growing urban landscape in these twentieth-century cities.[22]

This explains the radically different appearance of US cities from older and more compact European models. The desire to bring nature into the city, to create parks and open space, was a common sentiment on both sides of the Atlantic, and indeed in other parts of the world as well.[23] Yet the growth pattern of North American cities in the twentieth century, from Los Angeles to Sunbelt cities, as well as in new suburbs around existing older cities, was decisively new and different. Instead of bringing nature into the city, Americans took their cities out into nature, unleashing vast changes in the landscape of the nation. Yet this sprawling urbanism, which promised every family their own piece of nature, instead increasingly cut them off from nature. In a century in which rivers could be moved, and fossil fuels and nuclear power used to cool, heat, and light houses and buildings, cities could be built in inhospitable and unlikely places, from the Mojave and Sonoran Deserts to the swampy coastlines of Texas or Florida. Whether or not this was wise, or sustainable, was a different—and mostly unasked—question. It would lead millions to settle in regions of potential environmental hazards. Yet it also helped spur a new and reinvigorated environmental movement in the later twentieth century. The relationship between cities and nature, natural and built environments, and humans and the environment by the early twenty-first century was enormously complex, and urban and environmental historians are now exploring that complexity, to the benefit of both fields.

Surveying the Terrain: Early Efforts to Merge Nature and the City

Before historians began to contemplate the intersections of cities and nature, reformers were already doing so. Frederick Law Olmstead is remembered for Central Park, the Chicago streetcar suburb of Riverside, and many other examples or landscape architecture and urban planning. Yet contemporary Americans often completely misread the complexity of the places he created, as well as his intellectual and explicitly political worldview. As Anne Whiston Spirn has eloquently illustrated, we fundamentally misread the "naturalness" of his constructions. Park-goers perceive Central Park as a remnant of the natural landscape of Manhattan, when in fact it is a carefully designed and heavily engineered landscape. His design plan for the Boston Fens made them functioning wetlands that cleaned and filtered water from the city. When the Fens were restored in the late twentieth century, that restoration was purely aesthetic; their functionality had been forgotten. As an ardent abolitionist, Olmstead also wanted Central Park and other parks to function as a vital democratic meeting grounds, a social and recreational commons for all residents of the city. In reality, restrictions on picnics and alcohol steered away the Irish and German immigrants who had built the park, and it instead became a place for the wealthy to promenade, see, and be seen.[24]

Progressive reformers such as Jacob Riis also saw the need for nature in the city, particularly in the form of playgrounds. Chicago, and then Los Angeles, would be the first two cities in the United States to create city departments to oversee playgrounds. These agencies, however, were often as concerned about a bluntly forced "Americanization" as they were about childhood play. For that matter, in the era of a supposedly closing frontier, Jane Addams's Hull-House and other "settlement houses," which provided services (and more of that Americanization) in immigrant communities, were, at least symbolically, attempting to bring the vitality of nature into the city. Nor was this in any way limited to the United States. In Britain, Germany, Mexico, and elsewhere, nature was seen as the solution to the excesses of urbanization.[25]

In Britain, an important movement in the same time period was the Garden City Movement, spearheaded by Ebenezer Howard. Howard wanted new communities, whether they were new cities or suburbs of existing cities, that would be cleaner and safer than big cities such as London or Birmingham. He was inspired to design such places after reading Edward Bellamy's utopian novel, *Looking Backward* (1887). Meticulously planned, Garden Cities were intended to never become larger than perhaps 30,000 people, with greenbelts of open space separating one town from the next. These communities would ideally merge town and country into one livable form, with enough land to be agriculturally self-sufficient, and open spaces for recreation. Howard's model proved influential in Britain, and in other countries as well. In the United States, one famous example is Greenbelt, Maryland. Yet the degree of planning, zoning, and design control required under Howard's meticulous plans would find limited purchase in the US, where unrestricted development would lead suburban communities in decidedly different directions.

Architects also thought and wrote about the place of nature in the city. Frank Lloyd Wright, the most famous American architect of the twentieth century, is remembered for houses that responded to and reflected the surrounding landscape. Yet Wright also had a keen interest in urban planning. His most famous design in this regard was for "Broadacre City," a radically decentered and dispersed "urban" plan, where every home had its own farmland, and there was no downtown to be seen. Thomas Jefferson might have liked this plan, for it embodied the deep suspicion many Americans felt toward cities. Broadacre City remained only a model, but the sprawling patchwork development around Sunbelt cities such as Phoenix or Houston are indirect heirs of Wright's plan, and the broader cultural sentiments he embodied.

At least two writers and public intellectuals also deserve mention here: Lewis Mumford and Jane Jacobs. Mumford was the author of many books on cities, including *The City in History* (1961). This sprawling book connected the growth of cities to the rise of human civilization. From the perspective of the early twenty-first century, it is dated, yet nevertheless remains pertinent. Mumford was prone to grand pronouncements and generalizations. He clearly over-romanticized some cities and eras, such as his depiction of the medieval European city. He also was too quick to condemn the aspects of contemporary cities he did not like, from suburbs to overreliance on technology to the design of the Pentagon. Yet his concerns remain highly relevant. He believed that cities were connected to their surrounding landscape, that they were inhabited by human beings who were first and foremost social animals, and they were places where natural and built environments, nature, culture, and technology coexisted in a complex balance. His concern that urban redevelopment—whether in declining American downtowns, or in European cities destroyed by bombs—maintain a human scale seems especially farsighted, as architects following Le Corbusier's futuristic cityscapes of high tower apartment blocks and open spaces created dreary, unloved housing that demolished some existing neighborhoods, and in many cases were later demolished themselves, as failed models of urban design.[26]

A very different writer who likewise shared concerns about a rush to demolish "blighted" neighborhoods and replace them with the grand schemes of urban designers and developers was Jane Jacobs. She famously clashed with Robert Moses, who oversaw redevelopment projects in New York City. When he announced plans for a new freeway that would demolish Greenwich Village, Jacobs had had enough. Her book *The Death and Life of Great American Cities* (also published in 1961) was a manifesto in favor of the messy but vital living communities that planners such as Moses hoped to replace with a redesigned urban planner's utopia. Jacobs, who had no training in urban planning, had arguably written the most influential urban planning work in US history. Like Mumford, she called for the preservation of the human scale in cities, and wrote of the organic nature of lived urban life. Her book proved a stinging rebuke to midcentury urban renewal in the US and her native Canada.[27]

SCHOLARLY AVENUES INTO THE INTERSECTIONS OF NATURE AND THE CITY

Though neither Mumford nor Jacobs were trained as historians, their publications arrived at an opportune moment. Urban history and environmental history both emerged as distinct fields in the US beginning in the 1960s and 1970s. These two fields—which for a long time did not intersect—both emerged in a rapidly changing society, one being remade by the Civil Rights Movement, the women's movement, and a new wave of environmentalism that followed Rachel Carson's seminal book *Silent Spring* in 1962. Historical writing was changing as well. A new emphasis on social history, race, class, gender, quantitative history, and other fields and topics led historians away from an emphasis on top-down political histories of presidents or monarchs, and toward histories that looked at broader and more diverse swaths of society. Both urban and environmental history were written in response to contemporary concerns. Wracked by riots, with downtowns in decline and whites fleeing to the suburbs, cities in the

1960s were considered by many to be in a state of crisis. Likewise, environmentalists were haltingly moving from an environmentalism focused on preserving heroic landscapes to one rooted in ecology, worried about pollution, toxins, and, quite literally, the nature in one's backyard rather than only the nature in a national park.

Yet despite arising in the same era, and with some similar concerns, urban and environmental history remained not only distinct, but highly divergent. While urban history examined infrastructure, political machines, voting patterns, immigrant communities, and other topics, environmental analysis was largely absent. At the same time, environmental history eschewed urban topics. Roderick Nash's *Wilderness and the American Mind* (1967) contained only a single reference to cities—a mention that Victorians sometimes derisively termed a slum a "city wilderness." Alfred Runte's *The National Parks: The American Experience* (1979) certainly did not have an urban focus. Neither did Donald Worster's *The Dust Bowl* (1979), Richard White's *Land Use, Environment, and Social Change: The Shaping of Island County, Washington* (1979), or William Cronon's *Changes in the Land: Indians, Colonists, and the Ecology of New England* (1983). Boston makes a few brief appearances in Cronon's book, however, as when deforestation caused a fuel shortage in the settlement.[28]

In fact, it was urban history, or a few subsets within the field, that first began moving in an environmental direction. These were histories of infrastructure, pollution, and sanitation. Martin Melosi crafted connections between cities and the environment in multiple books, starting with *Garbage in the Cities: Refuse, Reform and the Environment, 1880–1980* (1981), *The Sanitary City: Urban Infrastructure in America from Colonial Times to the Present* (2000), and *Effluent America: Cities, Industry, Energy, and the Environment* (2001). Garbage and sewage were obvious contact points between cities and the larger land- and waterscapes they occupied and interacted with. Even so, Melosi's merging of urban and environmental history initially remained an outlier in both fields. Gradually, however, interest in the environmental footprint of cities grew. Urban pollution, for example, has received scrutiny on both sides of the Atlantic, examples being Joel A. Tarr's *The Search for the Ultimate Sink: Urban Pollution in Historical Perspective* (1996), and Frank Uekoetter's study of industrial pollution and regulation, *The Age of Smoke: Environmental Policy in Germany and the United States, 1880–1970* (2009).

The watershed moment in the merging of environmental and urban history, however, arrived in 1991. That year, William Cronon published *Nature's Metropolis: Chicago and the Great West.* Cronon's second monograph represented a quantum leap in the integration of urban and environmental history. It posed new theoretical understandings of natural environments, and ones remade by humans, which Cronon, borrowing from Hegel and Marx, termed "first" and "second" natures.[29] Through commerce, commodity flows, and the collection and processing of crops, animals, and timber, Cronon detailed in precise ways exactly how Chicago was related to and exploited its hinterland, the "Great West." He also showed how "Nature's Metropolis" profited directly from the industrialization of agriculture and the mechanization of meat processing. In all of these ways, *Nature's Metropolis* was, and remains, a landmark, and many subsequent works have borrowed in some way from it. If in retrospect it has any absences or omissions, the most evident one might be labor and laborers. In Cronon's book there are commodities, and cows, and logs, and pig-killing machines, but there are not that many people, at least not working people.

Nevertheless, Cronon's monograph proved hugely influential: studies influenced by his work include Ari Kelman's *A River and Its City: The Nature of Landscape in New Orleans* (2003), which traces the city's complex and troubled history with its environment, from floods to hurricanes to yellow fever; Coll Thrush's *Native Seattle: Histories from the Crossing-Over Place* (2008), which reinserts Native Americans into the urban history narrative; David Stradling's *Making Mountains: New York City and the Catskills* (2007), which traces the creation of a recreational hinterland for the city; and Matthew Klingle's *Emerald City: An Environmental History of Seattle* (2007), which demonstrates that a city that drew much of its identity from its spectacular surroundings had also profoundly altered that landscape, leaving environmental degradation and social inequality in its wake. Michael Rawson's *Eden on the Charles: The Making of Boston* (2010) carefully traces the evolving relationship between the city and its environment, continually remade by the actions and needs of the residents of the city. My own *The Frontier of Leisure: Southern California and the Shaping of Modern America* (2010) drew upon Cronon's model of an urban hinterland, in this case a recreational one that remade the Southwest from a supposed wasteland to a tourist wonderland, and used resorts such as Palm Springs to both promote Los Angeles, and, in the longer term, promote and propagate its distinctive form of urbanism.

Thematic Variations on Environmental Urban History: Disasters, Inequality, and Sprawl

In the more than two decades since *Nature's Metropolis*, studies mixing environmental and urban history have proliferated. Some have gone in very different directions from Cronon's example, or indeed from many of the earlier writings on the subject. To give just three examples, disasters, inequality, and sprawl have all received scholarly scrutiny in the last two decades. Some such studies built on earlier work, while others drew on current events as sources of inspiration. Kelman's *A River and Its City*, for example, was reissued with a new preface after the tragedy of Hurricane Katrina in 2005. Ted Steinberg's *Acts of God: The Unnatural History of Natural Disasters in America* (2000) recast the landscape of hurricane-prone Florida as a "Do-It-Yourself Deathscape." Other chapters in the same volume examined efforts to conceal the true damage of the San Francisco earthquake of 1906, and the less well-known Charleston South Carolina earthquake of 1886. Steinberg also details how the federal government subsidizes and encourages coastal development, and risky development in other regions.

Another city that could be a veritable poster child for disaster is Los Angeles. Mike Davis unpacks layer upon layer of disasters, real and imagined, in his *Ecology of Fear: Los Angeles and the Imagination of Disaster* (1998). According to Davis, Southern California's natural hazards, plus its rampant urban sprawl, created a particularly volatile playground for disaster, an "apocalypse theme park." Readers are taken on a terrifying tour of hazards, from killer quakes to mountain lions which have developed a taste for joggers. In "The Case for Letting Malibu Burn," he presents a damning picture of a region that expends vast sums to combat naturally occurring chaparral fires in über-rich Malibu, while doing nothing to prevent horrific fires in high-rise tenement buildings and low-rent hotels housing poor immigrants. He also asks a provocative question: why, in literature, television, and films, do people enjoy watching the destruction of Los Angeles? His argument is that the city represents the other: nonwhite, noncitizen, nonheterosexual. Its destruction is therefore to be cheered, in some of the darkest recesses of the national psyche.

A less flamboyant book, but one no less insightful, is Jared Orsi's *Hazardous Metropolis: Flooding and Urban Ecology in Los Angeles* (2004). In a city purported to exist in a desert, Orsi recounts the long history of disastrous floods in Los Angeles. He ties these directly to a lack of or obliviousness to environmental knowledge and awareness of environmental risk. The founders of the city in 1781 had to relocate the city after spring rains swelled the little Los Angeles River to a rampaging torrent, despite having been warned of this danger by local Indians. The same story would play out in the late nineteenth and early twentieth centuries, when city residents developed all the land along the river, only to see floods damage their properties. In response, the federal government turned the entire river and its major tributaries into a concrete drainage system. A city in desperate need of water now dumps almost all of the rainwater it receives into the Pacific Ocean in less than an hour.

If Los Angeles was a poster child for disaster, it was unfortunately also one for inequality. The poor residents of the city were likely to breathe more polluted air, to be exposed to more toxic chemicals and industrial waste, and to have far less access to recreational space. Los Angeles might have been an especially stark example, but variations of the same themes could be found in many cities, from New York to Chicago, Miami to Houston. While gentrification added some luster to city centers, and suburbs became more racially and socioeconomically diverse, American cities in the early twenty-first century remain places of clear inequality, and unequal risks and opportunities. Social justice and the combating of environmental racism are two of the great challenges of cities now, and in the future. These issues also represent a challenge for older, predominantly white environmental groups such as the Sierra Club, which have haltingly begun to pay more attention to urban environmental issues and inequalities.[30]

American cities also remain distinctive. Their sheer scale, at least since 1945, is perhaps their most unusual feature. American cities grew exponentially in the decades after World War II, devouring farmland and open space, making Americans ever more reliant on the automobile and petroleum, and segregating the nation's cities by race. Urban historians have capably analyzed the rise of mass suburbia, particularly in works such as Kenneth Jackson's *The Crabgrass Frontier: The Suburbanization of the United States* (1985), which touches upon the physical implications of this distinctive form of urbanism, but does not investigate it in great detail. A landmark book on the subject, Adam Rome's *The Bulldozer in the Countryside: Suburban Sprawl and the Rise of American Environmentalism* (2001), interrogates this distinctive American urbanism in promising ways. On the one hand, Rome traces the political and economic side of this history. He shows how industrial production from World War II was revamped to build houses, and how the nation's mortgage system was redesigned to make new housing affordable. He then traces its material history, from its growing reliance on fossil fuels, to its polluting septic tanks, and to its devouring of open space. Most provocatively, however, he then makes a compelling case for understanding how sprawl helped spur a newly energized environmental movement in postwar America. This history, with multiple foci and broad implications, illustrates how urban environmental history can help us better understand our cities, our environment, and the natural and human worlds which encompass them both.

LOOKING BACKWARD—AND FORWARD

After decades of generally ignoring each other, environmental and urban history are now engaged in one of the richest collaborative enterprises currently underway in any historical field. It is hard to know where this dialogue may go in the future. Transnational studies are still needed, as is much more on globalization. The rapid urbanization occurring on the US-Mexico border is one place to start, but only one. The twenty-first century is already a world of mega-cities, from Tokyo to Beijing, Mexico City to Mumbai, Lagos to Los Angeles. These big cities have big problems: current environmental hazards, future climate change, pandemics, pollution, and more. Historians must look at the cities of the past to see how they coped—or did not cope—with similar challenges.[31] In a world with half a billion people living outside their native nations, and hundreds of millions more pouring out of rural areas into cities, making cities livable will be one of the great challenges of this century. Another will be sustainability. In a world where everyone wants a local variation of the American good life, circa 1960, with refrigerators, cars, and air conditioning, societies will have to figure out how to spread material abundance more equitably, but also more sustainably. The United States had the supposed luxury of an abundant and sparsely populated continent. Now, however, we all live in an increasingly crowded world, with commodities of all kinds increasingly scarce, and increasingly expensive. We are going to have to learn to live together more peaceably, more sustainably, and to use less. Those are daunting challenges, but if they can be met, they will be met in cities. A city with a smaller environmental footprint is far less destructive than a human population sprawled far and wide. Environmentalists—and sometimes environmental historians—have seen cities as the other, the anti-nature, to be loathed, avoided, and feared. Yet it is in our cities that humanity must first solve its environmental problems, and find a way toward a sustainable future, and historians can and will inform that process.

In the United States, few places serve as a more potent symbol of nature than Yosemite National Park. Tourists flock to it from around the world, and American environmentalists associate it with John Muir and the creation of the Sierra Club in 1892. Yet those tourists, and most of those environmentalists, forget that Muir and his fellow members lived in San Francisco and other cities. It was those cities that created a hunger for nature, and were bound in an inextricable relationship with the natural world. Environmentalists and historians alike would do well to tend both to that spectacular valley and also to that dynamic city, born of the hunger for gold, built of wood, metal, and stone atop trembling earth, destroyed by earthquake and fire, and then rebuilt, booming anew during World War II, and then sprawling outward in one suburban development after another in the postwar era. This changeable city has most recently reinvented itself once more, this time not as a city of gold, but of silicon, as the fabled orchards of the Santa Clara Valley gave way to Apple and Silicon Valley. Rare earths and metals from around the world are assembled by workers in China and Mexico into phones, laptops, and tablets, with glowing pixels that reproduce in beautiful detail the tourist photos taken in Yosemite. The city and the valley are not opposites. San Francisco and Yosemite are inextricably bound, one to the other. If we can create a worldview that encompasses both—the natural environment and the built one, the human and the natural, the city and the valley—then we can see the world whole, perhaps for the first time.

NOTES

1. See Martin Melosi, *The Sanitary City: Urban Infrastructure in America from Colonial Times to the Present* (Baltimore: Johns Hopkins University Press, 2000), 2–8. Much of the introduction to *Sanitary City* overlaps with Melosi's "The Place of the City in Environmental History," *Environmental History Review* 17 (Spring 1993): 1–24. See also Joel Tarr, "The Metabolism of the Industrial City: The Case of Pittsburgh," *Journal of Urban History* 28 (July 2002): 511–545; and *The Search for the Ultimate Sink: Urban Pollution in Historical Perspective* (Akron, OH: University of Akron Press, 1996).
2. William Cronon, *Nature's Metropolis: Chicago and the Great West* (New York: W. W. Norton, 1991).
3. Andrew Hurley, *Environmental Inequalities: Race, Class, and Industrial Pollution in Gary, Indiana, 1945–1980* (Chapel Hill: University of North Carolina Press, 1995).
4. Charles C. Mann, *1491: New Revelations of the Americas before Columbus* (New York: Vintage, 2005); William M. Denevan, "The Pristine Myth: The Landscape of the Americas in 1492," *Annals of the Association of American Geographers*, 82 (September 1992): 369–385.
5. Cronon, *Nature's Metropolis*, 53. For the land survey, see John Opie, *The Law of the Land: Two Hundred Years of American Farmland Policy* (Lincoln: University of Nebraska Press, 1994).

6. See Richard C. Wade, *The Urban Frontier: The Rise of Western Cities, 1780–1830* (Cambridge: Harvard University Press, 1959).

7. Ted Steinberg details the travails and misapprehensions of early colonists in *Down to Earth: Nature's Role in American History* (New York: Oxford University Press, 2002), 21–27.

8. William Cronon traces environmental changes in New England in *Changes in the Land: Indians, Colonists, and the Ecology of New England* (New York: Hill and Wang, 1983). Gary B. Nash examines the region's urban history in *The Urban Crucible: The Northern Seaports and the Origins of the American Revolution* (Cambridge, MA: Harvard University Press, 1979). See also T. H. Breen, *The Marketplace of Revolution: How Consumer Politics Shaped American Independence* (New York: Oxford University Press, 2004) For the "Atlantic World," see Nicholas Canny and Philip Morgan, eds., *The Oxford Handbook of the Atlantic World, c.1450–c.1850* (New York: Oxford University Press, 2011).

9. J. R. McNeill examines the role that malaria and other diseases played in the spread of slavery and empire in the Atlantic World in *Mosquito Empires: Ecology and War in the Greater Caribbean, 1620–1914* (Cambridge: Cambridge University Press, 2010).

10. For more on the environmental requirements for cotton production, see Steinberg, *Down to Earth*, 81–87. For a broader environmental history of the South, see Jack Temple Kirby, *Mockingbird Song: Ecological Landscapes of the South* (Chapel Hill: University of North Carolina Press, 2006). For New Orleans, see Ari Kelman, *A River and Its City: The Nature of Landscape in New Orleans* (Berkeley: University of California Press, 2003).

11. Carol Sheriff, *The Artificial River: The Erie Canal and the Paradox of Progress, 1817–1862* (New York: Hill and Wang, 1996). Though not an environmental history, Christopher Clark examines the economic transformations in this region and era in *The Roots of Rural Capitalism: Western Massachusetts, 1780–1860* (Ithaca, NY: Cornell University Press, 1990).

12. Cronon, *Nature's Metropolis*.

13. Jared Farmer, *On Zion's Mount: Mormons, Indians, and the American Landscape* (Cambridge, MA: Harvard University Press, 2008).

14. Andrew C. Isenberg, *Mining California: An Ecological History* (New York: Hill and Wang, 2005); Gray Brechin, *Imperial San Francisco: Urban Power, Earthly Ruin* (Berkeley: University of California Press, 2006).

15. Joel A. Tarr, ed., *Devastation and Renewal: An Environmental History of Pittsburgh and Its Region* (Pittsburgh: University of Pittsburgh Press, 2005); Kathleen Brosnan, *Uniting Mountain and Plain: Cities, Law and Environmental Change along the Front Range* (Albuquerque: University of New Mexico Press, 2002); Thomas G. Andrews, *Killing for Coal: America's Deadliest Labor War* (Cambridge, MA: Harvard University Press, 2008); David W. Lewis, *Sloss Furnaces and the Rise of the Birmingham District: An Industrial Epic* (Tuscaloosa: University of Alabama Press, 2011).

16. Conevery Bolton Valencius explores folk beliefs about the environmental causes of illness in *The Health of the Country: How American Settlers Understood Themselves and Their Land* (New York: Basic Books, 2002). For the appeal of Los Angeles, see Lawrence Culver, *The Frontier of Leisure: Southern California and the Shaping of Modern America* (New York: Oxford University Press, 2010). The statistic on the total percentage of health migrants is derived from John E. Baur, *The Heath Seekers of Southern California, 1870–1900* (San Marino, CA: Huntington Library Press, 1959).

17. Culver, *Frontier of Leisure*.

18. Mike Davis takes a grim view of the rise of Los Angeles in *City of Quartz: Excavating the Future in Los Angeles* (New York: Verso, 1990). See also William Deverell and Greg Hise, *Land of Sunshine: An Environmental History of Metropolitan Los Angeles* (Pittsburgh: University of Pittsburgh Press, 2005). The Mulholland quote is from Marc Reisner, *Cadillac Desert: The American West and Its Disappearing Water* (New York: Penguin, 1986), 86.

19. Ted Steinberg posits the boll weevil's role in spurring the Great Migration in *Down to Earth*, 103–04.

20. For the history of Phoenix, see Philip VanderMeer, *Desert Visions and the Making of Phoenix, 1860-2009* (Albuquerque: University of New Mexico Press, 2010). For Las Vegas, see Hal Rothman, *Neon Metropolis: How Las Vegas Started the Twenty-First Century* (New York: Routledge, 2003).

21. The definitive history of the Sunbelt—environmental or otherwise—has yet to be written. For economic and political perspectives, see Lisbeth Cohen, *A Consumers' Republic: The Politics of Mass Consumption in Postwar America* (New York: Vintage, 2003).

22. See, for example, Harold L. Platt, *Shock Cities: The Environmental Transformation and Reform of Manchester and Chicago* (Chicago: University of Chicago Press, 2005).

23. See Dorothee Brantz and Sonja Dümpelmann, eds., *Greening the City: Urban Landscapes in the Twentieth Century* (Charlottesville: University of Virginia Press, 2011) for a comparative examination of efforts to bring nature into the city, from Mexico City to London to Sofia, Bulgaria.

24. Anne Whiston Spirn, "Constructing Nature: The Legacy of Frederick Law Olmstead," in *Uncommon Ground: Toward Reinventing Nature*, ed. William Cronon (New York: W.W. Norton, 1995), 91–113; Witold Rybczynski, *A Clearing in the Distance: Frederick Law Olmstead and America in the 19th Century* (New York: Touchstone, 1999).

25. Dorceta E. Taylor provides a broad overview of urban environmental reform in *The Environment and the People in American Cities, 1600s–1900s* (Durham, NC: Duke University Press, 2009). See also Dominick Cavallo, *Muscles and Morals: Organized Playgrounds and Urban Reform, 1880–1920* (Philadelphia: University of Pennsylvania Press, 1981). For an international and comparative perspective, see Brantz and Dümpelmann, eds., *Greening the City*.

26. Lewis Mumford, *The City in History: Its Origins, Its Transformations, and Its Prospects* (New York: Harcourt, Brace,1961).

27. Jane Jacobs, *The Death and Life of Great American Cities* (New York: Vintage, 1992).

28. Cronon, *Changes in the Land*, 25–26.

29. Cronon, *Nature's Metropolis*, xix.

30. See, for example, Julie Sze, *Noxious New York: The Racial Politics of Urban Health and Environmental Justice* (Cambridge, MA: MIT Press, 2007); Sylvia Hood Washington, *Packing Them In: An Archaeology of Environmental Racism in Chicago, 1865–1954* (New York: Lexington Books, 2004), and Don Mitchell, *The Right to the City: Social Justice and the Fight for Public Space* (New York: The Guilford Press, 2003).

31. Andrew Isenberg, ed., *The Nature of Cities* (Rochester: University of Rochester Press, 2006) offers one example of comparative international environmental and urban history, especially in its comparative examinations of cities in not just different regions, but different historical eras.

PART IV

ENTANGLING ALLIANCES

CHAPTER 20

....

RACE AND ETHNICITY IN ENVIRONMENTAL HISTORY

....

CONNIE Y. CHIANG

OVER the past three decades, historians have recognized environmental history's potential to illuminate the complex dynamics of human societies. William Cronon, Carolyn Merchant, Ted Steinberg, Richard White, and others have all called attention to the ways in which scholars, both within and outside the field, have explored and could further investigate the construction of social categories, like race and ethnicity, in tandem with changes to the natural world. By integrating social and environmental history, they could also address broader questions about power relations.[1] As Alan Taylor, a self-described social historian "who reads and admires environmental history," has noted, "we can discern and assess inequality by attending to differing experiences, past and present, with nature within our society."[2]

For some of these scholars, bringing social and environmental history together seemed logical, given the fields' similar roots and interests. Both fields gained momentum in the United States during the social and political movements of the 1960s and 1970s. "With society rent asunder by the civil rights movement, antiwar protests, and feminist demands," Alice Kessler-Harris explains, American historians "had difficulty reconciling myths of national progress and consensus with the tensions around them." These "new" social historians soon rejected studies of political leaders and began to explore "the dynamic interaction of a multiracial and multiethnic population." Likewise, by exposing widespread ecological changes, the environmental movement became "the engine that drove environmental history" and gave the field an audience. Both environmental and social history also drew intellectual inspiration from the French *Annales* School. While environmental historians were attracted to the School's attention to climate and geography as historical factors over the *longue dureé*, social historians emulated its commitment to exhaustive community studies and the *mentalités* of ordinary people.[3] Indeed, the fields also shared "a preoccupation with the common and the previously inconspicuous" and a commitment to the moral and political relevance of their work.[4]

Particularly since the late twentieth century, many environmental historians and scholars in allied fields, such as historical geography and the history of science and medicine, have recognized the compatibility of social and environmental history and have incorporated race and ethnicity into their studies. They have traced how diverse people integrated environmental activities into their cultural traditions and social identities, and have examined how these interactions with nature helped to define notions of racial and ethnic difference. Several scholars have also explored how those in power used theories of scientific racism and environmental determinism to maintain their hegemony over certain people and places. As historians have long argued, race and ethnicity are constructions; scholars of environmental history have added that nature has played a formative role in shaping those constructions.

From the city to the country, environmental historians have also illustrated how racial and ethnic divisions became literally inscribed on land and water. Environmental changes, from industrialization and the creation of parks to deforestation and devastating hurricanes, had profound social implications for certain racial and ethnic groups. They often lived in polluted neighborhoods, on low-lying land prone to flooding, or along the outskirts of conservation areas, and their rights to the pleasures and economic benefits of the natural world were routinely circumscribed. Indeed, certain individuals and groups tried to control access to nature in order to dominate those people they believed to be their social inferiors. Yet even as they faced unrelenting efforts to exclude and subordinate them, people of different racial and ethnic backgrounds used their connections to the natural world to fight oppressive regimes and policies. Environmental historians, in other words, have largely rejected a narrative of victimization, and instead have shown how marginalized groups made claims to nature and used it as a tool of resistance.

While studies of North America still dominate the scholarship, environmental historians of Africa, Latin America, and Asia have also explored how the categories of race and ethnicity came to bear on the physical world and vice versa. Collectively, their work has addressed numerous topics across space and time, from colonial encounters, slavery, and the development of extractive industries to conservation, outdoor recreation, and environmental justice. By telling compelling and sometimes violent stories, they have demonstrated nature's profound role in the emergence, reinforcement, and attempted eradication of racial and ethnic divisions. This is arguably one of the field's most important contributions, but it demands more depth and nuance. Environmental historians need to delve further into the flow of power between *and* within racial and ethnic groups, paying closer attention to class differences and intellectual traditions. They also need to analyze whiteness as a distinct racial identity in its own right, one that was also forged through interactions with nature. In these ways, environmental history can reinforce its centrality to understanding power relations involving racial and ethnic groups across the globe.

INDIGENOUS KNOWLEDGE AND COLONIAL ENCOUNTERS

Some of the first historical studies that addressed the environmental experiences of nonwhites focused on Native Americans. William Cronon, Richard White, Timothy

Silver, and Arthur F. McEvoy examined native transformations to the natural world prior to colonization, contrasting precolonial indigenous knowledge with the environmental changes that emerged after contact. They suggest that natives did not live in idealized harmony with the natural world, as many modern environmentalists often believed. Instead, they actively shaped the world around them. They burned forests to encourage the growth of plant species, facilitate the hunting of game, and sow their own crops. They hunted, fished, and gathered fruits, nuts, and roots. They adapted to seasonal changes, migrating to where they could find food at a given time.[5] Coll Thrush has built on this work by suggesting how place names reflected natives' intimate understanding of the natural world. Whereas European explorers named land and bodies of water after themselves or their patrons, natives named them in relation to their lived experiences. For instance, Indians around Puget Sound called a bend on the Duwamish River "Lots of Douglas Fir" because this tree was in abundant supply there.[6]

Subsistence patterns shifted dramatically when natives clashed with colonial powers. As Alfred Crosby argues, Europeans' settlement of North and South America, Australia, and New Zealand had a crucial ecological component that was just as important as military prowess. In a process dubbed the "Columbian exchange," Old World plants, animals, and microbes "Europeanized" these lands and served as tools of political and economic conquest. For example, Europeans brought *trébol*, a clover, which soon smothered native crops in Peru and North America. Crosby suggests that this weed performed the dual roles of "pioneer and *conquistador*." Of even greater importance were Old World pathogens. Diseases such as smallpox and influenza spread rapidly and devastated native peoples, because they had no acquired immunity. These virgin soil epidemics often wiped out entire villages, disrupted kinship networks, and wreaked havoc on social organization, weakening native resistance and helping to clear the way for European colonization.[7]

The natives who survived or avoided disease forged tenuous relationships with the colonists. Not surprisingly, these two groups understood and engaged with nature in very different ways. As William Cronon explains in his classic study, *Changes in the Land: Indians: Colonists, and the Ecology of New England* (1983), Indians in southern New England moved seasonally, planting crops in the spring, hunting and fishing in the summer, and harvesting and hunting again in the fall before migrating to winter sites. The English, on the other hand, sought permanent settlements with fields, pastures, and fences. Cronon explains, "English fixity sought to replace Indian mobility; here was the central conflict in the ways Indians and colonists interacted with their environments." These differences would have real consequences. Because the English viewed Native American activities—hunting, fishing, burning—as unproductive, they denied their claims to the land. "European perceptions of what constituted a proper use of the environment thus reinforced what became a European ideology of conquest," Cronon concludes.[8]

Nonetheless, Indians and colonists came to depend on and take advantage of each other for various goods. Eager to obtain prized beaver pelts, colonists traded brass and copper pots, woven fabrics, iron tools, and wampum, strings of white and purple beads that symbolized power and wealth. Indians grew more dependent on Europeans, but Europeans also became enmeshed in trade and even armed conflict with the Indians of Long Island Sound, who gathered the whelks and quahogs used to make wampum.[9] The political and economic implications of the fur trade were inseparable from the environmental ones. Because of the devastating impact of disease, natives participated in the fur trade to retain or obtain power. Trading animal pelts for metal goods or wampum was a route to prestige. But in the process, they put pressure on the very game populations on which they depended. Slow to reproduce, the beaver began to disappear from coastal regions of Massachusetts by 1640. Meanwhile, as colonists' settlements spread, other important game species, such as white-tailed deer, began to decline. The natives were the "real losers," as their earlier subsistence and settlement patterns became untenable.[10]

The ramifications of native-white relations and trade also reverberated on the Great Plains. Elliott West's eloquent study, *The Contested Plains: Indians, Goldseekers, and the Rush to Colorado* (1998), details the dramatic changes that unfolded after the Spanish reached the interior of North America in the sixteenth century. The story of the Cheyenne is particularly instructive. To obtain Spanish horses and tap into their potential power, they had to adopt a nomadic lifestyle, moving to the central plains where the trade nexus developed and the bison were abundant. Their involvement in the bison trade made them more dependent on whites, who were not always sympathetic to their needs. Internal unity also broke down, as they had to live in smaller groups because of the need for horse pasture and protection. By the 1840s, drought and declining bison populations exposed the environmental limits of their lifestyle. Like the Indians in New England, the Cheyenne became "reliant on the essentials they were helping to destroy." The discovery of gold in 1857 and the arrival of American settlers—along with market hunters who targeted bison herds—soon squeezed out the Cheyenne.[11]

The dramatic social and environmental impacts of colonization were not unique to native peoples of North America. Environmental historians of Asia, Latin America, and Africa have identified similar patterns and themes. Brett Walker's *The Conquest of Ainu Lands: Ecology and Culture in Japanese Expansion, 1590–1800* (2001) is an excellent example. Drawing on scholarship from the "New Western history," which focuses on the convergence of diverse peoples on the North American West, Walker suggests that the fate of the Ainu, the native people of Hokkaido, paralleled their indigenous counterparts across the Pacific.[12] Like European colonial endeavors in North America, Japanese expansion during the seventeenth and eighteenth centuries undermined native autonomy. Lured by trade goods that they could not produce themselves, like rice, the Ainu intensified their exploitation of their territory. Heightened competition for resources put pressure on the natural world and led to warfare among Ainu groups. Trade also spread diseases—smallpox, measles—that shrank the Ainu population and limited their ability to resist the Japanese. By the end of the eighteenth century, they had become integrated into Japan's commercial economy.[13]

These studies follow a similar narrative, in which colonization altered indigenous subsistence patterns and instigated widespread social, economic, and political changes, but other scholars, particularly in the history of medicine, suggest that scientific theories of race and environment also shaped imperial endeavors. As Joyce Chaplin argues, sixteenth-century English colonists in North America initially sought out native environmental knowledge. But by the early seventeenth century, attitudes had shifted as epidemic diseases annihilated native populations. The colonists pointed to these deaths as evidence of Indians' intrinsic inferiority, while construing their own ability to reproduce, in spite of their new surroundings, as evidence of their bodily superiority and fitness to establish dominion over North America. These notions of racial difference led the English to reject natives and exalt their own technology's ability to master nature.[14]

Espoused by eighteenth-century philosophers such as Charles de Montesquieu and Immanuel Kant, subsequent theories that linked racial difference to the environment, specifically climate, also proved influential to colonialism.[15] As Mark Harrison explains, "It was axiomatic in colonial medical topography that the superior attributes of the conquering race were derived from the wholesome climate of northern Europe, just as the 'degeneracy' of subject races owed much to their native environments and lifestyles." Places like the tropics, Warwick Anderson adds, were "No place for a white man, and yet just the place for white dominion over man and nature." From the eighteenth to the early nineteenth centuries, physicians tried to address "this medical conundrum of imperialism" by developing the study of acclimatization. They subscribed to a monogenist framework in which the human race originated from one source and variations resulted from environmental influences. People could thus adapt to new surroundings and settle in conquered places. But by the early nineteenth century, racial lines hardened with intensifying debates over slavery, and a polygenist framework gained favor, claiming that differences between races were fixed. These factors led physicians to assert that Europeans were incapable of adjusting to unfamiliar environments. As a result, they could colonize unknown places, but they could not become permanent inhabitants.[16]

While these theories restrained imperial ambitions to some extent, they did little to halt the social transformation of colonized people and the physical transformation of colonized places. Ultimately, colonial endeavors changed how native peoples interacted with the natural world and undermined previous strategies for procuring resources. The cohesion of native groups weakened, and they became more dependent on colonial powers. Historians have made it clear that in gaining control over the natural world, Europeans and Americans also obtained power over the people who stood in the way of their imperial goals.

SLAVERY AND NATURE

The history of slavery, like the history of colonialism, has often been told from a socioeconomic point of view, not through the lens of environmental history. Yet African and African American slaves radically transformed the environment in the name of profit and power for their masters. The physical changes that masters ordered and slaves carried out, under force and threats of punishment, were substantial. Before they could plant sugar, tobacco, or cotton, they had to cut down forests and remove stumps. In the Atlantic tidewater, they built hydraulic systems of canals and drains for rice plantations, moving massive amounts of swamp muck in the process. To ensure continued productivity, masters imposed discipline on their slaves. According to Mart Stewart, "The domination of nature and the domination of one group of humans by another evolved mutually."[17]

Masters affirmed slaves' suitability for forced labor by emphasizing their intrinsic connections to nature. For instance, they often equated their slaves to animals—both wild and domesticated—to justify their regime. In the case of British Honduras, a colonial official described a slave who spotted mahogany trees as an "instinctual creature" imprinted with knowledge of the jungle—not a rational human being who found trees through observation and reason.[18] Constructing slaves as animal-like also dehumanized them, and made the use of brute force appear acceptable and necessary to ensure submission. As Karl Jacoby explains, "The drive for control is so essential—and so similar, whether the object of control is a slave or a domestic animal—as to overwhelm most distinctions between humans and animals." In the minds of slaveholders, domestic animals and slaves followed a similar trajectory, from wild to tame. Slavery served to civilize slaves and had the potential to make them fully human.[19]

As with colonial ambitions, scientific theories that attributed racial difference to climate also contributed to the defense of slavery. Many masters insisted that Africans were better acclimated to hot environments and thus ideal for Southern plantation labor. Even when American scientists, most notably Samuel George Morton, George R. Gliddon, and Josiah C. Nott, challenged climatic and environmental determinism in the 1840s and 1850s and, in accordance with polygenist thought, argued that race was immutable and that Africans evolved separately, the link between climate and race remained. Africans could work in the summer heat, but not because they had long lived in tropical climates; rather, as Mart Stewart concludes, "they were created differently. Acclimation was now evidence of permanent and immutable differences, rather than a process of differentiation."[20]

With their suitability for plantation labor affirmed, African slaves toiled on the land, providing and acquiring invaluable knowledge of the natural world in the process. Stewart notes that they developed "a closer view of the cultivated environment... They were aware, from row to row, of the progress of the plants during the growing season... Their hands experienced crop cultures from dawn to dusk, from day to day, and from the ground up."[21] According to Judith Carney, those who came to the coast of South Carolina and Georgia from the rice-growing regions of West Africa even helped their masters re-engineer the land with similar systems of flooding and drainage as those used in the Upper Guinea Coast. By the mid-eighteenth century, South Carolina planters came to prefer slaves from specific ethnicities and regions known for rice cultivation, such as the Mandinka from Gambia.[22] Several scholars have repudiated Carney's findings, arguing instead that slaves had limited agency in rice culture. But

Paul Sutter suggests that the very debate "makes clear that the social relations of slavery took place within complex agroecosystems, and master-slave relations were constantly reinvented and recalibrated in relation to those systems."[23]

Slaves' skill and know-how gave them some autonomy and allowed them, in a small way, to resist their enslavement. Particularly for slaves who labored under the task system—most common on the rice plantations, where drainage ditches demarcated areas for measuring tasks—they could occasionally complete their work before dusk and pursue numerous activities in nature.[24] They gathered materials for handicrafts, weaving sea grass and bark into baskets and using plants for fabric dyes. They tended gardens and grew small crops for subsistence, to trade with other slaves, and to sell at market. To add more protein to their diet and to obtain more saleable and tradable goods, many slaves fished, hunted game, and kept livestock. Since slaves were prohibited from owning firearms, they learned to use fire, dogs, and homemade gear—from traps to watercraft—to catch their prey.[25]

These endeavors, in turn, acquired social and cultural meaning. Slaves incorporated the behaviors of favored game animals, such as squirrels and rabbits, into songs and stories. In addition to foodstuffs, many slaves cultivated medicinal plants to sustain their healing traditions. Property ownership—whether of livestock, farm equipment, or bushels of corn—reinforced kin networks; relatives helped with the family's productive activities, and children inherited their parents' property. Hunting, fishing, and gardening also created camaraderie among family and community members, while allowing slaves to assert their role as providers.[26] These activities even served subversive purposes. During their excursions, slaves plotted out escape routes and scouted places to hide and find food.[27]

For their part, masters were calculating in using the natural world to wield power over slaves. Some slaveholders encouraged hunting, fishing, and tending gardens and livestock to reduce the costs of food rations. They also saw these endeavors as a way to maintain control, as they could threaten to take away property and privileges if slaves did not obey orders. They even took advantage of slaves' desire to hunt game to rid the nearby forests of opossums and raccoons, which ravaged crops. Some white mistresses were also known to tell stories about dangerous creatures living in the woods to deter slaves from running away. As Scott Giltner explains, "southern fields, forests, and streams were contested arenas wherein key tensions in antebellum southern life were played out."[28]

Mark Fiege brings together much of this scholarship in his recent innovative study, *The Republic of Nature: An Environmental History of the United States* (2012). In his "King Cotton" chapter, he integrates the life cycle of the cotton plant with the life cycle of the slave. He explains, "The crop cycle of cotton was the point of struggle between masters and slaves, as masters tried to discipline slaves for agricultural work and as slaves resisted that domination." Indeed, masters drove their slaves the hardest at particular times during the cycle: when cotton-choking weeds and plants needed to be removed from May to July and during picking time beginning in late July. But cotton depended upon and competed with the life cycle of the slave. Slave bodies had a variety of needs—"food, water, rest, nurture, and reproduction"—all of which took away from the tending and harvesting of cotton. In the end, resistance mounted, and the Civil War "ruptured the crop cycle." King Cotton was no more.[29]

COMPETITION AND ADAPTATION IN EXTRACTIVE INDUSTRY

As Fiege suggests, the nature of the agricultural landscape helped to shape power relations between masters and slaves, and whites and nonwhites. In the extractive industries of the North American West, the dual labor system similarly relegated nonwhites to bottom-tier jobs—typically the least desirable, most arduous positions—and limited their upward mobility.[30] And like slavery, a "loose evolutionary theory" sometimes justified this regime. As Douglas Sackman argues in his study of the California citrus industry, early twentieth-century growers maintained that Mexican, Chinese, and other nonwhite workers were biologically suited for agricultural labor. They attributed their acquired skills to natural selection and construed them as "racial essences."[31] Environmental historians have also illustrated how extractive industries sparked competition over resources and initiated cultural conflicts between whites and nonwhites. Those in power tried to use the natural environment as a tool for social control, while marginalized nonwhite groups, much like slaves, learned to adapt to changing political and environmental conditions. In some cases, their resistance was unsuccessful. In other cases, their changes to the natural world became a way to evade discrimination and defend their autonomy.

One groundbreaking study that demonstrates how nonwhites devised strategies to sustain their environmental activities is Arthur McEvoy's *The Fisherman's Problem: Ecology and Law in the California Fisheries, 1850-1930*. He starts his study with the Yurok, Hupa, and Karok of the Klamath-Trinity watershed who, unlike other California Indians, were able to stave off European and American efforts to destroy their subsistence patterns. Several factors explain their success. Prior to colonization and settlement, they limited their use of the salmon fishery in order to share with groups upstream and to lower the risk of failure. This "nature banking" left them "a natural reserve on which to draw during hard times. Indians also protected their traditional fishing rights by exercising military skill and engaging in the market economy. These adaptations allowed them to continue fishing salmon, which "remained key to their sense of themselves as Indians," and retain much of their autonomy well into the twentieth century.[32]

Similarly, Chinese and Italian fishermen in late nineteenth-century California developed management strategies that shielded their fishing activities from outside interference. They formed producers' associations made up of fishers who came from a specific locality back in "the old country" and caught a specific species of fish.

These groups controlled and distributed the harvest to ensure equity among members. McEvoy explains that these extralegal policies were not necessarily environmentally benign, but their intent was to manage resources "so as to protect the stability and longevity of their communities." Since profit was less important than "the preservation of a traditional way of life," rapacious behavior was kept in check. However, protective groups that "resisted the demand of a market society" were short-lived. With increased ethnic diversity in the fisheries and the growing power of fish brokers and processors, class solidarity replaced ethnic solidarity by the end of the nineteenth century.[33]

While not explicitly an environmental history, Sucheng Chan's *This Bittersweet Soil: The Chinese in California Agriculture, 1860–1910* (1986) also demonstrates the development of effective environmental strategies for economic and cultural survival in the face of overt racism. Chinese immigrants saw the cultivation of land as a route to prosperity and stability, and their establishment of permanent rural communities was a testament to their ability to adapt to environmental conditions. This was evident in the San Joaquin Delta, where many Chinese potato farmers prospered. Because the region's peat soil encouraged the growth of a potato-ravaging fungus that necessitated crop rotation, white landowners rarely focused on this crop. As tenant farmers, the Chinese did not have to worry about this problem and could move from field to field. Chinese and other immigrant farmers were also willing to cultivate the swampy land behind levees. As Chan explains, the land "influenced the social ecology of a community where race, ethnicity, and class were all aligned according to the contours created by nature."[34]

But nonwhites' power to use nature as a tool of resistance was circumscribed. Constructions of racial and ethnic difference routinely defined these groups as inferior and expendable, and a variety of legal cases and government regulations, from immigration restrictions to conservation measures, often worked against them. In the case of the Chinese farmers in California, the Chinese Exclusion Act of 1882, which prohibited the immigration of Chinese laborers, reduced their presence in the delta. Nonetheless, racial and ethnic workers developed several strategies to respond to ongoing pressure to marginalize and eliminate them from the extractive economy. Many historians have succeeded in showing their flexibility as they negotiated their tenuous place in these landscapes.[35]

THE INEQUITIES OF ENVIRONMENTAL REFORM

By the second half of the nineteenth century, urbanization and the rapid development of extractive industries in the United States gave impetus to reform movements that focused on the conservation of natural resources and the protection of wilderness. In fact, conservation became a hallmark of the Progressive Era. Emphasizing efficiency and scientific expertise, government officials implemented policies that would ensure a regular supply of timber, soil, water, and other natural resources. At the same time, many Americans also took an interest in preserving wilderness through the creation of national parks and other reserves.[36] Several environmental historians have documented the impact of these reforms on racial and ethnic minorities, finding that they often compromised cultural traditions and economic livelihoods and justified expulsion. But rather than paint them as victims, scholars have shown that many individuals pushed back and tried to shape policies in ways that would benefit their own communities.

In the United States, Progressive Era reformers often targeted racial and ethnic minorities and adopted a sort of environmental determinism in which they claimed that improving the physical conditions of their neighborhoods and fostering connections to nature would bring assimilation and enhanced social conditions. For example, in his study of Jane Addams, Harold Platt finds that she was deeply concerned about the environment on Chicago's West Side, which was dominated by immigrants. Along with Florence Kelley and the other women of Hull-House, the settlement house that she cofounded in 1889, Addams began to survey housing, sanitation, and disease incidence. Their data challenged ethnic stereotypes and "demonstrated that land rents, not ethnicity, accounted for different conditions in these neighborhoods."[37] Although Addams and her colleagues dispelled the idea that immigrants were inherently unclean, they were not cultural pluralists. Their ultimate goal was mass assimilation of immigrants, which would be brought about by imposing "American" standards and values. According to Adam Rome, Progressive Era reformers also believed that nature study in the public schools could help immigrants become Americans. Through lessons on conservation principles and working in school gardens, immigrant children would develop a connection to the land and learn "civic virtues."[38]

While many immigrants and nonwhites embraced reforms, others maintained their own agendas. For instance, in response to the nature study curriculum, some immigrant children "challenged the assumption that they needed to be taught how to respect nature."[39] In the South, rural African American women participated in the Home Demonstration Service projects of the Negro Cooperative Extension Service. Employed as agents, they taught other black women the principles of "cleanliness, thriftiness, and management" and how to grow beautiful gardens that would improve their homes. But Dianne Glave argues that African American women did not abandon their own conservation strategies; rather, they joined their previous traditions with the techniques promulgated by reformers.[40]

For George Washington Carver, head of the Tuskegee Institute's agricultural school, it was necessary to develop his own brand of conservation to address racial oppression. As Mark Hersey details, he advocated sustainable agricultural techniques to poor black farmers throughout the south, insisting that "they could defend themselves against the economic and political vicissitudes they faced as a result of their race by turning to the natural environment." This was of utmost importance after the Civil War, when black tenants and sharecroppers went into great debt buying expensive fertilizers to increase cotton yields. Carver recommended that farmers diversify their crops and raise live-stock, abandoning cotton and store-bought goods. While he ultimately could not chal-lenge the South's entrenched racial politics and the growth of modern agribusiness, his vision of environmental reform addressed the links between racial inequities and environmental degradation.[41]

By defining a proper way of interacting with the natural world, other reform-ers attacked the strategies that local people had devised to sustain their families and communities. In many cases, their efforts took on a racist or nativist edge. Early twentieth-century wildlife management in Pennsylvania, for instance, estab-lished a closed season, required hunters to use only guns, and banned the hunting of songbirds. These game laws were especially detrimental to Italian immigrants, who hunted songbirds for subsistence. Since they were easy to identify—they lived in eth-nic enclaves and spoke a different language—the state game commission believed that Italian immigrants were responsible for most game law violations. As Louis Warren explains, state officials even created conservation laws, like the 1909 Alien Gun Law, that ostensibly protected the public good but really attacked Italians' "cultural reori-entation of local hunting practices."[42] Similarly, planters in the postbellum South sup-ported game and stock laws in the name of conservation, agricultural improvement, and private property, but these measures also allowed them to reassert authority over former slaves and lower-class whites.[43]

Such strategies moved easily from land to water, as government officials also used conservation measures to displace immigrants and nonwhites in the fisheries. As Joseph E. Taylor, Matthew Klingle, Andrew Fisher, and others have demonstrated, this was particularly pronounced in the Pacific Northwest. In the late nineteenth cen-tury, Taylor argues, "Salmon management increasingly became a political process of determining who should benefit from salmon and who should be excluded." With the marginalization of Indians and Asians, "the industrial salmon fishery became a white fishery."[44] In the early twentieth century, Washington State officials further codi-fied this development. Its first fisheries code of 1915 banned traditional gear, including spears and snares. This was followed by bans on purse seines and gill nets, popular gear among Indian and immigrant fishermen. Those subsequently charged with illegal fish-ing were mostly Indians and Italian, Greek, Slavic, or Japanese immigrants.[45]

Given the exclusionary effects of these policies, it is not surprising that some American conservationists were eugenicists who saw a connection between the con-servation of natural resources and the supremacy of the Anglo-Saxon race. According to Gray Brechin, the National Conservation Congresses, first convened by forester Gifford Pinchot in 1909, asserted these links. One attendant, Mrs. Matthew T. Scott of the Daughters of the American Revolution, argued, "the mothers of this genera-tion...have a right to insist upon the conserving not only of soil, forest, birds, minerals, fishes, waterways, in the interest of our future home-makers, but also the conserving of the supremacy of the Caucasian race in our land."[46] President Theodore Roosevelt also blended natural resource conservation and the prevention of "race suicide." He sup-ported the country life movement, in part, because he saw the farmer as "the exemplar of American racial character." To make farming more attractive, reformers empha-sized efficiency and provided education on water and soil conservation. Implementing conservation strategies, then, would bolster the family farm and stop white degenera-tion.[47] These ties between conservation and whiteness certainly merit further scholarly investigation.[48]

Other conservation measures, especially national parks, had dire consequences for native peoples, whose activities, and indeed very presence, were shunned. In the United States, an ahistorical vision of national parks as pristine wilderness devoid of humans and preserved for white tourists necessitated the removal and dispossession of native peoples. As Mark Spence argues, "uninhabited wilderness had to be created before it could be preserved, and this type of landscape became reified in the first national parks." Most Americans came to see national parks as places that displayed the most remarkable examples of the nation's natural wonders and scenic beauty, untouched by human hands. But this ideal was an invention predicated on the subordination and removal of Indian peoples.[49]

Yellowstone National Park exemplified this process. Park officials moved Crow, Shoshone, and Bannock to reservations outside park boundaries, but many Indians continued to hunt in the park, given the meager food rations on the reservations. Although treaties granted them the right to hunt on unoccupied public lands, park offi-cials and outdoor enthusiasts claimed that they were destroying the park's forests and wildlife. In *Ward v. Race Horse* (1896), the US Supreme Court affirmed the uninhabited park vision and dismissed the treaty provisions.[50] For Native Americans, Karl Jacoby explains, "conservation was but one piece of a larger process of colonization and state building in which Indian peoples were transformed (in theory, at least) from indepen-dent actors to dependent wards bound by government controls." It was also an attack on the "material underpinning of their existence" and the "spiritual and moral under-standings that gave their lives meaning."[51]

Despite these assaults on their culture and livelihoods, Indians who claimed lands that became national parks did not abandon their enduring connections to these places, and resisted efforts to prohibit their customary uses. The Blackfeet, for instance, hunted deep in the Glacier National Park backcountry and campaigned tirelessly for their right of access.[52] It was not until the late twentieth century that the federal govern-ment began to concede to Native American demands. In a departure from the unin-habited park vision, nine of ten national parks established in Alaska in 1980 allowed Alaskan natives (including whites) to retain their "customary and traditional" uses within park boundaries. Elsewhere, native people have continued to battle with the park service to restore access and occupancy rights.[53]

Outside the United States, conservation measures likewise called for the removal of indigenous peoples and the prohibition of their environmental practices. Several scholars of Africa, Asia, and Latin America have demonstrated how colonial regimes instituted conservation policies that criminalized and devalued indigenous environ-mental activities. Seeking to save resources for current and future white settlers and to force nonwhites into wage labor, forests, soil, watersheds, and game animals all became targets of reform.[54] The impact of national parks on indigenous peoples also parallels the United States example. In Namibia's Etosha National Park, Hai//om San (Bushmen) initially stayed in the park, which they believed to be ancestral land. However, because tourists and park administrators grew wary of their presence "in what some clearly wished was a 'pristine' wilderness," roughly six hundred Hai//om were evicted from the park in 1954, with most becoming landless farm laborers for white-owned commercial agricultural businesses.[55] Similarly, as Roderick Neumann demonstrates, the Meru were denied customary rights to grazing and building materials and access to religious and cultural sites in Tanzania's Arusha National Park.[56]

Like their American counterparts, these indigenous groups resisted state policies that prohibited and restricted their access. The Meru, for instance, continued to graze cattle and collect firewood within the park, even befriending and negotiating with park rangers to evade punishment. In Matopos National Park, remaining African residents protested their removal to tribal trust lands in 1962 by cutting park fences and poaching game, and they demanded access to the park to visit ancestral homes and shrines.[57] These conflicts, like those in the United States, demonstrated indigenous peoples' enduring ties to specific lands and activities in nature.

Marsha Weisiger's compelling book, *Dreaming of Sheep in Navajo Country* (2009), provides further nuance to this line of inquiry. Like other studies, she documents how conservation policies—in this case, a New Deal livestock reduction program—had a devastating impact on the Navajo/Diné. She also traces Diné resistance to these policies and dissects how they understood the problem. While New Dealers blamed Diné overstocking for degraded rangeland and erosion, many Diné denied the existence of a problem and performed ceremonies to restore balance. In the end, both perspectives were based on incomplete understandings of the environment. This is an important point that moves beyond a narrative of oppression and resistance. Weisiger also offers a powerful assessment of the major lesson from this episode: "When conservationists high-handedly imposed measures that were profoundly antithetical to Diné culture, they helped begin the process of their program's unraveling."[58] Environmental reform, in other words, needs to include, rather than ignore or ostracize, the people who use the lands and waters in question.

Outdoor Recreation and Its Discontents

The marginalization of racial and ethnic minorities extended to leisure experiences. Mark Spence documents how Yosemite Indians, who lived within the national park into the twentieth century, often worked for the government and participated in the tourist economy by acting as guides or selling baskets. In 1916, park officials and concessionaries even started "Indian Field Days" with horse races and a rodeo, forcing the

Yosemite to conform to "popular conceptions of what Indians supposedly did." While
these jobs and programs allowed them to remain in their ancestral home, they were
naturalized as part of the park landscape and objectified as exotic spectacles. When
they defied romantic images, they were punished or banished.[59]

Nonwhites were also excluded and segregated from many pleasure grounds,
leading to open conflict. As Colin Fisher details, African American reformers in
early twentieth-century Chicago encouraged black residents to escape the physical and
social ills of the city and vacation in the countryside or find refuge in city parks. In
a system of de facto segregation, they could retreat to the Lake Michigan beach near
Twenty-Fifth Street. In July 1919, a group of five African American boys paddled their
raft into the lake, crossing the racial line and prompting white beachgoers to throw
rocks at them. Eugene Williams was struck in the head and drowned. The other boys
identified the perpetrator, but a white policeman refused to arrest him. This event set
off the infamous Chicago race riots, which lasted four days and left thirty-eight people
dead. While many historians have argued that long-standing housing, labor, and political issues caused the riot, Fisher suggests that African American exclusion from urban
nature was another factor.[60]

As some African Americans grew more affluent in the 1910s and 1920s, they continued to seek leisure activities in nature. This impulse led to the establishment of
all-black resorts, such as Idlewild in Michigan. With private lots for sale, African
Americans built both modest and more lavish homes and escaped the cities during the
summer to hike, boat, swim, and fish along Lake Idlewild.[61] Elsewhere, substandard,
segregated facilities were the norm, even at the national parks. In accordance with
Jim Crow practices, Shenandoah and Great Smoky Mountains National Parks created
separate black campsites. In response to complaints, William J. Trent, Jr., Secretary
of Interior Harold Ickes's adviser for Negro Affairs, called for a change in park policy in 1939. According to Terence Young, Trent believed that national parks were for
all Americans and that segregated park facilities were "a contradiction in democratic
government." Ickes responded by desegregating a Shenandoah picnic ground and
upgrading a black-only facility. By 1942, racial segregation in the national parks came
to an end. African Americans could now join whites in the national parks where they
could "together attempt to reinforce their common American values and their shared
national identity."[62]

While Fisher, Young, and others have illuminated the importance of recreational
landscapes to civil rights protests, Andrew Kahrl's *This Land Was Ours: African
American Beaches from Jim Crow to the Sunbelt South* (2012) builds upon their work
by digging deeper into the political and economic institutions and environmental changes that contributed to the marginalization of black recreation.[63] According
to Kahrl, African Americans bought oceanfront property during the Jim Crow years
for leisure and business pursuits, creating a diverse array of resorts, from "high-toned
exclusive retreats" to "bawdy, raucous beaches" with music, alcohol, and gambling.
However, during the second half of the twentieth century, they steadily lost their holdings due to federal engineering projects that improved the coastline for economic

development but degraded their land in the process. They were also exposed to "new and more aggressive forms of predation" that dispossessed them. Throughout the book, Kahrl is careful not to paint African Americans as a unified group. Indeed, some wealthier individuals tried to participate in corporate coastal development, "often with devastating consequences for those fellow African Americans who found themselves torn from the land."[64] This attention to class divisions within racial and ethnic groups warrants further attention, as individuals sought leisure in nature for distinct and sometimes contradictory reasons.[65]

Whiteness is another analytical dimension that would enrich environmental histories of outdoor recreation. While Kahrl and others have shown how African Americans tried to made in-roads into recreational activities often understood to be white, they do not fully explain how these pursuits became constructed as white in the first place. Annie Gilbert Coleman has explored this line of inquiry in her work on skiing. First brought to the country by Scandinavian and European immigrants, the sport retained a close connection to white ethnicities. Advertisements featured handsome white men or beautiful blonde women clad in the latest ski haute couture. Coleman suggests that these "images of whiteness" made skiing "a potentially alienating experience" for nonwhites. While more affluent minorities have recently taken to the slopes, few resort promoters or equipment manufacturers have featured them in their ads. The non-Northern European faces at ski resorts are often Latin American, African, and Eastern European immigrants who work behind the scenes.[66] Scholars could build on this work and examine the racialization of other outdoor activities as white. It is crucial to detail this process in order to understand the struggles for access that followed.

ENVIRONMENTALISM AND ENVIRONMENTAL JUSTICE

Just as environmental historians have demonstrated the exclusionary tendencies of conservation, wilderness protection, and outdoor recreation, they have similarly deconstructed the most celebrated environmental reform movements: the American environmental movement. Emerging in the 1960s, it "questioned corporate capitalism's tendency to view nature solely as an instrument in the service of economic gain," as Ted Steinberg explains. Middle- and upper middle-class activists addressed a wide range of environmental problems, from wilderness and wildlife protection to air and water pollution and population growth.[67] Some of these issues took on a racist dimension. For instance, Zero Population Growth (ZPG), a group that formed in response to Paul Ehrlich's *The Population Bomb*, lamented the environmental implications of population growth and advocated voluntary sterilization and birth control. Opponents argued that such policies were racist, sexist, and open to abuse, as when two black women were sterilized involuntarily in Montgomery, Alabama in 1973. A people of color caucus even walked out of a Population and World Resources Conference in 1970 because the organizers, which included ZPG, did not address "the racial connotations of the issues under consideration." Because they wanted to curb Mexican immigration, ZPG and similar groups also clashed with Latino groups that advocated immigrant rights. As Robert Gottlieb notes, "the focus on immigration created concerns about racist assumptions within environmentalism."[68]

In addition to promoting potentially racist policies, the environmental movement did not adequately address the concerns of poor, working-class, and nonwhite communities, specifically their disproportionate exposure to toxics in their homes and workplaces. Many environmental historians have demonstrated that this problem had deep historical roots. For instance, Andrew Hurley's work on late nineteenth-century St. Louis suggests that city officials decided that the easiest solution to pollution and odor complaints was to locate noxious businesses in neighborhoods "where political opposition was weakest, property values were lowest, and residents were poorest." At the time, these parameters best described where German immigrants had congregated. Hurley concludes, "the late nineteenth-century approach to industrial pollution had less to do with abating emissions than with allocating social costs."[69]

Hurley developed these ideas further in *Environmental Inequalities: Class, Race, and Industrial Pollution in Gary, Indiana, 1945–1980* (1995). He argues that environmental changes had a "differential impact on society's members," depending on their social and economic status. Liberalism's failure to challenge property rights or industrial capitalism and its inability to balance "the imperatives of industrial growth and social welfare" explained these inequities. In Gary, middle-class whites moved to the suburbs to escape the industrial environment, but blue-collar workers could not leave degraded neighborhoods around the industrial sector. Likewise, African Americans found it difficult to avoid unhealthy working and living conditions due to housing covenants, intimidation, and discriminatory hiring practices. When all three groups began to protest industrial pollution, U.S. Steel threatened to eliminate jobs, "exploit[ing] the vulnerability of those at the lower level of Gary's social hierarchy for the purpose of maintaining [its] environmental hegemony." Subsequent industrial activities continued to concentrate where poor and nonwhite residents lived.[70] Several environmental historians, including Ellen Stroud, Dolores Greenberg, and Sylvia Hood Washington, have identified similar patterns elsewhere.[71]

Activists of color began to campaign for the eradication of these environmental inequalities beginning in the late 1970s. From Texas and California to North Carolina and the Rosebud Indian Reservation in South Dakota, African American, Latino, Native American, and Asian American organizers protested the siting of solid waste landfills, lead smelters, refineries, and waste incinerators in their backyards. Linda Nash and Laura Pulido have also detailed campaigns against pesticide use, which brought chronic health problems to Mexican farm workers. Activists used many tactics, from demonstrations and public hearings to fact-finding and lobbying, and targeted all levels of government. As sociologist Robert Bullard explains, this environmental justice movement "extend[ed] the quest for basic civil rights." He concludes,

"It is unlikely that this nation will ever achieve lasting solutions to its environmental problems unless it also addresses the system of racial injustice that helps sustain the existance [sic] of powerless communities forced to bear disproportionate environmental costs."[72]

Yet some historians have been quick to complicate the scholarship on environmental inequities. According to Hal Rothman, the environmental justice movement has obscured the importance of class. While the United Church of Christ's 1987 report, "Toxic Wastes in The United States: A National Report on the Racial and Socio-Economic Characteristics of Communities with Hazardous Waste Sites," documented that at least 80 percent of people of color lived in close proximity to hazardous and toxic waste facilities, Rothman countered that the "overwhelming majority" of people affected by toxics were working-class whites. Because some polluted neighborhoods were also inhabited by racial minorities, race and class could not be easily disentangled.[73] Monica Perales, moreover, points to the "multiple meanings of environmental justice." When the El Paso, Texas community of Smeltertown was targeted for demolition due to lead contamination in 1972, Mexican American residents protested because they wanted to save their cherished community. In short, they wanted "more than an expedient solution to an environmental problem." While they ultimately lost this battle, Perales suggests that this incident demonstrates that environmental justice was not solely about providing healthy work and living environments for all populations. It also included "understanding the needs of different communities and allowing them to play a role in determining fair and realistic solutions when faced with environmental crisis."[74]

Environmental injustices, moreover, were not confined to the United States and moved freely across international borders. Maquiladoras south of the US-Mexico border polluted nearby towns, causing health problem among workers and residents. Industrialized nations also targeted poor communities of color in the developing world for toxic waste disposal and high-risk technologies. This dumping policy created a scandal when a 1991 memo signed by Lawrence Summers, then chief economist of the World Bank, was made public. The memo justified dumping in Less Developed Countries (LDCs), noting that under-polluted areas in Africa had air quality that was "inefficiently low." Some multinational companies, most notably Royal Dutch Shell, appeared to concur with Summers's assessment. Beginning in 1956, it began to exploit billions of barrels of crude oil in the Niger Delta, which has contributed to both the degradation of the natural environment and the oppression local people.[75]

By the late twentieth century, mainstream environmental groups began to take a more active interest in environmental justice issues, but critics continue to charge that environmentalism is elitist. The reemergence of the immigration issue in Sierra Club politics, with an anti-immigration referendum in 1998 and anti-immigration board candidates in 2004, demonstrates these tensions. Fearing perceptions of being racist, Sierra Club leaders rejected claims that immigrants would strain natural resources. Both times, the anti-immigration forces were defeated, but as Robert Gottlieb suggests, the controversy "revealed the vulnerability of the national environmental groups like the Sierra Club on issues of race and justice" and their failure to make immigrants a constituency.[76] Even the environmental justice movement has failed to integrate poor, working-class, and nonwhite groups. For example, the Campo Indians of San Diego County campaigned to lease space on their reservation for a landfill, a move that troubled activists.[77] Environmental historians need to continue analyzing the unfolding complexities of the environmental justice movement and assess the extent to which it has served the interests of racial and ethnic groups, past and present.

FUTURE DIRECTIONS

Environmental historians and scholars in allied fields have produced many excellent studies that integrate race and ethnicity and demonstrate the reciprocal relationship between social and environmental change. They have blurred the natural and the social, and have provided many compelling narratives about nature's role in power relations. As Mark Fiege's *The Republic of Nature* (2012) suggests, many historical episodes warrant this kind of analysis. His brilliant interpretation of *Brown v. Board of Education* shows that even an event that seems firmly about race has an environmental component. In painstaking detail, he traces the environmental history of the color line in Topeka, documenting how African American neighborhoods were relegated to low-lying areas by the Kansas River and the Shunganunga Creek. Linda Brown and her family lived on the fringes of these enclaves, near the river and surrounded by industry. Rather than attend a white school close to her home, she had to walk along the railroad tracks to catch a bus to a black school. Her lawsuit to end school segregation, then, was part of a longer history in which the physical landscape of the city reinforced social inequalities.[78] As Fiege explains, "The color line ... was more than just a legal abstraction; it was a material practice grounded in the social experience of landscape and its physical properties."[79]

Fiege suggests the tremendous potential of using environmental history to examine episodes outside the field's usual purview, but the scholarship on race, ethnicity, and environment could expand in other areas as well. As this essay has suggested, there could be more attention to class, as most studies tend to focus on working-class or impoverished minorities and not their middle-class or elite counterparts. The latter individuals clearly pushed for more access to leisure areas, but how else did they experience, understand, or imagine the natural world? Future studies could also explore the role of the environment in the construction of whiteness, a racial identity that often made certain claims to nature, excluding and marginalizing others in the process. Last, more scholars could examine the environmental experiences of Latinos and Asian Americans, two groups who have received minimal attention from environmental historians.[80]

Environmental historians who want to continue analyzing race and ethnicity may also want to return to one of the roots of the field: intellectual history. Roderick Nash's *Wilderness and the American Mind*, first published in 1967, remains one of the classics of the field. By analyzing the writings of America's most famous environmental thinkers, from Henry David Thoreau and Ralph Waldo Emerson to John Muir and Aldo Leopold, Nash traces shifting attitudes toward wilderness and its meaning to American civilization. Yet the figures in Nash's study are mostly privileged, well-educated white men.[81] Few subsequent historians have examined the environmental thought of nonwhites, nor have they explored how racist and nativist attitudes influenced understandings of nature. Studies by Peter Coates and Miles Powell are notable exceptions.[82]

Did different racial and ethnic groups develop a distinct intellectual tradition with respect to the natural world? Did their racial and ethnic backgrounds help to shape a specific environmental ethos? Did they discuss the environment's role in their history and contemporary lives? How *did* nonwhite intellectuals write about nature, if at all? Kimberly K. Smith's study, *African American Environmental Thought: Foundations* (2007), suggests the potential of this line of inquiry. She begins her book with this question: "One might think that 250 years of slavery would have left black Americans permanently alienated from the American landscape…Locked in a struggle for social justice, what interest could they have in the claims of nature?" By analyzing leading black thinkers like Frederick Douglass, W. E. B. DuBois, and Booker T. Washington, Smith discovers "a rich tradition of black environmental thought," dating back to abolitionism, that informed the contemporary environmental justice movement.[83]

The intellectual traditions of racial and ethnic groups can be difficult to uncover, but their environmental thought can further complicate the narrative of victimization and inequality. While many environmental historians have demonstrated how members of racial and ethnic groups asserted control over their interactions with the natural world, the ways that they thought about nature and how these ideas shaped their culture and informed their political agendas also point to their autonomy and unwillingness to accept racial and ethnic constructions that limited their access to the environment. Given environmental historians' recent explorations of the transnational dimensions of environmental change, it is also important to analyze the extent to which these ideas of nature transgressed borders. As diverse people moved from region to region and nation to nation, their environmental views likely transferred as well, ultimately shaping the physical and cultural contours of landscapes around the world. Race and ethnicity, like nature, are historical categories that never stay still.

Notes

I would like to thank David Gordon, David Hecht, Drew Isenberg, Matthew Klingle, Rachel Sturman, an anonymous reviewer, and participants in the Temple University workshop in 2007 for their comments and feedback.

1. William Cronon, "Modes of Prophecy and Production: Placing Nature in History," *Journal of American History* 76, no. 4 (March 1990): 1122–1121; William Cronon, "Kennecott Journey: Paths out of Town," in *Under an Open Sky: Rethinking America's Western Past*, ed. William Cronon, George Miles, Jay Gitlin (New York: W.W. Norton, 1992), 28–51; Carolyn Merchant, "Shades of Darkness: Race and Environmental

History," *Environmental History* 8, no. 3 (July 2003): 380–394; Ted Steinberg, "Down to Earth: Nature, Agency, and Power in History," *American Historical Review* 107, no. 3 (June 2002): 798–820; Richard White, "Afterword Environmental History: Watching a Historical Field Mature," *Pacific Historical Review* 70, no. 1 (February 2001): 103–111. See also Paul S. Sutter, "The World with Us: The State of American Environmental History," *The Journal of American History* 100, no. 1 (June 2013): 94–119; Sarah T. Phillips, "Environmental History," in *American History Now*, ed. Eric Foner and Lisa McGirr (Philadelphia: Temple University Press, 2011), 285–313; Ted Steinberg, *Down to Earth: Nature's Role in American History* (New York: Oxford University Press, 2002); Mart A. Stewart, "Environmental History: Profile of a Developing Field," *History Teacher* 31, no. 3 (May 1998): 351–368; Stephen Mosley, "Common Ground: Integrating Social and Environmental History," *Journal of Social History* 39, no. 3 (Spring 2006): 915–933; Alan Taylor, "Unnatural Inequalities: Social and Environmental Histories," *Environmental History* 1, no. 4 (October 1996): 6–19; Martin V. Melosi, "Environmental Justice, Political Agenda Setting, and the Myths of History," *Journal of Policy History* 12, no. 1 (2000): 43–71.

2. Taylor, "Unnatural Inequalities," 16.

3. Alice Kessler-Harris, "Social History," in *The New American History*, rev. and expanded ed., ed. Eric Foner (Philadelphia: Temple University Press, 1997), 231–236; Alfred W. Crosby, "The Past and Present of Environmental History," *American Historical Review* 100, no. 4 (October 1995): 1184–1188.

4. Taylor, "Unnatural Inequalities," 8–9; Mosley, "Common Ground," 919–920.

5. William Cronon, *Changes in the Land: Indians, Colonists, and the Ecology of New England* (New York: Hill and Wang, 1983, 2003); Richard White, *Land Use, Environment, and Social Change: The Shaping of Island County, Washington* (Seattle: University of Washington Press, 1980); Timothy Silver, *A New Face on the Countryside: Indians, Colonists, and Slaves in South Atlantic Forests, 1500–1800* (New York: Cambridge University Press, 1990); Arthur F. McEvoy, *The Fisherman's Problem: Ecology and Law in the California Fisheries* (New York: Cambridge University Press, 1986). See also Shepard Krech, III, *The Ecological Indian: Myth and History* (New York: W. W. Norton, 1999); William M. Denevan, *Cultivated Landscapes of Native Amazonia and the Andes* (New York: Oxford University Press, 2001).

6. Coll Thrush, *Native Seattle: Histories from the Crossing-Over Place* (Seattle: University of Washington Press, 2007), 238.

7. Alfred W. Crosby, *Ecological Imperialism: The Biological Expansion of Europe, 900–1900* (New York: Cambridge University Press, 1986), 154–155. See also Alfred W. Crosby, *The Columbian Exchange: Biological and Cultural Consequences of 1492* (Westport, Conn.: Greenwood, 1972).

8. Cronon, *Changes in the Land*, 8.

9. Cronon, *Changes in the Land*, 90–97.

10. Cronon, *Changes in the Land*, 99–101. See also Calvin Martin, *Keepers of the Game: Indian-animal Relationships and the Fur Trade* (Berkeley: University of California Press, 1978).

11. Elliott West, *The Contested Plains: Indians, Goldseekers, and the Rush to Colorado* (Lawrence: University Press of Kansas, 1998), 33–93, esp. 55, 90. For similar patterns among other native groups, see Richard White, *The Roots of Dependency: Subsistence, Environment, and Social Change among the Choctaws, Pawnees, and Navajos* (Lincoln: University of Nebraska Press, 1983). For the decline of the bison, see Dan Flores,

"Bison Ecology and Bison Diplomacy: The Southern Plains from 1800 to 1850," *Journal of American History* 78, no. 2 (September 1991): 465–485 and Andrew Isenberg, *The Destruction of the Bison: An Environmental History, 1750–1920* (New York: Cambridge University Press, 2000).

12. For the "New Western History," see Patricia Nelson Limerick, *The Legacy of Conquest: The Unbroken Past of the American West* (New York: W. W. Norton, 1987) and Richard White, *"It's Your Misfortune and None of My Own": A New History of the American West* (Norman: University of Oklahoma Press, 1991).

13. Brett L. Walker, *The Conquest of Ainu Lands: Ecology and Culture in Japanese Expansion, 1590–1800* (Berkeley: University of California Press, 2001). See also Elinor G. K. Melville, *A Plague of Sheep: Environmental Consequences of the Conquest of Mexico* (New York: Cambridge University Press, 1994); Cynthia Radding Murrieta, *Wandering Peoples: Colonialism, Ethnic Spaces, and Ecological Frontiers in Northwestern Mexico, 1700–1850* (Durham, NC: Duke University Press, 1997); Warren Dean, *With Broadax and Firebrand: The Destruction of the Brazilian Atlantic Forest* (Berkeley: University of California Press, 1995); Richard Grove, Vinita Damodaran, Satpal Sangwan, eds., *Nature and the Orient: The Environmental History of South and Southeast Asia* (Delhi: Oxford University Press, 1998); Nancy J. Jacobs, *Environment, Power, and Injustice: A South African History* (New York: Cambridge University Press, 2003).

14. Joyce E. Chaplin, *Subject Matter: Technology, the Body, and Science on the Anglo-American Frontier, 1500–1676* (Cambridge, MA: Harvard University Press, 2001).

15. David N. Livingstone, "Race, Space, and Moral Climatology: Notes Toward a Genealogy," *Journal of Historical Geography* 28, no. 2 (2002): 159–180.

16. Mark Harrison, "'The Tender Frame of Man': Disease, Climate, and Racial Difference in India and the West Indies, 1760–1860," *Bulletin of the History of Medicine* 70, no. 1 (1996): 91; Warwick Anderson, "Disease, Race, and Empire," *Bulletin of the History of Medicine* 70, no. 1 (1996): 63. Both articles appeared in theme issue entitled "Race and Acclimatization in Colonial Medicine." For other studies of climate, race, and imperialism, see Warwick Anderson, "The Natures of Culture: Environment and Race in the Colonial Tropics," in *Nature in the Global South: Environmental Projects in South and Southeast Asia*, ed. Paul Greenough and Anna Lowenhaupt Tsing (Durham, NC: Duke University Press, 2003), 34–35; Warwick Anderson, *The Cultivation of Whiteness: Science, Health, and Racial Destiny in Australia* (Durham, NC: Duke University Press, 2006); Mark Harrison, *Climates and Constitutions: Health, Race, Environment and British Imperialism in India, 1600–1850* (New York: Oxford University Press, 1999); Conevery Bolton Valenčius, "Histories of Medical Geography," in *Medical Geography in Historical Perspective*, ed. Nicolaas A. Rupke (London: Wellcome Trust Centre for the History of Medicine at UCL, 2000), 15–19.

17. Mart A. Stewart, *"What Nature Suffers to Groe": Life, Labor, and Landscape on the Georgia Coast, 1680–1920* (Athens: University of Georgia Press, 1996), 148.

18. Melissa A. Johnson, "The Making of Race and Place in Nineteenth-Century British Honduras," *Environmental History* 8, no. 4 (October 2003): 604–606.

19. Karl Jacoby, "Slaves by Nature?: Domestic Animals and Human Slaves," *Slavery and Abolition* 15, no. 1 (April 1994): 89–99.

20. Stewart, *What Nature Suffers to Groe*, 63–64; Mart A. Stewart, "'Let Us Begin with the Weather?': Climate, Race, and Cultural Distinctiveness in the American South," in *Nature and Society in Historical Context*, ed. Mikuláš Teich, Roy Porter, and Bo Gustafsson (New York: Cambridge University Press, 1997), 248–249; George M. Fredrickson, *The*

Black Image in the White Mind: The Debate on Afro-American Character and Destiny, 1817–1914 (New York: Harper & Row, 1971), 71–96. For other work on scientific theories of race and environment, see Nancy J. Christie, "Environment and Race: Geography's Search for a Darwinian Synthesis," in *Darwin's Laboratory: Evolutionary Theory and Natural History in the Pacific*, ed. Roy MacLeod and Philip F. Rehbock (Honolulu: University of Hawai'i Press, 1994): 426–473; Nicholas Hudson, "From 'Nation' to 'Race'" The Origin of Racial Classification in Eighteenth-Century Thought," *Eighteenth-Century Studies* 29, no. 3 (1996): 247–264.

21. Mart A. Stewart, "Slavery and African American Environmentalism," in *"To Love the Wind and the Rain": African Americans and Environmental History*, ed. Dianne D. Glave and Mark Stoll (Pittsburgh: University of Pittsburgh Press, 2006), 11.

22. Judith A. Carney, "The African Antecedents of Uncle Ben in U.S. History," *Journal of Historical Geography* 29, no. 1 (January 2003): 1–21; Judith A. Carney, *Black Rice: The African Origins of Rice Cultivation in the Americas* (Cambridge, MA: Harvard University Press, 2001). Both Carney and Philip D. Morgan suggest that prior to the mid-eighteenth century, the percentage of South Carolina slaves from African rice-producing areas was low. See Philip D. Morgan, "Work and Culture: The Task System and the World of Lowcountry Blacks, 1700–1880," *William and Mary Quarterly*, 3d Ser., 39, no. 4 (October 1982): 567–568. For other studies of slaves and rice cultivation, see also Joyce E. Chaplin, "Tidal Rice Cultivation and the Problem of Slavery in South Carolina and Georgia, 1760–1815," *The William and Mary Quarterly*, 3rd Ser., 49, no. 1 (January 1992): 31, 54–56.

23. Sutter, "The World With Us," 107–108. For the debate over Carney's "black rice thesis," see David Eltis, Philip Morgan, and David Richardson, "Agency and Diaspora in Atlantic History: Reassessing the African Contribution to Rice Cultivation in the Americas," *American Historical Review* 112, no. 5 (December 2007): 1329–1358. There was also an exchange entitled "The Question of 'Black Rice'" in the February 2010 issue of the *American Historical Review*.

24. Morgan, "Work and Culture," 569; Chaplin, "Tidal Rice Cultivation," 56–57.

25. Stewart, "Slavery and African American Environmentalism," 11–13, 17; Scott Giltner, "Slave Hunting and Fishing in the Antebellum South," in *"To Love the Wind and the Rain,"* 25–30; Elizabeth D. Blum, "Power Danger, and Control: Slave Women's Perceptions of Wilderness in the Nineteenth Century," *Women's Studies* 31 (2002): 260–262. For slave property ownership, see Dylan Penningroth, "Slavery, Freedom, and Social Claims to Property among African Americans in Liberty County, Georgia, 1850–1880," *Journal of American History* 84, no. 2 (September 1997): 405–435.

26. Stewart, "Slavery and African American Environmentalism," 14–16; Blum, "Power, Danger, and Control," 256–260.

27. Giltner, "Slave Hunting and Fishing," 33–36; Blum, "Power, Danger, and Control," 250–251.

28. Giltner, "Slave Hunting and Fishing," 22–23; Blum, "Power, Danger, and Control," 255. See also Scott Giltner, *Hunting and Fishing in the New South: Black Labor and White Leisure after the Civil War* (Baltimore: The Johns Hopkins University Press, 2008). Even after the Civil War, white planters linked the control of nature with the control of their black workers. See James C. Giesen, "'The Truth About the Boll Weevil': The Nature of Planter Power in the Mississippi Delta," *Environmental History* 14 (October 2009): 683–704.

29. Mark Fiege, *The Republic of Nature: An Environmental History of the United States* (Seattle: University of Washington Press, 2012), 100–138, quotes on 105, 128, 137.

30. White, *"It's Your Misfortune and None of My Own,"* 282–285; Chris Friday, *Organizing Asian American Labor: The Pacific Coast Canned-Salmon Industry, 1870-1942* (Philadelphia: Temple University Press, 1994), 14–24.

31. Douglas Cazaux Sackman, *Orange Empire: California and the Fruits of Eden* (Berkeley: University of California Press, 2005), 128.

32. McEvoy, *Fisherman's Problem*, 53, 56, 62. For other examples of Indian maintenance of subsistence activities, see Cary C. Collins, "Subsistence and Survival: The Makah Indian Reservation, 1855–1933," *Pacific Northwest Quarterly* 87, no. 4 (Fall 1996): 180–193; Alexandra Harmon, *Indians in the Making: Ethnic Relations and Indian Identities around Puget Sound* (Berkeley: University of California Press, 1998), 218–244.

33. McEvoy, *Fisherman's Problem*, 98–99.

34. Sucheng Chan, *This Bittersweet Soil: The Chinese in California Agriculture, 1860–1910* (Berkeley: University of California Press, 1986), 224.

35. For the links between labor and environmental history, see Gunther Peck, "The Nature of Labor: Fault Lines and Common Ground in Environmental and Labor History," *Environmental History* 11 (April 2006): 212–238.

36. Samuel P. Hays, *Conservation and the Gospel of Efficiency: The Progressive Conservation Movement, 1890–1920* (Cambridge, MA: Harvard University Press, 1959); Roderick Nash, *Wilderness and the American Mind*, 4th ed. (New Haven, CT: Yale University Press, 2001). For national parks, see Alfred Runte, *National Parks: The American Experience*, 3d ed. (Lincoln: University of Nebraska Press, 1997).

37. Harold L. Platt, "Jane Addams and the Ward Boss Revisited: Class, Politics, and Public Health in Chicago, 1890–1930," *Environmental History* 5, no. 2 (April 2000): 194–222, esp. 201 and Harold L. Platt, *Shock Cities: The Environmental Transformation and Reform of Manchester and Chicago* (Chicago: University of Chicago Press, 2005).

38. Adam Rome, "Nature Wars, Culture Wars: Immigration and Environmental Reform in the Progressive Era," *Environmental History* 13 (July 2008): 445. Neil Maher argues that New Deal conservation policies were intended to have a similar impact on immigrants. See Neil M. Maher, *Nature's New Deal: The Civilian Conservation Corps and the Roots of the American Environmental Movement* (New York: Oxford University Press, 2008).

39. Rome, "Nature Wars, Culture Wars," 446.

40. Dianne D. Glave, "'A Garden so Brilliant With Colors, so Original in its Design': Rural African American Women, Gardening, Progressive Reform, and the Foundation of an African American Environmental Perspective," *Environmental History* 8, no. 3 (July 2003): 395–411. See also Jack Temple Kirby, *Mockingbird Song: Ecological Landscapes of the South* (Chapel Hill: University of North Carolina Press, 2006), 182–184.

41. Mark Hersey, "Hints and Suggestions to Farmers: George Washington Carver and Rural Conservation in the South," *Environmental History* 11, no. 2 (April 2006): 239–268.

42. Louis Warren, *The Hunter's Game: Poachers and Conservationists in Twentieth-Century America* (New Haven, CT: Yale University Press, 1997), 29, 45. See also Rome, "Nature Wars, Culture Wars," 434–445.

43. Steven Hahn, "Hunting, Fishing, and Foraging: Common Rights and Class Relations in the Postbellum South," *Radical History Review* 26 (1982): 36–64; Scott Giltner, *Hunting and Fishing in the New South: Black Labor and White Leisure after the Civil War* (Baltimore: The Johns Hopkins University Press, 2008).

44. Joseph E. Taylor, *Making Salmon: An Environmental History of the Northwest Fisheries Crisis* (Seattle: University of Washington Press, 1999), 137.

45. Matthew Klingle, *Emerald City: An Environmental History of Seattle* (New Haven, CT: Yale University Press, 2007), 174–176 and Andrew H. Fisher, *Shadow Tribe: The Making of Columbia River Indian Identity* (Seattle: University of Washington Press, 2010), 168–169. See also Roberta Ulrich, *Empty Nets: Indians, Dams, and the Columbia River* (Corvallis: Oregon State University Press, 1999); Katrine Barber, *Death of Celilo Falls* (Seattle: University of Washington Press, 2005); Bill Parenteau and James Kenny, "Survival, Resistance, and the Canadian State: The Transformation of New Brunswick's Native Economy, 1867–1930," *Journal of the Canadian Historical Association* 13 (2002): 49–71; Douglas C. Harris, *Fish, Law, and Colonialism: The Legal Capture of Salmon in British Columbia* (Toronto: University of Toronto Press, 2001).

46. Gray Brechin, "Conserving the Race: Natural Aristocracies, Eugenics, and the U.S. Conservation Movement," *Antipode* 28, no. 3 (1996): 238. See also Alexandra Minna Stern, *Eugenic Nation: Faults and Frontiers of Better Breeding in Modern America* (Berkeley and Los Angeles: University of California Press, 2005), 115–149.

47. Laura L. Lovett, *Conceiving the Future: Pronatalism, Reproduction, and the Family in the United States, 1890–1938* (Chapel Hill: University of North Carolina Press, 2007), 110–130.

48. For further development of this point, see Miles Powell, "Vanishing Species, Dying Races: A History of Extinction in America" (PhD diss., University of California, Davis, 2012).

49. Mark Spence, *Dispossessing the Wilderness: Indian Removal and the Making of the National Parks* (New York: Oxford University Press, 1999), 4.

50. Spence, *Dispossessing the Wilderness*, 50, 66–67. See also Karl Jacoby, *Crimes Against Nature: Squatters, Poachers, Thieves, and the Hidden History of American Conservation* (Berkeley: University of California Press, 2001), 81–146.

51. Jacoby, *Crimes Against Nature*, 151, 187.

52. Spence, *Dispossessing the Wilderness*, 83–100. Indian fishermen also resisted conservation efforts. See Klingle, *Emerald City*, 239–241; Fisher, *Shadow Tribe*; Douglas C. Harris, *Landing Native Fisheries: Indian Reserves and Fishing Rights in British Columbia, 1849–1925* (Vancouver: University of British Columbia Press, 2008).

53. Theodore Catton, *Inhabited Wilderness: Indians, Eskimos, and National Parks in Alaska* (Albuquerque: University of New Mexico Press, 1997); Spence, *Dispossessing the Wilderness*, 136–139. See also John Sandlos, *Hunters at the Margins: Native People and Wildlife Conservation in the Northwest Territories* (Vancouver: University of British Columbia Press, 2007).

54. For Africa, see William Beinart, "Introduction: The Politics of Colonial Conservation," *Journal of Southern African Studies* 15, no.2 (January 1989): 143–162; Richard Grove, "Scottish Missionaries, Evangelical Discourses and the Origins of Conservation Thinking in Southern Africa, 1820–1900," *Journal of Southern African Studies* 15, no.2 (January 1989): 163–187; Hans G. Schabel, "Tanganyika Forestry Under German Colonial Administration, 1891–1919," *Forest and Conservation History* 34, no. 3 (July 1990): 130–141; Jacob Tropp, "Dogs, Poison and the Meaning of Colonial Intervention in the Transkei, South Africa," *Journal of African History* 43 (2002): 451–472; Edward Steinhart, *Black Poachers, White Hunters: A Social History of Hunting in Colonial Kenya* (Athens, OH: Ohio University Press, 2006); Jacobs, *Environment, Power, and Injustice*. For Asia, see Ramachandra Guha, *The Unquiet Woods: Ecological Change and Peasant Resistance in the Himalaya* (Berkeley: University of California Press, 2000). For Latin America, see Richard H. Grove, *Green Imperialism: Colonial Expansion, Tropical Island*

Edens, and the Origins of Environmentalism (New York: Cambridge University Press, 1995), 264–308.

55. James Suzman, "Etosha Dreams: An Historical Account of the Hai//om Predicament," *Journal of Modern African Studies* 42, no. 2 (2004): 221–238.

56. Roderick P. Neumann, *Imposing Wilderness: Struggles over Livelihood and Nature Preservation in Africa* (Berkeley: University of California Press, 1998). See also Jane Carruthers, *The Kruger National Park: A Social and Political History* (Pietermaritzburg: University of Natal Press, 1995).

57. Neumann, *Imposing Wilderness*, 198–199; Terence Ranger, "Whose Heritage?: The Case of Matobo National Park," *Journal of Southern African Studies* 15, no. 2 (January 1989): 217–249; Terence O. Ranger, *Voices from the Rocks: Nature, Culture and History in the Matopos Hills of Zimbabwe* (Bloomington: University of Indiana Press, 1999).

58. Marsha Weisiger, *Dreaming of Sheep in Navajo Country* (Seattle: University of Washington Press, 2009), 242–243.

59. Spence, *Dispossessing the Wilderness*, 101–123, esp. 117.

60. Colin Fisher, "African Americans, Outdoor Recreation, and the 1919 Chicago Race Riot," in *"To Love the Wind and the Rain,"* 63–76. For the exclusionary nature of outdoor recreation, see also Klingle, *Emerald City*, 154–179 and Lawrence Culver, *The Frontier of Leisure: Southern California and the Shaping of Modern America* (New York: Oxford University Press, 2010), 52–82.

61. Mark S. Foster, "In the Face of 'Jim Crow': Prosperous Blacks and Vacations, Travel and Outdoor Leisure, 1890–1945," *Journal of Negro History* 84, no. 2 (Spring 1999): 130–149. See also Brian McCammack, "Recovering Green in Bronzeville: An Environmental and Cultural History of the African American Great Migration to Chicago, 1915–1940" (PhD diss., Harvard University, 2012); Cindy S. Aron, *Working at Play: A History of Vacations in the United States* (New York: Oxford University Press, 1999), 206–236; Culver, *Frontiers of Leisure*, 70–72.

62. Terence Young, "'A Contradiction in Democratic Government': W. J. Trent, Jr., and the Struggle to Desegregate National Park Campground," *Environmental History* 14 (October 2009): 651–682.

63. For beaches as sites for civil rights protests, see J. Michael Butler, "The Mississippi State Sovereignty Commission and Beach Integration, 1959–1963: A Cotton-Patch Gestapo?" *Journal of Southern History* 68, no. 1 (February 2002): 107–147, esp. 124; Lena Lenček and Gideon Bosker, *The Beach: The History of Paradise on Earth* (New York: Penguin Books, 1999), 242–245; Culver, *Frontiers of Leisure*, 68–82.

64. Andrew W. Kahrl, *This Land Was Ours: African American Beaches from Jim Crow to the Sunbelt South* (Cambridge, MA: Harvard University Press, 2012), 14, 17, 211.

65. Studies of the class dimensions of outdoor recreation often focus on white consumerism, not the experiences of racial and ethnic minorities. See James Morton Turner, "From Woodcraft to 'Leave No Trace': Wilderness, Consumerism, and Environmentalism in Twentieth-Century America," *Environmental History* 7, no. 3 (July 2002): 464–484.

66. Annie Gilbert Coleman, "The Unbearable Whiteness of Skiing," *Pacific Historical Review* 65, no. 4 (November 1996): 583–614, esp. 606, 608; Annie Gilbert Coleman, *Ski Style: Sport and Culture in the Rockies* (Lawrence: University Press of Kansas, 2004), 202. For whiteness and leisure, see also Culver, *Frontiers of Leisure*, 53, 76–82.

67. Steinberg, *Down to Earth*, 240; Adam Rome, "'Give Earth a Chance': The Environmental Movement and the Sixties," *Journal of American History* 90, no. 2 (2003): 525–554.

68. Robert Gottlieb, *Forcing the Spring: The Transformation of the American Environmental Movement*, rev. and updated ed. (Washington, DC: Island Press, 2005), 332–335. See also Powell, "Vanishing Species, Dying Races." He argues that this anti-immigration stance among environmental groups actually began much sooner.

69. Andrew Hurley, "Busby's Stink Boat and the Regulation of Nuisance Trades, 1865–1918," in *Common Fields: An Environmental History of St. Louis*, ed. Andrew Hurley (St. Louis: Missouri Historical Society, 1997), 145–147. See also Andrew Hurley, "Creating Ecological Wastelands: Oil Pollution in New York City, 1870–1900," *Journal of Urban History* 20 (May 1994): 340–361.

70. Andrew Hurley, *Environmental Inequalities: Class, Race, and Industrial Pollution in Gary, Indiana, 1945–1980* (Chapel Hill: University of North Carolina Press, 1995), 3, 13, 149.

71. Ellen Stroud, "Troubled Waters in Ecotopia: Environmental Racism in Portland, Oregon," *Radical History Review* 74 (1999): 65–95; Dolores Greenberg, "Reconstructing Race and Protest: Environmental Justice in New York City," *Environmental History* 5, no. 2 (April 2000): 223–250; Sylvia Hood Washington, *Packing Them In: An Archaeology of Environmental Racism in Chicago, 1865–1954* (New York: Lexington Books, 2005). See also Christopher G. Boone, "Zoning and Environmental Inequity in the Industrial East Side," in *Land of Sunshine: An Environmental History of Metropolitan Los Angeles*, ed. William Deverell and Greg Hise (Pittsburgh: University of Pittsburgh Press, 2006), 167–178.

72. Robert D. Bullard, "Anatomy of Environmental Racism and the Environmental Justice Movement," in ed. Robert D. Bullard, *Confronting Environmental Racism: Voices from the Grassroots* (Boston: South End Press, 1993), 15–39, esp. 22, 30; Eileen Maura McGurty, *Transforming Environmentalism: Warren County, PCBs, and the Origins of Environmental Justice* (New Brunswick: Rutgers University Press, 2007); Linda Nash, *Inescapable Ecologies: A History of Environment, Disease, and Knowledge* (Berkeley: University of California Press, 2006), 170–208; Laura Pulido, *Environmentalism and Economic Justice: Two Chicano Struggles in the Southwest* (Tucson: University of Arizona Press, 1996), 57–124.

73. Hal K. Rothman, *The Greening of a Nation: Environmentalism in the United States Since 1945* (New York: Harcourt Brace, 1998), 161–165. Hurley details a similar phenomenon among working-class whites in Gary. See Hurley, *Environmental Inequalities*, 77–110. For connections between race and class, see Martin V. Melosi, "Environmental Justice, Ecoracism, and Environmental History," in *"To Love the Wind and the Rain,"* 123; Melosi, "Environmental Justice, Political Agenda Setting, and the Myths of History"; Stroud, "Troubled Waters in Ecotopia," 68.

74. Monica Perales, "Fighting to Stay in Smeltertown: Lead Contamination and Environmental Justice in a Mexican American Community," *Western Historical Quarterly* 39 (Spring 2008): 41–69, quote on p. 63 and Monica Perales, *Smeltertown: Making and Remembering a Southwest Border Community* (Chapel Hill: University of North Carolina Press, 2010).

75. Bullard, "Anatomy of Environmental Racism," 19–20; Devon Peña, *The Terror of the Machine: Technology, Work, Gender, and Ecology of the U.S.-Mexico Border* (Austin: University of Texas Press, 1997); Ike Okonta and Oronto Douglas, *Where Vultures Feast: Shell, Human Rights, and Oil in the Niger Delta* (San Francisco: Sierra Club Books, 2001).

76. Gottlieb, *Forcing the Spring*, 24–25.

77. Dan McGovern, *The Campo Landfill War: The Fight for Gold in California's Garbage* (Norman: University of Oklahoma Press, 1995). A similar battle unfolded on the Skull Valley Goshute Reservation in Utah. An agreement with Private Fuel Storage (PFS) to store nuclear waste on the reservation divided the tribe, PFS recently asked the Nuclear Regulatory Commission to terminate its license. See Judy Fahys, "Utah N-Waste site backers call it quits," *Salt Lake Tribune*, December 21, 2012. See also Sierra M. Jefferies, "Environmental Justice and the Skull Valley Goshute Indians' Proposal to Store Nuclear Waste," *Journal of Land, Resources, and Environmental Law* 27, no. 2 (2007): 409–429.

78. Fiege, *The Republic of Nature*, 318–357.

79. Ann M. Little, "'Race and nature are at the heart of the story': Part I of my interview with *The Republic of Nature* author Mark Fiege," *Historiann Blog*, March 27, 2012, http://www.historiann.com/2012/03/27/race-and-nature-are-at-the-heart-of-the-story-part-i-of-my-interview-with-the-republic-of-nature-author-mark-fiege/.

80. White, "Afterword Environmental History," 105–108. For recent work on Asian Americans and environmental history, see Dorothy Fujita-Rony, "Water and Land: Asian Americans and the U.S. West," *Pacific Historical Review* 76, no. 4 (November 2007): 563–574 and Connie Y. Chiang, "Imprisoned Nature: Toward an Environmental History of the World War II Japanese American Incarceration," *Environmental History* 15, no. 2 (April 2010): 236–267.

81. Nash, *Wilderness and the American Mind*.

82. Peter Coates, *American Perceptions of Immigrant and Invasive Species: Strangers on the Land* (Berkeley: University of California Press, 2007) and Powell, "Vanishing Species, Dying Races."

83. Kimberly K. Smith, *African American Environmental Thought: Foundations* (Lawrence: University Press of Kansas, 2007), 1–3. See also Blum, "Power, Danger, and Control," 250. For a study of environmental thought in India, see also Ramachandra Guha, *How Much Should a Person Consume?: Environmentalism in India and the United States* (Berkeley: University of California Press, 2006).

CHAPTER 21

···

WOMEN AND GENDER

Useful Categories of Analysis in Environmental History

···

NANCY C. UNGER

IN 1990, Carolyn Merchant proposed, in a roundtable discussion published in *The Journal of American History*, that gender perspective be added to the conceptual frameworks in environmental history.[1] Her proposal was expanded by Melissa Leach and Cathy Green in the British journal *Environment and History* in 1997.[2] The ongoing need for broader and more thoughtful and analytic investigations into the powerful relationship between gender and the environment throughout history was confirmed in 2001 by Richard White and Vera Norwood in "Environmental History, Retrospect and Prospect," a forum in the *Pacific Historical Review*. Both Norwood, in her provocative contribution on environmental history for the twenty-first century, and White, in "Environmental History: Watching a Historical Field Mature," addressed the need for further work on gender. "Environmental history," Norwood noted, "is just beginning to integrate gender analyses into mainstream work."[3] That assessment was particularly striking coming, as it did, after Norwood described the kind of ongoing and damaging misperceptions concerning the role of diversity, including gender, within environmental history. White concurred with Norwood, observing that environmental history in the previous fifteen years had been "far more explicitly linked to larger trends in the writing of history," but he also issued a clear warning about the current trends in including the role of gender: "The danger ... is not that gendering will be ignored in environmental history but that it will become predictable—an endless rediscovery that humans have often made nature female. Gender has more work to do than that."[4] Indeed it does.

In 1992, the index to Carolyn Merchant's *The Columbia Guide to American Environmental History* included three subheadings under women. "Women and the egalitarian ideal" and "women and the environment" each had only a few entries. Most entries were listed under the third subheading, "activists and theorists," comprising seventeen names.[5] Nine years later Elizabeth Blum compiled "Linking American

Women's History and Environmental History," an online preliminary historiography revealing gaps as well as strengths in the field emerging "at the intersection of these two relatively new fields of study." At that time Blum noted that, with the exception of some scholarly interest being diverted to environmental justice movements and eco-feminism, "most environmental history has centered on elite male concerns; generally, women's involvement tends to be ignored or marginalized."[6]

One way to measure scholars' responses to the challenge articulated by Merchant, Norwood, White, and Blum, is to count the citations indexed under various topics in the Forest History Society's *Environmental History Bibliography*, a continuously updated electronic database citing books, articles, theses, and dissertations on a wide variety of topics including global climate change, sustainable development, environmental justice, and the depiction of nature in art, film, and literature.[7] The bibliography includes citations to 45,000 published works. In 2013, the search term "trees" generated 10,200 entries, "water" generated 6,087, and "men" generated 5,320. The search term "women" generated 1,675 entries, "gender" generated 268, and "sexuality" only 19. Some entries appeared in more than one category, but even when these multiple listings are included in the final tally, the total for the three latter terms constitutes less than 5 percent of this database. A more encouraging sign is the fact that women and gender are finding their way into major reference sources. One of the thirty-two chapters that make up the 2010 *A Companion to American Environmental History* is devoted to "Gender," and "Ecofeminism" is the subject of one of the thirty-six chapters in the *Companion to Environmental Philosophy*.[8] Although environmental scholars and historians are increasingly investigating topics involving gender, women, and sexuality, such research is still clearly in its infancy. In 2013, in a series of essays entitled "State of the Field: American Environmental History," the *Journal of American History* followed up on its 1990 roundtable. While several contributors note in passing that gender has become meaningful component in environmental history, not one of the eight essays addresses gender with any specificity beyond noting that feminist scholar Donna Haraway's studies illuminate "the particularly gendered and racialized ways science constructed its animal objects."[9] A survey of the existing literature, however, reveals that gendered approaches have the potential to contribute to a more genuinely comprehensive understanding of environmental history.

As differentiated from sex, which pertains to physiology, and sexuality, which focuses on sexual practices and sexual identity, gender refers to culturally defined and/or acquired characteristics. "Gender" has frequently been misused as a synonym for "women" in historical writing. Joan W. Scott's "Gender: A Useful Category of Historical Analysis" stresses that "gender" is "a constitutive element of social relationships based on perceived differences between the sexes" and "a primary way of signifying relationships of power." According to this definition, in which women and men are defined in terms of one another, "no understanding of either [can]...be achieved by entirely separate study."[10] Incorporating gender into environmental history is a complex undertaking, as it by definition does not occur in a vacuum. Race, ethnicity, and class—among other factors—help construct gender roles, and the culture that results can change dramatically over time. Despite these complexities, emerging studies reveal the differences that gender, sex, and sexual identity have made in shaping men's and women's attitudes toward, and relationships with, the environment and each other.

When the new women's history first emerged in the 1970s, many of the early studies emulated the male model of focusing on the "greats" of the past and provided accounts of female leadership in various traditionally male-dominated fields, such as politics, medicine, and literature. Environmental historians followed suit: many of the environmental history studies of the 1970s and 1980s that focused on women examined the contributions of individual female scientists (Alice Hamilton, founder of occupational medicine/industrial toxicology), conservationists (naturalist Caroline Dormon), and nature writers (Mary Austin, *The Land of Little Rain*), with Rachel Carson (*Silent Spring*) by far the most frequently cited and celebrated female environmentalist. However, just as women's history rapidly developed from a rather pale imitation of men's history into a vibrant, rich, and important field in its own right, environmental history broadened its focus to become a vast multidisciplinary field encompassing the entire globe and investigating time periods from the primordial to the present. Women—not just individual female "greats"—increasingly appear, and issues of masculinity and homosexuality are recognized as well. Incorporating this broader context, even the more recent works on individual women are less traditional in their subject matter and make broader contributions to environmental history. For example, Mary Joy Breton's *Women Pioneers for the Environment* (2000) profiles forty-two nineteenth- and twentieth-century women from around the world who broke with their prescribed subservient gender roles to become leading environmental activists.[11] Studies of women in environmental history have also broadened from an emphasis on women activists who consciously worked to protect the environment to include a myriad of gender-based environmental relationships.

The uniqueness and importance of women's roles cannot be fully appreciated, though, unless placed in the appropriate gendered context. Elizabeth Blum warns against the "lack of cross-field knowledge [that] has contributed to isolated, often ahistorical studies of women's involvement in the environment," a concern echoed by Leach and Green.[12] Most gendered analyses strive to give both sexes equal consideration—an effort still too infrequently made in the field as a whole. Because men's roles in environmental history have traditionally received greater scholarly attention overall, this chapter focuses primarily on women (but certainly not to the exclusion of men) in seeking to understand both women and gender as useful categories of analysis within environmental history.[13]

PREMODERN INDIGENOUS PEOPLES

Indigenous peoples serve as the "miners' canaries of the modern world."[14] Because, according to environmentalist Alan Durning, "When the Indians vanish, the rest will follow," much attention has been paid to premodern environmental practices, especially those proven to be sustainable.[15] Worldwide, many environmentalists study indigenous peoples who continue to live lifestyles relatively similar to those lived by their ancestors prior to contact with outside cultures and peoples. There is a growing sense of urgency surrounding such studies, as neighboring peoples' population growth, industrialization, and factory farming threaten those lifestyles by over-fishing, over-hunting, bulldozing, and otherwise depleting or destroying the resources vital to indigenous sustainability.

Although there have been a few attempts to survey the environmental histories of indigenous peoples at the global level, most studies focus on a particular continent, country, or region.[16] The environmental history of Native Americans is in many ways unique, yet some of the most vexing problems that plague its study hinder scholars of other parts of the world as well. Few places on earth remain untouched by modernity, and their environments have changed accordingly. Understanding the role of gender in early Native American environmental history, for example, proves difficult for several reasons, including the fact that in the centuries following the European invasion, virtually no tribe was able to consistently practice pre-contact ways. Moreover, in what is now the United States, there were hundreds of Indian nations whose resource strategies, whether based in fishing, hunting, gathering, or farming, were as diverse and as changeable as the different landscapes they inhabited.[17] It is as dangerous to generalize about Americans in the pre-Columbian period as it is about precolonial Africans, Australians, or any other indigenous peoples.[18] Another barrier to reclaiming the pre-Columbian environmental past based on only fragmentary evidence is that the earliest accounts of Native Americans were written by colonial observers (as was the case with other encounters around the world), who were almost exclusively male and whose own culture, especially their presumptions about gender, strongly colored their perspective on aboriginal ways. English observers, for whom hunting was a recreational activity for aristocrats, were disdainful of the considerable time and energy native men devoted to hunting. English male disapproval intensified when they witnessed activities that they imagined as suitable for men—planting and harvesting—performed primarily by women. Women, according to this view, remained within the local villages caring for the children and doing the "drudgery" appropriate to their subordinate social status, producing the finished goods (especially clothing and food that would not spoil) from the raw materials provided by the men and thereby having little direct impact on the environment.[19]

Women's historians, in tandem with historians of premodern societies, have worked to replace this stereotype with more nuanced understandings of indigenous peoples' gendered relationships with the environment across many cultures.[20] The "grunt" work many missionaries in North America misunderstood as indicative of women's inferior status was in fact a centuries-old attempt at an equitable division of labor, which made it possible for women to feed and care for children while carrying out various tasks communally. Such gendered divisions of labor were rarely rigid. While California Indian men, for example, were the primary hunters and fishers and the women the primary gatherers and food preparers, men sometimes aided in the gathering (such as knocking acorns off oak limbs) and women hunted, fished, and trapped small game.[21]

In more recent years, premodern ecologists have replaced the old stereotypes with the popular glorification of indigenous peoples living totally off the land, but having virtually no impact upon it. Yet in Indian societies throughout North America, men frequently manipulated the environment by burning, hunting, and fishing. Women too manipulated the environment as they provided, via gathering or farming, much of their communities' total food. In areas where Indians farmed, women were usually the primary distributors of the corn, beans, squash, and pumpkins they planted, weeded, and harvested.[22] In Southeastern New England, for example, since about the year 1000, the corn alone produced by women provided about 65 percent of their tribes' caloric input. By planting mixed crops, women shielded the soil from excessive sun and rain and cut down drastically on the amount of weeding that subsequent European farming methods would necessitate, thereby also slowing soil exhaustion.[23]

Like other indigenous peoples, Indians did not live in total harmony with nature.[24] Their resource strategies continued to be largely sustainable, however, even as the individual elements changed over time. In some areas of North America, as in other areas around the world, local conditions were sufficiently harsh to ensure a low population. Among more prosperous tribes threatened by overpopulation, the key to their ability to carry out what William Cronon calls "living richly by wanting little," was that they controlled their numbers.[25] In those tribes, Native American women's greatest environmental impact came not through their gathering, irrigation projects, horticulture, fishing, herding, or their ability to preserve foods. Instead, their greatest single impact came through their nearly universal practice of prolonged lactation. Breast feeding was very common for the first three years after childbirth, but among some tribes it lasted for four years and sometimes even longer. Certainly breast feeding in the first two years had enormous practical benefits, primarily convenience and mobility. It was also valued because it brought decreased fertility. Because Native American women actively sought to control their populations, they routinely nursed their babies past when children could easily thrive on solid foods, and frequently more than twice as long as their European contemporaries.[26]

Along with prolonged lactation, some Native American women, like their European counterparts, also practiced infanticide and abortion.[27] To guarantee population control, breast feeding was sometimes combined, as in the case of the Huron, the Cheyenne, and California's Ohlones, with sexual abstinence, a method also practiced by many indigenous peoples worldwide, including those who lived along the Amazon and within Africa's Congo basin.[28] By carefully controlling their populations, and keeping them below the land's "carrying capacity," Indian women made a crucial contribution to their peoples' ability to live relatively sustainable lifestyles. Indian populations were also periodically checked by other factors, including wars, droughts, and floods. In addition some endured "lean" winters, during which the stores of food intentionally limited by the tribe ensured that the weakest were winnowed out.[29] But these latter factors alone cannot account for the remarkably stable (although larger than previously believed) numbers of Indians estimated to have populated what is now the United States.[30] The contributions made by Indian women were crucial.

Europeans quickly changed the landscape of many of the places they came to domi-nate. European men did so partly based on their patriarchal beliefs that the Bible com-manded them to be the leaders and providers for their families and to "subdue the earth," but also in large part because they needed that landscape to increasingly meet "the demands of faraway markets for cattle, corn, fur, timber, and other goods whose 'values' became expressions of the colonists' socially determined 'needs.'" As early as 1653, the colonial historian Edward Johnson, considering the New England ecosystems, marveled at the fact "that this Wilderness should turn [into] a mart for Merchants in so short a space [with] Holland, France, Spain, and Portugal coming hither for trade."[31] European methods of farming and concepts of progress quickly supplanted the ways of life of those Indians across North American who managed to survive the vast waves of death brought by exposure to European diseases to which they had no immunities. In much of North America and in other colonized places around the globe, many tradi-tional links between gender, sexuality, and environmental sustainability were shattered.

Native peoples' environmental knowledge and skills nevertheless made them valu-able as guides (Sacajawea's work as guide and interpreter has produced a very large literature, especially for children) and as key contributors to the newcomers' ways of living on the land.[32] When Indians in the Great Lakes region were drawn into the fur trade, native men were often held in greater esteem than were native women, and not only because of the patriarchal traditions of the European invaders.[33] Men's traditional hunting skills were highly valued in the fur trading economy. (An exception was the Great Plains, where women's labor was crucial to transforming raw bison hides to mar-ketable commodities.)[34] Yet Indian women did not submit passively to the obliteration of their sustainable practices and the other incursions into their peoples' traditions, but instead practiced cultural self-determination in a number of powerful ways, including through their reproductive choices.

Susan Sleeper-Smith and Sylvia Van Kirk have done extensive work on gender in the fur trade.[35] Native women in the western Great Lakes region frequently married or were otherwise paired with French fur traders. Such marriages á la façon du pays (in the custom of the country) facilitated fur traders' commerce by enveloping it in native customs of reciprocal exchange among kin. These kin networks facilitated men's access to valuable pelts, fueling an industry so lucrative that the population of fur-bearing animals, including seals, otters, and beavers, was rapidly depleted all across North America, empowering women to negotiate positions of prominence. Women strove to maintain their Indian identities largely through the same extensive kin networks.

In western lands under mission control, some Indian girls and women were converted to Christianity and inculcated with European gender norms, but their skills as farmers and herders remained in demand. Severe conditions left native men and women with few options. Both provided forced labor, but they were not completely powerless.[36] Indian deaths outstripped births not only due to disease, inadequate food supplies, and over-work (especially when Indians were forced to grow crops in arid climates), but because women consciously limited their reproduction through sexual abstinence, abortion, and infanticide. In California in the second half of the nineteenth century, as Albert Hurtado has shown, large numbers of native women seemed to disappear from native populations, as they were increasingly drawn into Anglo society as domestic laborers.[37]

ENSLAVED AFRICANS

For centuries, many of the dramatic changes made to colonized environments were frequently carried out by the people the invaders brought with them as forced labor to mine (in Brazil, for example) or to clear or drain and plant land. As was the case in many slave-owning societies, slave owners in the Americas depended on trade rather than natural increase to maintain their supply of slaves, and most, valuing size and strength, preferred to buy men. Such was particularly the case in Brazil and in the sugar islands of the Caribbean, where the most lucrative plantations were located. Mainland North America, a relative backwater in the New World plantation economy in the eighteenth century, had a lower gender imbalance in the slave trade. Nevertheless, only as the overseas trade was legally phased out did the numbers of enslaved women begin to approach the number of enslaved men, with the sex ratio evening out around the 1740s.[38] Forced laborers of both sexes used environmental knowledge gained in both Africa and North America to not only ostensibly do the work their masters required of them, but also to improve the quality of their own lives.[39]

Enslaved women gathered naturally growing herbs, roots, and berries for both dietary and medicinal purposes, and sometimes joined men and boys in fishing and in trapping and hunting for small game.[40] Women who were granted garden patches grew food that was used to partially (sometimes nearly wholly) provide for their families' diet, and in some instances to sell or trade. The mixed crops women grew resulted in far less soil exhaustion than did the monocrops of their owners. In her study of female slaves' perceptions of wilderness, Elizabeth Blum explores how the ability to live off the land in local swamps and woods allowed runaway slaves, male and female, to survive before making their way north to freedom or, as happened far more frequently, returning to their owners either through resignation or coercion.[41]

The enslaved also used their environmental knowledge to subtly undermine the institution that bound them. Studies by Sharla Fett, Liese Perrin, and Marie Schwartz emphasize the knowledge of abortion and contraception enslaved peoples brought from Africa and the Caribbean.[42] Methods used previously to control their local populations to their own benefit were adapted in their new situations as forms of resistance to slavery. The demands of forced field labor precluded most enslaved women's ability to breastfeed with sufficient frequency to suppress ovulation. Instead, they used the environmental knowledge gained in their homelands concerning the use of a number of medicinal plants also available in the New World as abortifacients (especially cotton root). Such practices, which were severely punished when detected, not only reduced their masters' supplies of new generations of forced laborers, but also served as a kind of strike, since reproduction was considered a central function of enslaved women, contributing to higher prices for women considered to be promising "breeders."[43]

Enslaved women used their environmental knowledge concerning production to combat the injustice of slavery in other ways as well. Because most slave owners shared the gendered perception that skilled work was the natural domain of men, enslaved men and boys worked as carpenters, smiths, teamsters, and the like, leaving a disproportionate amount of field work to women and girls. Figures for individual farms and plantations vary widely, but in rough numbers, 13 percent of enslaved males compared to 4 percent of females carried out skilled labor in the 1740s. Fifty years later the percentage of female slaves performing skilled work had increased to 6 percent (and higher on larger estates), but the figure for males was 26 percent.[44] By 1850, 89 percent of female slaves worked in the fields, still outnumbering the 83 percent of males relegated to field work.[45] While slave owners may have considered the fieldwork carried out by women to be unskilled labor left to them by default, they nevertheless benefited from the gendered expertise of female field hands. Daniel Littlefield and Judith Carney show that women's agricultural expertise in rice, indigo, corn, and cotton production was a result of the specialized knowledge and hand tool experience garnered in their native lands. In turn, observes Carney, "subordinated peoples used their own knowledge systems of the environments they settled to reshape the terms of their domination."[46]

Agricultural experience and wisdom combined with gender roles to empower enslaved women. All family members were, of course, subject to the will of the master, but within the cabins of the enslaved, women generally enjoyed greater gender equity than did their white counterparts. Limiting their masters' supplies of new slaves was only one of many forms of resistance to white tyranny. Another was the passive refusal of field workers to fertilize increasingly depleted cotton fields or to terrace untilled hillsides. Field workers rarely refused outright to increase their masters' crop yields, but the expensive tools required for the work were ill used, forever breaking or disappearing mysteriously. Costly fertilizers were applied improperly. So widespread were these actions that most slave owners preferred to view them as further proof of their slaves' inherent laziness and stupidity rather than as calculated forms of resistance, and quickly abandoned terracing and fertilizing efforts.[47] Although some planters did rotate corn with cotton and others had some success with fertilizers, the actions of enslaved field workers, disproportionately female, hastened the necessity for the geographic expansion of slavery.[48] (A similar, although more flagrant refusal to terrace white-owned hillsides occurred in Kenya in 1948–1949. Known as "The Revolt of the Women," local workers undermined colonists' efforts to stem soil erosion.)[49]

MEN AND WOMEN ON THE AMERICAN FRONTIER

Carolyn Merchant's *The Death of Nature: Women, Ecology, and the Scientific Revolution* (1980) reveals the vast changes in perceptions of both women and nature brought about by the Scientific Revolution of the sixteenth and seventeenth centuries.[50] In the centuries that followed, gender, race, class, and ethnicity all played a role in Americans' relationships with frontier environments. Where wilderness posed a threat to survival as well as a resource to be exploited, strict gender lines often blurred, as the labor of both men and women was necessary to clear or drain land, plant and harvest, and tend to animals as well as to prepare and preserve food. Only when a certain level of settlement and prosperity was achieved could families afford to have women return to their prescribed domestic sphere within the home. Even on well-established farms, however, kitchen gardens remained the province of women and ensured that women spent considerable time outdoors attuned to the weather and the changing seasons.[51] Susan Scott Parrish has revealed colonial attitudes and perceptions of nature reflected in the writings of eighteenth-century American women interested in natural history, and Joan Jensen offers a particularly insightful study of farm women's contributions to the growth and development of the nation.[52]

Much of the research published in the 1970s and 1980s on gender on the American frontier focused on pioneers from the east, some recently migrated from Europe, who made the overland journey to settle on the Great Plains or in the Far West in the nineteenth century. The degree of influence held by the evolving prescribed gender spheres (the male worlds of outdoor activities, money, and politics described in E. Anthony Rotundo's *American Manhood* (1994) and Michael Kimmel's *Manhood in America* (2011); the female indoor worlds of domesticity discussed—and contested—by scholars including Barbara Welter, Gerda Lerner, Mary Ryan and Suzanne Lebsock) on the activities and attitudes of the newcomers to the west remains the subject of lively debate.[53] According to the prescribed ideals, men, the natural protectors and the material providers for their families, gloried on the frontier where they were made (or broken) by their ability to wrest a fortune, or at least a livelihood, from the land in the form of precious metals, crops, or livestock. The transformation of the land wrought by such endeavors was not seen by men as exploitative, destructive, or shortsighted, but rather as their right and their familial and even religious obligation. In the words of Reverend Thomas Starr King, in an 1862 address before the San Joaquin Valley Agricultural Society, "The true farmer is an artist. He brings out into fact an idea of God." Men in environments that did not naturally lend themselves to carrying out God's idea were not exempt from their obligation. Even the desert like valley floor of central California must be transformed because "[t]he *earth is not yet finished*...It was made for grains, for orchards, for the vine, for the comfort and luxuries of thrifty homes." How must such a transformation take place? "[T]hrough the educated, organized, and moral labor of men."[54]

Chad Montrie's "'Men Alone Cannot Settle a Country': Domesticating Nature in the Kansas-Nebraska Grasslands," traces the evolution of the relationship between gender and women's roles (especially outdoor versus indoor labor) during the transition from homesteading to settlement.[55] Some authors, including Lillian Schlissel, focus more on the women who accompanied their husbands to the frontier only reluctantly, and who were eager to recreate in the west the spheres of domestic influence they had maintained by virtue of their gender in the east. Others, including John Mack Faragher, highlight the similarities between men and women, and, like Sandra Myres, feature wives who were either eager to partner with their husbands in the business of homesteading, or, because the Homestead Act of 1862 was not gender specific, sought homesteads in their own right.[56]

Gender values prescribed that men were responsible for providing for their families materially, while women were charged with "civilizing" the frontier. Women insisted upon the establishment of schools and churches rather than saloons, and turned crude dwellings into homes.[57] Women on the western frontier commonly brought beauty and diversity to their homes and gardens through the exchange of seeds and cuttings with family members back east, transforming a strange environment into one more inviting and familiar, and creating a tangible tie with people and environments seemingly worlds away.[58] Richard Westmacott and Dianne Glave have studied the aesthetics and cultural significance of traditional gardening practices of rural African American women.[59] Black and white pioneer women also sought seeds and roots to augment local plants (including berries, barks, roots, and flowers) to be used for medicinal purposes, homeopathy being another art practiced by women within the domestic sphere centered on nurturance.

Sheryll Patterson-Black's *Western Women in History and Literature* (1978) notes the preponderance of literature focusing on homesteading families, including Willa Cather's *O Pioneers!* and *My Ántonia* as well as Ole Rolvaag's *Giants in the Earth*.[60] Perhaps the most influential has been Laura Ingalls Wilder's extremely popular *Little House* series, which reveals not only the huge divide between proper men's work and proper women's work, but also how frequently that divide was breached in order to secure the family's survival.[61] Memoirs including Hal Borland's *High, Wide, and Lonesome: Growing Up on the Colorado Frontier* reveal the hardships settlers faced when families attempted to transform western environments beyond what nature would allow.[62] Archival repositories all across the country contain letters and diaries of pioneering and rural settlers that continue to enhance understanding of how race, place, time, and ethnicity combined with gender to shape family members' relationships with each other and their environments.[63]

Glenda Riley's sweeping *Women and Nature: Saving the "Wild West"* (1999) looks more specifically at western women's unique, gender-based contributions to environmental protections.[64] Other recent studies look at western expansion's environmental and cultural impact on the peoples already on the contested terrain of "new" lands. Scholars who emphasize the role of race and ethnicity featuring Native American and Mexican American women include Vicki Ruiz, S. J. Kleinberg, Deborah Kanter, Paula Nelson, and Katherine Benton-Cohen.[65] Theda Perdue and Bruce White have produced particularly useful studies of the gendered roles played by Native Americans in the ongoing fur trade.[66]

Masculinity, Hunting, and Conservation in Empire

Indigenous people, plants, and animals throughout much of the world experienced profound change as the result of colonization.[67] During the Victorian and Edwardian eras, big game hunting was celebrated as excellent training for future soldiers of the British Empire because it fostered fearlessness, coolness under pressure, and aggression.[68] All around the world, including in North America, as Greg Gillespie demonstrates in *Hunting for Empire* (2008), these markers of masculinity made big game hunting an integral part of European colonial culture.[69] Elite colonists and their visitors who hunted lions and tigers in Africa and Asia depended on a large traveling party of local porters and attendants to see to their every comfort. These hunts sometimes occasioned grudging respect for local trackers, but mostly they affirmed the colonists' domination over the local people and non-human environment.[70] Many of these hunters also self-identified as scientists and students of natural history, donating numerous specimens to regional museums and, especially, the Natural History Museum of London.[71] Others hunted so indiscriminately as to be described as undermining the English ideal of masculinity. In the wake of concerns about the impact on hunters as well as alarm over the rapid reduction of big game, humanitarian concerns began to enter the public discourse on hunting, initiating a preservationist movement.[72]

THE ROLE OF GENDER IN NEWLY URBANIZED, INDUSTRIALIZED SOCIETIES

By the 1850s nearly a fifth of the American population was living in towns and cities. As the ranks of middle-class urbanites swelled prior to the Civil War, new notions of gender emerged. As Carolyn Merchant notes, "Men's identities as frontiersmen, explorers, fur traders, and soldiers truncated, while their employment as industrial laborers, mechanics, and businessmen expanded." Outdoor clubs for men formed throughout the latter half of the century to provide reassurances of masculinity through the kinds of tests of male strength and endurance that were no longer part of everyday life. Even the men who were not members of the Sierra Club (founded in 1892) or the Appalachian Mountain Club (1886) could have their masculine imaginations stirred and affirmed by the new outdoor journals, such as *American Sportsman* (1871) and *Forest and Stream* (1873).[73]

Changes in men's identities came to influence the way middle-class Americans defined "true womanhood," or woman's proper sphere. According to the emerging doctrine, commerce and wage work and the moral compromises that accompanied them were presumed to be the realm of men; women were to stay at home, where they could remain immune to the corruptions of urban life and create a healing domestic atmosphere. Women's prescribed sentimental, selfless, and nurturing natures presumably rendered them unfit to compete in the urban world, but ideally suited to the domestic concerns that preoccupied them.[74] These prescribed gendered spheres ignored the realities of race, class, and ethnicity that left many Americans unable and often unwilling to carry out these "natural" modes of living. In privileging the middle and upper classes, these constructed identities also in many ways limited the women in those classes to activities inside the home. They nonetheless ultimately encouraged the notion of women as uniquely qualified and obligated to seek environmental reform.

In a 1915 newsletter article entitled "The College Woman and Citizenship," the Syracuse University Alumni Club reminded its members that "The woman's place is in the home. But today, would she serve the home, she must go beyond the home. No longer is the home encompassed by four walls. Many of its important activities lie now involved in the bigger family of the city and the state."[75] During the Progressive Era (circa 1890–1917), many middle-class female reformers, primarily but not exclusively white, claimed that male domination of business and technology had resulted in a skewed value system.[76] Profit had replaced morality, they charged, as men focused on financial gain as the sole measurement of success, progress, and right. Men profited, for example, by selling impure food and drugs to an unsuspecting public. In the factories whose profits turned a few individuals into millionaires, working-class men, women, and children toiled long hours for low wages in unsafe conditions, only to go home to urban squalor. Non-renewable resources were exploited with no thought to their conservation.[77] In the face of so much gross injustice, environmental and otherwise, women, long prescribed to be the civilizers of men, staged protests and organized reform efforts. The nature of their proposed solutions, including resource conservation and wilderness preservation, reveal new insights into the power of gender in early industrialized society.

GENDER IN AMERICAN PROGRESSIVE-ERA WILDERNESS PRESERVATION AND RESOURCE CONSERVATION

According to Lydia Adams-Williams, who promoted herself in 1908 as the first woman lecturer and writer on conservation, "Man has been too busy building railroads, constructing ships, engineering great projects, and exploiting vast commercial enterprises" to consider his environmental impact.[78] Adams-Williams claimed that it fell to "woman in her power to educate public sentiment to save from rapacious waste and complete exhaustion the resources upon which depend the welfare of the home, the children, and the children's children."[79] Many women agreed that, in the words of environmental historian Carolyn Merchant, "Man the moneymaker had left it to woman the moneysaver to preserve resources."[80] Nature, in other words, had been denied nurture.

Some men risked being scorned for holding "unmanly" views by promoting resource preservation. Adam Rome frames "'Political Hermaphrodites': Gender and Environmental Reform in Progressive America," around a contemporary cartoon rendering wilderness preservation icon John Muir both impotent and feminine.[81] In the drawing, Muir is elaborately clothed in a dress, apron, and flowered bonnet as he fussily (and fruitlessly) attempts to sweep back the waters flooding Hetch Hetchy Valley. Gifford Pinchot, who became first chief of the Forestry Service, escaped such denunciations by making it clear that even as a conservationist with an abiding love for the outdoors, he still saw nature as in the service of men. "Wilderness is waste," he infamously proclaimed, "Trees are a crop, just like corn." He dedicated his agency to "the art of producing from the forest whatever it can yield for the service of man."[82] Theodore Roosevelt, too, framed his support for conservation in terms of benefitting human rather than non-human nature. In 1907 he addressed both houses of Congress to gain support for his administration's effort to "get our people to look ahead and to substitute a planned and orderly development of our resources in place of a haphazard striving for immediate profit."[83]

It is a testament to Roosevelt's hyper-masculine persona (detailed in *Rough Rider in the White House: Theodore Roosevelt and the Politics of Desire* [2003], by Sarah Watts), that he could so successfully sow the seeds of conservationism within a male population concerned about the encroachment of federal control into state sovereignty, and deeply suspicious of any argument even tinged with sentimentality.[84] Writer George L. Knapp was one of many men who remained unconvinced, terming the call for conservation "unadulterated humbug" and the dire prophecies "baseless vaporings." He preferred to celebrate the fruits of men's unregulated resource consumption: "The pine woods of Michigan have vanished to make the homes of Kansas; the coal and iron which we have failed—thank Heaven!—to 'conserve' have carried meat and wheat to the hungry hives of men and gladdened life with an abundance which no previous age could know." According to Knapp, men should be praised, not chastened, for turning "forests into villages, mines into ships and skyscrapers, scenery into work."[85] Such beliefs were reinforced in the press. The Houston *Post*, for example, declared, "Smoke stacks are a splendid sign of a city's prosperity," and the Chicago *Record Herald* reported that the Creator who made coal "knew that smoke would be a good thing for the world." Pittsburgh city fathers equated smoke with manly virtue and derided the "sentimentality and frivolity" of those who sought to limit industry out of baseless fear of the by-products it released into the air.[86]

The notion of a strict gender divide over the need for wilderness preservation and resource conservation is, however, belied by the number of male leaders in the nascent environmental movements in the early 1900s, with Roosevelt, Pinchot, and Muir topping the list. Hunting had been long linked with masculinity and, by the turn of the century, was increasingly the domain of the elite.[87] Roosevelt and Pinchot were both founding members of the American Bison Society in 1905, an organization composed almost entirely of men who associated the bison with the imagined qualities of frontier masculinity.[88] Similarly, other sportsmen lobbied for animal conservation laws in order to sustain a supply of game for sport hunting. Scholars including Sarah Watts, Peter Bayers, Tina Loo, and Gail Bederman all emphasize, to varying degrees, the importance of hunting in shoring up a waning sense of masculinity.[89] Andrea Smalley, however, points to evidence from sportsmen's periodicals to suggest that men during this period did not perceive the sport as an exclusively masculine past time. Smalley emphasizes women's frequent appearance in magazines and journals that popularized sport hunting's image, revealing how gender played a complex role in turn-of-the century "blood sports." She argues that elite sportsmen encouraged rather than shunned women's involvement "as a way to promote their political agenda and to legitimate their conception of 'correct' hunting." Women's involvement, as hunters and as hunting reformers, upheld rather than undermined Victorian notions of respectable recreation.[90]

Scholarship on the role of Progressive-Era women who dedicated themselves to wilderness protection and resource conservation is particularly rich, built on the foundation established by Suellen Hoy, Carolyn Merchant, Maureen Flanagan, and Nancy C. Unger.[91] Women, prohibited from voting and shut out from so much of the world of business and power, found an outlet for their energies in environmental activism. This was an arena in which their prescribed gender role was a credential rather than a handicap. Accounts of Progressive-Era scholars (including botanists and nature writers) as well as individual activists have been augmented by studies of groups, especially women's organizations and clubs, although women also played significant roles in some environmental agencies open to both sexes, such as the Audubon Society and the Sierra Club.[92]

THE CAMPAIGN TO SAVE BIRDS

Women's conservation efforts extended to the protection of both domesticated and wild animals, particularly birds. In 1886, nature writer Sarah Orne Jewett published the short story "A White Heron," an early, impassioned plea for plume-bird conservation.[93] By 1910, the activities of the Massachusetts Audubon Society (established in 1896 by Boston socialite Harriet Lawrence Hemenway in response to the slaughter of the heron exposed by Jewett) were augmented by those of the 250 women's clubs that were active nationwide, dedicated specifically to the protection of birds and plants. Resistance to the ongoing extermination of entire bird species, women's club leader Marion Crocker insisted, was vital to the preservation of the human race. Before the widespread use of insecticides following World War II, birds provided virtually the only check on the insect population that threatened crops prior to harvest. Warned Crocker, "If we do not follow the most scientific approved methods, the most modern discoveries of how to conserve and propagate and renew wherever possible those resources which Nature in her providence has given to man for his use but not abuse, the time will come when the world will not be able to support life and then we shall have no need of conservation of health, strength, or vital force, because we must have the things to support life or else everything else is useless."[94] Crocker's approach was unusual for a female Progressive-Era reformer, because although she appealed specifically to women, she chose not to play the maternal card. She stressed the necessity of birds in interrelated plant and animal kingdoms, reminding her listeners of the crucial roles birds played in agriculture and pest control. "This is not sentiment," she stated flatly. "It is pure economics."[95]

Speeches on the floor of the US Senate fueled Crocker's and Lydia Adams-Williams's assertions that men could not be trusted to carry out the crucial task of saving the birds and, ultimately, humanity. Missouri's James A. Reed responded to a 1913 bill introduced to protect migratory birds by asking: "Why should there be any sympathy or sentiment about a long-legged, long-beaked, long-necked bird that lives in swamps and eats tadpoles?" He urged, "Let humanity utilize this bird for the only purpose that the Lord made it for…so we could get aigrettes for the bonnet[s] of our beautiful ladies."[96] To the horror of those who saw clearly the crucial role that pest control provided by wild birds played in national and international economies and ecosystems, Reed dismissed the protection of birds as trivial, born out of "an overstrained, not to say maudlin sympathy for birds born and reared thousands of miles from our coast."[97]

The notion that women were especially suited to carry out the campaign to save the birds was reinforced by the photographs and stories of birds by popular novelist and nature author Gene Stratton Porter. Most of Porter's nine nature books featured her signature close-up photographs of birds in their natural habitat. Porter claimed that the female domestic sphere was responsible for the patience, sympathy, and attentiveness necessary to capture those photos. All Porter's books suggested women's special affinity with nature, reinforcing that it was their obligation to be at the vanguard of bird preservation.[98]

Women reformers sought legislation protecting the birds but took more immediate action as well. In describing the practice of using birds' bodies to decorate women's hats, the chair of the conservation committee of the Colorado Federation of Women's Clubs observed, "Each beautiful head, wing, or breast mutely declar[es] the cruel death of the bird—the mercenary spirit of men and the vanity of women."[99] Some bird protectionists deemed feather wearing to be the antithesis of true womanhood, for it "demoralized and degraded womankind and made a travesty of the better instincts of motherhood." Most admonitions appealed to maternalism. Aigrettes, for example, were "harvested" during the breeding season, when the feathers were at the height of their beauty, leaving the parents dead and the young to die of starvation. "Remember, ladies," urged a California Federation of Women's Clubs newsletter, "that every aigrette in your hat costs the life of a tender mother."[100]

The campaigns of various women conservationists ultimately led to a variety of legislative successes, indicated by the plea from a Colorado legislator to the president of the General Federation of Women's Clubs: "Call off your women. I'll vote for your bill."[101] In October 1913, a new Tariff Act outlawed the import of wild bird feathers into the United States, and in 1918 the Migratory Bird Treaty Act established protections for birds that migrated between the United States and Canada. Women continued to wear hats, but milliners throughout the United States and Europe bowed to the legal and societal pressures to dramatically reduce the use of feathers as primary decoration, although they did continue to use naturally shed feathers, particularly ostrich and peacock. Thus, prior to achieving suffrage, women were able to successfully wield legislative influence and, by preserving millions of birds, protect complex and vital environmental relationships from ruin by a powerful American industry. Using similar rhetoric and techniques, women worked to protect forests, lakes, rivers, and a host of other natural resources. Their language was not always conciliatory. Clara Bradley Burdette, the first president of the women's California Club, spoke plainly of the gendered divide that existed across the nation on issues of natural resource conservation: "While the women of New Jersey are saving the Palisades of the Hudson from utter destruction by men to whose greedy souls Mount Sinai is only a stone quarry...the word comes to women of California that men whose souls are gang-saws are mediating the turning of our world-famous Sequoias into planks and fencing worth so many dollars."[102]

WOMEN AS MUNICIPAL HOUSEKEEPERS

While the much celebrated "John Muirs, Gifford Pinchots, and Teddy Roosevelts of the conservation movement gave little mind to the quality of urban life," lesser-known and frequently female activists, including Alice Hamilton, Jane Addams, and Ellen Swallow Richards, "struggled with the blight of pollution, health hazards, and the physical degradation" undermining urban homes and work places.[103] Women of middle- to upper-class backgrounds were leaders in urban organizations promoting reforms including civic cleanliness and sanitation, smoke and noise abatement, and pure food and drugs, making clear the "absolute necessity of combating health hazards and pollution for the safety of all citizens."[104] Journalist Rheta Childe Dorr called community "Home," deemed city dwellers "the Family," and public schools "the Nursery," then added, "And badly do the Home and the Family and the Nursery need their mother."[105] "Women are by nature and training, housekeepers," asserted handbill author Susan Fitzgerald, urging, "Let them have a hand in the city's housekeeping, even if they introduce an occasional house-cleaning."[106]

Women's educational programs to promote public health ranged from persuading citizens not to spit on city sidewalks to alerting tenement dwellers to the dangers of lead poisoning. They also addressed concerns specific to women in economically oppressed neighborhoods, revealing the hidden environmental hazards in many women's occupations and promoting pure milk, healthful food preparation, and proper care of infants and children. Labor activist Rose Schneiderman railed against hazardous workplace environments where property was held so dear and human lives, especially the lives of "working girls," so cheap that tragedies like the Triangle Factory fire (which killed 146 women, mostly young and Jewish, in New York City) were commonplace. However, Schneiderman's immigrant and working-class origins, as well as her emphasis on corporate responsibility for urban suffering, set her apart from most of her sister reformers who were bent on eliminating urban environmental ills. While

Schneiderman defended the rights of working women, many middle-class female reformers expressed their concern for these working "girls" as the future mothers of the race, arguing that "the greatest danger was not to the 'girls' but to 'racial vitality' in the form of 'nervous exhaustion,'" ultimately resulting in "undervitalized" children.[107]

The middle- and upper-class status of white early urban environmental reformers frequently led them to not only ignore or neglect issues of class and race, but to openly exhibit hostility to various ethnic groups and people of color, judging them as inferiors who were as much the cause as the victims of disease and sanitation problems. However, the urban environmental historian Martin Melosi concludes, "The fairest assessment to make about the turn-of-the-century urban environmentalism is that it provides a partial legacy for modern environmental justice activists, rather than no legacy at all."[108] The role of gender and sexuality in toxic urban environments is evident in this early period, as women were recognized both as uniquely affected by urban environmental dangers (at home and in the workplace) and as uniquely qualified to offer relief from some of those burdens.

CHILDREN'S OUTDOOR ORGANIZATIONS AND THE GENDER DIVIDE

As Nancy C. Unger, Susan Schrepfer, and Ben Jordan have shown, children's organizations formalized early in the twentieth century codified a gendered approach to the values to be found in non-human nature.[109] The Boy Scouts of America began in 1910, augmenting the ranch camps in the Far West that promised to inculcate urban boys with the traditional masculine frontier values and skills that were quickly disappearing from American cities. When their daughters were excluded from such experiences, some women, themselves barred from membership in many outdoor associations, began creating outdoor activities and organizations for girls.

Charlotte Farnsworth, preceptor of New York City's prestigious Horace Mann School, helped to establish the Camp Fire Girls in 1911. Like many other youth group organizers, Farnsworth embraced outdoor activities because she believed they would reinforce, rather than challenge, the separate gender spheres. Farnsworth stated unequivocally that girls "are fundamentally different from boys in their instincts, interests, and ambitions." Luther Gulick, another key cofounder of the Camp Fire Girls, spent twenty years investigating anatomy, physiology, psychology, ethics, and religions in his effort to understand what it meant to be "manly" and "womanly," and saw the Camp Fire Girls as a "clearer vision of this question." He believed that to copy the Boy Scouts "would be utterly and fundamentally evil. ... We hate manly women and womanly men, but we all love to have a woman who is thoroughly womanly, and then adds to that a splendid ability of service to the state." Gulick put a new spin on the traditional perception of womanhood, claiming that "women had acquired the undesirable trait of independence because they had been sequestered in their homes" and therefore needed to be taught collective obedience to best prepare for gainful employment, efficient homemaking, and public service.[110] Until 1943, although women served on the board of directors of the Camp Fire Girls, all of the executive directors were men who shared some version of Gulick's basic belief that the organization's primary goal was to discourage female independence. They sought instead to imbue girls with the values and skills that would allow them to excel in women's traditional spheres. Members were routinely reminded in one signature song that "love is the joy of service so deep that self is forgotten."[111]

A year after the founding of the Camp Fire Girls, Juliette Gordon Low established the Girl Scouts of America. Low was convinced of the value of sports for women, the advantages of outdoor exercise, and the wisdom of preserving the environment. Boy Scouts and Girl Scouts were taught the value of routine, patriotism, and skills necessary to outdoor living, but they were expected to learn very different lessons from their activities and study of nature. According to the original handbook for the Girl Scouts, *How Girls Can Help Their Country*, "The Scout movement, so popular among boys, is unfitted for the needs of girls." A different system was needed to give "a more womanly training for both mind and body." A Girl Scout's first duty was to "Be Womanly," for "none of us like women who ape men." Scouting adhered to a strict gender divide: "For the boys it teaches *manliness*, but for the girls it all tends to *womanliness*. ... If character training and learning citizenship are necessary for boys, how much more important it is that these principles should be instilled into the minds of girls who are destined to be the mothers and guides of the next generation."[112]

To Boy Scouts the outdoors was a stage on which to rehearse manhood, while "one of the most important principles to be instilled" in Girl Scouts was "strict and prompt obedience to laws and orders." Where Boy Scouts were taught to be aggressive in order to become providers and fighters, Girl Scouts were told to go about their business "quietly and gently," to "never draw attention to themselves unnecessarily," to display "moral courage," and "to never marry a man unless he is in a position to support you and a family." The first wish of the Girl Scout was "to make others happy."[113] The task of the Girl Scout was to become a proper mother to the next generation of workers, fighters, and scouts. Even something as seemingly gender-neutral as the campfire, the heart of communal rituals for both sexes, held entirely different meanings for children in scouting organizations. Boys were told that fire stood for the camaraderie of the battlefield, factory, and office.[114] Girls learned that fire represented hearth and home; without the fire of domesticity brought by woman's "magic touch," a house is "dark, and bare and cold."[115]

Girls Scouts were told that they were the natural leaders in conservation: "Women and girls have it infinitely more in their power than men have to prevent waste. ... The real test of a good cook is how little food she wastes." Trained to think about future generations, Girl Scouts were urged to apply to natural resources the principles of conservation practiced at home, recognizing that "in this United States of ours we have cut down too many trees and our forests are fast following the buffalo."[116] After an initial emphasis on dominating the environment by routinely mandating activities like chopping down trees, Boy Scout leaders acknowledged that women were correct that nature needed to be conserved, but they stereotyped women as too sentimental and selfish toward nature to conserve it properly. In 1916, National Boy Scout commissioner Dan Beard celebrated the Boy Scouts' dedication to bird protection, contrasting it to the actions of women, who sought only the "upholstered skins of these poor birds as ornaments for their hats," ignoring the leadership women had provided in the Save the Birds campaign of the previous decade.[117]

While the Boy Scouts were taught that conservation was the rightful domain of the male sex, girls in outdoor organizations were learning about females' special powers, abilities, and rights. Girls' alleged innate qualities left them uniquely qualified—and obligated—to conserve, protect, and defend parks and forests. Camp Fire Girls earned special honors when they contributed to their community via street cleaning; beautifying yards; conserving streams, birds, trees or forests; or improving parks and playgrounds. Girls in scouting organizations eagerly joined other groups dedicated to nature appreciation, like hiking clubs and urban improvement societies promoting sanitation and health education. The natural nurturers, women and girls who contributed to the uplift of society and the protection of the natural world would themselves be nurtured. This message found ready adherents, and in its first ten years membership in the Camp Fire Girls jumped from 60,000 to 160,000. The Girl Scouts of America experienced a similar explosion, with membership rising from 70,000 at the end of the 1910s to 200,000 one decade later. For generations, both organizations (as well as a multitude of private and civic summer camps) gave girls outdoor experiences and fostered their environmental awareness.[118]

WOMEN AND NATURE WRITING

Carolyn Merchant argues that male and female conservationists bridged the gender divide between 1880 and 1905 through a "gendered dialectic," a kind of call and response that culminated in "men and women working together to form Audubon societies and to pass legislation to preserve avifauna."[119] This bridging of the gendered divide proved temporary, however. Although male environmentalists were gratified by the moral authority women's activism brought to conservation and preservation concerns, Adam Rome has traced the ultimate rejection of the female incursion into the world of masculine authority.[120] Men feared the homophobic rhetoric that linked the sentimentality and emotionalism associated with women with effeminacy.[121] As such smear tactics ultimately weakened many progressive reforms, including environmentalism, men eased or forced women out of positions of authority. Susan R. Schrepfer's *Nature's Altars: Mountains, Gender, and American Environmentalism* (2005) introduced many of the threads pursued by Rome. Schrepfer shows how, between the 1860s and the 1960s, American men and women, primarily but not exclusively of the white middle class, viewed mountains through the very gendered lenses prescribed by their particular era, and responded accordingly. She notes that, moving beyond the Progressive Era, women were not only pressured into resigning from the kind of leadership positions detailed in Rome's study, but into quitting various outdoor activities as well. Women's participation in mountaineering did not disappear entirely, but declined significantly in the late 1930s, as did their leadership positions in the Sierra Club and other alpine organizations. Some found acceptance for their outdoor activities by forming leagues and societies promoting wilderness preservation, especially the protection of wildflowers.[122]

Welcome in fewer and fewer branches of the conservation and preservation movements, women nevertheless continued to contribute, frequently through nature writing. Vera Norwood provides an excellent overview of women's predominantly literary contributions. Edited collections by Marcia Myers Bonta and by Lorraine Anderson and Thomas S. Edwards feature the writings of a wide range of American women, while Rachel Stein argues that even women authors not traditionally associated with "nature writing" reinterpreted nature, incorporating alternative conceptions of nature into their works.[123]

THE OUTDOORS AS THE SAVIOR OF MASCULINITY

In much of the industrialized world, technology and urbanization were perceived as alienating men from non-human nature. That change in perception, along with the expansion of women's political authority, posed a threat to traditional notions of power being based in the masculine body's mastery over the natural world. Bryant Simon argues that anxieties about cities, decadence, and the dissolution of manhood that pervaded the last two decades of the nineteenth century and the first two decades of the twentieth were confirmed in the 1930s by the Great Depression. In the United States, New Deal reformers seized on the Civilian Conservation Corps (CCC) "as a way to 'beef up' male bodies and strengthen the state" because "physically weak men, they believed, weakened the nation." Educated men had been building their minds rather than their bodies. Worse still, urban degeneracy had produced a vast urban poor: unwashed, uneducated, and physically unfit. The solution was to place them outdoors "in close communion with beneficent nature," the "wholesome, pure" source of male "virility and toughness." "The greatest achievement of the CCC," according to one administrator, "has not been the preservation of material things such as forests, timber-lands, etc., but the preservation of American Manhood." And as Simon points out, the type of male the CCC boasted of producing bears a striking resemblance to the ideal masculine images celebrated in Nazi Germany, the Stalinist Soviet Union, and other Western nations in the 1930s. The male body, its muscles honed by strenuous activity in the outdoors, represented strength and health—a welcome reassurance in a time of tension and uncertainty.[124]

At the insistence of First Lady Eleanor Roosevelt, residence camps were also created for unemployed women, but female campers were prohibited from reforestation and environmental projects and focused instead on learning and practicing housekeeping skills. As noted by scholar Heather Van Wormer, women could only stay for two to three months in the camps, whereas men were recruited to the CCC for a year. The CCC ultimately employed some three million men, while only 8,500 women experienced life in residence camps. In addition to their room and board, CCC men were paid thirty dollars a month as compensation for their labor, of which twenty-five dollars was automatically sent to their families. Women, seen more as charity cases despite their labor, were "given" an "allowance" of fifty cents a week.[125] Eleanor Roosevelt regretted that more extensive opportunities were not offered for women, but few in government shared her view that women should participate equally in voluntary service and education programs. Even within the Roosevelt administration, the women's camps were referred to derisively as "She She She," a parody of "CCC."[126]

HOMEMAKING, SUBURBIA, AND THE BOMB

Despite male efforts to minimize female power in formal organizations, even full-time homemakers, both rural and urban, continued to see environmental issues as part of their rightful sphere, and still included environmental activism in their various individual, club, and volunteer activities.[127] The shortage of male labor created by World War II brought women into jobs for which they had previously been declared unfit. In particular, work widely available for the first time in sawmills, in logging camps, and in forest management brought women new environmental insights, as well as new authority.[128]

The atomic bombings of Hiroshima and Nagasaki in 1945 and the threat of nuclear warfare in the Cold War that followed forged a new kind of global environmental consciousness. Within the United States the immediate postwar period featured stricter and more rigid gender prescriptions as patriarchy, Christianity, and especially the heterosexual nuclear family were prescribed as not only socially desirable, but politically necessary if the nation was to survive—and to triumph over—the communist menace.[129] The control of nature extended to activities that included developing bigger and more deadly chemical weapons and controlling pests and weeds through poisons.

Women had been gradually easing into the workplace since the beginning of the twentieth century, and moved into wage work in large numbers during the war, but with the end of the conflict, the return of servicemen, and new prosperity, many middle-class women returned to full-time domesticity in the rapidly developing suburbs. Nevertheless, the number of women taking on paid labor increased in the decades following the war, especially among those of the working class. Despite that fact, woman's prescribed proper and natural place was, once again, decidedly not in the workplace but within the home, where her role was to see to the health, happiness, and safety of her husband and children. Yet, as in previous decades, "home" extended into the public sphere. In the local community, women provided much of the unpaid labor in neighborhood schools and houses of worship. Women around the world also participated in far more political activities, including "Ban the Bomb" campaigns and demonstrations, frequently citing their status as mothers and homemakers as their most compelling credential.[130] Even as the Cold War subsided, women continued to lead protests against nuclear power plants.[131] Middle-class women in their thirties or forties who were raising children and were not employed outside the home were, claims one scholar, "naturals" for this protest work, because their role as the primary care-givers to their children had previously involved them in "broad humanistic/nurturing issues," their interactions with other activists were minimally contentious, and their lack of conventional power left them with little to lose.[132]

Some suburban women focused on other environmental hazards. Homemakers, repeatedly urged to conserve during the war years, were now encouraged to consume the many products their husbands' spent their days working to provide. By 1955 half of all American households owned a television—one of the most powerful tools promot-ing what was later termed derisively "growthmania," an obsession predicated on the assumption that "more [more goods, more living space, more people, more profits] is better." The gendering of consumption of the many newly available and heavily adver-tised consumer goods aligned suburban women in particular with environmentally harmful practices. New standards of cleanliness and appearance necessitated a range of chemical compounds inside each suburban home, garage, and tool shed.

Phosphates routinely found in detergents (as well as disinfectants and deodor-ants) unbalanced ecosystems by fostering dangerously prolific marine plant growth. Women encouraged to beautify the inside of their homes and keep them in pristine condition routinely used solvent-based paints, primers, and varnishes that emitted volatile organic compounds, contributing to the destruction of the stratospheric ozone layer and playing a significant role in the creation of the greenhouse effect. And the pesticides and herbicides touted as essential to women's beautifying their homes' exte-riors, especially through the cultivation of colorful flowerbeds, made their way into the groundwater. The result was serious health problems in humans, including disrup-tions in the endocrine system, cancer, infertility, and mutagenic effects.[133] A few crit-ics, including Elizabeth Dodson Gray, began to recognize that rampant consumerism was rapidly depleting natural resources and poisoning the environment, with women uniquely at risk. Gray warned that more chemicals were found in the average modern home than in chemical labs of the past, and that "many homemakers know little about these chemicals and even less about their toxic and polluting effects."[134]

Silent Spring

In 1962, the ecologist Rachel Carson dramatically challenged conventional notions of progress and celebrations of prosperity with *Silent Spring*, and she used some of the prevailing beliefs about gender to give credence to her message. Carson had spent many years working in the Fish and Wildlife Service where, despite her abilities as a biologist, she hit the so-called "glass ceiling," the level beyond which women were rarely promoted. She turned to nature writing, which both made her financially independent (her first book, *The Sea Around Us*, published in 1951, stayed on the *New York Times* bestseller list for eighty-six weeks) and gave her the platform to challenge authority—both the scientific authority that assured the public that chemical pesticides were safe, and, implicitly, the patriarchal authority that relegated women to second-tier positions in professional science. In 1954 Carson proclaimed women's "greater intuitive understanding" of the value of nature as she denounced a society "blinded by the dollar sign" that was allowing rampant "selfish materialism to destroy these things." She also announced, "I am not afraid of being thought a sentimentalist," and stressed the role of natural beauty in human spiritual growth at the same time that she defended the presence of emotion in science and nature writing.[135]

These "female" values were very much in evidence in *Silent Spring*. Chastening "man" for his "arrogant" talk of the "conquest of nature," Carson warned that men's power had not been tempered by wisdom. Specifically, she questioned the governmental fathers' decisions concerning industrial waste and their contributions to the nation's vast reliance upon pesticides, especially DDT, which was effective to a fault, remaining toxic for weeks and months. Following its initial surface application, DDT seeped into the soil and water and ultimately into the food chain, as it was ingested by birds and animals, including humans, in whose tissues it caused cancer and genetic damage. "Future historians may well be amazed by our distorted sense of proportion," Carson warned, asking "How could intelligent beings seek to control a few unwanted species by a method that contaminated the entire environment and brought the threat of disease and death even to their own kind?"[136]

Gender and sexuality influenced the critical response to *Silent Spring*.[137] The many denunciations within the popular press of Carson's work as overly emotional played to the stereotype of women as unscientific and inherently hysterical (a word derived from the Greek word *hystera*, "womb"). The popular press focused on Carson's marital status. She was variously described as "unmarried," "never married," and "a shy female bachelor." Such references served to "desex Carson and brand her as not-quite-woman" in an age in which marriage and motherhood were upheld as woman's highest calling and defined femininity.[138]

Carson's defenders, however, openly defied disparaging arguments that hinged on widely held perceptions of gender and sex. *Silent Spring*, which attacked the government's misplaced and ineffectual paternalism, appeared just one year before Betty Friedan's assault on patriarchy, *The Feminine Mystique*. Many of the women "awakened" by Freidan's work took their first steps toward finding a larger place in the world by responding to Carson's call to question authority. Friedan's urging that women throw off patriarchy contributed to Carson's message that they no longer assume "that someone was looking after things—that the spraying must be all right or it wouldn't be done."[139] One woman reader's praise for Carson denounced the highly touted postwar notion that "Father Knows Best": "'Papa' does not always know best. In this instance it seems that 'papa' is taking an arbitrary stand, and we, the people are just supposed to take it, and count the dead animals and birds."[140]

When President John F. Kennedy's Science Advisory Committee validated Carson's claims about pesticides, her emphasis on the interconnectedness of all life was no longer dismissed as feminine romanticism. The understanding that any disturbance to the web of life has consequences throughout was accepted, by most, as a scientific reality.[141] Through her challenge to authority, Carson made the public aware of attempts by the scientific-industrial complex to manipulate and control nature to the ultimate detriment of all.[142] Her critique of the country's dependence on chemical pesticides has since been widely recognized as one of the most influential books of the twentieth century.[143]

As environmental historian Adam Rome notes, "Carson cultivated a network of women supporters, and women eagerly championed her work."[144] Her message inspired the untold numbers of local grassroots groups and movements that continued to multiply in cities, suburbs, and on college campuses throughout the 1960s. Women cited *Silent Spring* in educational pamphlets and in their letters to editors and petitions to politicians. Individually and in groups, women stepped up their campaigns to ban the bomb, clean up rivers, save forests, and stop pollution.[145] Women's organizations particularly active in promoting environmental awareness and protection included the League of Women Voters, the American Association of University Women, the Federation of Women's Clubs, and the Garden Club of America.

Ecofeminisms

The feminist and environmental movements of the 1960s contributed significantly to the ecofeminist and environmental justice movements of subsequent decades. Although the concept of ecofeminism is grounded in the movements launched to no small degree by the writings of Carson and Friedan, its definition depends on which ecofeminist, scholar, or critic is asked.[146] Ecofeminism unites environmentalism and feminism into a global cause, holding that there is a relationship between the oppression of women and the degradation of nature throughout the world. Some argue that, because of that relationship, women are the best qualified to understand, and therefore to right, environmental wrongs.[147] In most parts of the world, women are the ones who are "closest to the earth," that is, the ones who gather the food and prepare it, who haul the water and search for the fuel with which to heat it. Everywhere, they are the ones who bear children, or in highly toxic areas, suffer miscarriages and stillbirths or raise damaged children. Brazilian ecofeminist Gizelda Castro echoed the sentiments expressed by Lydia Adams-Williams nearly a century earlier: by dedicating themselves to the pursuit of immediate profit, "Men have separated themselves from the ecosystem." Castro concluded that it therefore falls to women to fight for environmental justice and to save the earth.[148]

A variety of mutually exclusive forms of ecofeminism rival for dominance.[149] One branch emphasizes the power of goddess mythology. Scholar Vandana Shiva, for example, presents the precolonial period as India's golden age, when feminine, conservation, and ecological principles predominated, and the forests were "worshipped as Aranyani, the Goddess of the Forest, the primary source of life and fertility."[150] Practitioners of Goddess Spirituality seek to reclaim ancient traditions in which, they assert, a Mother Goddess (rather than a Holy Father) was revered as the great giver of life. Some argue that despite the efforts of the patriarchal Judeo-Christian tradition to eradicate this belief, all women, especially mothers, are the natural guardians of "Mother Earth."

Their horrified feminist rivals counter that these kinds of claims perpetuate old gendered stereotypes and are a violation of the egalitarianism of true feminism.[151] Moreover, such claims are insufficiently grounded in science to be compelling to the non-feminists who make up a majority of the population. In the words of Bella Abzug, congresswoman and cofounder in 1990 of the Women's Environment and Development Organization, "it's OK to show your emotion and come in as a mother...to say that this is going to hurt my children, but it's not good enough."[152] Moreover, this school of ecofeminists insists, nature should not be anthropomorphized into a mother to be protected, but must instead be respected as a non-human, non-gendered partner in the web of life. They argue that women and nature are mutually associated and devalued in Western culture. Because of this tradition of oppression, they argue, women are uniquely qualified to understand and empathize with the earth's plight and to more fairly distribute its resources. According to 1980s activist Donna Warnock, "The road to women's liberation lies not only in ousting patriarchy, but also in rejecting its inequitable and environmentally and socially disastrous production system which is based on man's dominion over women and the earth, and the illusion of infinite resources." Warnock represented many ecofeminists who see the anthropocentrism that is so damaging to the earth as just one strand in a web of unjust "isms" including ageism, sexism (including heterosexism), and racism, all of which must be destroyed in order to achieve a truly just world: "[T]he eco-system, the production system, the political/economic apparatus and the moral and psychological health of a people are all interconnected. Exploitation in any of these areas affects the whole package." The only hope for human survival, Warnock concludes, "lies in taking charge: building self-reliance, developing alternative political, economic, service and social structures, in which people can care for themselves...to promote nurturance of the earth and its peoples, rather than exploitation."[153] Catriona Sandilands-Mortimer, Catherine Kleiner, and Nancy C. Unger have examined the efforts of lesbians to create just such alternative communities.[154]

Melissa Leach and Cathy Green, however, offer pointed warnings about the dangers of romanticizing and oversimplifying women's relationships with nature, both in the past and in the present. They question the accuracy of "primordial harmony," and argue, moreover, that "by essentializing the relationship between women and nature, ecofeminist analyses have represented history in generalized ways which entrap women in static roles."[155] Their article "Gender and Environmental History: From Representation of Women and Nature to Gender Analysis of Ecology and Politics," focuses primarily on women's land use changes as a result of imperial and colonial policies and politics, but it has larger implications and applications as well. They reveal how a narrow ecofeminist focus on "Women, Environment, and Development" that emphasizes a special relationship between women and environment produces "generalized accounts" that "obscure rather than clarify linkages between changing gender relations, ecologies, and colonial science, ideology and policy, and they deploy history to suggest policies which could well prove to be detrimental to women." Specifically, they note the danger in speaking generically of "third world women," which ignores multitude of distinguishing factors such as age, race, marital status, class, caste, ethnic group, and local ecology. They also warn against regarding men and women as dichotomously separate, and urge recognition of the dynamics of gender, social stratification, and environmental change that were influenced by precolonial trade and commerce. They argue, in essence, for specificity concerning both gender and ecology, and the need to "highlight the variability in experiences of change which emerged as different ecological possibilities, relations of land and labor use, and dynamics of marriage and household formation interplayed with regional political issues." Leach and Green conclude by questioning "the extent to which Northern or elite Southern feminists," including themselves, can or even should set feminist research agendas on behalf of others. In view of their concerns about the politics of voice, they cite as a major challenge for future work the examination of "the production of diverse historical representations about a place, produced at different times and by different authors (local women and men, chiefs and commoners; colonial and modern anthropologists, colonial administrators) exploring how these accounts speak to and past each other, and how (as discourses) they had material effects."[156]

THE ENVIRONMENTAL JUSTICE MOVEMENT

The environmental justice movement frequently incorporates aspects of feminism in its efforts to enforce the right to a safe and healthy physical, social, political, and economic environment for all people. Issues of race, class, sexuality, and gender are regularly addressed in environmental justice studies as factors that frequently subject people to injustice, but have also served to unify and mobilize those same people in their struggles against that injustice. Beginning in the 1950s, for example, in a series of actions later denounced as "Plundering the Powerless," mining companies aggressively gutted lands held by Chicanos and especially by Native Americans for nuclear fuel.[157] Native American women established the national organization Women of All Red Nations (WARN) in 1978 to strengthen themselves and their families in the face of ongoing attacks on Indian culture, health, and lands, drawing attention to the fantastically high increase in miscarriages, birth defects, and deaths due to cancer on Indian reservations in areas of ongoing intense energy development including Nebraska, the Southwest, and western South Dakota.[158] WARN's emphasis on the drastic increase in childhood cancers of the reproductive organs (at least fifteen times the national average) made the demands for action by mothers particularly compelling; however, the involvement of many WARN members was motivated by a variety of factors in addition to maternal concerns, including property rights and values based in gendered traditions. Among the Navajo, for example, land often belonged to the women, since it could be passed down from father to daughter, uncle to niece.[159] In addition, many men had died as a result of their work (miners' risk of lung cancer increased by a factor of at least eighty-five), leaving their widows to band together to seek compensation.[160] WARN also worked to inform Native American women of their rights to resist an aggressive government-funded mass sterilization program WARN termed genocidal.[161] At a WARN sovereignty workshop Indian women were told they "must lead." Activists urged them, "Control your own reproduction: not only just the control of the reproduction of yourselves... but control of the reproduction of your own food supplies, your own food systems" to rebuild traditional native cultures and ways of living with the earth.[162]

People of color perpetually bring unique perspectives (evolving as well out of class, education, religion, and a host of other factors in addition to race) to ongoing issues concerning their environments. Toxic waste facilities, chemical emissions, and health risks from air pollution disparately affect economically poor communities disproportionately populated by people of color.[163] In the modern environmental justice movement in the United States, African American women in particular, frequently the heads of single-parent households, bring a legacy of assertiveness, leadership, and maternal concerns.[164] They play a prominent role in a number of community organizations, waging campaigns against environmental dangers in the workplace and the home, especially in areas known as "brown fields" because of their toxicity.[165] Latinas, too, emphasize their dual role as mothers and workers in combating environmental hazards.[166] In California, for example, they continue to build on a long legacy of struggle led by the United Farm Workers against various pesticides, particularly those affecting reproduction.

Much of the best scholarship on environmental justice is being carried out by sociologists, political scientists, and legal experts, but scholars are increasingly providing crucial historical and gendered viewpoints. Rachel Stein's edited collection *New Perspectives on Environmental Justice: Gender, Sexuality, and Activism* (2004) offers a wide-ranging and particularly valuable introduction to the history of efforts to achieve environmental justice in communities suffering from factors including poverty, racism, ethnocentrism, sexism, and homophobia.[167] Nancy C. Unger; Robert D. Bullard and Damu Smith; Andrea Simpson; and Giovanna Di Chiro have revealed the widely ranging role gender has played in environmental justice history.[168]

NEW TRENDS AND THE FUTURE

Virginia Scharff's 2003 edited collection, *Seeing Nature Through Gender*, was widely praised for placing sexuality and gender into environmental history.

Although scholarship at the intersection of gender studies and environmental history is still in its infancy, many exciting studies are emerging, including *Queer Ecologies: Sex, Nature, Biopolitics, and Desire* (2010), edited by Catriona Sandilands-Mortimer and Bruce Erickson, and Nancy C. Unger's *Beyond Nature's Housekeepers: American Women in Environmental History* (2012). One of the most important new trends is recognition of the importance of gender and environment across all times, cultures, and geographic boundaries. The huge outpouring of environmental histories by and about women from around the world is finally being appreciated and widely disseminated beyond their countries of origin. Women and gender are increasingly being recognized as useful categories of analysis within environmental history. From a plethora of studies about the Chipko environmental resistance movement by women in India in the 1970s, scholarship has grown to include the roles of gender and sexuality in the environmental histories of Africa, Asia, Canada, Europe, South America, Australia.[169] One of the most prolific and influential scholars is Bina Agarwal, whose investigations into the gendered ways in which women in poor rural households suffer uniquely from environmental degradation resulting from decreased resource access and control are centered primarily in India, but are frequently relevant to other parts of the developing world.[170] Of particular interest are her studies of women's activism in environmental protection and regeneration, including the gender dimensions of decision-making and property rights.[171]

Perhaps even more exciting than the single-area studies are those that compare environmental issues in two or more countries or regions, imbuing the material with crucial context. Carol MacCormack and Marilyn Strathern's *Nature, Culture, and Gender* (1980) culturally contextualizes perceptions of the relationship between nature and gender by examining eighteenth- and nineteenth-century concepts and practices in Bolivia, Papau New Guinea, Sierra Leone, and Europe.[172] Other excellent examples of this kind of vitally important work include Carolyn E. Sachs's *Gendered Fields: Rural Women, Agriculture, and Environment* (1996), which compares the role of women in rural agriculture in Africa, Asia, and the United States, and Glenda Riley's *Taking Land, Breaking Land: Women Colonizing in the American West and Kenya, 1840–1940* (2003).[173] Gender dimensions of the new transnational model of development are explored in William Robinson's "(Mal)Development in Central America: Globalization and Social Change" (1998) and Andres Serbin's "Transnational Relations and Regionalism in the Caribbean" (1994).[174] The effect of nongovernmental organizations (NGOs) on local gendered relationships with the environment are examined by Millie Thayer in *Making Transnational Feminism: Rural Women, NGO Activists, and Northern Donors in Brazil* (2009); Lisa Marie Aubrey's *The Politics of Development Co-operation: NGOs, Gender and Partnership in Kenya* (1997), and "Sovereign Limits and Regional Opportunities for Global Civil Society in Latin America" (2001) by Elisabeth Friedman, Kathryn Hochstetler, and Ann Marie Clark.[175]

Why a Gendered History of the Environment Matters in the Twenty-First Century

Both the physiological differences between the sexes and the vast array of culturally created differences (i.e., gender) have profoundly shaped the environmental past. The results in the present day are nevertheless often startling in view of the fact that men and women generally inhabit the same environments and usually share equally in their benefits and detriments. Susan Schrepfer argues that William Cronon's influential essay "The Trouble with Wilderness; Or, Getting Back to the Wrong Nature" (1997) is incomplete as a reparative, because Cronon targets only the masculine version or myth of wilderness. She maintains that "scholars working at the intersection of gender and environmental history need to continue to disentangle the strands of masculinity and femininity that have been wrought into the…all visions of nature over the course of American history."[176]

Differentiating and disentangling gender is not just an intellectual exercise. In the twenty-first century, men and women work together under the leadership of women as well as environmentalists like Bill McKibben and Al Gore, men who are raising awareness and providing vital direction. Yet a growing body of social science research indicates that "women rank values linked to environmental concern as more important than men do and see environmentalism as important to protecting themselves and their families." American women make up the bulk of a great many environmental organization memberships and are "less likely than men to support environmental spending cuts and are less sympathetic to business when it comes to environmental regulations." They are also more concerned about environmental risks to health, especially local ones. And this gendered difference is not limited to the United States. Throughout industrialized countries, women are "more likely to buy ecologically friendly and organic foods, more likely to recycle and more interested in efficient energy use."[177] In developing nations as well, women are often at the vanguard of environmental leadership, both locally and nationally.[178] Gender continues to influence how the environment is understood, used, abused, exploited, and healed.[179] If men and women are to be partners in protecting the environment, they need to recognize the factors, especially the socially constructed ideas and attitudes, which have influenced and sometimes divided them. In the decades to come, studies at the intersection of gender and environment will enrich both disciplines as they immeasurably expand understandings of the past, inform the present, and shape the future.

NOTES

Portions of this essay are drawn from works by Nancy C. Unger, including *Beyond Nature's Housekeepers: American Women in Environmental History* (Oxford University Press, 2012); "Gendered Approaches to Environmental Justice: An Historical Sampling," in *Echoes from the Poisoned Well: Global Memories of Environmental Injustice*, ed. Sylvia Hood Washington, Paul Rosier, and Heather Goodall (Rowman & Littlefield, 2006): 17–34; and "Women, Sexuality, and Environmental Justice in American History," in *New Perspectives in Environmental Justice: Gender, Sexuality, and Activism*, ed. Rachel Stein (Rutgers University Press, 2004), 45–60. Beth Bailey and Andrew Isenberg offered valuable suggestions to this essay, as did many of the contributors to this volume. Mary Whisner and Don Whitebread provided editing expertise.

1. Carolyn Merchant, "Gender and Environmental History," *The Journal of American History* 76 (1990): 1117–1121.
2. Melissa Leach and Cathy Green, "Gender and Environmental History: From Representation of Women and Nature to Gender Analysis of Ecology and Politics," *Environment and History* 3 (1997): 343–370. Subsequent quotes from this article have been altered to conform to American spellings.
3. Vera Norwood, "Disturbed Landscape/Disturbing Process: Environmental History for the Twenty-First Century," *Pacific Historical Review* (February 2001): 84.
4. Richard White, "Environmental History: Watching a Historical Field Mature," *Pacific Historical Review* (February 2001): 104; 109.
5. Carolyn Merchant, *The Columbia Guide to American Environmental History* (New York: Columbia University Press, 2002), 448.
6. Elizabeth Blum, *Linking American Women's History and Environmental History: A Preliminary Historiography*, http://www.h-net.org/~environ/historiography/uswomen.htm.
7. Forest History Society, *Environmental History Bibliography*, http://www.foresthistory.org/Research/biblio.html.
8. See Susan R. Schrepfer and Douglas Cazaux Sackman, "Gender," in *A Companion to American Environmental History*, ed. Douglas Cazaux Sackman (Malden, MA: Wiley-Blackwell, 2010), and Victoria Davion, "Ecofeminism," in *A Companion to Environmental Philosophy*, ed. Dale Jamieson (Malden, MA: Wiley-Blackwell, 2003), 233–248.
9. Linda Nash, "Furthering the Environmental Turn," *Journal of American History* 100 (June 2013): 132.
10. Joan W. Scott, "Gender: A Useful Category of Historical Analysis," *American Historical Review* 91 (December 1986): 1053–1075.
11. Mary Joy Breton, *Women Pioneers for the Environment* (Boston: Northeastern University Press, 1998).
12. Blum, "Linking American Women's History," and Leach and Green, "Gender and Environmental History."
13. See John Herron and Andrew Kirk, eds., *Human/Nature: Biology, Culture, and Environmental History* (Albuquerque: University of New Mexico Press, 1999), especially Virginia Scharff, "Man and Nature! Sex Secrets of Environmental History," 31–48; Vera Norwood, "Constructing Gender in Nature," 49–62.

14. Jose Barreiro, "Indigenous Peoples are the 'Miners' Canary' of the Human Family," in *Learning to Listen to the Land*, ed. William Willers (Washington, DC: Island Press, 1991), 199–201.

15. Alan Durning, "Worldwatch Paper #112: Guardians of the Land: Indigenous Peoples and the Health of the Earth," (Worldwatch Institute, 1992).

16. For a global overview, see J. Donald Hughes, *An Environmental History of the World: Humankind's Changing Role in the Community of Life* (New York: Routledge, 2002). In Washington, Rosier, Goodall, eds., *Echoes from the Poisoned Well,* thirteen chapters are devoted to the environmental practices of indigenous peoples from various parts of the world.

17. See Charles C. Mann, *1491: New Revelations of the America's Before Columbus* (New York: Vintage, 2006).

18. Leach and Green, "Gender and Environmental History," 352.

19. See, for example, Thomas Forsyth, "Manners and Customs of the Sauk and Fox Nations of Indians," 1827, Thomas Forsyth Papers, Lyman Draper Manuscripts, Volume 9T, 20, Wisconsin Historical Society, Madison, Wisconsin.

20. See Hermien Soselisa, "The Significance of Gender in the Fishing Economy of the Goram Islands, Maluku," in *Old World Places, New World Problems: Exploring Issues of Resource Management in Eastern Indonesia*, ed. Sandra Pannell and Franz von Benda-Beckmann (Center for Resource and Environmental Studies, ANU, 1988), 321–35.

21. See James J. Rawls and Walton Bean, *California: an Interpretive History*, 7th ed., (New York: McGraw-Hill, 1998), 11–13; Richard Dasmann, *California's Changing Environment* (San Francisco: Boyd and Fraser, 1981), 1–8; George Phillips, *The Enduring Struggle: Indians in California History* (San Francisco: Boyd and Fraser, 1981), 4–12.

22. See Morrill Marsten to The Reverend Jedediah Morse, November 1820, Thomas Forsyth Papers, Volume 1T, 65, Lyman Draper Manuscripts, Wisconsin Historical Society, Madison, Wisconsin.

23. Carolyn Merchant, *Earthcare: Women and the Environment* (New York: Routledge, 1995), 92–95.

24. See Mann, *1491*; Shepard Krech, *The Ecological Indian: History and Myth* (New York: W. W. Norton, 2000); Leach and Green, "Gender and Environmental History," 351.

25. William Cronon, *Changes in the Land: Indians, Colonists, and the Ecology of New England* (New York: Hill and Wang, 1983).

26. Forsyth, "Manners and Customs of the Sauk and Fox Nations of Indians."

27. Russell Thornton, American Indian Holocaust and Survival: A Population History Since 1492 (Norman: University of Oklahoma Press, 1987), 31. See also Demitri Shimkin, "Eastern Shoshone" in Handbook of North American Indians (Great Basin), ed. Warren L. D'Azevedo (Washington DC: Smithsonian Institution 1986), 330; Francis Riddell, "Maidu and Konkow," in Handbook of North American Indians (California), ed. Robert F. Heizer (Washington DC: Smithsonian Institution, 1978), 381; T. N. Campbell, ed., "Coahuiltecans and Their Neighbors," in Handbook of North American Indians (Southwest), ed. Alonso Ortiz (Washington DC: Smithsonian Institution, 1983), 352; Walter O'Meara, Daughters of the Country (New York: Harcourt, Brace and World, 1968), 84; John Demos, The Tried and the True: Native American Women Confronting Colonization (New York: Oxford University Press, 1995), 77.

28. Thornton, *American Indian Holocaust and Survival*, 31, Adam Hochschild, *King Leopold's Ghost* (Boston: Mariner Books, 1999), 73. See also Liese M. Perrin, "Resisting

Reproduction: Reconsidering Slave Contraception in the Old South," *Journal of American Studies* 35, no. 2 (2001): 258–259; 263; 266.

29. In a manuscript on tribal traditions, a US government agent noted, "All Indians are very fond of their children and a sick Indian is loth [*sic*] to leave this world if his children are young, but if [the children are] grown up and married they know they are a burden to their children and don't care how soon they die." Forsyth, "Manners and Customs," 15.

30. See Unger, "Women, Sexuality and Environmental Justice," 48.

31. Cronon, *Changes in the Land*, 167.

32. For works for adults on Sacajawea, see Thomas P. Slaughter, *Exploring Lewis and Clark: Reflection on Men and Wilderness* (New York: Vintage Books, 2003); Sally McBeth, "Memory, History, and Contested Pasts: Re-imagining Sacagawea/Sacajawea," *American Indian Culture and Research Journal* 27, no. 1 (2003): 1–32. For a case study outside North America, see Barbara Walker, "Engendering Ghana's Seascape: Fanti Fishtraders and Marine Property in Colonial History," *Society & Natural Resources* 15 (May-June 2002): 389–407.

33. See Virginia Bouvier, *Women and the Conquest of California, 1542–1840: Codes of Silence* (Tucson: University of Arizona Press, 2001).

34. See Andrew Isenberg, *The Destruction of the Bison: An Environmental History, 1750-1920* (Cambridge: Cambridge University Press, 2000).

35. Susan Sleeper-Smith, *Indian Women and French Men: Rethinking Cultural Encounter in the Western Great Lakes* (Amherst: University of Massachusetts Press, 2001); Sylvia Van Kirk, *Many Tender Ties: Women in Fur-Trade Society, 1670–1870* (Norman: University of Oklahoma Press, 1983).

36. See Miroslava Chavez-Garcia, *Negotiating Conquest: Gender and Power in California, 1770s–1880s* (Tucson: University of Arizona Press, 2004); Rose Marie Beebe and Robert Senkewicz, eds., *Lands of Promise and Despair* (Santa Clara, CA: Santa Clara University, 2001).

37. Albert Hurtado, "Sexuality in California's Franciscan Missions: Cultural Perceptions and Sad Realities," *California History* 72 (Fall 1992): 370–385; Hurtado, *Indian Survival on the California Frontier* (New Haven, CT: Yale University Press, 1988) 169–192.

38. Deborah Gray White, *Ar'N't I a Woman?: Female Slaves in the Plantation South* (New York: W. W. Norton, 1985), 67.

39. See Mart Stewart, "Slavery and the Origins of African American Environmentalism," in *"To Love the Wind and the Rain": African Americans and Environmental History*, ed. Dianne Glave and Mark Stoll (Pittsburgh, PA: University of Pittsburgh Press, 2006), 9–20; Whitney Battle, "A Yard to Sweep: Race, Gender, and the Enslaved Landscape" (PhD diss., University of Texas at Austin, 2004).

40. Scott Giltner, "Slave Hunting and Fishing in the Antebellum South," Glave and Stoll, eds., *"To Love the Wind and the Rain,"* 21–36.

41. Elizabeth Blum, "Power, Danger, and Control: Slave Women's Perception of Wilderness in the Nineteenth Century," *Women's Studies* 31, no. 2 (2002): 247–266.

42. See Sharla Fett, *Working Cures: Healing, Health and Power on Southern Slave Plantations* (Chapel Hill: University of North Carolina Press, 2002), 65; 176–177; Perrin, "Resisting Reproduction," 255–274; Marie Schwartz, *Birthing a Slave: Motherhood and Medicine in the Antebellum South* (Cambridge, MA: Harvard University Press, 2006).

43. Perrin, "Resisting Reproduction," 258–59; 263; 266.

44. Phillip D. Morgan, *Slave Counterpoint: Black Culture in the Eighteenth Century Chesapeake and the Low Country* (Chapel Hill: University of North Carolina Press, 1998).

45. S. Mintz, "Slavery Fact Sheet," *Digital History,* 2007, http://www.digitalhistory.uh.edu/historyonline/slav_fact.cfm.

46. Daniel Littlefield, *Rice and Slaves: Ethnicity and the Slave Trade in Colonial South Carolina* (Urbana: University of Illinois Press, 1991); Judith Carney, *Black Rice* (Cambridge, MA: Harvard University Press, 2001), 162. Carney's conclusion "that knowledge of rice cultivation enabled slaves arriving in South Carolina to enjoy greater autonomy from their owners than was possible for other crops" is questioned by David Eltis, Philip Morgan, and David Richardson in "Agency and Diaspora in Atlantic History: Reassessing the African Contribution to Rice Cultivation in the Americas," *American Historical Review* 112, no. 5 (December 2007): 1328–1358. The controversy was revisited in "*AHR* Exchange: The Question of 'Black Rice," *American Historical Review* 115, no. 1 (February 2010); 123–171.

47. See Eugene Genovese, "Cotton, Slavery, and Soil Exhaustion in the Old South," *Cotton History Review* 2, no. 1 (1961): 3–17.

48. See Frederick Law Olmsted, *The Cotton Kingdom: A Traveler's Observations on Cotton and Slavery in the American Slave States* (New York: Knopf, 1953), 410–411.

49. See Fiona Mackenzie, "Political Economy of the Environment, Gender and Resistance Under Colonialism: Murang'a District, Kenya, 1910–1950," *Canadian Journal of African Studies* 25, no. 2 (1991): 226–256.

50. Carolyn Merchant, *The Death of Nature: Women, Ecology, and the Scientific Revolution* (New York: Harper & Row, 1980). For an application of Merchant's theory to the history of South Africa, see Nancy Jacobs, "The Colonial Ecological Revolution in South Africa: The Case of Kurman," in *South Africa's Environmental History: Cases and Comparisons,* ed. Stephen Dovers, Ruth Edgecombe, Bill Guest (Athens: Ohio University Press, 2003), 19–33.

51. See Clarissa Dillon, "'A Large an [*sic*] Useful, and a Grateful Field': Eighteenth Century Kitchen Gardens in Southeastern Pennsylvania, the Uses of Plants, and Their Place in Women's Work" (PhD diss., Bryn Mawr College, 1986).

52. See Susan Scott Parrish, "Women's Nature: Curiosity, Pastoral, and the New Science in British America," *Early American Literature* 37, no. 2 (2002): 195–238; Joan Jensen, *Loosening the Bonds: Mid-Atlantic Farm Women, 1750–1850* (New Haven, CT: Yale University Press, 1988); and especially Jensen's document collection *With These Hands: Women Working on the Land* (New York: The Feminist Press, 1981). See also Laurel Thatcher Ulrich, *A Midwife's Tale: The Life of Martha Ballard, Based on Her Diary, 1785–1812* (New York: Vintage, 1991).

53. E. Anthony Rotundo, *American Manhood: Transformations From Revolution to the Modern Era* (New York: Basic Books, 1994); Michael Kimmel, *Manhood in America: A Cultural History* (New York: Oxford University Press, 2011). Although few historians dispute what was being prescribed, there is controversy over to what degree it was internalized and by whom. R. W. Connell, for example, emphasizes in *Masculinities,* 2d ed. (Berkeley: University of California Press, 2005) that there is never a single concept of masculinity. While some historians defend Barbara Welter's conclusions about the era's "Cult of True Womanhood," others object to the negative connotations of the word "cult," and criticize Welter's definition as both simplistic and overly rigid. More recent additions to the controversy include Mary Kelley, "Beyond the Boundaries," *Journal of the Early Republic* 21, no. 1 (2001): 73–78; Carol Lasser, "Beyond Separate Spheres: The

Power of Public Opinion," *Journal of the Early Republic* 21, no. 1 (2001): 115–123; Mary Cronin, "Redefining Woman's Sphere," *Journalism History* 25, no. 1 (1999): 13–25; Brian Gabrial, "A Woman's Place," *American Journalism* 25, no. 1 (2008): 7–29; Mary Louise Roberts, "True Womanhood Revisited," *Journal of Women's History* 14, no. 1 (2002): 150–155; Nancy A. Hewitt, "Taking the True Woman Hostage," *Journal of Women's History* 14, no. 1 (2002): 156–162.

54. Donald Worster, *Rivers of Empire: Water, Aridity, and the Growth of the American West* (New York: Oxford University Press, 1985), 97.

55. Chad Montrie, "'Men Alone Cannot Settle a Country': Domesticating Nature in the Kansas-Nebraska Grasslands," *Great Plains Quarterly* 25, no. 4 (2005): 245–258. See also Cynthia Prescott, "'Why She Didn't Marry Him': Love, Power, and Marital Choice on the Far Western Frontier," *Western Historical Quarterly* 38, no. 1 (2007): 25–45.

56. John Mack Faragher, *Men and Women on the Overland Trial*, 2d ed. (New Haven, CT: Yale University Press, 2001); Sandra Myres, *Westering Women and the Frontier Experience, 1800–1915* (Albuquerque: University of New Mexico Press, 1992); Julie Roy Jeffrey, *Frontier Women: "Civilizing" the West? 1840–1880*, 2d ed. (New York: Hill and Wang, 1998). A variety of books focus on pioneering women in general, including Linda Peavy and Ursula Smith's *Pioneer Women: the Lives of Women on the Frontier* (Norman: University of Oklahoma Press, 1998); Ruth Moynihan et al.'s edited collection *So Much to be Done: Women Settlers on the Mining and Ranching Frontiers* (Lincoln: University of Nebraska Press, 2nd ed., 1998), and Susan Roberson's edited collection *Women, America, and Movement: Narratives of Relocation* (Columbia: University of Missouri Press, 1998). Others take a more regional approach, such as Terri Baker and Connie Henshaw's edited collection *Women Who Pioneered Oklahoma: Stories From the WPA Narratives* (Norman: University of Oklahoma Press, 2007); JoAnn Levy's *They Saw the Elephant: Women in the California Gold Rush* (Norman: University of Oklahoma Press, 1992); and the section "Women on the Wisconsin Frontier, 1836–1848," in Genevieve McBride's edited collection *Women's Wisconsin* (Wisconsin Historical Society Press, 2005). These various accounts are filled with women gathering fuel along westward trails; building sod houses; clearing land; planting, harvesting, and preserving crops; herding animals; milking cows, and tending chickens.

57. See Unger, "Civilizing the Frontier," in *Beyond Nature's Housekeepers*, 64–70.

58. See Unger, "Beautifying the Environment," in *Beyond Nature's Housekeepers*, 70–72.

59. Richard Westmacott, *African-American Gardens and Yards in the Rural South* (Knoxville: University of Tennessee Press, 1992); Dianne Glave, "'A Garden So Brilliant with Colors, So Original in Its Design': Rural African American Women, Gardening, Progressive Reform, and the Foundation of an African American Environmental Perspective," *Environmental History* 8, no. 3 (2003): 395–411.

60. Sheryll Patterson-Black, *Western Women in History and Literature* (Crawford, NE: Cottonwood Press, 1978).

61. See Cary W. de Wit, "Women's Sense of Place on the American High Plains," *Great Plains Quarterly* 21, no. 1 (2001): 29–44. A particularly valuable Canadian study is Sandra Rollings-Magnusson's "Canada's Most Wanted: Pioneer Women on the Western Prairies," *Canadian Review of Sociology and Anthropology* 37, no. 2 (2000): 223–238.

62. Hal Borland, *High, Wide, and Lonesome: Growing Up on the Colorado Frontier* (Olympic Marketing Corp, 1984); Alma Lou Plunkett, *The Kansas-Colorado Line: Homesteading Tales of Several Families, A Memoir* (iUniverse, 2012). See also Jean Gray, *Homesteading*

Haxtun and the High Plains: Northeastern Colorado History (Charleston, SC: The History Press, 2013); David Laskin, *The Children's Blizzard* (New York: Harper Collins, 2004).

63. See, for example, Adrienne Caughfield, *True Women and Westward Expansion* (College Station: Texas A&M University Press, 2005), based on the diaries and letters of ninety Texas women.

64. Glenda Riley, *Women and Nature: Saving the "Wild" West* (Lincoln: University of Nebraska Press, 1999).

65. Vicki Ruiz, "Shaping Public Space/Enunciating Gender: A Multiracial Historiography of the Women's West, 1995–2000," *Frontiers* 22, no. 3 (2001): 22–25; S.J. Kleinberg, "Race, Region, and Gender in American History," *Journal of American Studies* 33, no. 1 (1999): 85–88; Deborah Kanter, "Native Female Land Tenure and Its Decline in Mexico, 1750–1900," *Ethnohistory* 42 (Fall 1995): 607–616; Paula Nelson, "Women and the American West: a Review Essay," *Annals of Iowa* 50, nos. 2-3 (1989): 269–273; Katherine Benton-Cohen, "Common Purposes, Worlds Apart: Mexican-American, Mormon, and Midwestern Women Homesteaders in Cochise County, Arizona," *Western Historical Quarterly* 36 (Winter 2005): 429–452.

66. Theda Perdue, *Cherokee Women: Gender and Culture Change, 1700–1835* (Lincoln: Bison Books, 1999), and Bruce White, "The Woman Who Married a Beaver: Trade Patterns and Gender Roles in the Ojibwa Fur Trade" *Ethnohistory* 46, no. 1 (1999): 109–147.

67. See William Beinart and Lotte Hughes, *Environment and Empire* (Oxford: Oxford University Press, 2009).

68. See William K. Story, "Big Cats and Imperialism: Lion and Tiger Hunting in Kenya and Northern India, 1898–1930," *Journal of World History* 2, no. 2 (June 1991): 135–173; J. A. Mangan and C. McKenzie, "'Duty unto Death'—the Sacrificial Warrior: English Middle Class Masculinity and Militarism in the Age of the New Imperialism," *International Journal of the History of Sport* 25, no. 9 (August 2008): 1080–1105; Mangan and McKenzie, "Martial Conditioning, Military Exemplars and Moral Certainties: Imperial Hunting as Preparation for War," *International Journal of the History of Sport* 25, no. 9 (August 2008): 1132–1167.

69. Greg Gillespie, *Hunting for Empire: Narratives of Sport in Rupert's Land, 1840–70* (Vancouver: University of British Columbia Press, 2008). See also Matt Carmill, *A View to a Death in the Morning: Hunting and Nature Through History* (Cambridge, MA: Harvard University Press, 1993).

70. See Joseph Sramek, "'Face Him Like a Briton'; Tiger Hunting, Imperialism, and British Masculinity in Colonial India, 1800–1875," *Victorian Studies* 48, no. 4 (Summer 2006): 659–680.

71. See J. A. Mangan, "Publicist and Proselytizer: The Officer-Hunter as Scientist and Naturalist," *International Journal of the History of Sport* 25, no. 9 (August 2008): 1189–1217; John M. MacKenzie, *The Empire of Nature: Hunting, Conservation and British Empire* (New York: St. Martin's Press, 1989).

72. See Harriet Ritvo, "Destroyers and Preservers," *History Today* 52, no. 1 (January 2002): 33–40; "Martial and Moral Complexities," *International Journal of the History of Sport* 25, no. 9 (August 2008): 1168–1188.

73. Carolyn Merchant, "George Bird Grinnell's Audobon Society: Bridging the Gender Divide in Conservation," *Environmental History* 15 (January 2010): 4–5.

74. See Allison Helper, *Women in Labor: Mothers, Medicine, and Occupational Health in the United States, 1890–1980* (Columbus: Ohio State University Press, 2000).

75. Clara Bradley Burdette, "The College Woman and Citizenship," *The Syracusan* (June 15, 1910): 3, Box 125, File 1, Clara Burdette Collection, Huntington Library, San Marino, California.

76. See Elizabeth Blum, "Women, Environmental Rationale, and Activism during the Progressive Era," in *To Love the Wind and the Rain*, 77–92.

77. Nancy C. Unger, *Fighting Bob La Follette: The Righteous Reformer*, 2d ed., (Madison: Wisconsin Historical Society Press, 2008), 86–87.

78. Unger, "Gendered Approaches to Environmental Justice," 86–87.

79. Ibid.

80. Merchant, "Women of the Progressive Conservation Movement: 1900-1916," *Environmental Review* 8, no. 1 (Spring 1984), 65.

81. Adam Rome, "'Political Hermaphrodites': Gender and Environmental Reform in Progressive America," *Environmental History* 11 (July 2006): 440–463.

82. See Robert V. Hine and John Mack Faragher, *Frontiers: A Short History of the American West* (New Haven, CT: Yale University Press, 2000), 176–190. For a nuanced biography of Pinchot, see Char Miller, *Gifford Pinchot and the Making of Modern Environmentalism* (Washington, DC: Island Press, 2004).

83. Theodore Roosevelt, *State of the Union Addresses by Theodore Roosevelt* (MacLean, VA: IndyPublish, 2007), 218.

84. Sarah Watts, *Rough Rider in the White House: Theodore Roosevelt and the Politics of Desire* (Chicago: University of Chicago Press, 2003).

85. "George L. Knapp Opposed Conservation," in *Major Problems in American Environmental History*, ed. Carolyn Merchant (Lexington: DC Heath, 1993), 322.

86. Elizabeth Blum, *Love Canal Revisited: Race, Class, and Gender in Environmental Activism* (Lawrence: University Press of Kansas, 2008), 28.

87. John F. Reiger, *American Sportsmen and the Origins of Conservation*, 3d rev. ed., (Corvallis: Oregon State University Press, 2000); Daniel Justin Herman, *Hunting and the American Imagination* (Washington, DC: Smithsonian Institution Press, 2001).

88. Isenberg, *Destruction of the Bison*, 169–170.

89. Watts, *Rough Rider in the White House*; Peter Bayers, "Frederick Cook, Mountaineering in the Alaskan Wilderness and the Regeneration of Progressive Era Masculinity," *Western American Literature* 38, no. 2 (Summer 2003): 170–193; Tina Loo, "Of Moose and Men: Hunting for Masculinities in British Columbia, 1880–1939," *Western Historical Quarterly* 32, no. 3 (Autumn 2001): 296–320; Gail Bederman, *Manliness and Civilization: A Cultural History of Gender and Race in the United States, 1880–1917* (Chicago: University of Chicago Press, 1996).

90. Andrea Smalley, "'Our Lady Sportsmen': Gender, Class, and Conservation in Sport Hunting Magazines, 1873–1920," *Journal of the Gilded Age and Progressive Era* 4, no. 4 (October 2005): 359; 364.

91. See Suellen Hoy, "'Municipal Housekeeping': The Role of Women in Improving Urban Sanitation Practices, 1880–1917," in *Pollution and Reform in American Cities, 1870–1930*, ed. Martin Melosi (Austin: University of Texas Press, 1980), 173–198; Maureen A. Flanagan, "The City Profitable, The City Livable: Environmental Policy, Gender, and Power in Chicago in the 1910s," *Journal of Urban History* 22 (January 1996): 163–190; Merchant, "Women of the Progressive Conservation Movement, 1900–1916," 55–85; Unger, "Nature's Housekeepers: Progressive-Era Women as Midwives to the Conservation Movement and Environmental Consciousness," *Beyond Nature's Housekeepers*, 75–104.

92. See Linda C. Forbes and John M. Jermier, "The Institutionalization of Bird Protection: Mabel Osgood Wright and the Early Audubon Movement," *Organization & Environment* 15 (December 2002): 458–474; Kathy S. Mason, "Out of Fashion: Harriet Hemenway and the Audubon Society, 1896–1905," *Historian* 65 (Fall 2002): 1–14; Jennifer Price, "Hats Off to Audubon," *Audubon* 106 (November-December 2004): 44–50; Polly Welts Kaufman, *National Parks and the Woman's Voice: A History* (Albuquerque: University of New Mexico Press, 1996).

93. Sarah Orne Jewett, *A White Heron and Other Stories* (Mineola: Dover Publications, 2012).

94. Merchant, "Women of the Progressive," 64–65.

95. "Mrs. Marion Crocker on the Conservation Imperative, 1912," in Merchant, *Major Problems in Environmental History*, 2d ed., 325.

96. Joseph Kastner, "Long Before Furs, It was Feathers That Stirred Reformist Ire," *Smithsonian* 25, no. 4 (July 1994): 100.

97. "Ill Advised Sarcasm," *La Follette's Magazine* 5, no. 35 (August 30, 1913): 3.

98. Porter practiced what scholar Amy Green terms a form of "muscular womanhood" grafted "onto the new womanhood ideal." Amy Green, "She Touched Fifty Million Lives," in *Seeing Nature Through Gender*, 235.

99. Wendy Keefover-Ring, "Municipal Housekeeping, Domestic Science, Animal Protection, and Conservation: Women's Political and Environmental Activism in Denver, Colorado, 1894–1912" (MA thesis, University of Colorado, 2002), 146.

100. Robin W. Doughty, *Feather Fashions and Bird Preservation: A Study in Nature Protection* (Berkeley: University of California Press, 1975), 63; 72.

101. Merchant, "Women of the Progressive," 68.

102. Ibid., 59.

103. Martin Melosi, "Environmental Justice, Political Agenda Setting, and the Myths of History," *Journal of Policy History* 12, no. 1 (January 2000): 53. See, for example, Claudi Clark, *Radium Girls: Women and Industrial Health Reform, 1910–1935* (Chapel Hill: University of North Carolina Press, 1997); Allison L. Hepler, *Women in Labor: Mothers, Medicine, and Occupational Health in the United States, 1890–1980* (Columbus: Ohio State University Press, 2000); Angela Gugliotta, "Class, Gender, and Coal Smoke: Gender Ideology and Environmental Injustice in Pittsburgh, 1868–1914," *Environmental History* 5 (April 2000): 165–193; Mark Speltz, "An Interest in Health and Happiness as Yet Untold: The Woman's Club of Madison, 1893– 1917," *Wisconsin Magazine of History* 89 (Spring 2006): 2–15.

104. Melosi, "Environmental Justice," 54.

105. Rheta Childe Dorr, *What Eight Million Women Want* (Boston: Small, Maynard, and Co., 1910), 327.

106. Susan Fitzgerald, "Women in the Home," in *One Half the People: The Fight for Women Suffrage*, ed. Anne Firor Scott and Andrew MacKay Scott (Urbana: University of Illinois Press, 1982), 115.

107. John McClymer, *The Triangle Strike and Fire* (Fort Worth, TX: Harcourt Brace, 1998), 4–5. See also Arthur McEvoy, "The Triangle Shirtwaist Factory Fire of 1911: Social Change, Industrial Accidents, and the Evolution of Common Sense Causality," *Law & Social Inquiry* 20 (July 2006): 621–651.

108. Melosi, "Environmental Justice," 55.

109. Unger, *Beyond Nature's Housekeepers*, 109–117; Susan Schrepfer, *Nature's Altars: Mountains, Gender, and American Environmentalism* (Lawrence: University

Press of Kansas, 2005), 152–158; Ben Jordan, "'Conservation of Boyhood': Boy Scouting's Modest Manliness and Natural Resource Conservation, 1910–1930," *Environmental History* 15 (October 2010): 612–642.

110. Schrepfer, *Nature's Altars*, 153–154.

111. *Camp Fire Girls* (1912, repr., Applewood, n.d), 25.

112. W.J. Hoxie, *How Girls Can Help Their Country: Handbook for Girl Scouts* (1913; repr., Applewood, n.d.), 25; 23.

113. Hoxie, *How Girls*, 2; 108–109; 16.

114. Schrepfer, *Nature's Altars*, 157.

115. *Camp Fire Girls*, frontispiece.

116. Hoxie, *How Girls*, 110; 80; 63.

117. Jordan, "Conservation of Boyhood," 614, 617, 628.

118. See Unger, *Beyond Nature's Housekeepers*, 112.

119. Merchant, "George Bird Grinnell's Audubon Society," 8.

120. Rome, "Political Hermaphrodites."

121. See Kevin P. Murphy, *Political Manhood: Red Bloods, Mollycoddles, and the Politics of Progressive Reform* (New York: Columbia University Press, 2008).

122. Schrepfer, *Nature's Altars*. See also Mary Leffler Cochran, *Fulfilling the Dream: The Story of National Garden Clubs, Inc., 1929–2004* (St. Louis, MO: National Garden Clubs, Inc., 2004); Cameron Binkley, "'No Better Heritage Than Living Trees': Women's Clubs and Early Conservation in Humboldt County," *Western Historical Quarterly* 33 (Summer 2002): 179–203.

123. Vera Norwood, *Made From this Earth: American Women and Nature* (Chapel Hill: University of North Carolina Press, 1993); Marcia Myers Bonta, *American Women Afield: Writings by Pioneering Women Naturalists* (College Station: Texas A&M University Press, 1995); Lorraine Anderson and Thomas S. Edwards, eds., *At Home on This Earth: Two Centuries of U.S. Women's Nature Writing* (Hanover, NH: University Press of New England, 2002). Rachel Stein, *Shifting the Ground: American Women Writers' Revisions of Nature, Gender, and Race* (Charlottesville: University of Virginia Press, 1997). Follow ups to Stein include Justyna Kostkowska, *Ecocriticism and Women Writers: Environmental Poetics of Virginia Woolf, Jeanette Winterson, and Ali Smith* (New York: Palgrave Macmillan, 2013); Karen Waldron and Rob Friedman, *Toward a Literary Ecology: Places and Spaces in American Literature* (Lanham, MD: Scarecrow Press, 2013).

124. Bryant Simon, "'New Men in Body and Soul': The Civilian Conservation Corps and the Transformation of Male Bodies and the Male Politic," in *Seeing Nature Through Gender*, ed. Virginia Scharff (Lawrence: University of Kansas Press, 2003), 82–83; 87–90; 95–97. See also Neil Maher, "Labor: Enrollee Work and the Body Politic," in *Nature's New Deal: The Civilian Conservation Corps and the Roots of the American Environmental Movement* (Oxford: Oxford University Press, 2009): 77–114.

125. Heather Van Wormer, "A New Deal for Gender: The Landscapes of the 1930s," in *Shared Spaces and Divided Places: Material Dimensions of Gender Relations and the American Historical Landscape*, ed. Deborah L. Rotman and Ellen-Rose Savulis (Knoxville: University of Tennessee Press, 2003): 219.

126. Blanche Weisen Cook, *Eleanor Roosevelt, vol. 2, 1933–1938* (New York: Viking, 1999): 88–91.

127. Nancy C. Unger, "Women for a Peaceful Christmas: Wisconsin Homemakers Seek to Remake American Culture," *Wisconsin Magazine of History* 93, no. 2 (Winter

2009-2010): 2–15; Ted Moore, "Democratizing the Air: The Salt Lake Women's Chamber of Commerce and Air Pollution, 1936–1945," *Environmental History* 12 (January 2007): 80–106. As Grey Osterud points out in *Putting the Barn before the House: Women and Family Farming in Early-Twentieth-Century New York* (Ithaca, NY: Cornell University Press, 2012), rural families often rejected urban gender models, yet more traditional gendered beliefs as well as the agency described by Osterud fueled rural women's sense of environmental authority and responsibility. See for example, Nancy C. Unger, "The We Say What We Think Club," *Wisconsin Magazine of History* 90, no. 1 (Autumn 2006): 16–27.

128. Jan C. Dawson, "'Lady Lookouts' in a 'Man's World' During World War II: A Reconsideration of American Women and Nature," *Journal of Women's History* 8 (Fall 1996): 99–113; Sarah Smith, "'They Sawed up a Storm,'" *Northern Logger & Timber Processor* 48 (April 2000): 8–10, 61–62; Roxane S. Palone, "Women in Forestry—Past and Present," *Pennsylvania Forests* 92 (Spring 2001): 7–10. For Canada, see Susanne Klausen, "The Plywood Girls: Women and Gender Ideology at the Port Alberni Plywood Plant, 1942–1991," *Labour/Le Travail* 41 (Spring 1998): 199–235.

129. See Elaine May, *Homeward Bound: American Families in the Cold War Era* (New York: Basic Books, 1988).

130. See Adam Rome, "'Give Earth a Chance': The Environmental Movement and the Sixties," *Journal of American History* 90, no. 2 (2003): 525–554.

131. See Jens Ivo Engels, "Gender Roles and German Anti-Nuclear Protest. The Women of Wyhl," In *Le Démon Moderne: La pollution dans le sociétés urbaines et industrielles d'Europe*, ed. Christoph Bernhardt and Geneviève Massard-Guilbaud (Clermont-Ferrand, France: Presses Universitaires Blaise-Pascal, 2002): 407–424; Nancy C. Unger, "Gendered Approaches to Environmental Justice," 17–34.

132. Virginia Kemp Fish, "'We Stopped the Monster': LAND in Retrospect," 1994, in LAND, Box 1, Folder 1, Wisconsin Historical Society, 5; and "Widening the Spectrum: The Oral History Technique and Its Use with LAND, a Grass-Roots Group," *Sociological Imagination*, 31, no. 2 (1994): 106.

133. See Nancy Langston, "Gender Transformed: Endocrine Disruptors in the Environment," in *Seeing Nature Through Gender*, ed. Virginia J. Scharff (Lawrence: University of Kansas Press, 2003), 129–66 and Langston, *Toxic Bodies: Hormone Disruptors and the Legacy of DES* (New Haven, CT: Yale University Press, 2010).

134. Donna Warnock, pamphlet "What Growthmania Does to Women and the Environment," circa 1985, Syracuse: Feminist Resources on Energy and Ecology, Box 14, File 6, Atlanta Feminist Lesbian Alliance, Rare Book, Manuscript, and Special Collections Library, Duke University.

135. Linda Lear, *Rachel Carson: Witness for Nature* (New York: Henry Holt, 1997), 259–60. See also Christopher Sellers, "Body, Place and the State: The Makings of an 'Environmentalist' Imaginary in the Post-World War II U.S.," *Radical History Review* 74 (Spring 1999): 50.

136. Rachel Carson, *Silent Spring*, 25th anniversary ed., (Boston: Houghton Mifflin, 1987), 8.

137. See Michael B. Smith, "'Silence, Miss Carson!': Science, Gender, and the Reception of *Silent Spring*," *Feminist Studies* 27, no. 3 (Fall 2001): 733–752; Maril Hazlett, "'Woman vs. Man vs Bugs,': Gender and Popular Ecology in Early Reactions to *Silent Spring*," *Environmental History* 9 (October 2004): 701–729.

138. Julia Corbett, "Women, Scientists, Agitators: Magazine Portrayal of Rachel Carson and Theo Colborn," *Journal of Communication* 51 (2001): 728–729.

139. Lear, *Rachel Carson*, 423.

140. Vera Norwood, "Rachel Carson," in *The American Radical,* ed. Mari Jo Buhle (New York: Routledge, 1994), 318. See also Maril Hazlett, "Voices from the Spring: *Silent Spring* and the Ecological Turn in American Health," in *Seeing Nature Through Gender,* 103–128.

141. Efforts continue to discredit her work. See Roger Meiners, Pierre Desrochers, and Andrew Morriss, eds., Silent Spring *at 50: The False Crises of Rachel Carson* (Washington, DC: Cato Institute, 2012).

142. See Mark Lytle, *The Gentle Subversive: Rachel Carson,* Silent Spring, *and the Rise of the Environmental Movement* (New York: Oxford University Press, 2007).

143. See Udall, "How Wilderness was Won," 98–105, and Norwood, "Carson," 313–15. Carson's life and contributions continue to attract serious study. See William Souder, *On a Farther Shore: The Life and Legacy of Rachel Carson, Author of* Silent Spring (New York: Broadway, 2013).

144. Rome, "Give Earth a Chance," 536–537.

145. See Scott Hamilton Dewey, "'Is This What We Came to Florida For?' Florida Women and the Fight against Air Pollution in the 1960s," in *Making Waves: Female Activists in Twentieth-Century Florida,* ed. Jack E. Davis and Kari Frederickson (Gainesville: University Press of Florida, 2003), 197–225; Catherine Howett, "Grounding Memory and Identity: Pioneering Garden Club Projects Documenting Historic Landscape Traditions of the American South," in *Design With Culture: Claiming America's Landscape Heritage,* ed. Charles A. Birnbaum and Mary V. Hughes (Charlottesville: University of Virginia Press, 2005), 19–38.

146. See Noel Sturgeon, *Ecofeminist Natures* (New York: Routledge, 1997); Karen Warren, ed., *Ecofeminism* (Bloomington: Indiana University Press, 1997); Mary Heather MacKinnon and Moni McIntyre, eds., *Readings in Ecology and Feminist Theology* (Kansas City, MO: Sheed and Ward, 1995); Barbara Epstein, "Ecofeminism and Grass-roots Environmentalism in the United States," in *Toxic Struggles: The Theory and Practice of Environmental Justice,* ed. Richard Hofrichter (Salt Lake City: University of Utah Press, 2002), 144–152; Mary Mellor, *Feminism & Ecology* (New York: New York University Press, 1997); Carolyn Merchant, *Earthcare: Women and the Environment* (New York: Routledge, 1995).

147. See Yonnette Fleming [aka Reign], "Earth Mother," *Sinister Wisdom* 63 (Winter 2004-2005): 88; Mona Domosh and Joni Seager, *Putting Women in Place; Feminist Geographers Make Sense of the World* (New York: Guilford Press, 2001).

148. Carolyn Merchant, *Radical Ecology* (New York: Routledge, 1992), 205.

149. Mary Ellen Shaw, "The Varieties of Goddess Experience: Feminist Pragmatism in the Study of Women's Spirituality" (PhD diss., University of Minnesota, 2001).

150. Vandana Shiva, quoted in Leach and Green, "Gender and Environmental History," 347.

151. See Melissa Leach, "Earth Mother Myths and Other Ecofeminist Fables: How a Strategic Notion Rose and Fell," *Development and Change* 38, no. 1 (January 2007): 67–85.

152. Suzanne Braun Levine and Mary Thom, *Bella Abzug* (New York: Farrar, Straus and Giroux, 2007), 61.

153. Warnock, "What Growthmania Does to Women." See also Chaone Mallory, "Ecofeminism and Forest Defense in Cascadia: Gender, Theory and Radical Activism," *Capitalism Nature Socialism* 17 (March 2006): 32–49.

154. Catriona Sandilands, "Lesbian Separatist Communities and the Experience of Nature," *Organization & Environment* 15 (June 2002): 131–163; Catherine Kleiner, "Nature's Lovers: The Erotics of Lesbian Land Communities, 1974–1984," in Scharff,

Seeing Nature, 242–262; Nancy C. Unger, "The Role of Nature in Lesbian Alternative Environments in the United States: From Jook Joints to Sisterspace," in *Queer Ecologies: Sex, Nature, Biopolitics and Desire*, ed. Catriona Sandilands-Mortimer (Bloomington: University of Indiana Press, 2010); Unger, "Women's Alternative Environments: Fostering Gender Identity by Striving to Remake the World," *Beyond Nature's Housekeepers*, 163–185.

155. Leach and Green, "Gender and Environmental History," 349, 366.

156. Ibid., 344; 350; 351; 353; 357; 365. Ecofeminism continues to be the subject of study and controversy. A sampling of the most recent works include Anne Stephens, *Ecofeminism and Systems Thinking* (New York: Routledge, 2013); Dianne Rocheleau, Barbara Thomas-Slayter, and Esther Wangari, eds., *Feminist Political Ecology: Global Issues and Local Experience* (New York: Routledge, 2013); Greta Gaard, Simon Estok, and Serpil Oppermann, eds., *International Perspectives in Feminist Ecocriticism* (New York: Routledge, 2013); Christine Cuomo, *Feminism and Ecological Communities* (New York: Routledge, 2012).

157. Gail Robinson, "Plundering the Powerless," *Environmental Action* (June 1979): 3. See Peter Eichstaedt, *If You Poison Us: Uranium and Native Americans* (Sante Fe: Red Crane Books, 1994); Doug Brugge, Timothy Benally, and Esther Yazzie-Lewis, *The Navajo People and Uranium Mining* (Albuquerque: University of New Mexico Press, 2007); Raye Ringholz, *Uranium Frenzy: Saga of the Nuclear West* (Logan: Utah State University Press, 2002); Judy Pasternak, *Yellow Dirt: An American Story of a Poisoned Land and a People Betrayed* (New York: Free Press, 2010).

158. Eda Gordon, "Health Study Exposes Water Contamination on Pine Ridge Reservation," *Black Hills Report* 5, no. 1 (March 1980): 1.

159. Loretta Schwartz, "Uranium Deaths at Crown Point," *Ms.* (October 1979): 81.

160. Tom Barry, "Bury my Lungs at Red Rock," *The Progressive* (February 1979): 25–28.

161. "The Women's Health and the Future Generations Workshop," WARN Report (June-December 1979): 33–42, LAND Papers, Box 7-file 14, Wisconsin Historical Society, Madison, Wisconsin. See also Unger, "Women, Sexuality, and Environmental Justice," and, for historical context, Mark A. Largent, *Breeding Contempt: The History of Forced Sterilization in the United States* (New Brunswick, NJ: Rutgers University Press, 2007).

162. WARN Report II, "The Sovereignty Workshop," (June-December 1979), LAND Papers, Box 7-file 14, Wisconsin Historical Society, Madison, Wisconsin.

163. See Dorceta Taylor, *Race, Gender, and American Environmentalism*, (U.S. Department of Agriculture, Forest Service, Pacific Northwest Research Station, 2002); Luke Cole and Sheila Foster, *From the Ground Up: Environmental Racism and the Rise of the Environmental Justice Movement* (New York: New York University Press, 2001).

164. In Stein, *New Perspectives*, see Valerie Ann Kaalund, "Witness to Truth: Black Women Heeding the Call for Environmental Justice," 78–92, and Julie Sze, "Gender, Asthma Politics, and Urban Environmental Justice Activism," 177–190. See also Robert D. Bullard, ed., *The Quest for Environmental Justice: Human Rights and the Politics of Pollution* (Berkeley: University of California Press, 2005); Andrea Simpson, "'Who Hears Their Cry?': African American Women and the Fight for Environmental Justice in Memphis, Tennessee," in *The Environmental Justice Reader: Politics, Poetics, and Pedagogy*, ed. Joni Adamson, Mei Mei Evans, and Rachel Stein (Tucson: University of Arizona Press, 2002), 82–104.

165. See Marie Bolton and Nancy C. Unger, "Pollution, Refineries, and People: Environmental Justice in Contra Costa County, 1980s" in *The Modern Demon: Pollution in Urban and Industrial Societies*, ed. Cristoph Bernhardt and Genevieve Massard-Guilbaud (Clermont-Ferrand: Presses Universitaires Blaise-Pascal, 2002), 425–437.

166. See Margaret Rose, "'Woman Power Will Stop Those Grapes': Chicana Organizers and Middle-class Female Supporters in the Farm Workers' Grape Boycott in Philadelphia, 1969–1970," *Journal of Women's History* 7, no. 4 (1995): 6–36; "From the Fields to the Picket Line: Huelga Women and the Boycott, 1965-1975," *California Labor History* 31, no. 3 (1990): 271–293; "Traditional and Nontraditional Patterns of Female Activism in the United Farm Workers of American, 1962–1980," *Frontiers* 11, no.1 (1990): 26–32.

167. Case studies of environmental racism abound. See Sylvia Hood Washington, *Packing Them In: An Archeology of Environmental Racism in Chicago, 1985-1954* (Lanham: Lexington Books, 2005); Melissa Checker, *Polluted Promises: Environmental Racism and the Search for Justice in a Southern Town* (New York: New York University Press, 2005); Eileen McGurty, *Transforming Environmentalism: Warren County; PCBs, and the Origins of Environmental Justice* (New Brunswick, NJ: Rutgers University Press, 2007). For a thoughtful investigation of how the role of race can be underappreciated even in well-known stories of environmental justice, see Blum, *Love Canal Revisited*.

168. Unger, "The Modern Environmental Justice Movement," in *Beyond Nature's Housekeepers*, 187–214; Unger, "Gendered Approaches to Environmental Justice," 17–34; Unger with Marie Bolton, "'Mother Nature is Getting Angrier': Turning Sacred Navajo Land into a Toxic Environment," *Environmental Crisis and Human Costs* (Alcala de Henares: Instituto Franklin, forthcoming 2014); Robert D. Bullard and Damu Smith, "Women Warriors of Color on the Front Line," in *The Quest for Environmental Justice: Human Rights and the Politics of Pollution*, ed. Robert Bullard (Berkeley: University of California Press, 2005), 62–84; Andrea Simpson, "Who Hears Their Cry?: African American Women and the Fight for Environmental Justice in Memphis, Tennessee," in *The Environmental Justice Reader: Politics, Poetics, and Pedagogy*, ed. Jodi Adamson, Mei Mei Evans, Rachel Stein (Tucson: University of Arizona Press, 2002); Giovanna Di Chiro, "Environmental Justice from the Grassroots: Reflections on History, Gender, and Expertise," in *The Struggle for Ecological Democracy: Environmental Justice Movements in the United States*, ed. Daniel Faber (New York: Guilford Press, 1998), 104–136. See also Barbara Epstein, "Grassroots Environmental Activism: The Toxics Movement and Directions for Social Change," in *Earth, Air, Fire, Water: Humanistic Studies of the Environment*, ed. Jill Kerr Conway, Kenneth Kiniston, and Leo Marx (Amherst: University of Massachusetts Press, 1999); Shannon Elizabeth Bell, *Our Roots Run Deep as Ironweed: Appalachian Women and the Fight for Environmental Justice* (Urbana: University of Illinois Press, 2013); Peggy Frankland and Susan Tucker, *Women Pioneers of the Louisiana Environmental Movement* (Jackson: University Press of Mississippi, 2013); Joyce Barry, *Standing Our Ground: Women, Environmental Justice, and the Fight to End Mountaintop Removal* (Athens: Ohio University Press, 2012).

169. For the Chipko movement, see Vandana Shiva, *Staying Alive: Women, Ecology and Survival in India* (New Delhi: Kali for Women, 1988); Willow Ann Sirch, *Eco-Women: Protectors of the Earth* (Golden, CO: Fulcrum Publishing, 1996); and Shobita Jain, "Standing Up for Trees: Women's Role in the Chipko Movement," *Unasylva* 36, no. 4 (1984): 12–20. Global studies include Louis Fortmann and Dianne Rocheleau, "Women and Agroforestry: Four Myths and Three Case Studies," *Women in Natural Resources* 9, no. 2

(1987): 35–44, 46, 51; Rosemary Ruether, ed., *Women Healing Earth: Third World Women on Ecology, Feminism, and Religion* (Maryknoll, NY: Orbis Books, 1996); William Beinart and Joann McGregor, *Social History & African Environments* (Athens: Ohio University Press, 2003); Jean Davidson, ed., *Agriculture, Women, and Land: The African Experience* (Boulder, CO: Westview Press, 1988); Anoja Wickramasinghe, *Deforestation, Women and Forestry: The Case of Sri Lanka* (Institute for Development Research Amsterdam by International Books, 1994); Richard A. Schroeder, *Shady Practices: Agroforestry and Gender Politics in the Gambia* (Berkeley: University of California Press, 1999); Allison Goebel, *Gender and Land Reform: The Zimbabwe Experience* (Montreal: McGill-Queen's University Press, 2006); Arun Agrawal and K. Sivaramakrishnan, eds., *Agrarian Environments: Resources, Representation, and Rule In India* (Durham, NC: Duke University Press, 2000); Antonia Finnane, "Water, Love, and Labor: Aspects of a Gendered Environment," in *Sediments of Time: Environment and Society in Chinese History*, ed. Mark Elvin and Ts'ui-jung Lui (Cambridge: Cambridge University Press, 1998): 657–690; Philippa Crawford and Philippa Maddern, eds., *Women as Australian Citizens: Underlying Histories* (Carlton South, Victoria, Australia: Melbourne University Publishing, 2001); Melody Hessing, Rebecca Raglon, Catriona Sandilands, eds., *This Elusive Land: Women and the Canadian Environment* (Vancouver: University of British Columbia Press, 2005); Engels, "Gender Roles and German Anti-Nuclear Protest," *Le Démon Modern*, 407–424; Constance E. Campbell, "On the Front Lines But Struggling for Voice: Women in the Rubber Tappers' Defence of the Amazon Forest," *Ecologist* 27 (March/April 1997): 46–54; Thomas Klubok, *Contested Communities: Class, Gender, and Politics in Chile's El Teniente Copper Mine, 1094–1951* (Durham, NC: Duke University Press, 1998).

170. See Bina Agarwal, "Re-sounding the Alert: Gender, Resources, and Community Action," *World Development* 25, no., 9 (1973): 1372–1381; Agarwal, "Gender and Forest Conservation: The Impact of Women's Participation in Community Forest Governance," *Ecological Economics* 68, no. 11 (September 2009): 2785–2799; Agarwal, "The Gender and Environment Debate: Lessons From India," *Feminist Studies* 18 no. 1 (Spring 1992): 119–159.

171. See Agarwal, "Does Women's Proportional Strength Affect Their Participation? Governing Local Forests in South Asia," *World Development* 38, no. 1 (January 2010): 98–112; Agarwal, "A Challenge for Ecofeminism: Gender, Greening, and Community Forestry in India," *Women and Environments International Magazine* 52/53 (Fall 2001): 12–16. See also Leach and Green, "Gender and Environmental History," 358–360.

172. Carol MacCormack and Marilyn Strathern, eds., *Nature, Culture, and Gender* (Cambridge: Cambridge University Press, 1980).

173. Carolyn E. Sachs, *Gendered Fields: Rural Women, Agriculture, and Environment* (Boulder, CO: Westview Press, 1996); Glenda Riley, *Taking Land, Breaking Land: Women Colonizing in the American West and Kenya, 1840–1940* (Albuquerque: University of New Mexico Press, 2003).

174. William Robinson, "(Mal)Development in Central America: Globalization and Social Change," *Development and Change* 29, no. 3 (July 1998): 467–498; Andres Serbin, "Transnational Relations and Regionalism in the Caribbean," *Annals of the American Academy of Political and Social Science* 533 (May 1994): 139–161.

175. Millie Thayer, *Making Transnational Feminism: Rural Women, NGO Activists, and Northern Donors in Brazil* (New York: Routledge, 2009); Lisa Marie Aubrey, *The Politics of Development Co-operation: NGOs, Gender and Partnership in Kenya*

(New York: Routledge, 1997); Elisabeth Friedman, Kathryn Hochstetler, and Ann Marie Clark, "Sovereign Limits and Regional Opportunities for Global Civil Society in Latin America," *Latin American Research Review* 36, no. 3 (2001): 7–35.

176. Considerable space is dedicated in Schrpefer and Sackman's "Gender" to a gendered critique of William Cronon's "The Trouble with Wilderness; Or, Getting Back to the Wrong Nature," in *Out of the Woods: Essays in Environmental History*, ed. Char Miller and Hal Rothman (Pittsburgh: University of Pittsburgh Press, 1997): 28–50.

177. Gary Polakovic, "Are Women Greener Than Men?" *Los Angeles Times*, June 13, 2012. See also Lynette C. Zelezny, Poh-Pheng Chua, and Christina Aldrich, "Elaborating on Gender Differences in Environmentalism," *Journal of Social Issues* 56, no. 2 (2000): 443–457; Steven Arnocky, "Gender Differences in Environmentalism: The Mediating Role of Emotional Empathy," *Current Research in Social Psychology*, February 19, 2011, http://www.uiowa.edu/~grpproc/crisp/crisp16_9.pdf ; Mark Somma and Sue Tolleson-Rinehart, "Tracking the Elusive Green Women: Sex, Environmentalism, and Feminism in the United States and Europe," *Political Research Quarterly* 50, no. 1 (March 1997): 153–169; Amy Caiazza and Allison Barrett, "Engaging Women in Environmental Activism: Recommendations for Rachel's Network," Institute for Women's Policy Research, September 2003, http://www.csu.edu/cerc/researchreports/documents/EngagingWomenInEnvironmentalActivism2003.pdf.

178. That list includes women like the late Wangari Maathai, Kenyan founder of the Green Belt Movement and winner of the 2004 Nobel Peace Prize.

179. See Kari Norgaard and Richard York, "Gender Equality and State Environmentalism," *Gender and Society* 19, no. 4 (August 2005): 506–522.

CHAPTER 22

....

CONQUEST TO CONVALESCENCE

Nature and Nation in United States History

....

WILLIAM DEVERELL

> How much has been lost in our short years as a nation, how much have
> we to be nostalgic about.
>
> —Donald Worster[1]

In the spring of 1850, the Philadelphia physician John Kearsley Mitchell made his first visit to Niagara Falls. A prominent antebellum medical figure, Mitchell helped further Philadelphia's reputation as the *urbs medica* of nineteenth-century America. Among a small contingent of other surgeons and researchers at institutions such as Jefferson College and the University of Pennsylvania, Mitchell carried forward the intimate relationship between medicine and the City of Brotherly Love in the generation following that of the prime mover in American medicine, Benjamin Rush.

A middling, sentimental poet, Mitchell wrote verse and parlor lyrics ("Oh! Fly to the Prairie") applauding the nature he imagined in the continent's western flat lands and forests.[2] But nothing prepared him for what he saw at the waterfall. Niagara stunned him. Dr. Mitchell's response was physical, emotional, and intellectual. Given to overwork, prone to exhaustion and bouts of what the next generation would call neurasthenia, Mitchell looked upon the falls as a kind of necessary shock—virtually electric and certainly sublime—which could right and rejuvenate his body, if not his very soul. Writing that evening to his son Weir, on his way to becoming a physician himself, the elder Mitchell nearly lost the ability to fathom, much less explain, what he saw and felt, and why it was important to him, though he certainly tried. "My son," he wrote, "this has been to us a day of unutterable sublimity." The experience at the waterfall had produced both "awe and dread."

Moments of the day lay beyond confusion, and the acoustic assault of the experience resonated well after. "Amidst the roar of the cataracts, and the tempestuous call of the rapids," Mitchell wrote, the "mighty voiced message from Erie to Ontario, to whose ceaseless summons from creation the lower lake has designed no answer." In purpled prose spot on for the time and for someone of Mitchell's standing and station, he wrote on as if diagnosing a patient tumbling toward suicide.

> The outpourings of a majestic lake of clear pure water tremble as they approach the surge of the giant cliff as if afraid of the leap which forever separates them from their gentle source. Long before they reach the precipice they whiten as if with terror and seem to cling to every rock that offers them a vain support. Despite their rough detention, they stagger downwards, singing a mournful song, from the deep roar of the great billow to the tiny voice of the rippling wavelet, the whole treasury of tone is opened, as if a world were in the agony of expression. Heedlessly hastening onward or reluctantly holding back, the waters seem to acquire at length the calmness of despair, as they prepare for the desperate leap. Smooth and green come they to the rounded verge and over go they, green, green, but soon, the pent up air between rock and water, forces itself thro' the sheet in most fanciful forms, like snow wreaths, until the whole sheet of liquid green becomes as white as the driven snow and with verdure above and snow caps below Niagara looks like an Alp upon its head, save that the eternal noise and ceaseless motion break in upon the delusion.

Mitchell concluded his virtually medical narrative with the era's obligatory confession as to the impossibility of accurate representation of what he had witnessed. "Description of the indescribable is ever a folly. To half know Niagara one must see it and then one feels as if the soul could not take it in entirely."[3]

Dr. Mitchell played to antebellum type here. Standing there at the cataracts, he helps us open our story. Seeing Niagara up close was *supposed* to provoke aphasia in word or text. Men and women were supposed to feel small and puny in the face of God and His awesome waterworks, and the sublime experience rendered words, or at least the right words, inaccessible. Another narrator struggled similarly: "Over this scene the cloud of foam mysteriously moved, rising upward, so as to spread itself partly on the face of the fall, and partly on the face of the sky: while over all were seen the beautiful and soft colors of the rainbow, forming almost an entire circle, and crowning it with celestial glory. But it is in vain. The power, the sublimity, the beauty, the bliss of that spot, of that hour—it cannot be told."[4] The greatness of God, the greatness of the falls and their roar, even the greatness—as potential yet fully realized—of the young nation all rolled together as tributaries of promise and portent, of awe, of grandeur, and of purpose.

Abraham Lincoln, contemplating Niagara at about the same time ("Niagara Falls!" he wrote exultantly), seemed no less perplexed than John Mitchell. The existence of the falls made self-evident geologic, if pedantic, sense to Lincoln: "If the water moving onward in a great river, reaches a point where there is a perpendicular jog, of a hundred feet in descent, in the bottom of the river,—it is plain the water will have a violent and continuous plunge at that point. It is also plain the water, thus plunging, will foam, and roar, and send up a mist, continuously, in which last, during sunshine, there will be perpetual rain-bows." The straightforward cause and effect interplay of water, gravity, even sunshine: "the mere physical of Niagara Falls is only this." Yet mystery overwhelmed empirical detachment. Facts were "really a very small part of that world's wonder."[5] The enduring power of Niagara and its "indefinite past" lay in inchoate realms flowing from its inimitable, indescribable ability to "excite reflection" and "emotion."[6]

Nor could paint find the way to truth. None but the most talented narrative or history painters of the age could hope to capture Niagara at all accurately, and the base of the falls is figuratively littered with canvases discarded in frustration. "Niagara," writes Bryan Wolf, "like an Old Testament God jealous of betrayal by a given image, mocks those who aspire to record it."[7]

Laboring at his writing desk, John Mitchell admitted as much in paternal confession to his son. But he then closed his letter with what might at first seem a curious association—one that brought *nation* directly to Mitchell's dialogue and confusion about *nature*. Describing Niagara was a fool's errand, even for someone peering through the mists of the falls. It was every bit as chimerical as any attempt to render a full and accurate 1850 rendering of the young nation's greatest human symbol. Father offered a lesson to son: both waterfall and George Washington (who had been dead for half a century) were simply "too great for our finiteness to comprehend." It was best to admit failure at the outset rather than to try with words inept and ill-suited.

In pre-Civil War America, no one embodied the young republic better than George Washington. Benjamin Franklin, Thomas Jefferson, other founders: they had their place and their roles to play. But none but Washington *loomed*. Already deeply enmeshed with the meaning of the nation by "father-of-his country" paternity and enshrined by way of commemorative coinage and place-name geography, Washington towered over all others as the representative man. The pedestal upon which the Early Republic so enthusiastically placed him was a well-burnished heroic trope prior to the Civil War. And there he would stand, indescribable in greatness, until the martyred Lincoln gently ushered him aside in 1865.

Mitchell's juxtaposed placement of Niagara and Washington as great and unable to be described thus ought not to be especially surprising. As icons and monuments—in life, death, and spectacle—they suggested to antebellum Americans that the republic could produce national and natural greatness. But as such greatness could be literally beyond description; it would likely remain difficult to grasp fully.[8] Seeing the nation—either as an infant polity or through its major symbol—as Niagara make allegorical sense to Americans who believed national destiny to be at once propitious, divinely inspired, a tad frightening, and ultimately difficult to translate.

Nature and Nation and History

Mitchell's flabbergasted wordplay opens the fascinating—and disarmingly complex—relationship between nature and nation in American history. To be sure, the doctor's confusion is but a hint of a topic that inspires its own version of analytical awe and conceptual dread. Note that Mitchell never blended the meanings of Washington and Niagara; he used one as example for the other. Describing Niagara was *like* describing Washington. The topic only grows infinitely more complex when we find, as we do in the American historical record, nature and nation intertwined so much so—in metaphor, ideology, iconography, and idealism—as to render unclear where one begins and the other ends. Suffice it to say that, as we leave Dr. Mitchell in exhilarated letter-writing confusion, we begin with an obligatory caveat. This chapter is but a preface to a gargantuan topic which sprawls uncontained across scholarly fields, among them intellectual history, cultural history, environmental history, and political history. It is worth consideration and mention here, too, that the association of nature and nation flirts with analytical pitfalls, as the historian Richard White has argued.[9] In the mere pages which follow, we can but touch the surface of a theme which might itself be labeled "too great for our finiteness to comprehend." But over the falls we go.

We plunge in superb company. *Many* scholars have contributed works and ideas that, either chiefly or as they steep over time, have contemplated intrinsic, special, deleterious, or spiritual magnetism between nature and nation in American history.[10] The environmental history of the United States, writ large or small, is of course concerned with the question of how nation, nationalism, and nationhood are all tied to environmental realities and environmental perspectives.[11] I might somewhat unhelpfully cite here a veritably comprehensive bibliography of works published in the last twenty years—a high water mark, as it were, of the subfield—which have helped make the environmental history of the United States required reading for Americanist specialists generally.[12]

Chief among our guides is Perry Miller who, in both *Errand into the Wilderness* (1956) and especially *Nature's Nation* (1967), took the measure of American regard for, and response to, nature as an indicator of a young nation's brewing intellectual and literary ferment. Miller remains of critical importance in the early twenty-first century simply because his best work has the kind of rare intellectual heft borne of great talent, incessant curiosity, and penetrating analysis; he remains an essential guide to anyone interested in the intellectual trajectories of nation and nature through American culture. *Errand into the Wilderness* was concerned with colonial thought in the seventeenth and eighteenth centuries; while I am concerned with a later period in this chapter, the volume is nonetheless critical for positioning nature ("Nature" in Miller's interpretive schema) as central to national self-conception. Miller's contribution is more fully drawn within the essays of his posthumously published *Nature's Nation*. Rooted in embryonic nationalism was belief in the environment as bearing fundamental integrity for the American experiment in the New World. In Miller's erudite reckoning, the American psyche holds—and continues to hold—a fundamental faith in the *virtue* of the North American landscape. As American romanticists saw it, this virtue is both reflective of, and vulnerable to, the virtue of the people upon it. And though Miller ("the best historical mind of his generation, perhaps of his century," in Edmund Morgan's reckoning) was especially attuned to the ways in which the vision fell short of ideal (and thus provided a dark, even Gothic stain to American cultures and letters), the connection of nature to nation remained clear.[13] A symbiotic and ideally mutually reinforcing relationship can thus describe natural history and political economy; as thrives American nature, so thrives the American polity and the American experience. It is here, in the idealized form of that connection, where we locate enthusiasm and exultation, even triumph, in Dr. John Mitchell's pen. The Republic could be reflected in the cascade of Niagara; the falls could be described as Washingtonian; all three were great, powerful, virtuous.

As a way to think of the early republic, then, the American response to landscape and nature can be categorized as a search for knowledge amid confidence and presumptions of eventual, if not inevitable (if not infinite), greatness. We have John Winthrop's ever-useful and anticipatory "City on a Hill" insistence about comingled natural vistas and extant national virtues to highlight this categorization. There is no straight line to progress, however. Miller's cautionary voice echoes in the early twenty-first century to remind us of a contradiction in the passion of the Early Republic's embrace of nature. If, as Miller and his Puritan, Romantic, and Transcendentalist informants insisted, regard for the sublime in American nature helped define America as exceptional, what would happen when that very same nature (and even its prelapsarian inhabitants) fell before the axe, the railroad, myriad antebellum "internal improvements"? The anxiety Miller read in the works of Emerson, Thoreau, and Melville foreshadowed his own. Progress, as a self-congratulatory state of mind and being, "had moved his nation from covenant to revival to a romantic destiny as nature's favor child, and the moral universe had contracted accordingly." [14] Scholar of jeremiad, scholar as Jeremiah.

Marked especially, but by no means exclusively, by the expedition of Lewis and Clark, the early republic, as regards nation and nature, might best be characterized as a quest of equal parts for empire and empirical knowledge, with the search for each an overlapping and reinforcing enterprise.[15] "Thinking about Niagara" helps drive home the point, as the young nation and its philosophers (natural and otherwise) searched for the meaning and the promise of the United States in physical laws and physical monuments revealed. Scientific curiosity about landscape, land forms, paleontology, and the like could be, and were often, civilian passions.[16] But the federal and military supervision of botanical, topographical, and cartographic vistas to the west in the two generations prior to the Civil War drives home the point: the extent of the nation, as well as its very promise, could be fathomed by way of scientific and cultural reckoning with North American nature.[17] That such reckoning included native peoples within nature, even the state of nature, is a given. Examples abound in ever-widening echoes as the nineteenth century deepened. We refer to one lengthy example here: Captain Zebulon Pike's instructions for his expeditionary journey into the far West and New Spain, commencing in 1805. Note the usual confluence of national objectives with the desire for empirical environmental readings and measurements:

> HAVING completed your equipments, you are to proceed up the Mississippi with all possible diligence, taking the following instructions for your general government, which are to *yield,* to your discretion in all cases of exigency. You will please to take the course of the river, and calculate distances by time, noting rivers, creeks,

highlands, prairies, islands, rapids, shoals, mines, quarries, timber, water, soil, Indian villages, and settlements, in a diary to comprehend reflections on the winds and weather. It is interesting to government to be informed of the population and residence of the several Indian nations, of the quantity and species of skins and furs they barter per annum, and their relative price to goods'; of the tracts of country on which they generally make their hunts, and the people with whom they trade. You will be pleased to examine strictly, for an intermediate point between this place and the prairie des Chiens, suitable for a military post, and also on the Ouisconsin [sic], near its mouth, for a similar establishment, and will obtain the-consent of the Indians for their erection, informing them that they are intended to increase their trade, and ameliorate their condition. You will please to proceed to ascend the main branch of the river until you reach the source of it, or the season may forbid your further progress without endangering your return before the waters are frozen up. You will endeavour to ascertain the latitude of the most remarkable places in your route, with the extent of the navigation, and the direction of the different rivers winch fall into the Mississippi; and you will not fail to procure specimens of whatever you may find curious, in the mineral, vegetable, or animaı kingdoms, to be rendered at this place. In your course you are to spare no pains to conciliate the Indians, and to attach them to the United States; and you may invite the great chiefs of such distant nations as have not been at this place, to pay me [General James Wilkinson] a visit.[18]

This outward, largely though not uniformly, western reach emanated from many places. The infant capital city of Washington was of course critical as the seat of government and the War Department. But the nation was stitched together by way of territorial acquisitiveness via many centers of power, opportunity, and influence. New York and Boston played critical roles in supplying adventurers and venture capital, as did other cities large and small in the antebellum period. None, though, were as important as Philadelphia. That cradle of American nationalism furthered the connection of nature and nation like no other locale. The early republic's center of publishing and learning—where Lewis and Clark stopped off for pre-expedition instrument training—might even be called the capital city of Manifest Destiny. It was Philadelphia that provided much of the ideological, intellectual, and financial adhesive to attach nation to nature prior to the Civil War. There Charles Willson Peale showed off the continent's natural history mysteries in his eponymous museum; numerous distinguished publishing houses brought forth the best of scientific and expeditionary writings about the North American landscape; and Philadelphia merchants, lawyers, and bankers helped keep the Manifest Destiny drumbeat going through speeches, speculative investment, and generally xenophobic rhetoric.[19] The antebellum zeitgeist demanded that nation envision and then quickly claim its national apogee through exuberant conquest and dominion—over land and landscape, over indigenous people, over other nations. That vision, as sanguinary as it was bold, was sprinkled with comforting prophecy and religious obligation. "That it is one of the designs of Providence that the United States shall go forward, develop new resources, acquire new possessions, and new power, and spread whatever they have, of truth, goodness and justice in their institutions over the earth, is very evident," insisted one proud commentator in the usual parlance of nationalizing inevitability.

> Nothing can prevent this. The growth of a nation cannot be stayed. Its march may be bloody—war, for long years, may be its condition, but nevertheless it advances. Republican Rome illustrated this over two thousand years ago. At the point of the sword, and on the wings of her eagles, and through conquest, she spread art, science, philosophy, and all the treasures of civilization over the entire earth, and thus prepared the way for Christianity.[20]

Way back when, as early as in the anxious writings of seventeenth-century Puritans, Perry Miller saw Gothic darkness amid all the triumphalism. He could see the coming transition from, as Joyce Appleby has phrased it, a "celebration of progress into the history of declension."[21] The attack upon nature lay at the center of this moral and other degeneration. The conquest of North America, its landscape, and its indigenous people (a linked enterprise, to be sure) threatened to obliterate any visions of an Edenic environmental tabula rasa upon which Americans could envision a perfectible future. Not only could the conquest of nature mean the end of a critical phase of history (without wilderness, there can no longer be any errand into it), but the land was forever bloodied, even saturated. Puritan guilt lingered, certainly enough to question occasionally, then more frequently, if destruction of native peoples had truly been called for in the divine scheme. And the conquest of the landscape only hastened the great rupture over how that "virgin land" was to be worked: by laborers enslaved or by laborers free? Perhaps the sectional clash was indeed what many a theologian insisted: retribution for getting it all so wrong in the first place.

Nature and Nation Postbellum

National aggrandizement—expressed as territorial cession and xenophobic bluster—of course did not end with some arbitrary Manifest Destiny sunset. Yet the rupture that was the Civil War did provoke, and even require, new visions of the nation's nature. It did so in the 1860s, and it ought to in our time as well for scholars tracing political, cultural, and environmental trajectories through American history. On the one hand, the war's onset was all about the implantation of one labor system versus another in the vast new territorial gains of the conquered far west and Southwest; decades of tug-of-war and the eventual capitulation to violence was an increasingly futile sectional debate about land, labor, and the future of the Union. The war's resolution, the maintenance of the Union, and the eradication of slavery fostered fundamental changes in the ways in which national destiny and even the meanings of American nationalism worked in relation to nature. And on a smaller scale, as historian Lisa Brady has recently noted, even the war's destruction, measured against an environmental metric, suggests ways in which the struggle wounded the land just as it killed and wounded people and animals.[22]

Even ecstatic Niagara could be viewed differently in the aftermath of the Civil War, its metaphorical power as a mysterious embodiment of nature's promise for America shaken in the postbellum new order of things. The best example of this rethinking comes not from an American. The historian Thomas Carlyle, stewing in bitter antagonism over reform efforts in Great Britain, brought out his long screed of a pamphlet "Shooting Niagara, and After?" in 1867. A crude polemic widely panned in the United States, Carlyle's rant insisted that the current of democracy—too much, too fast—in Great Britain (and by analogy and extension the United States) was about to shoot over the allegorical falls and plunge in destruction to the very bottom of the catch basin of history.[23]

To a particularly keen observer like Walt Whitman, Carlyle's metaphor fell far short of apt or accurate. Grabbing hold of Niagara as the way to illustrate the cacophonous excess of democracy was not only attacking something the Good, Gray Poet exalted, but made for poor 1860s geography and even worse timing. In taking on Carlyle and his environmental metaphor for political decline, Whitman celebrated the sad but profound sacrifice of the Union dead and wounded. It was they who had proved democracy's future and muscle, even if in doing so they grimly illustrated the nation's war-torn weakness. Whitman answered Carlyle with a different metaphor: "We have seen this race proved by wholesale by drearier, yet more fearful tests—the wound, the amputation, the shatter'd face or limb, the slow hot fever, long impatient anchorage in bed, and all the forms of maiming, operation and disease. Alas! America we have seen, though only in her early youth, already to the hospital brought."[24]

Here Whitman suggests something very important about the Civil War and its impact upon the nation's ideas about nature. Carlyle's attempt to draw Niagara into an admonishment of democracy's leveling ways was anachronistic, and likely outright wrong. What is more, as Whitman suggests, Carlyle insulted the honored place of Niagara in capturing the potential of the Early Republic. Democracy was alive, though wounded, by the sectional clash. The future depended on convalescence, so that the terrible obligation of warfare in the name of union and freedom might prove ultimately strengthening. The nation's people, even its democratic essence, were also moving—literally—away from the ease of Niagara allegories and symbols. The vertical movement of the cascade had been now met by the horizontal movement of westering Americans. The touchstone that had once been the upper reaches of the Niagara River at the US-Canada border had lost its illustrative power. It would not be long, Whitman wrote of post-Civil War America, before "the dominion-heart of America will be far inland, toward the West."[25]

That said, it still made sense to characterize nation and nature together, if bound now by sinews of different form, shape, and purpose. "It is certain to me that the United States, by virtue of that war and its results," Whitman wrote, "are now ready to enter...upon their genuine career in history as no more torn and divided in their spinal requisites, but a great homogeneous Nation...a moral and political unity in variety, such as Nature shows in her grandest physical works."[26] The momentous transition borne of fratricide's blood carried with it obligation. Were the hundred-year-old nation to survive, it would have to become "entirely adjusted to the West."

CONVALESCENCE IN NATURE

What did such adjustment entail? If we can, albeit somewhat crudely, characterize the relationship between nation and nature in the antebellum period as marked by confusion, awe, aggressive possession, and scientific curiosity, the post-Civil War era introduced a new vision of nature as a tonic to the psychological and physiological insults of war. Broken and bloodied, the nation and its hundreds of thousands of wounded forged a new vision of nature as convalescent. Immersion in nature offered the suggestion, if not the very promise, of healing, for both nation and individuals alike, following what was newly indescribable in America: the trauma of the Civil War.

The convalescent vision of the postwar nation seemed now to incorporate the west— its land, vistas, landscapes, and environment—as fundamental to healing the nation and its people. In this way, might we at least metaphorically expand the point by pointing out that the transcontinental railroad project completed just after the war was a suture laid across the land, tying North and South together in reunion, terminating in the hopeful west?

Imagine the figure of Frederick Law Olmsted. Eyewitness to the horrific destruction of the war as Secretary of the United States Sanitary Commission, Olmsted despaired as the nation became, in his memorable phrasing, a "Republic of suffering." Yet by 1864, relocated briefly to California, Olmsted envisioned a breathtaking Sierra Nevada valley as a convalescent, even redemptive, site for national healing. Thus was the movement to preserve Yosemite by federal protections furthered, and thus was the movement toward the creation of national parks inaugurated.[27] That the idea of a national park system—the *ideology* of a national park system—forever altered the relationship between nature and nation is a given (though undeniably important at that). But histories of the American national parks do the most service and most analytical good when they acknowledge the political, cultural, and convalescent contexts in which they were imagined and created. In other words, the later and tireless advocacy of John Muir on behalf of Yosemite and other wilderness spaces mattered. But so too, as Muir himself knew, did the trauma of the Civil War and an awakening to the belief that nature could be as a balm to the soul, psyche, or body. As we contemplate nature and nation, especially in the nineteenth century, we must reinsert the Civil War into the transition between manifest destiny and the rise of conservation; it is simply too important not to take its full measure and impact. In the postwar period (and even during the war, as seen with 1864's federal protection of Yosemite), a shattered, wishful nation and its wounded added convalescence to the list of benefits that nature offered.[28] The connective tissue between nature and nation had taken new form, and new urgency.

Ironically, I might best illustrate this transition to a convalescent nature by returning to Philadelphia and the orbit of John Kearsley Mitchell, the physician whose rhapsodic take on Niagara started this chapters's voyage. Recall that Mitchell looked upon Niagara as did legions of antebellum Americans: wild, infinite, sublime, and awesome, Niagara roared with the potential of the young nation. It might also help individuals like Dr. John Mitchell return to health, but it would do so as if by an electroshock of the sublime.

In the years following the war, hundreds of thousands lay wounded, as did the nation. What was needed was convalescence, on grand and simple scales. Might nature be viewed through a different prism, one that pushed aside the jolt of the sublime in favor of nature as more soothing, passive, and inviting of repose?

It is not merely coincidental that we use as illustration of our aims here the two Dr. Mitchells, the father John and the son Silas Weir. As the Civil War ebbed, the medical response to environment and nature is an especially important feature of the nation/nature history we are trying to tell. We are especially drawn to a circle in and around the Gilded Age medical department of the University of Pennsylvania, then the center of neurological research in the nation.[29]

The Gilded Age was, as many an historian and other scholar has suggested, an age of neurasthenia. Generally a nonspecific medical diagnosis for a grab-bag association of ills, neurasthenia ("nerve weakness") tended to be diagnosed in the upper class, seems to have been gendered, and was—not coincidentally—neatly bounded by the end of the Civil War and the early ascent of Freudianism.[30] This is a period marked by the postwar lifespan of a Civil War soldier who lived through the war and his wounds. For the purposes of establishing chronological brackets for this essay, I will also suggest that as the "convalescent period" of nature and nation wanes, it gives way to a "conservation period" in which the Progressive Era's attachment to preservation and conservation (linked, if often countervailing, ideas) rises to prominence at the close of the First World War.[31]

At the center of the neurasthenic world in the United States stood Philadelphia physician Silas Weir Mitchell. Along with Sir William Ohsler and Harvey Cushing, both of whom he knew well, Mitchell was among the best known physicians of the era. A lot has been written about Weir Mitchell. He is a charismatic historical figure: vain, smart, driven, and multifaceted, and no less an observer of the post-Civil War era than John Hay called him (and this was in praise) "the most impossible man of my generation."[32] In the felicitous phrasing of his best biographer, he was "almost a genius." For in addition to his medical practice and medical research, Mitchell found time and had the discipline to write novels, short stories, poetry, and historical nonfiction. He collected autographs, and bric-a-brac: an Inca arylballus and Zuni pottery. While perhaps tempting, it would not be fair to characterize Mitchell as a mere dilettante. Several novels, now largely forgotten, were bestsellers of the day. He was sought after for both literary and medical commentary by the salon, newspaper, and magazine world of his day, and, at the apex of his fame around the turn of the century, his was a recognizable name in international literary and medical circles. He has been called, with justification, "the father of American neurology." "No one ever knew how he did all this," is how one writer sums up Mitchell's proclivities and productiveness.[33]

A desultory student, Weir Mitchell nonetheless began college at fifteen at the University of Pennsylvania. He studied some and misbehaved as well. At sixteen, much as his father had also experienced as a young man, Mitchell began to spit blood. In the medical fashion of the day, as both faith in outdoor pursuits as well as a "it can't do any harm" kind of fatalism when it came to respiratory and pulmonary disorders, Mitchell, who had to withdraw from the university, was ordered to spend more time outdoors. This he did, thus crystallizing both a long life's passion for hiking, camping, and fishing and a faith in the physiological and psychological good that could come from such pursuits.

Weir Mitchell recovered. He studied medicine at Philadelphia's Jefferson Medical College, where his father was a faculty member, and graduated at twenty-one. He then went for a year to England and Paris, bought a microscope, and began to concentrate on his medical studies. Both parents continued to express concern for his health. "If you study too hard [and] exercise too little," his father cautioned him, "you may also have a typhoid exemplification of the same truth." Writing later with similar concern, Mitchell's mother admonished her son and insisted that he "come home fat."[34]

Bright, privileged, and interested in medicine in the *urbs medica* of nineteenth-century America, Weir Mitchell would have likely been able to carve out a fine, even distinguished, career had the Civil War not happened. But the war did happen. And by it and through it, Mitchell was transformed from "simply another family physician in the nineteenth century Philadelphia pattern" to something else entirely.[35] And within this personal transition, we can divine far greater change in the ways in which America and Americans looked to nature to help heal wounds of disunion.

When he returned to the United States in the 1850s, Mitchell threw himself into exacting medical research, much of it involving vivisection. He made careful, thorough, and pathbreaking studies of rattlesnake venom. Then the war arrived. As a young doctor, Silas Weir Mitchell became interested in nervous disease during the Civil War. Though offered a regular Army position as a Brigade Surgeon, Mitchell opted in 1862 for contract work with the Union Army so that he could stay in or near Philadelphia. He trained at several Civil War convalescent and surgical hospitals in Philadelphia (there were fifty or more). The city was, as historian Matthew Gallman notes, "close enough to the seat of war that Union soldiers were a familiar sight on its streets, and after each major eastern battle the city's hospitals filled with wounded."[36] The writer and Phildelphian Struthers Burt put it more graphically. The city was, he wrote, inundated by the "river of wounded" regularly washing in from the front.[37] With other young doctors and surgeons, several of whom, such as William Williams Keen and Jacob Mendes Da Costa, would become every bit as influential, Mitchell worked in Turner's Lane Hospital, the Union Army's 400-bed facility for diseases of the nervous system. Turner's Lane was the first neurological hospital in the United States, specifically designed, partly at Mitchell's suggestion, to treat injuries and disorders of the central and peripheral nervous system.[38] At Turner's Lane, Mitchell observed and treated patients with all manner of wounds and disease; he and his colleagues took thousands of pages of observational, preoperative, and postoperative notes. Mitchell was especially interested in what he called "causalgia," or the burning, often excruciating pain associated with gunshots. He and his colleagues experimented with a wide range of treatments: they massaged the limbs of their patients (or their stumps); they turned to "incessant use" of morphine (one patient received nearly 400 injections). Thoroughly a student of what he called "the therapeutics of despair," Mitchell took what his "vast experience" in the war taught him and eventually operated the neurological clinic at the Infirmary for Nervous Diseases, in addition to running an extensive private practice on aristocratic Walnut Street.[39]

I am interested here in Weir Mitchell primarily due to two of his preoccupations and influences: the Civil War and the American West, and how they (and he) helped to symbolize the refiguring of the nation/nature dyad in American culture. Mitchell never got over the Civil War and its ravages—it is the central theme of his fiction, appearing over and over again.[40] He and each of his three brothers served for the Union during the war, despite paternal ties to Virginia and initial indifference to the war, especially on Weir Mitchell's part.

In 1866, Edward Everett Hale, the editor of *Atlantic Monthly*, published Mitchell's anonymously authored short story "The Case of George Dedlow." The story is presented as an autobiography of the life of quadruple amputee Dr. George Dedlow, a fictional figure rendered so real in the story that Dedlow not only received mail in his name at Philadelphia's United States Army Hospital for Injuries and Diseases of the Nervous System (Turner's Lane) or at the facility colloquially known as "Stump Hospital" (Philadelphia's South Street Hospital); visitors also came to meet and console him.[41] Based entirely on first-hand knowledge of wartime amputees—if curiously autobiographical in some ways—the story prefigures a great deal of medical insight into the "phantom limb" phenomenon following amputation of an arm or leg.

Mitchell's most important medical work of the war itself, what amounts to the nonfictional companion published shortly after "George Dedlow," was a treatise on the treatment of wartime gunshot wounds published with two colleagues.[42] Later novels (a long span of nearly twenty years separates "Dedlow" from the 1884 publication of Mitchell's first novel *In War Time*) explore the fates and foibles of Civil War physicians.[43]

Near the end of the century, Mitchell fell increasingly under the spell of the Civil War, its trauma, and its continued hold upon him. An early student during the war itself of what would later be called battle fatigue, then shell shock, and, eventually post-traumatic stress disorder, Mitchell consulted with wartime commanders and inquired about their recollections of "the various phases of courage under fire" during the war.[44]

Working closely with his physician son John, Mitchell embarked on an important longitudinal study of Civil War wounded as they aged. He wrote to former patients, inquiring how their injuries affected them as they grew older. A long form letter was mailed to dozens of veterans who had been treated in Philadelphia during the war, asking them to reply to a series of questions about their wounds. The questionnaire was sophisticated and thorough, and, assisted by the War Department Surgeon General's office, the Mitchells consulted a master list of company and regimental rosters to find their informants.

As the fiftieth anniversary of Gettysburg approached, Mitchell chaired the Committee on Memorial to the Medical Officers Who Lost Their Lives in the Civil War. Months later, at the end of his life in 1914, Mitchell hallucinated that he was back at Gettysburg, treating the wounded in one of the battlefield tent hospitals. Mitchell insisted that the imaginary soldier he was attending to must undergo immediate amputation of a limb or he would surely die; as Robert Goler notes, "the traumas of bodily wounding during the Civil War continued to torture his soul."[45]

How to connect Mitchell with changing ideas of nature in the aftermath of the Civil War? One point of significance is simply to allow Mitchell's experience to serve as exclamation point for the enduring impact of the Civil War—it simply changed everything. At Mitchell's death, a cousin of his penned a long, poignant eulogy that speaks to the issue, setting Mitchell on the Civil War stage, forever to be shaped by it, never to leave it. "Not on soldiers only did the Civil War thus set its seal, but upon all patriotic men and women also. And so that generation saw a vision that we see not, and wrote with a pen that never can be ours." The eulogist continued: "To Dr. Mitchell came hundreds of the maimed and wounded from that conflict. Wounds, blood, agony, and noble courage were for him a daily sight. In hospitals he helped to save the men, or saw them die. Never could he be the same again. Whatever his thoughts thenceforth, deep down was that memory perpetual. . . . We can only mark, we cannot measure, the effect of the Civil War upon Dr. Mitchell."[46]

There is a companionate obsession (although perhaps "obsession" is too strong a word) to Weir Mitchell. He was almost, at a subconscious level, fixated on the Gilded Age American west, that very same west that Walt Whitman correctly insisted was coming to preoccupy the post-Civil War nation. The interest shines through Mitchell's fiction: an important character in a novel is Dr. Westerly and an entire novel is called Westways. The west figures into his poetry, and Mitchell took an important early trip to Yellowstone not long after its establishment as a national park in the 1870s, during which he dutifully recorded his travels, his awe, and his complaints, in a small leather bound diary.

It is instructive to spend some time with this journal. Middle-aged and successful, Mitchell took care to make certain that he was well outfitted for the trip. The journal includes his list of things to bring along: netting, gloves, shoes, rod, flies, breeches, cold cream. Throughout, the journal carefully notes the beauty of the far western landscape, and Mitchell was an especially observant student of color, even as he entered his thoughts in penciled shorthand. Near Salt Lake, for instance, where he started out on the last leg of the journey to Yellowstone, he noted the "amazing beauty of blue lake" and the "amazing yellows of dying grass on hills gray greens of sage velvety softness." He was equally interested in getting the language and colloquial dialects and descriptions of the Rocky Mountain West right. For example, he made sure to note that "to stretch your blanket" meant to exaggerate, and that "to be heeled well is to be armed."

While Mitchell seems to have been especially preoccupied with the trout fishing in Rocky Mountain streams –the fishing was "immense," even "absurd"—and lakes, he did note the beauty of the alpine landscape and "the pure lemon glow of sunset." At the Grand Tetons south of Yellowstone, he looked up from the valley floor near Jackson and saw the "3 splendid Tetons looking like the Matterhorn." His journal suggests that the trip meant a lot to him (and he would return to Yellowstone as an old man). One day was filled, for example, with "amazing variety and beauty not to be forgotten." Another, more odd comment (perhaps the far west was very exotic for the urbane Philadelphian?) in the journal is that, amid "unconceivable beauty," the "wonder is not the grotesqueness but the beauty." Ever the patrician, even while "roughing it" in the Rockies, Mitchell's companions included a party of "7 gentlemen" and a detachment of nearly two dozen soldiers. The mountain west would continue to play a role in Mitchell's life and, significantly, in his therapeutics.

Mitchell's rapidly ascending profile as a doctor for neurasthenics, as well as an innovative practitioner in the treatment of neurasthenia, insured a long list of patients, men and women alike, waiting to see him. Given neurasthenia's racial, gender, and class characteristics, as well as Mitchell's increasing fame, his patients included a number of the era's best-known people.

Mitchell divided his treatments for neurasthenic convalescence along sharp gender lines. Women he sent to bed, often for months at a time. The so-called "rest cure," which he more or less invented, excised all intellectual or physical activity from the patient; she was only to rest and gain weight. Weir Mitchell, a plaque in his honor in Philadelphia explains, "taught the use of rest for the nervous."[47]

Then there was the opposite prescription: Mitchell's "west cure." Weir Mitchell figured that nature had convalescent powers. As such, he sent his male patients outside and, as Americans were fast beginning to appreciate, the nation's big outdoors was the newly claimed American west, far from the theaters of the recent conflagration. For instance, Mitchell sent Thomas Eakins west to the Dakota Badlands to overcome neurasthenic exhaustion and socialize with cowboys, whom he later and memorably painted. The west was not the sole destination for Mitchell's nervous young man (camping, sailing, and canoeing along the Atlantic seaboard captured his attention as well), but, because of a single patient's experiences, the west became a fixture in the minds, if not the travel plans, of the generations brought low by the neurasthenic reaction to the Civil War.

NATURE, NATION, AND WEIR MITCHELL'S MOST INFLUENTIAL PATIENT

One such patient was Owen Wister, Weir Mitchell's cousin. It was Wister who so aptly eulogized Mitchell as molded by the Civil War. Presenting himself to his cousin as confused, aimless, and nervous, Wister sought expert advice in the 1880s. Weir Mitchell promptly ordered his anxious young cousin to Wyoming, and Owen Wister—called Dan by his family—dutifully headed to Jackson Hole. The rest was American cultural history. Much as Montana and Dakota Territory would invigorate myopic, asthmatic Teddy Roosevelt at the same time (a reinvigoration not at all unimportant to the future of nation and nature), Wyoming remade Owen Wister. The wilderness beauty of the landscape, where the Great Plains meet the Rocky Mountains, made him a writer, a regular visitor, and an eventual ranch owner, while influencing American literary and cultural history in ways germane to our discussion of nature and nation.

Dan Wister went west at the age of twenty-five. He followed in his cousin's 1870s footsteps, when the world's first national park had only recently been carved out of the rugged river and mountain country at the far northern corner of the territory. Wyoming was very far away the tree-named streets of aristocratic Philadelphia, far from the Wister family mansion where his famous grandmother, the actress Fanny Kemble, lived, far from the Harvard Yard, and far from the gaze of a mother in whose eyes he never quite seemed to measure up.

Perhaps some of Wister's problems were hereditary. Both his mother and father suffered from broken health at points in their lives. Owen J. Wister, a physician, was forced to suspend his medical practice for three years not long after the Civil War ended. Decades of overwork had shattered his health, and he suffered from migraines; a patient and friend wrote that he "has neuralgia & nervous weakness from overwork, cannot exert his brain *at all,* not even to read, & has given up his practice *entirely* for the present."[48] Neurasthenia by any other name. His wife Sarah, too, suffered bouts of ill health which prompted her to travel as an exercise in convalescence. The Civil War, too, brought about especially stressful circumstances in Sarah Butler Wister's life, given her mother's anti-slavery sentiments and the slave-keeping realities of her father, the Southerner Pierce Butler.[49] "Without a doubt," concludes physician and historian of medicine Steven J. Peitzman, "neurosis permeated this remarkable but troubled family."[50]

WEST FROM PHILADELPHIA

By the mid-1880s, the bulk of the nearly two thousand miles separating Owen Wister from Wyoming could be covered by train. Wister boarded one in Philadelphia and made his way to Wyoming Territory. Even before he made it to the Rockies, the pampered young man had already committed thoughts about the west in his diary. "One must come to the West to realize what one may have most probably believed all one's life long—that it is a very much bigger place than the East—and future America is just bubbling and seething in bare legs and pinafores here. I don't wonder a man never comes back after he has once been here for a few years."[51]

His cousin had provided an essential prescription: go outside and do things. Wister's daughter remembered that her father "planned to shoot big game, fish for trout, camp in the wilderness, and see the Indians." In this he joined legions of late nineteenth-century tourists. But, in his daughter's memory at least, the west painted for Owen Wister a different story. "Struck with wonder and delight," he "had the eye to see and the talent to portray the life unfolding in America." After what she called "six journeys for pleasure," but what we might call something more searching and hopefully therapeutic, Dan Wister quit his path to lawyering (he had graduated Harvard Law in 1888) and became a writer. In this he was both phenomenally successful and something of a literary meteor, shooting to rapid fame and then burning out in the second half of his life.

Not that Wister's western experiences took immediately. Dan Wister may have fallen hard for the west, for Wyoming, and for the Grand Tetons, but he cast about for a good while before inspiration struck after his twenties were behind him. The story is worth recounting here, as it ties Wister's clubby Philadelphia life to the trans-Rocky Mountain West in important and lasting ways. As Wister himself told it, he had been so enamored of his time in Wyoming, and believed it to be so regenerative of health and vigor, that he went back and forth a number of times throughout his late twenties. While the tourist genre had not yet fully congealed in the 1880s, Wister was dude ranching at the Tetons: learning about the West and "its wild glories," its people, its landscapes, history, and vistas, in carefully choreographed settings of comfort and ease. And he was clearly thinking about the nation and its future.

It worked, or at least for a while—Wister's health would again break in the early twentieth century. But that would be later. For the young Wister, the west had begun to mean a great deal in terms of both personal and professional redemption. Though he had already tried his hand at writing about the west, Wister later recalled a galvanizing moment. Returning to Philadelphia from one of his Wyoming sojourns in the fall of 1891, Dan Wister went, not surprisingly, to the club where he rented bachelor's quarters upstairs. There he met a friend: the well-to-do, profligate son of a highly regarded Shakespearean actor and nephew of famed architect Frank Furness. Over more than a few drinks, talk turned to the west. The west that Wister had encountered when Weir Mitchell first sent him there was, he could plainly see, fading away. He'd gone back and forth enough times, and for a half dozen years, to be able to notice the deep changes: more people, more tourists, more amenities, less nature. Why, Wister and his friend Walter Furness wondered, "wasn't some Kipling saving the sage-brush for American literature, before the sage-brush and all that it signified went the way of the California forty-niner, went the way of the Mississippi steam-boat, went the way of everything?"

It was not that these transitions had yet to receive comment. Wister remembered later thinking that his good friend Theodore Roosevelt "had seen the sage-brush true, had felt its poetry; and also Remington, who illustrated his articles so well.

> But what was fiction doing—fiction, the only thing that has always outlived fact? Must it be perpetual tea-cups? Was Alkali Ike in the comic papers the one figure which the jejune American imagination, always at full-cock to banter or to brag, could discern in that epic which was being lived at a gallop out in the sage-brush? "To hell with tea-cups and the great American laugh!" "Walter," Wister exclaimed

to his drinking companion, "I'm going to try it myself! I'm going to start this minute."[52]

And so, if this tale is to be believed, he did. He went upstairs to the library of the Philadelphia Club that night and wrote a good chunk of the short story "Hank's Woman." Within a few months, he'd written "How Lin McLean Went East." And with them, he was well on his way to inventing a genre of American literature: the modern western.

Did the therapies work? Did they cure Wister of his "nervous prostration?" Probably not. Letters home to his mother—with whom he shared an exceedingly difficult relationship—revealed him to be unhappy and anxious well into and throughout adulthood. Wister's daughter writes that her father was not happy until he married his cousin Mary Channing Wister in 1898; it was her mother's "steady nerves and health," Fanny Wister wrote years later, which "made all serene around her."[53] Children arrived quickly, eventually three boys and three girls. Wister hung Philadelphian Thomas Sully's oil portrait of his grandmother in the parlor of his house, and put two by Frederick Remington—who was now one of the illustrators for Wister's stories—in the front hall.

By the turn of the century, Wister was at work at his largest, most ambitious work. Admitting later the scale of his ambition, Wister wrote Oliver Wendell Holmes that he wished for his protagonist to be "a man of something like genius— the American genius—."[54] The result, named precisely for that protagonist, was *The Virginian: A Horseman of the Plains*, published in 1902.

With this novel (never out of print since), Weir Mitchell's cousin more or less knocked the good doctor off the best-seller lists of American letters. The book is many things, among them the most popular book by any Philadelphian save for Benjamin Franklin and his *Autobiography*.[55] And Owen Wister is certainly the most-read Philadelphian besides Franklin to this day. To be sure, the novel is at once a western and a tale of reunion and post-Civil War national redemption. It is all about the ways in which region and landscape act as postbellum convalescent institution (recall Whitman's observation of a wounded America "to the hospital brought"). The gist of the story, which was made into an important film just as the fiftieth anniversary of Gettysburg and *Birth of A Nation* approached, is of a white-hatted, white-horsed Confederate veteran who brings justice and civilization to the frontier West. There the unnamed hero (ever "the Virginian") meets and falls for Molly (Wister's real wife's nickname), a Yankee schoolteacher.[56] Eventually they marry, thus drawing North and South, the former sectional foes, together in matrimonial reunion. That is the first step. More significantly, with their many children who follow, they figuratively create or procreate a new nation in, of, and because of the west.

It would be difficult to exaggerate the cultural power of Wister's creative invention. The western, and the west it purports to describe and explore, is of surpassing importance in myriad facets of American life, and it has been thus since Wister released the Virginian into popular culture. I have attempted to sketch its genesis as inexorably tied to a period in which nature and nation, because of the searing impact of the Civil War, was refigured in American life and lives.

That convalescent moment, that convalescent *demand*, ought to be factored into the environmental history of the American nation as a transition between the bloody acquisitiveness of answering manifest destiny's divine or other obligations, and the early twentieth century's moves toward an exceptional regard for conservation. Only with convalescent nature—and especially a convalescent west—are we able to get from the early nineteenth century to the early twentieth. Only with convalescent nature are we able to get from Niagara Falls and the electric sublime to Yosemite and repose.

Wister himself anticipates the transition from convalescence to conservation. Recall his insistence that *The Virginian* had roots in his belief that no American Kipling was saving the Old West for posterity by way of narrative fiction. Though I believe convalescence (his own and the nation's) overwhelms other features of the novel and short stories, it is clear that Wister saw the [57]western landscape as rapidly evolving from "the Old West" to a lesser "New West."

The convalescent moment is fairly easily bracketed by the Civil War, given that conflagration's impact. An attempt to heal the nation by way of landscape does not cure all, nor does a belief in the healing power of landscape for individuals wrestling with all manner of disease. By the early twentieth century, as regards numerous regions all across (but not limited to) the west, it is as if the formerly fearful appraisals of miasmic landscapes had been replaced by uniformly optimistic assumptions of nature's healing powers.

With that comes this new fear, however, which Wister notes. Perhaps the west—a Turnerian west, a safety-valve west, a reinvigorating west, a wilderness west—was fast falling away? As scholars such as Turner fretted, might that mean the veritable loss of one side of the nature/nation binary? If nature were to go away, wouldn't nation suffer inevitably? If the nation had been forged from nature, if conquering nation had fostered American prominence, and if immersion in nature had healed a nation newly returned from war, what now if nature were no longer so robust?

Such concerns were central to the conservation moment of the early twentieth century. Perhaps such worries about finite natural resources and their fate did motivate Owen Wister within his factual search for personal redemption and fictional search for reunification of the nation. Wister claimed that he had to act, with his pen, before the west "went the way of everything." Such convictions of declension, and the price the health of the nation would pay for it, certainly did color the conservationist ethos of the latter nineteenth and early twentieth centuries, even to the point of echoing the earlier strands of guilt which Perry Miller so clearly saw and heard in the pages and words of Puritan claims of New World exceptionalism.

First set loose by convalescent hopes and needs, conservation and environmentalism had begun to remake the American relationship with nature by the early twentieth century. But the stakes had changed. While nature might no longer be able to make or heal the nation to the degree it once seemed to have boundless power to do, the nation could nonetheless add to its distinctiveness, and even distinction, by and through what it did to save nature. Prime mover positions had changed and flipped, and they had done so with remarkable historical velocity. Nation now clearly held dominion over nature, and amidst all the triumphalism of American power and might we can hear passages laced with guilt and anxiety. What the Puritans had worried about, what Perry Miller had both analyzed and amplified, seemed to have come to pass. Nation had beaten nature in a contest of the former's making, and what might be saved before all had gone away? Nature might have convalesced nation; but the moment had seemingly passed at which nation could return the gift.

In Donald Worster's brilliant characterization, modern American environmentalism is "that peculiar search for national atonement."[58] Americans yet may be Puritans all. Several things are clear. There is enough blame to spread guilt around. There is much to be guilty about.

NOTES

1. Donald Worster, *The Wealth of Nature: Environmental History and the Ecological Imagination* (New York: Oxford University Press, 1993), 4.
2. John Kearsley Mitchell, *Indecision, a Tale of the Far West; and Other Poems* (Philadelphia: E. L. Carey & A. Hart, 1839); Joseph Philip Knight, *Oh! Fly to the Prairie: A Song Dedicated to George S. Morris* (Philadelphia: Geo. W. Hewitt & Co, 1839) [words by Mitchell].
3. John Kearsley Mitchell to Silas Weir Mitchell, May 19, 1850; Silas Weir Mitchell Papers, College of Physicians and Surgeons, Philadelphia.
4. Horatio Adams Parsons, *Steele's book of Niagara Falls* (Oliver G. Steele, 1840), 105.
5. Abraham Lincoln (ca. late September, 1848), from Roy P. Basler, *The Collected Works of Abraham Lincoln*, 9 vols. (New Brunswick, NJ: Rutgers University Press, 1953): 2, 10–11.
6. Lincoln closed his unfinished paean to Niagara, which scholars date to the late 1840s, with praise for the infinite endurance of the falls: "Never dried, never froze, never slept, never rested." In cadence and reference to infinite temporality, the phrase seems to anticipate Lincoln's immortal Gettysburg address with its plea that "government of the people, by the people, for the people, shall not perish from the earth."
7. Bryan Wolf, "The Fall of Niagara," *Winterthur Portfolio* 22 (Spring 1987): 81–86. "The road to Niagara is paved with the work of artists who have failed to understand its secrets." See also Barbara Novak, *Nature and Culture: American Landscape and Painting, 1825–1875* (New York: Oxford University Press, 1980).
8. David Miller, among others, suggests that the Falls and Washington were actually in a bit of a contest over representation, with the former overtaking the other in popularity as the eighteenth century faded in favor of the nineteenth. See Miller, "Sublimity and the Civilizing Process," *American Quarterly* 38 (Winter 1986): 854–859.
9. In "The Nationalization of Nature," White offers a caution and a caveat regarding consideration of environment and environmental histories as necessarily bounded by political or national scales and/or perimeters. See White, "The Nationalization of Nature," *Journal of American History* 86 (December 1999): 976–986.
10. As but important primers, see, for example, Donald Worster, *Nature's Economy: A History of Ecological Ideas*, New ed., Studies in environment and history (Cambridge [Cambridgeshire]: Cambridge University Press, 1985); Donald Worster, *Dust Bowl: The*

Southern Plains in the 1930s (New York: Oxford University Press, 1979); Donald Worster, *Under Western Skies: Nature and History, and the American West* (New York: Oxford University Press, 1992); Annette Kolodny, *The Land Before Her: Fantasy and Experience of the American Frontiers, 1630–1860* (Chapel Hill: University of North Carolina Press, 1984); Annette Kolodny, *The Lay of the Land: Metaphor as Experience and History in American Life and Letters* (Chapel Hill: University of North Carolina Press, 1975); Frederick Jackson Turner, *The Frontier in American History* (New York: H. Holt and Company, 1920); William Cronon, ed., *Uncommon Ground: Rethinking the Human Place in Nature*, pbk. ed. (New York: W. W. Norton, 1996); William Cronon, *Nature's Metropolis: Chicago and the Great West* (New York: W. W. Norton, 1991); William Cronon, *Changes in the Land: Indians, Colonists, and the Ecology of New England*, 1st ed. (New York: Hill and Wang, 1983); Richard White, *The Organic Machine* (New York: Hill and Wang, 1995); Simon Schama, *Landscape and Memory* (New York: Knopf, 1995); Carolyn Merchant, *American Environmental History: An Introduction* (New York: Columbia University Press, 2007); Perry Miller, *Nature's Nation* (Cambridge, MA: Belknap Press of Harvard University Press, 1967); Perry Miller, *Errand into the Wilderness* (Cambridge, MA: Belknap Press of Harvard University Press, 1956). Two of Miller's most important successors as to the conception of nature's nation have been Leo Marx, who took the story forward past Transcendentalism well into the industrial age, and Roderick Nash, who connected Miller's Romantics with American conservationists and American environmentalism by the end of the nineteenth century (in a work first published the same year as *Nature's Nation*). See Marx, *The Machine in the Garden* (New York: Oxford University Press, 1964) and Nash, *Wilderness and the American Mind* (New Haven, CT: Yale University Press, 2001). For a slightly different analytical perspective, one that privileges geography and place over nature per se in the ongoing dialogue with nation and nationalism, see David Jacobson, *Place and Belonging in America* (Baltimore: Johns Hopkins University Press, 2002). Among the most important works addressing aspects of the nation and nature relationship in recent years is Mark Fiege, *The Republic of Nature* (Seattle: University of Washington Press, 2011).

11. Drawing upon Miller's work as inspiration and commemorative opportunity (twenty-five years after publication), a recent insightful compilation provides perspective and validation of Miller's continuing influence. See European Association for American Studies, *"Natures Nation" Revisited: American Concepts of Nature from Wonder to Ecological Crisis*, European contributions to American studies 4 (Amsterdam: VU University Press, 2003).

12. An excellent synthesis of American environmental history, to which this essay is indebted, is Douglas Cazaux Sackman, ed., *A Companion to American Environmental History* (Malden, MA: Blackwell-Wiley, 2010). Two newer textbook syntheses are helpful in that they both attempt to address the history of the nation with explicit attention to environmental history and, every bit as much, rest upon the insights of much recent scholarship. See Theodore Steinberg, *Down to Earth: Nature's Role in American History* (New York: Oxford University Press, 2002) and John Opie, *Nature's Nation: An Environmental History of the United States* (Fort Worth, TX: Harcourt Brace College Publishers, 1998). A thoughtful and more episodic treatment is Gunther Paul Barth, *Fleeting Moments: Nature and Culture in American History* (New York: Oxford University Press, 1990). For readings on the nature as commodity within American culture, see Alexander Wilson, *The Culture of Nature: North American Landscape from*

Disney to the Exxon Valdez (Cambridge, MA: Blackwell, 1992); Susan Davis, *Spectacular Nation: Corporate Culture and the Sea World Experience* (Berkeley: University of California Press, 1997).

13. Edmund Sears Morgan, *American Heroes: Profiles of Men and Women Who Shaped Early America* (New York: W. W. Norton, 2009), 251.

14. See, for example, the review by Neil Harris in *American Quarterly* 19 (Winter 1967): 725–730. Harris calls *Nature's Nation* "an epithet with ambiguous implications." "Moral universe" quote is from Joyce Appleby, "History as Art: Another View," *American Quarterly* 34, no. 1 (Spring 1982): 30.

15. For a concise assessment of eighteenth- and early-nineteenth-century visions of nature and nation, explored through genealogy and peregrinations throughout the Atlantic seaboard pre- and post-Revolution, see Thomas P. Slaughter, *The Natures of John and William Bartram*, 1st ed. (New York: Knopf, 1996).

16. The public dimension to scientific scrutiny of North America is a vast topic in antebellum American culture. Charles Willson Peale and Thomas Jefferson dominate the scene. First-rate introductions to the theme are David R. Brigham, *Public Culture in the Early Republic: Peale's Museum and Its Audience* (Washington: Smithsonian Institution Press, 1995); David C. Ward, *Charles Willson Peale: Art and Selfhood in the Early Republic* (Berkeley: University of California Press, 2004); Charles A. Miller, *Jefferson and Nature: An Interpretation* (Baltimore: Johns Hopkins University Press, 1988). Matthew Cordova Frankel, "Nature's Nation" Revisited: Citizenship and the Sublime in Thomas Jefferson's "Notes on the State of Virginia," *American Literature* 73, no. 4 (December 2001): 695–726.

17. Regardless of which edition, the handbook of exploration in the service of empire in the period under review remains William H. Goetzmann, *Army Exploration in the American West, 1803–1863*, Yale publications in American studies 4 (New Haven, CT: Yale University Press, 1959); see also Roger L. Nichols, *Stephen Long and American Frontier Exploration*, Oklahoma pbk. ed. (Norman: University of Oklahoma Press, 1995).

18. Zebulon Montgomery Pike, *Exploratory travels through the western territories of North America: comprising a voyage from St. Louis, on the Mississippi, to the source of that river, and a journey through the interior of Louisiana, and the north-eastern provinces of New Spain: Performed in the years 1805, 1806, 1807, by order of the government of the United States* (printed for Longman, Hurst, Rees, Orme, and Brown, 1811), xii-xiii.

19. See, generally, Russell F. Weigley, ed., *Philadelphia: A 300 Year History* (New York: W. W. Norton, 1982); see also Laura Rigal, *The American Manufactory: Art, Labor, and the World of Things in the Early Republic* (Princeton, NJ: Princeton University Press, 1998). And though it is perhaps stretching the elasticity of credulity more than is appropriate, might we view Peale's red curtain, which he pulls back in his famous 1822 self-portrait "The Artist in His Studio", as the very falls of Niagara? Beyond the venerable falls, to the West (and in Peale's displays), lay the natural resources and natural patrimony to remake America as the nineteenth century deepened.

20. *The Gazette of the union, golden rule, and odd fellows' family companion*, Saturday May 26,1849 (New York: Crampton and Clarke, 1849).

21. Appleby, "History as Art," 28.

22. Lisa Brady, *War upon the Land: Military Strategy and the Transformation of Southern Landscapes during the American Civil War* (Athens: University of Georgia Press, 2012).

23. Thomas Carlyle, *Shooting Niagara, and After?* (London: Chapman and Hall, 1867).

24. Walt Whitman, *Specimen Days and Collect...: Also Walt Whitman's Poems, Leaves of Grass* (Philadelphia: Rees Welsh & co, 1882), 217. Whitman's response to Carlyle softened some over time; see the helpful discussion in Brent E. Kinser, *The American Civil War in the Shaping of British Democracy* (Surrey, UK: 2011), 52–53.

25. Ibid., 222.

26. Ibid., 262.

27. Helpful here is the work of Lynn Ross-Bryant; see "Sacred Sites: Nature and Nation in the U.S. National Parks," *Religion and American Culture* 15 (Winter 2005): 31–62; Richard A Grusin, *Culture, Technology, and the Creation of America's National Parks*, Cambridge studies in American literature and culture 137 (Cambridge: Cambridge University Press, 2004). See also William Deverell, "Redemptive California? Re-Thinking the Post-Civil War," *Rethinking History* 11 (March 2007): 61–78.

28. See Donald Worster, *A Passion for Nature: The Life of John Muir* (Oxford: Oxford University Press, 2008). Faith in such protected landscapes as natural sites of individual and national reconstruction and healing remains deeply rooted in American culture. Following the attacks of September 11, 2001, the Secretary of the Interior invited Americans to visit the national parks for free over Veterans Day weekend: "What better places to begin [the] healing process than in our parks, where Americans can draw strength from national icons of freedom and peace from splendors of nature...and reconnect with the values that have made this nation great." Secretary of the Interior Gale Norton, October 4, 2001 news release, quoted (in epigraph) in Ross-Bryant, "Sacred Sites," p. 31.

29. See, for example, Francis Gosling, "Neurasthenia in Pennsylvania: A Perspective on the Origins of American Psychiatry, 1870–1910," *Journal of the History of Medicine and Allied Sciences, Inc.*, v. 40, 188–206.

30. In the words of one literary scholar, nineteenth-century neurasthenia encompassed a "nearly uninterpretably wide variety of symptoms." See Todd Robinson, "'There is not much Thrill about a Physiological Sin': Neurasthenia in Willa Cather's *The Professor's House, M/C Journal* 4, no. 3 (June 2001). While the symptoms and diagnosis were both widespread, Robinson perhaps reaches too far in suggesting that, within the first few decades of the twentieth century, "nearly every educated American had suffered under the rubric of 'neurasthenic.'"

31. As Marike Gijswijt-Hofstra has recently noted, "The First World War marked the more or less final retreat of neurasthenia." See Gijswijt-Hofstra, "Introduction: Cultures of Neurasthenia from Beard to the First World War," in Porter, 1–30.

32. Worster, *A Passion for Nature*.

33. The deftly stated confusion belongs to writer Struthers Burt from his book *Philadelphia: Holy Experiment* (Garden City, NY: Doubleday, 1945).

34. John K. Mitchell to Silas Weir Mitchell, July 5, 1851; Matilda Henry Mitchell to Silas Weir Mitchell, July 26, 1857; Silas Weir Mitchell Papers, College of Physicians and Surgeons, Philadelphia. Typhoid would carry away the elder Mitchell in 1858. The younger Mitchell would later note that "the British are too fat." S. Weir Mitchell, *Fat and Blood: And How to Make Them*, 2d ed. (Philadelphia: J. B. Lippincott & Co., 1879), 231.

35. Ernest Penney Earnest, *S. Weir Mitchell, Novelist and Physician* (Philadelphia: University of Pennsylvania Press, 1950), 45.

36. J. Matthew Gallman and ebrary, Inc., *Touched with Fire? Two Philadelphia Novelists Remember the Civil War*, Frank L. Klement lectures no. 11 (Milwaukee, WI: Marquette University Press, 2002), 10.

37. The writer Struthers Burt writes that Philadelphia's antebellum political and regional loyalties were split between North and South. But when "the great river of wounded came pouring back, all this changed." Philadelphia became a forthrightly Northern city. See Burt, *Philadelphia: Holy Experiment* (Garden City, NY: Doubleday, 1945).

38. See F. R. Freemon, "The first neurological research center: Turner's Lane Hospital during the American Civil War," *Journal of the History of the Neurosciences* 1 (135–142).

39. See Cerutti, "S. Weir Mitchell: The Early Years." As Tom Lutz points out, and Cerutti notes implicitly, the grievousness of Civil War wounds often meant that Mitchell had little to offer his patients but rest and food; this is likely what led him in later years to the 'rest cure' for neurasthenics, men and women alike: "since he had little to offer them than rest and nutriment, he had his first glimpse of their value by accident." See Lutz, "Varieties of Medical Experience: Doctors and Patients, Pysche and Soma in America," in Porter Neurasthenic special issue 51–76, quoted at 56.

40. Especially helpful in analyzing the Civil War's persistence in Mitchell's fiction is D. J. Canale, "S. Weir Mitchell's Prose and Poetry on the American Civil War," *Journal of the History of the Neurosciences* 13, no. 1 (2004): 7–21.

41. Robert J. Goler, "Loss and the Persistence of Memory: "The Case of George Dedlow" and Disabled Civil War Veterans," *Literature and Medicine* 23, no. 1 (Spring 2004): 160–183. See also Debra Journet, "Phantom Limbs and 'Body-Ego': S. Weir Mitchell's 'George Dedlow,'" *Mosaic* 23:1 (Winter 1990): 87–99. Mitchell clearly tied his Civil War experiences to his treatment of neurasthenia in the postwar period. As he noted in one of his medical works, amputation and the subsequent pain and trauma of a phantom limb could make "the strongest man ... scarcely less nervous than the most hysterical girl." See Bourke, "Silas Weir Mitchell's *The Case of George Dedlow*."

42. S. Weir Mitchell, *Gunshot Wounds, and Other Injuries of Nerves* (Philadelphia: J. B. Lippincott & Co., 1864).

43. Mitchell discusses the fictive use of the Civil War, especially in his novel *Westways*, in the *New York Times*, October 12, 1913; see "American History in Fiction." I do not think it coincidental that Mitchell returned to the war just as Gettysburg's fiftieth anniversary approached. *Westways* is published in the year of that half-century anniversary.

44. See copy of letter from "General Smith," n.d., to S. Weir Mitchell, Weir Mitchell Papers.

45. Robert J. Goler, "Loss and the Persistence of Memory: "The Case of George Dedlow" and Disabled Civil War Veterans," 181.

46. Perpetuity of memory continued to the edge of death. "During the last five days of his illness his wandering mind returned to those scenes; his wandering talk was of mutilation and bullets; he conversed and argued with that past. Therefore it is that when we read his tales and poems, no matter what be their subject, all come from a spirit over which had passed the great vision; every drop of ink is tinctured with the blood of the Civil War." Wister's eulogy, "S. Weir Mitchell: Man of Letters" was published in College of Physicians of Philadelphia, *S. Weir Mitchell: M.D., LL.D., F.R.S., 1829–1914: Memorial Addresses and Resolutions* (Philadelphia: s.n., 1914). See also, more generally, Canale, "S. Weir Mitchell's Prose and Poetry."

47. David M Rein, *S. Weir Mitchell as a Psychiatric Novelist* (New York: International Universities Press, 1952), 1.

48. Steven J. Peitzman, "'I Am Their Physician': Dr. Owen J. Wister of Germantown and His Too Many Patients," *Bulletin of the History of Medicine* 83 (2009): 261.

49. See, generally, Peitzman, "I Am Their Physician"; see also Catherine Clinton, *Fanny Kemble's Civil War* (New York: Oxford University Press, 2000).

50. Peitzman, "'I Am Their Physician,'" 270.

51. Owen Wister journal entry, July 2, 1885; Owen Wister Papers, Library of Congress.

52. For a fictionalized visit to that evening through vignette (one which utilizes Wister's journal entries), see John Lukacs, *A Thread of Years* (New Haven: Yale University Press, 1998), chapter one.

53. Owen Wister, *Owen Wister Out West: His Journals and Letters* (Chicago: University of Chicago Press, 1958), 14.

54. Ibid., 16. Holmes knew firsthand of the Civil War's traumas; he had been grievously wounded at Antietam (he was shot through the neck).

55. Nathaniel Burt, *The Perennial Philadelphians: The Anatomy of an American Aristocracy* (Philadelphia: University of Pennsylvania Press, 1999).

56. See Eric M. Augenstein, "Mary Channing Wister: An Unknown Legend," on the LaSalle University's Connelly Library website. See http://www.lasalle.edu/commun/history/articles/marychanningwister.htm.

57. For an insightful reading of this ethos, see David M. Wrobel, *The End of American Exceptionalism: Frontier Anxiety from the Old West to the New Deal* (Lawrence: University Press of Kansas, 1993).

58. Donald Worster, *The Wealth of Nature: Environmental History and the Ecological Imagination* (New York: Oxford University Press, 1993): 9.

CHAPTER 23

BOUNDLESS NATURE

Borders and the Environment in North America and Beyond

ANDREW R. GRAYBILL

IN January 2008, the *New York Times* ran a lengthy travel story on southern Arizona.[1] In the article, the reporter fondly recounts a visit to Organ Pipe Cactus National Monument, a drive across the Tohono O'odham Indian Reservation, and requisite details on where to find good meals, quaint accommodations, and colorful local residents. Notable for its absence, however, is much discussion of Mexico, surprising given the title of the piece ("An Arizona Road Trip on the Edge of America"), and especially given the fact that the landscape so captivating to the author lies right in the middle of the Sonoran Desert, a vast transnational ecosystem that stretches hundreds of miles above and below the border. Though he steals occasional glimpses of the nation to the south when crossing several mountain ranges, the writer scrupulously observes the international boundary—in body as well as in mind—even when it is marked only by a rickety barbed-wire fence or a lonely white obelisk.

Journalists, of course, are not alone in observing these borderlines. For most of the twentieth century, scholars of North America rarely wrote beyond the boundaries of the nation-state. But in recent years, students of Canada, the United States, and Mexico have developed new approaches to the study of regions that straddle these international divides, places like the American Southwest or the northern reaches of the Great Plains. The anthropologists, historians, and geographers—among others—associated with the emerging field of "borderlands" scholarship have investigated a range of common themes, including the politics of boundary creation, the implications of such processes for aboriginal peoples, and the role of borders in the formation and maintenance of national identity.[2]

Environmental historians and their colleagues in related disciplines, however, have been somewhat slower to embrace this concept.[3] As noted by Richard White in a 1999 essay, "Particularly in the United States, where environmental history is strongest, national boundaries have largely determined studies of the depletion of resources, efforts at conservation and preservation, and ecological change."[4] The reason for this strict adherence to national borders is simple enough and common throughout the historical profession: academic history developed in the late nineteenth century alongside the nation-state, which thus came to serve as the basic unit of measure. It is no surprise, then, that the work of many scholars remains coterminous with national boundaries.[5]

And yet no field seems better suited to a borderlands approach than environmental history, considering the extent to which political demarcations arbitrarily divide the natural world. Put another way, the environment is the historical actor seemingly least confined by the dictates of the nation-state, and as such it offers scholars an opportunity to think beyond national borders. Indeed, environmental historians have long produced studies that—in their attention to climate, natural catastrophes, or the movement of plants, animals, and microbes across continents—deemphasize the importance of national borders.[6] To offer but one example, consider how the 1815 eruption of the Tambora volcano on the Indonesian island of Sumbawa affected climate and harvests in North America and Western Europe for two years afterwards.[7]

Apart from these works, which, given their subjects, necessarily downplay the importance of boundaries, there is a small but growing body of historical literature concerned with borders and the environment. This scholarship—nearly all of which takes state sovereignty as its (unstated) foundation—falls into three broad categories, the first of which considers the struggle between national states for control of natural resources such as land, water, and wildlife, especially where boundaries are unclear or contested. A second group explores the significance of borders in determining the human impact upon ecosystems after the hardening of international divides. And a third collection of studies examines the role of landscape in shaping human perceptions of regions that defy and transcend the borders of the singular nation-state. This essay provides a brief survey of some of these works, and offers a handful of concluding thoughts on new methods and avenues of inquiry that scholars interested in such questions might pursue.

INTERNATIONAL RESOURCE COMPETITION AND CONSERVATION

The contest among nations for control of natural resources is, understandably, an extensively treated subject, given that such matters underscore the shared dependence of neighboring countries upon the same fragile ecosystems. In North American historiography, Michael F. Logan's study of the Santa Cruz River provides a wonderful illustration of this dilemma. In *The Lessening Stream* (2002), Logan traces the history of the Santa Cruz, which, he explains, "is the only river to originate within U.S. boundaries, flow out of the country, and then return to continue its course within the United States."[8] This unique circumstance has meant that residents of both Arizona (where the river begins and ends) and the Mexican state of Sonora (which enjoys thirty-two miles of the waterway) have vested economic interests in the health of the river, but this was not enough to prevent the flow of the Santa Cruz from dwindling to a mere trickle by the late twentieth century, largely because of intensive human use.[9]

More familiar as a source of friction between Mexico and the United States is the Río Grande. Accounting for three-fifths of the nearly two-thousand-mile border that divides the two nations, the river has served as a flashpoint for international rivalry since the 1846 outbreak of the Mexican-American War. Although the Treaty of Guadalupe Hidalgo established the Río Grande as the boundary between Texas and the states of northeastern Mexico, the river itself did not always abide by the dictates of federal officials in distant capital cities. Indeed, as political scientists Alan C. Lamborn and Stephen P. Mumme explain in *Statecraft, Domestic Politics, and Foreign Policy Making* (1988), the most significant territorial dispute between the countries during the twentieth century arose from a series of natural shifts in the river's course between 1852 and 1868, which displaced 650 acres of land to the US side of the line. The matter languished until 1911, when a Canadian arbiter attempted to resolve the conflict by restoring part of the area to Mexico, a solution that the United States rejected out of hand. More than a half-century passed before presidents John F. Kennedy and Adolfo Lopez-Mateos signed the Chamizal Treaty, which awarded two-thirds of the tract to Mexico and provided for redirecting portions of the river.[10]

While these riparian matters have been thorny issues in the US-Mexico borderlands, diplomatic skirmishes between Canada and the United States involving environmental concerns have often turned on access to fish and wildlife.[11] The existence of such tensions challenges the idea that the continent's northern border has experienced little of the turmoil usually associated with its southern counterpart. Take, for example, the case of the Fraser River sockeye salmon, the source of a bitter dispute between American and Canadian fishers and their government advocates that dates to the 1890s and remains unresolved even today. The lifecycle of the sockeye neatly demonstrates the problems inherent in managing a migratory species in the borderlands, as explained by historian Joseph E. Taylor III. From its natal waters in British Columbia, the sockeye travel to the Pacific Ocean for a period of one to five years, before returning to the Fraser River to spawn. On this return voyage, however, the fish pass through American waters, where—especially in the early twentieth century—US fishermen intercepted them at places like Point Roberts, the tip of a tiny peninsula connected to the Canadian mainland but which nevertheless belongs to the United States because it extends below the forty-ninth parallel. After years of wrangling and against the backdrop of declining harvests, officials in Ottawa and Washington, DC finally ratified the Salmon Convention of 1937, though as Taylor explains this agreement has not stopped fishers from both nations from "leapfrogging" one another by capturing the sockeye further out in the ocean, so that "a local conflict metastasized into a cancerous blight that has enveloped most of the northeastern Pacific."[12]

Lissa Wadewitz adds another dimension to this story in her essay "The Scales of Salmon: Diplomacy and Conservation in the Western Canada-U.S. Borderlands" (2010).

As her research reveals, the lengthy and costly delay that preceded the 1937 agreement on managing the Puget Sound fishery owed in part to local resistance in British Columbia and especially Washington State. As Wadewitz explains, American fishers and canners pressed their state representatives in Olympia to modify or reject compromise measures suggested by US federal officials, complaining that the health of the local economy depended on the salmon harvest. Sometimes the dissidents even invoked the doctrine of states' rights, which had a legal basis given that US states exercise jurisdiction over their fisheries within three miles of the coast (though they are prohibited from negotiating directly with other nations). Across the line, meanwhile, where Ottawa retains control of Canadian waters through the Department of Marine and Fisheries, federal agents in British Columbia began to relax the enforcement of their own conservation initiatives, as largely unchecked American fishing depleted the sockeye runs and eroded the size of the Canadian catch. In the end, it took the worsening harvests described by Taylor—as well as political changes in Washington State, including the creation of a fisheries board and the election of an amenable governor—to smooth the way for an international agreement, however ineffective it has since proven to be.[13]

And yet, as Rachel St. John has shown, local actors on either side of international divides were capable also of helping to preserve—rather than annihilate—transborder spaces for animal resources, even in the most unlikely contexts. Whereas the typical narrative about cattle ranching in the nineteenth-century US-Mexico borderlands emphasizes the extraordinary violence of cross-border raiding, particularly on the Río Grande frontier, St. John's work notes that farther west along the line, ranchers in places like Arizona and Sonora actually challenged federal efforts to strengthen the border between the nations. Such measures—which, as St. John explains, included the construction of the first border fence between the countries—aimed to address official concerns about land ownership, the enforcement of customs laws, and the spread of bovine diseases. But for their part, both Mexican and American ranchers insisted that excessive border control would rupture the integrated landscape upon which they ranged their herds, and create unnecessary economic hardship. The resistance of these cattlemen forced US and Mexican officials to implement a less onerous regulatory regime, with the result—according to St. John—that "borderlands ranching flourished within the context of negotiated, if not necessarily compromised, [national] sovereignty."[14]

At times the federal governments of North America have sought to attach a "public face" to these attempts at forging international harmony on resource issues, of which the most visible (and studied) are transborder peace parks.[15] To be sure, as Stephen P. Mumme explains, just such a place exists in downtown Ciudad Juarez, Chihuahua, where a fifty-acre section of the land awarded to Mexico in the 1963 Chamizal Treaty was set aside as the Chamizal International Peace Park, intended to commemorate the amicable resolution of the dispute.[16] Other, more ambitious ventures in the US-Mexico borderlands have been less fruitful, like the failure (at least thus far) to create a transnational park in the Sonoran Desert, which has foundered in part because of US concerns about illegal immigration and national security.[17] But as environmental policy specialist Charles C. Chester points out, not-for-profit corporations like the International Sonoran Desert Alliance (a group of indigenous and non-native residents from Mexico and the United States) have emerged to promote "conservation across borders," underscoring once more the vital role of local actors in shaping environmental developments in the borderlands.[18]

Despite its potential to build and enforce shared conservation regimes between the nations of North America, the peace park concept is not without its shortcomings, as historian Catriona Mortimer-Sandilands has suggested in a recent essay about Waterton-Glacier International Peace Park. Established in 1932 by the efforts of Rotary Club members from both sides of the border, the 195-square-mile reserve links southern Alberta's Waterton Lakes National Park (formed in 1895) with northern Montana's Glacier National Park (created in 1910), and was intended to offer a hopeful model of international agreement in the wake of World War I (even if the two nations were allies, and not combatants, in that conflict). Ultimately, however, the park's significance was almost entirely symbolic, as officials in both countries retained exclusive control over their own halves of the reserve and implemented conservation and land management policies that, as Mortimer-Sandilands argues, were driven by national imperatives. In other words, proactive attempts by Canadian and US officials to manage—or even merely to acknowledge—ecological interdependence have suffered from the same nation-based focus that characterizes the competition over resources, like the salmon of Puget Sound. Moreover, as Mortimer-Sandilands observes, the governments have maintained a clear-cut swath through the forest along the forty-ninth parallel meant to draw attention to the fact that the reserve is a place of international meeting and agreement, but with the ironic result that "the artificial border was actually made even more visible than it had been."[19]

Indeed, at least initially, most borderlands preserves were conceived as assertions of national sovereignty, rather than as opportunities for international cooperation. Consider two of the best-known such places in Argentina: Los Alerces National Park, a 2600-square-kilometer preserve along the mountainous border with Chile; and Iguazú National Park, on Argentina's northeastern border with Brazil. The landscapes of both parks are iconic: Los Alerces contains the world's largest alerce forest (alerce is an evergreen species that, like the North American sequoia, lives for thousands of years); Iguazú is home to Iguazú Falls, a series of almost three hundred cascades including the horseshoe-shaped, eighty-two-meter-tall *Garganta del Diablo* (Devil's Throat). When Argentina established both preserves in the 1930s, the border regions encompassing the parks were remote and the international boundaries were imprecise. Neighboring states had competing claims to the area (in fact, Brazil's Iguaçu National Park is directly opposite its Argentine counterpart).[20] Only in recent years has a veneer of internationalism been imposed upon Iguazú: it is now a UNESCO World Heritage Site.

HUMAN IMPACT UPON ECOSYSTEMS

A second cluster of works concentrates less on international competition for resources than on the role of boundaries themselves in determining and revealing human impact upon ecosystems. To put it another way, these studies highlight the environmental divergences between neighboring states that developed in the wake of border formation. Take, for example, the work of geographers Conrad J. Bahre and Charles F. Hutchinson, which investigates changes in the plant life along the US-Mexico border during the twentieth century. In proving that these developments were anthropogenic—and not caused by shifting climatic conditions, as others have argued—Bahre and Hutchinson use a series of photographs taken in the 1890s by the International Boundary Commission that show the 258 obelisks used to mark the border between Mexico and the United States west of El Paso, Texas (the same structures observed by the *New York Times* travel writer mentioned above). Taken from multiple directions, these pictures show that prior to the imposition of the border, vegetation in the region was basically identical north and south of the line (as one would expect). But when comparing these earlier photographs to images of the same markers taken in the 1960s, 1970s, and 1990s, major changes are evident, especially in the denuded landscape of Sonora. With land use practices the only significant variable, Bahre and Hutchinson attribute the differences to intensive and unregulated Mexican farming and ranching south of the border, in contrast to tight US Forest Service control north of the line.[21]

Geographer Simon M. Evans reaches a similar conclusion about the significance of the Canada-US divide in his extensive work on the ranching industry in the Alberta-Montana borderlands. While conventional wisdom on the subject long held that the cattle business in the Canadian West was merely a northward diffusion of the system that spread throughout the American Great Plains, Evans argues that there were, in fact, critical differences that distinguished the Canadian example from its US counterpart, including a much tighter transatlantic connection between Canada and Great Britain, borne of the enduring imperial relationship between London and Ottawa. Even more important in creating this "institutional fault line" along the forty-ninth parallel were the divergent land policies of the United States and Canada. Whereas US federal officials, for the most part, failed to codify regulations governing western ranges (a situation which produced endemic conflict between ranchers and farmers throughout much of the late nineteenth century), Dominion authorities retained jurisdiction over the ranching country of southern Alberta by leasing large tracts (up to 100,000 acres) to individuals or syndicates and investing lessees with the authority to expel interlopers. While such differences may not have produced landscape contrasts as visible to the naked eye as the changes in vegetation along the US-Mexico border, the impact of the Canadian policy was clear enough, as farmers did not make significant inroads into southern Alberta (as they had across the line in Montana) until the late nineteenth century.[22]

Other studies that consider the influence of people in altering borderlands ecosystems focus on trans-boundary pollution, which is the subject of Mark Cioc's examination of the Rhine River during the nineteenth and twentieth centuries. Originating in Switzerland, the Rhine follows a northwesterly course, flowing through or alongside Austria, Liechtenstein, Germany, France, Belgium, and Luxembourg before emptying into the North Sea near the Dutch port of Rotterdam. Though celebrated for the lovely countryside through which it passes, heavy industrialization trans-formed the Rhine into "Europe's romantic sewer," as the many factories that sprang up along its banks dumped their wastes into the river. The nadir came in 1986, when a Swiss chemical plant caught fire, and in the ensuing attempts to extinguish the flames, firefighters washed tons of chemical products into the river, killing off huge numbers of fish downstream. As Cioc explains, given the Rhine's importance to interstate transportation and the generation of hydroelectric power, contemporary restoration efforts have focused rather narrowly on limiting contamination, rather than recovering habitat.[23]

Transborder pollution has also been a concern in North America since the massive industrial expansion of the late nineteenth century and the attendant transfer (acci-dental or otherwise) of waste between nations. Such problems accelerated in the period after World War II, so that in recent years, as policy specialist Richard Kiy and histo-rian John D. Wirth explain, Canada, Mexico, and the United States have each become embroiled in controversies with their neighboring nations over environmental deg-radation.[24] One particularly contentious issue between Canada and the United States has centered on the acid rain that falls periodically on both sides of the boundary in the northeastern borderlands. Direct negotiations between the two countries began in the late 1970s, in the wake of a joint international report attributing some 70 to 80 per-cent of the trans-boundary pollution to the United States. While US President Jimmy Carter actively sought a diplomatic solution, his successor, Ronald Reagan, evinced lit-tle interest in solving (or even acknowledging) the problem. American recalcitrance in the early 1980s led frustrated Canadian officials (at the national as well as the provincial levels) to cast aside "a long tradition of noninterference in United States domestic pol-icy making," up to and including "directly lobbying the United States public, media, nongovernmental organizations, and Congress."[25] Their efforts culminated in the sign-ing by both nations of the Air Quality Accord in March 1991.[26]

Attempts to address other, less famous disputes between North America's mem-ber nations have not been as successful. For instance, in a report issued by the North American Commission on Environmental Cooperation in 2000, lead author and prominent environmentalist Barry Commoner asserted that the high levels of dioxin (a carcinogenic chemical compound that enters the food chain through the air) found in parts of the northern Canadian territory of Nunavut came primarily from sources in the Upper Midwestern United States. However, one of the biggest polluters identi-fied by the report—a municipal power plant in Ames, Iowa—refused even to test the levels of dioxin contained in its discharges.[27] Likewise, sulfur emissions from coal-fired power plants in Texas and especially northern Mexico have led to the sharp deteriora-tion of air quality at Big Bend National Park in West Texas, so that on some days Big Bend has the worst visibility of any park in the trans-Mississippi United States.[28] These conditions have generated tensions between government actors and regulatory agen-cies on both sides of the boundary.

A similar set of conditions prevailed for much of the twentieth century in the so-called "Sulfuric Triangle" of Eastern Europe. This area—located between Germany,

Poland, and the Czech Republic—sits stop enormous deposits of brown coal (lignite), known for high concentrations of sulfur and ash. These reserves helped sustain Nazi industrial production during World War II, and then fueled manufacturing in the communist states that emerged after the conflict. By the 1970s, the region's endemic air pollution (especially in southern Poland, which—courtesy of wind patterns—absorbed many of the contaminants belched by German and Czech power plants) led to widespread deforestation and a spike in birth defects and other medical troubles. As explained by historian J. R. McNeill, in the end, pervasive frustration with state-sanctioned pollution played a role in toppling the region's communist regimes in the late 1980s and early 1990s. Clean-up efforts have been uneven in the two decades since.[29]

Of particular interest to scholars are the environmental implications of the North American Free Trade Agreement (NAFTA), which, when implemented on January 1, 1994, integrated the continent's economies (and by extension, its ecosystems) to an unprecedented degree. In fact, worries that NAFTA might precipitate the relaxation of environmental standards led to the passage of a side-treaty, the North American Agreement on Environmental Cooperation (NAAEC), requiring the three nations to enforce their own regulatory measures. To be sure, this seems somewhat less important along the forty-ninth parallel, which marks the meeting point between two postindustrial societies, than it does on the US-Mexico border, which, in the words of one scholar, offers "the strongest contrasts in the entire world in terms of economic differences from one side of the boundary to the other."[30] Indeed, this financial disparity fueled the development of Mexican *maquiladoras*, factories located just across the border that import raw materials duty free and then produce finished products for export. While these plants first appeared in the 1960s, they experienced explosive growth in the 1980s and early 1990s, even as many observers worried that foreign companies (especially those in the United States) were shifting production to such locales in order to exploit Mexico's lax labor and environmental standards.[31]

However, as John D. Wirth explains, the continent's borderlands are not merely the sites of outsourced and amplified environmental degradation. From his study of the mining and smelting industry, Wirth concludes that on occasion North America's border regions have also served as possible models for international compromise on environmental matters. As he notes in his book *Smelter Smoke in North America* (2000), beginning in the late nineteenth century "a shared industrial culture developed early on in the continental space from British Columbia to Sonora," one characterized by the fluid exchange of proven financial and technological practices throughout the transnational Rocky Mountain West.[32] Thus, when disputes over international air pollution erupted during the twentieth century, there was already a framework in place for negotiating solutions, with the result that lessons learned in the (imperfect) resolution of one such conflict along the boundary separating British Columbia and Washington State between 1927 and 1941 were successfully applied to a similar disagreement in the 1980s in the Arizona-Sonora borderlands.

Of course smelter smoke was not the only boundary-crossing contaminant that North American nations attempted to manage. Epidemic disease, especially illnesses carried across the line by non-white peoples, were a source of considerable anxiety for borderlands residents as well as government officials, who sought to maintain the supposed purity of their Anglo populations in the face of rising immigration and cross-border travel. This was particularly true in the United States, where— beginning in the later decades of the nineteenth century—border-making took on pronounced racial dimensions along both the forty-ninth parallel and the Río Grande. As historian Jennifer D. Seltz explains in her essay "Epidemics, Indians, and Border-Making in the Nineteenth-Century Pacific Northwest" (2010), white residents of Washington State's Olympic Peninsula attempted to prevent the crossing of various aboriginal peoples from Canada into the United States during the 1870s and after, insisting that these "northern Indians" were more susceptible to diseases (by virtue of both the unhealthy environments in which they lived as well as their perceived racial inferiority) which they might then communicate to whites. These early efforts often were fruitless, however, because—prior to the turn of the century— there was not sufficient state and federal support for such injunctions, due in part to the demands of the regional economy (in which migratory Indian laborers played an indispensable role).[33]

Such relative permeability had disappeared by the early decades of the twentieth century, at least along the Río Grande. In her article "Buildings, Boundaries, and Blood: Medicalization and Nation-Building on the U.S.-Mexico Border, 1910-1919" (1999), historian Alexandra Minna Stern notes that state and federal officials in El Paso, Texas (and elsewhere along the boundary) began to apply rigorous standards to Mexicans seeking to cross into the United States, subjecting them to degrading health inspections as well as disinfection and even occasional quarantines. While such measures were meant, ostensibly, to limit the transmission of diseases such as typhus and smallpox, these measures also had the effect of marking Mexicans as "others"— a group of potentially threatening non-whites who endangered the health as well as the racial integrity of Anglo-Americans. In fixing the physical boundary between the two nations, this process of medicalization also hardened the symbolic division as well, casting Mexico "as a totally foreign land."[34]

IMAGINING TRANSNATIONAL LANDSCAPES

A final collection of studies examines how human perceptions of, and interactions with, the natural world sometimes give rise to regional visions that transcend political boundaries. Such ideas hold obvious appeal when considered from multiple locations, such as the borderline between North Dakota and Saskatchewan or the mid-point of the Río Grande as it cuts through West Texas and the northern Mexican state of Chihuahua. In such places, borders seem especially permeable and the grip of the state more fleeting. Nevertheless, despite the apparent sameness of the landscape and the absence of imposing physical boundaries, many observers remain blinkered by the centripetal pull of the nation, like the *New York Times* travel writer wandering about southern Arizona but still largely oblivious to the presence of Mexico, lying just a few miles away.

Cultural geographer Peter Morris offers an intriguing glimpse of this phenomenon in his essay, "Regional Ideas and the Montana-Alberta Borderlands" (1999). In the piece, Morris explores the obstacles that have impeded the development of a cross-border regional identity on the northwestern plains, an area—if indeed such a place exists in North America—ideally suited to the evolution of a transnational consciousness. After all, for much of the nineteenth century, the boundary along this portion of the forty-ninth parallel was marked only by mounds of dirt, and until the terrorist attacks of September 11, 2001 it was an especially porous section of the so-called "longest unde-fended border in the world." And yet, as Morris argues, rather than binding together the Canadian and American residents of the area, the environment itself has at times actually divided them, which he demonstrates in his study of the chinook, a warm dry, wind that, during the winter months, removes fantastic quantities of snow from the ground. In southern Alberta, the significance of the chinook is reflected in the numer-ous place and business names that dot the landscape, most likely because of its histori-cal importance to area ranchers in exposing winter feed to grazing cattle. In Montana, by contrast, where ranching was of secondary importance to agriculture (the cultural bedrock of American civilization, according to Morris), the chinook was viewed far less favorably, for it siphoned off much-needed moisture. Thus, the word appears almost nowhere in the state outside of Chinook, Montana, which lies right on the bor-der. As Morris concludes, it is the national context that shapes the regional reception of this weather pattern.[35]

Still, other scholars have insisted that the landscape can serve to unite borderland peoples, creating pockets of regional unity that supersede national boundaries. Take, for example, historian Molly P. Rozum's "Grasslands Grown" (2001), a study of the northern reaches of the Great Plains during the late nineteenth and early twentieth centuries. In contrast to Morris, Rozum argues that, despite the unmistakable presence of the forty-ninth parallel, residents of Canada's Prairie Provinces and the neighbor-ing US states of Montana and North and South Dakota developed a shared sense of place based upon their common encounter with the regional landscape. Among the formative and transnational experiences that Rozum describes are interaction with the region's flora and fauna; exposure to climatic extremes of broiling summers and brutal winters; patterns of rural settlement (and later out-migration); agricultural labor; and economic hardship in the face of environmental catastrophe (such as the Dust Bowl of the 1930s). While Rozum does not ignore the significance of the border—in fact, she acknowledges that it was made increasingly manifest in the nationalist policies and cultural initiatives that emanated from both Ottawa and Washington, DC after about 1950—she maintains that, in the end, such efforts could never entirely obliterate a sense of place shared by the inhabitants of this unbroken landscape.[36]

If the experience of a common ecosystem could lead some borderlanders to envision transnational spaces, so, too, could the distribution of natural resources and the economic possibilities that they offered. Consider the case of "the temperate woods tradition" that—as geographer Victor Konrad describes—characterized the transborder lumbering industry of the nineteenth century that stretched westward along the Canada-US boundary from the Gulf of Maine to Puget Sound (and even north to Alaska and the Yukon). As Konrad explains, the border meant very little to the woodsmen who worked in these forest regions, as they migrated back and forth across the boundary to take trees where they could find them, employing similar equipment and techniques. In fact, Konrad argues, their efforts served actually to obscure the border, as "Tracts of stumps amid an interlacing of curvilinear woods roads created a landscape of arboreal devastation and resource extraction that obliterated evidence of the international boundary." Though the industry has since given way to agricultural and residential settlement, some vestiges of the lumbering past remain, seen especially in similar home building styles found in cross-border areas such as Michigan, Ontario, Washington State, and British Columbia.[37]

Historian Samuel Truett makes a similar argument about the role of copper in forging a transnational space linking southern Arizona and northern Sonora during the late nineteenth and early twentieth centuries. In his book *Fugitive Landscapes* (2006), Truett explains how the joint efforts of corporate titans and political figures from both sides of the US-Mexico divide built an industry that transcended the international boundary. Multiple networks held this borderland together, from labor migrations and capital flows to the most obvious (and perhaps most important) manifestation of all: railroads that ran north and south and which met at the border, facilitating the movement of people and minerals. But just as the timber operations of the Canada-US borderlands gave out in the period after World War I, the survival of the Arizona-Sonora copper industry proved similarly elusive. Labor unrest and social turmoil had long threatened to tear apart the carefully constructed regional economy, but it was the eruption of the Mexican Revolution in 1911 that permanently rearranged the cross-border nature of the business, as "A land transformed by the captains of transnational industry passed increasingly under the shadow of the Mexican state." The legacy of these borderland ties lives on, however, in the Spanish names of many communities in southern Arizona, as well as in the transportation corridors that now facilitate the passage of illegal immigrants.[38]

In other cases, a borderlands identity predicated on resource commodification grew from efforts by residents and boosters after the fact of settlement and national incorporation, as historian Katherine G. Morrissey shows in her book *Mental Territories* (1997). Focusing on a rural portion of the Pacific Northwest that radiated outward from Spokane to encompass portions of eastern Washington, northern Idaho, and southeastern British Columbia, Morrissey argues that area inhabitants (followed later by land and railroad speculators) pushed the idea of an "Inland Empire" that held the promise of enormous wealth for those willing to move their families—or at least their money—to it. These riches, as described by the region's promoters, consisted of fertile

agricultural land suited to the cultivation of wheat, abundant orchards, and even scattered mineral deposits ripe for exploitation. While these efforts met with mixed success—in part because later visitors, like the writer Zane Grey, celebrated the area for its
relative inaccessibility, thus casting aspersions on its economic potential—Morrissey
explains that the idea of the Inland Empire proved far more durable, echoing Rozum's
notion that transnational regions are sometimes rooted more firmly in the mind than
on the ground.[39]

Future Directions

Given that environmental history is—relatively speaking—a young area of inquiry
and that the rigorous study of North America's borderlands is still emerging, there
are multiple paths that scholars working at the intersection of these two fields might
follow. Perhaps the most obvious course involves the adoption of a truly continental
perspective that addresses similar ecological developments along both northern and
southern boundaries. Some scholars have already begun calling for such an approach
more generally in the study of North America's border regions, noting that—while students of the Canada-US and US-Mexico borders are asking increasingly similar questions and investigating parallel historical actors—they usually work in personal and
intellectual isolation from one another. As such, fresh perspectives and methodologies
applied to the study of one borderland rarely inform research conducted on the other,
even though a range of groups—illegal immigrants, for example—adapted the knowledge gained along one boundary to the challenges posed by its northern or southern
counterpart.[40]

That said, a small number of works have addressed both borders simultaneously,
including a book-length study examining Chinese immigrants and US border security,
a volume on the exclusionary practices of American authorities directed at undocumented immigrants during the late nineteenth and early twentieth centuries, and a
consideration of the efforts by Japanese immigrants to circumvent measures aimed
at limiting their entry to and mobility within North America between the 1880s and
1920s.[41] Scholars of the environment would do well to follow suit, and there is already a
serviceable model at their disposal.

John D. Wirth's *Smelter Smoke in North America* is the rare work of environmental
history that is continental in scope, and it offers a solid template for such investigations,
with its tight focus on a common, transnational problem and its grounding in Mexican,
Canadian, and American sources. There is a need for more studies of this type and the
possibilities abound, as suggested by even a quick consideration of the three general
research areas outlined above. In terms of natural resource competition and conservation, a sustained comparison of the peace park concept in both borderlands could be a
useful frame with which to consider the stark diplomatic contrasts that characterize
the northern and southern boundaries. Likewise, a focused inquiry into water-borne

pollution in Puget Sound and the Río Grande could offer valuable insight into the efficacy of cooperative regulatory regimes in ameliorating environmental degradation
in the borderlands. And a study of Anglo settlement in remote stretches along both
boundaries might highlight the ways in which visions of a unified landscape were actually contingent upon perceived cross-border racial sameness.

It follows that a second approach might involve the comparison of borderlands at
the global level. At first blush, this may seem overly ambitious, in light of the previously
acknowledged reluctance of scholars to think even on a continental scale, not to mention that the study of two borderlands implies passing familiarity with (at least) four
national histories. However, such a method offers the same promises ascribed to the
internationalization of environmental history more generally, including the challenge
to (North) American exceptionalism as well as a forced engagement with the burgeoning environmental historiography that considers the world beyond the friendly confines of the United States and Canada (and Mexico, to a lesser degree).[42] Imagine the
payoffs of a study comparing the struggle over water resources in the US-Mexico borderlands with similar conflicts in desert regions, such as the area around Lake Chad,
which is used heavily by residents of four different North African nations.[43] Likewise, a
comparison of Waterton-Glacier International Peace Park with Iguazú National Park
could test whether the empty rhetoric of trans-boundary conservation in the North
American context is the rule or the exception in other parts of the world. Moreover,
scholars and policy makers interested in the history of the acid rain dispute between
Ottawa and Washington might find it profitable to consider analogous situations in
Europe, where trans-boundary pollutants emanating primarily from Germany and
the United Kingdom have caused widespread deforestation in portions of southern
Scandinavia.[44] And an examination of Russian colonization in Eurasia juxtaposed
with frontier absorption in North America could illuminate the multiple contexts in
which the possibilities of natural resource extraction influenced political expansion.[45]

A third—and particularly innovative—path could produce works that abandon the
state-based inquiry altogether, framing questions that transcend national space and
time.[46] In some sense, this seems an especially difficult proposition when writing environmental histories of the borderlands, because so many of the driving questions (as
indicated in the foregoing discussion) emerge from the competition—and occasional
cooperation—between states. As Richard White has explained, in the main it is only
colonialists who have attempted to write histories of North America (and beyond) that
chart broad developments across what later becomes bounded national space.[47]

A small number of environmental scholars have worked in this milieu, exploring
questions rooted in stateless locales throughout North America, sometimes focusing
on indigenous peoples and their changing relationships to the landscape.[48] Those looking to escape state-centered environmental studies might also consider the example
offered by various works on cattle ranching in what later became the US Southwest.
These studies note that the ranges of Texas and California were stocked with animals interbred from Mexican and Anglo cattle, were owned by Mexican and Anglo
ranchers, and were managed by Mexican and Anglo drovers using hybridized herding

techniques—all of which neatly illustrate the cultural accommodation and borrowing that often flourishes in borderland areas.[49] Indeed, human-animal interactions—the fur trade in both North American borderlands being another—seem promising subjects for scholars looking to transcend the nation, and perhaps it is from their inquiries that satisfying environmental histories of regions (rather than bifurcated international locales) will emerge.[50]

Regardless of the directions that these new studies may take, one thing seems certain: the intersection of borders and the environment will remain a fruitful and vitally important field of inquiry, as suggested by the contemporary debates in the United States about the construction of a fence along the entire US-Mexico border. Although the wall is intended to keep out illegal immigrants as well as potential terrorists, it will also cut right through multiple biotic communities which straddle the international divide. One proposed stretch of the fence near Brownsville, Texas, will threaten the plant and animal species of a one-thousand-acre preserve along the Río Grande, leading its manager to exclaim: "I'm all for border protection...But here's the question— at what price?"[51] The answers to such dilemmas promise to keep government officials, policy specialists, scientists, and environmental historians busy for some time.

NOTES

1. *New York Times*, 11 January 2008.
2. For a discussion of these trends and some of the most significant scholarship, see Benjamin H. Johnson, and Andrew R. Graybill, "Borders and Their Historians in North America," in *Bridging National Borders in North America: Transnational and Comparative Histories*, ed. Benjamin H. Johnson and Andrew R. Graybill (Durham, NC: Duke University Press, 2010), 1–29. See also Pekka Hämäläinen and Samuel Truett, "On Borderlands," *Journal of American History* 98, no. 2 (September 2011): 338–361.
3. It is worth noting that in 2009, a group of scholars convened a month-long National Endowment for the Humanities Summer Institute for University and College Teachers with the title "Nature and History at the Nation's Edge: Field Institute in Environmental & Borderlands History." And some historians have indeed thought in broad conceptual terms about the relationship between environment and the nation-state. See, for instance, William G. Robbins, "Bioregional and Cultural Meaning: The Problem with the Pacific Northwest," *Oregon Historical Quarterly* 103, no. 4 (Winter 2002): 419–427; Joseph Taylor, "Boundary Terminology," *Environmental History* 13, no. 3 (July 2008): 454–481; Ian R. Tyrrell, "America's National Parks: The Transnational Creation of National Space in the Progressive Era," *Journal of American Studies* 46, no. 1 (February 2012): 1–21; and Donald Worster, "World Without Borders: The Internationalizing of Environmental History," *Environmental Review* 6, no. 2 (Autumn 1982): 8–13. Finally, a handful of essays pick up such themes in Douglas Cazaux Sackman, ed., *A Companion to American Environmental History* (Malden, MA: Blackwell, 2010).
4. Richard White, "The Nationalization of Nature," *Journal of American History* 86, no. 3 (December 1999): 976. White's point is underscored by a recent special issue on Canada published by the leading journal in the field, in which none of the seven articles probes

the region bisected by the forty-ninth parallel (although two shorter "reflections" do address the boundary). See *Environmental History* 12, no. 4 (October 2007).

5. For a helpful introduction to this sort of thinking and its implications for the writing of history, see Charles S. Maier, "Consigning the Twentieth Century to History: Alternative Narratives for the Modern Era," *American Historical Review* 105, no. 3 (June 2000): 807–831.

6. For the migrations of plant and animal species and the attendant effects, see Alfred W. Crosby, *Ecological Imperialism: The Biological Expansion of Europe, 900–1900* (Cambridge: Cambridge University Press, 1986), and William H. McNeil, *The Global Condition: Conquerors, Catastrophes, and Community* (Princeton, NJ: Princeton University Press, 1992). On climate, see the work of H. H. Lamb, especially *Climate, History and the Modern World* (New York: Routledge, 1995).

7. See Henry M. Stommel and Elizabeth Stommel, *Volcano Weather: The Story of 1816, the Year Without a Summer* (Newport, RI: Seven Seas Press, 1983).

8. Michael F. Logan, *The Lessening Stream: An Environmental History of the Santa Cruz River* (Tucson: University of Arizona Press, 2002), 5.

9. For another study that places the contest over shared water resources at the heart of the complex ecological relationship between Mexico and the United States, see Evan R. Ward, *Border Oasis: Water and the Political Ecology of the Colorado River Delta, 1940–1975* (Tucson: University of Arizona Press, 2003).

10. Alan C. Lamborn and Stephen P. Mumme, *Statecraft, Domestic Politics, and Foreign Policy Making: The El Chamizal Dispute* (Boulder, CO: Westview Press, 1988). See also Stephen P. Mumme, "The International Boundary and Water Commission, United States and Mexico, and the Municipal Development of El Paso, Texas and Ciudad Juarez, Chihuahua," *Journal of the West* 44, no. 3 (Summer 2005): 38–43, and Jeffrey M. Schulze, "The Chamizal Blues: El Paso, the Wayward River, and the Peoples in Between," *Western Historical Quarterly* 43, no. 3 (Autumn 2012): 301–322. Seemingly less contentious was the Mexican government's mid-twentieth-century effort to develop cotton agriculture in the Río Grande Valley, using Mexican-American workers repatriated from Texas. See Casey Walsh, *Building the Borderlands: A Transnational History of Irrigated Cotton Along the Mexico-Texas Border* (College Station: Texas A&M University Press, 2008).

11. See Kurkpatrick M. Dorsey, *The Dawn of Conservation Diplomacy: U.S.-Canadian Wildlife Protection Treaties in the Progressive Era* (Seattle: University of Washington Press, 1998). It is worth noting that the United States and Canada also have a complex and at times antagonistic—if also much less studied—history of shared water practices. See for example Daniel Macfarlane, *Negotiating a River: Canada, the US, and the Creation of the St. Lawrence Seaway* (Vancouver: University of British Columbia Press, 2014), and Lynn Heasley and Daniel Macfarlane, eds., *Border Flows: A Century of Canadian-American Water Relations* (Calgary: University of Calgary Press, forthcoming).

12. Joseph E. Taylor III, "The Historical Roots of the Canadian-American Salmon Wars," in *Parallel Destinies: Canadian-American Relations West of the Rockies*, ed. John M. Findlay and Ken S. Coates (Seattle: University of Washington Press, 2002), 155–180, quote p. 175. For more information on the sockeye, see Taylor's *Making Salmon: An Environmental History of the Northwest Fisheries Crisis* (Seattle: University of Washington Press, 1999), and Matthew D. Evenden, *Fish Versus Power: An Environmental History of the Fraser River* (New York: Cambridge University Press, 2004).

13. Lissa Wadewitz, "The Scales of Salmon: Diplomacy and Conservation in the Western Canada-U.S. Borderlands," in *Bridging National Borders*, 141–164. For a wider discussion of these matters, see Lissa K. Wadewitz, *The Nature of Borders: Salmon, Boundaries, and Bandits on the Salish Sea* (Seattle: University of Washington Press, 2012).

14. Rachel St. John, "Divided Ranges: Trans-border Ranches and the Creation of National Space Along the Western Mexico-U.S. Border," in *Bridging National Borders*, 116–140, quote p. 118.

15. See Charles C. Chester, *Conservation Across Borders: Biodiversity in an Interdependent World* (Washington, DC: Island Press, 2006), 20–23.

16. Mumme, "The International Boundary and Water Commission," 40.

17. Former Secretary of the Interior Stewart Udall explained in a 1997 interview that he urged President Lyndon Johnson to create such a park, but that Johnson balked (for reasons that remain unclear). See "Stewart Udall: Sonoran Desert National Park," *Journal of the Southwest* 39, nos. 3 and 4 (Autumn-Winter, 1997): 315–320. The special double issue in which the interview appears—titled "Dry Borders: Binational Sonoran Desert Reserves"—contains essays by a range of specialists on the region. For more on the ecosystem, see Steven J. Phillips and Patricia Wentworth Comus, *A Natural History of the Sonoran Desert* (Berkeley: University of California Press, 2000). Moreover, as Emily Wakild has explained, Mexican officials have expressed their own concerns on the idea of joint parks, grounded in their different conservation priorities. See her article, "Border Chasm: International Boundary Parks and Mexican Conservation, 1935–1945," *Environmental History* 14, no. 3 (July 2009): 453–475. For more on the national parks concept in Mexico, see Emily Wakild, *Revolutionary Parks: Conservation, Social Justice, and Mexico's National Parks, 1910–1940* (Tucson: University of Arizona Press, 2011).

18. Chester, *Conservation Across Borders*, 53–133.

19. See Catriona Mortimer-Sandilands, "'The Geology Recognizes No Boundaries': Shifting Borders in Waterton Lakes National Park," in *The Borderlands of the American and Canadian Wests: Essays on the Regional History of the 49th Parallel*, ed. Sterling Evans (Lincoln: University of Nebraska Press, 2006), 309–333, quote p. 326. It could be fruitful to juxtapose this example with its closest counterpart in the East, Niagara Falls. As this chapter was going to press, one very promising study emerged which compares national parks on both sets of North American borders: see Neel G. Baumgardner, "Bordering North America: Constructing Wilderness Along the Periphery of Canada, Mexico, and the United States" (PhD diss., University of Texas, 2013).

20. Like Iguazú, Victoria Falls—on the Zambezi River in Southern Africa—is located between two countries, Zambia and Zimbabwe, which, as in the South American example, have each established national parks on their respective sides of the cascade. See JoAnn McGregor, "The Victoria Falls 1900–1940: Landscape, Tourism and the Geographical Imagination," *Journal of Southern African Studies* 29, no. 3 (September 2003): 717–737.

21. Conrad J. Bahre and Charles F. Hutchinson, "Historic Vegetation Change in La Frontera West of the Rio Grande," in *Changing Plant Life of La Frontera: Observations on Vegetation in the United States/Mexico Borderlands*, ed. Grady L. Webster and Conrad J. Bahre (Albuquerque: University of New Mexico Press, 2001), 67–83. See also Conrad J. Bahre, *A Legacy of Change: Historic Human Impact on Vegetation in the Arizona Borderlands* (Tucson: University of Arizona Press, 1991). Other scholars have used photographic evidence to track such developments. See two works by James R. Hastings and Raymond M. Turner: *The Changing Mile: An Ecological Study of Vegetation Change with Time in*

the Lower Mile of an Arid and Semiarid Region (Tucson: University of Arizona Press, 1965); and *The Changing Mile Revisited: An Ecological Study of Vegetation Change with Time in the Lower Mile of an Arid and Semiarid Region* (Tucson: University of Arizona Press, 2003). For more on anthropogenic change in the region, see Henry F. Dobyns, *From Fire to Flood: Historic Human Destruction of Sonoran Desert Riverine Oases* (Socorro, NM: Ballena Press, 1981). For a broader survey of the region's plant and animal life, see Frederick R. Gehlbach, *Mountain Islands and Desert Seas: A Natural History of the U.S-Mexican Borderlands* (College Station: Texas A&M University Press, 1993 [1981]). Jared Diamond makes a similar argument about the contrast between Haiti and the Dominican Republic (which share the island of Hispaniola) in *Collapse: How Societies Choose to Fail or Succeed* (New York: Penguin, 2005), 329–357.

22. See Simon M. Evans, "The Origin of Ranching in Western Canada: American Diffusion or Victorian Transplant?" *Great Plains Quarterly* 3, no. 2 (Spring 1983): 79–91. For his part, geographer Terry G. Jordan-Bychkov acknowledges the differing land policies between Canada and the United States but maintains that, "as far as cattle ranching is concerned, the border does not matter." See his essay "Does the Border Matter? Cattle Ranching and the 49th Parallel," in *Cowboys, Ranchers, and the Cattle Business: Cross-Border Perspectives on Ranching History*, ed. Simon Evans, Sarah Carter, and Bill Yeo (Calgary: University of Calgary Press, 2000), 1–10. For additional comparisons between the Canadian and US ranching economies, see David H. Breen, *The Canadian Prairie West and the Ranching Frontier, 1874–1924* (Toronto: University of Toronto Press, 1983), and Andrew R. Graybill, *Policing the Great Plains: Rangers, Mounties, and the North American Frontier, 1875–1910* (Lincoln: University of Nebraska Press, 2007), especially 110–157.

23. Mark Cioc, *The Rhine: An Eco-Biography, 1815–2000* (Seattle: University of Washington Press, 2002). For a broader consideration of these issues in the context of Asia, see Piers M. Blaikie and Joshua S. S. Muldavin, "Upstream, Downstream, China, India: The Politics of Environment in the Himalayan Region," *Annals of the Association of American Geographers* 94, no. 3 (September 2004): 520–548.

24. Richard Kiy and John D. Wirth, "Introduction," in *Environmental Management on North America's Borders*, ed. Richard Kiy and John D. Wirth (College Station: Texas A&M University Press, 1998), 5–31.

25. See Leslie R. Alm, *Crossing Borders, Crossing Boundaries: The Role of Scientists in the U.S. Acid Rain Debate* (Westport, CT: Praeger, 2000), quote p. 51. For more on the issue, see Kathryn Harrison and George Hoberg, *Risk, Science, and Politics: Regulating Toxic Substances in Canada and the United States* (Montreal: McGill-Queen's University Press, 1994); John Herd Thompson and Stephen Randall, *Canada and the United States: Ambivalent Allies* (Montreal: McGill-Queen's University Press, 1994); and Jurgen Schmandt, Judith Clarkson, and Hilliard Roderick, *Acid Rain and Friendly Neighbors: The Policy Dispute Between Canada and the United States* (Durham, NC: Duke University Press, 1988).

26. The formal name for this pact is *The Agreement Between the Government of Canada and the Government of the United States of America on Air Quality*. See Alm, *Crossing Borders*, 26.

27. Michael S. Carolan and Michael M. Bell, "No Fence Can Stop It: Debating Dioxin Drift from a Small U.S. Town to Arctic Canada," in *Science and Politics in the International Environment*, ed. Neil E. Harrison and Gary C. Bryner (New York: Rowman & Littlefield, 2004), 271–295.

28. Meanwhile, Americans in the Puget Sound area have complained bitterly about the dumping of partially treated sewage by the city of Victoria, British Columbia, which then contaminates shared waterways. For more on these matters, see R. Anthony Hodge and Paul R. West, "Achieving Progress in the Great Lakes Basin Ecosystem and the Georgia Basin-Puget Sound Bioregion," in *Environmental Management*, 72–107; and Mary Kelly, "Carbón I/II: An Unresolved Binational Challenge," in *Environmental Management*, 189–207. For a social history of a borderland community beset by airborne environmental contamination, see Monica Perales, *Smeltertown: Making and Remembering a Southwest Border Community* (Chapel Hill: University of North Carolina Press, 2010). And for the story of a small but successful attempt to limit cross-border industrial contamination in the Southwest, see Allen Blackman and Geoffrey J. Bannister, "Pollution Control in the Informal Sector: The Ciudad Juárez Brickmakers' Project," (Washington, DC: Resources for the Future, 1998).

29. J. R. McNeill, *Something New Under the Sun: An Environmental History of the Twentieth-Century World* (New York: W. W. Norton, 2000), 89–92.

30. Paul Ganster, "The United States-Mexico Border Region and Growing Transborder Interdependence," in *NAFTA in Transition*, ed. Stephen J. Randall and Herman W. Konrad (Calgary: University of Calgary Press, 1995), 141. See also John J. Audley, *Green Politics and Global Trade: NAFTA and the Future of Environmental Politics* (Washington, DC: Georgetown University Press, 1997); and Tom Barry, *The Challenge of Cross-Border Environmentalism: The U.S.-Mexico Case* (Albuquerque, NM: Resource Center Press, 1994).

31. For an examination of the contemporary *maquiladora*, see Devon G. Peña, *The Terror of the Machine: Technology, Work, Gender, and Ecology on the U.S.-Mexico Border* (Austin: University of Texas Press, 1997).

32. John D. Wirth, *Smelter Smoke in North America: The Politics of Transborder Pollution* (Lawrence: University Press of Kansas, 2000), 201. A helpful (if anonymous) reviewer of this chapter points out that "Wirth's sanguine assessment about the outcome in British Columbia was at least premature. Industrial pollution continues to pour into the Columbia River and down into Lake Roosevelt and the Coleville Indian Reservation," while noting also that suits filed against one of the biggest polluters "continue to stumble" through the court system. For more on this long-running dispute and attempts at its resolution, see D. H. Dinwoodie, "The Politics of International Pollution Control: The Trail Smelter Case," *International Journal* 27, no. 2 (Spring 1972): 219–235; and especially Rebecca M. Bratspies and Russell A. Miller, eds., *Transboundary Harm in International Law: Lessons from the* Trail Smelter *Arbitration* (New York: Cambridge University Press, 2006).

33. Jennifer D. Seltz, "Epidemics, Indians, and Border-Making in the Nineteenth-Century Pacific Northwest," in *Bridging National Borders*, 91–115. See also Seltz, "Embodying Nature: Health, Place, and Identity in Nineteenth-Century America," (PhD diss., University of Washington, 2005), and John Lutz, "Work, Sex, and Death on the Great Thoroughfare: Annual Migrations of 'Canadian Indians' to the American Pacific Northwest," in *Parallel Destinies*, 80–103.

34. Alexandra Minna Stern, "Buildings, Boundaries, and Blood: Medicalization and Nation-Building on the U.S.-Mexico Border, 1910–1919," *Hispanic American Historical Review* 79, no. 1 (February 1999): 41–81, quote p. 81. See also Stern, *Eugenic Nation: Faults and Frontiers of Better Breeding in Modern America* (Berkeley: University of California

Press, 2005); David Dorado Romo, *Ringside Seat to a Revolution: An Underground Cultural History of El Paso and Juárez, 1893–1923* (El Paso, TX: Cinco Puntos Press, 2005); and Miguel Antonio Levario, *Militarizing the Border: When Mexicans Became the Enemy* (College Station: Texas A&M University Press, 2012), 53–66. For a contemporary perspective on such matters, see Giovanna Di Chiro, "'Living is for Everyone': Border Crossings for Community, Environment, and Health," *Osiris* 19 (2004): 112–130.

35. Peter S. Morris, "Regional Ideas and the Montana-Alberta Borderlands," *Geographical Review* 89, no. 4 (October 1999): 469–490.

36. Molly Patrick Rozum, "Grasslands Grown: A Twentieth-Century Sense of Place on North America's Northern Prairies and Plains," (PhD diss., University of North Carolina, 2001).

37. Victor Konrad, "Borderlines and Borderlands in the Geography of Canada-United States Relations," in *NAFTA in Transition*, 179–192, quote p. 187.

38. Samuel Truett, *Fugitive Landscapes: The Forgotten History of the U.S.-Mexico Borderlands* (New Haven, CT: Yale University Press, 2006), quote p. 179. See also D. W. Meinig, *The Shaping of America: A Geographical Perspective on 500 Years of History, vol. 3: Transcontinental America, 1850–1915* (New Haven, CT: Yale University Press, 1998), 152–157.

39. Katherine G. Morrissey, *Mental Territories: Mapping the Inland Empire* (Ithaca, NY: Cornell University Press, 1997).

40. Johnson and Graybill address such matters in "Borders and Their Historians," in *Bridging National Borders*, 1–29.

41. Erika Lee, *At America's Gates: Chinese Immigration During the Exclusion Era, 1882–1943* (Chapel Hill: University of North Carolina Press, 2003); Patrick Ettinger, *Imaginary Lines: Border Enforcement and the Origins of Undocumented Immigration, 1882–1930* (Austin: University of Texas Press, 2009); and Andrea Geiger, *Subverting Exclusion: Transpacific Encounters with Race, Caste, and Borders, 1885–1928* (New Haven, CT: Yale University Press, 2011). For another continental study, see Dominique Brégent-Heald, "Projecting the In-Between: Cinematic Representations of Borderlands and Borders in North America, 1908–1940," in *Bridging National Borders*, 249–274.

42. Richard White discusses these trends in "Afterword Environmental History: Watching a Historical Field Mature," *Pacific Historical Review* 70, no. 1 (February 2001): 103–111.

43. In terms of competition for scare natural resources, it could also be fruitful to compare lumbering operations along the Canada-US border with examples drawn from elsewhere, such as Finland's eastern boundary with Russia. See Ismo Björn, "Life in the Borderland Forests: The Takeover of Nature and its Social Organization in North Karelia," in *Encountering the Past in Nature: Essays in Environmental History*, ed. Timo Myllyntaus and Mikko Saikku (Athens: Ohio University Press, 2001), 49–73.

44. Kenneth E. Wilkening, "Localizing Universal Science: Acid Rain Science and Policy in Europe, North America, and East Asia," in *Science and Politics*, 209–240.

45. See Nicholas B. Breyfogle, Abby Schrader, and Willard Sunderland, eds., *Peopling the Russian Periphery: Borderland Colonization in Eurasian History* (New York: Routledge, 2007).

46. The work of political scientist James C. Scott offers an excellent model for such scholarship. See especially his *Seeing Like a State: How Certain Schemes to Improve the Human Condition Have Failed* (New Haven, CT: Yale University Press, 1998), and *The Art of Not Being Governed: An Anarchist History of Upland Southeast Asia* (New Haven, CT: Yale University Press, 2009).

47. White, "The Nationalization of Nature."

48. See Theodore Binnema, *Common and Contested Ground: A Human and Environmental History of the Northwestern Plains* (Norman: University of Oklahoma Press, 2001), and Cynthia Radding, *Wandering Peoples: Colonialism, Ethnic Spaces, and Ecological Frontiers in Northwestern Mexico, 1700–1850* (Durham, NC: Duke University Press, 1997).

49. See, for example, Terry G. Jordan, *North American Cattle Ranching Frontiers: Origins, Diffusion, and Differentiation* (Albuquerque: University of New Mexico Press, 1993); and Andrew C. Isenberg, *Mining California: An Ecological History* (New York: Hill & Wang, 2005), 103–130.

50. Bethel Saler and Carolyn Podruchny explain why fur trade histories very rarely cross borders in their essay, "'Glass Curtains and Storied Landscapes': The Fur Trade, National Borders, and Historians," in *Bridging National Borders*, 275–302. Studies of the fur trade, moreover, allow scholars to investigate questions of sovereignty and its implications for the environment in the pre-state era. See, for example, Jennifer Ott, "'Ruining' the Rivers in the Snake Country: The Hudson's Bay Company's Fur Desert Policy," *Oregon Historical Quarterly* 104, no. 2 (Summer 2003): 166–195.

51. *San Antonio Express-News*, 14 May 2008. For more on such dilemmas elsewhere along the US-Mexico border, see the story on fence construction in the Yuma-area wetlands in *Arizona Republic*, December 24, 2007.

CHAPTER 24

CROSSING BOUNDARIES

The Environment in International Relations

KURK DORSEY

IN 1992, at the United Nations Conference on Human Development meeting in Rio de Janeiro, delegations from most of the governments of the world completed two explicitly environmental conventions and established three broad sets of goals. The two conventions dealt with protecting biodiversity and curbing anthropogenic climate change, while the common goals included a host of environmental practices, such as sustainable forestry. Despite the sense that the Earth Summit, as the UN meeting was known, marked something new under the sun—the culmination of a twenty-year process begun at the UN Conference on the Human Environment in Stockholm in 1972—these agreements were only the latest chapter in the long history of environmental diplomacy. As early as 1713, British and French diplomats had worked out a plan to divide the cod fisheries off the coast of modern-day Canada, and in 1892 Canada and the United States had introduced the idea of scientifically driven conservation to their discussion of the fisheries in their boundary waters.[1] In 1909, Canada and the United States had also added transborder pollution to the diplomatic record with their Boundary Waters Treaty. In particular, the work of the Earth Summit fit into a few trends that have evolved over the last century. The most important of these trends has been the shift from bilateral agreements dealing with relatively small problems, to multilateral agreements dealing with regional problems, to global agreements that deal with planetary problems. The first explicitly environmental treaties frequently were negotiated to regulate access to resources, rather than to conserve any particular resource. Only since the early twentieth century, as the agreements have moved beyond fisheries deals, has explicit environmental protection become an important goal. The growth in the scope of environmental diplomacy reflects both the ways in which big conventions build upon the success of smaller treaties, and the ways in which science and conservationism have allowed people to think beyond the local and see problems at the ecosystem level. It would have been impossible to address some of the global problems—such as trade in endangered species or damage to the ozone layer—without precedents that showed what might be feasible, as well as a growing belief that people have something akin to ownership rights for shared global resources. As more people around the world have adopted environmentalism, they have joined non-governmental organizations (NGOs) that have frequently been an important force for pressuring governments into partaking in treaty negotiations. Between concern for the environment and a general (if not consistent) faith in science, it has become possible since the early twentieth century for nations to address problems such as pollution and extinction of species while they are still manageable—although that ability has not always led to success.

General patterns of diplomacy have had profound influence on the environment. In recent history, the Cold War and the Vietnam War each shaped the way people used and thought about the environment; more subtly, each was shaped by assumptions about the environment. Much the same could be said about the competition between France and England in North America or Japanese expansionism in the 1930s. Even the name "Earth Summit" implied a similarity to Cold War summit conferences, and the post-Cold War context of the Earth Summit was significant. Scholarship on these overlaps is still rare, but what has been done demonstrates the important connections between the environment and traditional diplomacy. More important, the work on the environment in international relations suggests the limits of trying to write environmental history about any one nation.

All diplomacy is limited by the difficulty of creating a permanent structure that can deal with shifting conditions. An arms-control treaty might be undermined by new technology, a trade treaty can be made obsolete by new economic patterns, and a treaty of alliance might no longer be suitable when one party changes governments. Environmental treaties have those potential problems, as well as the unique problem that they attempt to regulate an ecosystem that is constantly changing, frequently in ways that humans cannot understand or even perceive. Often, they create static institutions to govern invariably fluctuating situations: species migrate, markets change, technology becomes more efficient, peoples' perception of the resource shifts. Even those institutions created to be flexible rarely have the kind of nimbleness necessary to respond to what can be quick shifts in use of a resource. Perhaps that is inevitable, because these organizations ultimately rely for their power on public support for member government policies, and neither the policies nor the support can be pulled together quickly. Over the last century, most successful environmental treaties have had a permanent commission of some sort that has the authority to respond to changing conditions, so the major challenge has been to balance the commission's need for power with the member states' desire to retain as much sovereignty as possible.

The global approach to environmental diplomacy has grown from the need to bring together all of the relevant stakeholders if lasting solutions are to be found—and since the end of World War II there has been a new idea that all people, not just the immediate users, have a stake in resource questions. Expanding demand for a higher standard of living around the world has inevitably increased the pressure on natural resources, and the response has wavered between a rush to get them while one can and steps to close off all access. The diplomatic response to the competition for resources had been at first to set up systems of regulated access as a way to reduce the chance for diplomatic crises, followed by a search for means to create the maximum sustainable yield. More recently, pressure has grown to preserve resources from exploitation altogether, particularly in the cases of some specific types of charismatic megafauna. The trend toward stricter global treaties has depended on the greater influence of science. The role of science is frequently contested, between those who look to science for a rational way of attacking problems and those who see science as insufficiently connected to the ethical quandaries of life on the planet and not nearly as impartial as its proponents claim.

The grassroots debate about the meaning of science reflects another important force in environmental diplomacy: public opinion. Political scientist Lynton K. Caldwell has argued that most such diplomacy came about because citizens pressured reluctant governments to act, which seems more accurate in the period since 1970 than the one before. In his view, the 1972 UN Conference on the Human Environment at Stockholm was a turning point because it galvanized public opinion behind international environmental cooperation.[2] Beyond applying pressure through NGOs, citizens have organized across borders and sometimes imagined themselves citizens of the planet more than citizens of any one country. In many cases they have very astutely used public relations, science, sentiment, and other forces to persuade governments to respond to their concerns, although at times NGOs have found themselves working at cross purposes. The effort to regulate whaling has shown both the skill of citizens in bringing pressure on governments and the potential problem when environmental organizations bring different interests to the table—in this case those opposed to all whaling struggling against those who supported aboriginal whaling rights.

Finally, one must note that the willingness of states to resort to environmental diplomacy often depends on their inability to handle a crisis themselves. States are naturally hesitant to hand over even a small slice of their sovereignty to an international agency unless something compels them to, so they resort to environmental diplomacy reluctantly, and sometimes even see it as an admission of defeat. Of course, many important environmental amenities, whether fisheries, clean air, or the earth's climate, do not recognize political boundaries. Governments need to cooperate or risk the diminishment of those resources, but sometimes they are unwilling to admit their inability to solve problems unilaterally.

The diplomacy of the environment can be broken into three types. The first set of treaties was, mostly, bilateral attempts to regulate access to fisheries as a means of defusing potential conflict. The second consists of, mostly, multilateral treaties meant to apply conservationist principles to fish and other wildlife, culminating in some of the global wildlife treaties of the 1970s and 1990s. The third type deals with pollution control, ranging from specific bilateral accords of nearly a century ago to the global conventions of the last twenty-five years. At the same time, the environment has been a powerful force in international relations, from old imperial rivalries through the Cold War to recent free trade agreements. Alfred Crosby and Jared Diamond have long since become household names by connecting environmental forces to broad imperial outcomes.[3]

It is worth remembering that, in addition to formal environmental diplomacy, political scientists have been examining informal international exchanges as well. Kate O'Neill, in *The Environment and International Relations* (2009), observes that the "pervasiveness of collective action problems in a world of territorially sovereign nation states where environmental problems refuse to stay within national boundaries" has led to non-state actors playing a larger role. While there have been more than seven hundred formal agreements, protocols, and amendments dealing with the environment signed in the last century, they have left enough gaps that non-governmental organizations have sprung up to address some of the problems.[4] Historians have barely noticed these grassroots organizations with an international focus, but O'Neill's book is a good starting point for considering how political scientists have studied their effectiveness. Environmental historians have long recognized that nature does not confine itself to political boundaries, but only a few have crossed the disciplinary or national boundaries in order to write about environmental diplomacy or the role of the environment in shaping international relations. In fact, the number of historians who have examined the place of the environment in international relations is small enough (but growing) that there are hardly any historiographical debates, even though political scientists have been happily constructing and analyzing competing models of environmental diplomacy for at least three decades.

Regulating Access to the Seas

Diplomats have been wrestling with fisheries problems for centuries—long before there was any concern about scarcity or the possibility of overfishing. Beginning with treaties that temporarily reconciled colonial problems in the Western Hemisphere and continuing into the twentieth century, fisheries were usually on the agenda to prevent small diplomatic problems from becoming big ones. The sea seemed inexhaustible, so the central challenge was to ensure clear political control over the best fishing grounds as a means of eliminating misunderstandings and codifying encroachments. Conservation of the resource appears not to have occurred to anyone until the latter half of the nineteenth century, and even then the idea was applied only to very limited fisheries along specific borders. But before that century ended, fisheries diplomacy became a matter of using science to manage the resource in a utilitarian way. At the center of most of the diplomacy in this period was the cod.

The earliest treaty to address access to fisheries came with Great Britain's victory in the War of Spanish Succession, or Queen Anne's War. The 1713 Treaty of Utrecht relieved France of Newfoundland and Nova Scotia, relegating most French power in North America to the St. Lawrence River valley in modern-day Quebec, extending out to Prince Edward Island and Cape Breton Island. The French may have lost the war, but they valued the fisheries enough to bargain hard for continued access to the cod around Newfoundland. The historian J. M. Bumsted notes that, by coincidence, in 1714 fishermen began to perfect off-shore fishing for cod, which was actually cheaper and more productive than setting up in a sheltered cove, and was probably the first step in significantly reducing the cod population.[5] Thus the first attempt to regulate use of a resource was undercut within a year by a change in fishermen's behavior and technology, which would be a recurring theme in environmental diplomacy. The frequent wars in North America finally led to the inevitable conquest of the less populous French colonies in the Seven Years' War. In the Treaty of Paris of 1763, once huge New France was reduced to two small islands south of Newfoundland, St. Pierre and Miquelon, which were used as fishing stations. France also retained the right to dry fish on the shore of Newfoundland. French subjects held the right to fish beyond three leagues (about nine miles) from the coast of Newfoundland, and fifteen leagues from Cape Breton Island. Neither side expressed any concern for conservation of the apparently illimitable horde of cod; rather, they focused on dividing the sea into French and British zones.[6]

The newly formed United States also found itself negotiating over access to fisheries. John Adams, appointed by the Continental Congress to a leadership position on the United States delegation to peace negotiations in Paris in 1781, was determined to protect New Englanders' privileges throughout the northwest Atlantic. He worked to ensure that the fishermen from Massachusetts (including the part that became Maine in 1820) in particular had access to the Grand Banks off Newfoundland, which remained a British colony. These shoal waters offered some of the best cod fishing in the world, and the citizens of the new United States did not want something as petty as a revolution to disrupt the old patterns that made towns like Gloucester wealthy. The Treaty of Paris of 1783 consequently allowed New England's fishermen to continue as they had before the American Revolution began.[7] Looking back at the various treaties that regulated access to the cod, it is worth noting that they frequently were part of a larger context of imperial struggles over natural resources, such as precious metals, valuable pelts, or simply arable land. The decisions to go to war, attack or defend certain places, and keep or return conquered lands often depended on perceptions of natural value.

In the next century, the Grand Banks remained a point of contention, as the War of 1812 undermined the Treaty of Paris. London was not in an especially charitable mood during the negotiations for the Treaty of Ghent in 1814, so US fishing privileges were not resumed and left for future negotiations. The Convention of 1818 was just the first of several attempts to establish a permanent framework for bilateral use of the cod grounds, as British ability to control access to the fish was balanced by the US ability to close its market to Canadian fishermen. At times, as in the 1880s, the biggest news from the fisheries was aggressive Canadian efforts to enforce the agreements by confiscating Gloucester boats, but more important was the shift in technology that made the old agreements obsolete. Refrigeration, in particular, removed the need for US fishermen to come to shore to acquire bait and to salt their catch, which meant that the British had less to offer as bargaining chips and the Americans had more flexibility. As the fishery underwent these transitions, there were still enough outstanding issues of access to the Grand Banks so that the Empire and the United States submitted their positions to an international arbitration board in 1909, and they finally implemented the arbitrators' decision with a treaty in 1912—it was the fifteenth agreement between them on cod, and it finally resolved the issue, though it did not protect the cod stocks.[8]

Europeans were also working among themselves in the nineteenth century on regulating fisheries on the other side of the Atlantic, but not in the cauldron of war that shaped the Anglo-American negotiations on cod. In 1843, France and Britain agreed to split access to the North Sea, set rules for use, and give themselves each the power to monitor the other's activities at the fishing grounds—a system which the scholar A. P. Daggett called "unscientific, clumsy, and of value only in not being observed."[9] A six-power agreement in 1882 dealt with the same issues, again hinting at conservationist ideas, as well as inspection and enforcement provisions. They followed that up in 1885 with an agreement to protect the Rhine River's salmon population, presumably a recognition that the 1882 agreement was not sufficient to deal with an anadramous fish.[10] Of all of the loosely environmental treaties that Europeans concocted in the late nineteenth and early twentieth centuries, 1882's was the most influential in creating standards for registering and regulating fishing vessels.

European and North American efforts to regulate the cod paralleled each other more closely when nations began to extend their jurisdiction over the continental shelf out to two hundred miles. Canada and the United States, which wanted to control access to the cod of much of the Northwest Atlantic, eventually forced most foreigners out of their zones. Despite evidence that the fishery was stretched too far by the 1960s, both governments subsidized their fishermen and ignored any evidence of overfishing. By the 1990s they had managed to destroy their cod stocks, and early in the twenty-first century there is little evidence of a recovery.[11] With fisheries in the Atlantic under great pressure, governments increasingly turned to force to regulate access. Iceland engaged in a series of small Cod Wars with Britain and, to a lesser extent, West Germany, between the 1950s and 1970s to protect the fisheries in its zone. The struggle among NATO allies featured hard bargaining and occasional violence, but tiny Iceland prevailed in its attempts to regulate the fisheries in its two-hundred-mile zone, and fisheries ecologist Roy Hilborn concluded that the cod fisheries in the eastern North Atlantic had weathered overfishing better than on the other side of the ocean.[12] Canada, too, became increasingly militant in the 1990s. After the collapse of the cod fishery, the Canadian government took the unusual step of extending its protection for turbot, a distantly related but suddenly valuable fish, beyond the two-hundred-mile zone, on the theory that it had an obligation to protect the species, which migrated through its waters, from overharvesting. Canadian coast guard vessels used machine guns to deter Spanish fishing boats from pursuing their quarry. Canadians argued that their action was based on conservation, but Spaniards saw old-fashioned access control behind Canada's actions.[13]

It did appear that conservation was a useful smokescreen to hide the more likely goal of simply controlling access when governments extended their exclusive economic zones out to two hundred miles. Iceland is one of the few countries to succeed in conserving fisheries by cracking down on its own fishermen. The US and Canadian experiences in the Atlantic Ocean are more common—kicking out the foreigners then subsidizing citizens to such an extent that they almost cannot help but overfish. There was so much faith in the sea's ability to produce an unlimited supply of fish that few thoughtful people worried about fish stocks until a crash forced the suspension of most fishing. As historian Jeff Bolster noted, the sea was mortal long before anyone recognized that reality. Fishing became of interest to the diplomatic community only when fishermen threatened enough violence to destabilize otherwise peaceful diplomatic relations. But with few citizens organizing for tighter restrictions on fishing, and scientific advice often politically unpalatable, governments have rarely tried to conserve fish before a fishery hit crisis state.

CONSERVING FISH AND WILDLIFE

Despite the failure to develop a conservationist structure for the cod fishery, conservation of fisheries has been a common goal over the last century. The idea that fisheries have no limits was first discarded when the sea in question was reasonably small or the fish in question were actually mammals, particularly seals and whales, which often fell under the catch-all term of "fisheries" in any effort to extract a living resource from the sea. Even in these specialized conditions, though, promoting conservation could be very difficult, because of differing ideas of what it meant to conserve a resource, or even about whether a certain resource was in need of protection.

The first agreement to use scientific inquiry to regulate a disputed fishery before it was destroyed was the 1892 agreement between the Canadian and American governments to have a joint commission study the fisheries, pollution, and obstructions of their boundary waters. The Great Lakes had been the scene of increasing tension between Canadian and American fishermen, who chased the same fish but operated under different rules—and sometimes the same erroneous beliefs, such as the idea that three miles from shore they entered international waters and the rules fell away. In addition to the Great Lakes, there were smaller disputes on the lakes and rivers from the Atlantic Coast to the Pacific, especially in the salmon fisheries of Puget Sound. The worst problems came where competition was the fiercest, of course, and fishermen complained about their neighbors enough to hinder smooth diplomatic relations. To solve the problem, Canada and the United States each appointed a scientist to a joint committee with the task of studying all of the shared problems and producing a set of recommendations to solve them. The boundary fisheries problem became entangled with other US-Canadian disputes, and the issue lingered until a similar joint study began in 1908. That effort led to a proposed law for each country, but the United States Congress failed to pass the law recommended by American ichthyologist and president of Stanford University, David Starr Jordan. The Great Lakes fisheries in particular suffered from the subsequent overfishing, but faith in conservation diplomacy remained in place.[14]

Most of the reason for continued optimism came from the successful effort to conserve the North Pacific Fur Seal, which breeds mainly on Alaska's Pribilof Islands. In the eighteenth century, Russia took control of the region and paved the way for harvesting seals for their thick fur coats, an industry that was passed on to Americans when the United States purchased Alaska in 1867. Seals spend about three-quarters of their time in the water, mostly well off their rookery islands, so before long Canadians and Americans were catching seals on the high seas. US officials sought various ways of ending this hunting, which they compared to piracy, leading to the Paris Arbitration of 1893. Neutral arbitrators crafted a policy that prevented pelagic (or high seas) sealing within a sixty-mile radius of the islands, which represented a melding of the diplomatic arguments with scientific evidence about the role of pelagic sealing in the decline of the seal population. In particular, scientists began to accumulate evidence that pelagic sealing took breeding females disproportionately. A herd that had once numbered more than two million was well below one million in 1893 and heading for a low of about 150,000 in 1910. From that arbitration forward, science became not only an important tool in the discussions between the United States and Canada but also a point of contention, as it has remained ever since. Attempts to work out a consensus on the impact of land sealing versus pelagic sealing failed, both within the US scientific community and between US and Canadian government scientists. As would become apparent so often in the future, science could yield different conclusions depending on the interests of the people asking the questions.[15]

The United States government found that scientific research generally supported its diplomatic efforts to end pelagic sealing by Canadians and later Japanese. It ultimately offered the Canadian and Japanese enough material concessions to win their support for the recommendations from American scientists that sealing should be restricted to taking a few thousand young males on land. The final agreement in the North Pacific Fur Seal Convention of 1911 left the US government with the power to decide how many seals to kill, and that power would be managed by a commission of experts who were confident that they knew everything there was to know about fur seal biology. Their management succeeded: after a short hiatus, the herd yielded revenue from tens of thousands of skins each year, and the population rebounded to more than two million before World War II. Moreover, in later discussions the fur seal convention became a model for US diplomats which proved that scientifically driven conservation could provide appropriate management of shared resources—even if the US government had been reluctant to define them as shared.

The fur seal treaty served as a bridge between negotiations intended to divide resources to allow orderly use and those that aimed to protect a resource from unsustainable use. Canada and the United States followed it up with the Migratory Bird Treaty of 1916 and the Pacific Halibut Agreement of 1923, both of which relied on scientific expertise for insights on how to manage migratory resources that people were using unsustainably. The US Supreme Court case *Missouri v. Holland* in 1920, which cemented the bird treaty, served as the foundation for federal intervention into interstate environmental matters. Justice Oliver Wendell Holmes, writing for the majority, held that federal treaties trumped any state laws about resource use, opening the way for more use of the treaty power in environmental protection, with the United States and Great Britain usually being the leaders, or the nations whose support was necessary for progress. The halibut agreement set up an International Fisheries Commission to handle management, and the two nations have revised it periodically. The halibut convention was cleaner because it dealt with one species, rather than the several hundred species of birds that crossed the forty-ninth parallel, but the bird treaty was more important both for catching public attention and for generating a new precedent.[16]

The success that the United States enjoyed when working with Canada was not paralleled in its work with Mexico, which suggests that nations with different conceptions of the value of nature may have a difficult time reaching an accommodation on environmental matters. As early as 1913, US conservationists proposed negotiating a bird protection treaty with Mexico, but that treaty was not completed until 1936. Rather than a sign of a comity of interests between the two states, historian Keri Lewis has shown that the 1936 convention reflected Mexican willingness to placate the United States in order to win concessions on the more important issue: namely, Pacific fisheries. California fishermen had fought off Mexican attempts to tax and regulate their operations off Mexico's coast since at least 1922, and they continued to fight hard into the 1930s. In fact, they derailed the fishing negotiations so that Mexico received nothing in return for its concession on the bird treaty. The California fishermen were not persuaded of the need to conserve the tuna fisheries they were working, in part because the Mexican government seemed more interested in generating tax revenue from US fishermen than in promoting conservation. Few Americans believed that the Mexican government took conservation that seriously, and the US government was not terribly sympathetic to closing off access for American fishermen just to benefit someone else.[17] It is worth noting that the more recent record of cooperation in North America, through the creation of transboundary civil society, shows promise in a way that conservationists of a century ago would envy.[18]

The 1936 bird treaty was well known to US conservationists, but it was a mere footnote in a decade of big conventions. In 1933, several European nations signed a deal to protect African big game from overhunting; in 1931 and 1937 nations from several continents signed conventions to regulate and limit whaling; and in 1940 most of the states of the Western Hemisphere signed the Convention on Nature Protection in the Western Hemisphere to combat extinction and habitat loss. Although historian Mark Cioc argued that these conventions were little more than hunting agreements that missed opportunities to advance conservation of resources, together they marked the transition from narrowly conceived regulations to broader agreements that emphasized complexity and science, as well as sustainable business practices. In the years after World War II, diplomats launched a new era with the International Convention for the Regulation of Whaling (ICRW), which combined the stringent regulations of some of the smaller treaties with the recognition of the complexity of the multilateral conventions. The ICRW became a model for postwar conservation treaties for more than twenty years, until its flaws, particularly in the areas of amending the rules and enforcement, showed the way to new ideas.

The 1933 convention was an attempt to revive a similar failed effort from 1900, signed in London. The first agreement had united the European colonial powers in an effort to protect some of the best-known examples of charismatic megafauna, like gorillas and giraffes, from extinction. The 1900 agreement listed as many species that were targeted for reduction (because they competed with colonial hunters) as there were species to be protected. The convention did not come into force, but it apparently influenced policy in a few of the empires. The follow-up in 1933, the Convention Relative to the Preservation of Fauna and Flora in their Natural State, included most of the same states, as well as South Africa and Sudan, which were dominions of the British Empire. In addition to hunting regulations, the 1933 London Convention featured a requirement that members establish national parks as soon as possible for the protection of certain species. The convention was superseded in 1968 by a similar one adhered to by many of the newly independent states of the African continent, which again moved beyond its predecessor by adding material on protection of soil and water. The African agreement is one of the most ambitious of the efforts to establish common standards of protection, and it also demonstrates how such conventions frequently cannot be fully implemented by states with limited means.[19]

In 1940, most of the states in the Western Hemisphere followed the lead of the London Convention with the Convention on Nature Protection in the Western Hemisphere (CNP). Like its African predecessor, most of the ideas and motivation came from the Northern Hemisphere, and it was up to each member state to take steps to protect endangered species and establish the various parks and reserves required. Two key differences, though, were that the US scientists who conceived of the CNP drew from their own conservation experience in the United States, and they were careful to work through the Pan-American Union to make sure that the convention was not tainted with the appearance of being imposed on the Latin American states. Both the African conventions and the CNP had many far-sighted provisions, but they were also crippled by a lack of any central authority to enforce those provisions. As the various treaties on whaling would prove, few states would be willing to abide by any convention that had any real central power. Even a secretariat with the power to provide scientific advice would have to tread carefully to avoid the impression that northerners were telling southerners what to do, or likewise non-whalers and whalers.[20]

The 1933 and 1940 conventions left unaddressed what might happen to people living in the areas to be designated national parks. Mark David Spence, Mark Cioc, and other scholars have shown how the creation of national parks often has meant relocation of people in the name of luring ecotourists and fulfilling their desire for pristine nature. Conservation treaties involving the United States have generally protected aboriginal hunting and fishing rights, but it appears that few governments even considered the implications of displacement of aboriginal peoples or more recent inhabitants in the process of establishing national parks. The African and Western Hemisphere conventions left to individual states any decision about aboriginal rights to the land itself. In the United States, aboriginal hunting rights were often preserved, even as the national parks excluded the aborigines from their lands.[21]

The 1931 and 1937 whaling conventions, which have drawn attention from Tønnessen and Jonssen, Mark Cioc, Graham Burnett, and myself, were significantly more complex because they dealt with the right to harvest resources on the high seas, which meant that they ran counter to long traditions of freedom of the seas; therefore, they had to deal with complex issues of enforcement, while creating numerous opportunities for scientific research. Whales were particularly valuable for their fat, which could be rendered into oil and refined into margarine. A powerful international whaling industry had developed by the 1920s, and there was no consensus on what sustainable harvesting of whales meant. At the urging of the International Council for the Exploration of the Sea, the League of Nations helped formulate an agreement in 1931 to set basic standards for whaling signed by twenty-six nations (a remarkable number given that there were only about fifty recognized states in the world at the time). The main breakthrough was the requirement to report accurate statistics to the International Bureau for Whaling Statistics (IBWS) in Norway, so that there would be a common basis for analyzing the data, and hence making recommendations in the future based on scientific data.[22]

The Great Depression disrupted the industry, and whalers tried to regulate themselves by forming a private quota system, but still it was evident that the industry was taking more whales than was prudent. The nine states that signed a new agreement in 1937 accepted a closed season, size limits, and standards for processing the carcass completely, all with an eye to making the industry sustainable. The convention was notable for its reliance on scientists to help frame the terms and the difficulty the delegates had in dealing with the Japanese decision to whale outside of the convention framework. That decision had a host of long-range consequences, particularly the creation of an image of the Japanese as anti-conservationists, but in the short term their behavior allowed the whalers to evade attempts at cracking down. Cioc sees the conventions of the 1930s as missed opportunities for protection of whales, but the chance for real conservation depended on Japanese, and to a lesser extent German, policy. The outbreak of war in Europe in 1939 ended attempts to modify the 1937 convention to make it more effective.[23]

In 1946, the United States put together a proposal for a new convention that featured ideas cribbed from the Norwegians and British, reinforced by Americans' notion that there needed to be a global system that would facilitate trade and generate peace and prosperity for the world, and hence the United States. This International Convention for the Regulation of Whaling resembled some of the previous conventions, but it marked a breakthrough in two key ways. Most important, it established a permanent commission of member states charged with updating the regulations contained in the convention. These amendments were to be based on scientific evidence, which would have to be very convincing to earn the three-quarters vote necessary to be accepted. Second, the commission's powers were supposed to be balanced among whalers, scientists, and fisheries bureaucrats, each with its own goals and methods.[24]

The International Whaling Commission is not generally thought of as a success, given that the whaling industry was not wisely developed and the largest species of whales were steadily reduced over the next forty years. At its peak of public attention, from roughly 1973 to 1983, most people thought it was a disaster, a whalers' club that never had any pretense of conservation. Environmentalists working through NGOs such as Project Jonah and Greenpeace gained influence in the commission and succeeded in adopting a moratorium on commercial whaling, which has served to focus attention even more on Japan, as it allows some whaling to continue under the banner of scientific research. The most powerful enforcement mechanism that the IWC has is American legislation, namely the Pelly and Packwood-Magnuson amendments, which allowed the United States to reduce seafood imports from and close its fisheries to nations that undermined the working of international conservation organizations, such as the IWC.[25] In its wild swing from an agency that endorsed whaling in its founding document to one that was being used to end almost all whaling, the IWC reflected the trend in environmental diplomacy in the 1970s, from diplomacy rooted in the US idea of progressive conservationism—faith in experts and efficient long-term use—to diplomacy rooted in the more globalist, modern environmentalism that is deeply suspicious of business and has a much larger list of species or things that need to be preserved.

But before that brand of environmentalism emerged, the IWC served as a working model of international cooperation to conserve a useful resource for the long term. After 1946, there was an explosion of fisheries diplomacy, much of which involved bilateral treaties or small conventions about cooperation among political allies, like the 1978 agreement between revolutionary Angola and Eastern Bloc Bulgaria (strange bedfellows otherwise), or the 1985 Pacific Salmon agreement between Canada and the United States. The more important treaties, however, were the multilateral ones, such as the 1948 Northwest Atlantic Fisheries Convention, which followed the ICRW's lead in creating a quasi-sovereign commission to improve the management of a particular resource. That pact was revised in the 1970s and also followed by others for the Northeast Atlantic, the Northern Pacific, and the Southeastern Pacific. It would be nearly unthinkable now to draft such a convention without a commission driven by science and obligated to manage the fisheries in question with an eye toward sustainability.[26]

At the same time, as Carmel Finley has shown, the idea of sustainable management of fisheries has a complex history that is more about politics than science. The idea of maximum sustainable yield (MSY) gained favor in the 1930s among scientists working with the International Council for the Exploration of the Sea and among people trying to regulate whaling, some of whom were the same people. Finley argues that, after the war, the idea of MSY as policy gained traction as much because it was endorsed by American diplomacy as because of scientific developments. Particularly in negotiations in the North Pacific between Japan, Canada, and the United States, US political leaders elevated maximum sustainable yield from an idea to an unassailable goal, even though the science has still not caught up to the hopes of the diplomats in the twenty-first century. In short, MSY served as a smokescreen, casting political goals as scientifically valid.[27]

While the Northwest Atlantic convention suggested one way of regulating the resource, via cooperative management, coastal states were simultaneously beginning to exert more control over the open sea. Previously, states had claimed three-, or occasionally even twelve-mile zones of sovereignty or control, and most parties agreed that the rest of the seas beyond those zones were free to anyone who wanted to use them. In 1945, President Harry Truman issued a proclamation that extended US jurisdiction over natural resources on the continental shelf off the US coast and a second one establishing fisheries conservation zones that could stretch for miles from the coast. In 1952, three of the South American Pacific states—Chile, Peru, Ecuador—pursued a more radical solution: extension of sovereignty for two hundred miles into the ocean. As other states followed suit, the extension of jurisdiction was almost always tied to the continental shelf, which was itself a fairly new idea and a feature that had only recently been mapped to a large enough extent to be worth noting. The 1952 announcement triggered complex negotiations about the law of the sea, which will be addressed later in this essay.

Focused treaties with just a few member states, such as the 1973 Polar Bear protection pact among various northern governments, continue to be negotiated, but after 1946 most of the important work on animals was done in four global conventions: the 1971 Ramsar agreement on wetlands; the 1973 Convention on Trade in Endangered Species (CITES); the 1979 Bonn Convention on Migratory Species (CMS); and the 1992 Biodiversity convention from Rio. In many ways, the Bonn and Ramsar agreements follow in the footsteps of the 1940 CNP, in that member states agree to undertake a series of domestic reforms that have international implications, and therefore each agreement gives environmentalists a tool with which to ensure compliance from their governments. But ultimately, these conventions are more statements of principle than anything that can be enforced.

The most important of the three early 1970s agreements, therefore, was CITES, because it was the only one that included built-in enforcement mechanisms. The core idea—that trade in wildlife was leading to the extinction of certain species of charismatic megafauna—came from the International Union for the Conservation of Nature (IUCN) in 1963, which called for a diplomatic agreement to regulate such trade. The US Endangered Species Act of 1969 endorsed such a convention, as did the delegates at Stockholm in 1972.[28] It was one of the last great environmental conventions to feature prominent US leadership, as the final agreement was worked out at the Pentagon, of all places. Simon Lyster calls CITES "perhaps the most successful of all international treaties concerned with the conservation of wildlife," because its principles are broadly acceptable and it is more easily enforced than others.[29]

Under CITES, member states have an obligation to list endangered species that exist within their territory and to respect the lists of endangered species that other nations compile. There are two lists that matter: the so-called "black list" of species that should not be traded, and the "grey list" of species that can be traded only with a permit from the exporting and importing governments. As of 2013, the 178 member states had placed about 925 species on the black list and 33,000 on the grey. Like the CMS and Ramsar, CITES has a permanent governing body, but unlike them, its governing body has some enforcement authority, particularly in its power to call for an embargo on trading listed species, and sometimes all trade, to specific countries. In the early twenty-first century, several African states lost their CITES trading privileges for failing to complete a questionnaire on elephants, reflecting widespread concern about fraudulent commerce in ivory. Most famously, CITES punished the United Arab Emirates (UAE) in 2001 for a series of flagrant violations regarding falcons, caviar, and cheetahs. The UAE changed policy the next year to conform with the convention, rather than face the continued embarrassment of being sanctioned, which suggests that the ultimate power of these broad conventions is to focus attention on unacceptable behavior. In fact, the secretariat of the Convention on Migratory Species suggests that a major benefit of joining the convention is to focus domestic conservation efforts, and there is no mention of enforcement powers. Given that, it is not entirely clear why a government would join the Bonn or Ramsar agreements. Any government willing to join such an organization presumably is already willing to take the legal steps necessary to protect wetlands or migratory species, and if that government loses public support for such protections, it can simply stop participating, with no international repercussions. The CITES case against the UAE proved that it takes an egregious affront to the rules just to get the machinery creaking slowly forward, in part because organizations cannot dilute the power of sanctions by applying them for every infraction. Shame is frequently just as effective, and it can be more easily conjured up by NGOs than by governments that might wish to maintain cordial relations in general. On the other hand, NGOs are usually not especially effective at arranging boycotts, so economic sanctions have to come from states.[30]

Of course, the central problem of any convention with enforcing power is that very few countries would want to join it. In recent years, the United States has relinquished its position as a strong supporter of diplomatic efforts to protect the environment, in large part because of a fear that any convention with teeth is likely to bite American practices. In part, that fear is based on the US roles as the leading consumer of just about everything in the world, and a major exporter of liquid, solid, and gaseous pollution. The drop-off in leadership is also a product of changing political realities. Each small and weak state not only gets the same vote as the United States in these various international agencies, but is more likely to see the United States as a cause of the problem rather than a leader to be emulated in general. Some US leaders have argued that the United States in many cases already has environmental laws that are more restrictive than those in these conventions. The extensive network of wildlife refuges in the United States far surpassed any requirements in the Ramsar treaty, for example, and US endangered species legislation provides more protection to biodiversity than any current convention.

The various wildlife treaties came to a logical culmination in 1992's Convention on Biological Diversity (CBD), drafted at the Earth Summit. The convention built upon years of work by groups such as the IUCN, which had been arguing for forty years that humans were causing a wave of extinction. In particular, concern about the tropical rainforest had reached nearly the same level as the Save the Whales fervor of the late 1970s. Most invertebrate species in these forests, as well as many vertebrates and plants, had not been catalogued, and it was obvious that large, but unknowable, numbers were being driven to extinction. CITES had addressed trade in some of these species, but it had become obvious that habitat destruction was a far greater threat to a larger range of species than trade, so a broader agreement was necessary. The problem to overcome was that the countries harboring the most species in danger of extinction through habitat destruction were relatively poor, which meant that they lacked the resources to set aside and monitor habitat and, frequently, lacked the motivation to prioritize wild-life protection over economic development. The CBD required that signatory nations take steps to preserve biodiversity and set up a mechanism for wealthier countries to assist poorer countries in that task. More controversially, the CBD took steps to assign intellectual property rights for biotechnology to the state that hosted the species upon which a product was based, rather than to the companies that developed the product. While 192 of the world's countries have ratified the convention, the United States, because of its objection to the biotechnology sections, has not taken that crucial step.[31]

In less than a century, wildlife diplomacy had developed from a focus on specific migratory species of high aesthetic appeal, through a stage of targeting certain habitat or trade practices, to finally a broad inclusion of numerous species and landscapes. Each nation, then, should have a small say in the protection of the Amazon basin or mammals of the high seas, particularly if each was willing to pay a part of the price of protection. Species were no longer just part of a nation's larder, but were now aspects of humanity's heritage—and obligation. At the same time, people advocating for diplomacy to protect the environment were making choices about what was important enough to fight for. Whales drew public attention, but fish did not, even though many fish species, such as the cod, had been as depleted as whales. Protecting the rain forest was framed as saving species that either might be useful (for example, in the production of anti-cancer drugs), or had some unique quality (such as immense age or flashing colors). These choices reflected both the reality that campaigners cannot work on everything at once, and also that it is very hard to mount a public pressure campaign on the back of many species of beetles or a fish species with no charisma at all. Once protection of the environment became the goal, the decisions about what to protect first were more often reflections of popularity than scientific or ecological necessity.

POLLUTION CONTROL

As with the formal efforts to conserve fish and wildlife, diplomatic efforts to deal with transborder pollution are now about a century old. Like their counterparts, pollution treaties began with bilateral efforts to solve specific problems, grew to include multinational efforts to address regional issues, and have now come to the stage of global agreements designed to curb problems that may destabilize the biosphere, such as damage to the ozone layer or anthropogenic climate change. In many ways, these are the most controversial agreements because they face the challenge of finding agreement among three groups of states with divergent interests: industrialized societies that create most of the pollution; poorer states that would rather industrialize to raise standards of living than reduce pollution output; and vulnerable states that will be unable to respond to the consequences of that pollution. Addressing pollution, in most people's eyes, means curbing the economic practices that make societies wealthy, and many people have a hard time choosing the possibility of long-term protection if it seems likely to undermine short-term prosperity. Of course, the divide between rich polluting nations and poor recipient nations is hardly exact. China, a country that is poor by standards of per capita GDP, has become the world's largest source of pollution, while Canada and the United States, two of the wealthiest countries, find themselves downwind of each other's pollution.

Canada and the United States, in fact, took the first steps toward transborder pollution control in 1909, when they established the International Joint Commission, which is still in existence. The IJC was geared especially toward water pollution and navigation issues, but over time it made its mark with air pollution issues as well. Either of the two governments could refer something to the IJC, which would then assemble an impartial commission of experts to examine the problem and devise a solution. The IJC spent most of its early years looking at the mess that the two countries were creating in the Great Lakes and their tributaries, but the political difficulties of addressing such pollution were too great until the environmental movement in the 1970s emphasized the need for seeing the lakes, their tributary streams, and watersheds as a whole, interwoven system. In the 1920s, the IJC tried to resolve the Trail Smelter case, but failure led Canada and the US to an arbitration tribunal. The smelting operation in Trail, British Columbia, has long been a notorious polluter in the surrounding farm region, and the farmers of northeastern Washington have been dissatisfied with the Canadian firm's compensation offers. The US and Canadian governments agreed to arbitration in 1935, and the final report in 1941 left a contested legacy. The tribunal assigned financial liability and developed a plan to eliminate damage across the border; more important, it created the precedent that a nation should treat the citizens of another country the way it would treat its own citizens. Alexandre Kiss wrote that "it is difficult to overestimate the importance" of the ruling, because it established the twin principles of cooperative regulation and a state's responsibility for acts within its borders. On the other hand, John Wirth argued that the Trail Smelter case set back pollution control thirty years by focusing on monetary damages from one plant, rather than addressing transborder pollution more broadly. That it took nearly fifty years for the United States to cooperate with Canada in the reduction of acid rain from the American Midwest suggests that the Trail Smelter dispute failed to serve as much of a precedent, despite the thousands of hours put into resolving it.[32]

Before World War II, there were just a few other attempts to regulate transborder pollution. This lack of action was not surprising, given that most states did not try to regulate the pollution inside their borders. Industrialization brought prosperity and pollution, and any attempt to limit one would hinder the other. After the Second World War and the subsequent rise of reformist governments that focused on centralized democratic power to address the inequities of the 1930s, most of the industrialized states began taking baby steps toward pollution control. The United States Congress made its first attempt at clean water legislation in 1948, for example, but it was years before national governments began to provide the huge amount of funding necessary to address such basic problems as local sewage abatement. Shared waterways, whether the Rhine River or the Great Lakes/St. Lawrence River, usually flowed in such a way that some countries received significantly more pollution than they produced. This inequity frequently made it difficult to address localized pollution, because the benefits of a clean-up would not necessarily accrue to the nation that paid for it. The most successful pollution abatement treaties have been those that have brought together a range of states, featured some attempt at balancing costs and benefits, and were based on faith that the majority of scientists correctly had identified the problem.[33]

The first of the large-scale postwar efforts to deal with pollution came in 1954 with the International Convention for the Prevention of Pollution of the Sea by Oil, which built on similar but failed efforts from the 1920s by the United States and 1930s by the League of Nations. Tankers from signatory states were prohibited from purposefully discharging oil, particularly in the form of oily ballast water, in coastal zones. The regulations were gradually tightened into the early 1970s, when marine pollution suddenly became the subject of several regional and global meetings. North Sea states met in Oslo in 1972 to curtail pollution from ships and planes crossing that sea. In that same year, diplomats concluded the Convention on the Prevention of Marine Pollution by Dumping of Wastes and Other Matter (MARPOL), which was intended to stop the practice of using the sea as the final resting place for radioactive waste and other toxins, while regulating the deposit of less noxious substances. The next year a global meeting in London crafted the International Convention on the Prevention of Pollution from Ships, which finally entered into force in 1981, just before the signing of the final Law of the Sea (commonly abbreviated as UNCLOS) agreement, which is the most comprehensive pollution control convention for the oceans.[34]

The dumping of radioactive waste in the sea was one of the clearest examples that most people thought of the ocean as a bottomless pit, a place where out-of-sight, out-of-mind thinking was applicable and appropriate. Jacob Darwin Hamblin demonstrated how the nuclear nations of Europe came to see the oceans as the place to get rid of waste that was too controversial to store in their own countries, even as they paid lip service to the idea of diplomatic consultations about limiting such waste. Governments determined what kind of waste they had and then worked to outlaw the other kinds, or, in the Soviet case, just dumped whatever they had no matter the rules. Dramatic moments when fishermen hauled up waste items that were supposed to be lost forever forced the public to confront what Hamblin calls "the poison in the well." But this left governments with a conundrum: how could they reconcile their desire to keep using nuclear power without an acceptable way of disposing of the waste? Most governments had the good fortune either to be able to cover up their actions, or sacrifice some of the least important sources of waste. As was often the case, public pressure could force governments to act, but those actions did not always address the biggest problems.[35]

UNCLOS negotiations were originally the product of concern about extension of jurisdiction over the continental shelf, but they evolved into a broad discussion of topics, including pollution. After nine years, the third set of negotiations concluded in 1982. Historian Ralph Levering and his mother Miriam, who was a member of the private agency known as the Neptune Group, argued that that group was critical in closing the gap between the industrial states and the less-developed countries, known as the Group of 77, by emphasizing scientific accuracy. The convention defined pollution broadly, including standard ideas such as threats to human health and living resources, but also adding in introduced species as a type of marine pollution to be combated. States were required to act against pollution from ships and from land (which accounts for as much as 60 percent of marine pollution), including airborne sources and that which fouled estuaries. The convention spelled out in some detail how states could deal with polluting vessels as well as what each state's responsibilities were in terms of controlling ships that flew its flag, and the pollution emanating from its soil.[36]

UNCLOS has been quite successful and widely accepted. Each year since 1982, a few more governments have ratified it, so that, as of 2013, the number stands at 165, and many have moved beyond the convention by signing regional protocols to address local conditions in places such as the Baltic Sea. In 1996, negotiators created an addendum that specifically targets fisheries that straddle borders. The United States waited several years to sign the original convention and has still not ratified it (although the government accepts it as customary law), in part because the convention appeared to undermine the ability of high-tech companies to profit from the exploitation of seabed mineral deposits, and in part because of ideological concerns about US sovereignty. The Law of the Sea convention and MARPOL have left the oceans with a general code for pollution control, which, while inadequate in some ways, is more thorough than any other sector of environmental diplomacy, including atmospheric pollution.[37]

The first broad attempt at atmospheric pollution control—protecting the ozone layer—required a combination of faith in science and evenly distributed costs and benefits. In the 1970s, scientists began to fear that an apparently benign and very useful group of chemicals, chlorofluorocarbons (CFCs) and halons, could break down ozone (O_3) molecules in the stratosphere. The ozone layer protects the planet from excessive solar radiation. CFCs had been used for more than 40 years in refrigeration and air conditioning, as well as aerosol spray cans, and they were widely regarded as safe. The threat to the ozone layer was both existential and still hypothetical when diplomats convened in 1985 in Vienna. There, twenty governments out of the forty-three present agreed in principle that CFCs should be reduced, but opposition from industry and scientific uncertainty combined to prevent significant progress. It is also worth noting that no environmental NGOs attended the Vienna meeting.[38]

In the next two years, though, the United States government accepted the scientific argument that CFCs could seriously endanger human life, and American chemical companies embraced the idea that they could create safe replacements. Other industrialized states were split on the science, and there was no consensus on how to deal with the less industrialized countries that wanted the benefits of CFCs. At a subsequent meeting in Montreal in 1987, delegates from sixty nations agreed on a 50 percent cut in CFC production within twelve years. On one hand, this was a cautious response based on the inability of some industries to deal with a rapid phase-out, but it was also a remarkable agreement given that there was still no hard evidence that CFCs were actually eroding the ozone layer. The negotiations were helped along by reports of a hole in the ozone layer above Antarctica, but the cause of that hole was not clear. As the evidence piled up to prove that the theories were accurate, the signatories met in Helsinki in 1989 and agreed to ban all CFCs within a few years, as well as to provide for technology transfers so that industrializing countries could meet its terms and still have refrigeration. And yet even as ozone diplomacy is justly hailed as an impressive accomplishment, CFCs that were released legally years ago are still slowly wafting up to the stratosphere and disrupting ozone molecules. It may be decades before the damage to the ozone layer is halted.[39]

Of course, it is impossible, early in the twenty-first century, to mention long-term consequences of inaction without thinking about the impact of greenhouse gases on the planet. In the late 1970s, the widespread theory that the planet was headed toward a mini Ice Age (or even the slightly later theory that a world war would bring a nuclear winter which would destroy human civilization) began to give way to the concern that people could be causing the planet to heat up. Scientists had long understood that carbon dioxide, water vapor, and other gases in the earth's atmosphere trapped the sun's energy well enough to make life possible. In the last quarter of the twentieth century, they began to study how the burning of fossil fuels, such as coal and oil, might change the atmosphere by adding to the natural store of greenhouse gases. As early as 1977, a panel from the National Academy of Sciences was trying to predict the impact of doubling the amount of CO_2 in the earth's atmosphere. The next year an international group of scientists was predicting at least a tripling or quadrupling of CO_2, with consequences that on balance would be bad for humanity.[40] The validity of the science of climate change has since become the subject of intense debate. The vast majority of scientists studying the atmosphere have come to the conclusion that humans are changing the atmosphere through the burning of fossil fuels, and, although there is a large range of possible scenarios, most agree that unchecked use of fossil fuels will cause serious disruptions, from more flooding of coastal communities to catastrophic collapses of ocean current systems. And yet, unlike the ozone layer discussion, the undercurrent of dissent has been powerful enough to disrupt diplomatic efforts to respond to the theories, with a small but persistent group of scientists who generally argue that the observed changes over the past few decades fit within a pattern of natural variation. The doubters are bolstered by the leaps of faith that some of the computer-based climate models require, and they feed into a political debate about whether the costs of human-induced climate change will be worse than the costs of weaning industrialized societies off fossil fuels.[41] In their widely hailed book, Naomi Oreskes and Erik Conway concluded that the root of climate-change skepticism can be found in the well-funded activities of industry scientists, who have generally placed politics ahead of science.[42] Debates about environmental science have increasingly included the phrase "junk science," indicating the depths of animosity that have built up over everything from climate change to DDT to migrating shorebird populations.

The scientific consensus was complete enough by 1992 to allow most delegates at Rio to sign the Framework Convention on Climate Change. That convention acknowledged that human activity was a cause of climate change and that nations should take steps to reduce their impact. But specific goals were postponed, largely at the behest of the United States and a few allied nations, until a subsequent meeting in 1997. The sides generally were reversed from the 1980s discussions about ozone, when Japan and several of the Western European states were unwilling to accept the theories about CFCs; they now were among the leaders on climate change, while the United States was the main obstacle to turning scientific theories into policy. The subsequent Kyoto Protocol in 1997 is probably the most controversial environmental agreement in history. The signatories agreed to reduce their greenhouse gas emissions by various amounts by 2012. The United States was supposed to reduce its emissions by 7 percent from 1990's number. In addition, already industrialized states were supposed to share their technology with poorer countries that desperately wanted to industrialize to improve their standards of living. Recognizing that desire for improved standards of living, the negotiators did not set any goals for industrializing states like India and China. That decision gave Kyoto's opponents in the United States and a few other countries three points of attack: the science behind global warming was insufficient; the costs of compliance would be out of line with the costs to be avoided; and the burden was not fairly distributed.[43] The Kyoto Protocol was ratified by the necessary fifty-five states, which produced, in aggregate, at least 55 percent of the world's carbon output, but the United States rejected the treaty in 2001, which meant that it was essentially symbolic. If the United States, China, and India expand their emissions without serious limits, they will more than compensate for anything that the other industrial states do to cut theirs.

The odd state of the diplomacy of global warming was reflected in the meeting of the G8 industrial democracies in 2008. The leaders agreed they would achieve a 50 percent reduction of greenhouse gases by 2050 and that India and China would also be expected to reduce their emissions. They neglected to determine whether the baseline for the reduction would be 2008 or 1990 (as in Kyoto)—a fairly important distinction, which only served to magnify the lack of seriousness behind delaying a solution for forty years. But at the same time, United States President George W. Bush, who had largely been unsympathetic to the idea that the United States should take precautions against climate change, agreed in principle that people were causing climate change in a way that ought to be addressed. In the meantime, between the great recession of 2008 and a switch to natural gas, the United States actually witnessed a decline in its greenhouse gas emissions over the course of several years, although it was still nowhere close to meeting the target assigned by Kyoto, which only underlined the ways in which a diplomatic framework can become obsolete quickly. European states reported that they generally were going to meet their targets. Meanwhile, China has surpassed the United States as the leading producer of greenhouse gases in raw terms and as the fastest growing producer overall. In 2012, China formally agreed that it would meet binding targets, perhaps as early as 2020.[44]

The Kyoto Protocol and subsequent climate negotiations face three fundamental problems that ultimately afflict most environmental diplomacy: difficulty in reconciling the different expectations of countries with very different attitudes toward nature; the inability of scientists and diplomats to think in the same terms; and the complexity of creating a system that is properly both rigid in some places and flexible in others. The problem of divergent national expectations will be surmountable only when India and China see themselves as polluters of the environment, rather than as players rushing to catch up in the exploitation of natural resources. By that point, the technology of energy production or our understanding of the atmosphere may have changed so much that the old diplomatic framework will be obsolete. By then, perhaps the United States will be aligned with the other industrial democracies in pursuing a precautionary approach to climate change, or at least will see profitable possibilities in changing technologies; its shift from leadership to obstructionism in environmental diplomacy has done more than any other factor to disrupt the completion of conventions. Even if the major polluting states reach some sort of relative similarity, they will still experience internal dissension and confusion, because the scientists and diplomats will not agree on the true effects and costs of climate change.

ENVIRONMENTAL RAMIFICATIONS OF TRADITIONAL DIPLOMACY

Almost every decision that people make or policy that governments enact has some ramification for the environment and is based, at least in part, on deeply held assumptions about the best use of resources. So it follows that diplomatic policies usually have deep roots in environmental assumptions, and in turn have unintended environmental consequences. Environmental and diplomatic historians are just beginning to explore the intersection of environmental forces and great power diplomacy, although military historians have long been aware of the role of food supplies and weather in shaping the outcome of campaigns. Five recent edited collections show the potential for crossing the boundaries between environmental history and the traditional fields of military and diplomatic history: Charles Clossmann's *War and the Environment* (2009), John McNeill and Corinna Unger's *Environmental Histories of the Cold War* (2010), Erika Bsumek, David Kinkela, and Mark Lawrence's *Nation-States and the Global Environment* (2013), Richard Tucker and Edmund Russell's *Natural Enemy, Natural*

Ally (2004), and Kurk Dorsey and Mark Lytle's special issue of *Diplomatic History* on environmental issues (2008).[45] While these collections have an array of authors and topics, their purviews are wide enough that there has been almost no disagreement among them. Instead, they have pointed the way to a handful of more fully developed monographs and a number of possibilities for future work.

Environmental forces have played a critical role in the shaping of ongoing struggles between cultures and societies. The long-running wrestling match between France and England for control of North America was a test of wills in European capitals, but it was also a test of the ability of each colonial population to thrive in its setting, which in turn was a product of decisions about the value of the landscape—whether it was best used to produce food or furs, for example. The struggle for dominance of the Great Plains between Native American nations and the United States was probably destined to end with a defeat of the less numerous Indians, but it also seems likely that the timing of the US government's victory depended in part on cyclical environmental forces that influenced bison populations.[46]

Historian John McNeill has shown how the course of empires in the Caribbean in the years before 1914 was set as much by mosquitoes as by bewigged men in Western capitals. From Columbus's arrival in 1492, European heads of state struggled with one another for four centuries to control valuable ports, sugar production, and slave populations in the Caribbean. They came to realize that diseases would have to be accounted for when making battle plans, controlling the time of the campaign season and even assumptions about the number of casualties an attacking army would take. Generals and common soldiers alike grew to understand the risk of disease, even if they could not comprehend how the diseases spread. McNeill has skillfully demonstrated the complex intertwining of diplomatic and environmental forces, reaching even to small touches, such as how the introduction of clay pots in the sugar milling process created unlikely habitat for disease-carrying mosquitoes, which made some islands more hazardous than others. As late as the Spanish-American War in 1898, disease was shaping the role of armed forces in Cuba.[47]

Historian Greg Cushman has made similar connections using the guano islands on the other side of South America. Guano, or bird excrement, has long been an important source of fertilizers, and the rocky coasts of Peru and islands farther out to sea provided enormous amounts of fertilizer for decades. The guano harvested in Peru, Cushman argues, helped to spur development throughout the Pacific world, and contributed to a war in 1879 among the Pacific nations of South America. It was also a product of ecological cycles and a product vulnerable to competition for fish. As human fisheries grew in importance, guano-producing birds lost out. Eventually, guano birds found themselves not only outcompeted, but occupying perilous perches in international relations, up against everything from shifting commodity prices to nuclear testing. Ironically, it may be that demand for organic food worldwide will lead to management of marine resources that encourages a return of the guano-producing species. The seemingly insignificant rocks fit only for bird nests were clear evidence of the interlocking relationship between nature and international politics.[48]

In a parallel vein, once the United States emerged as the dominant power in the Western Hemisphere, its industrial appetites demanded resources from the tropics. In some cases, tropical goods were nearly unique, such as rubber, sugar, and bananas. If Americans wanted tires for their cars and sugar and tropical fruit for their meals, they would have to control the environments that produced the goods. In addition, even some things like beef became easier to access via tropical landscapes. The logical conclusion, which historian Richard Tucker ties to larger forces in US diplomacy, was to promote easier exchange of goods and finances across international borders, so Americans could at least take the edge off their insatiable appetites for tropical products. In the process of building this system in the late nineteenth century and extending it well into the next, US officials unleashed a cascade of environmental and social consequences, such as moving people to marginal land so that the most productive could be used for export agriculture.[49]

Tucker's book is one of a handful that allows us to begin making direct connections between environment and international relations broadly. The very idea of development—transferring ways of thinking about people and resources from wealthy countries to poor ones—is based on assumptions about the environment. Going back at least to 1948, when Fairfield Osborn and William Vogt published their warnings about the global catastrophes inherent in overpopulation, some voices warned that foreign policy would have to take into account resource depletion. Historian Thomas Robertson has unpacked the complex history of post-World War II anxiety about population, arguing that Americans had a hard time reconciling the warnings with their desire to use more resources. A country happily enjoying the fruits of the Baby Boom and worrying about Communists everywhere struggled to accept messages from Vogt, Osborn, Aldo Leopold, and Paul Ehrlich.[50]

But even as they struggled, the US government launched or participated in grand schemes to limit population, bring economic development, and improve agriculture in poorer countries around the world. Poverty might not be conquered, but it could be combatted, and fundamentally the battleground would be nature: cycles of fertility, patterns of weather, resources waiting to be tapped. The very idea of prosperity through development depended on a host of assumptions about the environment, such as the best use for a river (to generate power) or an agricultural system (to generate export crops). The Green Revolution and US support for intertwining of national markets were both based on those kinds of assumptions, and they had enormous environmental ramifications. The flaws in development theory would become more obvious and more hotly debated later, but in the 1960s in particular there was great faith that the industrialized Western states, using programs such as the Peace Corps, could help to modernize poorer countries through projects like dam building. Whether scientists were working to control the weather, improve strains of staple crops, or introduce new fertilizers, they were part of a grand struggle not just with nature but also with the Communist world, in an effort to show that the capitalist system provided more hope for the world's poor. Historians Nicholas Cullather, writing on the Green Revolution, and Matthew Connelly, writing on population control, have demonstrated the ways in which faith in modern science in the name of winning hearts and minds often led to unintended consequences.[51] The history of efforts to control pollution intersected with the history of using the environment to gain an advantage in great power struggles.

Early in the Cold War, each side began experimenting with ways to use natural forces as weapons. Of course, the energy released by the decaying radioactive isotope was one of the most powerful natural forces to be adapted for military purposes. But as Jacob Hamblin and other historians of science have shown, the high-level experiments got underway to harness weather, starting with timing of rains to help crops grow and later expanding to hinder use of the Ho Chi Minh trail, and then they moved to weaponizing phenomena in the earth's high atmosphere, including processes that might create global warming to undermine Communism. When John F. Kennedy spoke of the long twilight struggle, perhaps he was actually suggesting using the twilight as a weapon to blind Communist armies.[52]

The place where push came to shove in the competition between ideologies was in Vietnam, and David Biggs has shown that the Vietnam War is as ripe for environmental analysis as anything in the history of international relations. Westerners brought with them a host of assumptions about the best ways to use and overcome nature, and those assumptions were linked to other assumptions about the best way to fight Ho Chi Minh's followers. The ease with which US forces turned to defoliating chemicals, forest fires, and napalm to gain the upper hand suggests a disdain for Vietnam's nature along the lines of the general US ambivalence about their Vietnamese allies. US attempts to remake the Mekong River Delta showed not only an ignorance of history but also a lack of appreciation for the very ground (and water) in the delta. Western models of development were supposed to win the hearts and minds of the peasants. For their part, those peasants were also influenced by long traditions of living off the land, whether they enjoyed those traditions or not. In large part, they wanted to control which parts of modern technology and ideology they got to use.[53]

David Zierler and Ed Martini have each delved into one particular issue from Vietnam: the use of the defoliant Agent Orange. Early in the war, US commanders decided that the foliage of South Vietnam was as much their enemy as the actual soldiers whom they opposed. They not only saw the forests as hiding places for guerrilla forces, but they also saw agricultural fields as sources of food for the South Vietnamese rebels. With a deep faith in technology, they launched a war on South Vietnam's ecosystems. A rainbow of herbicidal agents rained down on the land, and in some cases Americans used fires to fight the forest. The herbicides were frequently laced with poisonous and long-lasting byproducts, such as dioxins, meaning that their impact would be for generations, not just a few growing seasons. Martini demonstrates that the use of Agent Orange was an international phenomenon, and attempts to determine its short- and long-term impact have been clouded by uncertainty over such things as how much was used, how much dioxin was in any batch, or who was actually exposed. Zierler argues specifically that the backlash against Agent Orange was critical in the creation of modern environmentalism, particularly the fear that technology would lead to catastrophic ecocide.[54]

No doubt other wars and struggles could have similar histories written. The Mexican-American War might have had a different outcome if more Americans had viewed northern Mexico as salubrious, or if yellow fever had done more damage to Winfield Scott's army. Every invasion, real or contemplated, of Russia has been shaped by assumptions about the value of the steppes and planning for winter warfare. Certainly, the course of World War I would have been different if the influenza epidemic had broken out on a different time frame or if North American food production had not been so abundant.

People have long understood that the environment crossed political borders, whether in the form of migrating animals, flowing waters, widely distributed natural resources, or dispersing pollution. Despite that recognition, scholars have tended to write about the environment within nation-states, rather than across them. Environmental historians, diplomatic historians, and historians of science have slowly started to cross their disciplinary borders to broaden our understanding of the place of the environment in international relations. The space for further work is as broad as the planet itself.

NOTES

1. A. P. Daggett, a specialist in international law, refers to a Franco-Spanish treaty of 1683 that regulated access to citizens of both nations at the mouth of a boundary river, the Bidassoa, which might be the first treaty to specifically address fishing rights. A. P. Daggett, "The Regulation of Marine Fisheries by Treaty," *American Journal of International Law* 28, no. 4 (October 1934): 697.
2. Lynton K. Caldwell, *International Environmental Policy: Emergence and Dimensions*, 2d ed. (Durham, NC: Duke University Press, 1990), 2, 29.
3. Alfred Crosby, *Ecological Imperialism: The Biological Expansion of Europe, 900–1900*, 2d ed. (Cambridge: Cambridge University Press, 2004); Jared Diamond, *Guns, Germs, and Steel: The Fates of Human Societies* (New York: W. W. Norton, 1999).
4. Kate O'Neill, *The Environment and International Relations* (Cambridge: Cambridge University Press, 2009), 5, 199.
5. J. M. Bumsted, *The Peoples of Canada: A Pre-Confederation History*, vol. I (Toronto: Oxford University Press, 1992), 120.
6. A detailed history of the northwest Atlantic cod fishery can be found in Harold A. Innis, *The Cod Fisheries: The History of An International Economy*, rev. ed (Toronto: Ryerson Press, 1954). Mark Kurlansky's *Cod: A Biography of the Fish that Changed the World* (New York: Walker, 1997) is shorter, more accessible, and broader in scope in that it brings the story up to the late twentieth century and covers European politics.
7. A concise history of US negotiations with Great Britain can be found in Chandler P. Anderson, "The Final Outcome of the Fisheries Arbitration," *The American Journal of International Law* 7, no. 1 (January 1913): 1–16.
8. Brian Payne, *Fishing a Borderless Sea: Environmental Territorialism in the North Atlantic, 1818–1910* (East Lansing: Michigan State University Press, 2010) is the best brief history of the international nature of the fishery; see also W. Jeffrey Bolster's *The Mortal Sea: Fishing the Atlantic in the Age of Sail* (Cambridge, MA: Harvard University Press, 2012) for a comprehensive account of how early fishing affected fish populations.

9. Daggett, "The Regulation of Marine Fisheries by Treaty," 714.

10. Daggett suggests that the first Franco-British agreement was in 1828, 699, 703.

11. Kurlansky has a good summary of the late-twentieth-century cod disputes; for the possible permanence of the decline, see P. Neubauer, O. P. Jensen, J. A. Hutchings, and J. K. Baum. "Resilience and recovery of overexploited marine populations." *Science* 340 (2013): 347–349.

12. Roy Hilborn, *Overfishing: What Everyone Needs to Know* (Oxford: Oxford University Press, 2012), 8–10.

13. Kurlansky, *Cod*, 161–73; "Spanish Stirred by 'War' Over a Fish They Don't Eat," *New York Times*, April 15, 1995.

14. Kurkpatrick Dorsey, *The Dawn of Conservation Diplomacy: U.S.-Canadian Wildlife Protection Treaties in the Progressive Era* (Seattle: University of Washington Press, 1998), 40–48.

15. The paragraphs on fur seal diplomacy are drawn from Dorsey, *The Dawn of Conservation Diplomacy*, chapters 4 and 5.

16. For more on the bird treaty and *Missouri v. Holland*, see Dorsey, *The Dawn of Conservation Diplomacy*, chapters 6–8, especially pages 234–235. The halibut treaty is discussed in John Q. Adams, "The Pacific Halibut Fishery," *Economic Geography* 11, no. 3 (July 1935): 247–257.

17. Keri E. Lewis, "Negotiating for Nature: Conservation Diplomacy and the Convention on Nature Protection in the Western Hemisphere, 1929–1976," (PhD diss., University of New Hampshire, 2008); chapter 5 deals with US-Mexican fisheries.

18. Charles Chester, *Conservation Across Boundaries: Biodiversity in an Interdependent World* (Washington, DC: Island Press, 2006).

19. Mark Cioc, *The Game of Conservation: International Treaties to Protect the World's Migratory Animals* (Athens: Ohio University Press, 2009), chapter 1; Simon Lyster, *International Wildlife Law: Emergence and Dimensions* (Cambridge: Cambridge University Press, 1987), chapter 7; P. van Heijnsbergen, *International Legal Protection of Wild Fauna and Flora* (Amsterdam: IOS Press, 1997) has several short sections on Africa, including 16–17, 24–25, and 152.

20. Lewis, "Negotiating for Nature," chapter 1, explains the U.S. origins of the convention; subsequent chapters focus on the response of Mexico, Venezuela, Argentina, and Costa Rica to its terms. See also Lyster, *International Wildlife Law*, chapter 6.

21. Cioc, *The Game of Conservation*, chapter 1, is even called "Africa's Apartheid Parks"; Mark David Spence, *Dispossessing the Wilderness: Indian Removal and the Making of the National Parks* (Oxford: Oxford University Press, 2000).

22. This and the following paragraphs on whaling are taken largely from J. N. Tønnessen and A. O. Johnsen, *The History of Modern Whaling* (Berkeley: University of California Press, 1982). Specifics on the 1931 convention can be found at Lyster, *International Wildlife Law*, 17–18; see also Cioc, *The Game of Conservation*, chapter 3; D. Graham Burnett, *The Sounding of the Whale: Science and Cetaceans in the Twentieth Century* (Chicago: University of Chicago Press, 2012); and Kurkpatrick Dorsey, *Whales and Nations: Environmental Diplomacy on the High Seas* (Seattle: University of Washington Press, 2013).

23. Tønnessen and Johnsen, *The History of Modern Whaling*, 452–456.

24. Tønnessen and Johnsen, *The History of Modern Whaling*, 499–506, and Dorsey, *Whales and Nations*, chapter 3.

25. Scott Barrett, *Environment and Statecraft: The Strategy of Environmental Treaty-making* (Oxford: Oxford University Press, 2003), 69–71; Dorsey, *Whales and Nations*, chapter 6.

26. The text of the Northwest Atlantic treaty can be found at www.nafo.int, and dozens of other fisheries and environmental agreements can be found at http://iea.uoregon.edu.

27. Carmel Finley, *All the Fish in the Sea: Maximum Sustainable Yield and the Failure of Fisheries Management* (Chicago: University of Chicago Press, 2011).

28. Thomas R. Dunlap, *Saving America's Wildlife: Ecology and the American Mind, 1850–1990* (Princeton, NJ: Princeton University Press 1988), 145–146.

29. Lyster, *International Wildlife Law*, chapter 12, especially page 239.

30. Information about CITES can be found at www.cites.org; for specifics on CITES sanctions, see also Rosalind Reeve, "Wildlife Trade, Sanctions and Compliance: Lessons from the CITES Regime," *International Affairs*, 82, no. 5 (September 2006): 881–897.

31. Van Heijnsbergen provides a brief overview of the history of such protection and then focuses on modern material, such as CITES.

32. Alexandre Kiss and Dinah Shelton, *International Environmental Law* (Ardsley-on-Hudson, NY: Transnational Publishers, 1991), 122–126; Caldwell, *International Environmental Policy*, 122–123; for an overview of the IJC, see Chirakaikaran J. Chacko, *The International Joint Commission between the United States of America and the Dominion of Canada* (New York: Columbia University Press, 1932). The historian who has written the most about the Trail Smelter is John Wirth, whose main book is *Smelter Smoke in North America: The Politics of Transborder Pollution* (Lawrence, KS: University Press of Kansas, 1999).

33. The best history of a river is Mark Cioc's *The Rhine: An Eco-biography, 1815–2000* (Seattle: University of Washington Press, 2006).

34. Kiss and Shelton, *International Environmental Law*, 162–168,175–178; Caldwell, *International Environmental Policy*, 294. The efforts of the League of Nations on oil pollution have been studied by Anna-Katarina Woebse, "Oil on Troubled Waters? Environmental Diplomacy in the League of Nations," *Diplomatic History*, 32, no. 4 (September, 2008): 519–538.

35. Jacob Hamblin, *Poison in the Well: Radioactive Waste in the Oceans at the Dawn of the Nuclear Age* (New Brunswick, NJ: Rutgers University Press, 2009).

36. Kiss and Shelton, *International Environmental Law*, 169–174; Ralph Levering and Miriam Levering, *Citizens' Action for Global Change: The Neptune Group and Law of the Sea* (Syracuse, NY: Syracuse University Press, 1999).

37. The UNCLOS website can be found at http://www.un.org/depts/los/index.htm.

38. A firsthand account of the negotiation of the Vienna and Montreal agreements is in Richard Benedick, *Ozone Diplomacy: New Directions in Safeguarding the Planet* (Cambridge, MA: Harvard University Press, 1991). For the Vienna convention see pages 44–47.

39. Benedick, *Ozone Diplomacy*, 51–55, 77–97; the most recent summary of ozone depletion can be found at the EPA's website, http://www.epa.gov/ozone/science/

40. Caldwell, *International Environmental Policy*, 263–264.

41. Barrett, *Environment and Statecraft*, 374–380.

42. Naomi Oreskes and Erik Conway, *Merchants of Doubt: How a Handful of Scientists Obscured the Truth on Issues from Tobacco Smoke to Global Warming* (New York: Bloomsbury Press, 2010); chapter six focuses on global climate change.

43. See Barrett, *Environment and Statecraft*, 359–399, for a good summary of the negotiations at Kyoto and subsequent meetings, the flaws in the Kyoto system, and a suggested alternative path.

44. "Clouds Part Slowly in Climate Change Diplomacy," *The Guardian,* July 8, 2008, http://www.theguardian.com/world/2008/jul/08/g8.climatechange; "The East is Grey," *The Economist,* August 10, 2013.

45. Edmund Russell and Richard Tucker, eds., *Natural Enemy, Natural Ally: Toward and Environmental History of Warfare* (Corvallis: Oregon State University Press, 2004); Erica Bsumek et al., *Nation-States and the Global Environment: New Approaches to International Environmental History* (Oxford: Oxford University Press, 2013); J. R. McNeill and Corinna Unger, *Environmental Histories of the Cold War* (Cambridge: Cambridge University Press, 2010); Charles E. Closmann, *War and the Environment: Military Destruction in the Modern Age* (College Station: Texas A & M University Press, 2009); Kurk Dorsey and Mark Lytle, "Forum: New Directions in Diplomatic and Environmental History," *Diplomatic History* 32, no. 4 (September 2008): 517–646.

46. Andrew Isenberg, *The Destruction of the Bison* (Cambridge: Cambridge University Press, 1999).

47. J. R. McNeill, *Mosquito Empires: Ecology and War in the Greater Caribbean, 1620–1914* (Cambridge: Cambridge University Press, 2010).

48. Gregory Cushman, *Guano and the Opening of the Pacific World: A Global Ecological History* (Cambridge: Cambridge University Press, 2013).

49. Richard Tucker, *Insatiable Appetite: The United States and the Ecological Degradation of the Tropical World* (Berkeley: University of California Press, 2010).

50. This paragraph and the next one are drawn from Thomas Robertson, *The Malthusian Moment: Global Population Growth and the Birth of American Environmentalism* (New Brunswick, NJ: Rutgers University Press, 2012); see also William Vogt, *Road to Survival* (New York: W. Sloane Associates, 1948), Fairfield Osborn, *Our Plundered Planet* (Boston: Little, Brown, 1948), Aldo Leopold, *A Sand County Almanac* (New York: Oxford University Press, 1949), Paul Ehrlich, *The Population Bomb* (New York: Ballatine Books, 1968).

51. Nick Cullather, *The Hungry World: America's Cold War Battle against Poverty in Asia* (Cambridge, MA: Harvard University Press, 2008); Mathew Connelly, *Fatal Misconception: The Struggle to Control World Population* (Cambridge, MA: Harvard University Press, 2008).

52. Jacob Darwin Hamblin, *Arming Mother Nature: The Birth of Catastrophic Environmentalism* (Oxford: Oxford University Press, 2013).

53. David Biggs, *Quagmire: Nation-building and Nature in the Mekong Delta* (Seattle: University of Washington Press, 2010).

54. David Zierler, *The Invention of Ecocide: Agent Orange, Vietnam, and the Scientists Who Changed the Way We Think about the Environment* (Athens: University of Georgia Press, 2011); Edwin Martini, *Agent Orange: History, Science, and the Politics of Uncertainty* (Amherst: University of Massachusetts Press, 2012).

CHAPTER 25

....

THE POLITICS OF NATURE

....

FRANK ZELKO

ANYONE with even a vague awareness of contemporary public affairs knows that environmental issues are near the top of the political agenda, at least rhetorically, in most Western democracies. Those with a deeper historical interest might argue that the roots of environmental politics lie in the nineteenth century, when various conservation and preservation movements began to emerge in North America and Europe. But can we go back even further? Is it possible to study the environmental politics of ancient Egypt, Ming China, and medieval Europe? Theoretically, the answer is yes: politics, after all, is an omnipresent phenomenon in all societies, and nature—whether conceived as a sacred entity, a natural resource, or something in between—has always been the subject of political contestation. Nevertheless, the potential scope of "politics" is so vast, that an overview of the subject requires some limits and a clear focus.[1]

Given that the vast majority of historians concern themselves with the most recent phase of our species' history—the period characterized by populous, hierarchical societies rather than the small hunter-gatherer bands that constituted 99 percent of our history—it makes sense to focus on how elites have attempted to use their power to shape society's relationship with nature and how others, in turn, have accepted or contested their policies and pronouncements. From this perspective, the state is the locus of politics. Traditionally, scholarly studies of the state have focused on issues such as the maintenance of security, sovereignty, and social order. Environmental historians have added another vital dimension: the management of environmental affairs. Adam Rome has labeled this branch of state activity the "environmental-management state". Although Rome was employing the concept with reference to twentieth-century US history, it is capacious enough to embrace virtually every populous, hierarchical society that has ever existed. It can therefore function as an overarching conceptual framework for studying environmental politics throughout history.[2]

THE NATURE OF POLITICS

....

Modern Venice is not exactly a city that people associate with forests. In fact, the exquisite urban confection of the Adriatic Sea—a marvel of stone, water, and human ingenuity—is almost bereft of trees. Apart from a few hardy weeds poking through the cracks in the centuries-old cobblestones, most of the city's plant life is confined to window boxes and rooftop terraces. But of course, Venice has not always been just a tourist Mecca, a kind of Italianate Disneyland for adults. It was arguably the most important European city-state of the late medieval era, and it controlled large tracts of forest in its rural hinterland. In fact, these forests literally constitute the foundations of the city, which rests upon innumerable piles of larch and alder. The merchant fleet that plied the Mediterranean trade routes, bringing silver, spices, and slaves into the city and making Venice into Europe's greatest wholesale marketplace, also required vast quantities of timber. This heavy reliance on wood forced the Venetian Senate to gradually assert the city's influence over the forests of the northern Adriatic, a process that culminated, in 1476, in a number of legal provisions prescribing how the forests could be exploited. In short, the city's economic needs prompted a series of political decisions that determined how the region's forests could be used, both by the Venetian state and the people within its extended political orbit.[3]

Fast forward to early twentieth-century Prussia, where we find the Ministry of Education busy establishing an Institute for the Care of Natural Monuments (*Staatliche Stelle für Naturedenkmalpflege*). Prussia may forever be associated with Junkers, *Pickelhauben*, and blood-and-iron militarism, but this interest in nature protection, which prefigured later developments in German environmental policy, might well be its most enduring legacy. The institute, created in 1906, was the first government organ that dealt exclusively with nature protection, and it marked the beginning of state-sponsored environmental protection in Germany, initiating an institutional continuity that has lasted right up to the creation of the current *Bundesamt für Naturschutz*. The Institute was a political expression of the fear and nostalgia that accompanied Germany's rapid modernization. As traditional, quintessentially "Germanic" landscapes were sacrificed to industry and large-scale agriculture, various voices, mostly among the bourgeoisie and nobility, urged the state to act in order to protect the precious remnants of premodern Prussian nature.[4]

Skip ahead to January 1, 1970, a momentous day in the history of environmental policy. On that day, Richard Nixon, a man whose interest in environmental issues was shallow but whose instinct for political opportunity was razor sharp, signed the National Environmental Policy Act (NEPA), which required federal agencies to conduct environmental impact assessments before embarking on development projects. In the following years, Nixon put his stamp of approval on a host of seminal environmental acts, including the Endangered Species Act and the Marine Mammal Protection Act. Furthermore, promoting environmental awareness overseas became a key platform of Nixon's foreign policy, largely through the agency of Russell Train, the chairman of the newly created Council on Environmental Quality. All of this came about after more than a decade of environmental activism on the part of various non-governmental organizations and citizen's groups, as well as the actions of politicians at all levels of government.[5]

These three cases—from fifteenth-century Venice, early twentieth-century Germany, and the late twentieth-century United States—are all instances of what might broadly be defined as "environmental politics." They are "political" in the sense that they deal with the exercise of power and the resolution of conflict among individuals and groups pursuing their interests; and they are "environmental" insofar as they constitute examples of human interaction with the non-human world.[6] Certainly, there are important differences in the goals and ideological underpinnings that led to them: in the case of Venice, the goal was conserving a vital natural resource; in Prussia it was maintaining what many saw as desirable elements of the Prussian landscape; and in the US, state policy had numerous motivations, including resource conservation, protecting human health, and protecting various ecosystems and species. Nevertheless, all were efforts on the part of the state to manage certain environmental problems, and all no doubt involved multiple constituencies. The concept of the environmental-management state is therefore at least as useful as some of the more traditional areas of state activity studied by political historians, such as the national security state or welfare state.[7]

The environmental-management state, one can argue, has existed since the rise of the first agrarian civilizations. We do not have much direct evidence of the quotidian politics that preoccupied the rulers and denizens of ancient Sumer, but we can be fairly certain that they involved issues such as water engineering. The same was no doubt true for the numerous other hydraulic civilizations that bloomed in the arid zones of southwest Asia and northern Africa. Over time, ideas about "nature" changed considerably, as did the human capacity to exploit it. Such changing perceptions have altered the range of actions open to the environmental-management state and those who try to influence it. For example, the Venetian forest provisions of 1476 and NEPA both involve state environmental management, but the two states' conceptions of nature—its structure, its boundaries, and its relationship to humans—differed significantly. For late twentieth-century Americans, "nature" is laden with two hundred years of post-Enlightenment history. Its meaning has been shaped by the insights of evolutionary theory and ecology and by the experience of industrialization, which in turn led to the development of concepts such as interconnectedness and finitude. Fifteenth-century Venetians may have understood that there was a connection between deforestation and the siltation of Venice's lagoon, but the policies revolving around such issues took place within a different epistemological and moral framework.

Therefore, although the environmental-management state concept is capacious— perhaps *because* it is so capacious—it must be used with some caution and nuance. In particular, historians must avoid conflating "environmental politics" with "environmentalism." The former describes a broad concept applicable across time and space, while the latter is a recent worldview representing a momentous cultural shift in humanity's understanding of its place in nature. In Michael Bess's words, environmentalism "reinvent[ed] nature as a bounded and fragile space, requiring intensive human nurturing and protection." While the roots of this shift date back to the eighteenth century, its manifestation as a mass phenomenon, "a central feature in the cultural landscape of modernity," occurred only after the Second World War.[8] Strictly speaking, therefore, the term "environmentalism" is anachronistic when used to describe pre-1960s environmental politics.[9]

The rest of this chapter will examine how various historians have studied environmental politics in specific times and places. Although few scholars have used the term "environmental-management state" explicitly, many implicitly employ the concept. Given the voluminous nature of this literature, such a review will be necessarily selective. The focus will mostly be on Europe and North America since the fifteenth century. This overview will elucidate how the environmental-management state has developed in response to changes in technology and resource use, the opening and closing of political opportunity structures, and other factors. It will examine familiar developments such as conservation, preservation, sanitation reform, nature romanticism, and environmentalism. It will also touch on how state efforts to manage the environment have been affected by other broad historical developments, including imperialism, capitalism, democracratization, totalitarianism, and the rise of the modern nation-state. Some of these studies focus on elites; others show how various actors—from bureaucrats, to workers, to peasants—have at various times been both victims of the environmental-management state and active agents in its operation.

Environmental Management and Politics in the Preindustrial Era

Our modern world is inconceivable without fossil fuels, but prior to the mid-eighteenth century, they played little role in human life. Coal was difficult and expensive to mine, oil more so. By and large, early modern Europeans harnessed nature's energy in much the same way as their ancestors had been doing for centuries: animals ploughed fields, wind and water mills ground grain, and wood fuelled the hearths that blazed in numerous smitheries and warmed countless homes. The same was true for the rest of the world, although the influence of geography and climate helped ensure that there was considerable variation in the techniques and methods of food procurement and energy extraction. In much of the world, life revolved around agricultural villages, where the benighted inhabitants were vulnerable to the whims of nature, the pecuniary demands of urban powers, and the opportunistic raids of nomadic pastoralists. Civilizations rose and fell. The Sumerians built an intricate network of canals and irrigation channels and made the desert bloom, until evaporation finally rendered their fields barren with salt, the poisonous residue of virtually all subsequent hydraulic schemes. Water shortages were rarely a problem in central Europe. Instead, dense forests constituted the backdrop to people's lives and the backbone of much economic activity. In both cases, heavy dependence on a particular natural resource—be it water or wood—provided abundant opportunities for conflict and the concentration of power and wealth. Not surprisingly, therefore, political life frequently revolved around such natural resource disputes.[10]

In much of preindustrial Europe, forests constituted the lifeblood of society. Unsurprisingly, therefore, they were major sites of political contestation. To the nobility, for whom sustenance was rarely a problem, the forest's main function was as a hunting preserve, and throughout the Middle Ages an increasingly powerful ruling class cordoned off vast forest acreages for its recreational pleasure. Naturally, this conflicted with the subsistence needs of peasants, for whom the forests constituted, first and foremost, a vital supply of firewood, as well as a source of meat, fruit, and nuts. Not surprisingly, therefore, the royal dominion over forests provoked a great deal of seething resentment. In England, for example, royal dominion was imposed at the time of the Norman Conquest in the eleventh century, and was enforced through the threat of gruesome punishments such as castration and blinding.[11]

Throughout Europe, Joachim Radkau notes, medieval princes also discovered that "forest protection (was) a first-rate instrument of political power." This protection was justified as a traditional hunting and mining prerogative. However, as public opinion became increasingly important as a result of the print revolution and the Reformation, rulers had to convince the population that restricting access to forests was for the common good, rather than merely for the nobility's recreational pleasure. Thus the political discourse shifted from *preservation* for the sake of the elite few, to *conservation* for the benefit of the masses. Henceforth, restrictions were increasingly justified on the grounds that wood shortages were becoming more and more severe, thereby threatening the entire populace. The result was a veritable avalanche of forest legislation throughout northern and central Europe, leaving behind a trail of archival records that constitute an excellent source of primary materials for environmental historians who are interested in forest politics and state formation.[12]

Several historians have followed this trail, but few have done so with the assiduousness of Paul Warde. In his dense and detailed study of forest politics and economics in early modern Württemberg, Warde not only tracks the development of a "moral economy" centered around peasant access to forests, but also convincingly argues that the efforts to extend and improve administrative control over forests was a vital factor in the formation of the early modern state. Warde describes a "wooden world" almost completely dependent on forests, which provided firewood for heating and manufacturing, revenue for the dukes, and fodder for pigs and other livestock on which people depended for protein. Like any good environmental historian, Warde examines ecological change, as well as the politics and economics of forestry, and demonstrates the connection between them all. He shows how peasants and local communities, in a process that was both voluntary and subtly coercive, gradually ceded their customary usage rights to ducal authority on the understanding that central administration offered better protection and superior management. Thus both parties could agree that they were acting in defense of the common wealth, or *Gemeinnutz*, although ultimately, it was the duke who was in the position to define the interests of the community—usually, not surprisingly, in a way that maximized his power and wealth.[13]

The forests in the Franche-Comté, a couple of hundred kilometers southwest of Württemberg, underwent a similar political transformation, although the reaction

of the region's citizens was more rambunctious than that of their German counterparts. Successive political upheavals in France throughout the late eighteenth and
early nineteenth centuries catalyzed resistance to the state's efforts to control the
region's forests—a resistance, Kieko Matteson argues, that produced some tangible
benefits. Nevertheless, as in Württemberg and most of Europe, forest conservation in France had less to do with farsighted environmental protection than with
suppressing sedition, expanding state power, and substituting commercial exploitation for communal utility.[14] Matteson joins a number of other scholars, particularly those studying Africa and Asia, who argue that conservation practices were
frequently imperialistic and despotic in nature. By ignoring local knowledge and
criminalizing customary usage, state forest policies not only restricted peasant
access to woodlands, but also damaged their ecological integrity by gradually converting them into timber plantations.[15] This is a form of politics that might best be
described as *dirigiste conservation*: the process whereby political elites, with varying degrees of coercion, attempt to gain greater control over natural resources in
order to benefit the state.[16]

A similar process was occurring south of the Alps. Karl Appuhn shows how the
Venetian state expanded its bureaucratic reach in order to secure its timber supply.
The state appropriated technical knowledge and rationalized forest management,
but whether this created an effective system of rational forestry depends to some
extent, Appuhn argues, on the ideological sympathies of the historian. If one takes
an anti-statist approach of the kind adopted by James Scott, Venetian forestry practices were ultimately detrimental to both the forests and the state, since they failed to
take adequate account of people's local knowledge, or what Scott refers to as *metis*, the
"wide array of practical skills and acquired intelligence in responding to a constantly
changing natural and human environment."[17] On the other hand, one could argue that
Venice developed a reasonably effective set of rational tools for controlling its timber
resources and that it managed its forests better than its early modern counterparts in
England and France.[18] In this case, the state's political management of nature could
once again be described as *dirigiste conservation*.

Several scholars have tried to link early modern European conservation efforts to
twentieth-century environmentalism, though none have been as vociferous as Richard
Grove. In a work of prodigious research and remarkable erudition, Grove examines the
encounter between colonial officials and exotic nature on tropical islands in the seventeenth and eighteenth centuries. The compact scale of islands allowed various English
naturalists and French physiocrats to observe the ecological consequences of intensive
resource exploitation and agriculture in places such as Mauritius and the Bahamas.
This led many to lobby, with some success, for various forms of conservation, thereby
establishing the political and intellectual basis for twentieth-century environmentalism.[19] While Grove convincingly describes the formation of a kind of incipient conservationist worldview, most commentators agree that he pushes his argument too far
when he tries to portray colonial scientists as progenitors of twentieth-century environmentalists. As Michael Bess notes, Grove

cannot show any real link of continuity between these men and the green movements of the present. The chasm is simply too great between the mass politics of an
information age, animated by visions of ozone holes and global warming, and the
relatively isolated (and soon forgotten) work of these pioneering colonial officials.[20]

Emma Spary, historian of science, has covered similar territory, without making Grove's contentious claims about the link between eighteenth-century science
and twentieth-century environmentalism. In fact, unlike Grove, she uses the word
"environment" sparingly and never suggests a direct link between her subjects and
twentieth-century environmentalists. Spary's study of French naturalists and the
Muséum d'Histoire Naturelle (known during the *Ancien Régime* as the *Jardin du Roi*)
examines the changing attitudes toward nature reflected in the Muséum's collection.
She argues that the work of categorizing, analyzing, and displaying natural objects was
hardly a neutral scientific project; rather, it dovetailed with the state's broader political
aims. During the French Revolution, the culture of aristocratic patronage that characterized the *Jardin* gave way to a kind of republican naturalism. Nature, like humanity,
could be improved, and the goal of an enlightened society should be to strive toward
the perfection of both. The Muséum and its naturalists collected species from all over
the world and seemed particularly obsessed with how well exotic species could acclimatize to new surroundings. The notion of improving individual species for the bet-
terment of both nature and mankind was then applied to entire ecosystems, with some
naturalists, such as Buffon, even asserting that humans could alter the world's climate
in a positive way. For the leaders of the new republic, society was malleable and improv-
able. Naturalists created a model of nature that reflected this political view, and the
Muséum d'Histoire Naturelle's collections and exhibitions embodied this ideology
and put it on display for the pedagogical benefit of the republic's citizens. Here was an
example of an environmental-management state that was interested in more than just
securing access to natural resources; it was also consciously trying to shape its citizens'
ideas about nature in a way that would serve the broader interests of the Republic.[21]

While dirigiste conservation might be an apt description for state environmental-
management throughout much of history, not all states practiced it, at least not everywhere and all the time. For one thing, enforcement of state policy became more difficult and expensive the farther one moved from the center of power, although as
Matteson shows, the Parisian bureaucracy could certainly cause problems in distant
Franche-Comté.[22] The Ottomans, however, took a different approach. After conquering Egypt in the late seventeenth century, Ottoman rulers, according to Alan Mikhail,
were content to give Egyptian peasants "near-absolute authority over the function and
repair of irrigation works because they were the ones with the most specialized and
longest experience of those irrigation features and of the environments they served."[23]
This bottom-up practice remained in place for over a century and only ended, somewhat ironically, when the Ottomans retreated from Egypt and power devolved back
to Cairo in the early nineteenth century. Until then, the powerful Ottoman state felt
it was useful to exploit the local knowledge of Egyptian peasants, who grew much of

the empire's food. In turn, Egyptian peasants maintained a sense of independence and pride in their ability to run and maintain a complex irrigation network and highly productive agricultural system, and were thus less inclined to feel resentful toward their rulers in Istanbul.

THE INDUSTRIAL AGE

The nineteenth and twentieth centuries saw all kinds of changes in the ways humans exploited their environment. In a remarkably short time, scientists and engineers developed ways to extract vast quantities of energy from the bowels of the earth, while a new class of entrepreneurs harnessed it to produce cornucopian quantities of food, clothing, and the various knick-knacks demanded by a burgeoning population that was undergoing a rapid metamorphosis from peasants to proletarians to fully fledged consumers. Liberal capitalism was the locomotive of the early stage of economic modernization, but it was not essential to its momentum. Socialist, communist, and fascist states were as fanatically committed to the central tenets of modernity as the pioneer industrial capitalists of Manchester and Pittsburgh. All desired constant economic growth and expansion, sharing an ideological commitment to what twentieth-century critics would label "productivism."[24] The ecological impact of human activity—never negligible—began to produce levels of environmental change that in the past could only be caused by cosmic cataclysms such as meteor strikes, volcanic eruptions, or global climate change. Industrial effluent choked rivers and streams, clouds of coal smoke shrouded grimy cities, and ancient forests and grasslands were sacrificed to make way for crops and cattle.[25]

Not surprisingly, this level of destruction, whether creative or not, spurred significant cultural and political reactions throughout the world, thereby altering the context of state environmental management. For simplicity's sake, we can classify these reactions into three broad, and frequently overlapping, categories. The first of these is *nature romanticism*, a cultural current that flowed, and continues to flow, beneath the sweeping tide of industrial modernity. In Europe and North America, science and reason promoted views of nature that became increasingly rational and instrumental. The grid and the straight line defined the worldview of the scientific rationalist, and efficiency and maximum yield became the mantra of the industrialist and the forest manager alike. In a remarkably short space of time, this mode of thinking was inscribed in the landscape: "inefficient" rivers were straightened, "useless" wetlands reclaimed, and "chaotic" forests were remade for the efficient production of timber. Such developments, combined with a dramatic increase in air and water pollution, damaged or destroyed the appearance of "nature" that people were accustomed to, even if that nature was itself the product of many centuries of human industry and exploitation. For many people, such changes came to be seen as signs of progress and prosperity. However, for a small but vocal minority, they were deplorable evidence of humankind's greed and destructiveness. Throughout the nineteenth century, writers on both sides of the Atlantic condemned the trail of social, environmental, and aesthetic destruction that industrialization left in its wake, frequently fetishizing preindustrial landscapes in the process.

In Britain and Europe, industrialization was accompanied by a wave of nostalgia for the rural landscapes and livelihoods it contaminated. William Wordsworth, John Ruskin, and Edward Carpenter exemplified this yearning for the preindustrial countryside, while the socialist writer William Morris argued that successful nature protection required the restriction of private property rights.[26] In the United States, transcendentalists such as Ralph Waldo Emerson and Henry David Thoreau glorified what they saw as the untamed nature of the American frontier, with Thoreau famously proclaiming "in wildness is the preservation of the world." Painters such as Albert Bierstadt and Thomas Moran offered monumental renderings of sublime wilderness, while John Muir, the son of deeply religious Scottish immigrants, took Thoreau's declaration to heart, roaming the mountains of the American west and becoming the great evangelist of wilderness preservation.[27] Nature romanticism spawned numerous movements and organizations dedicated to preserving various versions of preindustrial nature, be it the pastures and hedgerows of the English and French countryside or the sublime landscapes of the American frontier.[28] In the United Kingdom, the Back to Nature movement of the late nineteenth century advocated a rural utopia; idyllic English villages full of healthy, self-sufficient communities of educated citizens with time to read and reflect on their good fortune. Germany, too, had its back-to-nature cults, some of which had a pronounced tendency toward mysticism and paganism, including groups such as the *Naturmenschen*, as well as the broader *Lebensreform* movement. Nature romanticism also spawned various overtly political organizations, no more so than in the United States, where John Muir's Sierra Club emerged as the major lobby group for the preservation of the sublime landscapes of the west.[29]

There is, not surprisingly, a wealth of literature dealing with various forms of nature romanticism around the world. Relatively few of these studies, however, explicitly explore the linkages between this cultural development and its political manifestations. A notable exception is Peter Gould's work on the impact that various "green" ideas had on the evolution of socialism in late nineteenth-century Britain. Gould's study demonstrates how ideas about nature informed politics in the pre-environmentalist era. Influenced by the romanticism of writers such as Ruskin, William Morris and other socialists advocated lifestyles that were grounded in the rural rhythms and pastoral aesthetic of nature, or at least, the particularly English version of nature they favored. Such ideas spawned various "Back to Nature" and "Back to the Land" movements that sought to temper the squalor of urban-industrial life by encouraging the preservation of open space and more dispersed, semi-rural settlements. Ironically, such aims dovetailed rather nicely with conservative concerns about the immorality of the new industrial proletariat, a condition best purged by the cleansing air of the English countryside. Depending on one's politics, therefore, nature could be viewed as either a promoter of freedom or a means of social control.[30]

In Germany at roughly the same time, nature romanticism fostered various preservation movements among middle-class citizens anxious about the impact of
industrialization on German landscapes, and by extension, national identity. As
Thomas Lekan notes in his study of the cultural and political discourse of nature in
late nineteenth-century Germany, "Preservationists' efforts to set aside natural monuments from industrialization did not merely prevent environmental destruction but
produced new political identities by naturalizing modern political borders and school-
ing local citizens in the aesthetic beauties of nature."[31] In the United States, according
to Donald Worster, the nature romanticism exemplified by John Muir was "an integral part of the great modern movement toward democracy and social equality, which
has led to the pulling down of so many oppressive hierarchies that once plagued the
world."[32]

Romantics such as Muir and Wordsworth were able to wield considerable cultural
and political influence within their respective societies. By problematizing industrialization and economic growth, they persuaded an influential segment of the bourgeoisie to attempt to reshape the state's environmental management practices to include
the preservation of scenic landscapes, a position that frequently ran counter to the
more traditional state practices of natural resource exploitation and forest conservation. An early example occurred in the 1870s, when the city of Manchester attempted
to meet its burgeoning water needs by damming Lake Thirlmere in England's scenic
and much-loved Lake District. John Ruskin was among those who strongly opposed
the plans, arguing that it would be an act of desecration, not only of a beautiful landscape but also of a vital part of England's cultural heritage. Opponents of the dam also
frequently quoted Wordsworth's lovingly poetic descriptions of the region.[33] Although
unsuccessful, the Thirlmere campaign and others like it would thereafter complicate
the task of state environmental management, as various elites attempted to harness
the power of the state to pursue environmental policies that were often in opposition
to one another. Over the next century, particularly in Western democracies, interest
groups urging the protection of wildlife and preservation of landscapes would become
increasingly adept at convincing the state to act on their behalf.

A second major reaction to the environmental impact of industrial modernity
was the intellectual and political movements we can group together under the heading *scientific conservation*. An outgrowth of European forest conservation, scientific
conservation in the nineteenth century became an ideology of "doom and resurrection,"[34] which predicted that new forms of agriculture and industry, combined with
laissez-faire resource extraction, would decimate nature and leave few resources for
future generations. The answer was state management of resource extraction and agricultural practices in order to encourage the greatest degree of efficiency through the
application of modern science. The most influential model for this attitude toward
natural resources arose from German forestry practices of the nineteenth century, in
which forests were managed with mathematical precision by foresters and bureaucrats
for the greater good of the Prussian, and after 1870, German, state. German silvicul-
ture techniques influenced British and French practices, thereby spreading scientific

forestry to the far-flung corners of their respective empires, and German forestry schools trained some of the most influential foresters in North America, including Gifford Pinchot, who would become the leading proponent of scientific conservation in the United States.[35]

Politically, conservation manifested itself in multiple debates and power struggles throughout the world. In the United States, for example, leading conservationists such as Gifford Pinchot stood in opposition to both the proponents of laissez-faire resource exploitation—and by extension, the dominant political discourse of small government—and the preservationists represented by John Muir. Aligning themselves with the broader progressive movement, scientific conservationists tended to be well-educated members of the white Anglo-Saxon Protestant upper-middle class. Their attitudes toward nature frequently conflicted with the interests of rural people and immigrants. As Louis Warren has shown, conservationists' ideas of what constituted "proper" hunting practices differed considerably from those of Italian immigrants in Pennsylvania or the Blackfeet in Montana. Elite sportsmen such as Teddy Roosevelt, who combined Ivy League refinement with frontier machismo, successfully lobbied for hunting regulations that severely curtailed the subsistence and market hunting practices of rural hunters.[36] Similarly, Benjamin Heber Johnson's study of mining strikes in early twentieth-century Minnesota reveals how mine owners and merchants sought to enforce conservation laws in order to deprive workers of the fruits of subsistence hunting, thereby exerting greater pressure on them to return to work.[37] Colonial conservation policies frequently possessed a similar political dynamic. In East Africa, they overturned traditional hunting practices in favor of laws that suited European big game hunters, while in South and Southeast Asia, scientific conservation undermined customary forest rights, depriving people of resources they had used sustainably for centuries.[38] Such concerns have continued to animate scholarship in the Global South, where the policies of the environmental-management state frequently conflict with the needs of people trying to live according to long-standing natural resource practices. Much of this scholarship falls under the broadly defined field of political ecology, and is vigorously critical of state policy while championing the rights of the poor and oppressed. One of the more influential theories to emerge from such work is Arun Agrawal's notion of "environmentality," which examines how the environmental-management state inculcates its subjects with the state's environmental ethos. The term was inspired by Foucault's notion of "governmentality," and its goal, according to Agrawal, "is to understand and describe how modern forms of power and regulation achieve their full effects not by forcing people toward state-mandated goals but by turning them into accomplices."[39] When a state such as India—the primary site of Agrawal's research—attempts to impose environmental reforms (frequently those advocated by Western governments and NGOs) such as forest conservation upon people who rely on the forest for sustenance, it is usually met with noncooperation or determined opposition. However, when power, knowledge, and regulation are dispersed in such a way that local people feel they have a stake in—and an understanding of the goal of—environmental management, they are likely to not only cooperate with the state, but to help spread and enforce its policies. The historical nature of Agrawal's work—which is based on a long-term analysis of the Indian forest councils created by the colonial state in the 1930s—suggests that the environmentality concept, especially the idea of the formation of environmental "subjects," could have broader historical applicability, and not just in the Global South.

Industrialization and technological modernity also gave rise to a series of movements that were not so much concerned with the preservation of landscapes or the efficient exploitation of resources as they were with protecting human health. Various reformers, particularly in rapidly growing industrial cities, stressed the importance of proper waste disposal and pollution abatement as ways of improving the health of urban citizens. As Radkau notes, "(m)ore so even than today, urban environmental concerns in the nineteenth century were part of contexts that seemed to pose an immediate threat. The dirt and stench of industrial cities formed a continuum with the danger of epidemics and socially explosive problems," thereby tying urban environmental problems to various social pathologies.[40] Like scientific conservationists, anti-pollution advocates and urban health reformers in Europe and North America—members of what has broadly been labeled the *sanitation movement*—were mostly upper-middle-class reformers.[41]

The largely urban concerns of the sanitation movement, combined with the fact that it coincided with important developments in federal and municipal politics, has ensured that the various branches of sanitation reform have received considerable historical coverage. Among American environmental historians, Martin Melosi and Joel Tarr have led the way. Melosi's magnum opus, *The Sanitary City* (2000), offers the most complete and sophisticated overview of the history of waste disposal in the United States. Melosi organizes his study around three major biological paradigms: the age of miasmas (the early to mid nineteenth-century); the bacteriological era (late nineteenth to mid twentieth-centuries); and the post-WWII period in which ecology became highly influential. As well as documenting the impact of technological and engineering developments on urban environments, Melosi's narrative charts the political struggles that were a vital part of the process of creating the "sanitary city." Scientists, engineers, and health advocates needed to persuade skeptical machine politicians and barely literate immigrants of the benefits of installing expensive sewage treatment facilities, while also trying to regulate the waste disposal habits of both industry and the general public.[42]

Since the sanitation movement, like the Progressive Era in general, was a transatlantic if not global phenomenon, there is plenty of room for comparative and transnational studies.[43] Frank Uekötter, for example, has compared how American and German cities coped with the problem of industrial smoke. His study elucidates some key political differences between the two nations in the arena of environmental reform. In the United States, with its long tradition of grass-roots movements and relatively weak central government, numerous local groups pressured factory owners to disperse and reduce smoke emissions. In Germany, with its more centralized and bureaucratic political structure, people relied more on the bureaucracy and the judicial system to control air pollution. The result, according to Uekötter, was that American reformers were more successful in individual battles, particularly where smoke was highly visible and posed a threat to a company's image. In Germany, and Europe in general, such problems were usually dealt with through various bureaucratic arms of the state, leading to more uniform outcomes, with fewer spectacular victories.[44] By the early twentieth century, as Radkau notes, the sanitation movement could look back on several decades of political success. While nature romantics had only limited success in preserving various wild and cultural landscapes, urban reformers, who tended to embrace both modern technology and interest-group politics, could celebrate half a century of rapid improvement in sewage treatment, smoke abatement, and overall urban environmental health. More so than most other branches of environmental reform, the sanitation movement, with its emphasis on human health and urban environments, was highly successful at mobilizing the state.[45]

The first half of the twentieth century was characterized by unprecedented global warfare and the sudden emergence of powerful totalitarian states that controlled vast territories. The rise to power of such regimes—the Soviets in Russia, the Nazis in Germany, and Mao's communists in China are the three most prominent examples— was accompanied by bloody warfare, massive socioeconomic disruption, and brutal repression. There was little political space and few resources for the type of environmental management that was occurring in Western democracies. The Soviet and Chinese states were mostly concerned with wringing natural resources and food from the land as quickly as possible, with little thought for conservation, let alone nature preservation, wildlife protection, or concerns about pollution. In the Soviet Union under Stalin, a few environmentally minded scientists managed to eke out what Douglas Weiner called "a little corner of freedom": a sliver of political space in which they could discuss ideas about nature protection and sustainable agriculture. The situation in China was similarly bleak, with Mao's "Man must conquer Nature" slogan setting the tone. In both these states, environmental management, if one can call it that, was devoted to intensive environmental exploitation regardless of the ecological and social costs.[46]

The German case, however, was somewhat different. One of the key questions occupying twentieth-century German environmental historians, and the title of a major publication on the subject, is "How green were the Nazis?" The sometimes acrimonious debate on this question was sparked by Anna Bramwell, an iconoclastic British historian who has argued that the Nazis, and the anti-humanist Romantic traditions that inspired them, played an important role in the evolution of twentieth-century green politics. Bramwell displays some curious sentiments: she is sympathetic toward Hitler's agriculture minister, Walther Darré, whom she views as a prophet for the organic farming movement and a reluctant Nazi. Today's Greens, on the other hand, "bear more resemblance to those pre- and proto-Nazi groups that sprang up during the 1920s in an outbreak of quasi-religious prophecy and radicalism."[47] Her argument overemphasizes the Nazi's commitment to environmental issues, while also deliberately attempting to taint green political parties with a Nazi lineage.[48]

Historians who have studied the period with more care have largely dismissed Bramwell's analysis. While a few high-ranking Nazis expressed some sympathy for organic farming and the preservation of what they saw as quintessentially German landscapes, such views were always subordinated to the regime's military and economic aims. And while some leading conservationists cooperated with the Nazis, it was mostly in the spirit of thoughtless pragmatism. As Frank Uekötter notes: "Without much thought to universal principles such as democracy and human rights, the German conservation movement acted on the basis of an exceedingly simple political philosophy: any legal provision, and any alliance with the Nazi regime, is fine as long as it helps our cause." There was, in short, nothing inherently fascist about German nature protection movements. Furthermore, Green Parties of the late twentieth century have very little in common with prewar conservationists, Nazi or otherwise. On the contrary, they are just as devoted to participatory democracy and social justice as they are committed to environmental protection, although like any political movement, they draw their fair share of utopian fanatics.[49]

THE AGE OF ENVIRONMENTALISM

As far as the politics of nature is concerned, the era since the Second World War, at least in North America, Western Europe, and Australia and New Zealand, has been the age of environmentalism. It was the period in which environmental concerns became a mass phenomenon and, to repeat Michael Bess's apt description, "a central feature in the landscape of modernity."[50] It is the period in which various strands of environmental politics from the previous era merged together, while also developing a new sense of political urgency and visibility. And it saw the rise of a relatively new science—ecology— which provided people with a discourse of both doom and redemption. Environmentalism has generated a huge literature across many disciplines, much of which deals with overtly political topics. This review will be confined to a handful of the more prominent studies, with a particular focus on the rise of environmental activism and the emergence of Green political parties. The former challenged the practices of the environmental-management state, while the latter attempted to gain a foothold within it.

In recent years, numerous books have purported to explain the emergence of American environmentalism, their subtitles promising readers new insights into the origin of the phenomenon. The list of influences is long and varied: the New Deal; automobile culture; suburban sprawl; scientists; forestry; and consumerism, among other things, have all been touted as significant, if not primary, factors in the shaping of modern environmentalism.[51] Nearly all of these authors make a compelling case for their own chosen topic, but after reading them, one is left with the feeling that virtually every significant twentieth-century social, economic, intellectual, and political development was a major factor in the rise of postwar environmentalism. A few authors, however, have attempted to offer more totalizing explanations of environmentalism, locating it in the broader cultural currents of modern history.

In addition to unleashing the atomic bomb, the Second World War also brought an unprecedented degree of economic growth and material prosperity to the United States. In his exhaustive study of postwar environmental politics, Samuel Hays has argued that modern environmentalism is largely a product of these broad socioeconomic forces. While an interest in environmental quality has nearly always existed among the well-to-do, according to Hays, it was not until the mid-twentieth century that the general level of prosperity reached a stage where "the search for amenities became a normal expectation on the part of most Americans."[52] Hays views environmentalism as an expression of human wants and needs along the same lines as the desire for better housing, improved recreational opportunities, and better health. These are fundamental notions of "progress" in American society and do not need to be explained away in other terms. "An interest in the environmental quality of life," Hays writes, "is to be understood simply as an integral part of the drives inherent in persistent human aspiration and achievement."[53] The prewar conservation movement, in other words, was primarily aimed at fostering a more efficient use of public resources, while postwar environmentalism was marked by an effort on the part of consumers to secure and maintain a more livable environment. Hays documents his argument with numerous opinion polls showing a growing environmental awareness that correlates with rising levels of affluence, as well as by analyzing dozens of newly formed citizen's organizations and the rapid growth of prewar environmental groups such as the Sierra Club and Wilderness Society. The result was not only a general growth of environmental awareness and wilderness appreciation, but also a loosening of the grip that professionals and policymakers had previously held on issues such as wilderness management and urban health.[54]

There is much merit in Hays's study. He makes a convincing argument that the socioeconomic changes of the postwar era were important in the development of a more environmentally conscious population. In general, this explanation of the broad material factors that underlay the rise of environmentalism makes sense. However, it does not adequately explain the rise of the more radical ideas and organizations that form an important part of modern environmentalism. As Donald Worster has noted, groups such as Greenpeace, with their campaigns to stop whaling, cannot meaningfully be described as products of an "evolving consumer value"; the actions of people risking their lives to save whales or rainforests cannot merely be explained away by rising levels of affluence. Furthermore, Worster points out, greater affluence and education do not necessarily correspond to a more receptive milieu for environmentalism: "There are too many wealthy Americans opposed to toxic waste regulation, too many Ivy League graduates who want to get rid of the EPA, for that explanation to get us very far."[55]

According to Worster, the catalyst for the rise of modern environmentalism was the atom bomb. Apart from injecting a greater sense of urgency into the environmental movement, the development of nuclear weapons was also an important factor in the popularization of ecology, a heretofore-obscure academic discipline that rapidly assumed an iconic status among environmentalists. Ecology, at least in its popular form, echoed the sentiments of John Muir and Rachel Carson, raising critical questions about the costs of unfettered scientific and economic "progress" and calling for a more respectful, humble, and holistic view of nature and humanity's place within it. Popular ecology merged easily with other holistic worldviews, such as Eastern religion, Native American spirituality, and the counterculture of the 1960s, thereby providing a common denominator for dissenting groups of various political and cultural stripes.[56]

By the late 1960s, ecology, as popularized by writers such as Rachel Carson, had become a metaphor for a certain way of viewing the natural world and the place of human beings within it. This was an unusual development for a branch of biology that evolved and was discussed mostly in the rarified air of university science departments. As Worster notes, few other academic disciplines have entered the public lexicon as catchwords denoting a particular worldview or political party. To those who fully embraced the new holistic ecological worldview proffered by the likes of Rachel Carson and Aldo Leopold, nature contained intrinsic wonders that offered a wholly satisfying form of "rational enchantment" that was entirely compatible with—indeed, was in large part derived from—a science-based cosmology.[57]

A critical turning point in the history of postwar American environmental politics occurred in the early 1950s, when the Bureau of Reclamation was busy planning a series of dams along the Colorado River. One of these projects, the Echo Park Dam, would have flooded the Dinosaur National Monument, a remote two-hundred-thousand-acre wilderness and recreation area on the Utah-Colorado border. Echo Park served as a rallying point for the postwar wilderness movement. Prominent figures such as David Brower, the executive director of the Sierra Club, and his Wilderness Society counterpart, Howard Zahniser, injected wilderness preservation with a newfound degree of urgency and activism. In addition to its lobbying activities in Washington, Brower's Sierra Club bombarded the public with direct-mail pamphlets asking: "What is Your Stake in Dinosaur?" and "Will you DAM the Scenic Wildlands of Our National Parks System?" The Club published advertisements in the national press, as well as arranging for the production of a film on the issue and publishing several impressive coffee table books. In the end, the Bureau bowed to the pressure from conservationists, and the Echo Park Dam project was shelved. The Sierra Club and other conservation groups, however, paid a high price for the victory. In return for the Bureau's decision to spare Dinosaur Monument, conservationists had agreed not to oppose other dam projects. One of these, the Glen Canyon Dam, flooded a magnificent gorge on the lower Colorado River, in the process creating a martyred landscape that to this day remains one of the most powerful symbols of ecological destruction among American environmentalists.[58]

Such political activism was a hallmark of postwar environmentalism. Groups such as the Sierra Club and Wilderness Society became part of an increasingly influential environmentalist lobby in Washington, and they grew skilled at using the mass media to influence public opinion and put pressure on legislators.[59] Not that all politicians necessarily needed convincing: quite a few were sympathetic from the beginning, becoming leading advocates for environmental reform. Tom McCall, the Republican governor of Oregon during the late 1960s and early 1970s, fought for a series of environmental reforms—from a bottle bill to various measures to curb urban sprawl—which made Oregon one of the most environmentally progressive states in the nation. At the federal level, Gaylord Nelson, a Democratic senator from Wisconsin, helped organize Earth Day in 1970, the event that did more than any other to launch environmentalism into the political mainstream.[60] A combination of lobbying, grass-roots activism, and the political leadership of figures such as Nelson and Stuart Udall, the Interior Secretary during the Kennedy and Johnson administrations, led to a veritable avalanche of environmental legislation, culminating in the National Environmental Policy Act of 1970.[61]

As in the United States, opposition to large-scale hydraulic engineering projects was also pivotal in radicalizing environmentalists in Australia and New Zealand. In 1972, opponents of large dam projects in Tasmania and on New Zealand's South Island formed what are widely considered to be the first green political parties in the world. These wilderness activists were soon joined by members of the anti-nuclear movement, as well as various anti-pollution and social justice groups, thereby forming the sort of coalition that would come to typify Green political parties throughout the world.[62] In West Germany, the Greens began as a loose coalition of ideologically diverse actors. In addition to the familiar New Left and countercultural activists, some of the most prominent early members were political conservatives, such as the well-known author and Christian Democrat, Herbert Gruhl. The more conservative members hoped to build a party that would transcend traditional ideological divides and concentrate on ecological issues. However, they soon found themselves outnumbered and overruled by people who saw environmentalism as but one plank of a broader agenda, which included feminism, grass-roots democracy, human rights, and anti-militarism, along with a critical attitude toward industrial capitalism.[63]

The United States has also had a Green Party, though it has never achieved the organizational unity, let alone the electoral success, of its European counterparts. In part, this is due to an electoral system that makes it extremely difficult for a third party to break the two-party stranglehold on the voting process. The electoral systems in Australia (preferential voting), the United Kingdom, and Canada (both first-past-the-post) also constitute significant barriers to smaller parties.[64] In the German version of proportional representation, by contrast, 5 percent of the vote is enough to gain a party representation in the Bundestag. Similar proportional systems throughout central and northern Europe have enabled Green Parties to gain substantial parliamentary representation, despite the fact that they receive a similar proportion of the vote as their Australian and Canadian counterparts.[65] Many European Greens, therefore, are the beneficiaries of the comprehensive political reforms that swept through the non-communist parts of the continent after the Second World War, creating electoral systems that are in many respects more "modern" than those in the so-called "New World."[66]

An increase in political activism and the rise of Green parties are two of the major traits that have characterized environmental politics in the age of environmentalism. Another notable characteristic has been the radicalization within sectors of the movement. Since the 1970s, certain activists have been increasingly willing to risk their lives in order to protect non-human life, whether specific species, such as whales and seals, or entire ecosystems, such as old-growth forests. The pioneer of this more direct-action oriented environmentalism was Greenpeace, a Vancouver organization founded in the late 1960s by a broad coalition of American peace activists and environmentalists and Canadian hippies and radicals. Blending the nonviolent, direct-action tactics used by the peace and civil rights movements with the rhetoric of popular ecology, Greenpeace activists sailed into US and French nuclear testing zones in the Pacific. Later in the 1970s, they placed themselves between whales and harpoons, daring the whalers to shoot straight over their heads, which they frequently did. The early members of Greenpeace, several of whom were prominent Canadian journalists, were followers of Marshall McLuhan's ideas about how the media could be used as a vehicle to change mass consciousness. Mass media expanded the Quaker notion of "bearing witness" to injustice to include the whole world, or at least those parts of it that had access to television and newspapers. The fact that Greenpeace actions frequently occurred on the high seas meant that the group had an international outlook from the very beginning—an outlook which fit well with the increasingly common idea that nature knew no borders. More than any other organization, Greenpeace pioneered both direct action ecological activism and international environmentalism.[67]

Greenpeace helped fashion what Paul Wapner has called "world civic politics": a level of politics where the promotion of broad cultural sensibilities represents a mechanism of authority that is able to shape human behavior. High-profile direct actions "sting" governments and corporations with an ecological sensibility, thereby "representing a mechanism of authority that is able to shape human behavior."[68] Some other organizations, however, have taken direct action a step further, insisting that violence—against property, if not humans—is sometimes necessary to stop what they view as ecological sacrilege. Two of the more prominent organizations in this category are Earth First! and the Sea Shepherd Conservation Society. The latter was founded by Paul Watson, a self-styled "eco-warrior" who was expelled from Greenpeace in the late 1970s. Sea Shepherd has rammed whalers and illegal fishing boats on the high seas. It has installed a sharp steel I-beam, known as the "can opener," into the starboard hull of one of its ships so that it can tear gashes in the hulls of its adversaries. Watson has been labeled an "eco-terrorist" and threatened with prosecution and jail by those he torments. However, few have followed through on these threats. This, Watson argues, is because his targets know they are engaging in acts that are illegal and morally repugnant, and they do not wish to draw further attention to their activities through high-profile international court cases. Watson, or "Captain Watson," as he prefers to be known, has cultivated a larger-than-life personality, and counts among his supporters well-known figures in the world of media and entertainment, including Mick Jagger, Pierce Brosnan, and Martin Sheen, as well as the Princeton philosopher and animal rights activist, Peter Singer. It is difficult to measure Watson's political effectiveness. Does his quasi-military style influence governments to any significant degree, or does it merely bolster his own fame? The Japanese government certainly is not impressed and shows no sign of abandoning its stubborn and disingenuous "scientific whaling" program, despite thirty years of harassment by both Sea Shepherd and Greenpeace. However, one could argue that without the likes of Sea Shepherd, the Japanese would feel embold-ened to hunt whales with impunity. And in some cases, groups like Sea Shepherd act as a kind of diplomatic buffer, enabling governments to indirectly express their anger and frustration with each other. For example, by not repudiating Sea Shepherd and allowing its ships to refuel in Hobart or Melbourne, the Australian government is able to demonstrate its people's strong opposition to Japanese whaling, but without directly assaulting Japanese whalers or insulting the Japanese government.[69]

What, then, has environmentalism wrought? How has it changed the way that humans organize their societies and practice politics? For skeptics and pessimists, the answer is obvious: very little. The forces of ecological despoliation—rampant con-sumerism, unrestrained population growth, and unchecked exploitation of the com-mons—continue apace, and economic and national security issues remain uppermost on most political agendas. Cyclical downturns notwithstanding, where is the evidence that Western capitalism has lost its momentum? At most, Western societies might be in for a brief Keynesian interlude—hardly a boon for environmentalists given Keynes' commitment to consumer spending—before the inevitable return of neoliberal capi-talism, with its "creative destruction" and gross inequalities.[70] Nevertheless, even the most hardened realist cannot look back on the past century without admitting that some things have changed considerably. As Michael Bess points out, in the early twen-tieth century, advertisers portrayed factories with belching smokestacks as quintes-sential symbols of progress and prosperity. But within half a century, "that symbol's emotional valence had undergone a complete reversal, from celebration to oppro-brium; the sky above the factory had been reinvented as a vulnerable, delicate space, and the factory recast as the aggressor." What occurred in France and the rest of the industrial West was "the partial greening of the mainstream...in which a whole new complex of discourses and institutions...came into being." This, according to Bess, was a "half-revolution," resulting in the development of a "light-green society."[71]

In the light-green society, the politics of nature is framed by the discourse of envi-ronmentalism, a discourse that was not fully formed until the latter part of the twen-tieth century. In earlier times, people's relationship with non-human nature was embedded in different worldviews. Societies that lacked a modern understanding of ecology and which had relatively limited ecological power could not imagine the kind of catastrophic scenarios that form the worldview of the nuclear age. Not sur-prisingly, the parameters of environmental politics have shifted considerably over the past few centuries as the stakes have gotten higher. Fifteenth-century Venetians could certainly cause considerable environmental change on a local and, to a lesser extent, regional scale, but their actions seem paltry when compared to the global ecological power of the United States in the nuclear age. The increasingly widespread recognition that we live in the Anthropocene—an era in which humans have altered the environment on a global scale—will almost certainly offer numerous new challenges for the environmental-management state. It is unlikely that past practices, which concentrated primarily on more efficient resource extraction and relatively minor attempts at ecological protection and amelioration, will be up to the job.

Notes

1. Among political scientists and philosophers, there is no widely agreed upon definition of "politics." A popular political science textbook begins with the disheartening observation that "even respected authorities cannot agree what the subject is about." Andrew Heywood, *Politics*, 2d ed. (Basingstoke, UK: Palgrave, 2002), 4.
2. Adam Rome, "What Really Matters in History: Environmental Perspectives on Modern America," *Environmental History*, 7 (April 2002): 303–318.
3. Karl Appuhn, "Inventing Nature: Forests, Forestry, and State Power in Renaissance Venice," *Journal of Modern History* 72 (December 2000): 861–889.
4. Friedmann Schmoll, "*Schönheit, Vielfalt, und Eigenart. Die Formierung des Naturschutzes um 1900, seine Leitbilder und ihre Geschichte*" in *Natur und Staat: Staatlicher Naturschutz in Deutschland 1906–2006*, eds. Hans-Werner Frohn and Friedmann Schmoll (Münster: Landwirtschaftsverlag, 2006). For more background on German nature protection during this period, see Thomas Lekan, *Imagining the Nation in Nature: Landscape Preservation and German Identity* (Cambridge, MA: Harvard University Press, 2004); William H. Rollins, *A Greener Vision of Home: Cultural Politics and Environmental Reform in the German Heimatschutz Movement, 1904–1918* (Ann Arbor: University of Michigan Press, 1997); and Raymond H. Dominick III, *The Environmental Movement in Germany: Prophets and Pioneers, 1871–1971* (Bloomington: Indiana University Press, 1992).
5. J. Brooks Flippen, *Nixon and the Environment* (Albuquerque: University of New Mexico Press, 2000). For Train's role in fostering international environmental policy, see Flippen, "Richard Nixon, Russell Train, and the Birth of Modern American Environmental Diplomacy," *Diplomatic History* 32, no. 4 (September 2008): 613–638.
6. Terms such as "environment" and "nature" are notoriously difficult to nail down. Environmental historians, like the population as a whole, tend to use them interchangeably. For a useful discussion, see Tim Ingold, *The Perception of the Environment: Essays in Livelihood, Dwelling, and Skill* (London: Routledge, 2000), especially 20.
7. For a summary of literature that deals with the environmental-management state in the US, see Paul Sutter, "The World with Us: The State of American Environmental History," *Journal of American History* 100, no. 1 (June 2013): 100–105. The political historian, Bruce Schulman, makes a similar case, arguing that the "resource management state" in the early twentieth century US was "much more extensive than the rudimentary welfare state." See "Governing Nature, Nurturing Government: Resource Management and the Development of the American State, 1900–1912," *Journal of Policy History* 17 (October 2005): 306.
8. Michael Bess, *The Light-Green Society: Ecology and Technological Modernity in France, 1960–2000* (Chicago: University of Chicago Press, 2003), 58.

9. An example of such anachronism is Richard Grove's otherwise brilliant *Green Imperialism: Colonial Expansion, Tropical Island Edens and the Origins of Environmentalism, 1600–1860* (Cambridge: Cambridge University Press, 1995).

10. There are a growing number of books that attempt to examine environmental history from a global perspective over the longue durée. Examples include: Joachim Radkau, *Nature and Power: A Global History of the Environment* (Cambridge: Cambridge University Press, 2008); I. G. Simmons, *Global Environmental History* (Chicago: University of Chicago Press, 2008); J. Donald Hughes, *An Environmental History of the World: Humankind's Changing Role in the Community of Life* (London: Routledge, 2002); Clive Ponting, *A New Green History of the World: The Environment and the Collapse of Great Civilizations* (London: Penguin, 2007). Though not explicitly an environmental history, J. R. McNeill and William McNeill's *The Human Web: A Bird's-Eye View of World History* (New York: W.W. Norton, 2003) is also useful. In scale and ambition, all of these works are dwarfed by David Christian's *Maps of Time: An Introduction to Big History* (Berkeley: University of California Press, 2004) which, as William McNeill notes on the dust jacket, "unites natural history and human history in a single, grand, and intelligible narrative."

11. The leading scholar of English forests is Oliver Rackham, whose works include: *Trees and Woodland in the British Landscape: The Complete History of Britain's Trees, Woods, and Hedgerows* (London: Phoenix Press, 2001) and *Woodlands* (London: Collins, 2006). For a more general history of English attitudes toward nature in the early modern period, see Keith Thomas, *Man and the Natural World: Changing Attitudes in England, 1500–1800* (Oxford: Oxford University Press, 1983). For a global overview of forest history, see Michael Williams, *Deforesting the Earth: From Prehistory to Global Crisis* (Chicago: University of Chicago Press, 2001) and Joachim Radkau, *Holz: Wie Ein Naturstoff Geschichte Schreibt* (Munich: Oekom, 2007).

12. Radkau, *Nature and Power*, 139.

13. Paul Warde, *Ecology, Economy, and State Formation in Early Modern Germany* (New York: Cambridge University Press, 2006). It is worth contrasting Warde's work with that of Dorothea Hauff, who covered similar political and economic territory, but without the ecological dimension. See *Zur Geschichte der Forstgesetzgebung und Forstorganisation des Herzogtums Württemberg im 16. Jahrhundert* (Stuttgart: Schriftenreihe der Landesforstverwaltung Baden-Württemberg, 1977). David Martin Luebke's *His Majesty's Rebels: Communities, Factions, and Rural Revolt in the Black Forest, 1725–1745* (Ithaca, NY: Cornell University Press, 1997) demonstrates how the process of centralization could spawn more violent resistance than that reported by Warde.

14. Carol Kieko Matteson, *Forests in Revolutionary France: Conservation, Community, and Conflict, 1669–1848* (New York: Cambridge University Press, 2015).

15. Examples of this "anti-conservation" literature include: K. Sivaramakrishnan, *Modern Forests: Statemaking and Environmental Change in Colonial Eastern India* (Stanford: Stanford University Press, 1999); Nancy Lee Paluso, "Coercing Conservation?: The Politics of State Resource Control," *Global Environmental Change* 3, no. 2 (1993): 199–218; Charles Zerner, *People, Plants, and Justice: The Politics of Nature Conservation* (New Haven, CT: Yale University Press, 1998); Mark Dowie, *Conservation Refugees: The Hundred-Year Conflict between Global Conservation and Native Peoples* (Cambridge, MA: MIT Press, 2009).

16. A more extreme version of this is what Douglas Weiner refers to as the "predatory tribute-taking state." In Russia, Weiner argues, the long-term natural resource policies of the state have been dominated by a form of institutionalized plunder: "Unbounded by the rule of law (although constrained somewhat by custom), these regimes saw the population and the land over which they ruled as a trove of resources to be mined for the rulers' purposes." Weiner, "The Predatory Tribute-Taking State: A Framework for Understanding Russian Environmental History," in *The Environment and World History*, ed. Edmund Burke III and Kenneth Pomeranz (Berkeley: University of California Press, 2009), 277.

17. James Scott, *Seeing Like a State: How Certain Schemes to Improve the Human Condition Have Failed* (New Haven, CT: Yale University Press, 1998), 311.

18. Appuhn, "Inventing Nature," 888. For an expanded version of Appuhn's influential *Journal of Modern History* article, see his book, *A Forest on the Sea: Environmental Expertise in Renaissance Venice* (Baltimore: Johns Hopkins University Press, 2009).

19. Grove, *Green Imperialism*.

20. Bess, *Light-Green Society*, 307.

21. E. C. Spary, *Utopia's Garden: French Natural History from Old Regime to Revolution* (Chicago: University of Chicago Press, 2000).

22. Matteson, *Forests in Revolutionary France*.

23. Alan Mikhail, *Nature and Empire in Ottoman Egypt: An Environmental History* (New York: Cambridge University Press, 2011), 3.

24. Some of the more prominent postwar "green" critics included Barry Commoner, E. F. Schumacher, André Gorz, Jacques Ellul, and Ivan Illich. For a summary, see Bess, *Light-Green Society*, ch. 5.

25. Important studies of the environmental impact of nineteenth-century industrialization include: William Cronon, *Nature's Metropolis: Chicago and the Great West* (New York: W. W. Norton, 1991); Theodore Steinberg, *Nature Incorporated: Industrialization and the Waters of New England* (New York: Cambridge University Press, 1991); Andrew C. Isenberg, *Mining California: An Ecological History* (New York: Hill & Wang, 2005); Mark Cioc, *The Rhine: An Eco-Biography, 1815–2000* (Seattle: University of Washington Press, 2002).

26. Ramachandra Guha, *Environmentalism: A Global History* (New York: Longman, 2000), 10–17; Peter C. Gould, *Early Green Politics: Back to Nature, Back to Land, and Socialism in Britain, 1880–1900* (New York: St. Martin's Press, 1988), 15–20.

27. However, as Donald Worster points out, Muir's fervent nature worship dissipated in middle age as he came to be increasingly concerned with the more mundane affairs of business and family. *A Passion for Nature: The Life of John Muir* (New York: Oxford University Press, 2008).

28. William Cronon offers a cogent analysis of the concept of the "sublime" as applied to wilderness. See "The Trouble with Wilderness; or, Getting Back to the Wrong Nature," in *Uncommon Ground: Rethinking the Human Place in Nature*, ed. Cronon (New York: W. W. Norton, 1995).

29. John Alexander Williams, *Turning to Nature in Germany: Hiking, Nudism, and Conservation, 1900–1940* (Stanford: Stanford University Press, 2007); Michael P. Cohen, *The History of the Sierra Club, 1892–1970* (New York: Random House, 1988).

30. Gould, *Early Green Politics*.

31. Lekan, *Imagining the Nation in Nature*, 4.

32. Worster, "John Muir and the Modern Passion for Nature," *Environmental History* 10, no. 1 (January 2005): 17.

33. Ritvo, *The Dawn of Green: Manchester, Thirlmere, and Modern Environmentalism* (Chicago: University of Chicago Press, 2009).

34. The phrase is William Beinart's. See "Soil Erosion, Conservationism and Ideas about Development: A Southern African Exploration, 1900–1960," *Journal of Southern African Studies* 11, no. 1 (October 1984): 59.

35. For more on German forestry practices, see Radkau, *Holz*, and Michael Imort, "A Sylvan People: Wilhelmine Forestry and the Forest as a Symbol of Germandom," in *Germany's Nature: Cultural Landscapes and Environmental History*, ed. Thomas Lekan and Thomas Zeller (New Brunswick, NJ: Rutgers University Press, 2005). Char Miller's *Gifford Pinchot and the Making of Modern Environmentalism* (Washington DC: Island Press, 2001) is the most comprehensive biography of Pinchot.

36. Louis S. Warren, *The Hunter's Game: Poachers and Conservationists in Twentieth Century America* (New Haven: Yale University Press, 1999).

37. Benjamin Heber Johnson, "Conservation, Subsistence, and Class at the Birth of the Superior National Forest," *Environmental History* 4, no. 1 (January 1999): 80–99.

38. David Anderson and Richard Grove, eds., *Conservation in Africa: People, Parks and Priorities* (Cambridge: Cambridge University Press, 1987); Mahesh Rangarajan, *Fencing the Forest: Conservation and Ecological Change in India's Central Provinces, 1860–1914* (New Delhi: Oxford University Press, 1996); Madhav Gadgil and Ramachandra Guha, *This Fissured Land: An Ecological History of India* (Berkeley: University of California Press, 1993); Nancy Peluso, *Rich Forests, Poor People: Resource Control and Resistance in Java* (Berkeley: University of California Press, 1992).

39. Arun Agrawal, *Environmentality: Technologies of Government and the Making of Subjects* (Durham, NC: Duke University Press, 2005), 216–217.

40. Radkau, *Nature and Power*, 240.

41. Unlike most other areas of environmental reform, women played a significant role in the sanitation movement. See Suellen M. Hoy, "'Municipal Housekeeping': The Role of Women in Improving Urban Sanitation Practices, 1880–1917," in *Pollution and Reform in American Cities, 1870–1930*, ed. Martin V. Melosi (Austin: University of Texas Press, 1980); Nancy Tomes, *The Gospel of Germs: Men, Women, and the Microbe in American Life* (Cambridge, MA: Harvard University Press, 1998); Carolyn Merchant, "Women of the Progressive Conservation Movement: 1900–1916," *Environmental Review* 8 (Spring 1984): 57–85.

42. Melosi, *The Sanitary City: Urban Infrastructure in America from Colonial Times to the Present* (Baltimore: Johns Hopkins University Press, 2000). Several of Tarr's most important essays are collected in *The Search for the Ultimate Sink: Urban Pollution in Historical Perspective* (Akron, OH: University of Akron Press, 1996).

43. For comparative studies, see Daniel T. Rogers, *Atlantic Crossings: Social Politics in a Progressive Age* (Cambridge, MA: Harvard University Press, 1998); E. P. Hennock, "The Urban Sanitary Movement in England and Germany, 1838–1914: A Comparison," *Continuity and Change* 15, no. 2 (September 2000): 269–296.

44. Frank Uekötter, *The Age of Smoke: Environmental Policy in Germany and the United States, 1880–1970* (Pittsburgh: University of Pittsburgh Press, 2009). For a similar study which focuses on Britain, see Peter Thorsheim, *Inventing Pollution: Coal, Smoke and Culture in Britain since 1800* (Athens: Ohio University Press, 2006).

45. Radkau, *Nature and Power*, 247.

46. Douglas Weiner, *A Little Corner of Freedom: Russian Nature Protection from Stalin to Gorbachev* (Berkeley: University of California Press, 1999); Judith Shapiro, *Mao's War Against Nature: Politics and the Environment in Revolutionary China* (New York: Cambridge University Press, 2001).

47. Anna Bramwell, "Was this Man the Father of the Greens?" *History Today* 34 (September 1984): 13.

48. Bramwell, *Blood and Soil: Walther Darré and Hitler's Green Party* (Abbotsbrook, UK: Kensal Press, 1985) and *Ecology in the Twentieth Century: A History* (New Haven, CT: Yale University Press, 1989). For a thorough critique, see Piers H. G. Stephens, "Blood, Not Soil: Anna Bramwell and the Myth of 'Hitler's Green Party,'" *Organization and Environment* 14, no. 2 (2001): 173–187.

49. Uekötter, *The Green and the Brown: A History of Conservation in Nazi Germany* (New York: Cambridge University Press, 2006), 16. See also Franz-Josef Brüggemeier, Mark Cioc, and Thomas Zeller, eds., *How Green were the Nazis?: Nature, Environment, and Nation in the Third Reich* (Athens: Ohio University Press, 2006). Janet Biehl and Peter Staudenmaier, members of Murray Bookchin's Institute for Social Ecology, are wary of the more extremist tendencies among some in the ecology movement—tendencies which they ascribe to the excesses of nineteenth-century German Romanticism and its anti-humanism. They arrive at similar conclusions to Bramwell, albeit from the other end of the political spectrum. See *Ecofascism: Lessons from the German Experience* (Edinburgh: AK Press, 1995). Also available online at: http://www.spunk.org/texts/places/germany/sp001630/ecofasc.html.

50. Bess, *Light-Green Society*, 58.

51. Maher, *Nature's New Deal: The Civilian Conservation Corps and the Roots of the American Environmental Movement* (New York: Oxford University Press, 2007); Paul S. Sutter, *Driven Wild: How the Fight Against Automobiles Launched the Modern Wilderness Movement* (Seattle: University of Washington Press, 2002); Adam Rome, *The Bulldozer in the Countryside: Suburban Sprawl and the Rise of American Environmentalism* (New York: Cambridge University Press, 2001); Michael Egan, *Barry Commoner and the Science of Survival: The Remaking of American Environmentalism* (Cambridge, MA: MIT Press, 2007); Gregory A. Barton, *Empire Forestry and the Origins of Environmentalism* (New York: Cambridge University Press, 2002); Gregory Summers, *Consuming Nature: Environmentalism in the Fox River Valley, 1850–1950* (Lawrence: University of Kansas Press, 2006).

52. Samuel P. Hays and Barbara D. Hays, *Beauty, Health, and Permanence: Environmental Politics in the United States, 1955–85* (New York: Cambridge University Press, 1987), 35.

53. Ibid., 5.

54. Hays, "The Structure of Environmental Politics since World War II," *Journal of Social History* 14 (1980–81): 724. Recent works have challenged, or at least complicated, Hays's assertions that postwar environmentalism was primarily a consumer-driven movement. In a study of the impact that suburban sprawl had on the environmental movement, Adam Rome has shown that consumers repeatedly made choices that contributed to environmental despoliation: "The consumer perspective was myopic," Rome concludes, and "Homeowners only cared about certain elements of the environment." See *The Bulldozer in the Countryside*, 13. Hal Rothman similarly concluded that Americans "embrace environmentalism when it is convenient and inexpensive, but when it challenges the

comforts to which they are accustomed, they ignore or avoid it." Rothman, *The Greening of a Nation?: Environmentalism in the United States since 1945* (Fort Worth, TX: Harcourt Brace, 1998), xii.

55. Worster, "Comment: Hays and Nash," in *Environmental History: Critical Issues in Comparative Perspective*, ed. Kendall E. Bailes (Lanham, MD: University Press of America, 1985), 259–260.

56. Donald Worster, *Nature's Economy: A History of Ecological Ideas*, 2d ed. (Cambridge: Cambridge University Press, 1994); Frank Zelko, "Challenging Modernity: The Origins of Postwar Environmentalism in the United States," in *Shades of Green: Environmental Activism Around the Globe*, ed. Christof Mauch, Nathan Stoltzfus, and Douglas R. Weiner (Lanham, MD: Rowman & Littlefield, 2006).

57. Worster, *Nature's Economy*, 360; Frank Zelko, "'A Flower is Your Brother!': Holism, Nature, and the (Non-Ironic) Enchantment of Modernity," *Intellectual History Review* 23, no. 4 (2013): 517–536.

58. The quotes are from Rothman, *Greening of a Nation?*, 41. Capitalization in original. The most comprehensive analysis of the Echo Park Dam affair is Mark W. T. Harvey, *A Symbol of Wilderness: Echo Park and the American Conservation Movement* (Albuquerque: University of New Mexico Press, 1994). Jared Farmer examines how Glen Canyon has come to stand for many competing visions of nature. See *Glen Canyon Damned: Inventing Lake Powell and the Canyon Country* (Tucson: University of Arizona Press, 1999). Brower expresses his regrets over the sacrifice of Glen Canyon in David Brower, ed., *The Place No One Knew: Glen Canyon on the Colorado* (San Francisco: Sierra Club Books, 1968). The Glen Canyon Dam was also the stage on which the radical environmental organization, Earth First!, introduced itself to the world.

59. As James Morton Turner convincingly argues, American wilderness advocates "adopted a political strategy grounded in 'reform liberalism,' which emphasized the role of government in protecting the public interest," thereby allying wilderness advocates with the Democratic Party and hitching "the success of environmental reform to an expansion of the powers and responsibilities of the federal government." See "'The Specter of Environmentalism': Wilderness, Environmental Politics, and the Evolution of the New Right," *Journal of American History* (June 2009): 125. For a more expansive treatment of the politics of wilderness in the US since the 1960s, see Turner, *The Promise of Wilderness: American Environmental Politics Since 1964* (Seattle: University of Washington Press, 2012). For a case study of interest-group politics at the local level, see Andrew Hurley, *Environmental Inequalities: Class, Race, and Industrial Pollution in Gary, Indiana, 1945–1980* (Chapel Hill: University of North Carolina Press, 1995). The book also examines the roots of the environmental justice movement that emerged in the 1980s.

60. Brent Walth, *Fire at Eden's Gate: Tom McCall and the Oregon Story* (Portland: Oregon Historical Society Press, 1994). Bill Christofferson, *The Man from Clear Lake: Earth Day Founder Senator Gaylord Nelson* (Madison: University of Wisconsin Press, 2004); Adam Rome, *The Genius of Earth Day: How a 1970 Teach-In Unexpectedly Made the First Green Generation* (New York: Hill and Wang, 2013).

61. The best summary is Hays, *Beauty, Health, and Permanence*. For more on environmentalism as a product of the 1960s, see Adam Rome, "'Give Earth a Chance': The Environmental Movement and the Sixties," *Journal of American History* 90, no. 2 (September 2003): 525–554.

62. Drew Hutton and Libby Connors, *A History of the Australian Environmental Movement* (Melbourne: Cambridge University Press, 1999). There is as yet no similar single volume history for the environmental movement in New Zealand. For a summary of the postwar era, and a guide to the existing literature, see Don Garden, *Australia, New Zealand, and the Pacific: An Environmental History* (Santa Barbara, CA: ABC CLIO, 2005), 174–180. On the Australian Greens, see Amanda Lohry, "Groundswell: The Rise of the Greens," *Quarterly Essay* 8 (2002) and Frank Zelko, "The Tasmanian Crucible: Bob Brown and the Australian Greens," in *Green Parties: Reflections on the First Three Decades*, ed. Frank Zelko and Carolin Brinkmann (Washington, DC: Heinrich Böll Foundation, 2006).

63. There is a voluminous literature on the German Greens. For a succinct English language overview, see Jürgen Hoffmann, "From Cooperation to Confrontation: The Greens and the Ecology Movement in Germany," in *The Culture of German Environmentalism*, ed. Axel Goodbody (New York: Berghahn Books, 2002). For a more detailed account of how the Greens emerged from the various strands of the West German left, see Andrei S. Markovits and Philip S. Gorski, *The German Left: Red, Green, and Beyond* (New York: Oxford University Press, 1993). For a truly exhaustive study in German, see Joachim Raschke's monumental *Die Grünen: Wie sie wurden, was sie sind* (Cologne: Bund Verlag, 1993). In recent years, the party has regained some of its earlier conservatism and pragmatism. In 2008, for example, the Hamburg Greens agreed to form a coalition government with the Christian Democrats, a development that could one day be replicated at the federal level.

64. The Australian Senate, however, is based on proportional voting. After the 2010 federal election, there were nine Greens in the Senate (out of a total of seventy-six senators).

65. Percentage of vote received by Green parties in various countries in previous federal election (as of January 2009): Australia: 9%; Austria, 10.4%; Canada: 6.8%; Germany: 8.3%; New Zealand: 6.7%; Sweden: 5.3%; Switzerland: 9.6%; United Kingdom: 1%; USA: 0.12%.

66. For a comparative study of European Green parties, see Michael O'Neill, *Green Parties and Political Change in Contemporary Europe: New Politics, Old Predicaments* (Aldershot, UK: Ashgate, 1997). For broader coverage, see Herbert Kitschelt, "The Green Phenomenon in Western Party Systems," in *Environmental Politics in the International Arena*, ed. S. Kamieniecki, G. A. Gonzalez, and R. O. Voss (Albany: SUNY Press, 1993). Also useful is the three-volume *Green Politics* series edited by Wolfgang Rüdig and published by Edinburgh University Press.

67. Frank Zelko, *Make it a Green Peace!: The Rise of Countercultural Environmentalism* (New York: Oxford University Press, 2013).

68. Paul Wapner, *Environmental Activism and World Civic Politics* (Albany, NY: State University of New York Press, 1996), 65. For detailed sociological studies of major environmental groups in the US and Germany, see Robert Brulle, *Agency, Democracy, and Nature: The US Environmental Movement from a Critical Theory Perspective* (Cambridge, MA: MIT Press, 2000) and William T. Markham, *Environmental Organizations in Germany: Hardy Survivors in the Twentieth Century and Beyond* (New York: Berghahn Books, 2008).

69. For an excellent profile of Watson, see Raffi Khatchadourian, "Neptune's Navy: Paul Watson's Wild Crusade to Save the Oceans," *New Yorker*, November 5, 2007. For a sympathetic account of Watson and other radical environmentalists, see Christopher Manes, *Green Rage: Radical Environmentalism and the Unmaking of Civilization* (Boston: Little,

Brown, 1990). On Earth First!, see Susan Zakin, *Coyotes and Town Dogs: Earth First! and the Environmental Movement* (Tucson: University of Arizona Press, 2002).

70. Skeptics of environmentalism's efficacy, at least as expressed through Green parties and environmental organizations, include: Anna Bramwell, *The Fading of the Greens: The Decline of Environmental Politics in the West* (New Haven, CT: Yale University Press, 1994); Grant Jordan and William Maloney, *The Protest Business?: Mobilizing Campaign Groups* (Manchester: Manchester University Press, 1997); Mark Dowie, *Losing Ground: American Environmentalism at the Close of the Twentieth Century* (Cambridge, MA: MIT Press, 1995), and most infamously, Bjørn Lomborg, *The Skeptical Environmentalist: Measuring the Real State of the World* (Cambridge: Cambridge University Press, 2001).

71. Bess, *Light-Green Society*, 58, 4.

INDEX

Italic page numbers indicate figures and illustrations.

Black Death, 1, 82, 86–87, 102n26, 133–134, 377
Black Rice (Carney), 432
Blighted neighborhoods, 563
Bloch, Marc, 4
"Blood and soil," 340–341, 351
Blum, Elizabeth, 600–601, 602, 606
Blumenbach, Johann Friedrich, 183
Bocking, Stephen, 215
Bodin, Jean, 115, 127n57
Body of Liberties (England 1641), 516
Boers. *See* South Africa
Bolster, W. Jeffrey, 219–220, 694
Bonn Convention on Migratory Species
    (CMS 1979), 700
Bonnifield, Matthew Paul, 135
Bonpland, Aimé, 186–187
Bonta, Marcia Myers, 619
Boone, Daniel, 519
Borah, Woodrow, 79–80
Borderlands (North America), 289, 668–687
    chinook wind on Canadian-US border, 677
    copper industry and, 678
    future of, 679–681
    human impact upon ecosystems of,
        672–676
    imagining transnational landscapes of,
        676–679
    interest in, 668–669
    international resource competition and
        conservation, 669–672
    lumber industry and, 678
    transborder peace parks, 671–672
Borland, Hal, 609
Boston, 554, 555, 561, 564, 649
Botanical gardens, 186
Botanists and tropics, 190–191
Boutmy, Emile, 554
Bovill, Edward, 120
Bovine spongiform encephalopathy, 94
Boy Scouts of America, 616–618
BPA, 493
Brady, Lisa, 650–651
Bramwell, Anna, 728, 729, 739n49
Brand, Stewart, 478
Braudel, Fernand, 4, 369, 371, 382–384,
    394n28, 395n33, 481
Brazer, Sheila, 138

Brazil
    chattel slavery in, 409
    drought in, 34, 35
    Escola Tropicalista Bahiana, 193
    forests in, 159–160, 232
    history of slavery and environment in, 435
Brazilwood, 160–161
*Breathing Space: How Allergies Shape Our
    Lives and Landscapes* (Mitman), 89, 240
Brechin, Gray, 583
Brenner, Robert, 378
Breton, Mary Joy, 602
Britain
    Back to Nature movement, 724
    big game hunting as desirable sport for,
        609–610
    chemicals and cancer in, 272
    climate and culture in, 39
    Cod Wars, 693
    common lands, enclosure of, 376–380
    common law, 515, 517
    coppicing trees in, 159
    dogs in Victorian England, 58–59
    early capitalism in, 381
    industrialization and nostalgia for
        preindustrial countryside in, 724
    international fishing rights of, 693
    migrating birds convention, 539–540
    royal forests in, 163, 173n24
    sanitary movement in, 89–90
    shipbuilding and need for timber in, 163
    TB epidemic in, 87
    Thirlmere campaign protesting damming
        as violation of nature, 725
    water and doctrine of natural flow, 525
    weather modification in, 30
British Garden Cities Movement, 553, 562
British Petroleum (BP), 495
Brock, Emily, 11, 154
Brockington, Dan, 296
Brooks, Karl, 536, 538
Brosnan, Kathleen, 12, 513
Brower, David, 731
Brown, Jessica, 299
Brown, John Croumbie, 117, 128n67
Brown, Linda, 590
Brown, Phil, 92

Franco-British treaty (1828), 712n10
Franco-Spanish treaty (1683), 712n1
Franklin, Benjamin, 646, 660
Freyfogle, Eric, 423n77, 522, 531
Freyre, Gilberto, 193
Friedan, Betty, 622, 623
Friedman, Elisabeth, 627
Friedman, Milton, 398
Friedrich, Caspar David, 335
"From Wilderness to Hybrid Landscapes:
    The Cultural Turn in Environmental
    History" (White), 349
*The Frontier of Leisure: Southern California
    and the Shaping of Modern America*
    (Culver), 565
Fudge, Erica, 54
Fuel. *See also* Coal mining; Fossil fuels;
    Fuelwood
  capitalism and, 388, 391n8, 396n44
  consumerism and, 485–487
  labor and, 440
Fuelwood, 159, 719
The Fundamental Constitutions of Carolina
    (Locke), 380
Furness, Frank, 659
Fur trade, 484, 576, 605, 609, 687n50

Gabler, Robert, 138
Gadgil, Madhav, 208–209
Gaia hypothesis, 213
Galapagos, 157
Galbraith, John Kenneth, 481
Galen, 181
Gallman, Matthew, 654
Galveston hurricane (1900), 30–31
*Garbage in the Cities: Refuse, Reform and the
    Environment, 1880–1980* (Melosi), 564
Garden Cities. *See* British Garden Cities
    Movement
Gary, Indiana, 12, 448–449, 588, 598n73
Gates, Paul, 520, 521–522, 532–533, 537
Geiling, E. M. K., 261
Gender. *See* Masculinity; Women and gender
*Gendered Fields: Rural Women, Agriculture,
    and Environment* (Sachs), 627
General Motors, 487
Genetic technology, 416

Genghis Khan, 141
Geography, 10, 195, 332. *See also* Cultural
    geography
  as field of study, 338
  historical geography, 234, 332, 355
Geography (Strabo), 180
Geopolitics, 339–340
Georgia
  Coast, 433
  Red Hills on Florida border, 434–435
Germany. *See also* National Socialism (Nazis)
  back-to-nature cults in, 724
  conservation history in, 717
  consumerism in, 476–477
  environmentalism in Nazi Germany, 728
  environmentalism in West Germany, 732,
    741n63
  forest management in, 720, 725–726
  forestry as profession in, 164
  geography as field of study in, 338
  geopolitics in, 339–340
  industrial culture in, 340–341, 727–728
  landscape study in, 338–339, 351
  nationalist meaning of landscapes of,
    334–337
  nature romanticism in, 725
  new generation of environmental
    historians in, 352
  renaturing in, 312
Germ theory, 78, 89, 261
GI Bill, 404, 559
Gilbert, Scott, 274
Gilded Age, 653
Gillespie, Greg, 610
Giltner, Scott, 579
Girl Scouts of America, 617–618
Glacier National Park, 285, 295, 415, 672
Glaciers, 28, 29, 38
Glacken, Clarence, 4, 115, 126n33, 181
Glave, Dianne, 582, 609
Gleason, Henry, 223n12
Glen Canyon Dam, 731, 740n58
Gliddon, George R., 578
Globalization, 234, 471, 689–690
  world civic politics and, 733
Global South, 726–727
Global warming, 25, 26, 29, 37, 370, 711

CPSIA information can be obtained
at www.ICGtesting.com
Printed in the USA
BVHW01s0023310518
517788BV00003B/26/P